The Acheulian Site of Gesher Benot Ya'aqov Volume IV

Vertebrate Paleobiology and Paleoanthropology Series

Edited by

Eric Delson
Vertebrate Paleontology, American Museum of Natural History
New York, NY 10024, USA
delson@amnh.org

Eric J. Sargis
Anthropology, Yale University
New Haven, CT 06520, USA
eric.sargis@yale.edu

Focal topics for volumes in the series will include systematic paleontology of all vertebrates (from agnathans to humans), phylogeny reconstruction, functional morphology, Paleolithic archaeology, taphonomy, geochronology, historical biogeography, and biostratigraphy. Other fields (e.g., paleoclimatology, paleoecology, ancient DNA, total organismal community structure) may be considered if the volume theme emphasizes paleobiology (or archaeology). Fields such as modeling of physical processes, genetic methodology, nonvertebrates or neontology are out of our scope.

Volumes in the series may either be monographic treatments (including unpublished but fully revised dissertations) or edited collections, especially those focusing on problem-oriented issues, with multidisciplinary coverage where possible.

Editorial Advisory Board
Ross D.E. MacPhee (American Museum of Natural History), **Peter Makovicky** (The Field Museum), **Sally McBrearty** (University of Connecticut), **Jin Meng** (American Museum of Natural History), **Tom Plummer** (Queens College/CUNY).

More information about this series at http://www.springer.com/series/6978

The Acheulian Site of Gesher Benot Ya'aqov Volume IV

The Lithic Assemblages

Naama Goren-Inbar

Institute of Archaeology, The Hebrew University of Jerusalem, Jerusalem, Israel

Nira Alperson-Afil

Martin (Szusz) Department of Land of Israel Studies and Archaeology, Bar-Ilan University, Ramat-Gan, Israel

Gonen Sharon

Prehistory Laboratory, Multidisciplinary Studies, East Campus, Tel Hai College, Upper Galilee, Israel

Gadi Herzlinger

Institute of Archaeology, The Hebrew University of Jerusalem, Jerusalem, Israel

With contributions by Tamar Lotan and Shmuel Belitzky

Naama Goren-Inbar
Institute of Archaeology
The Hebrew University of Jerusalem
Jerusalem
Israel

Nira Alperson-Afil
Martin (Szusz) Department of Land
of Israel Studies and Archaeology
Bar-Ilan University
Ramat-Gan
Israel

Gonen Sharon
Prehistory Laboratory, Multidisciplinary Studies
East Campus
Tel Hai College
Upper Galilee
Israel

Gadi Herzlinger
Institute of Archaeology
The Hebrew University of Jerusalem
Jerusalem
Israel

ISSN 1877-9077　　　　　ISSN 1877-9085　(electronic)
Vertebrate Paleobiology and Paleoanthropology Series
ISBN 978-3-319-74050-8　　ISBN 978-3-319-74051-5　(eBook)
https://doi.org/10.1007/978-3-319-74051-5

Library of Congress Control Number: 2018932200

© Springer International Publishing AG 2018

This work is subject to copyright. All rights are reserved by the Publisher, whether the whole or part of the material is concerned, specifically the rights of translation, reprinting, reuse of illustrations, recitation, broadcasting, reproduction on microfilms or in any other physical way, and transmission or information storage and retrieval, electronic adaptation, computer software, or by similar or dissimilar methodology now known or hereafter developed.

The use of general descriptive names, registered names, trademarks, service marks, etc. in this publication does not imply, even in the absence of a specific statement, that such names are exempt from the relevant protective laws and regulations and therefore free for general use.

The publisher, the authors and the editors are safe to assume that the advice and information in this book are believed to be true and accurate at the date of publication. Neither the publisher nor the authors or the editors give a warranty, express or implied, with respect to the material contained herein or for any errors or omissions that may have been made. The publisher remains neutral with regard to jurisdictional claims in published maps and institutional affiliations.

Cover illustration: The knapping of Giant Cores on the shore of the Paleo-Lake Hula. This image is based on studies of the lithic assemblages of Gesher Benot Ya'aqov, the essence of this volume. Artist: Guy Hivroni

Printed on acid-free paper

This Springer imprint is published by Springer Nature
The registered company is Springer International Publishing AG
The registered company address is: Gewerbestrasse 11, 6330 Cham, Switzerland

The Acheulian Site of Gesher Benot Ya'aqov

Coordinated by

Naama Goren-Inbar
Institute of Archaeology, The Hebrew University of Jerusalem

Foreword

Gesher Benot Ya'aqov is a site which can properly be described as central to our understanding of the Pleistocene record and of the course of human evolution in the last million years. The kind invitation from Naama Goren-Inbar to write this foreword gives me the chance to set its context in the broader picture, to justify the words above, and to touch on the meaning of its artefacts and analyses which this volume so clearly shows.

The Acheulean is the longest tradition in our human past, and geographically the most widespread. Naturally then, our knowledge depends on many sites, a thin scatter through time and space, telling a largely consistent if somewhat puzzling story: but very few of these sites provide a coherent picture of early human activities. Gesher Benot Ya'aqov stands out as one of the few sites around the world that can be a standard bearer of our knowledge. Is this because it was more important than other places in the past? That can hardly be so: its significance is rather for us, now, made by its archaeology.

Certainly, some favoured places drew a human return again and again, and GBY is one of those. Factors of resources – plants, animals and water – can make such a place focal for a local community. It would be hard to argue that one early human hunter gatherer community is more important than another, whether it is in the Levant, Europe or Africa. Even so, from the point of view of archaeologists, some areas give a better barometer of movement and dispersal than others, and the Middle East is crucial for an understanding of movements out of Africa, the first globalisations.

What makes GBY stand out further are the factors of deposition continuing steadily over a long period, of the order of hundred thousand years; the subsequent geological events which have created exposures, and then the breadth, scope and intensity of archaeological investigation. This combination is very rarely available.

Part of the strength of GBY comes from the 'bandwidth' of its preservation. If we ask what is central or critical to the Acheulean, the handaxe is certainly its hallmark, but field archaeology has moved beyond it, to the full range of evidence. The Acheulean is polythetic, in the sense that multiple features characterise it, but often they are not all present. The team at GBY recognised this early on, with studies of the 'visible and invisible' in the Acheulean which explored the difficulties of distinguishing between true absence and mere absence in a particular locality or preservational circumstances. Although the handaxe has proved as useful to study at GBY as anywhere, the outstanding value of the investigations is that they emphasise a far broader picture: not just categories such as small tools, wood, or butchery, but most especially exploration of the dynamics of activity – not just an abstract concept, but the actions of early humans - plain to discern in this volume.

Paradoxically, a new generation of scholarship with a psychological emphasis concentrates largely on the handaxe – often separated from archaeological context – and then risks missing the point clearly brought out in this research that it is one tool in a toolkit or even set of toolkits. In contradistinction, and over a long period, Naama Goren-Inbar and her colleagues have worked to construct from the multiple lines of evidence a sense of complete, functioning and competent human beings. Through that cognitive approach we see Homo erectus as effective problem solvers, competent and even comfortable in their world.

The Acheulean as global tradition: its reshaping in recent years

Views of the Acheulean have changed significantly in recent years. The tradition has grown yet longer, with dates pushing toward 1.8 million years for its beginnings. The idea of a link with *Homo erectus* remains. The Acheulean is usually seen as originating in Africa, and an early 'Out of Africa' has long been the dominant paradigm for modelling dispersals – one seriously considered in the GBY interpretations.

In this frame, starting dates at least as early as one million years are now possible in both Asia and Europe, although rarely supported by solid age determinations. Gesher Benot Ya'aqov, however, is well dated to periods just after the Brunhes-Matuyama palaeomagnetic boundary of 790,000 years ago. This places it at the head of the second half of the Acheulean, rather older than any handaxe site in Europe that we can see clearly, and older than most dated sites in India. This position makes the site able to stand as a benchmark for looking at the early development of the Acheulean across these areas.

The pivotal position of GBY, and the depth of the research, allow comparisons with other Acheulean industries, in the Levant and around the world, which are only just beginning – and which are made far more feasible by the content of this volume. The great majority of variability studies have been internal to each continent. From its stone technology, GBY seems to be closely allied with the African 'middle' Acheulean – with somewhat similar material occurring in India and parts of Europe such as Spain.

The record of stone

Lithics, the main subject of this volume, still provide the backbone of Palaeolithic studies. GBY shows handaxes and cleavers that are typical of the classic Acheulean idea. Broadly it has been known for a long time that such bifaces are made by striking large flakes from giant cores – but in the European record these are rare, and their global importance is often missed or even ignored. In Africa, the phenomenon is always noted, yet rarely treated with more than a few words. Our colleagues in Israel have provided a great service by exploring the issues, through systematic experiments, and then by systematic survey (Sharon 2009 and this volume).

If the GBY bifaces are very like those in Africa made of similar lavas, it is noticeable from the figures that they almost lack thick forms such as picks, even though these occur on the older sites of 'Ubeidiya. This is one aspect of the variability that is so prevalent in the Acheulean: that in almost any respect two sites close together can be as different as two sites far apart. Similarly, GBY shows echoes of techniques otherwise known several thousands of kilometres away. The careful working of large cores is one element of this ("The size and morphology of large flakes were pre-planned" p 376), and if anyone should doubt this, some of the proof lies in biface flakes made by the fascinating Kombewa technique, described perhaps for the first time outside Africa. In this by great skill the maker produces biface blanks with two bulbar faces. The technique occurs at La Kamoa in Congo, and in the Sahara, but is scarcely present on any of the major East African sites – and yet it crops up again at Gesher Benot Ya'aqov, not commonly but consistently (and probably this is its oldest securely dated occurrence).

Choice and cognition

Also of interest are the stone raw material choices. Over a period, here and on other sites too, Naama Goren-Inbar and her colleagues have shown the cognitive implications of selections which were evidently made to find the best properties for carrying out a particular task. Basalt, flint and limestone were preferred in different circumstances. The operational sequences so carefully traced by Goren-Inbar and colleagues show clear preferences of particular materials. Such selection can rarely be studied to the same degree on African sites, where flint-like materials are rare, or on most European sites where, barring the presence of quartzite, flint reigned supreme.

The research described here also reflects other new centres of interest – for example studies of percussion that now extend in a comparative frame to primate tool using practices. In these developments making links with primatology the GBY research has been at the forefront from an early stage. Again the studies have explored the dynamics of action, and there is the interlinking of operational chains – in this case the fashioning of stone, and the processing of nuts. From the circumstances of sedimentation and preservation GBY does not make available very large excavation surfaces, but they do allow the chains of activities to be seen in detail, especially in the case of association of artefacts and butchery.

Aspects of GBY revisited – bone, plants, wood and fire

The organic remains of GBY have been described in other volumes and papers, but it is important to allude to them here. In the later flowering of technology fire use and the hearth create a nexus for intersecting operational chains. It is more than possible that such interactions had already begun by 700,000 years ago. It is not coincidence that both wood and fire evidence are preserved at GBY. There are parallels at Kalambo Falls, and to some extent Schöningen. For physical and chemical reasons burnt material has little chance of being preserved, wood even less so. But where wood is preserved, burnt material is also likely to be found, almost without exception.

'WYSWTW' – what you see is what there was – has too often been the complacent mantra of Palaeolithic archaeology. There is far more than this. Recent approaches to social cognition make plain that we see little more than the tip of the iceberg. The co-preservation of wood and burning points towards complex behavioural patterns which we can only dimly see, but which are convincingly there.

GBY offers many lessons for the archaeology of the Acheulean. The most particular one is to see the value of intensive archaeology, carried out through with a total approach. That is a lesson carried forward in a tradition from the investigations of sites such as Pincevent, and applying equally to the Somme, Boxgrove, or Terra Amata. But GBY is older than these sites, and on older sites it bcomes harder to 'explicate' anything like a full 'palaeoethnology', as Andre Leroi-Gourhan attempted at Pincevent. The Hulah lake is gone, the beds have been tilted, the exposures in and around the present day river Jordan are limited. And yet, this research has emerged successfully in a way that must draw admiration, each question addressed methodically, and carried through in the most painstaking process from field to laboratory.

This volume, and the others which have already appeared, together make up a contribution marking the summation of year upon year of unrelenting hard work. Naama Goren-Inbar has inspired a team of collaborators, but it is to her above all that we owe this work. It is fitting that she has been recognised and honoured in various ways: the Gesher Benot Ya'aqov volumes give us an invaluable insight into the Acheulean and its makers, and will also give an enduring reward to all those whom Glynn Isaac termed the 'aficionados' of the handaxe tradition.

John Gowlett

Preface

The prehistoric site of Gesher Benot Ya'aqov (GBY) has been known since the mid 1930s. Three eminent archaeologists, D. A. E. Garrod, M. Stekelis, and D. Gilead, were responsible for its introduction to the academic community and beyond. Although each represented a different scholarly world, their studies shared much common ground. The location of the site, and its particular geographical, geological, and geopolitical characteristics, prevented these eminent prehistorians and their successors from conducting long-term investigations, or more precisely excavations, at the site. The efforts and achievements of this particular period are detailed elsewhere in this volume (Chapters 1–3; see also Appendix 1 for the earlier history of the site).

It was in the early 1980s that I was approached by a young geologist, Shmuel Belitzky, who asked my opinion about an isolated sedimentary outcrop located south of the known extent of the GBY Acheulian prehistoric site, in an area that had not previously been investigated for prehistoric finds. The outcrop, at the southern end of the Hula Valley, is located on the slopes of the Crusader fortress of *Vadum Iacob* and forms its sloping protective bank or glacis. This area, the northern extension of the "Korazim Saddle", is known for its extensive sequence of paleo-lake formations representing the very long depositional history of the northern Jordan Valley. As it was beyond the limits of my expertise to assign the exposure to a precise age within the Pliocene and Pleistocene deposits of the Upper Jordan Valley, we decided to examine the outcrop together. The two of us, joined by a small group of students who were working with me that summer (1985) at the Mousterian site of Quneitra, went to take a closer look at the edge of the fortress.

We could see that a series of eastward-tilted (70°) beds composed of sands, pebbles, fossil mammal bone fragments, and occasional boulders and cobbles form the glacis of the fortress. We then decided to excavate a protruding piece of flint from the cemented sediments of the outcrop, trying to minimize the surface damage in order to avert clandestine excavations in the future. After a few minutes we had exposed a flint handaxe and next to it a fossil gastropod (*Viviparus apameae*), the type fossil of the Middle Pleistocene Hula deposits; both finds are yardsticks for assigning the exposure to the Middle Pleistocene and to the Acheulian Technocomplex. Further survey of these beds resulted in the discovery of three superbly fashioned flint cleavers, which were left in place to be collected later, but were never found again.

The results of this short field trip testified to the extension of the Benot Ya'akov Formation south of the Benot Ya'aqov Bridges. They also inspired our ongoing study of this formation and its extremely rich assemblages, which has continued for over two decades.

In due course, and following a survey of the Jordan River banks south of the bridges, we initiated the excavations that are reported in this volume. They were located on the left bank of the Jordan River on a piece of land devoid of large cobbles and boulders, bordered on the west by the river and on the east by a dirt road with an extensive minefield beyond it. Although we clearly enjoy better geopolitical conditions than those faced by the previous researchers of the GBY site, we were still challenged by some similar difficulties.

The first field season was an extraordinary educational experience for us all, from the director to the most junior participant. It was then that it dawned on us, when we saw and understood the full extent of the tectonically deformed strata, that the site was extremely rich in artifacts and paleontological remains, and that its waterlogging had preserved wood, bark, fruits, and seeds. Concurrent with this understanding was my conviction that excavating this site demanded a commitment that went far beyond that required for a normative site. This was a goal that necessitated tremendous amounts of time, energy, dedication, strength, and perseverance, a certain amount of obstinacy, and masses of optimism.

During the early years of excavation at 'Ubeidiya, while standing in front of an excavated pit in Layer I-15, we jokingly quoted Prof. Moshe Stekelis, who referred to the pit as a "window into the Pleistocene". My feeling at GBY was that here we did indeed have an opportunity to see through a very small window into the Pleistocene. It was a window not only into the Pleistocene of the northern segment of the Dead Sea Rift, but also into an archive of information about the paleoenvironment, paleoecology, and hominin behavior in the eastern Mediterranean, the Levantine Corridor. The full understanding of this archive demanded a commitment that prevented me for many years from getting involved in studies, excavations, and analyses of any other site. It simply was, and still is, too large a job to be a part-time one.

There was no routine, in the usual meaning of the word, during the excavations, and I moved between excitement and despair. At other times there were absurd moments. An illustration of these was the fact that we needed to pump out the water that accumulated in the trenches and the excavation areas; on the other hand, the exposed organic materials, primarily the wood and bark fragments, had to be kept wet at all times to protect them from deterioration – in short, to the outside observer, an example of a *perpetuum mobile*. Another example is the extremely poor state of preservation of some of the most exquisite basalt bifacial tools (handaxes and cleavers). Many of these artifacts looked extremely impressive when they were taken out of the ground but, as we learned very quickly from bitter experience, they simply crumbled into mud (clay) if they were not immersed immediately in water. Despite all the care and conservation procedures that were applied to some of these items, some continued to disintegrate in the laboratory, ruling out their detailed analysis.

Tilted strata were familiar to me from my previous experience in the excavation of 'Ubeidiya. One aspect, and a very frustrating one, is the exposed surface of the excavated horizons, which usually resembles a strip of minimal height. The same frustration, but perhaps even more overwhelming, derived from the smaller extent of the excavation at GBY. We were able to expose an elephant skull and there were several lines of evidence indicating that it was not greatly disturbed taphonomically, and yet we could not continue the exposure of that particular level (Layer II-6 Level 1) because of the steep dip of the layer.

From the very first field season I fully grasped the enormous potential of the site. Step by step with the exposure of the archaeological horizons, the importance of the discoveries dawned on me. I still remember the tremendous excitement that I shared with Idit Saragusti when we looked at the exposed artifact horizon of Layer II-6 Level 4 – a tilted pavement of basalt bifacial tools. Both handaxes and cleavers were fully exposed on the limited surface of the dig, a thin layer one or two artifacts thick, minimally associated with other classes of finds and visibly unaffected by postdepositional processes. "Just like in Africa", we murmured in astonishment.

Such a complex project as that of GBY could not be run single-handed. It was and still is a team effort. During the last two decades I have been fortunate to be able to train a superb crew – all students of the Institute of Archaeology, The Hebrew University of Jerusalem. It was indeed a privilege to be with them, and eventually to learn from them. They are studious, serious, and joyful, and their uncompromising and long-lasting commitment was an unsurpassed experience for me. Each of the senior students (field and laboratory directors) has been working on the project for more than ten years, and all of them participate in the research effort to try to make sense of it all. There is no doubt in my mind that the GBY project has continued and has been so successful because of the support and enthusiasm of the team.

In the archives of the British Mandate, the Israel Department of Antiquities, and the Israel Antiquities Authority there are many letters documenting the importance that the presidents of the

Hebrew University and other high officials have attached to the study of the site. Alas, despite my attempts to raise awareness of the site as a national heritage locality of great importance, I have been unsuccessful. Furthermore, despite all my warnings, in December 1999, only two years after the end of the last excavation season, the site was deliberately ravaged and partially destroyed by drainage works.

One of the greatest responsibilities of running a project, well known and experienced by all my colleagues, is finding the resources to support the project. This task became harder and harder over the years and, despite my efforts, I had only partial success. For example, the expedition was never able to afford an advanced system of documentation. I feel that a project like GBY deserves the best of means with which to document such a unique field experience. Fortunately, the recent years of research have benefited from much better support.

The exceptional nature of the site (its waterlogged location right on the banks of the Jordan River) does not allow for its conservation in the usual sense. However, its scientific value and hence its importance go far beyond a signboard and some forgotten documents. It should be protected by declaring it a national archaeological heritage site and its natural and archaeological surroundings (including all the archaeological and historical monuments) should be designated a national reserve.

The field and laboratory work enabled me to expand my very limited background in many of the disciplines, and to learn about fields that I never thought would be an integral part of the project. There were outstanding moments of discovery when looking at newly assembled results, and there were other projects that could have been scientific gems and resulted in fascinating studies but were doomed by the human factor, a well-known stumbling block in multidisciplinary projects. And yet, above all this is the treasure of which we have so far exposed only very little. The site extends over 3.5 km on a north-south axis along the main fault line (a plate boundary). In a core drilled on the right bank immediately north of the northern bridge, penetrating to a depth of 80 m, we were able to retrieve a flint artifact from a depth of 54 m below the surface, demonstrating that the Benot Ya'akov Formation (whose base was not reached in this series of drillings) is far more extensive than what meets the eye on the surface, and that our understanding of it is still in its infancy. To illustrate this, I need only mention that the overall excavated surface of our expedition was limited to 246.84 m^2 and the excavated volume to 40.72 m^3. I give these figures simply to illustrate the enormity of the site and the unimaginable treasures buried under the Jordan River and its floodplain, and to attempt to convey the respect that I feel for the site and its research potential. It is also meant as a word of warning for those who may wish, at some time in the future, to conduct their own excavations at the site and to realize its potential. The task will be enormous!

This volume contains nine chapters: introduction (Chapter 1), the history of research (Chapter 2), the site of Gesher Benot Ya'aqov (Chapter 3), methodology (Chapter 4), a description of all layers and their detailed lithic content (Chapter 5), three chapters dedicated to each of the three raw materials used by the Acheulians – flint (Chapter 6), basalt (Chapter 7), and limestone (Chapter 8) – and finally a chapter dedicated to discussion and summary (Chapter 9).

The final chapter attempts to present some of the insights gained by the Gesher Benot Ya'aqov team through study of the material culture. It seems crucially important to me to point out that the contribution of the project goes far beyond simple description of the excavated stone artifacts. Rather, its most important contribution lies in our interpretations of the various behavioral patterns that emerge from the analyses, whether concerned with the cognitive abilities of the hominins, their other abilities, or arrays of hominin behavioral patterns that derive from our findings. We have acquired ample data on some of the main behavioral traits of the ancient hominins: their tool kits and the production of tools from the stage of acquisition of raw material through the transportation of the objects to the paleolandscape and finally their discard. Much has been gained by comparative study of the various archaeological horizons, and the emerging picture testifies to a multitude of different activities that were all carried out on the lake margin of the freshwater paleo-Lake Hula, in a lush environment of rich biomass. It is this variety of tasks – the monumental behavioral complexity of the Early–Middle

Pleistocene inhabitants of the Hula Valley region – that should be the focus of future work at the site and elsewhere.

This volume ends with our perspective on two general issues: hominin behavior at GBY and their cognitive abilities. If we have stretched the yarn too far, I take full responsibility.

N. G.-I.

Acknowledgements

This volume is the result of more than 25 years of continuous archaeological work in the field and in the laboratory. During these years many individuals and institutions supported and contributed to the project in numerous and various ways. There is no simple way for us to express our gratitude to them all. There has been much good will and interest in the project, and the following lines cannot fully express our appreciation. The realization of this volume, a very difficult task that could not have been accomplished without the help of many individuals, is the result of the prolonged efforts of a small permanent staff and of many students, among others, during an enormous number of working hours.

Foundation support

The fieldwork reported here was carried out with grants provided by the L.S.B. Leakey Foundation, the National Geographic Society (USA), the Binational Science Foundation USA-Israel, and the Irene Levi-Sala Care Archeological Foundation (several grants, all in support of conservation of organic materials). Other granting agencies provided the means for laboratory work and its different analyses: the Israel Science Foundation, which supported the project through several grants (grants nos. 152/92, 854/96; 782/99; 886/02; 1542/04; 300/06; 113/09; 858/09; 1838/11; 27/12), and the German–Israeli Foundation (GIF; grants nos. I-657-20.4/2000; I-896–208.4/2005). The Faculty of Humanities of the Hebrew University of Jerusalem, a special grant from the Israel Science Foundation, and the Ruth Amiran Fund of the Institute of Archaeology all provided support and were instrumental in the preparation and production of the current volume.

Special thanks are due to the members of Kibbutz Gadot, who lodged the expedition and made our stay in their home a most pleasant one. They provided outstanding hospitality and shared our enthusiasm; we felt we were at home and were supported, helped, and visited by many members of the kibbutz. Of particular help were Yossi Arbel and Haim Milrad, who constantly provided advice and physical help in many ways and for many years. Others, whose names are too numerous to be mentioned here, contributed to the present volume.

We thank the staff of the Israel Antiquities Authority, especially our colleagues of the Department of Prehistory, for their continuous support and help throughout many years. Their involvement covered diverse aspects and was always characterized by goodwill. We have profited immensely from many and varied discussions with our advanced students. They always showed a lively interest in our study and were helpful in many ways, contributing both their knowledge and their experience (particularly Arik Malinsky-Buller and Yossi Zaidner).

We express our thanks to Yuval Goren, who served as the treasurer of the Israel Prehistoric Society during the years of the project, and to the administrative staff of the Institute of Archaeology of the Hebrew University, Frida Lederman, Benny Sekay, and Smadar Pustilnik, who, usually unseen and working behind the scenes, greatly assisted us in completing the present study. Thanks are also due to the staffs of the Authority for Research and Development of the Hebrew University and of the Israel Science Foundation, who were always cooperative and helpful.

N.G.-I. wishes to thank the following institutions for continuous help and for providing a most pleasant working environment throughout the many years of research: the Israel Antiquities Authority (particularly Silvia Krapiwko, Ofer Marder, and Hamoudi Khalaily), the Institute for Advanced Studies of the Hebrew University, the Department of Biological Anthropology (and the Leverhulme Centre for Human Evolutionary Studies), and Clare Hall College at Cambridge University, UK.

The team

Idit Saragusti, Rivka Rabinovich, Gonen Sharon, and Nira Alperson-Afil formed the core of the GBY team. They all participated in fieldwork and took part in preparation, excavation, decision-making, and caring for the participants. Special thanks are due to Idit Saragusti, who was capable of turning some of my bizarre ideas into reality. Her command of the fieldwork and powers of observation, as well as her patience and calm, were a vital contribution to the success of the fieldwork. Rivka Rabinovich was in charge of the paleontological aspects in the field and the laboratory and was a most cheerful member of the team, always ready to take on any task. Gonen Sharon and Nira Alperson-Afil first participated as young undergraduate students and grew into prehistorians, each contributing many years and experience as well as understanding to the project. Many of their studies of the material culture of the site have been published in reputable journals, and their continuous involvement made this publication possible. We could not wish for better students, colleagues, and friends than these two.

Many other individuals took an active part in the excavations, whether colleagues belonging to other contributing disciplines or students, friends, and many who arrived from abroad, all full of enthusiasm and contributing to a most joyful atmosphere. We thank the following individuals for their involvement during the fieldwork: John Berg, Scott Brande, Mario Chech, Orna Cohen, Pepa Enamorado-Rivera, David Gordon, Noam Goren, Yuval Goren, Guy Hivroni, Erella Hovers, Simcha Lev-Yadun, Ofer Marder, Charles McNutt III, Dan Reshef-Mishmar, Hagar Taub, Eyal Vadai, Kenneth Verosub, Tzachi Vitelzon, Steve Weiner (for his contributions during the 1990 field season), Marcy Wiseman, and finally Dani Even-Zur (for his brilliant idea of utilizing anchors used in agriculture and hence providing unmatched precision of measurements throughout the different field seasons).

Lithic analysis

The lithic analysis of the GBY assemblages is an extremely difficult task, due to the various raw materials and their different weathering modes as well as the lack of standardization in some artifacts and tool classes. The entire analysis has been a prolonged process. Due to these constraints and many others, it was decided that only a small number of individuals would carry out the actual analyses (see Chapter 4). Those who conducted the analyses during the entire period are Idit Saragusti, Ofer Marder, Hagar Taub, Gonen Sharon, Josette Mimoun Sarel, Nira Alperson-Afil, Ron Lavi, Zinovi Matzkevich, Chagai Cohen, and Eyal Vadai.

We would also like to thank Bo Madsen, one of the most skilled and experienced modern knappers, particularly of bifacial tools, for his immense contribution to our understanding of technological aspects of Acheulian lithic technology. It was his experiments with the volcanic rocks from the vicinity of the site and his structured methodology that advanced our understanding of these aspects of the archaeological material.

The building and maintenance of the growing database became a central activity within the project over the years. Programs were changed and updating the files was a routine procedure. In the first years the team was joined by Ester Sivan, who was later replaced by Irena Laschiver. Special thanks are due to our team member Irena Laschiver, who worked with us in the laboratory for many years. She participated first in the sorting of sediments and then in that of the microartifacts. After this she was in charge of digitizing the full record of field maps and later on, for many years, of the GBY database. Her commitment, dedication, and permanent good humor made a huge contribution to the working environment.

The youngest person who made his mark in database management was Gadi Herzlinger, who added to it a programming aspect and as a result contributed much to the efficiency of the work, particularly in the last few years. His work was so valuable and his input so immense that he became a co-author of this volume.

The authors wish to thank the Gesher Benot Ya'aqov staff and students at the Institute of Archaeology of the Hebrew University for sorting of sediments and microartifacts over the years (1989–2007): Tchiya Alon, Doron Ben-Ami, Rachel Berman, Orly Boyum, Keren Braverman, Anat Cohen, Taufik Daa'dalla, Doron Dag, Hila Debono, Olga Dobovski, Michal Dovrat, Tomer Dror, Irit Eckshtein, Caroline Felus, Keren Finkelshtein, Talia Goldman, Emma Goldshmit, Natalya Gubenko, Debby Hochman, Ronit Israeli, Sharon Kaplan, Keren Katzir, Karni Klein, Christopher Konlan, Irena Laschiver, Liora Levy, Ravit Lin, Amit Marari, Maya Margalit, Efrat Mashiach, Noa Mashiach, Zinovi Matzkevitch, Sigalit Mazugi, Michal Meir, Uri Milshtein, Hadas Motro, Tali Nevo, Boaz Ofek, Dan Oriyan, Oded Ramati, Dan Reshef, Menachem Rogel, Keren Rosenberg, Karmit Rubin, Eli Shabo, Reut Shalib, Vivian Shalom, Ruthy Shimron, Iris Shmila, Esther Sivan, Guy Stiebel, Hagar Taub, Lilli-Shantal Tepber, Sigalit Tzaidi, Smadar Tzaidi, Eyal Vadai, and Yulya Volodosky.

Illustrations and photographs

Several individuals were responsible for drafting maps, cross-sections, and an array of other types of illustrations. Drawing in the field was a far from simple procedure, as it involved complicated geometric calculations and precision in laying out the boundaries of the drafted areas. Needless to say, apart from the searing summer temperatures, the tilted archaeological horizons and the dark colors of the muddy sediments presented many difficulties and necessitated inventions and innovations on the part of the drafting staff. These included Tzachi Vitelzon, Johanna Warshaw, Cathy D'Ouzil (Fig. 4.18), Raquel Rojas (Figs. 7.210a, b), Nira Alperson-Afil, Pepa Enamorado-Rivera, and Charles McNutt III. From time to time other talented participants joined in and produced excellent drawings.

Artifact illustrations were carried out by the following artists: Daliya Enoch, Yuliya Moskovich (Figs. 6.3, 6.14, 6.25, 6.30, 6.49, 6.50, 6.51, 6.53, 6.58, 6.76, 7.29, 7.30, 7.36-7.38), Leonid Zeiger (Figs. 6.5, 6.11, 6.46, 6.52, 6.54, 6.60, 6.65, 6.59, 6.69, 6.72, 6.76, 7.1, 7.5-7.15, 7.20, 7.22, 7.31, 7.34, 7.35, 7.143-7.145, 7.163, 7.165, 7.198, 7.199, 7.200, 8.1-8.14), Tzachi Vitelzon, Cathy Douzil, Paolo Giunti (Paolito) (Figs. 6.2, 6.17, 6.46, 6.48, 6.49, 6.55, 6.56, 6.65, 6.76, 7.2, 7.17, 7.23, 7.26, 7.27, 7.33, 7.39, 7.50, 7.56, 7.57, 7.86, 7.118, 7.121, 7.136, 7.141, 7.148, 7.157, 7.159, 7.160-7.162, 7.164, 7.166, 7.167, 7.170, 7.177, 7.178, 7.182, 7.185, 7.188, 7.190, 7.196, 7.197, 7.208). Yaron Hazan (Figs. 7.131, 7.133), Chagai Cohen, Guy Hivroni, and Noah Lichtinger contributed their expertise in the domain of computerized graphics.

For many years Gabi Laron, the former director of the photography laboratory at the Institute of Archaeology of the Hebrew University, invested many long working days in trying to achieve, and indeed achieving, most impressive and excellent results from very demanding and difficult objects. His quest for perfection is well documented in an array of superb photographs of very unimpressive artifacts that were extremely difficult to document. The following are his photographs: Figs. 3.3b & c, 3.19, 3.20, 3.21, 5.24, 5.36h & g, 6.78, 7.103, 7.104, 7.105, 7.106, 7.111, 7.112, 7.114, 7.115, 7.130, 7.134, 7.137, 7.138, 7.139, 7.140, 7.171, 7.172, 7.173, 7.186, 7.187, 7.192, 7.193, 7.194, 7.195, 7.213 (#7699), 8.4, 8.5, 8.6. The aerial photographs were taken by Arik Baltinester (Figs. 2.22 and 3.4).

In recent years great progress has been made in the field of digitized scanning (photography). All artifact illustrations that were not drawn by the above-mentioned artists were produced by 3D digitized scanning. Pioneers in this field, to whom we are extremely grateful, are Uzi Smilansky and particularly Leore Grosman of the Computerized Archaeology Laboratory at the Institute of Archaeology, who were attentive to our exacting demands and searched unknown territory in order to produce the digitized images of the artifacts. It is due to their collaboration that we are able to present many of these images in this report. In addition, Gadi Herzlinger produced many of the scans that appear in this volume, and a few were produced by Maya Oron.

Geologists

There is no doubt that we were able to decipher the complex geological structure of the site only thanks to Shmulik Belitzky's long years of involvement in the project. His involvement began with a survey on foot and by boat and included the decision as to where precisely the excavation should be located and, of prime importance, where and how to place the trenches (I–VI) that were the key to any stratigraphic, chronological, cultural, or other understanding of the site. Rarely does one find such an agreeable and talented colleague, with whom it is a pleasure to work. Shmulik's modesty and vast knowledge and experience should be a model for other geoscientists and scholars of other disciplines. We are greatly indebted to him and were privileged to work hand in hand with him.

We thank Craig Feibel for taking part in the last three field seasons and for sharing with us his understanding of the sedimentological aspects of the site. His involvement was crucial for gaining insight into the composition of the sediments, their origin, and their cyclic nature. His (so far) preliminary interpretations of the paleoclimate and the chronological location of the exposed sediments in the study area enabled us, both prehistorians and others, to provide better and more precise interpretations of the assemblages recovered during the excavation, and improved our understanding of hominin behavior in the Upper Jordan Valley.

Those who have passed away

Three colleagues and friends who were actively involved in our project passed away during the years of excavation and analysis. The late Shaul Sopher contributed his entomological knowledge and enabled us to enjoy the fieldwork without being bothered by insects (particularly wasps). The late Giora Parnos was an enthusiastic student in the first field seasons and became a much-appreciated colleague in the field of archaeology. His untimely death shocked us all, teachers and friends, and his loss is a great tragedy. The late veterinarian Dr. Eli Lotan, to whom we dedicated GBY Volume III, came for one field season and stayed for the entire project. He was hardworking and always ready to volunteer, and we were all jealous of his energies, strength, and manifold contributions to the team.

Others

The documentation of the large amount of activities causing destruction to the GBY site deserved a particular place in this volume. Data collection for parts of Appendix 1 was carried out by Tomer Dror, who familiarized himself with the different archives and zealously collected essential information. We thank Tamar Nachmias-Lotan for integrating Tomer's catalogue and archive into Appendix 1, covering the history of the Jordan River at the location of the site. Ofir Kelty helped in proofreading the volume.

We thank the staff of the Biological Database of the Israel Nature and Parks Authority for providing us with important data on the fauna of the Hula Valley.

Sue Gorodetsky edited this volume and helped immensely in making it clear and coherent in her meticulous and efficient way, for which we are most grateful. Noah Lichtinger made a great contribution to many of the computerized graphics and produced the book creatively and in her usual skillfull and professional manner.

Many thanks are due to Eric Delson and Eric Sargis, editors of the Vertebrate Paleobiology and Paleoanthropology book series, who generously accepted our study for publication and improved it with expertise and dedication. We thank the three anonymous reviewers for their comments and the editorial staff at Springer, particularly Sherestha Saini and Shobana Ramamurthy for their constant interest and efficiency.

Contents

1	**Introduction**		1
1.1	The Hula Valley		3
	1.1.1	Geography	3
	1.1.2	Lake Hula	3
	1.1.3	Climate	3
	1.1.4	Vegetation	5
	1.1.5	Fauna	5
	1.1.6	Geology	5
	1.1.7	The Hula Valley in the Past	6
2	**History of Research**		7
2.1	The History of the Benot Ya'aqov Bridges and Their Area		7
2.2	The History of Archaeological Research		8
	2.2.1	1934–1937	9
	2.2.2	1951–1968	12
	2.2.3	1981–1989	15
	2.2.4	The Study Area	16
		2.2.4.1 The Bar	16
	2.2.5	The Excavations	18
	2.2.6	December 1999	18
	2.2.7	2005, 2009–	19
2.3	The Archaeological Analyses		19
3	**The Site of Gesher Benot Ya'aqov**		21
3.1	Geographical Location		21
	3.1.1	Geographical Landmarks	23
3.2	Geology		24
3.3	Stratigraphy		28
	3.3.1	North of the Benot Ya'aqov Bridges	29
	3.3.2	The Crusader Fortress	29
	3.3.3	The Rosh Pinnah River	29
	3.3.4	The Study Area	29
		3.3.4.1 The Bar	29
		3.3.4.2 The Excavations	32
3.4	Chronology and Dating		34
	3.4.1	History of the Age and Date of the BYF	34
	3.4.2	Radiometric Dating and Paleomagnetism of the BYF and Its Environs	36
3.5	Raw Materials		37
	3.5.1	Basalt	37
		3.5.1.1 Origin	37

		3.5.2	Flint and Limestone	38
			3.5.2.1 Origin	38
		3.5.3	Weathering of Raw Materials	39
			3.5.3.2 Patination	41

4 Methodology — 43

- 4.1 Field Methodology — 43
  - 4.1.1 Grid and Shading — 43
  - 4.1.2 The Study Area and Trenches — 44
  - 4.1.3 The Excavations — 45
    - 4.1.3.1 Datum and Elevations — 46
    - 4.1.3.2 Excavation Difficulties and Techniques — 47
  - 4.1.4 Maps and Drafting — 49
    - 4.1.4.1 Maps — 49
    - 4.1.4.2 Cross-Sections (Profiles) — 51
  - 4.1.5 Sieving — 51
  - 4.1.6 Conservation — 52
- 4.2 Laboratory Methodology — 53
  - 4.2.1 Sediment Sorting and Analysis — 53
  - 4.2.2 Attribute Lists and Lithic Analysis — 54
    - 4.2.2.1 Flakes and Flake Tools — 55
    - 4.2.2.2 Cores and Core Tools — 55
    - 4.2.2.3 Bifaces: Handaxes and Cleavers — 55
    - 4.2.2.4 Natural Nodules — 57
  - 4.2.3 Digitization of Maps and Cross-Sections — 57
- 4.3 This Volume — 57
  - 4.3.1 Illustrating Lithic Artifacts — 58

5 The Lithic Assemblages in Context — 61

- 5.1 The Bar — 62
- 5.2 Area C — 62
  - 5.2.1 Layer V-1 (Unconformity C) — 65
  - 5.2.2 Layers V-2, V-3, and V-4 — 65
  - 5.2.3 Layer V-5 — 65
  - 5.2.4 Layer V-6 — 69
  - 5.2.5 Jordan Bank (JB) — 69
  - 5.2.6 The Enclosure (Jordan River) — 73
- 5.3 Area A — 74
  - 5.3.1 Layer I-4 — 75
  - 5.3.2 Layer I-5 — 75
- 5.4 Area B — 78
  - 5.4.1 Layer II-1 (Unconformity B) — 81
  - 5.4.2 Layer II-2 — 82
  - 5.4.3 Layer II-2/3 — 83
  - 5.4.4 Layer II-3 — 84
  - 5.4.5 Layer II-3/4 — 85
  - 5.4.6 Layer II-4 — 85
  - 5.4.7 Layer II-5 — 86
  - 5.4.8 Layer II-5/6 — 87
  - 5.4.9 Layer II-6 — 87
  - 5.4.10 Layer II-6/7 — 103
  - 5.4.11 Layer II-7 (Layer IV-7) — 103
  - 5.4.12 Layers II-7/8 and II-8 — 108
  - 5.4.13 Layer II-12 — 113

| | | 5.4.14 | Trench II. | 114 |
| | | 5.4.15 | Layer IV-25 | 115 |

6 The Flint Component … 119

6.1 Inventory and Taphonomy of the Flint Component … 119
6.1.1 Inventory of the Flint Component … 119
6.1.2 Taphonomy of the Flint Component … 119

6.2 Typology of Flint Flake Tools and Core Tools … 121
6.2.1 Typology of Flake Tools … 121
6.2.1.1 Discussion: Typology and Characteristics of Flake Tools … 162
6.2.2 Typology of Core Tools … 181

6.3 Technology of Flint Flakes … 185
6.3.1 Unretouched Flint Flakes … 185
6.3.1.1 *Éclats de Taille de Biface* … 190
6.3.1.2 Dorsally Plain (Kombewa) Flakes … 195
6.3.1.3 Ventral Removals … 200
6.3.2 Core Management Pieces … 201

6.4 Classification and Technology of Cores and Core Tools … 201
6.4.1 Cores … 203
6.4.2 Diverse Waste Artifacts … 219

6.5 Discussion and Summary of the Flint Component … 223
6.5.1 The Flint Reduction Sequences … 223
6.5.1.1 Percussor Techniques … 225
6.5.1.2 Cores … 227
6.5.1.3 The Levallois Method … 230
6.5.1.4 Tools … 232

7 The Basalt Component … 237

7.1 Inventory and Taphonomy of the Basalt Component … 237
7.1.1 Inventory of the Basalt Component … 237
7.1.2 Taphonomy of the Basalt Component … 237

7.2 Typology of Basalt Flake Tools … 241

7.3 Technology of Basalt Flakes … 250
7.3.1 Unretouched Flakes … 251
7.3.1.1 Large Flakes … 251
7.3.1.2 Small Flakes … 255
7.3.2 Unretouched Special Flakes … 259
7.3.2.1 Siret Flakes (Brut) … 260
7.3.2.2 Kombewa and Dorsally Plain Flakes (DPF) … 263
7.3.2.3 Biface Sharpening Flakes … 269
7.3.2.4 *Éclats de Taille de Biface* … 269
7.3.2.5 Flakes with Ventral Removals … 272
7.3.2.6 Core Management Pieces … 274
7.3.3 Giant Core Special Waste Products … 274
7.3.3.1 Shoulder/Corner Flakes … 276
7.3.3.2 Slice Flakes … 281
7.3.3.3 Wedge-Like Flakes … 284
7.3.3.4 Giant Flakes … 286
7.3.3.5 Waste of Giant Cores … 286
7.3.3.6 Flakes of Biface Morphology … 287

7.4 Typology and Technology of Basalt Cores and Core Tools … 298
7.4.1 Percussive Tools: Percussors, Pitted Stones, and Thin Anvils … 298
7.4.1.1 Percussors … 298

	7.4.2	Pitted Stones	306
	7.4.3	Thin Anvils	313
	7.4.4	Modified Artifacts	319
	7.4.5	Core Tools	320
		7.4.5.1 Chopping Tools	320
		7.4.5.2 Other Basalt Core Tools	320
	7.4.6	Angular Fragments	321
	7.4.7	Basalt Cores	323
		7.4.7.1 Cores	323
		7.4.7.2 Giant Cores	323
7.5	Bifacial Tools		331
	7.5.1	Methodology of Biface Analyses	332
	7.5.2	Raw Materials and Taphonomic Characteristics	332
		7.5.2.1 Raw Materials	332
		7.5.2.2 Preservation and Patination	333
		7.5.2.3 Breakage	335
	7.5.3	Technological Characteristics	337
		7.5.3.1 Size	337
		7.5.3.2 Blank Type	338
		7.5.3.3 Direction of Blow	341
		7.5.3.4 Striking Platforms	342
	7.5.4	Modification and Design	346
		7.5.4.1 Type of Retouch	348
		7.5.4.2 Quantity of Retouch	351
		7.5.4.3 Location of Retouch	352
	7.5.5	Morphology of Bifaces	353
		7.5.5.1 Morphology of Basalt Handaxes	355
		7.5.5.2 Morphology of Flint Handaxes	360
		7.5.5.3 Morphology of Basalt Cleavers	362
	7.5.6	Typology of Bifaces	365
		7.5.6.1 Typology of Basalt Handaxes	366
		7.5.6.2 Typology of Flint Handaxes	369
		7.5.6.3 Typology of Cleavers	370
		7.5.6.4 Predetermination of Cleavers	370
7.6	Discussion and Summary of the Basalt Component		371
	7.6.1	Reduction Sequences of Basalt Percussive Tools	373
	7.6.2	The Biface Reduction Sequence	374
		7.6.2.1 Reduction Strategies of Giant Cores	374
		7.6.2.2 Biface Design Strategies	376
		7.6.2.3 Biface Modification and Morphology	377

8 The Limestone Component — 379

8.1	Inventory and Taphonomy of the Limestone Component		379
	8.1.1	Inventory of the Limestone Component	379
	8.1.2	Taphonomy of the Limestone Component	380
8.2	Typology and Technology of the Limestone Flakes and Flake Tools		381
8.3	Typology and Technology of the Limestone Cores and Core Tools		385
	8.3.1	Percussors	385
	8.3.2	Modified Artifacts	391
	8.3.3	Chopping Tools	392
	8.3.4	Pitted Stones	392
	8.3.5	Cores and Other Core Tools	393
8.4	Discussion and Summery of the Limestone Component		394

9	**Discussion and Conclusions**		**397**
9.1	The GBY Lithic Assemblages		398
	9.1.1	The Different Raw Materials and Their Reduction Sequences	399
		9.1.1.1 Flint	399
		9.1.1.2 Basalt	401
		9.1.1.3 Limestone	402
		9.1.1.4 A Comparative View	402
	9.1.2	Typological and Technological Characteristics of the GBY Lithic Assemblages	403
		9.1.2.1 Typological Characteristics	403
		9.1.2.2 Percussive Techniques	405
		9.1.2.3 Core Reduction Methods	406
		9.1.2.4 Removal of Striking Platform	407
	9.1.3	Cultural Variability and Conservatism	408
		9.1.3.1 Variability	408
		9.1.3.2 Conservatism	412
9.2	Hominin Behavior at GBY		413
	9.2.1	Mobility	413
		9.2.1.1 Lithics	413
		9.2.1.2 Plant and Animal Resources	415
	9.2.2	Social Aspects	416
		9.2.2.1 Division of Labor	416
		9.2.2.2 Group Size	417
	9.2.3	Home Base – The Preferred Locale	420
		9.2.3.1 GBY as a Preferred Locale	420
		9.2.3.2 The Diversified Nature of Home	420
	9.2.4	Cognition	421
9.3	Final Words		424

References 427
Appendix 1: The Hula Reclamation Project and Its Implications for the Gesher Benot Ya'aqov Site 435
Appendix 2: Typological List 443
Appendix 3: List of Attributes for Flakes and Flake Tools 445
Appendix 4: List of Attributes for Cores and Core Tools 447
Appendix 5: List of Attributes for Handaxes 449
Appendix 6: List of Attributes for Cleavers 451
Appendix 7: A Comparison of Length of Flint Flake Tools 453

Index 455

List of Tables

Table 4.1	Lithic categories	58
Table 4.2	Typo-technological breakdown of the lithic inventory and the associated typologies and lithic categories.	58
Table 5.1	Excavation of areas and trenches by field season	61
Table 5.2	Data on excavated surface (m^2) and volume (m^3) of layers exposed in Area C	65
Table 5.3	Counts and frequencies (%) of the lithic assemblage of Layer V-1	65
Table 5.4	Counts and frequencies (%) of the typological composition of Layer V-1	65
Table 5.5	Counts and frequencies (%) of the lithic assemblages of Layers V-2–V-4/5	66
Table 5.6	Counts and frequencies (%) of the typological composition of Layers V-2–V-4/5	66
Table 5.7	Counts and frequencies (%) of the lithic assemblage of Layer V-5	67
Table 5.8	Counts and frequencies (%) of the typological composition of Layer V-5	67
Table 5.9	Counts and frequencies (%) of the lithic assemblage of Layer V-6	69
Table 5.10	Counts and frequencies (%) of the typological composition of Layer V-6	69
Table 5.11	Counts and frequencies (%) of the lithic assemblage of JB	73
Table 5.12	Counts and frequencies (%) of the typological composition of JB	73
Table 5.13	Data on excavated surface (m^2) and volume (m^3) of layers exposed in Area A	74
Table 5.14	Counts and frequencies (%) of the lithic assemblage of Layer I-4	77
Table 5.15	Counts of the typological composition of Layer I-4	77
Table 5.16	Counts and frequencies (%) of the lithic assemblage of Layer I-5	78
Table 5.17	Counts of the typological composition of Layer I-5	78
Table 5.18	Excavation of layers, levels, and trenches of Area B by field season	80
Table 5.19	Data on excavated surface (m^2) and volume (m^3) of archaeological horizons exposed in Area B	81
Table 5.20	Counts and frequencies (%) of the lithic assemblage of Layer II-1	81
Table 5.21	Counts and frequencies (%) of the typological composition of Layer II-1	82
Table 5.22	Counts and frequencies (%) of the lithic assemblage of Layer II-2	83
Table 5.23	Counts and frequencies (%) of the typological composition of Layer II-2	83
Table 5.24	Counts and frequencies (%) of the lithic assemblage of Layer II-2/3	84
Table 5.25	Counts and frequencies (%) of the typological composition of Layer II-2/3	84
Table 5.26	Counts and frequencies (%) of the lithic assemblage of Layer II-3	85
Table 5.27	Counts and frequencies (%) of the typological composition of Layer II-3	85
Table 5.28	Counts and frequencies (%) of the lithic assemblage of Layer II-4	86
Table 5.29	Counts and frequencies (%) of the typological composition of Layer II-4	86
Table 5.30	Counts and frequencies (%) of the lithic assemblage of Layer II-5	86
Table 5.31	Counts and frequencies (%) of the typological composition of Layer II-5	86
Table 5.32	Counts and frequencies (%) of the lithic assemblage of Layer II-5/6	87

Table 5.33	Counts and frequencies (%) of the typological composition of Layer II-5/6	87
Table 5.34	Counts and frequencies (%) of the lithic assemblage of Layer II-6 Level 1	91
Table 5.35	Counts and frequencies (%) of the typological composition of Layer II-6 Level 1	91
Table 5.36	Counts and frequencies (%) of the lithic assemblage of Layer VI-14	95
Table 5.37	Counts and frequencies (%) of the typological composition of Layer VI-14	95
Table 5.38	Counts and frequencies (%) of the lithic assemblage of Layer II-6 Level 2	96
Table 5.39	Counts and frequencies (%) of the typological composition of Layer II-6 Level 2	96
Table 5.40	Counts and frequencies (%) of the lithic assemblage of Layer II-6 Level 3	98
Table 5.41	Counts and frequencies (%) of the typological composition of Layer II-6 Level 3	98
Table 5.42	Counts and frequencies (%) of the lithic assemblage of Layer II-6 Level 4	100
Table 5.43	Counts and frequencies (%) of the typological composition of Layer II-6 Level 4	102
Table 5.44	Counts and frequencies (%) of the lithic assemblage of Layer II-6 Level 4b	103
Table 5.45	Counts and frequencies (%) of the typological composition of Layer II-6 Level 4b	103
Table 5.46	Counts and frequencies (%) of the lithic assemblage of Layer II-6 Level 5	106
Table 5.47	Counts and frequencies (%) of the typological composition of Layer II-6 Level 5	106
Table 5.48	Counts and frequencies (%) of the lithic assemblage of Layer II-6 Level 6	108
Table 5.49	Counts and frequencies (%) of the typological composition of Layer II-6 Level 6	108
Table 5.50	Counts and frequencies (%) of the lithic assemblage of Layer II-6 Level 7	111
Table 5.51	Counts and frequencies (%) of the typological composition of Layer II-6 Level 7	111
Table 5.52	Counts and frequencies (%) of the lithic assemblage of Layer II-6/7	113
Table 5.53	Counts and frequencies (%) of the typological composition of Layer II-6/7	113
Table 5.54	Counts and frequencies (%) of the lithic assemblage of Layer II-7	113
Table 5.55	Counts and frequencies (%) of the typological composition of Layer II-7	113
Table 5.56	Counts and frequencies (%) of the lithic assemblage of Trench II	114
Table 5.57	Counts and frequencies (%) of the typological composition of Trench II	114
Table 5.58	Counts and frequencies (%) of the lithic assemblage of Trench II green	115
Table 5.59	Counts and frequencies (%) of the typological composition of Trench II green	115
Table 5.60	Counts and frequencies (%) of the typological composition of Layer IV-25	117
Table 5.61	Descriptive statistics of flint artifacts of Layer IV-25 (mm)	117
Table 6.1	Counts of flint artifacts and microartifacts throughout the GBY stratigraphic sequence	120
Table 6.2	Frequencies (%) of state of preservation of flint CCT and FFT	121
Table 6.3	Frequencies (%) of breakage and patination of flint CCT and FFT	123
Table 6.4	Frequencies (%) of breakage patterns of flint FFT	124

List of Tables

Table 6.5	Typological frequencies (%) of flint flake tools by layer and level	124
Table 6.6	Counts and frequencies (%) of characteristics of flint notches	127
Table 6.7	Descriptive statistics of flint notches (size in mm; angle in degrees)	128
Table 6.8	Counts and frequencies (%) of retouch characteristics of flint notches	129
Table 6.9	Counts of multiple tools on flint notches	129
Table 6.10	Counts and frequencies (%) of characteristics of flint denticulates	131
Table 6.11	Counts and frequencies (%) of retouch characteristics of flint denticulates	132
Table 6.12	Descriptive statistics of flint denticulates (size in mm; angle in degrees)	133
Table 6.13	Counts and frequencies (%) of characteristics of flint borers	137
Table 6.14	Counts and frequencies (%) of retouch characteristics of flint borers	138
Table 6.15	Descriptive statistics of flint borers (size in mm; angle in degrees)	139
Table 6.16	Counts of multiple tools on flint borers	139
Table 6.17	Counts and frequencies (%) of characteristics of flint side-scrapers	142
Table 6.18	Descriptive statistics of flint side-scrapers (size in mm; angle in degrees)	143
Table 6.19	Counts and typological frequencies (%) of flint side-scrapers	144
Table 6.20	Counts and frequencies (%) of retouch characteristics of flint side-scrapers	149
Table 6.21	Counts and frequencies (%) of multiple (double) artifacts on flint side-scrapers	151
Table 6.22	List of multiple (triple) artifacts on flint sidescrapers (3 typological classifications)	154
Table 6.23	Counts and frequencies (%) of characteristics of flint end-scrapers	156
Table 6.24	Descriptive statistics of flint end-scrapers (size in mm; angle in degrees)	157
Table 6.25	Counts and frequencies (%) of retouch characteristics of flint end-scrapers	157
Table 6.26	Counts of multiple tools on flint end-scrapers	159
Table 6.27	Counts and frequencies (%) of characteristics of flint truncations	160
Table 6.28	Counts and frequencies (%) of retouch characteristics of flint truncations	161
Table 6.29	Counts and frequencies (%) of characteristics of retouched flint flakes	169
Table 6.30	Counts and frequencies (%) of retouch characteristics of retouched flint flakes	170
Table 6.31	Descriptive statistics of retouched flint flakes (size in mm; angle in degrees)	172
Table 6.32	Counts and frequencies (%) of characteristics of different flint flake tool categories and of unretouched flint flakes	173
Table 6.33	Counts and frequencies (%) of technological characteristics of different flint flake tool categories and of unretouched flint flakes	174
Table 6.34	Descriptive statistics of different flint flake tool categories and of unretouched flint flakes (complete artifacts only; size in mm)	179
Table 6.35	Counts and frequencies (%) of retouch characteristics of different flint flake tool categories	179
Table 6.36	Counts and typological frequencies (%) of flint core tools	182
Table 6.37	Counts and frequencies (%) of characteristics of flint chopping tools	183
Table 6.38	Descriptive statistics of flint chopping tools (size in mm)	184
Table 6.39	Counts and frequencies (%) of characteristics of flint core tools	184
Table 6.40	Descriptive statistics of flint core tools (size in mm)	185
Table 6.41	Counts and frequencies (%) of unretouched flint flakes by blank type	187
Table 6.42	Descriptive statistics of unretouched flint flakes (size in mm; angle in degrees)	188
Table 6.43	Counts and frequencies (%) of characteristics of unretouched flint flakes	191
Table 6.44	Counts and frequencies (%) of dorsal scar pattern and descriptive statistics of number of dorsal scars on unretouched flint flakes	192

Table 6.45	Counts and frequencies (%) of striking platform types of unretouched flint flakes.	193
Table 6.46	Counts of technological features associated with percussor type on flint FFT	193
Table 6.47	Descriptive statistics of flint *éclats de taille de biface* and unretouched flint flakes of Layers V-5–6 and Layer II-6 (Levels 1–7) (size in mm; angle in degrees)	199
Table 6.48	Counts and frequencies (%) of dorsal scar pattern of flint *éclats de taille de biface* and unretouched flint flakes of Layers V-5–6 and Layer II-6 (Levels 1–7)	200
Table 6.49	Counts and frequencies (%) of striking platform types of flint *éclats de taille de biface* and unretouched flint flakes of Layers V-5–6 and Layer II-6 (Levels 1–7)	200
Table 6.50	Counts and frequencies (%) of technological observations on flint *éclats de taille de biface* and unretouched flint flakes of Layers V-5–6 and Layer II-6 (Levels 1–7)	200
Table 6.51	Counts and frequencies (%) of lipped striking platform on flint *éclats de taille de biface* and unretouched flint flakes of Layers V-5–6 and Layer II-6 (Levels 1–7)	201
Table 6.52	Counts and frequencies (%) of dorsally plain flakes on flint FFT	203
Table 6.53	Counts and frequencies (%) of ventral removals on flint FFT	203
Table 6.54	Counts of core management pieces on flint FFT	205
Table 6.55	Counts and typological frequencies (%) of flint CCT	206
Table 6.56	Counts of Levallois flint cores throughout the GBY sequence	207
Table 6.57	Counts and frequencies (%) of characteristics of Levallois flint cores	212
Table 6.58	Descriptive statistics of Levallois flint cores (size in mm; angle in degrees)	214
Table 6.59	Counts and frequencies (%) of identifiable scar pattern of Levallois flint cores	214
Table 6.60	Counts and frequencies (%) of characteristics of flint cores on flakes of Layer II-6	216
Table 6.61	Descriptive statistics of flint cores on flakes of Layer II-6 (size in mm; angle in degrees)	217
Table 6.62	Counts and frequencies (%) of characteristics of globular and discoidal flint cores	217
Table 6.63	Descriptive statistics of globular and discoidal flint cores (size in mm; angle in degrees)	218
Table 6.64	Counts and frequencies (%) of technological characteristics of globular and discoidal flint cores	218
Table 6.65	Counts and frequencies (%) of characteristics of varia and formless flint cores of Layer II-6	225
Table 6.66	Descriptive statistics of varia and formless flint cores of Layer II-6 (size in mm; angle in degrees)	226
Table 6.67	Counts and frequencies (%) of technological characteristics of varia and formless flint cores of Layer II-6	227
Table 6.68	Counts and frequencies (%) of characteristics of flint angular fragments of Layer II-6	227
Table 6.69	Descriptive statistics of flint angular fragments of Layer II-6 (size in mm)	227
Table 6.70	Counts and frequencies (%) of characteristics of flint modified artifacts of Layer II-6	228
Table 6.71	Descriptive statistics of flint modified artifacts of Layer II-6 (size in mm)	228
Table 6.72	Counts and frequencies (%) of characteristics of flint core waste and amorphous waste	228

Table 6.73	Descriptive statistics of flint core waste and amorphous waste (size in mm)	229
Table 6.74	Descriptive statistics of flint percussors (size in mm)	229
Table 6.75	Counts and frequencies (%) of retouch/signs of use on flint CCT	229
Table 6.76	Counts and frequencies (%) of different typological groups and of artifacts with removed striking platforms (calculated out of the entire FFT component)	233
Table 6.77	Counts and frequencies (%) of scar pattern by striking platform and mean number of dorsal scars of flint FFT	234
Table 6.78	Counts and typological frequencies (%) of flint FFT by striking platform (removed/not removed)	235
Table 6.79	Counts and frequencies (%) of scar pattern of flint FFT by striking platform (removed/not removed)	235
Table 6.80	Counts and frequencies (%) of ventral removals on flint FFT by striking platform (removed/not removed)	235
Table 6.81	Counts and frequencies (%) of breakage pattern of flint FFT by striking platform (removed/not removed)	236
Table 7.1	Counts of basalt artifacts and microartifacts throughout the GBY cultural sequence	238
Table 7.2	Frequencies (%) of state of preservation of basalt artifacts	239
Table 7.3	Frequencies (%) of breakage and patination of basalt artifacts	240
Table 7.4	Frequencies (%) of breakage patterns of basalt FFT	241
Table 7.5	Typological frequencies (%) of basalt flake tools of Layer II-6	241
Table 7.6	Counts and frequencies (%) of taphonomic and technological characteristics of basalt flake tools of Layer II-6	242
Table 7.7	Counts and frequencies (%) of retouch characteristics of basalt flake tools of Layer II-6	243
Table 7.8	Descriptive statistics of different basalt flake tool categories of Layer II-6	244
Table 7.9	Counts of multiple tools on basalt flake tools	250
Table 7.10	Descriptive statistics of large basalt flakes (size in mm; angle in degrees) of Layer II-6	252
Table 7.11	Counts and frequencies (%) of characteristics of large basalt flakes of Layer II-6	253
Table 7.12	Counts and frequencies (%) of technological characteristics of large basalt flakes of Layer II-6	254
Table 7.13	Counts and frequencies (%) of special technological features of large and small basalt flakes	255
Table 7.14	Descriptive statistics of small basalt flakes (size in mm; angle in degrees)	256
Table 7.15	Counts and frequencies (%) of characteristics of small basalt flakes	257
Table 7.16	Counts and frequencies (%) of technological characteristics of small basalt flakes	258
Table 7.17	Counts and frequencies of unretouched special basalt flakes	259
Table 7.18	Counts and frequencies (%) of basalt Siret flakes on tools and blanks	260
Table 7.19	Descriptive statistics of basalt Siret flakes of Layer II-6 (size in mm; angle in degrees)	261
Table 7.20	Counts and frequencies (%) of taphonomic and technological characteristics of basalt Siret flakes of Layer II-6	262
Table 7.21	Counts and frequencies (%) of technological characteristics of basalt Siret flakes of Layer II-6	263
Table 7.22	Counts and frequencies (%) of basalt dorsally plain flake categories of Layer II-6	264
Table 7.23	Descriptive statistics of basalt dorsally plain flakes of Layer II-6 (size in mm; angle in degrees)	265

Table 7.24	Counts and frequencies (%) of taphonomic and technological characteristics of basalt dorsally plain flakes of Layer II-6	266
Table 7.25	Typological counts and frequencies (%) of basalt dorsally plain flakes on tools and blanks of Layer II-6	267
Table 7.26	Typological counts and frequencies (%) of Kombewa flakes	267
Table 7.27	Counts and frequencies (%) of special large basalt flake categories on Kombewa flakes	267
Table 7.28	Counts and frequencies (%) of different categories of basalt Kombewa flakes of Layer II-6	268
Table 7.29	Descriptive statistics of basalt Kombewa flakes of Layer II-6 (size in mm; angle in degrees)	270
Table 7.30	Counts and frequencies (%) of taphonomic and technological characteristics of basalt Kombewa flakes of Layer II-6	271
Table 7.31	Typological counts and frequencies (%) of biface sharpening flakes	272
Table 7.32	Descriptive statistics of basalt biface sharpening flakes of Layer II-6 (size in mm; angle in degrees)	275
Table 7.33	Counts and frequencies (%) of taphonomic and technological characteristics of basalt biface sharpening flakes of Layer II-6	276
Table 7.34	Counts and frequencies (%) of technological characteristics of basalt biface trimming flakes of Layer II-6	277
Table 7.35	Descriptive statistics of basalt *éclats de taille de biface* of Area C and Layer II-6 (size in mm; angle in degrees)	279
Table 7.36	Counts and frequencies (%) of taphonomic and technological characteristics of basalt *éclats de taille de biface* of Area C and Layer II-6	280
Table 7.37	Typological counts and frequencies (%) of basalt flakes and tools with ventral removals	281
Table 7.38	Counts and frequencies (%) of types of striking platforms on basalt flakes with ventral removals	281
Table 7.39	Descriptive statistics of basalt core management pieces of Layer II-6 (size in mm; angle in degrees)	282
Table 7.40	Descriptive statistics of basalt shoulder/corner flakes of Layer II-6 (size in mm; angle in degrees)	283
Table 7.41	Counts and frequencies (%) of taphonomic and technological characteristics of shoulder/corner flakes of Layer II-6	284
Table 7.42	Counts and frequencies (%) of technological characteristics of basalt shoulder/corner flakes of Layer II-6	285
Table 7.43	Descriptive statistics of basalt wedge-like flakes of Layer II-6 (size in mm; angle in degrees)	288
Table 7.44	Counts and frequencies (%) of taphonomic and technological characteristics of wedge-like flakes of Layer II-6	289
Table 7.45	Counts and frequencies (%) of taphonomic and technological characteristics of wedge-like flakes of Layer II-6	290
Table 7.46	Descriptive statistics of basalt waste of giant cores of Layer II-6 (size in mm; angle in degrees)	291
Table 7.47	Counts and frequencies (%) of the typology of basalt waste of giant cores	293
Table 7.48	Counts and frequencies of taphonomic characteristics of basalt waste of giant cores of Layer II-6	294
Table 7.49	Counts and frequencies of technological characteristics of basalt waste of giant cores of Layer II-6	294
Table 7.50	Descriptive statistics of basalt large flakes of biface morphology (size in mm; angle in degrees)	297
Table 7.51	Counts and frequencies (%) of taphonomic and technological characteristics of basalt large flakes of biface morphology	297

Table 7.52	Counts and frequencies (%) of technological characteristics of basalt large flakes of biface morphology	298
Table 7.53	Typological counts of basalt CCT by layer.	299
Table 7.54	Counts and frequencies (%) of basalt percussors and split percussors	300
Table 7.55	Counts and frequencies (%) of breakage patterns and cortical coverage of basalt percussors	304
Table 7.56	Descriptive statistics of basalt percussors (size in mm; weight in g)	305
Table 7.57	Counts and frequencies (%) of taphonomic and technological characteristics of basalt split percussors	307
Table 7.58	Descriptive statistics of basalt split percussors of Layer II-6 (size in mm)	308
Table 7.59	Counts of unmodified cobbles (potential percussors) with maximum size greater than 65 mm and roundness value greater than or equal to 7	308
Table 7.60	Counts and frequencies (%) of basalt pitted stones on CCT and FFT	310
Table 7.61	Counts and frequencies of co-occurrence of basalt pitted stones and other typological categories	310
Table 7.62	Counts and frequencies (%) of breakage patterns and cortical coverage of basalt CCT pitted stones of Layer II-6	311
Table 7.63	Counts and frequencies (%) of breakage patterns and corticas coverage of basalt FFT pitted stones of Layer II-6	312
Table 7.64	Descriptive statistics of basalt CCT pitted stones (size in mm) of Layer II-6	313
Table 7.65	Descriptive statistics of basalt FFT pitted stones (size in mm; angle in degrees) of Layer II-6	314
Table 7.66	Counts and frequencies (%) of pit characteristics of basalt pitted stones	315
Table 7.67	Counts and frequencies (%) of spatial location of pits of basalt pitted stones	315
Table 7.68	Counts and frequencies (%) of shape of pits of basalt pitted stones	315
Table 7.69	Counts and frequencies of basalt thin anvils	316
Table 7.70	Descriptive statistics of basalt thin anvils (N=42) (size in mm; weight in g)	318
Table 7.71	Damage types and location on basalt thin anvils	318
Table 7.72	Location and intensity of pitting damage on basalt thin anvils	318
Table 7.73	Type and direction of flaking damage on basalt thin anvils	319
Table 7.74	Counts and frequencies (%) of basalt modified artifacts	319
Table 7.75	Counts and frequencies (%) of breakage patterns and cortical coverage characteristics of basalt modified artifacts of Layer II-6	319
Table 7.76	Descriptive statistics of basalt modified artifacts of Layer II-6 (size in mm)	320
Table 7.77	Counts and frequencies (%) of basalt angular fragments	321
Table 7.78	Counts and frequencies (%) of breakage patterns and cortical coverage of basalt angular fragments of Layer II-6	321
Table 7.79	Descriptive statistics of basalt CCT angular fragments of Layer II-6 (size in mm)	322
Table 7.80	Typological counts and frequencies of basalt cores	323
Table 7.81	Descriptive statistics of basalt cores (N=31)	324
Table 7.82	Counts of groups of basalt giant cores	326
Table 7.83	Descriptive statistics of groups of basalt giant cores (size in mm)	327
Table 7.84	Counts and frequencies (%) of handaxes and cleavers by raw material	333
Table 7.85	Counts and frequencies (%) of preservation of basalt handaxes and cleavers of Layer II-6	335
Table 7.86	Counts and frequencies (%) of breakage patterns of basalt handaxes of Layer II-6	336
Table 7.87	Counts and frequencies (%) of breakage patterns of basalt cleavers of Layer II-6	336

Table 7.88	Counts and frequencies (%) of breakage patterns of flint and limestone handaxes of Layer II-6	337
Table 7.89	Descriptive statistics of basalt handaxes of Layer II-6 (size in mm)	338
Table 7.90	Descriptive statistics of basalt cleavers of Layer II-6 (size in mm)	339
Table 7.91	Descriptive statistics of flint and limestone handaxes of Layer II-6 (size in mm)	340
Table 7.92	Counts and frequencies (%) of blank type of bifaces of Layer II-6 by raw material	340
Table 7.93	Counts and frequencies (%) of direction of blow of basalt handaxes of Layer II-6	342
Table 7.94	Counts and frequencies (%) of direction of blow of basalt cleavers of Layer II-6	342
Table 7.95	Counts and frequencies (%) of striking platform type on basalt handaxes of Layer II-6	347
Table 7.96	Counts and frequencies (%) of striking platform type on basalt cleavers of Layer II-6	347
Table 7.97	Counts and frequencies (%) of type of retouch on basalt handaxes of Layer II-6	349
Table 7.98	Counts and frequencies (%) of type of retouch on both faces of basalt handaxes of Layer II-6	351
Table 7.99	Counts and frequencies (%) of type of retouch on basalt cleavers of Layer II-6	352
Table 7.100	Counts and frequencies (%) of quantity of retouch on basalt handaxes of Layer II-6	353
Table 7.101	Counts and frequencies (%) of quantity of retouch on basalt cleavers of Layer II-6	353
Table 7.102	Descriptive statistics of number of scars on basalt bifaces of Layer II-6	355
Table 7.103	Counts and frequencies (%) of location of retouch on both faces of basalt handaxes of Layer II-6	356
Table 7.104	Counts and frequencies (%) of location of retouch on both faces of basalt cleavers of Layer II-6	356
Table 7.105	Bordes' morphological ratios and their calculations	357
Table 7.106	Descriptive statistics of the morphological ratios of basalt handaxes of Layer II-6	357
Table 7.107	Distribution of cross-section shapes of basalt handaxes of Layer II-6	359
Table 7.108	Distribution of complete flint handaxes by layer	362
Table 7.109	Morphological characteristics of complete flint handaxes	362
Table 7.110	Morphological characteristics of basalt cleavers of Layer II-6	363
Table 7.111	Morphological ratios and threshold values	367
Table 7.112	Bordes' shape zones and corresponding typology	367
Table 7.113	Flatness class and elongation distribution of basalt handaxes of Layer II-6	367
Table 7.114	Shape zone distribution of basalt handaxes of Layer II-6	368
Table 7.115	Typological distribution of basalt handaxes of Layer II-6	368
Table 7.116	Typological classification data of flint handaxes	369
Table 8.1	Counts of limestone artifacts and microartifacts throughout the GBY sequence	379
Table 8.2	Counts and frequencies (%) of state of preservation of limestone CCT and FFT	380
Table 8.3	Counts and frequencies (%) of breakage and patination of limestone CCT and FFT	381
Table 8.4	Counts and typological frequencies (%) of limestone FFT	381
Table 8.5	Counts and frequencies (%) of characteristics of limestone FFT	382
Table 8.6	Descriptive statistics of limestone FFT (size in mm; angle in degrees)	384

List of Tables

Table 8.7	Counts and frequencies (%) of technological characteristics and descriptive statistics of number of dorsal scars of limestone FFT	386
Table 8.8	Counts and typological frequencies (%) of limestone CCT	387
Table 8.9	Counts of multiple typologies of limestone CCT	387
Table 8.10	Counts and frequencies (%) of characteristics and descriptive statistics of number of dorsal scars of limestone CCT from the entire cultural sequence of GBY by typological category	387
Table 8.11	Descriptive statistics of limestone CCT from the entire cultural sequence of GBY by typological category	388
Table 8.12	Counts and frequencies (%) of characteristics and descriptive statistics of number of dorsal scars of limestone percussors from the entire stratigraphic sequence of GBY by type	390
Table 8.13	Descriptive statistics of limestone percussors from the entire stratigraphic sequence of GBY by type	390
Table 9.1	Counts, frequencies (%)#, and and ratios of cores to different artifact categories in Layer II-6 by raw material	403
Table 9.2	Counts and scar counts of basalt bifaces and FFT and estimated counts of missing flakes throughout the cultural sequence	414
Table 9.3	Suggested generalized division of labor at GBY	417
Table 9.4	Activities recognized at GBY and their suggested cognitive traits	423

List of Figures

Fig. 1.1	Location map of the upper Jordan Valley (the Hula Valley and vicinity) . . .	4
Fig. 2.1	The Mamluk stone bridge with three arches: a) general view ca. 1895 (Central Zionist Archives), b) close-up view ca. 1912 (Central Zionist Archives) . . .	8
Fig. 2.2	The Mamluk bridge with four arches: a) general view showing the Mamluk caravanserai and the Lower Customs House ca. 1895 (courtesy École Biblique), b) close-up view (SRF_76; courtesy IAA)	8
Fig. 2.3	Australian soldiers repairing the Mamluk bridge during World War I ca. 1918 (SRF_76; courtesy IAA)	8
Fig. 2.4	The new British bridge: a) general view (1949), b) close-up view showing the erosion caused by the deepening of the Jordan River's bed by one meter by 1951, causing further damage to the Acheulian site (KKL Archives)	9
Fig. 2.5	View of the Jordan River's right bank and the remains of the destroyed Mamluk stone bridge (looking north), photographed by D.A.E. Garrod in 1935 (courtesy Pitt Rivers Museum)	9
Fig. 2.6	View of the Jordan River's right bank, the remains of the Mamluk stone bridge, and the team of D.A.E. Garrod during the excavation of trenches (showing Miss Gardner, the geologist), photographed by D.A.E. Garrod in 1935 (courtesy Pitt Rivers Museum)	10
Fig. 2.7	Garrod's letter to the Mandatory Department of Antiquities notifying her intention to end the project (16/4/1935 ATQ/75/6; courtesy IAA)	10
Fig. 2.8	Topographic map of the excavations at Gesher Benot Ya'aqov produced by the expedition of M. Stekelis (Stekelis et al. 1937), the black rectangle marking the location of D.A.E. Garrod's excavations; note the spoils along the right bank of the river and the disturbed area at the western end of the bridge (SRF_76; courtesy IAA)	11
Fig. 2.9	Excavation of an elephant tusk in the waterlogged sediments, photographed by the expedition of M. Stekelis (SRF_76; courtesy IAA).	12
Fig. 2.10	The Jordan River after the sluice was let down (1936); note the exposed boulders (SRF_76; courtesy IAA)	12
Fig. 2.11	Map sketched by Stekelis and Picard (1936) illustrating the localities excavated in the bed of the Jordan River (Stekelis' report of activities in 1936/7; SRF_76; courtesy IAA).	12
Fig. 2.12	Stratigraphy observed on the Jordan River's right bank (Stekelis 1960:66–67, fig. 4)	12
Fig. 2.13	View looking south of Stekelis' excavated pit (foreground) and the British bridge (background) (SRF_76; courtesy IAA).	12
Fig. 2.14	View of Stekelis' excavated pit (SRF_76; courtesy IAA)	13
Fig. 2.15	View of Stekelis' excavated pit: a) note the coarse sediments in the section, b) the pit during excavation (SRF_76; courtesy IAA)	14

xxxv

Fig. 2.16	Stekelis' final stratigraphic sequence (from Stekelis 1960: 68–69, fig. 5)	15
Fig. 2.17	The spoils of drained sediment originating from the bed of the Jordan River and its banks (after 1967).	15
Fig. 2.18	Map showing the location of previous excavations (1–3), surface collections made by D. Ben-Ami, and the study area	16
Fig. 2.19	A flint handaxe and the gastropod *Viviparus apameae* bedded in the tilted sediments of the eastern flanks of the Crusader fortress (the glacis), which are built of the Benot Ya'akov Formation	16
Fig. 2.20	Aerial view of the Jordan River in the vicinity of the Benot Ya'aqov Bridges with the location of the study area: a) January 1945, b) January 2007 (orthophoto courtesy Survey of Israel)	17
Fig. 2.21	The Bar's tilted layers and stratigraphy during a low water stand of the Jordan River in 1990	18
Fig. 2.22	Aerial view of the destruction caused by drainage activity in December 1999	18
Fig. 3.1	Map showing the area of the GBY site with the major geographical landmarks and archaeological sites mentioned in the text	21
Fig. 3.2	Exposures of the BYF: a) Jordan River, right bank north of the bridges (currently under water); note fossil mammal bone protruding from the sediments, b) right bank north of the bridges, c) right bank south of the bridges, d) and e) exposures in the Jordan River south of the bridges, f) small-scale excavations on the left bank of the Jordan River (NBA in Fig. 3.1).	22
Fig. 3.3	a) Aerial photo of the GBY area with locations of the most prominent landmarks (Crusader fortress, Customs House, Mamluk caravanserai, Mamluk stone bridge, b) the Crusader fortress (view from the east) and its slopes (glacis), c) view of the western slopes of the Golan Heights: the Mamluk caravanserai, the Lower Customs House, the ruins of the Mamluk stone bridge, and the stony spoils of the continuous drainage operations	23
Fig. 3.4	Aerial photograph showing the spoils made by the continuous drainage of the river on both banks of the river (January 2000)	24
Fig. 3.5a	Composite stratigraphic section of the rock units exposed within the Korazim Saddle (modified from Heimann and Ron 1993).	24
Fig. 3.5b	Geological map of the Rosh Pinnah area (courtesy of the Geological Survey of Israel and the GIS center of the Hebrew University)	25
Fig. 3.6	Generalized map of faults within the northern Dead Sea Rift (modified from Heimann and Ron 1993). Latitudes and longitudes are shown on the inset map, whereas the fault map uses the Israel grid system (from Belitzky 2002: fig. 1).	26
Fig. 3.7	Geological maps of the GBY area, including a) general and b) enlarged perspectives (from Belitzky 2002: fig. 6a)	27
Fig. 3.8	Geological map of the study area (from Goren-Inbar et al. 2002b: fig. 7)	28
Fig. 3.9	a) and b): the tilted BYF sediments exposed on the eastern flanks of the Crusader fortress	28
Fig. 3.10	The southernmost exposure of the BYF, the outlet of Nahal Rosh Pinnah to the Jordan River: a) and b) left bank of the river, c) right bank of the river	30
Fig. 3.11	The Bar: a) and b) view of the Bar sediments, c) tilted cemented sediments, d) close-up of the sediments, e) basalt handaxe *in situ* in the Bar sediments	31
Fig. 3.12	GBY study area: map of excavation areas and trenches	32
Fig. 3.13	Stratigraphic correlation chart of the GBY trenches' sedimentological sequences	33
Fig. 3.14a	Stratigraphic sections from GBY. Particle size and character is represented on the horizontal axis and with patterns	34
Fig. 3.14b	GBY composite section	35

List of Figures

Fig. 3.15	Aerial view of the GBY area and the location of the two drillings: GBY #2 and Eshel Ya'aqov	36
Fig. 3.16	The terraces of Nahal Rosh Pinnah in the southernmost exposure area of the BYF (1980s); note the bedded limestone cobbles *in situ*.	39
Fig. 3.17	A fresh basalt tool (# 8156 Layer II-6 Level 5)	39
Fig. 3.18	A giant basalt core (#10479 Layer II-6 Level 1): a) *in situ* and b) after removal; note the crust of basalt that remains on the sediment	40
Fig. 3.19	The distal edge of a basalt cleaver (#2255 Trench II green); note the development of exfoliation on the periphery.	40
Fig. 3.20	Basalt artifact (#2254 Trench II) that was not conserved and is consequently exfoliated	40
Fig. 3.21	Basalt handaxe (not catalogued Trench II) that was not conserved and is consequently exfoliated.	41
Fig. 4.1	The grid laid on top of the horizontal surface of the excavated area	44
Fig. 4.2	The grid laid on top of the horizontal surface of the excavated area; note the differences in elevation between the grid and the tilted excavated surface	44
Fig. 4.3	The freestanding grid system.	45
Fig. 4.4	The freestanding grid system.	45
Fig. 4.5	Plumb line marking a corner of a grid square on the tilted excavated horizon	46
Fig. 4.6	The movable awning located above the grid	46
Fig. 4.7	Backhoe trenching in the study area	46
Fig. 4.8	The GBY study area: map of trenches and excavation areas.	47
Fig. 4.9	View of the northern face of Trench II	47
Fig. 4.10	Excavation of Layer II-1 (the Area B Unconformity)	47
Fig. 4.11	Measuring elevations during excavation	48
Fig. 4.12	Segment of the field map of Layer II-6 Level 4	48
Fig. 4.13	Excavation of Area B; note the white strings laid on the Unconformity (Layer II-1) marking the strike of different levels of Layer II-6	48
Fig. 4.14	Two excavation methods: a) along the strike and dip, b) from top to bottom	49
Fig. 4.15	Field drawing of an exposed surface; note the aluminum drawing frame laid on the excavated surface	50
Fig. 4.16	Field drawing of an exposed surface; note the aluminum drawing frame laid on the excavated surface and the white strings laid on the Unconformity (Layer II-1) marking the strike of different levels of Layer II-6	50
Fig. 4.17	The method of field drafting; note the drawing frame in the foreground, where the interior part of the frame is aligned with the string marking the strike of the level	50
Fig. 4.18	Schematic illustration of different methodologies used during field drafting; in the foreground a drawing of an exposed surface and in the background a drawing of cross-sections perpendicular to the layer; note the freestanding grid and the strings marking the strike of the levels	51
Fig. 4.19	Wet sieving in the Jordan River	52
Fig. 4.20	Protecting the exposed archaeological horizons with wet geotextile	52
Fig. 4.21	Example of positioning of CCT: a flint core (#9689 Layer V-6) proximally positioned with the dominant working edge and the dominant debitage face on the left	55
Fig. 4.22	Flint handaxe with bifacial retouch (#6175 Layer II-6 Level 1), Face 1 on the left	56
Fig. 4.23	Basalt cleaver (#223 Layer II-6 Level 4); note the morphology of the proximal end of the ventral face, where thinning has resulted in a visible hinged scar	56

Fig. 4.24	Basalt handaxe with visible direction of blow on the ventral face (#5791 Layer II-6 Level 4b), Face 1 on the left	56
Fig. 4.25	Basalt cleaver with visible direction of blow on the ventral face (#5448 Layer II-6 Level 1), Face 1 on the left	57
Fig. 5.1	General view of the southern sector of the study area	62
Fig. 5.2	General view of Area C looking north	62
Fig. 5.3	General view of Area C looking south	63
Fig. 5.4	Map of excavated areas and geological trenches; the Area C excavations and the adjacent Trench V are highlighted	63
Fig. 5.5	Isometric view of the excavations and cross-sections in Area C, view to north	64
Fig. 5.6	The northern cross-section of the excavations of Area C; visible are Layers V-5 and V-6.	67
Fig. 5.7	Field map of Layer V-5: a) in plan view, view to north, b) in its original tilted position; view to southeast	68
Fig. 5.8	Layer V-6: a scatter of bones and lithics	69
Fig. 5.9	Field map of Layer V-6: a) in plan view, view to north, b) in its original tilted position; view to southeast	70
Fig. 5.10	Map of excavated areas and geological trenches; the Jordan Bank excavations are highlighted.	71
Fig. 5.11	Close-up view of the Jordan Bank coquina (Layer V-5)	71
Fig. 5.12	Jordan Bank: excavation of an elephant (*Palaeoloxodon antiquus*) tooth in Layer V-5; note the *Viviparus apameae* in the matrix of the tooth	72
Fig. 5.13	Excavation on the Jordan Bank.	72
Fig. 5.14	Excavation on the Jordan Bank.	72
Fig. 5.15	Jordan Bank, southern sector: examples of field maps (1996)	73
Fig. 5.16	Excavation of the elephant vertebra in the Enclosure	74
Fig. 5.17	Map of excavated areas and geological trenches; the Area A excavations and the adjacent Trench I are highlighted	74
Fig. 5.18	General view of Area A looking north	75
Fig. 5.19	General view of Area A looking south	75
Fig. 5.20	Isometric views of the excavations and cross-sections in Area A: a) view to southeast, b) view to northwest.	76
Fig. 5.21	Wood fragments bedded in the sediments of Layer I-5 (scale 7 cm)	76
Fig. 5.22	A bird bone bedded in the sediments of Layer I-5	77
Fig. 5.23	The stratigraphic position of Layers I-4 and I-5 as seen in a corner of an excavated square	77
Fig. 5.24	General view of Area B looking east at the beginning of excavation	78
Fig. 5.25	Map of excavated areas and geological trenches; the Area B excavations and the adjacent Trenches II, IV, and VI are highlighted	79
Fig. 5.26	Isometric view of the excavations and cross-sections in Area B, view to northeast	80
Fig. 5.27	General view of Area B looking east	81
Fig. 5.28	General view of Area B looking southeast	82
Fig. 5.29	General view of Area B looking northeast	82
Fig. 5.30	General view of Area B looking east.	83
Fig. 5.31	Close-up view of Layer II-2/3	84
Fig. 5.32	Close-up view of the archaeological horizons of Layers II-2 and II-3/4 (W=wood); in the background Layer II-6 Level 4	85
Fig. 5.33	Field maps of Layer II-6 Levels 1–7: a) in plan view, b) in its original tilted position; view to southeast	88
Fig. 5.34	Close-up view of Layer II-6 Level 1	89
Fig. 5.35	Field map of Layer II-6 Level 1: a) in plan view, b) in its original tilted position; view to southeast	90

List of Figures   xxxix

Fig. 5.36	The elephant skull and its associated artifacts from Layer II-6 Level 1: a) detailed view of the premaxillary part (ventral face) of the elephant skull (scale 1 m), b) a fragment of the premaxillary bone (conjoined with the skull), c) the elephant skull and associated finds; note the heavily fractured bone mass (scale 1 m), d) the wooden log and associated lithic artifacts, e) the wooden log overlying flint artifacts, f) the elephant skull entirely exposed and ready for plaster of Paris casting; note that a small part of a giant core is visible, the rest underlies the skull (scale 0.5 m), g) the log after final exposure and cleaning, h) the elephant skull after total exposure with close-up view of the trunk foramen (scale 20 cm)	92
Fig. 5.37	General view of Trench VI: a) looking west, b) looking east with Layer VI-14 in view	94
Fig. 5.38	Close-up view of Layer VI-14	95
Fig. 5.39	Close-up view of Layer II-6 Levels 1–4	95
Fig. 5.40	Close-up view of Layer II-6 Level 2 looking northeast	96
Fig. 5.41	Field map of Layer II-6 Level 2: a) in plan view, b) in its original tilted position; view to southeast	97
Fig. 5.42	Close-up view of Layer II-6 Level 3	98
Fig. 5.43	Field map of Layer II-6 Level 3: a) in plan view, b) in its original tilted position; view to southeast	99
Fig. 5.44	View of Layer II-6 Level 4	100
Fig. 5.45	Close-up view of Layer II-6 Level 4	100
Fig. 5.46	Close-up view of Layer II-6 Level 4	101
Fig. 5.47	Close-up view of Layer II-6 Level 4	101
Fig. 5.48	Close-up view of Layer II-6 Level 4	102
Fig. 5.49	Field map of Layer II-6 Level 4: a) in plan view, b) in its original tilted position; view to southeast	104
Fig. 5.50	Field map of Layer II-6 Level 4b: a) in plan view, b) in its original tilted position; view to southeast	105
Fig. 5.51	Close-up view of Layer II-6 Level 5	106
Fig. 5.52	Field map of Layer II-6 Level 5: a) in plan view, b) in its original tilted position; view to southeast	107
Fig. 5.53	Close-up view of Layer II-6 Level 6	108
Fig. 5.54	Field map of Layer II-6 Level 6: a) in plan view, b) in its original tilted position; view to southeast	109
Fig. 5.55	Close-up view of Layer II-6 Level 7; S=south test pit, N=north test pit	110
Fig. 5.56	Close-up view of Layer II-6 Level 7; S=south test pit, N=north test pit	110
Fig. 5.57	Close-up view of the north test pit of Layer II-6 Level 7: a) lower part (gravel), b) upper part (sand)	110
Fig. 5.58	Close-up view of the south test pit of Layer II-6 Level 7: a) lower part (gravel), b) upper part in section (sand)	111
Fig. 5.59	Field map of Layer II-6 Level 7 (excluding the test pits): a) in plan view, b) in its original tilted position; view to southeast	112
Fig. 5.60	General view of Trench II; the arrow indicates the location of a large bone	114
Fig. 5.61	Views of the hippopotamus pelvis in Trench II: a) general, b) close-up	114
Fig. 5.62	Flint double convex-concave side-scraper (#17356 Layer IV-25)	115
Fig. 5.63	Flint side-scraper on ventral face (#17357 Layer IV-25)	115
Fig. 5.64	Flint offset scraper (#17354 Layer IV-25)	116
Fig. 5.65	Flint atypical end-scraper (#17359 Layer IV-25)	116
Fig. 5.66	Flint typical borer **1** multiple views **2** detailed view of borer (#17358 Layer IV-25)	116
Fig. 5.67	Flint typical borer **1** multiple views **2** detailed view of borer (#17360 Layer IV-25)	116

Fig. 5.68	Flint notch (#17362 Layer IV-25)	116
Fig. 5.69	Flint *éclat de taille de biface* (#17361 Layer IV-25)	116
Fig. 5.70	Flint *éclat de taille de biface* (#17365 Layer IV-25)	117
Fig. 5.71	Flint burin spall (#17364 Layer IV-25)	117
Fig. 6.1	Length and preservation state of flint CCT and FFT	122
Fig. 6.2	Flint notch (#4716 Layer V-5)	125
Fig. 6.3	Flint notches (#938 Layer II-6 Level 1, #11730 Layer II-6 Level 7, #761 Layer II-5)	126
Fig. 6.4	Flint notches (#17235 Layer II-6 Level 1, #13866 Layer II-6 Level 7, **1** note bulb of percussion on dorsal face **2** detailed view of notch)	126
Fig. 6.5	Flint notch (#6136 Layer II-6 Level 1)	126
Fig. 6.6	Flint multiple tool: notch and denticulate (#9186 Layer II-6)	129
Fig. 6.7	Flint multiple tool: notch, denticulate and retouched flake with removed striking platform **1** multiple views **2** detailed view of notch (#5942 Layer II-6 Level 1)	130
Fig. 6.8	Flint denticulates (#5300 Layer V-5 with signs of utilization, #548 Layer II-3 with removed striking platform, #13640 Layer II-6/7 on biface sharpening flake)	130
Fig. 6.9	Flint denticulated points (#621 Layer II-2/3, #7343 Layer II-6 Level 4, #14962 Layer II-7)	134
Fig. 6.10	Flint borers **1** multiple views **2** detailed view of borer (#4526 Layer V-5, #11688 Layer II-6 Level 1 with signs of utilization, #11857 Layer II-6 Level 1, #13857 Layer II-6 Level 1 with signs of utilization, #6496 Layer II-6 Level 1 with signs of utilization, #6410 Layer II-6 Level 1, #9864 Layer II-6 Level 7, #17288 Layer II-6 Level 6 on *débordant* flake)	135
Fig. 6.11	Flint borers (#581 Layer II-2/3 with removed striking platform, #6068 Layer II-6 Level 1 with signs of utilization)	136
Fig. 6.12	Flint borers (#14319 Layer II-6 Level 7 **1** multiple views **2** detailed view of borer, #13939 Layer II-6 Level 7 with removed striking platform)	136
Fig. 6.13	Flint borers **1** multiple views **2** detailed view of borer (#737 Layer II-5, #1675 Layer II-6 Level 4b)	136
Fig. 6.14	Flint multiple tools: borer and a retouched flake (#4500 Layer V-4 **1** multiple views **2** detailed view of borer, #994 Layer II-5, note ventral removal)	140
Fig. 6.15	Flint multiple tools **1** multiple views **2** detailed view of borer (#4506 Layer V-4 borer and notch, #5301 Layer V-5 borer and notch, #11064 Layer II-6 Level 1 borer and end-notch)	140
Fig. 6.16	Flint borers **1** multiple views **2** detailed view of borer (#9803 Layer II-6 Level 7 multiple tool borer and single straight side-scraper, #4346 Layer II-6 Level 5)	141
Fig. 6.17	Flint side-scrapers **1** multiple views **2** detailed view of working edge (#11974 JB single straight side-scraper and burin, note removed striking platform, #4685 Layer V-5 single convex side-scraper on *débordant* flake, #6273 Layer II-6 Level 1 single convex side-scraper with signs of utilization, #3151 Layer II-6 Level 4 single convex side-scraper with signs of utilization, #3335 JB single convex side-scraper with signs of utilization)	145
Fig. 6.18	Flint abruptly retouched side-scraper **1** multiple views **2** detailed view of scraper edge (#565 Layer II-6 Level 1)	146
Fig. 6.19	Flint abruptly retouched side-scraper **1** multiple views **2** detailed view of scraper edge (#13007 Layer II-6 Level 1 with removed striking platform)	146
Fig. 6.20	Flint side-scrapers on ventral face **1** multiple views **2** detailed view of scraper edge (#5966 Layer II-6 Level 1 with signs of utilization, #14417 Layer II-6 Level 7, #13767 Layer II-6 Level 7, #17304 Layer II-6 Level 7 on biface sharpening flake, #13322 Layer II-6 Level 7 on hinged flake	

List of Figures   xli

	with signs of utilization)	147
Fig. 6.21	Flint side-scrapers **1** multiple views **2** detailed view of scraper edge (#12507 Layer II-6/7 convergent-convex side-scraper with ventral removal, note battering marks on cortex, #11958 Layer II-6 Level 6 convex transversal scraper, #9393 Layer II-6 Level 7 convex transversal scraper, #14386 Layer II-6 Level 7 alternately retouched side-scraper, #11914 Layer II-6 Level 7 convex transversal scraper, #9568 Layer II-6 Level 7 convex transversal scraper with removed striking platform)	148
Fig. 6.22	Flint double straight side-scraper **1** multiple views **2** detailed view of scraper edge (#13632 Layer II-6 Level 7)	149
Fig. 6.23	Flint multiple tools **1** multiple views **2** detailed view of scraper edge (#569 Layer II-5 side-scraper on ventral face and notch, #13635 Layer II-6 Level 7 single concave side-scraper and notch on a hinged flake, #504 Layer II-5 single concave side-scraper and retouched flake with signs of utilization, #14090 Layer II-6 Level 7 single straight side-scraper and retouched flake on hinged flake)	152
Fig. 6.24	Flint multiple tools **1** multiple views **2** detailed view of scraper edge (#11116 Layer II-6 Level 1 single convex side-scraper and borer, #9847 Layer II-6 Level 7 abruptly retouched side-scraper and borer with removed striking platform, #13773 Layer II-6 Level 7 offset side-scraper and side-scraper on ventral face, #13519 Layer II-6 Level 7 offset side-scraper and side-scraper on ventral face)	153
Fig. 6.25	Flint multiple (triple) tools (#3298 JB single straight side-scraper, burin, and truncation, #936 Layer II-5 single straight side-scraper, burin, and denticulate)	154
Fig. 6.26	Flint end-scrapers **1** multiple views **2** detailed view of scraper edge (#14897 Layer II-6 Level 3, #4261 Layer II-6 Level 6, #11740 Layer II-6 Level 7, #14014 Layer II-6 Level 7)	155
Fig. 6.27	Flint multiple tools (#727 Layer II-6 Level 4 end-scraper and retouched flake, #11846 Layer II-5/6 end-scraper and borer on *débordant* flake, #6077 Layer II-6 Level 1 end-scraper, notch, and retouched flake **1** multiple views **2** detailed view of retouched edge)	159
Fig. 6.28	Flint multiple tools (#2221 Trench II green truncation and denticulate with removed striking platform, #1317 Layer II-6 Level 4 truncation and denticulate)	162
Fig. 6.29	Flint backed knives (#4720 Layer V-5 backed knife with signs of utilization, #4726 Layer V-5 backed knife with signs of utilization **1** multiple views **2** detailed view of backed edge, #9610 JB naturally backed knife, #5920 V-5 naturally backed knife and denticulate with signs of utilization, note steps on proximal end)	163
Fig. 6.30	Flint multiple tool (#3336 JB backed knife with bipolar retouch and retouched flake with removed striking platform)	164
Fig. 6.31	Flint burins (#3956 JB burin on ventral face, #377 Layer II-6 Level 4 burin on ventral face with signs of utilization, note steps on proximal end, #1162 Layer II-6 Level 4b flint multiple tool: burin on ventral face and truncation, #3928 Layer II-6 Level 4 transversal burin, **1** multiple views **2** detailed view of burin, #19166 Layer V-5 burin spall)	164
Fig. 6.32	Flint burins (#9174 Layer II-6 Level 6 multiple tool: burin and retouched flake, #16142 Trench II burin on truncation, #13760 Layer II-6 Level 7 burin with signs of utilization **1** multiple views **2** detailed view of burin) . . .	165
Fig. 6.33	Flint varia tools (#4339 Layer II-6 Level 6 alternately retouched bec (beak) with signs of utilization and ventral removals, #6570 Layer II-6 Level 1 limace with removed striking platform, #6641 Layer II-6 Level 2 varia tool,	

	#9564 Layer II-6 Level 2 varia tool with ventral removals and removed striking platform, #6900 Layer II-6 Level 2 varia tool with signs of utilization, #1581 Layer II-6 Level 3 varia tool with Nahr Ibrahim technique, signs of utilization and removed striking platform)	166
Fig. 6.34	Flint varia tools (#14092 Layer II-6 Level 7 alternately retouched bec (beak) with signs of utilization **1** multiple views **2** detailed view of retouched edge, #14961 Layer IV-7 varia tool with ventral removals and removed striking platform)	167
Fig. 6.35	Flint retouched flakes (#14847 Layer II-6 Level 3 on *outrepassé* flake with removed striking platform, #3924 Layer II-6 Level 4 with Nahr Ibrahim technique, #4231 Layer II-6 Level 6)	168
Fig. 6.36	Flint flakes with removed striking platforms **1** multiple views **2** detailed view of proximal end (#9805 Layer II-6 Level 7 tanged tool and retouched flake, #5314 Layer V-5 denticulate)	175
Fig. 6.37	Flint flakes with removed striking platforms (#4343 Layer II-6 Level 6 single straight side-scraper, end-scraper, and borer, #13933 Layer II-6 Level 7 borer, single convex side-scraper, and notch, #13837 Layer II-6 Level 7 varia tool)	175
Fig. 6.38	Flint flakes with removed striking platforms (#6361 Layer II-6 Level 1 single convex side-scraper, side-scraper on ventral face, and notch, #6610 Layer II-6 Level 6 varia tool limace)	176
Fig. 6.39	Flint flakes with removed striking platforms **1** multiple views **2** detailed view of proximal end (#5307 Layer V-5 with signs of utilization, #3776 Layer II-6 Level 3 end-scraper, #2621 Layer II-6 Level 4 retouched flake with signs of utilization and ventral removals, #13580 Layer II-6 Level 7 naturally backed knife with signs of utilization, #14003 Layer II-6 Level 7 end-notch and denticulate, #6946 Layer II-6 Level 2 side-scraper on ventral face and retouched flake)	177
Fig. 6.40	Flint flakes with removed striking platforms **1** multiple views **2** detailed view of proximal end (#9220 Layer II-6 Level 6 side-scraper on ventral face with signs of utilization, #9216 Layer II-6 Level 6 abruptly retouched side-scraper with signs of utilization and ventral removals, #3116 Layer II-6 Level 4 denticulate)	178
Fig. 6.41	Flint flakes with removed striking platforms **1** multiple views **2** detailed view of proximal end (#4039 Layer II-6 Level 5 retouched flake, #9176 Layer II-6 Level 6 truncation and notch, #7622 Layer II-6 Level 7 notch with ventral removals)	178
Fig. 6.42	Flint chopping tools (#9688 Layer V-6, #10489 Layer II-6 Level 1 with signs of utilization, #708 Layer II-6 Level 4 with signs of utilization, #3176 Layer II-6 Level 4)	183
Fig. 6.43	Flint core tools (#10513 Layer II-6 Level 1 borer on core on flake, #3680 Layer II-6 Level 4 **1** burin on modified artifact **2** detailed view of burin, #17437 Layer II-6 Level 6 borer on modified artifact)	184
Fig. 6.44	Flint transversal scraper on nodule (#378 Layer II-6 Level 4)	184
Fig. 6.45	Flint blades with signs of utilization (#4408 Layer V-5, #4691 Layer V-5, #4723 Layer V-5 borer and *éclat de taille de biface*)	186
Fig. 6.46	Flint blades with signs of utilization (#7659 Layer V-6 *éclat de taille de biface*, #6561 Layer II-6 Level 1)	187
Fig. 6.47	Length and width of complete and broken unretouched flakes from Layer: V-6, II-6/L1, and II-6/L7	190
Fig. 6.48	Flint *éclats de taille de biface* (#5349 Layer V-5 borer with signs of utilization, note steps on proximal end, #5225 Layer V-5, #5246 Layer V-5, #7635 Layer V-5, note steps on proximal end, #7656 Layer V-5, #7644	

		Layer V-6, #7675 Layer V-6, #7633 Layer V-6, note steps on proximal end) 194
Fig. 6.49		Flint *éclats de taille de biface* (#7677 Layer V-6, #5485 Layer V-6, note steps on proximal end, #4000 JB retouched flake) 195
Fig. 6.50		Flint *éclats de taille de biface*, note steps on proximal end (#4014 JB, note crushing on proximal end, #3974 JB, #4005 JB, note crushing on proximal end, #3998 JB, #3997 JB, #4001 JB *débordant* flake, #4013 JB) 196
Fig. 6.51		Flint *éclats de taille de biface* with signs of utilization (#3339 JB, #3341 JB, note steps on proximal end, #3347 JB, #3360 JB, #3348 JB, note steps on proximal end, #3356 JB) 197
Fig. 6.52		Flint *éclats de taille de biface* (#3334 JB with signs of utilization, note steps on proximal end, #3351 JB, note steps on proximal end, #3355 JB with signs of utilization, #3315 JB, #3318 JB, #3311 JB with signs of utilization, #3296 JB hinged flake, note steps on proximal end, #3295 JB hinged flake) 198
Fig. 6.53		Flint *éclats de taille de biface* with signs of utilization (#3294 JB, #3330 JB) 199
Fig. 6.54		Flint *éclats de taille de biface* (#6576 Layer II-6 Level 1, #1240 Layer II-6 Level 4 with signs of utilization) 199
Fig. 6.55		Flint *éclats de taille de biface*: finishing flakes struck perpendicular to biface edge (#5321 Layer V-5, #8266 Layer V-5, #5408 Layer V-6, #4711 Layer V-5) 201
Fig. 6.56		Flint *éclats de taille de biface* (finishing flakes struck obliquely to biface edge: #5265 Layer V-5, #5241 Layer V-5, #5501 Layer V-6, #5203 Layer V-5; finishing flakes struck perpendicularly to biface edge: #5407 Layer V-6, #5281 Layer V-5) 202
Fig. 6.57		Flint dorsally plain flake: *débordant* flake (#19185 Layer V-5) 203
Fig. 6.58		Flint flakes with ventral removals and signs of utilization (#3329 JB, #11425 Layer II-6 Level 1 borer with removed striking platform) 203
Fig. 6.59		Flint core trimming elements (#16553 Layer II-5 side-scraper on ventral face with removed striking platform, #710 Layer II-6 Level 1, #14509 Layer II-6 Level 2 with signs of utilization, #9845 Layer II-6 Level 7 with signs of utilization, note two cones) 204
Fig. 6.60		Flint *débordant* flakes (#9502 Layer II-6 Level 3 notch, #1525 Layer II-6 Level 4 with signs of utilization, #6591 Layer II-6 Level 6 with signs of utilization, #13622 Layer II-6 Level 7 with signs of utilization). 205
Fig. 6.61		Flint Levallois cores (#9689 Layer V-6, #378 Layer II-5, #9408 Layer II-6 Level, #1503 Layer II-6 Level 4, #3646 Layer II-6 Level 4, #2056 Layer II-6 Level 4) 208
Fig. 6.62		Flint Levallois cores (#2326 Layer II-6 Level 4, #3566 Layer II-6 Level 4, #891 Layer II-6 Level 4) 209
Fig. 6.63		Flint Levallois cores (#1106 Layer II-6 Level 4b, #6374 Layer II-6 Level 5, #6478 Layer II-6 Level 6, #9347 Layer II-6 Level 7, #9372 Layer II-6 Level 7, #9565 Layer II-6 Level 7). 210
Fig. 6.64		Flint Levallois cores (#9570 Layer II-6 Level 7, #9291 Layer II-6 Level 7, #2175 Trench II, #2242 Trench II, #2177 Trench II **1** multiple views **2** detailed view of working edge). 211
Fig. 6.65		Flint cores: Levallois cores (#4732 Layer V-5, #1272 Layer II-6 Level 4); cores on flakes (#9692 Layer V-6, #747 Layer II-5, #10650 Layer II-6 Level 1, #9504 Layer II-6 Level 3). 213
Fig. 6.66		Flint cores on flakes (#8339 Layer II-6 Level 3 on biface sharpening flake, #3001 Layer II-6 Level 4 on biface sharpening flake, #791 Layer II-6 Level 4 with Nahr Ibrahim technique, #391 Layer II-6

	Level 4 with retouch, #922 Layer II-5/6 on biface sharpening flake)	215
Fig. 6.67	Flint cores on flakes (#822 Layer II-6 Level 4 notch on both edges, #2384 Layer II-6 Level 4, #2832 Layer II-6 Level 4)	216
Fig. 6.68	Flint core on flake (#734 Layer II-6 Level 4)	216
Fig. 6.69	Flint discoidal core (#6613 Layer II-6 Level 1)	217
Fig. 6.70	Flint cores (#11849 Layer II-6 Level 3 prismatic core, #9457 Layer II-6 Level 7 pyramidal core, #429 Layer II-6 Level 4 core on broken biface)	219
Fig. 6.71	Flint varia cores (#5142 Layer V-5, #10570 Layer II-6 Level 1, #11183 Layer II-6 Level 1, #10752 Layer II-6 Level 1)	220
Fig. 6.72	Flint varia cores (#738 Layer II-5, #915 Layer II-6 Level 1, #557 Layer II-6 Level 1, #963 Layer II-6 Level 1)	221
Fig. 6.73	Flint varia cores (#2382 Layer II-6 Level 1, #8348 Layer II-6 Level 3, #15596 Layer II-6 Level 3, #860 Layer II-6 Level 4)	222
Fig. 6.74	Flint varia cores (#2926 Layer II-6 Level 4, #9360 Layer II-6 Level 7)	223
Fig. 6.75	Flint varia cores: with a single convex side-scraper (#1259 Layer II-6 Level 4), cores on broken bifaces (#9283 Layer II-6 Level 7, #1373 Layer II-6 Level 4), cores on biface sharpening flakes (#3960 JB, #8990 Layer II-6 Level 6, #9435 Layer II-6 Level 7)	224
Fig. 6.76	Flint cores: varia cores (#9408 Layer II-6 Level 3, #2172 Layer II-6 Level 4), formless core (#4748 Layer V-5)	225
Fig. 6.77	Proposed scheme for the flint reduction sequence	230
Fig. 6.78	A damaged cervid (*Dama* sp.) antler from GBY (left) and an experimental antler percussor that was used to replicate basalt bifacial tools (right)	230
Fig. 6.79	Volume and cortex residue of the entire flint core assemblage (N=524) (volume in mm^3=length x width x thickness)	230
Fig. 6.80	Flint Levallois cores: on biface sharpening flakes (#9033 Layer II-6 Level 6, #328 Layer II-5/6) and on a biface (#9687 Layer V-6)	231
Fig. 6.81	Flint flakes with partial modification of the striking platforms (#8428 Layer II-6 Level 3 convex transversal scraper and a borer with a tang-like proximal end, #735 Layer II-6 Level 4 abruptly retouched side-scraper with narrowing of the lateral edges)	235
Fig. 7.1	Basalt massive scraper and a notch (#5384 Layer II-6 Level 1)	245
Fig. 7.2	Basalt end-notch on Kombewa flake, biface thinning flake (#155 Layer II-6 Level 4)	245
Fig. 7.3	Basalt denticulate with ventral removals on dorsally plain flake (#5510 Layer II-6 Level 1)	245
Fig. 7.4	Basalt denticulate (#5570 Layer II-6 Level 1)	245
Fig. 7.5	Basalt single convex side-scraper (#5545 Layer II-6 Level 1)	246
Fig. 7.6	Basalt side-scraper on ventral face (#5586 Layer II-6 Level 1)	246
Fig. 7.7	Basalt thinned back side-scraper with ventral removals (#5532 Layer II-6 Level 1)	246
Fig. 7.8	Basalt thinned back side-scraper (#5572 Layer II-6 Level 1)	246
Fig. 7.9	Basalt thinned back side-scraper (#5523 Layer II-6 Level 1)	246
Fig. 7.10	Basalt massive scraper (#5682 Layer II-6 Level 1)	247
Fig. 7.11	Basalt massive scraper (#5338 Layer II-6 Level 1)	247
Fig. 7.12	Basalt massive scraper with ventral retouch (#5852 Layer II-6 Level 1)	247
Fig. 7.13	Basalt massive scraper with ventral retouch (#5542 Layer II-6 Level 1)	248
Fig. 7.14	Basalt massive scraper (#5587 Layer II-6 Level 1)	248
Fig. 7.15	Basalt denticulated massive scraper (#5453 Layer II-6 Level 1)	248
Fig. 7.16	Planform of selected of basalt massive scrapers from GBY. Lines mark the location of a retouched edge (a continuous line marks retouch on the dorsal face; a broken line marks retouch on the ventral face). (1) #13399, Layer II-6 Level 1; (2) #13411, Layer II-6 Level 4; (3) #13538, Layer II-6	

		Level 2; (4) #14190, Layer II-6 Level 2; (5) #14482, Layer II-6 Level 2; (6) #14739, Layer II-6 Level 2; (7) #16138, Trench II; (8) #2029, Layer II-6 Level 1; (9) #2032, Layer II-6 Level 1; (10) #2065, Layer II-6 Level 1; (11) #2077, Layer II-6 Level 1; (12) #29, Layer II-6 Level 4; (13) #4852, Layer II-5 Level 4; (14) # 4854, Layer II-6 Level 4; (15) #4887, Layer II-6 Level 4b; (16) #4913, Layer II-6 Level 4b; (17) #4920, Layer II-6 Level 4b; (18) #4943, II-6 Level 4; (19) #4959, Layer II-6 Level 4; (20) #5004, Layer II-6 Level 4; (21) #5338, Layer II-6 Level 1; (22) #5486, Layer V-6; (23) #5594, Layer II-6 Level 1; (24) #5608, Layer II-6 Level 4b; (25) #5671, Layer II-6 Level 6; (26) #5704, Layer II-6 Level 6; (27) #5726, Layer II-6 Level 4b; (28) #5739, Layer II-6 Level 4b; (29) #5838, Layer II-6 Level 2; (30) #8609, Layer II-6 Level 2; (31) #9247, Layer II-6 Level 6; (32) #4998, Layer II-6 Level 4	249
Fig. 7.17	Basalt retouched flake (#330 Layer II-6 Level 4)	250
Fig. 7.18	Basalt retouched wedge flake (#256 Layer II-6 Level 4)	250
Fig. 7.19	Flow chart illustrating the order of presentation of different classes of basalt flakes section 7.3.	251
Fig. 7.20	Basalt *outrepassé* flake (#5740 Layer II-6 Level 1)	255
Fig. 7.21	Basalt flake with hinge (#437 Layer II-6 Level 4)	255
Fig. 7.22	Basalt flake with hinge (#5583 Layer II-6 Level 1)	255
Fig. 7.23	Basalt Siret flake with steps (#5047 Layer II-6 Level 4)	255
Fig. 7.24	Basalt Siret flake (#2029 Layer II-6 Level 1)	260
Fig. 7.25	Basalt Siret flake (#13252 Layer II-6 Level 1)	260
Fig. 7.26	Basalt Siret flake (#854 Layer II-6 Level 4)	260
Fig. 7.27	Basalt Siret flake (#4960 Layer II-6 Level 4)	260
Fig. 7.28	Basalt Siret flake (#5575 Layer II-6 Level 4b)	260
Fig. 7.29	Basalt dorsally plain flake (#5474 Layer II-6 Level 1)	267
Fig. 7.30	Basalt dorsally plain flake (#5516 Layer II-6 Level 1)	267
Fig. 7.31	Basalt massive scraper on Kombewa biface thinning flake (#25 Layer II-6 Level 4)	268
Fig. 7.32	Basalt Kombewa biface thinning flake (#5573 Layer II-6 Level 1)	268
Fig. 7.33	Basalt Kombewa biface thinning flake (#721 Layer II-6 Level 4)	268
Fig. 7.34	Basalt Kombewa flake (#5314 Layer II-6 Level 1)	268
Fig. 7.35	Basalt Kombewa flake (#5544 Layer II-6 Level 1)	268
Fig. 7.36	Basalt Kombewa flake (#5525 Layer II-6 Level 1)	268
Fig. 7.37	Basalt Kombewa flake (#3332 JB)	269
Fig. 7.38	Basalt handaxe-shaped Kombewa flake (#5362 Layer II-6 Level 1)	269
Fig. 7.39	Basalt handaxe-shaped Kombewa flake (#107 Layer II-6 Level 4)	269
Fig. 7.40	Basalt cleaver-shaped Kombewa flake (#4850 Layer II-6 Level 4)	269
Fig. 7.41	Basalt biface sharpening flake (#8064 Layer II-6 Level 1)	273
Fig. 7.42	Basalt biface sharpening flake (#2095 Layer II-6 Level 1)	273
Fig. 7.43	Basalt biface sharpening flake (#2102 Layer II-6 Level 1)	273
Fig. 7.44	Basalt biface sharpening flake (#5732 Layer II-6 Level 1)	273
Fig. 7.45	Basalt biface sharpening flake (#5565 Layer II-6 Level 1)	273
Fig. 7.46	Basalt biface sharpening flake (#5589 Layer II-6 Level 1)	273
Fig. 7.47	Basalt biface sharpening flake (#7492 Layer II-6 Level 1)	274
Fig. 7.48	Basalt biface sharpening flake (#5555 Layer II-6 Level 1)	274
Fig. 7.49	Basalt biface sharpening flake (#147 Layer II-6 Level 4)	274
Fig. 7.50	Basalt biface sharpening flake (#783 Layer II-6 Level 4)	274
Fig. 7.51	Basalt *éclat de taille de biface* (#687 Layer II-6 Level 1)	278
Fig. 7.52	Basalt *éclat de taille de biface* (#5876 Layer II-6 Level 1)	278
Fig. 7.53	Basalt *éclat de taille de biface* (#6446 Layer II-6 Level 1)	278
Fig. 7.54	Basalt *éclat de taille de biface* (#12956 Layer II-6 Level 1)	278

Fig. 7.55	Basalt *éclat de taille de biface* (#13253 Layer II-6 Level 1)	278
Fig. 7.56	Basalt biface thinning flake with ventral removals (#18 Layer II-6 Level 4)	281
Fig. 7.57	Basalt shoulder flake with ventral removals (#170 Layer II-6 Level 4)	281
Fig. 7.58	Basalt shoulder flake (#5379 Layer II-6 Level 1)	282
Fig. 7.59	Basalt shoulder flake (#5841 Layer II-6 Level 1)	282
Fig. 7.60	Basalt shoulder flake (#5075 Layer II-6 Level 4)	282
Fig. 7.61	Basalt shoulder flake (#9343 Layer II-6 Level 7)	282
Fig. 7.62	Basalt slice flake (#9230 Layer II-6 Level 5)	285
Fig. 7.63	Basalt slice flake (#9154 Layer II-6 Level 6)	285
Fig. 7.64	Basalt wedge flake (#5366 Layer II-6 Level 1)	286
Fig. 7.65	Basalt wedge flake (#5588 Layer II-6 Level 1)	286
Fig. 7.66	Basalt wedge flake (#4818 Layer II-6 Level 4)	286
Fig. 7.67	Basalt wedge flake (#4948 Layer II-6 Level 4)	287
Fig. 7.68	Basalt wedge flake of handaxe morphology (#4998 Layer II-6 Level 4)	287
Fig. 7.69	Basalt wedge flake of handaxe morphology (#194 Layer II-6 Level 4)	287
Fig. 7.70	Basalt giant flake (#13397 Layer II-6 Level 2)	290
Fig. 7.71	Basalt waste of giant core (#5704 Layer II- 6 Level 1)	292
Fig. 7.72	Basalt waste of giant core (#5765 Layer II-6 Level 1)	292
Fig. 7.73	Basalt waste of giant core (#5639 Layer II-6 Level 4)	292
Fig. 7.74	Basalt waste of giant core (#13413 Layer II-6 Level 4)	292
Fig. 7.75	Basalt waste of giant core (#126 Layer II-6 Level 4)	292
Fig. 7.76	Basalt waste of giant core (#268 Layer II-6 Level 4)	292
Fig. 7.77	Basalt waste of giant core (#2014 Layer II-6 Level 1)	293
Fig. 7.78	Basalt waste of giant core (#5301 Layer II-6 Level 4)	293
Fig. 7.79	Basalt waste of giant core (#56 Layer II-6 Level 4)	293
Fig. 7.80	Basalt flake of handaxe morphology (#2120 Layer II-6 Level 1)	295
Fig. 7.81	Basalt flake of handaxe morphology (#5450 Layer II-6 Level 1)	295
Fig. 7.82	Basalt flake of handaxe morphology (#5540 Layer II-6 Level 1)	296
Fig. 7.83	Basalt flake of handaxe morphology (#5004 Layer II-6 Level 4)	296
Fig. 7.84	Basalt flake of handaxe morphology (#5125 Layer II-6 Level 4)	296
Fig. 7.85	Basalt flake of handaxe morphology (#5726 Layer II-6 Level 4)	296
Fig. 7.86	Basalt flake of cleaver morphology (#282 Layer II-6 Level 4)	298
Fig. 7.87	Basalt percussor (#16326 Layer II-2/3); note battering on surface and damage on both edges	300
Fig. 7.88	Basalt percussor (#5717 Layer II-6 Level 1); note spontaneous flaking on surface and damage on both edges	300
Fig. 7.89	Basalt percussor (#8073 Layer II-6 Level 1); note spontaneous flaking and battering damage on the same surface	301
Fig. 7.90	Basalt percussor (#831 Layer II-6 Level 1)	301
Fig. 7.91	Basalt percussor (#7030 Layer II-6 Level 2); note battering and pitting damage on the flat surface	301
Fig. 7.92	Basalt percussor (#13706 Layer II-6 Level 2)	301
Fig. 7.93	Basalt percussor (#14887 Layer II-6 Level 3)	302
Fig. 7.94	Basalt percussor (#5064 Layer II-6 Level 4); note battering and pitting damage on the flat surface and lateral edge	302
Fig. 7.95	Basalt experimental percussor (#P01); note battering and breakage damage on proximal edge	302
Fig. 7.96	Basalt experimental percussor (#P02)	302
Fig. 7.97	Basalt experimental percussor (#P03); **1** note breakage on surface **2** detailed view of percussive damage	303
Fig. 7.98	Basalt very large percussor; 2.81 kg (#5278 Layer II-6 Level 1)	306
Fig. 7.99	Basalt very large percussor; 2.18 kg (#2312 Layer II-6 Level 1)	306
Fig. 7.100	Basalt split percussor (#6877 Layer II-6 Level 2)	306

Fig. 7.101	Roundness index for GBY natural objects (after Krumbein 1941)	307
Fig. 7.102	Length, width, and thickness (bubble size) of Layer II-6 basalt complete percussors, split percussors, and potential percussors	308
Fig. 7.103	Basalt pitted stone (#13525 Layer II-6 Level 2)	309
Fig. 7.104	Basalt pitted stone (#13387 Layer II-6 Level 3)	309
Fig. 7.105	Basalt pitted stone (#5520 Layer V-5)	309
Fig. 7.106	Basalt pitted stone (#5508 Layer II-6 Level 6)	309
Fig. 7.107	Basalt pitted stone (#8907 II-6 Level 4b)	309
Fig. 7.108	Length, width, and thickness (bubble size) of Layer II-6 basalt pitted stones on FFT and on CCT	315
Fig. 7.109	Basalt giant core with a cluster of pits (#14177 Layer II-6 Level 2)	315
Fig. 7.110	Basalt thin anvil (#7695 Layer II-6 Level 6)	316
Fig. 7.111	Broken basalt thin anvil (#5889 Layer II-6 Level 2)	316
Fig. 7.112	Broken basalt thin anvil (#5883 Layer II-6 Level 2)	316
Fig. 7.113	Positioning method of basalt thin anvils and their A, B, and C planes (#5571 Layer II-6 Level 1)	316
Fig. 7.114	Basalt thin anvil (#16123 Layer II-6 Level 2): a) Plane A with view of the pitted surface, b) Plane C – a break, c) another view of Planes A and C	317
Fig. 7.115	Basalt thin anvil (#7498 Layer II-6 Level 1); note damage in the form of flake scars	317
Fig. 7.116	Location and direction of types of flaking damage on basalt thin anvils	317
Fig. 7.117	Length, width, and thickness (bubble size) of Layer II-6 basalt thin anvils, giant cores, and pitted stones	318
Fig. 7.118	Basalt chopping tool (#677 Layer II-6 Level 4)	320
Fig. 7.119	Basalt core (#7531 Layer V-6)	323
Fig. 7.120	Basalt core on flake (#2119 Layer II-5)	324
Fig. 7.121	Basalt core on flake (#268 Layer II-6 Level 4)	324
Fig. 7.122	Basalt giant cores showing notches, possibly the result of quarrying; (#5447 Layer II-6 Level 1; #5896 Layer II-6 Level 2; two faces of #7696 Layer II-6 Level 7)	325
Fig. 7.123	Cross-sections of basalt giant cores from Layer II-6 (obtained by 3-D scanning) indicative of their slab origin	325
Fig. 7.124	Cross-sections of basalt giant cores from Layer II-6 and Layer II-2/3 and their probable location within the basalt slab morphology (#5446 Level 1, #7701 Level 6, #3561 Layer II-2/3, #14213 Level 2, #10479 Level 1, #10478 Level 1)	326
Fig. 7.125	Basalt exhausted giant core (#3562 Layer II-6 Level 4)	328
Fig. 7.126	Basalt exhausted giant core (#9580 Layer II-6 Level 7)	328
Fig. 7.127	Basalt exhausted giant core (#14213 Layer II-6 Level 2)	328
Fig. 7.128	Basalt giant core of the slab slicing method (#10479 Layer II-6 Level 1)	328
Fig. 7.129	Hypothetical reconstruction of the slab slicing method from top to bottom according to the reduction sequence: stage 1) basalt slab; stages 2 and 3) removal of two opening shoulder flakes; stage 4) removal of a wedge flake creating a suitable debitage surface. Following this stage, two strategies are possible. The first (stages 5a and 6a) is to detach flakes that remove the entire thickness of the slab, resulting in flakes that have an inherent cleaver shape. The second strategy (stages 5 and 6) is to detach flakes that remove only half or more of the slab's thickness, creating a backed knife flake (after Goren-Inbar et al. 2011)	329
Fig. 7.130	Basalt giant core of the bifacial method (#5447 Layer II-6 Level 1)	329
Fig. 7.131	Hypothetical reconstruction of basalt giant core reduction sequence of the bifacial method	330
Fig. 7.132	Basalt giant core of the Kombewa method (#7704 Layer II-6 Level 7)	330

Fig. 7.133	Hypothetical reconstruction of basalt giant core reduction sequence of the Kombewa method (after Goren-Inbar et al. 2011)	330
Fig. 7.134	Basalt giant core of the Levallois method (#10478 Layer II-6 Level 1)	331
Fig. 7.135	Basalt slab fragment (5897# Layer II-6 Level 2)	331
Fig. 7.136	Limestone cleaver (#7717 Layer II-6 Level 6)	334
Fig. 7.137	Limestone handaxe (#469 Layer II-6 Level 1)	334
Fig. 7.138	Limestone handaxe (#6609 Layer II-6 Level 1)	334
Fig. 7.139	Limestone handaxe (#1124 Layer II-6 Level 1)	334
Fig. 7.140	Flint handaxe (#8169 Layer II-6 Level 2)	334
Fig. 7.141	Flint handaxe (#5785 Layer II-6 Level 4b)	334
Fig. 7.142	Basalt cleaver, possibly Kombewa (#5904 Layer V-5)	340
Fig. 7.143	Basalt cleaver, Kombewa (#223 Layer II-6 Level 4)	340
Fig. 7.144	Basalt cleaver, Kombewa (#331 Layer II-6 Level 4)	340
Fig. 7.145	Basalt cleaver, Kombewa (#121 Layer II-6 Level 4)	341
Fig. 7.146	Basalt cleaver, Kombewa (#11961 Layer II-6 Level 5)	341
Fig. 7.147	Limestone handaxe on cobble (#6204 Layer II-6 Level 1)	341
Fig. 7.148	Limestone handaxe on cobble (#5750 Layer II-6 Level 4b)	341
Fig. 7.149	Basalt handaxe (#8620 Layer II-6 Level 2)	343
Fig. 7.150	Basalt handaxe (#5672 Layer II-6 Level 2)	343
Fig. 7.151	Basalt handaxe (#7747 Layer II-6 Level 5)	343
Fig. 7.152	Basalt handaxe (#7724 Layer II-6 Level 7)	343
Fig. 7.153	Basalt handaxe (#2006 Layer II-6 Level 1)	344
Fig. 7.154	Basalt handaxe (#2126 Layer II-6 Level 1)	344
Fig. 7.155	Basalt handaxe (#5655 Layer II-6 Level 2)	344
Fig. 7.156	Basalt handaxe (#8614 Layer II-6 Level 2)	344
Fig. 7.157	Basalt handaxe on side-struck flake (#13764 Layer II-6 Level 1)	345
Fig. 7.158	Basalt handaxe on side-struck flake (#8623 Layer II-6 Level 2)	345
Fig. 7.159	Basalt handaxe on side-struck flake (#7719 Layer II-6 Level 7)	345
Fig. 7.160	Basalt handaxe on side-struck flake (#7720 Layer II-6 Level 6)	345
Fig. 7.161	Basalt cleaver on side-struck flake (#8666 Layer II-6 Level 3)	345
Fig. 7.162	Basalt cleaver on side-struck flake (#5778 Layer II-6 Level 4b)	345
Fig. 7.163	Basalt cleaver on end-struck flake (#180 Layer II-6 Level 4)	346
Fig. 7.164	Basalt cleaver on end-struck flake (#7737 Layer II-6 Level 7)	346
Fig. 7.165	Basalt cleaver on special side-struck flake (#5974 Layer V-6)	346
Fig. 7.166	Basalt cleaver on special side-struck flake (#5780 Layer II-6 Level 4b)	346
Fig. 7.167	Basalt cleaver on special side-struck flake (#5792 Layer II-6 Level)	346
Fig. 7.168	Basalt handaxe with removed striking platform (#5776 Layer II-6 Level 4b)	347
Fig. 7.169	Basalt cleaver with removed striking platform (#5908 Layer II-6 Level 3)	347
Fig. 7.170	Basalt cleaver with removed striking platform (#13762 Layer II-6 Level 3)	348
Fig. 7.171	Basalt handaxe with plain striking platform (#152 Layer II-6 Level 4)	348
Fig. 7.171	Basalt handaxe with plain striking platform (#152 Layer II-6 Level 4)	348
Fig. 7.173	Basalt handaxe with plain striking platform (#179 Layer II-6 Level 4)	348
Fig. 7.174	Basalt cleaver with plain striking platform (#13765 Layer II-6 Level 1)	349
Fig. 7.175	Basalt cleaver with plain striking platform (#5906 Layer II-6 Level 3)	349
Fig. 7.176	Basalt handaxe with bifacial retouch on both faces (#7202 Layer II-6 Level 1)	350
Fig. 7.177	Basalt handaxe with thinning retouch on ventral face (#5787 Layer II-6 Level 6)	350
Fig. 7.178	Basalt handaxe with thinning retouch on ventral face (#1099 Layer II-6 Level 4b)	350
Fig. 7.179	Basalt handaxe with mixed retouch types (bifacial and thinning) (#8057 Layer II-6 Level 1)	350

List of Figures

Fig. 7.180	Basalt handaxe with mixed retouch types (bifacial and thinning) (#5673 Layer II-6 Level 2)	350
Fig. 7.181	Basalt handaxe with mixed retouch types (bifacial and thinning) (#5867 Layer II-6 Level 4b)	351
Fig. 7.182	Basalt cleaver with scraper-like retouch (#13761 Layer II-6 Level 2)	352
Fig. 7.183	Basalt handaxe with intensively flaked Face 1 (#7727 Layer II-6 Level 7)	354
Fig. 7.184	Basalt handaxe with minimal modification (#14195 Layer II-6 Level 1)	354
Fig. 7.185	Basalt handaxe with minimal modification (#5782 Layer II-6 Level 4b)	354
Fig. 7.186	Basalt handaxe with cleaver-like distal end (#244 Layer II-6 Level 4)	358
Fig. 7.187	Basalt handaxe with cleaver-like distal end (#264 Layer II-6 Level 4)	358
Fig. 7.188	Basalt handaxe with cleaver-like distal end (#5772 Layer II-6 Level 4b)	359
Fig. 7.189	Basalt handaxe with cleaver-like distal end (#193 Layer II-6 Level 4)	359
Fig. 7.190	Basalt handaxe with triangular cross-section (#7743 Layer II-6 Level 6)	360
Fig. 7.191	A scatter plot presenting handaxe shape variability at GBY following Roe's method (1964)	360
Fig. 7.192	Basalt handaxe with pointed distal end (#1023 Layer II-6 Level 3)	361
Fig. 7.193	Basalt handaxe with pointed distal end (#5907 Layer II-6 Level 3)	361
Fig. 7.194	Basalt handaxe with pointed distal end (#81 Layer II-6 Level 4)	361
Fig. 7.195	Basalt handaxe with pointed distal end (#144 Layer II-6 Level 4)	361
Fig. 7.196	Basalt cleaver of hourglass form (#1102 Layer II-6 Level 4b)	364
Fig. 7.197	Basalt cleaver of hourglass form (#7783 Layer II-6 Level 7)	364
Fig. 7.198	Basalt cleaver with proximally pointed base (#269 Layer II-6 Level 4)	364
Fig. 7.199	Basalt cleaver with straight working edge (#220 Layer II-6 Level 4)	364
Fig. 7.200	Basalt cleaver with diagonal working edge (#5977 Layer V-6)	364
Fig. 7.201	Basalt cleaver with diagonal working edge (#5670 Layer II-6 Level 1)	365
Fig. 7.202	Basalt cleaver with diagonal working edge (#5835 Layer II-6 Level 5)	365
Fig. 7.203	Basalt cleaver with convex working edge (#176 JB)	365
Fig. 7.204	Basalt cleaver (#5976 Layer V-5)	366
Fig. 7.205	Basalt cleaver (#9896 JB)	366
Fig. 7.206	Basalt cleaver (#8646 Layer II-6 Level 3)	366
Fig. 7.207	Basalt cleaver (#13763 Layer II-6 Level 3)	366
Fig. 7.208	Basalt cleaver (#7784 Layer II-6 Level 7)	366
Fig. 7.209	A heat-map presenting the GBY morphological attributes of cleavers' planform	367
Fig. 7.210	A scatterplot presenting the distribution of GBY handaxes according to Bordes' (1961) shape zones	369
Fig. 7.210a	Schematic illustration of cleaver modifications; working edge modification after detachment of the predetermined cleaver flake	370
Fig. 7.210b	Schematic illustration of cleaver modifications; hypothetical case of cleaver modification with a predetermined working edge (a scar of the giant core) and a series of flakes struck after the detachment of the predetermined cleaver flake	371
Fig. 7.211	Proposed scheme for the basalt reduction sequences.	372
Fig. 7.212	Length, width, and thickness (bubble size) of basalt percussors, pitted stones, and thin anvils from Layer II-6.	373
Fig. 7.213	Visible bedding structure (marked by arrows) in basalt artifacts (#3562 and #9580 Layer II-6 Level 2; #7699 Layer II-6 Level 7)	375
Fig. 7.214	Length, width, and thickness (bubble size) of Layer II-6 basalt handaxe and handaxe-shaped flakes	376
Fig. 7.215	Length, width, and thickness (bubble size) of Layer II-6 basalt cleavers and cleaver-shaped flakes	376
Fig. 8.1	Limestone cortical flake (#2100 Layer II-6 Level 4)	383
Fig. 8.2	Limestone cortical flake (#9171 Layer II-6 Level 3)	383

Fig. 8.3	Limestone cortical flake (#3440 Layer II-6 Level 4b)	383
Fig. 8.4	Limestone flakes: a) small non-cortical limestone flake (#9633 Layer II-6 Level 5), b) limestone elongated flake (#13311 Layer II-6 Level 3), c) limestone elongated flake (#1693 Layer II-6 Level 4b), d) limestone cortical flake (#14322 Layer II-6 Level 7), e) small non-cortical limestone flake (#4329 Layer II-6 Level 6)	385
Fig. 8.5	Limestone knapping percussor (#705 Layer II-6 Level 4)	389
Fig. 8.6	Limestone knapping percussor (#413 Layer II-6 Level 4)	389
Fig. 8.7	Limestone knapping percussor (#844 Layer II-6 Level 4)	389
Fig. 8.8	Limestone knapping percussor (#2971 Layer I-5)	389
Fig. 8.9	Limestone *percuteur de concassage* (#9545 Layer II-6 Level 7)	391
Fig. 8.10	Limestone *percuteur de concassage* (#7330 Layer II-6 Level 1)	391
Fig. 8.11	Limestone modified artifact with scar resulting from spontaneous shattering (#7354 Trench II)	391
Fig. 8.12	Limestone chopping tool; note pecking/battering signs on the proximal end (#6038 Layer II-6 Level 4b)	392
Fig. 8.13	Limestone pitted stone on massive scraper (#663 Layer II-6 Level 1)	393
Fig. 8.14	Limestone core varia (#891 Layer II-6 Level 4)	393
Fig. 8.15	Length, width, and thickness of limestone artifacts (nodules, CCT, split hammers, FFT)	395
Fig. 8.16	Proposed scheme for the limestone reduction sequence	395
Fig. 9.1	Schematic illustration of the reduction sequences of a) flint, b) basalt, and c) limestone	400
Fig. 9.2	Frequencies of different artifact categories in Layer II-6 by raw material	403
Fig. 9.3	Frequencies of single, double, and triple tools on flint, basalt, and limestone	404
Fig. 9.4	Frequencies of FFT, CCT, and bifaces throughout the cultural sequence by raw material	409
Fig. 9.5	Frequencies of different artifact categories throughout the cultural sequence by raw material	410
Fig. 9.6	Frequencies of particular types of flint flake tools throughout the cultural sequence	411
Fig. 9.7	Frequencies of flint FFT with removal of striking platform throughout the cultural sequence	411
Fig. 9.8	Frequencies of main lithic categories of all raw materials throughout the cultural sequence	412
Fig. 9.9	Weights (kg) of different categories of basalt artifacts throughout the cultural sequence	414
Fig. 9.10	Ratios of microartifacts to artifacts throughout the cultural sequence by raw material	415

List of Abbreviations Used in the Text and the Tables

ATQ	Israel Antiquity Authority Archives: British Mandate Files
BF	bifaces
CCT	cores and core tools
BYF	Benot Ya'aqov Formation
CL	cleavers
CZA	The Central Zionist Archives
édtdb	*éclat de taille de biface*
DPF	dorsally plain flakes
EY	Eshel Ya'aqov
FFT	flakes and flake tools
FTIR	Fourier transform infrared spectroscopy
GBY	Gesher Benot Ya'aqov
IAA	Israel Antiquities Authority
IAAA	Israel Antiquity Authority Archives
IDF	Israel Defense Forces
Indet.	indeterminate
JB	Jordan Bank
KKL	Keren Kayemeth LeIsrael – Jewish National Fund Archive
LCT	large cutting tools
LFA	Large Flake Acheulian
msl	mean sea level
NBA	North of Benot Ya'akov Bridge (Acheulian site)
Std dev	standard deviation
Std err	standard error

Measures:

g	gram
ka	thousand years
kg	kilogram
km	kilometer
km^2	square kilometer
Ma	million years
m	meter
m^2	square meter
m^3	cubic meter
mm	millimeter

Chapter 1
Introduction

Abstract Chapter 1 presents the aims and contents of this volume together with the general geographical background of the study. The volume's main objective is to present a comprehensive description of the lithic component of Gesher Benot Ya'aqov. Such a description enables discussion of different cognitive, behavioral, and cultural aspects of the hominins who occupied the site. The long occupational sequence excavated at the site provides an exceptional opportunity to study the Acheulian material culture through time and in a unique geographical position on the route out of Africa.

In addition, this chapter provides the geographical background of the study and discusses the climate, vegetation, and fauna of Lake Hula and the Hula Valley in the past and present. There appears to be continuity in the habitats and their associated biological elements from Early Pleistocene to Recent times.

Keywords climate • fauna • flora • Gesher Benot Ya'aqov • Hula Valley • Lake Hula

This volume, the fourth in the Gesher Benot Ya'aqov (GBY) monograph series, is dedicated to the Acheulian lithic assemblages. These were excavated during seven field seasons (1989–1997a, b) under the direction of Naama Goren-Inbar of the Department of Prehistory in the Institute of Archaeology of the Hebrew University of Jerusalem. It provides the reader with a detailed account of all the lithic finds of the site, their stratigraphic location, and their typotechnological description. The significance of these assemblages derives not so much from their typological and technological characteristics but more from the fact that they originate from a unique setting. The excavations at GBY unearthed a long, successive, and well-delineated occupational record in which individual cultural levels are preserved and have been excavated with an exceptionally high resolution.

Such circumstances are unique among Early and Middle Pleistocene sites, since Levantine open-air sites seldom preserve long stratigraphic sequences. These are more frequently recorded in cave sites, where it is often impossible to distinguish between individual occupational events. GBY is an open-air site in which lake-margin occupations were preserved throughout a long stratigraphic sequence, each preserving botanical, faunal, and lithic remains in mint condition. This situation is unmatched in the Levantine record of Early and Middle Pleistocene archaeological sites.

Furthermore, the chronological and geographical position of GBY places it at the center of a crucial process in human evolution. Its location in the Levantine corridor, one of the diffusion routes out of Africa and into Eurasia, offers a unique opportunity to understand evolutionary processes by examining the behavior and cognition of hominin populations in the midst of diffusion, adaptation, and colonization.

These processes of diffusion and adaptation introduced the Acheulian culture, identified by its characteristic large cutting tools (i.e. handaxes and cleavers), into GBY and the Levant in general. The Acheulian technocomplex first appeared in East Africa at ca. 1.76 Ma (e.g. Lepre et al. 2011) and persisted until 0.3–0.25 Ma over a wide geographical range that includes Africa, the Iberian Peninsula, Europe, Asia, and the Levant. The key figure initiating these diffusion processes was *Homo erectus* (*sensu lato*), who introduced African traditions, some clearly present at the site of GBY, into the Levantine Corridor.

To arrive at a better understanding of these processes, our aim has been to acquire knowledge of the ability of the GBY hominins to adapt to changing environments and habitats, an ability that facilitated these processes and eventually enabled the colonization of Eurasia. The position of GBY, both geographically and chronologically, makes it a most promising place to examine these processes. Recognizing the capabilities of the GBY hominins can provide the background for understanding of the tempo and variability of cultural change during the diffusion processes of the Early and Middle Pleistocene.

GBY is a waterlogged Acheulian site located in the northern Jordan Valley (Israel), a segment of the Great African Rift

© Springer International Publishing AG 2018
N. Goren-Inbar et al., *The Acheulian Site of Gesher Benot Ya'aqov Volume IV*, Vertebrate Paleobiology and Paleoanthropology,
https://doi.org/10.1007/978-3-319-74051-5_1

Valley. This location is of great importance for scholarly attempts to understand the Acheulian material culture and hominin behavioral patterns in the Levant and the Old World in general. Fingerprints of the African Acheulian were recognized at GBY during early stages of archaeological research. The discovery of artifacts made of volcanic rock (in this case basalt) and the occurrence of a particular morphotype (the cleaver) led Stekelis to emphasize how these finds differed from those known at that time from other Lower Paleolithic sites in Israel and Europe (Stekelis 1960; Chapter 2). It was only years later, with the advancement of research, that the "Out of Africa" route and the passage through the Levantine Corridor became a widely discussed model. GBY and some other Lower Paleolithic sites in the Levant have made an enormous contribution to our understanding of the evolution and spread of mankind and the colonization of Eurasia (Goren-Inbar and Speth 2004).

To date, the records of the two earliest Levantine Lower Paleolithic Acheulian sites, 'Ubeidiya (Bar-Yosef and Goren-Inbar 1993) and GBY, provide milestones in the quest to understand the mechanisms of survival, adaptation, and behavior of Early and early Middle Pleistocene hominins. These two sites in two adjacent basins of the Dead Sea Rift, both comprising multiple stratigraphically superimposed Acheulian occupations in a lake-margin environment, have furnished an unprecedented wealth of stone artifacts and fossil bones, representing long-term exploitation of the landscape during some 100 ka (Feibel 2004). 'Ubeidiya and GBY have also furnished excellent data on the cultural evolution of the Acheulian along a time trajectory, representing two distinct, chronologically separated entities, 'Ubeidiya at around 1.6–1.2 Ma (e.g. Martínez-Navarro et al. 2012) and GBY at around 0.8 Ma, assigned to MIS 20–18 (Goren-Inbar et al. 2000; Feibel 2004).

The lithic assemblages of GBY, originating in well-defined strata, are first recorded a few meters above the Matuyama/Brunhes chron boundary (0.79 Ma; Goren-Inbar et al. 2000) and are present throughout the depositional sequence (Chapter 3). In some strata the numbers of items are small, but in others thousands to tens of thousands of stone artifacts were found in association with paleobotanical (Melamed 1997, 2003) and paleontological assemblages (Rabinovich and Biton 2011; Rabinovich et al. 2012). Among the archaeological horizons, 14 yielded large enough assemblages for detailed analysis. As they are located stratigraphically one on top of the other and spatially in the same locality, the paleo-Lake Hula margin, they should be viewed as 14 independent, chronologically separated archaeological occupations (sites). Thus, the record of the GBY site should actually be considered an accumulation of several sites. The estimated duration of the entire depositional column is 100 ka (Feibel 2004) and that of the cultural sequence is estimated to be in the order of 50 ka (Feibel personal communication 2009).

The GBY site is unique within the record of Acheulian sites in the Levant. It belongs to the Large Flake Acheulian (LFA) industries (Sharon 2007) and is unmatched among the 365 Acheulian find spots and sites that have been registered in Israel (Alperson-Afil and Goren-Inbar n.d.). It is also the only waterlogged Acheulian site known in southwestern Asia, thus contributing unique finds of organic material that add much to our ability to reconstruct, among others, the environment of paleo-Lake Hula and adjacent areas and the diet of the Acheulians who lived there. Farther away from the eastern Mediterranean zone, the sites that display the closest similarities to GBY are African sites such as Olduvai Gorge (Leakey and Roe 1994), Olorgesailie (Isaac 1977), Kilombe (Gowlett 1991), Isenia (Roche and Texier 1991), Kalambo Falls (Clark 2001), Melka Kunture (Piperno 2001), Gadeb (de la Torre 2011), Konso (Beyene et al. 2013), and Kokiselei (Lepre et al. 2011), to mention but a few.

Although this volume focuses on the lithic assemblages, the lithics are only one facet of the much larger cultural realm of the GBY hominins. Our knowledge of these hominins, acquired through long years of research on the GBY site, sheds light on various cultural, behavioral, and cognitive aspects. The location of the site on the shore of paleo-Lake Hula provided the GBY hominins with a favorable setting in which recurrent occupations took place. The hominins' ability to exploit their natural surroundings, including both wet and dry habitats, furnished a wide dietary spectrum. This involved both animal (medium-sized and large mammals, fish, and possibly crabs, reptiles, and birds) and plant (fruits, nuts, seeds, and underground storage organs) resources, which were brought to the site and processed there. In addition, the use of fire is recorded in each of the occupational layers, suggesting that the hominins of GBY possessed knowledge of fire making. A more detailed review of the various behavioral aspects of the GBY hominins is provided in the closing section of Chapter 2 and in the discussion of this volume (Chapter 9). The geology, sedimentology, and dating of the site are described in detail in Chapter 3. The methodology applied at GBY, both in the field and in the analyses of the lithic inventory, are detailed in Chapter 4.

In this volume, it is our aim to provide the reader with a detailed account of all the lithics excavated at the site. It is in our interest to add to the small number of reports an additional contribution that will provide data showing that the biface, the *fossile directeur* of the Acheulian, is only one aspect of a long and multifacetted cultural tradition. We hope to demonstrate that behavioral patterns can be discerned from other tool types, even from what are frequently considered waste products.

We consider the provision of a comprehensive description of the assemblages to be a most important aspect of our study. The lithic content of each of the artifact-bearing strata is given according to its location in the depositional sequence of the site (Chapter 5). Following this is a breakdown of the lithic assemblages into three distinct groups of raw materials: flint (Chapter 6), basalt (Chapter 7), and limestone (Chapter 8). The marked differences among these materials (in hardness, preservation, fracture mechanics, etc.) dictated a separate analysis for each. While we have published several studies on the lithic assemblages of GBY over the long years of research, this volume integrates them all and adds many and diverse studies, particularly in the domain of the flint flakes, tools, cores, and core tools, but also in those of the basalt flakes and cores and the limestone assemblage, most previously unpublished.

In our view, understanding of the cultural signature can derive only from study of the entire lithic assemblage. Hence, this volume includes a discussion (Chapter 9) that takes into account all three raw materials and addresses different cognitive, behavioral, and cultural aspects of the GBY hominins.

In view of the richness of the site and the extent of the various analyses, it is only natural that this volume focuses on the site itself and does not include detailed comparisons with other contemporaneous or non-contemporaneous lithic assemblages originating in the Levant or elsewhere. Clearly, such comparisons are both important and complementary and will be pursued in future studies. In the near future we intend to provide the academic community with a 3-D digitized image archive of lithics from GBY, which will include different categories of tools and waste and will enable access and new analyses, in addition to those presented here. We hope that in this way we shall contribute to the ongoing research of the Acheulian material culture.

1.1 The Hula Valley

1.1.1 Geography

The site of GBY is located in the northern sector of the Dead Sea Rift (Horowitz 2001) within the northern part of the Jordan Valley – the Hula Valley. This valley (Fig. 1) is bordered by rift escarpments on the east (the Golan Heights) and west (the Naftali Mountains), which rise to 500–1,000 m above mean sea level (msl). Two landform barriers delimit the Hula Valley within the rift: the Iyoun Valley and the flanking Hermon range (up to 2,800 m above msl) in the north, and the elevated basaltic block of Korazim (up to 300 m above msl) in the west and south. The Hula Valley is a long, narrow and flat region extending over 177 km^2 (25 km long and 6–8 km wide), with an average elevation of 70 m above msl. The southern sector was occupied by the shallow Lake Hula (14 km^2) until it was drained in the 1950s and the northern sector by swamps (Dimentman et al. 1992; Karmon 1960). The central part of the valley is covered by peat and lignite deposits and soils resulting from the rich organic material growing in the swamps (31 km^2) (Karmon 1960). The northern part, slightly more elevated, is covered by alluvial soils and crossed by several streams that supply water to the whole of the Jordan Valley (Belitzky 1987; Horowitz 1979; Karmon 1953). Springs are located mainly on the western edge of the valley. An extensive surface in the northern part of the valley was once covered by swamps, on which dense stands of papyrus grew; an extremely small surface, a small spot, appears in the southeastern corner of the lake and the valley and was mapped as early as the mid-nineteenth century (Dimentman et al. 1992: fig. 2). The Jordan River flows out of the Hula Valley (previously the outlet of Lake Hula) and through the basalts of the Korazim Saddle, forming a deep canyon for some 18 km until it drains into the Sea of Galilee.

In recent years and subsequently to the reclamation of the lake, two water bodies were established in the Hula Valley: the Hula Nature Reserve and Lake Agamon (the Hula Nature Park and Bird Sanctuary), both providing small-scale reconstructed ecological niches and habitats mimicking the conditions that prevailed during the times when Lake Hula existed (for evaluation of the drainage and the contribution of the newly established freshwater bodies, see Inbar 2002).

1.1.2 Lake Hula

Until the mid-1950s a shallow freshwater lake, Lake Hula, existed in the southern parts of the Hula Valley. The catchment area of the Hula Valley is 1,470 km^2, from which the lake received average precipitation in the order of 740 million m^3 annually. The lake also received three quarters of the flow of the Jordan River, which crossed the northern marshes and reached the lake; the river was fed by three streams, Hermon, Snir, and Dan (Dimentman et al. 1992). Lake Hula was 5.3 km long and 4.4 km wide, extending over 12–14 km^2. It was a pear-shaped freshwater body, shallow (1.5–4 m deep) and fluctuating seasonally from 21 km^2 in summer to some 60 km^2 during winter, with extensive seasonal and inter-annual fluctuations of water level (Dimentman et al. 1992). The lake was characterized by floating, submerged, and emergent vegetation, which furnished rich habitats for varied fauna, with some endemic species. It was this unique habitat that enabled the survival of the rich flora and fauna of the lake and its surroundings and provided migratory birds with a key feeding station on their route between Europe and Africa – a situation that still prevails today. Dimentman (Dimentman et al. 1992) suggests that Lake Hula contained the richest diversity of aquatic biota in the Levant south of Lake Amiq in Turkey, which was also drained in recent times (Ashkenazi 2004).

1.1.3 Climate

The climate of the Hula Valley is a warm Mediterranean one, due to its geographical setting and relatively low elevation. It

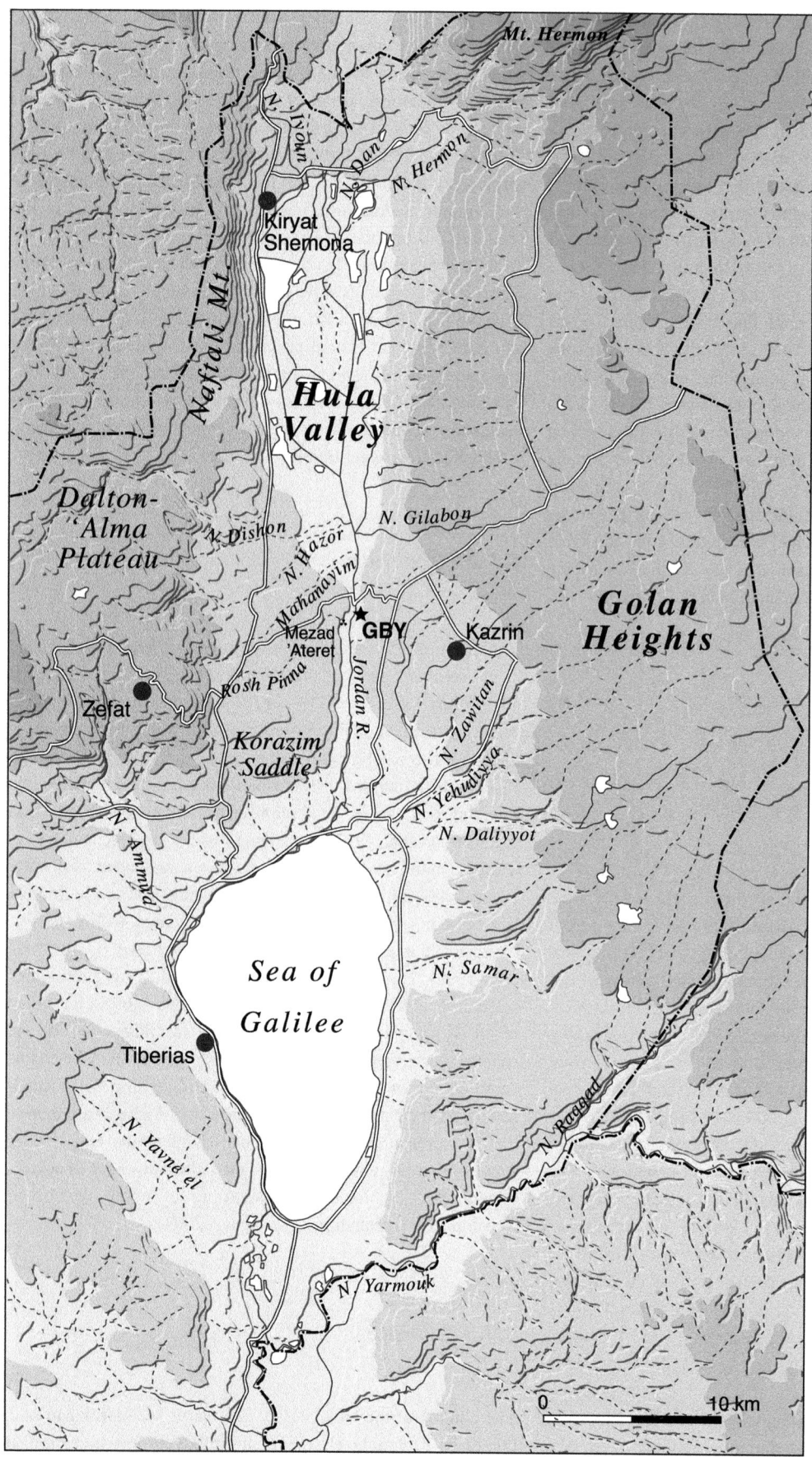

Fig. 1.1 Location map of the upper Jordan Valley (the Hula Valley and vicinity)

is characterized by hot dry summers and by cool rainy winters. The maximum summer temperature is 40°C, while during winter the temperature may drop below 0°C. The mean annual temperature is 21°C (Gat and Paster 1974). Annual precipitation in the valley ranges from 400 to 800 mm (Karmon 1960; Gat and Paster 1974).

1.1.4 Vegetation

The Hula Valley lies within the Mediterranean phytogeographical zone (Zohary 1959, 1962). Today most of the region comprises cultivated and fallow land. The wild vegetation accordingly consists mainly of segetal and ruderal plants. Among the segetal vegetation on the alluvial plain are scattered dwarf shrubs of *Prosopis farcta* (screw bean) (Zohary and Orshansky 1947). Along water margins grow trees of *Salix* (willow) and scattered trees of *Fraxinus syriaca* (ash). On dry land there are scattered trees of *Quercus ithaburensis* (Tabor oak) and *Pistacia atlantica* (Atlantic terebinth). On the border of the plain *Ziziphus spina-christi* (jujube) and *Z. lotus* (lotus thorn) are quite common. The original climax flora of the border was most likely the *Quercus ithaburensis–Pistacia atlantica* association (Zohary and Orshansky 1947). *Fraxinus syriaca* is concentrated today mainly along the Dan Valley in the north, together with *Salix acmophylla* (common willow). *Quercus ithaburensis* and *Pistacia atlantica* grow on the slopes of the Golan Heights adjacent to the Hula Valley in the east, in the Dan Valley, and in Upper and Lower Galilee. *Quercus ithaburensis* also grows in stands on the basalt of Korazim in the west and south. *Ulmus canescens* (hairy elm) grows in shady creeks by water in the Lower Galilee (Zohary 1966). An association of *Quercus calliprinos* (Kermes oak) and *Pistacia palaestina* (Palestine terebinth) is dominant in the mountains of Galilee west of the valley (Zohary 1960).

1.1.5 Fauna

The eastern Mediterranean, and in particular the Levantine Corridor, are regions of extreme biogeographical heterogeneity. The present-day fauna are the outcome of processes that took place during the Neogene and Pleistocene, and hence the Hula Valley fauna reflect a past biogeographical heterogeneity of elements of Palaearctic, Palaeotropic, Saharo-Arabian, and Ethiopian origin (Tchernov and Yom-Tov 1988). The extant taxa that originate in these distant biotic provinces are present all along the Dead Sea Rift and in its northern segment, the Hula Valley. The unique biogeographic setting of the Levant was described by Tchernov as "…characterized by patchy pattern of distribution and high disjunctiveness of many species, rather than obvious biogeographical gradient, and by scattered distribution of older relicts… small regions where evolving endemics are found, and amazing intermingling of species originating from very different geographic regions" (Tchernov and Yom-Tov 1988:239). While the Hula Valley is mainly a Mediterranean zone, it is a meeting point where taxa of African origin (e.g. Ethiopian: *Procavia capensis*, *Chamaeleo chamaeleon*) coexist with those of Euro-Siberian origin (e.g. *Pseudepidalea viridis*). These are associated with European carnivores, such as a species of otter (*Lutra lutra*), and among others with Asiatic species such as the mole-rat (*Spalax ehrenbergi*) and Gunthers's vole (Tchernov and Yom-Tov 1988). Before the lake's drainage, some 260 species of insects, 95 crustaceans, 30 snails and clams, 21 fishes, 7 amphibians and reptiles, 131 birds, and 3 mammals were recorded within the lake and in its immediate surroundings (Dimentman et al. 1992).

Recent sightings in Lake Agamon (Eshel 2008; Barnea 2009) include 222 taxa of birds and 7 species of mammals (excluding bats and micromammals): *Lutra lutra*, *Felis chaus*, *Vulpes vulpes*, *Sus scrofa*, and *Lepus capensis*. Faunal sightings in the Hula Valley beyond the wet habitats include 22 species of Actinopterygii, 273 species of birds, 26 species of mammals, and 12 species of reptiles (including introduced species) (Oron 2012). Human population growth, agricultural cultivation, animal husbandry, and the extensive use of pesticides have drastically changed the ecosystem and caused far-reaching changes in the faunal species and their abundance in the Hula Valley (details in Yom-Tov and Mendelssohn 1988). Yet, Lake Agamon and the Hula Nature Reserve have made a great contribution to the preservation of the Hula Valley fauna, particularly its birds.

1.1.6 Geology

The northern sector of the Dead Sea Rift, where the Hula Valley is located, is a segment of the Afro-Arabian Rift system (Prodehl et al. 1997; Horowitz 2001). This system is a complex rhomb-shaped graben structure bordered by internal faults, formed structurally by the 1,000 km long transform fault boundary of a plate convergence zone (Garfunkel 1981; Horowitz 2001; Belitzky 2002; Schattner and Weinberger 2008). The geological structure of this taphrogenic depression developed rapidly, with an estimated subsidence of some 2,500 m during the last 4 Ma (Heimann 1990; Belitzky 2002). According to Horowitz (1973), freshwater lakes and marshes have existed nearly continuously in the basin. The formation of the basin and its subsidence caused accumulation of lacustrine, paludal, and fluvial sediments with some basaltic intercalations (Horowitz 1973, 2001). Horowitz further suggested that intense downsinking took place between 1.5 and 1.25 Ma and during the last 0.7 Ma (Horowitz 1989); this has been refined with the study of GBY and is now estimated to be younger than 0.658 Ma (Sharon et al. 2010).

The sedimentary sequence of the Benot Ya'akov Formation (BYF) was defined by Horowitz in the Emeq Hula borehole (Horowitz 1973). Its thickness there is ca. 185 m and it is composed of a series of sediments, mainly of limnic chalk rich in malecofauna, overlain by the Hulata Formation and unconformably overlying the Yarda basalt (Horowitz 1979:135–136).

Outcrops along the Jordan River from its outlet from the Hula Valley to the Benot Ya'aqov bridge were studied by Picard (1963). His "Old Jordan Terrace" was 8 m thick with a marked middle horizon, marly and rich in shells of the endemic fossil gastropod *Viviparus apameae*; Picard named this horizon "Viviparus beds". These beds were named the Benot Ya'akov Formation by Horowitz (1973) and have undergone multidisciplinary investigations from the mid-1930s until today (Goren-Inbar and Belitzky 1989). The series of archaeological assemblages that originate in the excavations of GBY are bedded in these "Viviparus beds" of the BYF (Chapter 3).

1.1.7 The Hula Valley in the Past

The multidisciplinary research of GBY and its results have made remarkable scientific progress since the days when Stekelis discovered that the finds actually derived from strata *in situ* (Stekelis 1960). Analyses of the floral, faunal, sedimentological, and lithic assemblages originating from the renewed excavations have yielded a wealth of evidence on the prehistoric landscape of the Hula Valley.

Analysis of 1,568 wood macrobotanical remains has resulted in the identification of 26 taxa of trees, bushes, and climbers. Only one wood taxon is unknown in the Mediterranean today and remains to be identified. All 26 identified wood taxa are typical of the eastern Mediterranean flora and most grow today in the Hula Valley and adjacent areas (Goren-Inbar et al. 2002b). Analysis of some 100,000 additional macrobotanical remains yielded 107 identified taxa of nuts, fruits, vegetables, and plants producing underground storage organs, of which 14 are extinct wet-habitat species and over 55 species are edible (Melamed et al. 2011; Melamed et al. 2016).

Fossil faunal assemblages from other archaeological excavations documenting the last million years have provided insight into the history and evolution of the Hula Valley fauna. The excavations of the Acheulian GBY, the Mousterian Nahal Mahanaim Outlet (NMO), the Kebaran site north of GBY, and the Natufian Eynan have all furnished outstanding faunal data for the entire Pleistocene of the Hula Valley. The GBY assemblages illustrate the antiquity of many of the valley's extant species. These include mollusks (Mienis and Ashkenazi 2006, 2011), crustacea (Ashkenazi et al. 2005), fish (Alperson-Afil et al. 2009; Zohar and Biton 2011), amphibians (Rabinovich and Biton 2011), reptiles (Hartman 2004; Rabinovich and Biton 2011), birds (Simmons 2004), and terrestrial mammals (Rabinovich et al. 2008, 2012; Rabinovich and Biton 2011). Comparison between the present taxa of the Hula Valley and those of the prehistoric record shows that this ecological niche has retained many of its biotic elements throughout the Pleistocene and the Holocene. While most of the extinctions have been caused by human intervention, some of the species demonstrate resilience (e.g. the Hula painted frog *Latonia nigriventer*; Biton et al. 2013) and genetic continuity (e.g. the turtle; Hartman 2004).

In summary, the ecological characteristics of Lake Hula and its margins suggest overall continuity from Early Pleistocene to Recent times. The present-day habitat of the Hula preserves floral and faunal components that are recorded from some 0.78 Ma at the site of GBY. This setting provides a unique opportunity to examine the cultural and behavioral characteristics of the GBY hominins in their natural environment. To achieve better understanding of the material culture of GBY and the behavioral and cognitive aspects derived from it, we conducted in-depth analyses of the lithic assemblages from the site. This volume presents the results of our analyses of the entire lithic inventory of the site in an attempt to achieve insight into the Acheulian culture in general and the behavioral patterns and abilities of the GBY hominins in particular.

Chapter 2
History of Research

Abstract Chapter 2 presents the history of the archaeological discoveries and excavations at Gesher Benot Ya'aqov, from as early as the 1930s to very recently. Throughout these years, the area of the site was subjected to recurrent activities of construction, destruction, and drainage. The chapter provides a detailed account of the history of research on Gesher Benot Ya'aqov and the Benot Ya'akov Formation by scholars such as D. Garrod, M. Stekelis, and D. Gilead. Furthermore, as this volume is dedicated to the lithic assemblages originating from recent (1989 onwards) archaeological excavations at the site, an account of this phase of archaeological research is provided, as well as a summary of the main scientific achievements derived from it.

Keywords excavations · excavators · Benot Ya'aqov bridges · history of research · Hula drainage · surveys

The history of research at Gesher Benot Ya'aqov (GBY) involves three main topics. The first concerns the history of the construction, maintenance, and demolition of bridges that were built at the shallowest and hence the easiest crossing of the Jordan River, where the Acheulian site is located. Understanding of these activities provides the background to the archaeological discoveries and the importance of the area of the bridges to the prehistoric and paleobiological studies.

The second topic is the history of the various attempts to drain Lake Hula, which resulted in the prolonged destruction of the landscape and the removal of over a million cubic meters of Pleistocene sediments containing prehistoric remains. These events all touch on the area of the bridges, as this area was considered to be the core of the Hula Basin's drainage problems due to the presence of a massive volcanic flow that topographically "blocked" the drainage of the Jordan River from the Hula. The events associated with the drainage of the Hula, though relevant to the understanding of the general history of the GBY site, do not directly involve the precise location where the prehistoric excavations took place. Thus a detailed account of this topic is provided in Appendix 1.

The third topic is the history of prehistoric research at the site from the first identification of lithic artifacts in the 1930s until the most recent excavations, which are the subject of this volume.

2.1 The History of the Benot Ya'aqov Bridges and Their Area

The discovery and repeated destructions of the Acheulian site of GBY are associated with the different bridges built on the crossing of the Jordan River and coinciding with the location of the Acheulian site. For a better understanding of the site and its history of research, we will describe the history of the bridges, which are founded on the sediments of the Benot Ya'akov Formation (BYF). As the site extends for about 3.5 km (Belitzky 2002), many of the research activities and points of reference have involved the area of the bridges.

The bridge of Benot Ya'aqov on the route from Egypt to Syria is positioned about 15 km north of Lake Kinneret on the Jordan River, south of the river's outlet from Lake Hula, at about 70 m above msl. The location of the bridge, which possibly dates from Roman times, is considered the best crossing point (Karmon 1973:170). In about 1260 AD Sultan Baibars built an impressive basalt bridge, 18 m long; it had three arches and a tower at its western end, and lacked a parapet (Fig. 2.1) (Tepper and Tepper 2003). In 1887 a fourth arch was added to the existing bridge during an attempt to drain and reclaim the area of the Hula Basin by deepening the river and thus improving its flow capacity (Fig. 2.2). This basalt bridge was damaged in 1918 during World War I by the retreating army of the Ottoman Empire, and was then rebuilt by Australian soldiers (Fig. 2.3).

In 1930 the stone bridge was purposely demolished by the British and it was replaced by a steel "Bailey" bridge a few meters to the south. The building activities caused further deepening of

© Springer International Publishing AG 2018
N. Goren-Inbar et al., *The Acheulian Site of Gesher Benot Ya'aqov Volume IV*, Vertebrate Paleobiology and Paleoanthropology, https://doi.org/10.1007/978-3-319-74051-5_2

Fig. 2.1 The Mamluk stone bridge with three arches: a) general view ca. 1895 (Central Zionist Archives), b) close-up view ca. 1912 (Central Zionist Archives)

Fig. 2.2 The Mamluk bridge with four arches: a) general view showing the Mamluk caravanserai and the Lower Customs House ca. 1895 (courtesy École Biblique), b) close-up view (SRF_76; courtesy IAA)

the Jordan River's bed, which lowered the water level of Lake Hula (Fig. 2.4). This new bridge was severely damaged in 1946 during a military operation of the Haganah against the British Mandatory authorities, known as *Leil Hagsharim* ("the Night of the Bridges"), involving the demolition of several other bridges. It was once again severely damaged in 1948 during the War of Independence. During the Six Days' War in 1967 another "Bailey" bridge was erected by the IDF Engineering Corps south of the British "Bailey" bridge. These two bridges were in use, one in each direction, until 2007, when the southern one was removed and replaced by a concrete bridge, which today carries all traffic across the river in both directions (to and from the Golan Heights).

Fig. 2.3 Australian soldiers repairing the Mamluk bridge during World War I ca. 1918 (SRF_76; courtesy IAA)

2.2 The History of Archaeological Research

The prehistoric site of GBY has been known for some eighty years, during which many changes have taken place geographically, geopolitically, and scientifically. Within the area studied, names of Quaternary formations have been changed, marker beds have been identified and ruined, and, above all, changes induced by agricultural/industrial activities have destroyed the ancient extent of the Acheulian site. The history of research will be described here chronologically.

2.2 The History of Archaeological Research

Fig. 2.4 The new British bridge: a) general view (1949), b) close-up view showing the erosion caused by the deepening of the Jordan River's bed by one meter by 1951, causing further damage to the Acheulian site (KKL Archives)

2.2.1 1934–1937

The GBY archaeological site (north of the bridges) was discovered due to drainage and bridge construction activities. These two are intermingled, as described in Appendix 1. Because of the concept that the bridges were an obstacle to better drainage of the Hula Valley, the British Mandatory authorities decided to demolish the stone "Old Bridge" of the Mamluk period, a task that was carried out on October 4th, 1934. Two days later a report was written by Inspector N. Makhouly (6/10/1934): "Upon examination it was found that the river area on both sides of the Old Bridge is very rich in stone age flint implements, an evidence that the Stone Age Man (prehistoric) was dwelling on both banks of the river" (Israel Antiquities Authority Archive, henceforth ATQ; ATQ/512). In addition, the prehistoric site of GBY became known in that year because of the construction of a new bridge over the Jordan River, south of the old stone one. The construction of this bridge involved major drainage activities that included the lowering of the riverbed by two meters. This caused much damage to both river banks, where the dredged material was dumped in heaps, thus masking the visibility of sedimentological and volcanic outcrops and consequently the stratigraphy along the river bank.

The first to observe the prehistoric remains were the bridge constructors, though they were incapable of evaluating the meaning of the finds (mainly fossil mammal bones). Outstanding was the conduct of Mr. Horowitz, manager of the "Hula and Merom Sea Concession", who collected mammal bones from the earth dumps and sent the material to the then President of the Hebrew University, Dr. J.L. Magnes, who in turn gave it to Prof. G. Haas (1935), who identified fossil bones of elephants (remains of two molars and a single elephant tusk) (Stekelis and Picard 1936:214).

From 1934 onward, these heaps and additional ones (see below), all resulting from drainage and bridge construction activities, were the main source of archaeological and paleontological finds. During the same period D.A.E. Garrod and E.W. Gardner were the first to observe and report the presence of handaxes as well as splinters of fossil bones (Stekelis 1960). Garrod approached the Mandatory Department of Antiquities to request a permit to excavate some soundings (2/4/1935); she received the permit (no. 248) immediately and conducted her archaeological investigation in April 1935 (Figs. 2.5–2.6). During the same month she

Fig. 2.5 View of the Jordan River's right bank and the remains of the destroyed Mamluk stone bridge (looking north), photographed by D.A.E. Garrod in 1935 (courtesy Pitt Rivers Museum)

Fig. 2.6 View of the Jordan River's right bank, the remains of the Mamluk stone bridge, and the team of D.A.E. Garrod during the excavation of trenches (showing Miss Gardner, the geologist), photographed by D.A.E. Garrod in 1935 (courtesy Pitt Rivers Museum)

day both Garrod and E.W. Gardner collected fragments of fossil elephant bones (a molar and tusks) north of the trenches and also near the abutment of the "Old Bridge". There is no exact reference to the location of the two trenches and the pit, and only a single spot is illustrated in the map of the site that was produced by Stekelis (Fig. 2.8).

A few months later, an additional discovery was made in the vicinity of the bridge. This time, under four meters of overburden, a mass of fossil bones was exposed during the excavation of a percolating pit serving the stables of the British Military Mounted Police post on the Syro-Palestine border. The bones, bedded in a yellow clay deposit, were excavated and sent to Damascus by the contractors and were subsequently lost (Stekelis 1960). In two consecutive reports (Stekelis and Picard 1936, 1937), geological and archaeological findings originating from a sounding next to the ruins of the "Old Bridge" and from the riverbed were detailed. These summarized the stratigraphy and the archaeological remains. The 1937 report for the second season gives the geological and archaeological findings of the sounding carried out on the left bank of the river near the ruins of the "Old Bridge". It reported the identification of four strata: "A. Upper gravel of the old Jordan bed with few flint implements of Levallois technique. B. Yellowish clay of the transition one with basalt pebbles; molars and few broken bones of elephants; flint

informed R. Hamilton (16/4/1935), then the director of the Department of Antiquities, that she was not interested in continuing her activities there "Should Dr. Picard & M. Stekelis apply for a permit to excavate at this spot, this will be done as a result of discussion with me, as I consider it better that Dr. Picard, who has made the Jordan valley his special field, should investigate this matter" (ATQ/75/6) (Fig. 2.7). In Garrod's diary (Garrod archive, Saint Germaine-en-Laye, France: Notebook 33431, p. 10, April 8th, 1935), she described the landscape in the location of her activity (on the right bank of the Jordan River and between the old and new bridges) and observed abundant fresh flint and the first ever sighting of a biface: "A very fine hand-axe lay edgeways in a shelly grit, pointing in the direction of the stream, only one edge being exposed. It was totally unabraded." According to her diary, she excavated a small trench to a depth of ca. 0.50 m but could not continue due to the presence of large basalt blocks at the bottom. The stratigraphy included (from top to bottom): "… yellowish material (a flint scraper was found at its base) passing down into decomposing basalt, blue-black in colour, & in places bright green". A second trench was excavated north of the first, penetrating only as deep as the yellow sediment; it was found to be archaeologically sterile. In addition to these two trenches, an attempt was made to clear an existing sounding, located north of the two trenches, which was considered to be three meters deep, but work could not continue due to the presence of water. During the same

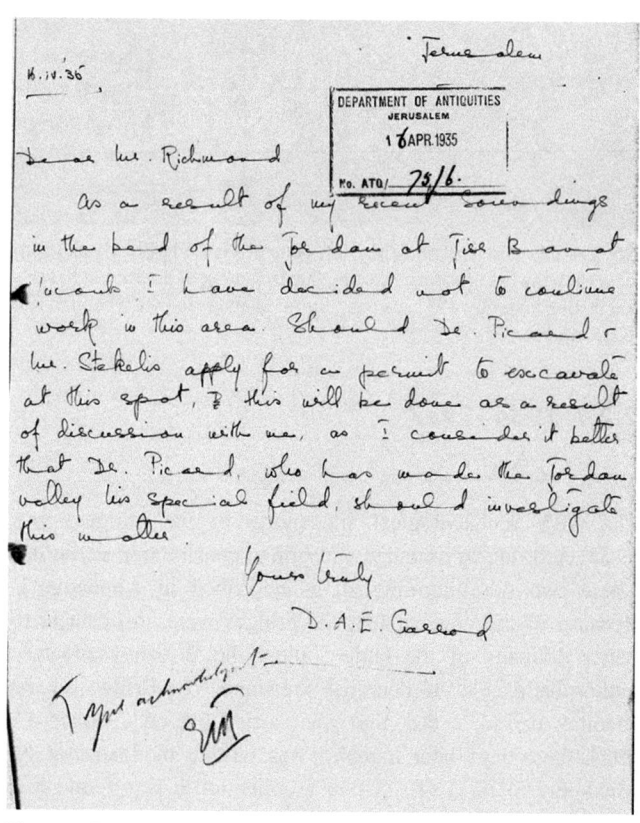

Fig. 2.7 Garrod's letter to the Mandatory Department of Antiquities notifying her intention to end the project (16/4/1935 ATQ/75/6; courtesy IAA)

2.2 The History of Archaeological Research

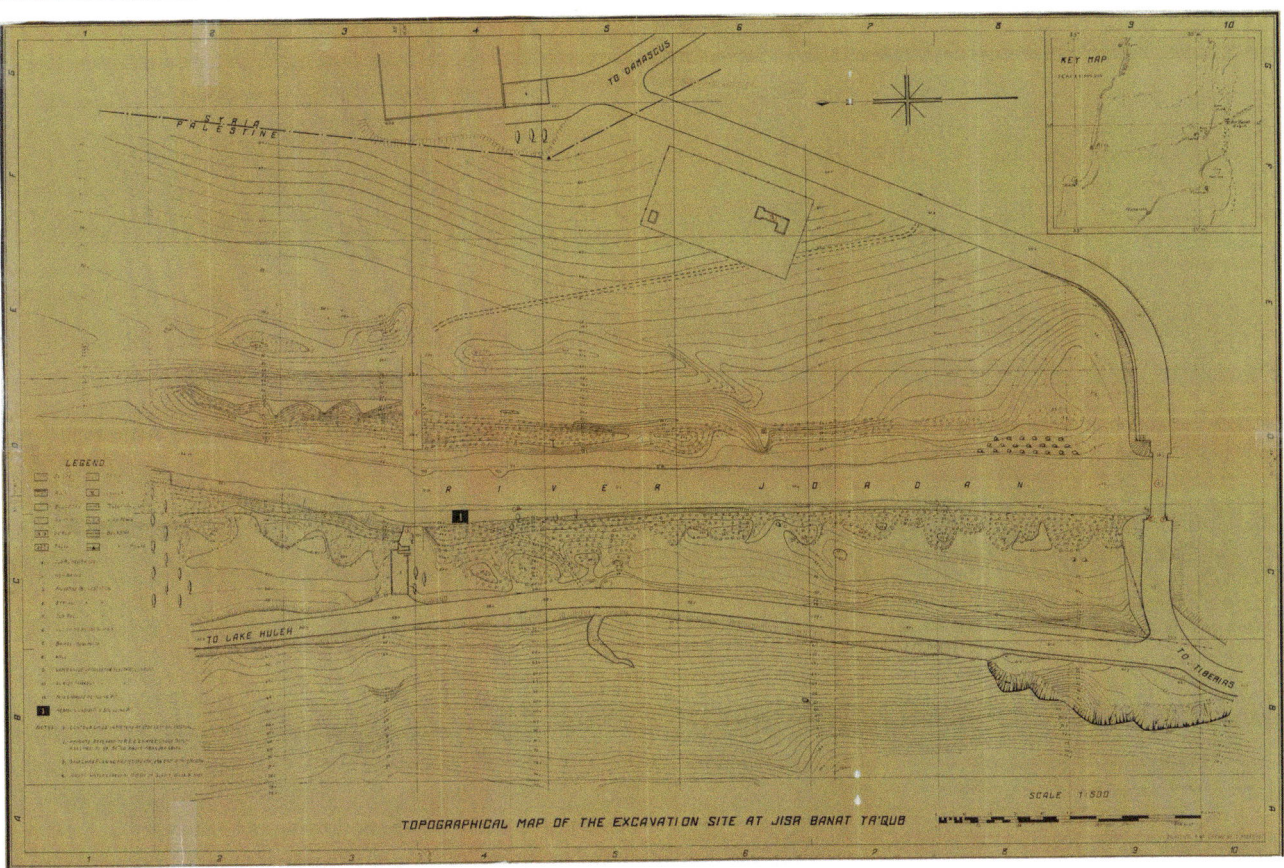

Fig. 2.8 Topographic map of the excavations at Gesher Benot Ya'aqov produced by the expedition of M. Stekelis (Stekelis et al. 1937), the black rectangle marking the location of D.A.E. Garrod's excavations; note the spoils along the right bank of the river and the disturbed area at the western end of the bridge (SRF_76; courtesy IAA)

implements of Acheulian technique. C. Green clay with basalt pebbles; molars and bones of elephants and other animals; flint implements of the same type as in Layer B. D. Lower black soil bed with basalt, flint, and limestone boulders, containing abundant fauna of mollusca (*Melanopsis* sp. and *Vivipara* sp.), elephant tusks (Fig. 2.9), molars, and parts of skeletons; flint and basalt implements of lower Acheulian technique" (Stekelis and Picard 1937:45).

During these and the following years, Stekelis conducted many field excursions to the vicinity of the bridge in order to examine the disturbed deposits. During one of these trips he observed the first basalt handaxe found up to then in Palestine (Stekelis 1960:61), but it was the discovery of a fossil cervid jaw, which he found on the shore of Sea of Galilee, that led him to think of its original context and eventually to conclude that it had been eroded by the Jordan River during a heavy storm and had been transported to where it was found. This realization led him to conduct the first large-scale archaeological survey in the Jordan River, since the task could not be carried out on the banks due to the disturbances caused by the repeated drainage activities described above. For proper investigation of the site, it was essential to survey the riverbed itself. And indeed: "On March 27, 1936 at midnight the sluice was let down and by 8.30 next morning no more than 0.30 m of water remained in the river and we were able to walk along and explore the river bed and its sections and look for fossilized bones and human artifacts" (Stekelis 1960:64) (Fig. 2.10). These special circumstances enabled Stekelis to survey the riverbed between the sluice (north) and the "New Bridge" (south) along ca. 3.5 km, reaching as far south as 200 m from the bridge. All along the riverbed, stone artifacts and fossil bones were found; the richest concentrations were observed in seven localities, four of them north of the bridge and three south of it (Fig. 2.11). Three of the localities situated to the north of the bridge, between it and the eucalyptus grove (marked on his map), were the most interesting; the details of each locality's finds were given in his report (Stekelis 1960:64). He further concluded that the artifacts and bones were not found *in situ* and summarized his stratigraphic observations of the river's right bank (Stekelis 1960).

In March 1937 Stekelis renewed his investigations. This time a pit measuring 7x7 m was dug on the right bank of the river south of the "Old Bridge" and its stratigraphy was described in detail (Stekelis 1960:66, fig. 4) (Fig. 2.12). No archaeological activities could subsequently be carried out due to logistical and geopolitical problems.

Fig. 2.9 Excavation of an elephant tusk in the waterlogged sediments, photographed by the expedition of M. Stekelis (SRF_76; courtesy IAA)

Fig. 2.10 The Jordan River after the sluice was let down (1936); note the exposed boulders (SRF_76; courtesy IAA)

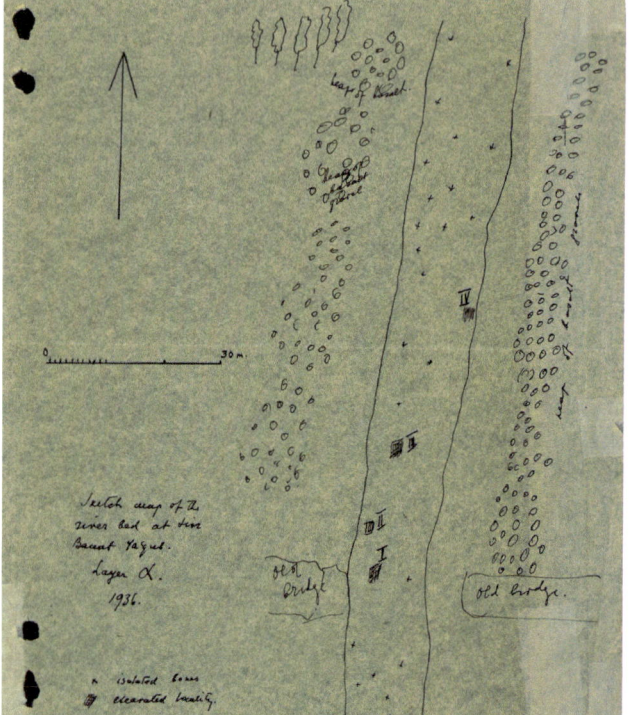

Fig. 2.11 Map sketched by Stekelis and Picard (1936) illustrating the localities excavated in the bed of the Jordan River (Stekelis' report of activities in 1936/7; SRF_76; courtesy IAA)

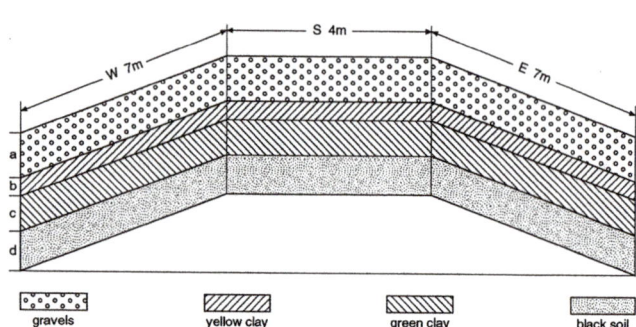

Fig. 2.12 Stratigraphy observed on the Jordan River's right bank (Stekelis 1960:66–67, fig. 4)

2.2.2 1951–1968

Stekelis' final field work and archaeological observations took place after 1951 (Figs. 2.13–2.15) following further drainage of Lake Hula by the Israeli authorities. The riverbed (already an artificial one) was deepened by more than 6 m and it became possible to modify some of the previous stratigraphic observations of the exposed sequence. The new sequence is presented in Figure 2.16. Following is a summary of Stekelis' description (1960:68, fig. 5) (from top to bottom):

Bed I: Gravel composed of flint and basalt pebbles (0.15x0.27 m), 0.42 m thick. Flint artifacts including Levallois technique.

Bed II: Basalt pebbles (0.55x0.07 m) grading into yellow clay, 18 cm thick. Handaxes (chert) and fossil mammal bones.

Bed III: Green clay with a few basalt pebbles and boulders (0.40x0.42 m), 0.30 m thick. Handaxes (chert) and fossil mammal bones.

Bed IV: Black soil including limestone, flint, and basalt

Fig. 2.13 View looking south of Stekelis' excavated pit (foreground) and the British bridge (background) (SRF_76; courtesy IAA)

2.2 The History of Archaeological Research

Fig. 2.14 View of Stekelis' excavated pit (SRF_76; courtesy IAA)

pebbles and boulders, 1.80 m thick. Abundant malecofauna (*Melanopsis* and *Vivipara*) and fossil mammal bones. Handaxes (basalt and chert).

Bed V: Hard black soil, 1.70 m thick. Abundant malecological fauna (*Melanopsis* and *Vivipara*) and fossil mammal bones. Many basalt artifacts, including fresh basalt handaxes and cleavers.

Bed VI: Gravel composed of flint limestone and basalt pebbles and boulders, 2.10 m thick. Abraded handaxes and cleavers.

Stekelis (1960:68–70) subdivided this sequence into upper (I–IV) and lower (V–VI) layers. He observed the use of chert and flint for artifact manufacture in the upper layers, and he also considered the very abraded basalt artifacts of these layers to be intrusive and not *in situ*.

Stekelis' publication of 1960 was a summary report of his archaeological activities and conclusions. In this report the main focus was placed on a detailed description of bifaces – handaxes and cleavers (Stekelis 1960). Summarizing the results of his research on the Acheulian site of GBY, Stekelis emphasized two main contributions: the first was the multidisciplinary approach (geological, paleontological, and archaeological) that was employed to study the site, the first of its kind during the preliminary stage of investigation of a site, and the second was the identification, for the first time ever in Palestine, of an open-air Acheulian site *in situ* that included both lithic and paleontological assemblages (Stekelis 1960).

The paleontological discoveries were studied and published during the same years by Bate and Hooijer (Stekelis et al. 1937; Hooijer 1959, 1960) and geological contributions were later provided by Picard (1952, 1963, 1965).

In 1967 two events took place that enabled the renewal of research at the site. The first was a geopolitical change in the border configuration, which allowed new research initiatives along the Jordan River. The second, associated with the first, was the renewal of drainage works along the river, which lasted for five months. D. Gilead accompanied these works by inspecting the deposits disturbed by heavy mechanical excavators during a period of four months. In April 1967 he carried out excavations of several test trenches north of Stekelis' pit of 1937 and identified the following stratigraphy some 20 m north of the "Old Bridge" (from bottom to top):

1) A layer of basalt boulders and pebbles over 2 m thick with flint tools and flakes and several fragments of animal bones.
2) A layer of dark clay rich in mollusks and 0.80 m thick, with a few flint flakes and a few fragments of animal bones.
3) An upper complex composed of lighter sandy clay rich in mollusks, 10–20 cm thick. Overlying it is a greenish-gray silt layer some 50 cm thick and a dark gray layer (ca. 40 cm thick) that contained no lithics or paleontological remains. The rest of the overlying layers were not examined.

According to Gilead's report, the objective of these excavations was to attempt to correlate the stratified sequences observed by Picard, Stekelis, and himself. This attempt was unsuccessful, as Gilead could not explain the following anomaly: "According to Stekelis' map... the upper part of the excavated stratigraphic complex appeared at 52 m above sea level and the excavation penetrated several meters lower. According to Picard's section (1963: fig. 3), the upper part of the same layers is observed at an elevation of 74 m above sea level and terminates at an elevation of 62 m above sea level. According to measurements recently taken, this deposit is located at an elevation of 62 m above sea level. Another difficulty is that while Picard remarks that the deposit is horizontal, in both Stekelis' section (1960: fig. 5) and the latest tests, this deposit was tilted toward the north or northwest" (ATQ/428/1–2; translated from Hebrew by N. G.-I.).

In addition to the above-mentioned activities, Gilead also surveyed sectors of the Jordan River banks along 4 km (mostly north of the bridge, though the exact area was not specified) and reported the presence of prehistoric finds on both banks. The right bank was found to be more interesting in view of its higher concentrations of artifacts. He also reported the presence

Fig. 2.15 View of Stekelis' excavated pit: a) note the coarse sediments in the section, b) the pit during excavation (SRF_76; courtesy IAA)

of Mousterian-like artifacts on the right bank of the Jordan some 300 m south of the new bridge.

In conclusion, Gilead described the enormous damage caused to the prehistoric site and doubted, in view of the difficulties mentioned above, the value of additional archaeological activities.

The most detailed lithic analyses of the GBY Acheulian material from both old surveys and excavations (Stekelis and Gilead) is included in Gilead's PhD thesis, which is based on four collections: 1) the Rockefeller Museum, 2) the Hebrew University, 3) the Department of Antiquities (currently the Israel Antiquities Authority), and 4) a collection made by Gilead (Gilead 1970). His analyses incorporated more than 380 artifacts of basalt and flint. Stekelis had assumed that the flint artifacts derived from Layers II–IV and the basalt ones from Layer V (Stekelis 1960). This view was adopted by Gilead (with appropriate reservations due to the need for further research), resulting in the division of the content of all collections according to their raw materials (basalt and flint); he even indicated that some of the flint bifaces were made "… with an elastic striker" (Gilead 1970:83). Gilead further indicated: "However, there are some indications that either there is a lower layer with both basalt and flint implements or that some flint implements do occur in Layer V. This must be verified by further research" (Gilead 1970:77). In the conclusion of his study he assigned GBY to the Middle Acheulian (Gilead 1970:320–321) and stated, comparing the assemblage with other Acheulian assemblages of the Old World: "There can be little doubt that in general the North African and the Near Eastern Acheulian compare rather closely. The resemblance in the main successive phases is quite striking… The 4,500 km separating the Near East from the Magreb apparently led to local variants…" (Gilead 1970:343). Emphasizing the importance of the Levant as a land bridge, he concluded: "There is, therefore, the highly interesting possibility of examination of the effects of cultural contacts and isolated development in our region" (Gilead 1970:352).

Gilead studied the lithics of both GBY and 'Ubeidiya long before the "Out of Africa" model was proposed and became the favored interpretation. In terms of 'Ubeidiya he wrote: "I would personally favor the hypothesis that it arrived from

2.2 The History of Archaeological Research

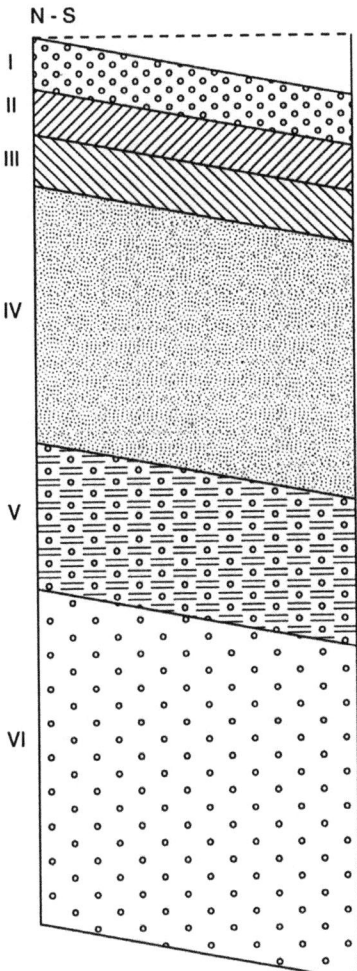

Fig. 2.16 Stekelis' final stratigraphic sequence (from Stekelis 1960: 68–69, fig. 5)

Fig. 2.17 The spoils of drained sediment originating from the bed of the Jordan River and its banks (after 1967)

Africa..." (Gilead 1970:316), he attributed more complexity to GBY, most probably due to the separation of the assemblages according to raw materials and the similarity observed between the flint handaxes derived from the upper units and those from Ma'ayan Baruch and other sites (Gilead 1970).

Heaps of gravels (spoils) originating from drainage activities in the course of the river and hence from the Acheulian site were available on both banks of the Jordan north of the present bridges and on the right bank south of it (Fig. 2.17). These heaps were a source of great interest to amateurs and collectors, but they also furnished finds for the various paleontological collections. The only discovery of hominin bones was made in the laboratory during a major paleontological revision carried out by Tchernov (1986), when he identified two hominin femur bones mixed with other paleontological remains of unknown context (Geraads and Tchernov 1983). An attempt to date these remains by U-Series radiometric method failed due to insufficient uranium content in the bones (Christophe Falguères personal communication 1994).

During these years at least three private lithic collections, which derive from the "surface" finds originating in the GBY Acheulian site, were formed. One of these was donated by the collector, D. Ben-Ami, to the Department of Prehistory at the Hebrew University. Of this collection only the cleavers have been studied (Goren-Inbar et al. 1991a; Zohar 1993), while the handaxes remained to be investigated. Much was gained from the cleavers study, especially with respect to technological aspects. These aspects of the data will be used for comparison with the data originating in the recently excavated material (see below). Ben-Ami's awareness of the details of the area where he collected the artifacts enabled us to use the material and to assign the artifacts to a specific area on the left bank of the river north of the bridges (Fig. 2.18).

Most of the amateurs' collections, by their nature, remain unknown. A recent change in the policy of the Israel Antiquities Authority may yield better access to these collections and their retrieval will enable us to widen the scope of comparison between the different areas of the site.

2.2.3 1981–1989

The geological survey and mapping conducted in the area of the bridge in the early 1980s resulted in the simultaneous discovery of freshwater tilted beds (64° E) by both Belitsky (Belitzky 1987, 2002) and Bar (Harash and Bar 1988) on the eastern flanks of the Crusader fortress of 'Ateret (*Vadum Iacob*). During a visit to the exposure, then an unassigned sedimentary deposit, in 1981, we scraped a minimal surface to reveal the presence of a flint handaxe bedded next to the gastropod *Viviparus apameae*, a guide fossil of the Hula Middle Pleistocene (Moshkovitz and Magaritz 1987) (Fig. 2.19). In additional visits to this exposure, other Acheulian artifacts (flint cleavers) were found, as well as mammal bones, including a fragment of an elephant tusk. Similar deposits were encountered repeatedly in very small exposures confined to the bed of the Jordan River and both banks during a geological survey conducted by Goren-Inbar and Belitzky between 1981 and 1989 (Goren-Inbar and Belitzky 1989). During this survey all visible units were re-mapped, resulting in a revised geological map (scale 1:50,000), presented and discussed in detail elsewhere (Chapter 3.2; Belitzky 2002). This fieldwork also included sampling of sedimentary formations for determination of paleomagnetism and age and sampling of volcanic formations for radiometric dating.

During the late 1980s, a plan proposed by Kibbutz Kfar HaNasi to construct a hydroelectric power station fed by an artificial lake on the Jordan River necessitated a geological survey in order to prevent possible destruction of the Pleistocene archaeological deposits. In July 1989 this survey was conducted by means of excavation of a series of pits with heavy machinery. The pits were arranged in a linear manner along three transects with the addition of two isolated pits. Four pits were dug along 73 m in the northernmost transect, exposing topsoil and the Gadot Formation. South of it, six pits were excavated along a transect of 117 m, exposing topsoil and basalt boulders. The sediments of the BYF were not exposed. The interpretation of this sequence is fully discussed in Chapter 3.

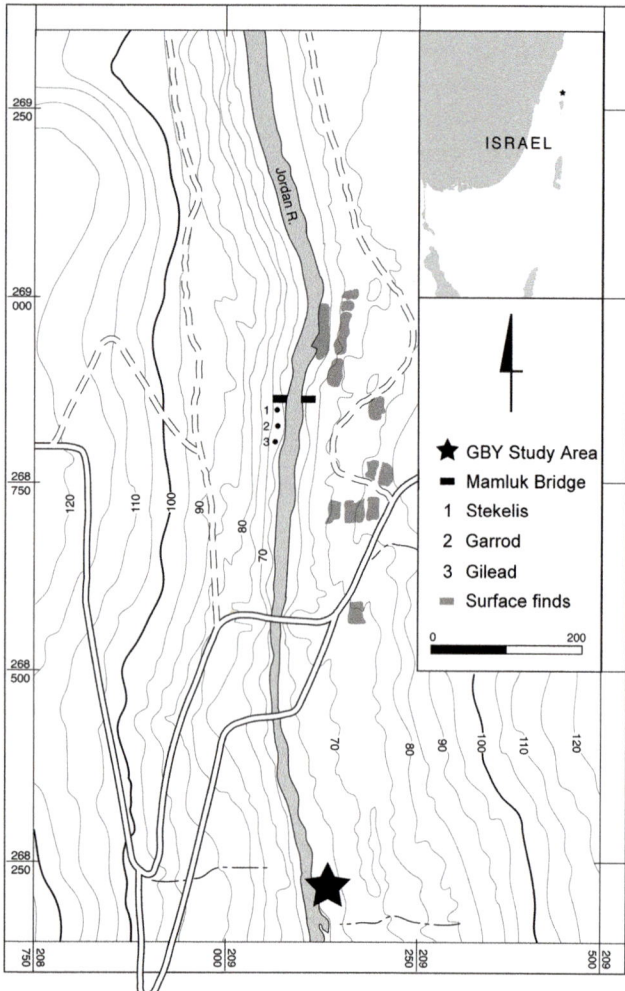

Fig. 2.18 Map showing the location of previous excavations (1–3), surface collections made by D. Ben-Ami, and the study area

2.2.4 The Study Area

In 1989 an area south of the bridges and on the left bank of the Jordan River was chosen for excavation (Fig. 2.20). This locality was selected for several reasons that included the structural position of the layers of the Bar (see below), which is situated immediately south of the study area, and the minimal exposure of what later became known as Layers V-5 and V-6, which are being eroded by the river and consequently fossil bones, organic material (wood and bark), and lithic artifacts have been washed away from their original environment of deposition. In this area, situated south of all previous archaeological activities, the Middle Pleistocene deposits are not exposed on the surface, since they are covered by the Jordan floodplain.

2.2.4.1 The Bar

As a result of an exceptionally low discharge of the Jordan River during the summers of 1989 and 1990, an extensive area, usually under water, was exposed on the left bank of the river. This area is built of limnic-fluvial sediments, which form a bar-like structure (Fig. 2.21). This structure, a component of the BYF, is built of mollusk-rich sediments and very large basalt boulders and was found to be composed of four lithofacial

Fig. 2.19 A flint handaxe and the gastropod *Viviparus apameae* bedded in the tilted sediments of the eastern flanks of the Crusader fortress (the glacis), which are built of the Benot Ya'akov Formation

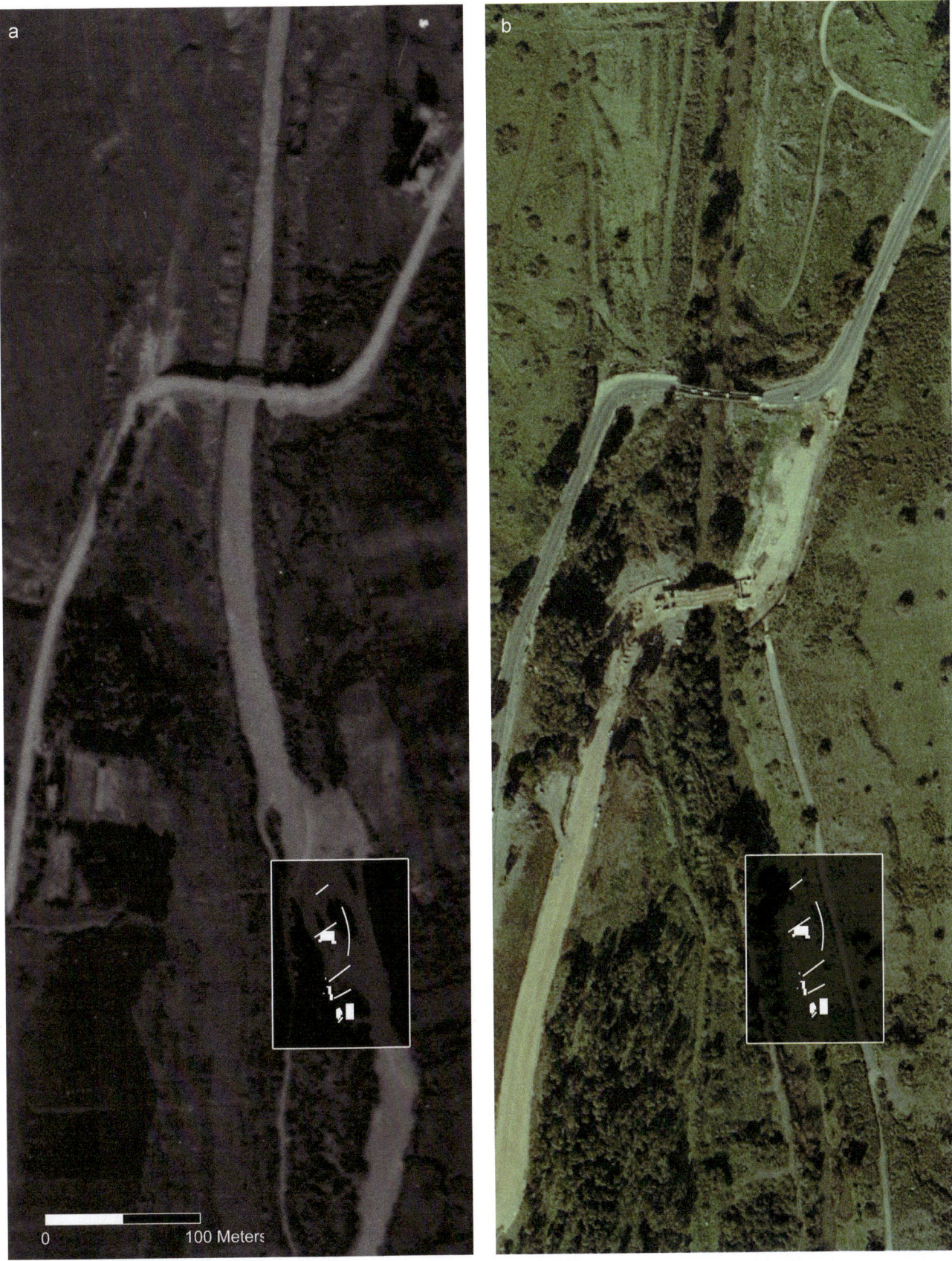

Fig. 2.20 Aerial view of the Jordan River in the vicinity of the Benot Ya'aqov Bridges with the location of the study area: a) January 1945, b) January 2007 (orthophoto courtesy Survey of Israel)

2.2.6 December 1999

In December 1999, working during the night, the Kinneret Drainage Authority undertook major drainage activity that caused one of the worst events of destruction of an archaeological site ever recorded, affecting both the GBY site and younger, previously unknown sites (Mousterian and Epi-Paleolithic sites) along the Jordan River. All along the river, from the Pkak Bridge to the Crusader fortress, the riverbed was dredged and huge amounts of sediments, mollusks, fossil bones, and lithic artifacts were piled on both banks (south of the bridges and only on the left bank) (Fig. 2.22). Some of these were flattened and others were piled up to a height of 2 m, while enormous quantities of spoils were later removed and deposited in fields for later cultivation, all east of the Jordan River (Sharon et al. 2002) and north of the study area. Irreversible damage was caused to the Acheulian site in the study area (and beyond), where thousands of cubic meters of *in situ* deposits containing a wealth of finds were destroyed. As the morphology of the banks and the Bar along the river was built from the BYF, the cutting, scraping, and mechanical excavation led in subsequent years to accelerated erosion of the Early–Middle Pleistocene strata. Geological and prehistoric surveys followed these events and are detailed elsewhere (Sharon et al. 2002, 2010b). These included additional surveys (on foot and by boat), a series of geological drillings each 10 m deep, and a small-scale excavation (2x2 m) of an Acheulian site denoted North of the Bridges (NBA), where collection of Acheulian artifacts and bones took place among other archaeological activities (Sharon et al. 2010). An Ar/Ar date of 658±15 ka was obtained on a fresh basalt flow (4 m thick) from a drilled core situated 5 m from the small excavation mentioned above; this flow immediately underlies the NBA horizon and is considered an intrusion within the BYF (Sharon et al. 2010).

Fig. 2.21 The Bar's tilted layers and stratigraphy during a low water stand of the Jordan River in 1990

fluvial and limnic units *in situ* but was never excavated. These are tilted in the same direction as that of the strike of BYF strata described below. The archaeological and paleontological finds and the characteristics of the sediments are discussed in detail elsewhere (Goren-Inbar et al. 1992). The abundance of the finds bedded in this complex advanced the decision to excavate in this area in order to ascertain the nature and spatial characteristics of these assemblages.

2.2.5 The Excavations

Seven field seasons took place in July between 1989 and 1997 under the direction of Naama Goren-Inbar on behalf of the Hebrew University of Jerusalem (with a second season in 1997 in September). In order to expose what were then considered the Middle Pleistocene strata, two trenches (Trenches I and II) were dug perpendicular to their strike during the first field season. These trenches enabled observation of the otherwise unseen stratigraphic sequence. Adjacent to these trenches two areas (A and B) were chosen for excavation, which took place in both areas during the first season (1989). From the field seasons of 1990 onwards excavation was limited to Area B (Goren-Inbar et al. 1992). In 1995 a third excavation area (C) was selected. Further details of field methodology and excavation approach are presented in Chapter 4.

Fig. 2.22 Aerial view of the destruction caused by drainage activity in December 1999

2.2.7 2005, 2009–

In addition to the long series of geological and archaeological activities described above, two drill cores were carried out in the BYF. The first was drilled during 2005 (with the help of the Israel Antiquities Authority) at the eastern pier of the southern bridge in preparation to its rebuilding to allow traffic in both directions. As we were aware of the fact that the bridge is built on the BYF we requested a drilling, which was carried out to a depth of 50 m. The core turned out to be of extreme importance, yielding data on the chronology and sedimentology and a wealth of paleontological finds, including outstanding paleobotanical remains. The second drilling (summer 2009) was carried out in Area C at the site. This drilling penetrated through 127 m of sediments and series of lava flows. The radiometric dates, paleomagnetic data, sedimentological and volcanic record, and paleontological remains from the two core drillings all furnish new information on the landscape, geological evolution, and environmental changes at the southern end of paleo-Lake Hula (Chapter 3). Furthermore, the data are complementary to those obtained from the analyses of the various assemblages deriving from the excavations. The record supplied by these cores close some gaps within the evolution of the Dead Sea Rift and the communities of prehistoric hunter-gatherers that occupied it during the Early and Middle Pleistocene.

2.3 The Archaeological Analyses

The excavations at GBY have yielded a wealth of prehistoric material and raised scientific questions that are relevant to a variety of disciplines. Over the long years of research, different research questions were postulated and multidisciplinary analyses were carried out. The analyses that are concerned with geology, sedimentology, and dating are described in detail in Chapter 3. The following is a brief review of the many and varied studies (excluding lithic studies) that have been carried out over the years, contributing data on the environment and behavior of the GBY hominins.

The subsistence of the GBY hominins was clearly dependent on the floral and faunal resources of paleo-Lake Hula and its environs. Analyses of the floral assemblages suggest that edible plants are numerous (over 50 species) and originate both in the wet habitats of the valley and in its higher topographical borders. Nuts, including two nutritionally important water nuts, *Euryale ferox* and *Trapa natans* (Goren-Inbar et al. 2002a; Melamed et al. 2011; Goren-Inbar et al. 2014), fruits, and plants producing underground storage organs were an important component of the varied diet at the site (Goren-Inbar et al. 2002a; Melamed et al. 2011; Melamed et al. 2016). Clearly, control of fire enlarged the scope of edible plant species and improved the efficiency of other components of the diet. Fire, and the ability to make and control it, was a major element in the subsistence of the GBY hominins and was an integral part of the Acheulian cultural realm at the site. Concentrations of burned flint microartifacts delineate the location of (phantom) hearths, which repeatedly occur through all of the archaeological horizons (Alperson-Afil 2008; Alperson-Afil and Goren-Inbar 2010).

The location of fire strongly influenced the spatial distribution of the lithic artifacts, wood, plants, and bones around it, revealing spatial patterning of activities similar to that known from more recent periods (Alperson-Afil et al. 2009).

As for the faunal analyses, fallow deer was preferred among the mammals (Rabinovich et al. 2008), but consumption of fish (carp) is also well documented (Zohar and Biton 2011; Zohar et al. 2014). Bone taphonomy studies point to extensive processing of meat and marrow, as documented by cut marks, hack marks, and extremely high frequencies of bone fragmentation (Rabinovich et al. 2012).

The different studies suggest that the hominins of GBY possessed developed cognitive abilities enabling procedural cognition, planning and working memory, cooperative provisioning, technological flexibility, spatial cognition, long-term memory, large-scale spatial thinking, contingency planning, and more (e.g. Goren-Inbar 2011).

On the background of these developed cultural and cognitive abilities, we can suggest that the hominins of GBY, who occupied the area for a duration of at least 100 ka, were fully adapted to the Mediterranean environment, its diverse ecological niches, and its rich biological resources. The great cultural similarities between GBY and the African LFA sites show that the process of adaptation to a new environment was a long and continuous one. The GBY record provides insight into the sophisticated adaptive abilities of the Acheulians in Early–Middle Pleistocene conditions in the Levantine Corridor, throwing light on models of the adaptations that enabled the process of dispersals into Eurasia.

Chapter 3
The Site of Gesher Benot Ya'aqov

Abstract Chapter 3 is concerned with the geography, geology, stratigraphy, and chronology of the site of Gesher Benot Ya'aqov. The geographical landmarks of the study area are presented in detail and the structural, morphotectonical, and geomorphological processes of the area are discussed. The different exposures of the Benot Ya'akov Formation, in which the archaeological material is bedded, are described. The stratigraphic sequence of the study area is provided, together with a composite section of 34 meters. In addition, the various attempts to determine the age of the site are discussed, followed by a description of the different types of raw material found at the site.

Keywords chronology · dating · excavations · Gesher Benot Ya'aqov · geography · geology · raw materials · stratigraphy · study area

The archaeological site of Gesher Benot Ya'aqov (GBY) is named after a bridge that crosses the Jordan River, connecting the Korazim Saddle in the west with the Golan Heights in the east. This area is well known as an important crossing of the river and is mentioned in historical documents from the 12th century onward (Chapter 2). This river crossing was mentioned by many early travelers, who gave descriptions of landmarks or their encounters or impressions (e.g. in Schur 2002). In historical times this geographical location has been given different names, of which the most common are the Arabic *Jisr Banat Yaqub* and the Hebrew *Gesher* (=bridge) *Benot* (=daughters of) *Ya'aqov* (Jacob).

3.1 Geographical Location

The site of GBY stretches along some 3.5 km of the bed and banks of the Jordan River (Fig. 3.1). The site is practically invisible in the landscape, as the Benot Ya'akov Formation (BYF), in which it is bedded, is minimally exposed; the exposures are limited mainly to the banks of the river and to a lesser extent to its bed (Fig. 3.2). The sediment packages of tilted layers that include the archaeological horizons are

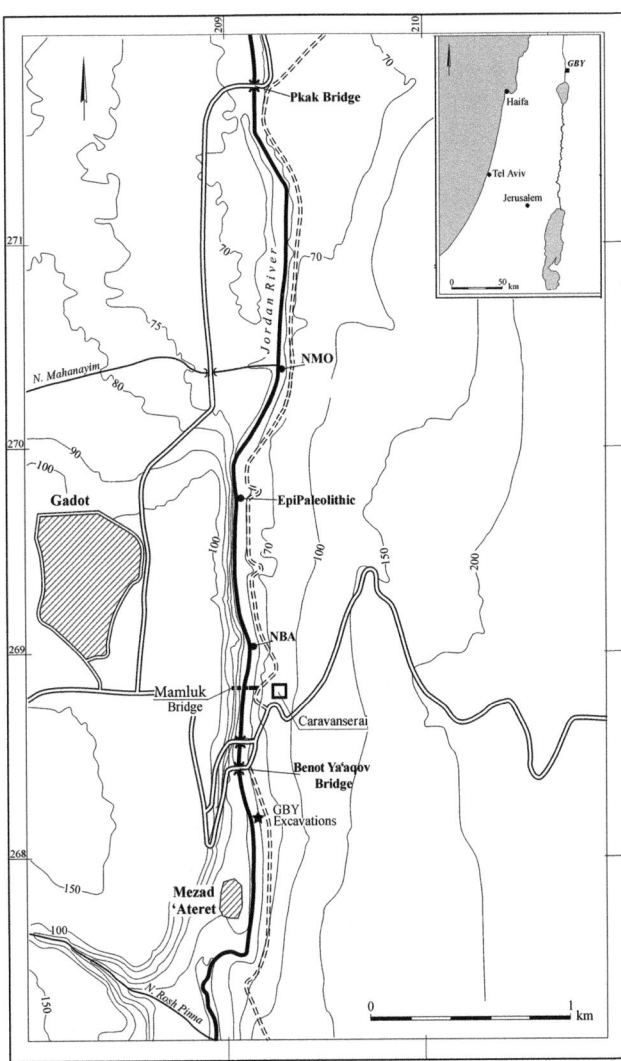

Fig. 3.1 Map showing the area of the GBY site with the major geographical landmarks and archaeological sites mentioned in the text

Fig. 3.2 Exposures of the BYF: a) Jordan River, right bank north of the bridges (currently under water); note fossil mammal bone protruding from the sediments, b) right bank north of the bridges, c) right bank south of the bridges, d) and e) exposures in the Jordan River south of the bridges, f) small-scale excavations on the left bank of the Jordan River (NBA in Fig. 3.1)

covered by the floodplain of the river (built mainly of sand, gravel, pebbles, and rarely boulders), which is stratified unconformably on top of the Early–Middle Pleistocene strata. The banks' exposures in the study area (the locality of the excavations in 1989–1997) are limited to two archaeological layers (V-5 and V-6), to which Volume III of the GBY series is dedicated (Rabinovich et al. 2012).

The structural complexity of the Hula Valley in general, and the area south of it where the site is located in particular, is responsible for the hidden nature of the site (Section 3.2). Although the course of the Jordan River, which follows a major fault line oriented north–south, is very narrow (Belitzky 2002), there is no correlation between the banks. Furthermore, the BYF is less visible on the right bank of the river (Belitzky 2002). Not only are exposures rare, but those that are visible are all tilted and not continuous (Fig. 3.2).

3.1.1 Geographical Landmarks

a) **Bridges:** The lack of visibility or continuous exposures has led both amateurs and excavators to refer to prominent landmarks. The most prominent feature is the stone bridge. From Garrod's and Stekelis' times, this landmark is mentioned and sometimes appears in various illustrations (e.g. Fig. 2.11). References to the bridge are very problematic, since as early as the 1930s there were two bridges, a Mamluk stone bridge and a British Mandatory steel "Bailey" bridge. Since then, and until today, there have been two bridges in this area, differing in structure and location (Chapter 2, Appendix 1).

b) **The Crusader fortress:** A southern landmark is the Crusader fortress of *Vadum Iacob* (Ellenblum et al. 1998), whose glacis facing east consists of a package of BYF beds tilted eastward (Goren-Inbar and Belitzky 1989) (Fig. 3.3). It was built in the 12th century AD (Marco et al. 1997) and was constructed on a long morphotectonic hill formed by a detached basalt block oriented north–south (Belitzky 1987), the northernmost in a row of similar blocks along the north–south Jordan Fault Line (Harash and Bar 1988).

c) **The caravanserai and the Customs House:** Between the earlier bridges and the Crusader fortress are a *caravanserai* (Mamluk) and a building known as the Lower Customs House (19th century). The fortress, the *caravanserai* and the Customs House are the only fixed points in the landscape in an area that has witnessed massive construction, resulting through the generations in devastating environmental and archaeological destruction (Fig. 3.3c).

d) **The proximity to the Syrian border:** The site is located in an area that was extremely sensitive from a geopolitical perspective during 1935–1967. Work (mainly surveys) was possible only during very short episodes, and the proximity to the Syrian border (1948–1967), located a few meters east of the Jordan River, rendered any research activity impossible (see details in Chapter 2).

The area adjacent to the site, from the dirt road situated east of the site and all along the north–south axis of the site, has not been investigated, as it has been declared a single large minefield. During the years of dredging carried out in the Jordan River, spoils that include many finds (bones, artifacts, and

Fig. 3.3 a) Aerial photo of the GBY area with locations of the most prominent landmarks (Crusader fortress, Customs House, Mamluk caravanserai, Mamluk stone bridge, b) the Crusader fortress (view from the east) and its slopes (glacis), c) view of the western slopes of the Golan Heights: the Mamluk caravanserai, the Lower Customs House, the ruins of the Mamluk stone bridge, and the stony spoils of the continuous drainage operations

Fig. 3.4 Aerial photograph showing the spoils made by the continuous drainage of the river on both banks of the river (January 2000)

organic materials) have been dumped in this area, but access to them is impossible (Fig. 3.4).

3.2 Geology

The archaeological site of GBY is located in the southern Hula Valley in the Dead Sea Rift, a segment of the African Great Rift System.

From the air, the southern Hula Valley looks like an intricate mosaic made up of blocks that differ in size and morphology, forming swells and troughs. The GBY archaeological sites are situated along the major fault zone that has formed many of the valley's peculiar traits, making it geologically and structurally a highly complex area, despite its modest size. Tectonic activity along an intricate net of faults has caused uplift, subsidence, and sinistral displacement of numerous structural features. The area has been affected by volcanic activity and by swift changes in the depositional environment. Because of this, scholars who have conducted research in the area have published geological cross-sections of their own, which differ substantially from each other. Under normal circumstances this phenomenon could be interpreted as one scholar being right and others wrong, but in fact no one was wrong. The reason lies in the extremely limited size of the exposures and the great complexity of the area (Fig. 3.5). Hence, a cross-section that supposedly represents an area in the order of hundreds of meters or more usually represents only the geological picture adjacent to the location where it was made. The exceptional complexity of the GBY area means that each cross-section is relevant only to a few meters. This situation makes GBY a most challenging area, despite its small size.

Quaternary tectonic activity has produced a complex local setting that controls the surrounding landscape of the site. The area of GBY was affected by young faulting and folding that shaped the regional morphology and affected the ongoing geomorphic activity in the area (Figs. 3.5b–3.7). In the Rift adjacent to GBY there are several discrete structural domains: the Korazim Saddle, the Ayyelet Hashahar area, the Hula Valley Basin, the Golan Heights, and the faulted margins (the Yehudiya Block) (Fig. 1.1).

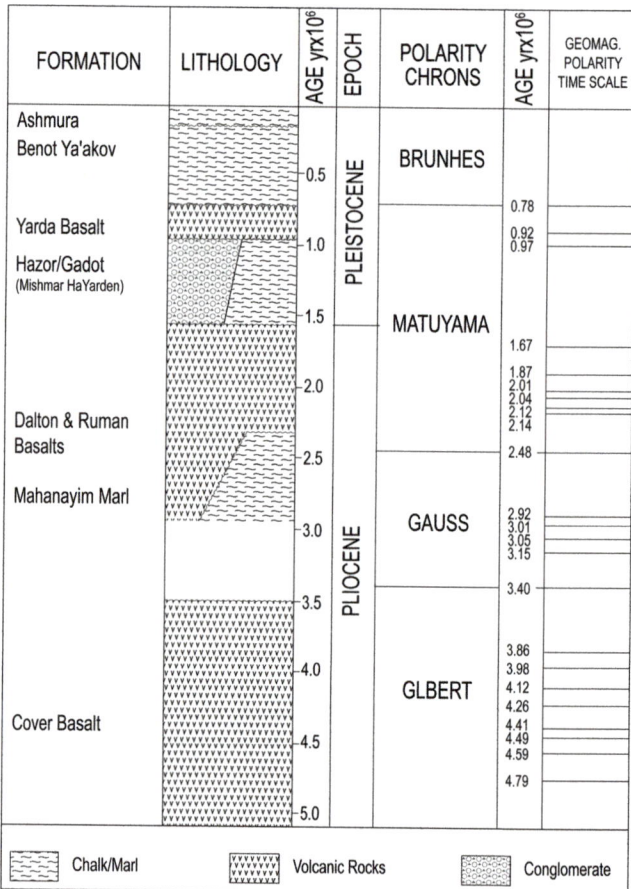

Fig. 3.5a Composite stratigraphic section of the rock units exposed within the Korazim Saddle (modified from Heimann and Ron 1993)

3.2 Geology

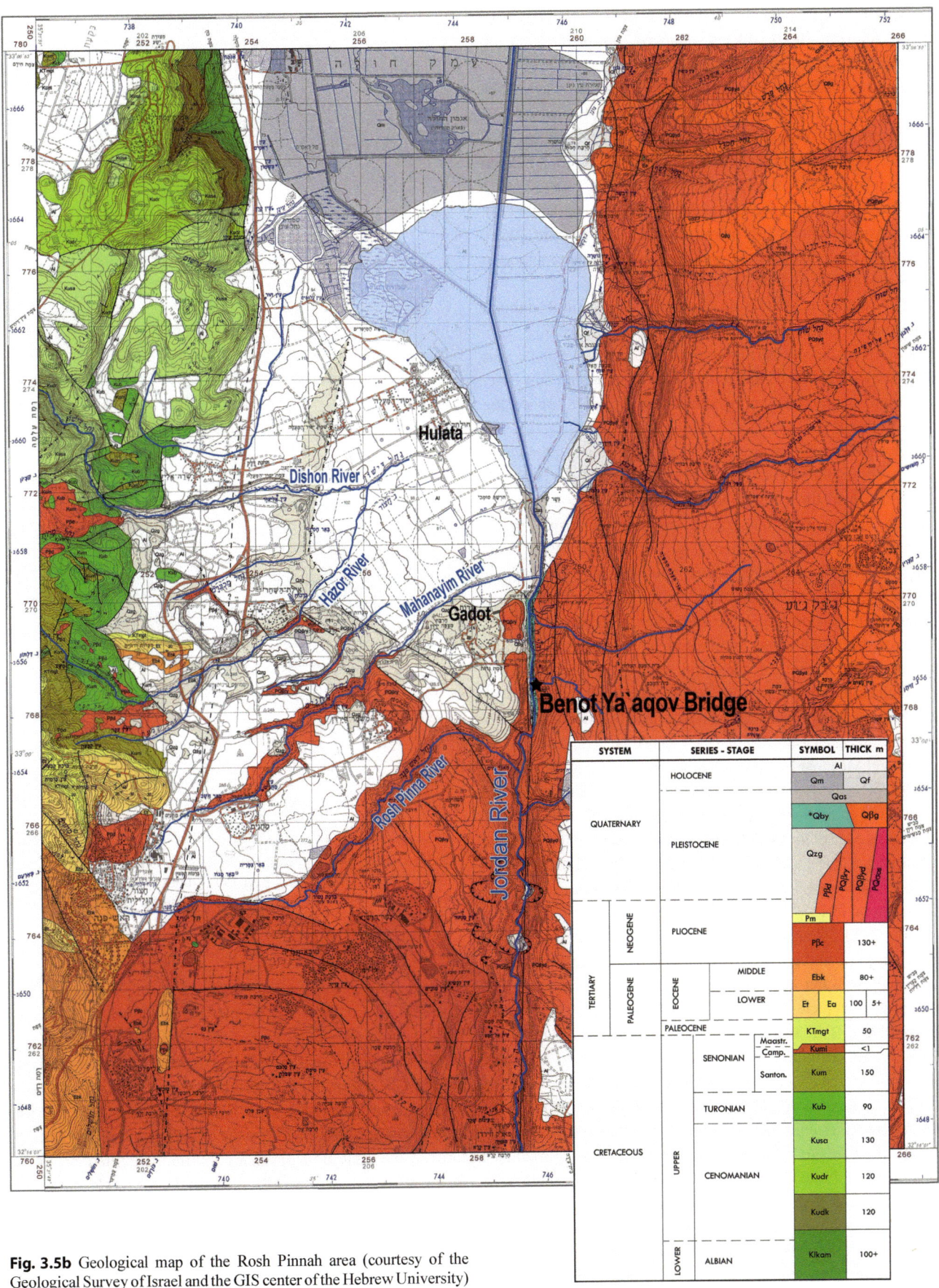

Fig. 3.5b Geological map of the Rosh Pinnah area (courtesy of the Geological Survey of Israel and the GIS center of the Hebrew University)

The Hula Valley Basin is a complex graben structure (pull-apart, Heimann 1990) with an estimated subsidence of 2,500 m during the last 4 Ma (Heimann 1990). The geological and structural subsidence data of the Hula Valley Basin have been recorded through a series of drill holes (Horowitz 1973; Heimann 1990). *Viviparus apameae*, a guide fossil of the BYF (Fig. 2.19), was found in the central part of the basin at a depth of about 500 m (Moskowitz and Magaritz 1987) just above a basalt flow that was dated to 1.13 Ma by Ar/Ar (Heimann 1990). This fossil disappears just below the Main Peat (Picard 1963), which has been dated by the uranium disequilibrium method to 222 ka (Magaritz and Moskowitz 1987).

The Korazim Saddle is an uplifted structural domain delimited on the east and west by left-lateral faults (Fig. 3.6). The uplift of the northern portion of the Korazim Saddle was formed by a rightward bend of the sinistral Korazim Fault Line. The uplift is accompanied by subsidence of blocks toward the Hula Valley Basin.

Morphotectonic, sedimentological, and geomorphological processes in the GBY area are still affected by tectonic deformation of three major active fault lines, Korazim to the west and Jordan and Almagor to the east, and their displacement has a major sinistral component (Fig. 3.6) (Belitzky 2002).

Late Pleistocene tectonic activity caused folding and faulting of the BYF. Several tens of meters of the BYF sediments were deposited in a morphotectonic depression, the Benot Ya'aqov Embayment, an attractive focal point in the landscape that is rich in water, fauna, and flora. The embayment was formed at the southeastern tip of the Hula Valley Basin, due to the folding and faulting along north–south directed plate border faults, which affected the older (mostly Pliocene) lacustrine sediments of the Gadot and Mishmar HaYarden Formations, as well as the Ruman and Yarda basalts (Goren-Inbar and Belitzky 1989). The depression is bordered on the west by the Korazim Saddle and on the east by the faulted edge of the Golan Heights basaltic plateau (the Yehudia Block (Fig. 3.6; Belitzky 2002). The lacustrine–fluvial sediments of the BYF have been deposited in the embayment and are composed of the littoral facies of the basin fill (Horowitz 1973).

The BYF strata include three dominant lithofacies: 1) coquina, sand, and gravel of a beach facies; 2) calcareous mud of a shallow-water lacustrine facies; and 3) conglomerate of a fluvial channel facies (Feibel 2001, 2004). The lacustrine–fluvial strata of the BYF that include Acheulian artifacts (Horowitz 1973) were described by Picard (1963) as *Viviparus* beds, following the presence of *Viviparus apameae*, a guide fossil of the BYF. The outcrops of the BYF extend about 3.5 km along the Jordan River until they disappear at the Jordan River gorge.

North of the GBY site, the strata are either almost horizontal or uplifted and tilted (up to 20–25° to the northwest, southeast, and northeast) (Figs. 3.2, 3.7). The strata form an anticline that is likely faulted in the hinge area, where the deposits are tilted 80° to the west-southwest. The anticline plunges slightly to the south. The trend of the hinge is south-southwest in the

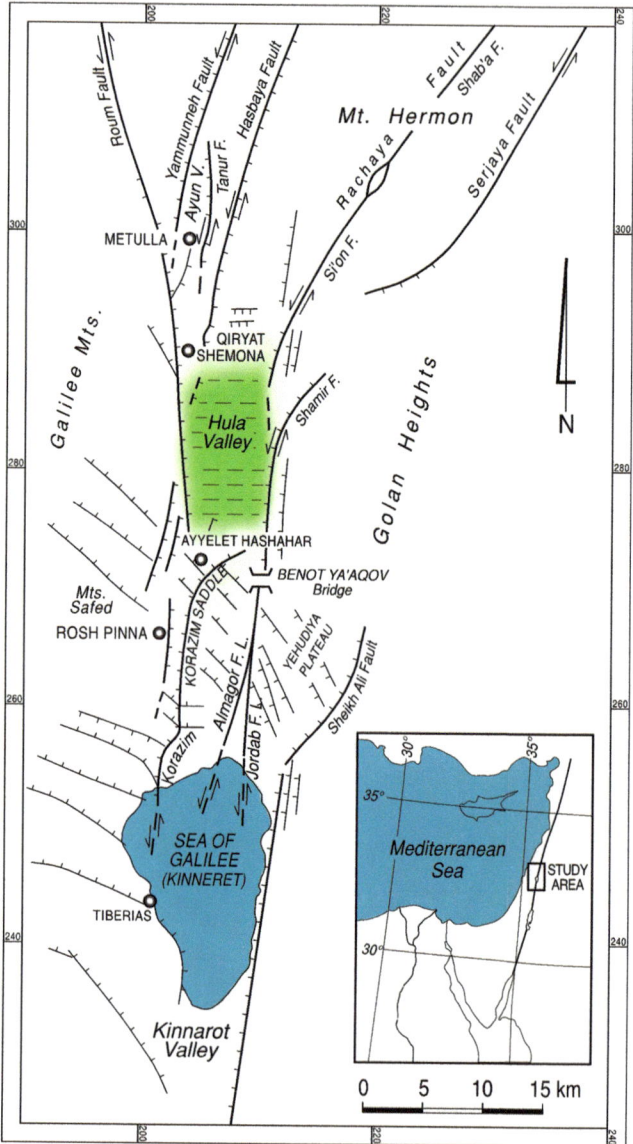

Fig. 3.6 Generalized map of faults within the northern Dead Sea Rift (modified from Heimann and Ron 1993). Latitudes and longitudes are shown on the inset map, whereas the fault map uses the Israel grid system (from Belitzky 2002: fig. 1)

north, gradually changing southward to south-southwest. The erosional terrace formed by the modern floodplain of the Jordan River truncates the upper part of the anticline.

In the study area, on the right bank of the Jordan River, six geological trenches revealed the tilted BYF strata and their stratigraphy (Goren-Inbar et al. 1992b; Feibel 2001, 2004). On the left bank of the river and on the eastern slope of the Crusader fortress, the same BYF strata are uplifted and tilted 70° to the east-southeast (Figs. 3.8–3.9), forming the fortress's glacis.

A small-scale basalt flow that came from the east yielded a date of 658±15 ka by Ar/Ar (Sharon et al. 2010). This flow probably blocked the embayment and stopped the sedimentological deposition of the BYF in the southern part of the embayment.

3.2 Geology

The uplift of the embayment area ended the deposition of the lacustrine sediments of paleo- Lake Hula and caused erosion and vertical incision by the Jordan River. The incision caused the capture of the Rosh Pinnah River and the exposure of the BYF strata. These strata are found in different structural positions to the north and south of the GBY site.

There is evidence for very recent tectonic activity expressed by sinistral displacements along the Jordan Fault Line. This movement is in the order of 2.1 m and has caused an offset of archaeological remains; this offset is visible in the walls of the Crusader fortress and has occurred since 1179 when the castle was conquered by Saladin (Ellenblum et al. 1998). Evidence of ongoing uplift of the Korazim Saddle was also detected by a geodetic survey of the area (Karcz 1995).

Recent developments have shed additional light on the geological structure and the stratigraphy of the BYF in the study area. These derive from two drillings initiated by the archaeological expedition. The first, GBY #2, was located at the east end of the current southern bridge and reached 50 m below the surface. The second, Eshel Ya'aqov (EY), was drilled in the study area of the site to a depth of 125 m below the surface. The results of both indicate that the tilted block continues to a depth of at least 125 m and that the BYF is laid on a thick series of basalt flows tilted 40° to the southwest. The dates and the location of the Matuyama-Brunhes chron boundary that were obtained from both drillings yield invaluable structural and depositional information and contribute much to the understanding of the

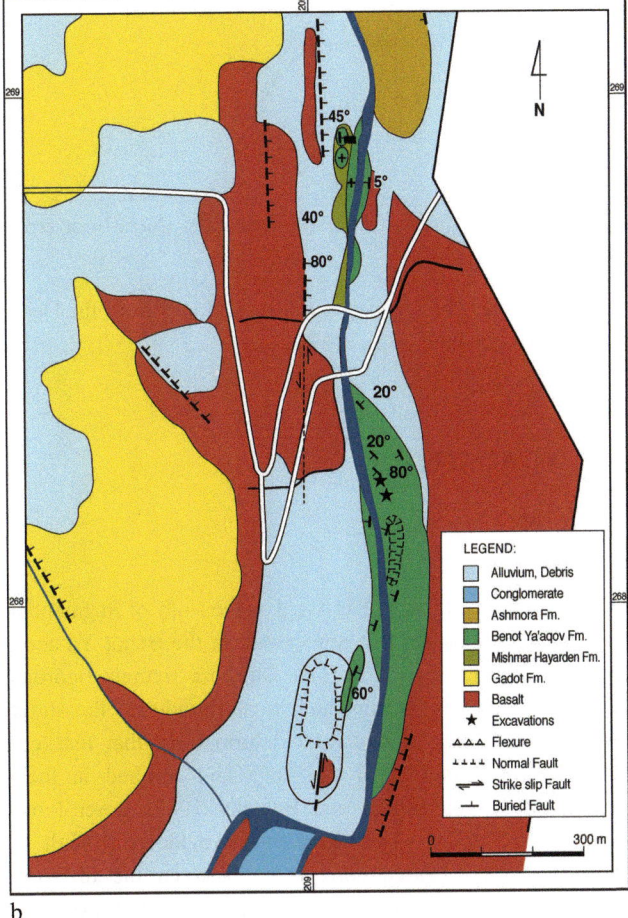

Fig. 3.7 Geological maps of the GBY area, including a) general and b) enlarged perspectives (from Belitzky 2002: fig. 6a)

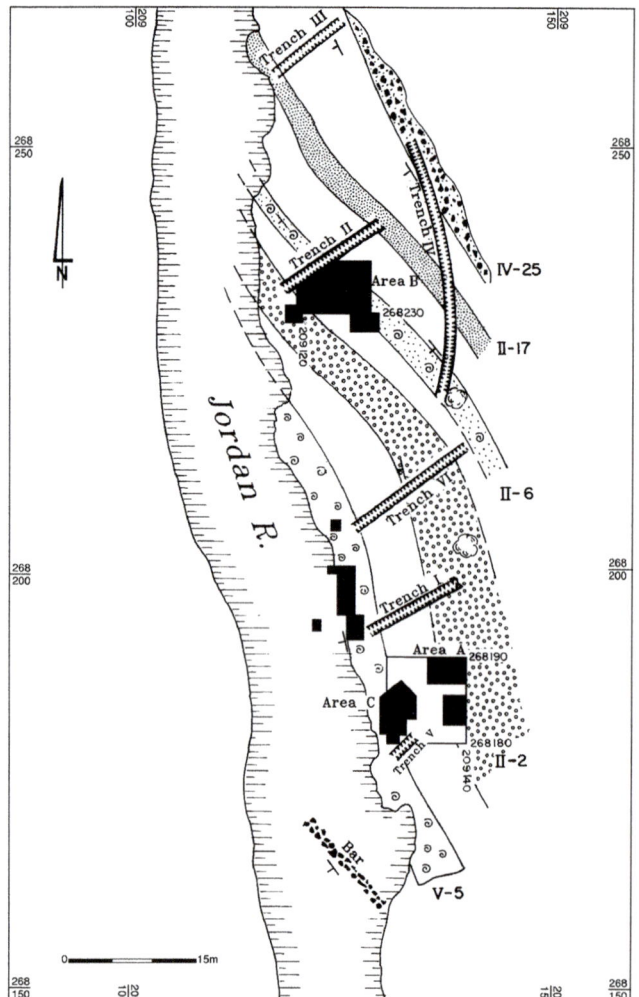

Fig. 3.8 Geological map of the study area (from Goren-Inbar et al. 2002b: fig. 7)

evolution of the Hula Valley and the development of the Dead Sea Rift (Goren-Inbar et al. 2012).

3.3 Stratigraphy

The stratigraphy of the BYF is based on the study of fragmented and isolated exposures that are located in the Benot Ya'aqov Embayment (Chapter 1; Fig. 3.5). Due to extremely rigorous tectonics in the embayment and its surroundings, the strata underwent a series of structural deformations that masked, deformed, and fragmented the strata and resulted in their discontinuity (Goren-Inbar and Belitzky 1989; Goren-Inbar et al. 1992a, 1992b). This situation has made it difficult to establish strongly based correlations between the different exposures. We therefore describe each individual exposure in detail (Section 3.1–3.4) and only after completion of the

Fig. 3.9 a) and b): the tilted BYF sediments exposed on the eastern flanks of the Crusader fortress

3.3 Stratigraphy

stratigraphic/sedimentological analyses do we attempt a general structural reconstruction of the BYF and the particulars of the Benot Ya'aqov Embayment.

3.3.1 North of the Benot Ya'aqov Bridges

These sediments are currently under water in the Jordan River. They were briefly exposed during the drought of July 1990 and were surveyed on foot and by boat, revealing a ca. 23 cm thick deposition of the BYF on the right bank, overlying a basalt flow. These BYF sediments are rich in mollusks, including *Viviparus apameae* and comprise artifacts as well as organic material and fossil mammal bone(s) (Fig. 3.2a, c, f).

Following this, and particularly after the devastating drainage episode in December 1999 (Sharon et al. 2002; Sharon et al. 2010), visibility was much better and stratification of sand, silts, and coquinas of the BYF was observed to interfinger with basalt flows of varying thickness. These basalt flows created the Jordan River bottleneck at its Hula Valley outlet and dictate the shallow nature of the water there. Several find spots were identified on both banks of the river, some 500 m north of the bridges. The detailed stratigraphy of this area, not yet precisely anchored to the study area, was obtained through a series of short drillings (each 10 m long), several sections, and a small-scale excavation that all took place on the left bank of the Jordan River (Sharon et al. 2010) (Fig. 3.2f). Overlying the BYF deposits is a conglomerate composed mainly of basalt artifacts, primarily handaxes and cleavers (Sharon et al. 2002; Sharon et al. 2010).

3.3.2 The Crusader Fortress

The tilted fluvio-lacustrine deposits observed along the eastern side of the Crusader fortress (*Chastellet, Vadum Iacob*, 'Ateret) are estimated to be 15–20 m thick (Fig. 3.9). They comprise basalt boulders, flint nodules, mammal bones, mollusks, and flint artifacts (including handaxes and cleavers). These exposures were not excavated, as the deposit is highly cemented and strongly tilted. The sedimentological composition of these exposures is described elsewhere (Belitzky 1987).

3.3.3 The Rosh Pinnah River

This is the southernmost presently known exposure of the BYF. It is located at the outlet of the Rosh Pinnah River to the Jordan River. These sediments are observed on the left bank of the Jordan River as well as its riverbed. The exposure on the left bank of the Jordan is composed of the following stratigraphic units, from top to bottom: a) an unconsolidated conglomerate ca. 2 m thick, including basaltic components of various sizes; in this conglomerate are bedded flint and basalt artifacts; b) a layer of white (sometimes reddish) sediments including mollusks (*Melanopsis, Melania, Theodoxus, Unio*, etc., but no *Viviparus*), ca. 4 m thick (Fig. 3.10); c) a thick green clay unit, of unknown thickness, exposed both on the bank and in the riverbed under water; and d) a black lignite layer exposed along the right bank of the river and including fragments of *Viviparus*. Units b–d are assigned to the Middle Pleistocene, while unit a is younger, most probably of Recent age.

3.3.4 The Study Area

The thickness of the stratigraphic sequence in the study area is the greatest currently known when compared to all the above-mentioned localities, reaching 34 m (Feibel 2001, 2004; Goren-Inbar et al. 2000) (Section 3.2). It comprises surface exposures of sediment packages as well as a series of beds that were artificially exposed as a result of the excavations. Some of the exposed surfaces, such as the Bar (Goren-Inbar et al. 1992b), are submerged under the waters of the Jordan during most of the year and are exposed only during months of very low discharge or during severe droughts. The sedimentary layers, together with the artifacts, fauna, and flora included in them, have been waterlogged since their deposition. Paleo-Lake Hula was one factor contributing to the high water table. More recently, the proximity of the perennial Jordan River, as well as the abundant springs that issue from between basalt flows of the Golan, have kept the site wet. The persistence of saturated conditions at the site has resulted in unusually good preservation conditions for some of the lithic artifacts and the organic remains reported elsewhere (Goren-Inbar et al. 2002a, b).

3.3.4.1 The Bar

This area, located immediately south of the study area on the left bank of the Jordan River, was studied because it covers a more extensive surface than other exposures of the BYF. It is composed of limnic-fluvial sediments that form a bar-like structure (hence its name) and include *Viviparus apameae*. When exposed, the dimensions of the Bar (before the 1999 destruction) are about 220 m long and up to 6 m wide. The strata are tilted in the northern part by some 20° to the WSW and in the southern part by some 25° to the WNW (Fig. 3.11). The exposed

Fig. 3.10 The southernmost exposure of the BYF, the outlet of Nahal Rosh Pinnah to the Jordan River: a) and b) left bank of the river, c) right bank of the river

strata, some 3.2 m thick, were divided into four lithofacies units (from top to bottom):

Unit 1: Basalt pebbles and boulders with marly matrix. A few artifacts and bones were found within this unit. It caps the more resistant underlying Unit 2.

Unit 2: Basaltic pebbles and boulders with marly matrix, highly cemented and very resistant to erosion. Morphologically, it is the dominant unit in the structure of the Bar. Artifacts and bones were found within this unit.

Unit 3: Marly rock with numerous basaltic pebbles and some well-preserved artifacts. This unit is more resistant to erosion than the underlying Unit 4 and structurally forms the lower part of the Bar.

Unit 4: Clay and marl with small basalt particles usually up to several millimeters in size. This unit is rich in molluska, bones, and artifacts. It is soft and erodes more easily than the overlying stratigraphic units, which form the Bar. This characteristic causes the formation of small elongated topographic depressions (Fig. 3.11b).

Stratigraphically, the Bar sediments are the youngest in the

3.3 Stratigraphy

Fig. 3.11 The Bar: a) and b) view of the Bar sediments, c) tilted cemented sediments, d) close-up of the sediments, e) basalt handaxe *in situ* in the Bar sediments

study area (Goren-Inbar et al. 2002b; Fig. 3.8). They contain many stone artifacts that are described elsewhere (Goren-Inbar et al. 1992b).

3.3.4.2 The Excavations

Six trenches and four excavation areas provided the background for the study area's stratigraphy. Trenches I, II, III, V, and VI were dug perpendicularly to the strike of the tilted block on which excavation took place. Areas excavated were B, C, A, and JB (Jordan Bank), in descending order of excavated surface and volume (Fig. 3.12). The cross-sections of the areas and the trenches (Chapter 4) provide extended exposure of the BYF sediments and information on their structural position. Sedimentological characteristics used in the definition of the GBY strata include grain size, color, porosity and compactness, thickness, orientation, content of organic material, continuity and structural position, mollusks, archaeological artifacts, and paleontological remains, including among others microfossils.

A composite stratigraphic section of the exposed walls of the trenches was produced, illustrating the sedimentological history and extent of the block on which the Acheulian activity took place. The sediments, an array of lake-margin deposits ranging from clay to boulders, include faunal and botanical remains and provide the background to hominin activity at the site. Stratigraphic criteria for differentiating between the different layers (geological and archaeological) were based on both sediment type and archaeological context. Several of the archaeological horizons were defined as stratigraphic units, as their finds formed a bedded entity of a thickness usually dictated by the thickest items. These horizons were first identified in Trench II and served as stratigraphic markers, while later they were sometimes used as the basis for the division between layers as well as levels (as in the case of Layer II-6 Levels 1–7).

Fig. 3.13 presents the stratigraphic correlation chart of the trenches with a detailed stratigraphic scheme. It presents the deposits of each trench and provides a correlation between the different strata observed. Fig. 3.14 is the composite section based on the above stratigraphic correlation, showing the entire 34 m deposit observed in the study area. It is bracketed between two coarse-grained fluvial depositional events; the uppermost is the Bar, discussed above (both illustrations are modified after Feibel 2001, 2004). A further volume in this series will be dedicated to the GBY sediments and their stratigraphy, which will not be described further here.

The sedimentological analyses have demonstrated a pattern of five second-order cycles of changing lacustrine facies recorded within a single first-order cycle that can likely be attributed to the ca. 20,000-year and 100,000-year Milankovitch cycles of global climate (Fig. 3.14b). (Goren-Inbar et al. 2000,

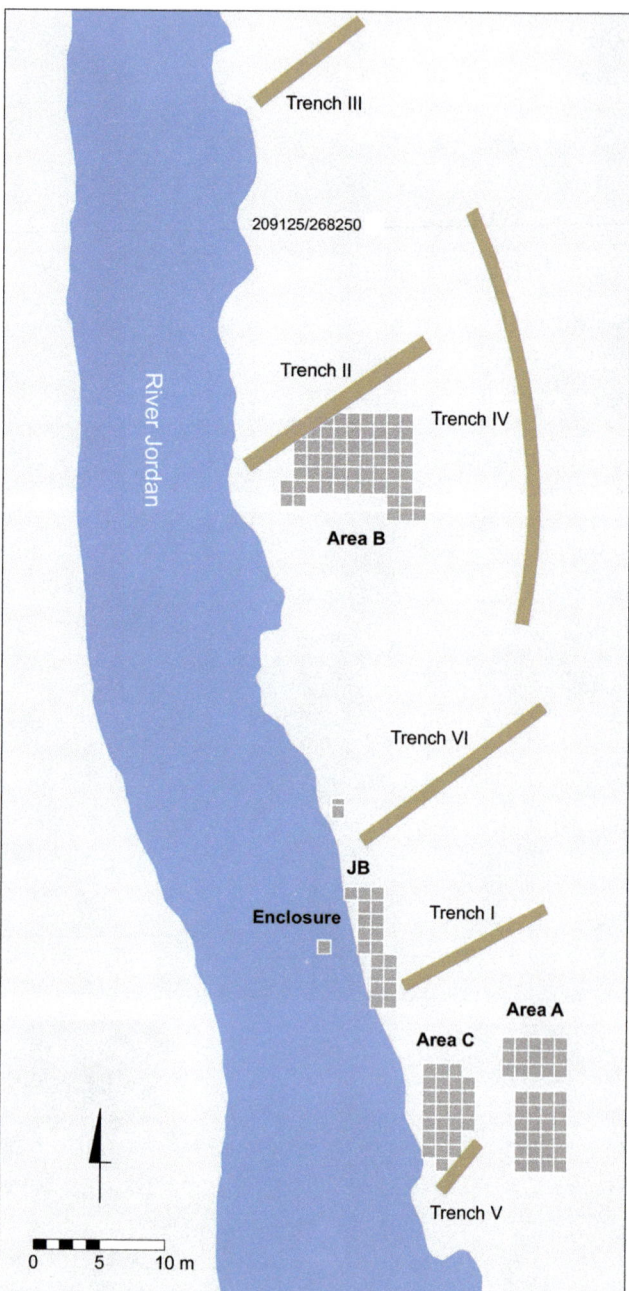

Fig. 3.12 GBY study area: map of excavation areas and trenches

2002b; Feibel 2001). The duration of the entire depositional sequence at GBY is thus estimated as ca. 100,000 years and it is assigned to MIS 18–20. Considering the paleomagnetic results (see below), the estimated date of the base of the sequence is 0.8 Ma and that of the top of the sequence is 0.7 Ma.

In the study area the horizontally laid sediments of the Recent Jordan River floodplain unconformably overlie the truncated Early–Middle Pleistocene tilted deposits, defined in this volume as "the Unconformity" (Chapters 4 and 5). Wherever these sediments were excavated in order to penetrate to the Pleistocene strata, they were denoted Layer 1 (e.g. Layer

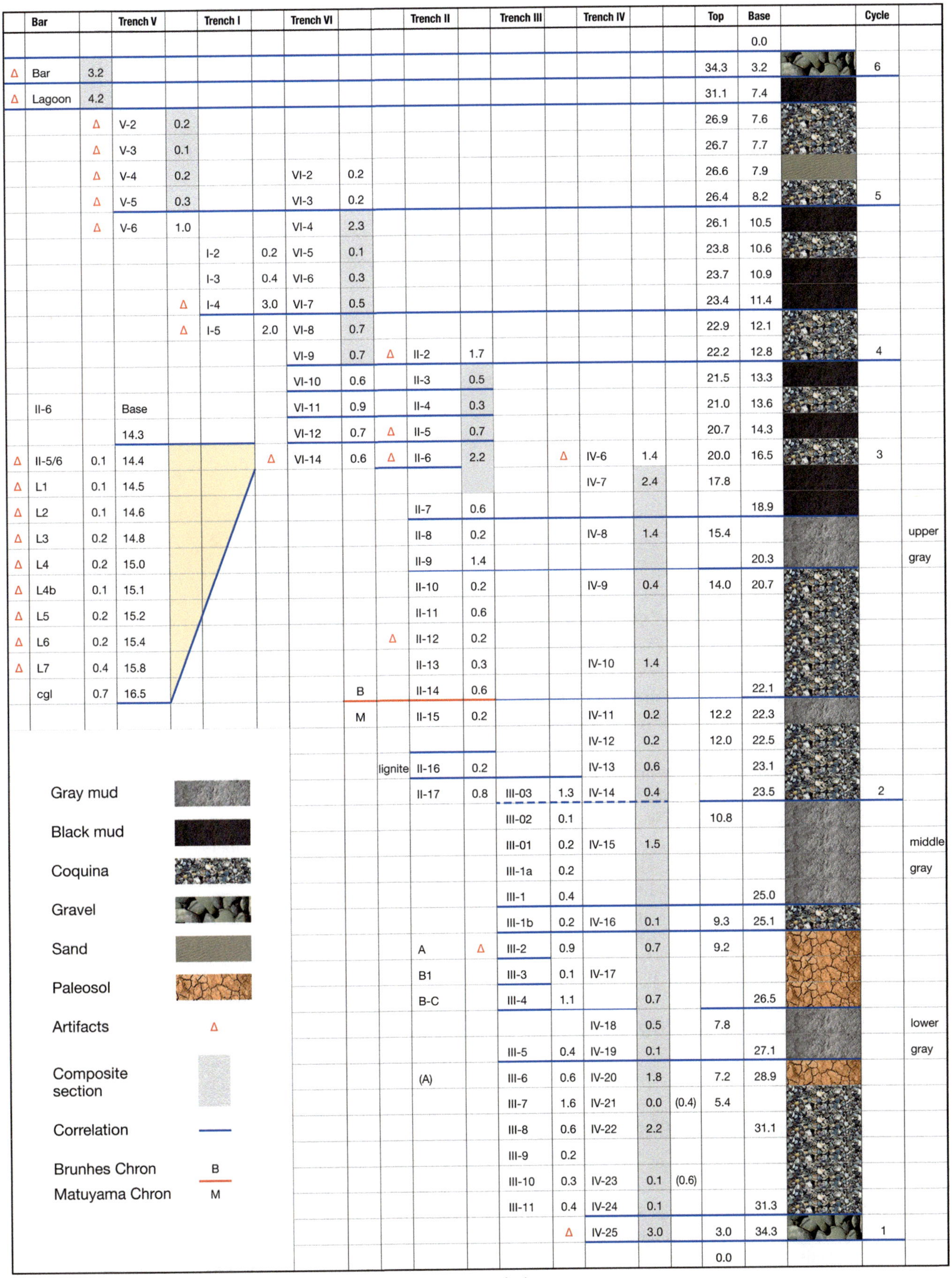

3.3 Stratigraphy

Fig. 3.13 Stratigraphic correlation chart of the GBY trenches' sedimentological sequences

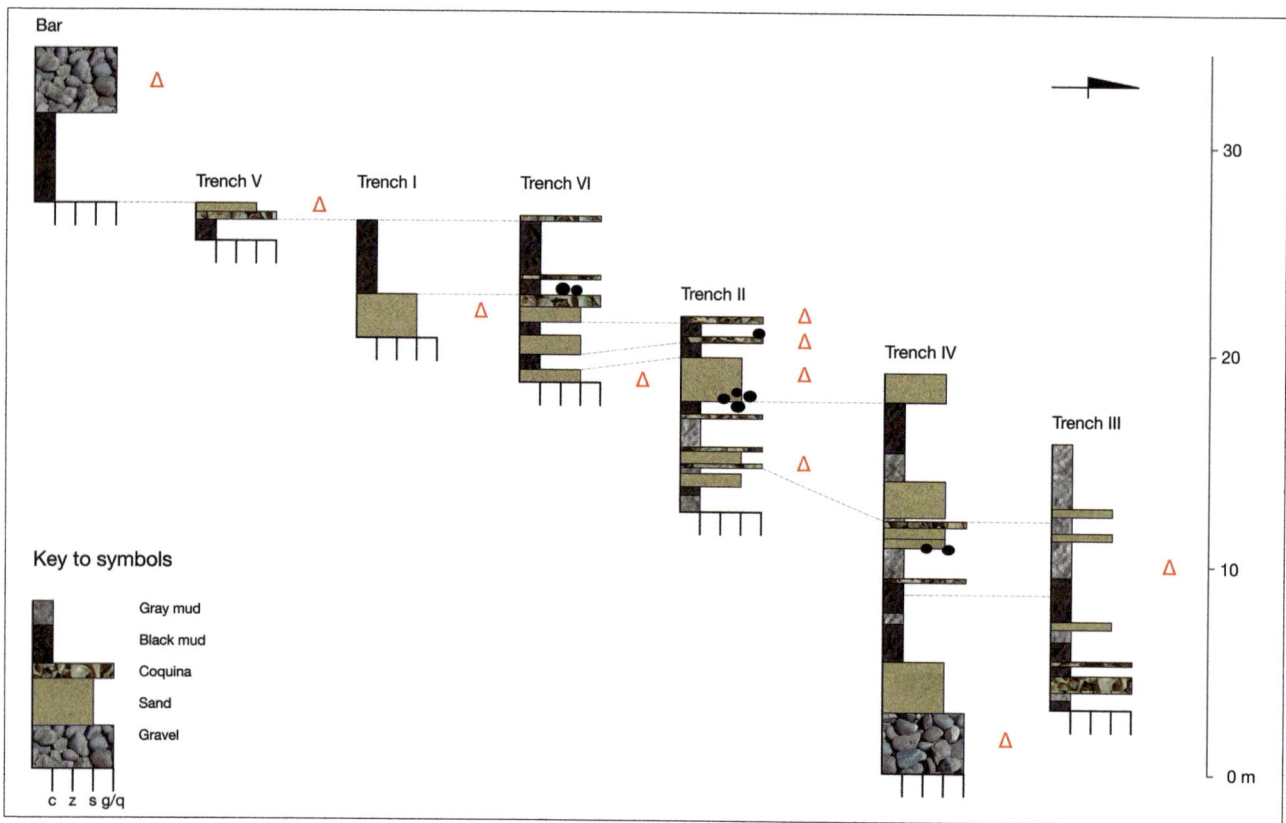

Fig. 3.14a Stratigraphic sections from GBY. Particle size and character is represented on the horizontal axis and with patterns

II-1, I-1, V-1, etc.). Directly below it, without intermediate material, are the BYF tilted strata; this makes it impossible to estimate the duration of the gap, when it was formed, and what was the order of sediment erosion.

New insight into the stratigraphic position of the excavated BYF deposits in the study area was gained from two drillings, GBY #2 and EY (described above) (Fig. 3.15). Their dating and sedimentological/stratigraphic results show that the BYF was deposited on top of a series of basalt flows and that the entire package (BYF and the basalts) is tilted similarly to that encountered in the study area, illustrating the enormous power of the young (intra-Pleistocene) tectonic activity in the region.

3.4 Chronology and Dating

3.4.1 History of the Age and Date of the BYF

The antiquity of the BYF was established as early as the 1930s following the discovery of large fossil mammal bones and stone tools (Stekelis 1960; Goren-Inbar and Belitzky 1989). These preliminary finds initiated studies of the geology of the area and later resulted in the identification of the Pleistocene outcrops of rich *Viviparus* beds and their assignment to the "Old Jordan Terrace" by Picard (1963). Picard (1965:345, table 1) assigned these sediments to the Middle Pleistocene, while Horowitz (1973:136) considered the BYF, which overlays the Yarda basalt dated to 640,000 BP, as Middle Pleistocene as well. Lacking radiometric dating, Picard (1963,1965), Hooijer (1959, 1960), Horowitz (Horowitz 1973), and Tchernov (1996), using different criteria, each assigned the BYF to a different glacial or interglacial period (Goren-Inbar and Belitzky 1989). Instrumental in these assignments were field stratigraphic observations of the formation, where findings such as mollusks, mammal paleontology, and pollen were recorded and analyzed.

The chronological assignment of the BYF was based from the very first discoveries on the biodiversity and biochronology of the faunal remains. Mammal assemblages were used to establish a date by Bate (Stekelis et al. 1937), Hooijer (1959, 1960), and Geraads and Tchernov (1983), and in a much later research phase by Rabinovich et al. (2012) and Martínez-Navarro and Rabinovich (2011). Micromammals provided a general chronological framework (Tchernov 1996 and in Goren-Inbar et al. 2000; Goldstein 2011). Further biochronological indications were gained from analyses of mollusks (Tchernov 1973; Ashkenazi and Mienis 2005). The presence of the extinct gastropod *Viviparus apameae galileae* remained for many years a yardstick for a *post quem* age, as according to

3.4 Chronology and Dating

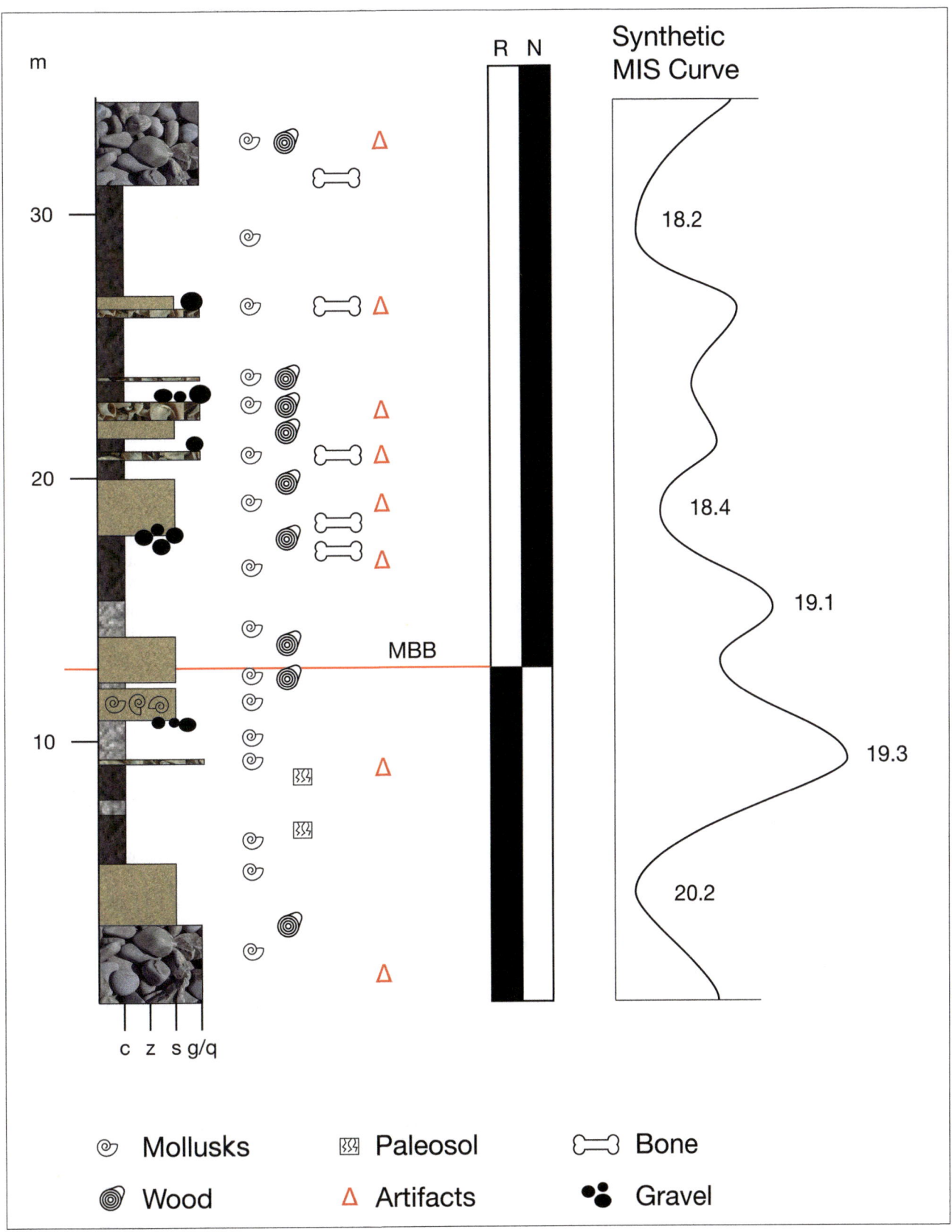

Fig. 3.14b GBY composite section

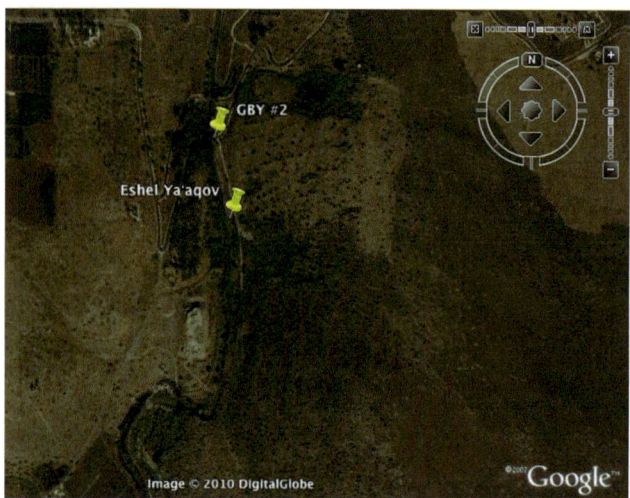

Fig. 3.15 Aerial view of the GBY area and the location of the two drillings: GBY #2 and Eshel Ya'aqov

Moshkovitz and Magaritz (1987) it became extinct sometime around 240 ka (Ashekenazi et al. 2010). Of all the above, it was the micromammals that provided an indication that the site's antiquity was greater than believed and that it in fact dates from the Early–Middle Pleistocene transition (Goren-Inbar et al. 2000). In addition to the above chronological considerations, and as early as the 1930s, the findings of Acheulian tools (handaxes and cleavers) provided another indication of the antiquity of the BYF; however, before the 1980s no dating for this culture in the Levant was available (Goren 1981). The similarities of the lithics to East and South African Acheulian entities drew the attention of Stekelis (1960) and later of Gilead (1970a), but Middle Pleistocene and earlier dates for these sites were not established until the 1980s. In sum, the different attempts that predate the most recent excavations generally assigned the BYF to the Middle Pleistocene.

3.4.2 Radiometric Dating and Paleomagnetism of the BYF and its Environs

Prior to the excavations in the study area, several radiometric datings became available from the site's vicinity, mainly from the Korazim Saddle and adjacent areas (Heimann et al. 1987; Heimann and Ron 1993). While the dates were very informative with regard to the buildup and evolution of volcanic landforms, they added no precision to the dating of the BYF. From the 1980s onwards, with the development of the field of radiometric dating, several attempts were made to obtain direct dating of the BYF in the study area. Among these were C14, OSL, ESR, uranium series, T/L, K/Ar, Ar/Ar, and cosmogenic nuclide exposure dating. Despite extensive efforts, a very small number of dates were obtained. The OSL dating attempts failed due to inconsistencies in the ages obtained in relation to the stratigraphy of the samples (N. Porat, personal communication). T/L samples were collected, including the dosimetry of their immediate surroundings, but results have never been obtained. The ESR and uranium series results (Rink and Schwarcz 2005) were unsatisfactory. The cosmogenic nuclide exposure dating method was unsuccessful as well.

A series of basalt samples was retrieved from the BYF in an attempt to achieve a more precise chronology by K/Ar and Ar/Ar dating (Geological Survey of Israel). The samples were selected from two different localities: 1) Layer II-25: a large boulder conglomerate exposed at the base of the located depositional sequence in the study area (it is also visible in the bed of the Jordan River, a few meters north and west of Trench III); 2) a basalt flow located north of the present bridges, which is exposed (as a result of quarrying and drainage of the river) on both banks. The dating results of these samples were unsatisfactory.

Organic material (charcoal) was sampled from the Unconformity in order to establish scientific proof for the assumption that these deposits (Layers I-1, II-1, III-1, IV-1, V-1, and VI-1) were all deposited by the Jordan River during Recent times. The two C14 dates obtained (Research Laboratory for Archaeology, Radiocarbon Accelerator Unit, Oxford) verified the Recent date and the non-homogeneous age of these deposits. In addition, after excavation it became evident that the Unconformity is indeed a very recent sediment package, consisting of floodplain material of the Jordan River. During its excavation in Areas A and B, aluminum foil, cigarette butts, plastic sandals, shells of modern Pistachio nuts, potshards, and an old Syrian communication radio set were found.

The most recent dating attempts were carried out on two drilled cores (GBY #2 and EY, described above). The Matuyama/Brunhes chron boundary was identified in the record of both cores, providing an overall correlation between the three stratigraphic sections (the excavations in the study area, GBY #2 and EY). Furthermore, basalt flows that underlie the BYF in both cores yielded similar dates (ca. 1.10 Ma by Ar/Ar) (Goren-Inbar et al. 2012). In addition, an Ar/Ar date of ca. 0.66 Ma was obtained for a basalt flow interfingering the BYF sediments of core GBY #2, providing a means of correlating with the BYF dated to 658±15 ka by Ar/Ar north of the bridges (Sharon et al. 2010). In view of the similarity in the thickness of the sediments in the three records, these geochronological correlations are considered reliable (Goren-Inbar et al. 2012).

Another attempt to obtain a chronological assignment for the depositional sequence in the study area was obtained by an extensive paleomagnetic study carried out by K. Verosub (Goren-Inbar et al. 2000), in which some 155 samples were analyzed and a shift from normal (top) to reversed (bottom) magnetic orientation was detected in Trench II (Fig. 3.14b), at the contact of Layers II-14 and II-15. This transition has been interpreted as the Matuyama-Brunhes chron boundary (0.79

Ma), relying on the age of the surrounding basalts, the structural geology of the site, and its location in the general geology of the region (as above). The paleomagnetic results enabled better precision for the assignment of the excavated sequence to both the Early Pleistocene (Layers II-25 to II-15) and the Middle Pleistocene (Layer II-14 to the Bar layers at the top of the sequence).

3.5 Raw Materials

Three different types of raw material were selected by hominins at GBY for the production and modification of stone artifacts and are the most abundant category of evidence in the archaeological horizons at the site. These raw materials are basalt, flint, and limestone. We present here the geographical and geological origins of the raw materials, some of their properties, and their susceptibility to chemical and mechanical weathering.

3.5.1 Basalt

The GBY site is located in a volcanic environment in which basalts form a major part of the landscape (Fig. 3.7). The Pliocene and Pleistocene volcanic rocks that dominate the site's vicinity are alkali and basanitic basalts (Weinstein et al. 2006) that usually display a landscape of weathered and exfoliated rock surfaces and basaltic soils. The structural aspects and evolution of the basalts were described by Picard (1952, 1963, 1965), Horowitz (1973), and Belitzky (1987, 2002; Goren-Inbar and Belitzky 1989) and were dated mainly by Heimann (Heimann et al. 1987; Heimann and Steinitz 1988; Heimann 1990; Heimann and Ron 1993).

3.5.1.1 Origin

Basalts are known from all around the site on both banks of the Jordan River, although along its course they are exposed artificially. The proximity of basalt is also extensively documented in the sediment record of the BYF, as revealed by the excavations carried out at the site and the various drillings carried out in the course and on both banks of the Jordan River along the 3.5 km axis of the site. This evidence is recorded in the form of small pebbles and sand-sized clasts, as well as the basalt-derived minerals that are known from the entire depositional sequence of the site. The size of the basalt component in the sediments is very different from that of the basalt artifacts, a fact that necessitates discussion of the possible origin of the latter. Since the details of the paleo-landscape in the vicinity of the site during depositional times are unknown, the amplitude of the geomorphologic and tectonic changes that have taken place since the time of the site's formation cannot be measured. Consequently, we rely on analysis of the present-day topography, the geological record, and to a lesser extent the dates available for the basalts.

From the geological and geomorphologic analysis (Belitzky 2002) it is evident that the volcanic landscape of the western slopes of the Golan Heights, situated east of the site, could not have been the source of basalt blanks of suitable size for the production of artifacts at the site. The reason for this is the lack of an erosive relief in the form of watercourses in the vicinity of the site. Well-developed systems do exist north of the site all along the eastern flanks of the Hula Basin (Fig. 1.1), but they drained into the paleo-Lake Hula to the north of the study area. On the west, the site is bordered by the volcanic structure of the Korazim Saddle (Goren-Inbar et al. 2002b), which geologically forms the (western) boundary of the Benot Ya'aqov Embayment (Belitzky 2002); it too lacks any visible present-day drainage courses in the immediate vicinity of the site.

Four major fluvial systems drained into the paleo-Lake Hula from northwest to southeast and are currently represented by the river systems draining Upper Galilee and the Safad Mountains, which include Nahal Rosh Pinnah, N. Mahanayim, N. Ayelet HaShachar and N. Dishon (Fig. 3.1). The Pleistocene rivers had somewhat different courses from the present-day ones, and the terrains that they drained were mainly sedimentary and to a lesser extent volcanic (Belitzky 2002; Goren-Inbar et al. 2002b). Although these fluvial systems could have contributed basalts to the site, the large basalt clasts that were found in the GBY sedimentological record were not transported by them. A wealth of drillings in the Hula Valley have furnished a most significant archive for the understanding of the evolution and history of the Plio-Pleistocene Lake Hula. It is also evident that the area to the north of the lake has been continuously flooded and poorly drained for a long duration, a phenomenon that resulted in a long record of stratified swamps (Picard 1952, 1963, 1965; Horowitz 1973; Heimann 1990). The paleo-geographical reconstructions of the GBY paleo-lake and its vicinity (Goren-Inbar et al. 2002b) rule out natural and repeated transportation of large basaltic clasts that would be suitable for production of artifacts (for further details see Feibel 2001, 2004). Within the studied depositional record of the site, which primarily documents an oscillating freshwater lake, there are only three episodes that are characterized by sediments reflecting the presence of a high-velocity fluvial system. These three layers are represented by basaltic gravel of similar chemistry to other basalts in the region. They are located at the top and bottom (the Bar and Layer IV-25 respectively) of the studied sequence

(Goren-Inbar et al. 2000; Feibel 2004), as well as at the base of Layer II-6, more precisely at its bottom, in Level 7 (Figs. 3.13–3.14). The stratigraphic positions of the Bar and Layer IV-25 rule out the option of viewing these conglomerates as the source of any of the lithic artifacts excavated at the site. The stratigraphic position of Layer II-6 Level 7 is such that it had already been deeply buried (by lake margin deposits and by sediments of the rising lake) and thus was not available to the occupants of the archaeological horizons that postdate it.

Considering the above, we conclude that the hominins must have selected and collected basalt raw material from exposures that are not necessarily visible at present and transported the selected objects to the site. Particular collection and transportation modes will be dealt with in other chapters of this volume (Chapters 6–9). Clearly, substantial changes have taken place since the deposition of the basalt objects at the site, particularly those relating to the freshness of the visible basalt flows in the landscape, which was a much younger volcanic terrain at the time of the site's occupation.

The basalt found within the excavated archaeological horizons is characterized by different morphology and size. Two main morphologies of basalt raw material are encountered: items that have slab morphology, either extremely large or small and thin, and those that are generally smaller and are characterized by a rounded shape of pebble morphology. The latter, which were either percussors or potential parent material for the production of cores, were most probably collected in the vicinity of the site from the wadi gravels that are so common in and around the Hula Valley. In contrast, the giant slabs and thin slabs must have been imported from particular fresh exposures of basalt flows where slabs were formed in the middle section of the flow (Chapter 7 and details therein).

It is important to note that tens of thousands of basalt artifacts and natural cobbles were excavated at the site. Most of these, regardless of their classification (modified or unmodified), are of compact and relatively fine-grained basalt. Less than a hundred items (all natural basalt pebbles) were found to be porous or vesicle-rich. Considering this, and the fact that artifacts cannot be made on vesicular basalt, the absence of the latter supports the suggestion of intentional selection and transportation of good-quality basalt items to the site.

During the field season of 1995 Prof. S. Weiner of the Weizmann Institute of Science carried out a preliminary examination of several samples of basalt artifacts originating in the archaeological horizons. The reason behind this attempt was the accumulation of field observations that signaled an extensive variability in the state of preservation of the basalt artifacts, as well as their color and texture. While some artifacts were extremely fresh and looked recently knapped, others were fissured, crumbling, and exfoliated as a result of exposure to atmospheric conditions (see below); a variety of intermediate states were also observed. The examination of the basalt samples was carried out by a portable FTIR machine in the field, and the results indicated that the sampled basalt artifacts differed chemically from each other and thus had a variety of origins.

There followed an attempt to qualify and quantify the variability observed in the basalt landscape and artifacts with the aim of gaining original data relating to the source of the basalt artifacts. A detailed geochemical study was then carried out (Light et al. 1999, Light 2001), applying a non-destructive XRF analysis to basalt samples from the landscape in the vicinity of the site, as well as to a selection of basalt artifacts originating from various archaeological horizons. The analysis yielded a most interesting and complex pattern of basalt exploitation, but regrettably it was never published (Verosub 2001 personal communication).

3.5.2 Flint and Limestone

3.5.2.1 Origin

Flint and limestone artifacts are described here together due to their shared origin in sedimentary rocks. The sedimentary Gadot and Mishmar HaYarden Formations (Horowitz 1979; Belitzky 2002) are both in the vicinity of the site and are both built of limnic sequences of freshwater lakes; however, they are devoid of large, hard limestone clasts that would have been suitable for the production of stone artifacts (Fig. 3.5). Flint and limestone clasts could have become available in this part of the lake margin through drainage systems that are located to the northwest and west of the site; they could have been transported by the four fluvial systems described above, which today, as in antiquity, carry their drained loads from high elevations (up to 1,208 m above msl at Mt. Meron) of the Upper Galilee and the Safad Mountains into the Rift Valley (Fig. 3.1). Indications of such transportation of sedimentary clasts into the Rift Valley in antiquity are documented in drillings of the Hula Basin (Belitzky 1987, 2002) as well as at the site. Although geomorphological indications suggest that the courses of these four fluvial systems have changed since Pleistocene times (Belitzky 1987, 2002; Feibel 2004), their antiquity is well established as the source of all sedimentary particles of the sedimentary sequence at the site.

Artifacts modified on flint and limestone that were excavated at the site are much larger in size than the clasts that are an integral part of the sediments. Therefore, as with basalt, the flint and limestone items that were modified into artifacts were selected and transported to the site by the hominins.

Extensive landscape destruction has taken place in the vicinity of the site (Chapter 2), particularly west of the river. During the early 1980s the terraces of Nahal Rosh Pinnah were still visible around its confluence with the Jordan River

3.5 Raw Materials

Fig. 3.16 The terraces of Nahal Rosh Pinnah in the southernmost exposure area of the BYF (1980s); note the bedded limestone cobbles *in situ*

in the southernmost exposure area of the BYF. These were photographed at the time with the aim of documenting the bedded limestone cobbles *in situ* (Fig. 3.16). Since then the entire landscape of this area has been destroyed as a result of the construction of the hydroelectric power station of Kibbutz Kfar HaNasi (Chapter 2). It is likely that some of these ancient terraces were visible to the Pleistocene occupants of the site and served as the source of sedimentary rock raw materials.

The extent of the raw material contributed by the four fluvial systems must remain unknown, as only a remnant of their histories is documented in the drillings carried out in the Hula Valley. It is evident that during the time span represented at the site (ca. 100 ka) there were sizable geomorphic changes in the vicinity of the site. Nevertheless, on the basis of diverse structural and geomorphological considerations, it seems reasonable to assume that the drainage regime was the main source of the limestone and flint raw materials at the site (Belitzky 2002). This assumption does not rule out transportation of sedimentary raw materials to the site by hominins during the time of the Acheulian occupation. Flint is available in the sedimentary formations of the eastern and northern parts of Galilee. In an attempt to locate possible sources of flint and limestone, an excursion to Nahal Dishon was carried out by expedition members, aimed at the procurement of cobbles of appropriate size for knapping. We found abundant nodules of suitable size. As this river is only one of the four options described above, and not the closest, it is only a possibility and cannot be considered the actual place of acquisition. A glimpse at such a possibility was furnished by the identification of a cortical flint type unknown from the Hula Valley but similar to a flint type widely known from the Central Jordan Valley in the vicinity of Kibbutz Sha'ar HaGolan (Delage 2007).

3.5.3 Weathering of Raw Materials

The chemical and mechanical properties of the three raw materials that were used at GBY for the modification of artifacts differ greatly from each other. These differences are best manifested archaeologically with regard to postdepositional processes, primarily preservation and patination. Evidently, the waterlogged environment of the site had a significant impact on the buried lithic artifacts in different ways. While the freshness of the flint and limestone artifacts was only partially impaired, some of the basalt artifacts were drastically affected and some had completely disintegrated. However, the destructive forces did not influence all of the basalts equally; one possible reason for this may be associated with the basalts of different origins that coexist at the site (see discussion above and Chapter 7).

Some of the basalt artifacts look recently knapped (Fig. 3.17), while others display different states of weathering. The most frequent preservation state is that of abraded objects that are characterized by rounding/smoothing of the surfaces, which makes the identification of particular features ("cortical" surfaces, direction of blow, conchoidal waves, cones, etc.) extremely difficult. Most basalts lack any distinctive cortex or rind that differentiates the exterior surface from the interior. The weathering of basalt was more understandable because of our careful method of excavation, which entailed the exposure of artifacts following by their careful removal. Looking at the undisturbed surface underneath the artifact, it was possible to see a thin crust that had detached from the artifact and adhered to the deposit. Fig. 3.18 illustrates a particular case encountered in Area B, where a basalt giant core was excavated in Layer II-6

Fig. 3.17 A fresh basalt tool (# 8156 Layer II-6 Level 5)

Fig. 3.19 The distal edge of a basalt cleaver (#2255 Trench II green); note the development of exfoliation on the periphery

process was augmented by exposure to atmospheric conditions and time, causing the artifacts to exfoliate and resulting in a smaller artifact with a rounded morphology and a harder core, lacking any visible traces of knapping (Figs. 3.20–3.21). Finally, some artifacts had reached total disintegration of the basalt into clay crust.

Fig. 3.18 A giant basalt core (#10479 Layer II-6 Level 1): a) *in situ* and b) after removal; note the crust of basalt that remains on the sediment

Level 1. After its removal an imprint was left in the deposit, where the surface of the sediment in which it was bedded consisted of a thin crust of basalt. Artifacts that are more severely damaged exhibit a development of fissures beginning from their extremities (Fig. 3.19). In cases where these artifacts were not conserved during fieldwork (immersed in water, Chapter 4), this

Fig. 3.20 Basalt artifact (#2254 Trench II) that was not conserved and is consequently exfoliated

3.5 Raw Materials

Fig. 3.21 Basalt handaxe (not catalogued Trench II) that was not conserved and is consequently exfoliated

Limestone artifacts seem to have been exposed to weathering processes that were most probably induced by the waterlogged nature of the sediments. Although they were made on hard rocks, all the limestone artifacts look somewhat rolled and smooth, and all edges and ridges are rounded. Thus, among the limestone artifacts there is a marked paucity of fresh items (Chapter 8). The weathered limestone artifacts are associated with an additional phenomenon in the form of clusters of striations visible (after some experience) to the naked eye. This is clearly a taphonomic, postdepositional phenomenon caused by an as yet uninvestigated agent. Although much progress has been achieved with regard to the taphonomic history of the bones at the site, particularly in connection with breakage and striation (Rabinovich et al. 2008, 2012), the striations on the limestone still await further study.

The preservation of flint displays a much greater variability than that of limestone. Some of the flint artifacts are fresh, but all degrees of weathering are observed as well. Although the flint weathering mechanism at GBY has not been investigated, it seems to us that the weathering of the artifacts was caused *in situ* by chemical weathering. What meets the eye is some sort of decalcification that primarily attacks the ridges of the artifacts. Clearly, the issue of flint weathering, as with the other raw materials, necessitates a future in-depth multidisciplinary study.

3.5.3.2 Patination

While patina is observed on most of the artifacts, it differs in nature among the different raw materials. The patina on basalt and limestone seems to be some sort of coloration resulting from the adherence of small-grained particles to the surfaces of the artifacts. In contrast, the patination of flint artifacts at GBY result from two different processes. The first is the adherence of particles from the surrounding sediments whose color dictates the color of the patina; in black mud environments rich in organic material, the developed rind is characteristically black. The second process is a biogenic development (Friedman et al. 1995; Gorbushina et al. 1996) that does not seem to be related to the color of the sediment. For some of the flint artifacts of GBY, longer life histories are expressed in double patina (Chapter 6).

Summing up the weathering history of lithics at GBY, our impression is that the depositional environment played a major role in the weathering of the artifacts.

In summary, this chapter has presented the geography, geology, stratigraphy, chronology, and raw materials of the site of GBY. A major part of the chapter is dedicated to the exposures of the BYF and their characteristics. The sedimentary package that was investigated in the study area (0.7–0.8 Ma), which combined drillings with the most extensive exposure of the BYF on land, is discussed in detail. These discussions provide the background for the following chapters, which deal with the methodology, inventory, and analyses of the lithic material from GBY.

Chapter 4
Methodology

Abstract Chapter 4 presents a detailed account of the field and laboratory methodologies used at Gesher Benot Ya'aqov. It provides a comprehensive description of the excavation methods and techniques pertaining to the grid, elevations, coordinates, sieving, sediment sampling, drafting of maps and cross-sections, and conservation. Laboratory methods include detailed descriptions of sediment sorting and lithic analysis, the latter including attribute analysis that incorporates in-depth descriptions of typological, technological, and morphological characteristics. This chapter is accompanied by five appendices presenting the typological list and the attributes and attribute states recorded for flakes and flake tools, cores and core tools, handaxes, and cleavers.

Keywords attribute analysis · cross-sections · excavation methods · Gesher Benot Ya'aqov · laboratory methodology · lithic analysis · maps · sieving and sorting · technology · typology

Chapter 4 is concerned with various aspects of the methodologies used in the Gesher Benot Ya'aqov (GBY) project. It comprises two main sections, the first presenting the field methodologies applied during excavation and the second describing the laboratory methodologies used for the analysis of the lithic assemblages. Other methodological aspects that are related to this volume are discussed as well.

4.1 Field Methodology

The field methodologies employed in the GBY project were based on cumulative experience and on the particular conditions and constraints dictated by the nature of the site. The site of GBY differs from most prehistoric sites in two major aspects: its waterlogged nature and its tectonically disturbed deposits. While the former necessitated excavation, registration, and conservation methodologies that have been employed elsewhere, the second dictated the development and application of unique methods, frequently based on trial and error and the accumulation of specific field experience.

4.1.1 Grid and Shading

During the survey of the Benot Ya'akov Formation (BYF) it became evident that its exposures are extremely small in size and are situated structurally in many and varied positions (Belitzky 1989, 2002; Goren-Inbar et al. 2000). These circumstances ruled out the possibility of applying a local coordinate system to the exposed surface, as was previously done at the site of 'Ubeidiya' where the grid was laid out directly on the tilted surface along the strike and dip of the surface intended for excavation (Bar-Yosef and Goren-Inbar 1993). As the BYF is fractured into isolated blocks with different orientations, the application of a similar method to GBY would have resulted in the coexistence of many coordinate systems, leading to confusion. Consequently, at GBY the Israeli Transverse Mercator (ITM) geographic coordinate system was applied to all excavation areas. The ITM, the coordinate system most commonly used for mapping in Israel, uses the transverse Mercator projection and meters as grid units; it also enables easy reference to unexcavated exposures and prominent landmarks (e.g. bridges). The ITM was used to position and orient the physical grid in the study area, serving as a reference system for the spatial location of all finds originating in the excavations.

The initial establishment of the grid was complicated by the presence of a minefield east and southeast of the study area and the fact that several map-marked triangulation points could not be located and had apparently disappeared from the vicinity of the site.

During the first two field seasons (1989, 1990) the grid was simply laid down on the horizontal surface of the excavated area

Fig. 4.1 The grid laid on top of the horizontal surface of the excavated area

Fig. 4.2 The grid laid on top of the horizontal surface of the excavated area; note the differences in elevation between the grid and the tilted excavated surface

(Fig. 4.1). At the end of the second season it became evident that pronounced differences in elevation (caused by exposure of the sediments along the strike and dip) were jeopardizing the accuracy of the grid (Fig. 4.2). Thus, starting with the field season of 1991, a suspended grid system was applied to the entire excavated area, a system that proved to be efficient and highly accurate throughout the field seasons. The grid system formed units of 1 m^2, each equally subdivided into four sub-squares. The southwestern corner of each square was the zero point for readings of x and y coordinates.

The suspended grid comprised a rectangular metal frame to which parallel steel cables were attached at intervals of 1 m. Strings with fishing weights attached to their ends (plumb lines) were tied at the crossing points of the cables, thus forming the four corners of each square meter (Figs. 4.3, 4.4). In this way the plumb lines marked, on the tilted horizon, the boundaries of the grid's square meters (Fig. 4.5).

Above the grid frame, a movable awning was installed. The awning rolled on metal cables and was deployed or retracted according to the light conditions and the requirements of photography (Fig. 4.6).

4.1.2 The Study Area and Trenches

On completion of the survey of the BYF (Chapter 3), a specific area on the left bank of the river was selected for excavation (1989). The area, in the Recent sandy flood plain of the Jordan River, was very flat and devoid of basalt boulders. As the floodplain deposits completely covered the Pleistocene deposits in this area, during the first field season two trenches were mechanically excavated by a backhoe, perpendicularly to the strike of the bedding, to provide observations of the stratigraphic sequence (Figs. 4.7, 4.8). Eventually, six trenches in all were excavated at GBY (Chapter 5). The northern face of each trench was drawn and the observed layers were numbered from

Fig. 4.3 The freestanding grid system

Fig. 4.4 The freestanding grid system

west to east (Fig. 4.9). Thus, each of the smallest recognizable stratigraphic units (layers) is named after the number of the trench in which it was observed (Trench I, II, etc.), with the addition of an Arabic number to refer to its location within the sequence of that trench. Hence, the numbering of the depositional sequence provides a relative chronology from younger (west) to older (east) layers. Trench IV was excavated during the 1995 season and was oriented, unlike the other trenches, in a roughly north-south curving line. This trench was dug in order to drain the main excavated area (Area B) by diverting the subsurface ground water originating in the Golan Heights and flowing from east to west. Although its original aim was simply to ease the conditions of excavation, the trench yielded a wealth of important information and was as thoroughly investigated as the other trenches.

Three areas (A, B, C) adjacent to the three main trenches were selected for excavation during the different field seasons (Fig. 4.8). In addition, excavations were carried out along the exposed layers on the Jordan River bank (referred to as JB) (Fig. 4.8). Limited excavation also took place in the Jordan River itself after the discovery of large bones that were exposed due to a very low water level (the Enclosure, Fig. 4.8).

4.1.3 The Excavations

In each of the areas excavation started with the removal of the Recent deposits (the Unconformity), which include redeposited Pleistocene material underlying Recent sediments (Chapters 3, 5), in order to expose the Lower–Middle Pleistocene bedding. The excavation of each of the Unconformities was carried out by conventional archaeological methods, from the surface downwards. The surface was drawn and the finds (lithic artifacts only) were registered (Fig. 4.10). Due to the taphonomically disturbed nature of these deposits, no sieving of the Unconformity sediments was carried out. The horizontal

Fig. 4.5 Plumb line marking a corner of a grid square on the tilted excavated horizon

Fig. 4.6 The movable awning located above the grid

Fig. 4.7 Backhoe trenching in the study area

(plan view) section of the underlying Pleistocene deposits was drawn to document the stratigraphy.

On exposure of the Pleistocene sediments and their stratigraphy, the excavation of the deposits began with the exposure of the tilted beds (each observed, for enhanced control, in the adjacent trench as well). The archaeological horizons were all exposed along the strike and dip of the layer/level. On completion of the lateral exposure of each artifact-bearing horizon, the finds were mapped and photographed and their spatial provenance was recorded before removal.

While the grid system was horizontally suspended above the excavation area, the archaeological horizons were tilted. The corners of each grid square were marked on the tilted surface by the hanging strings and plumb lines. The area excavated (on the tilted layers) under the 1 m² grid was thus a *projection* of 1 m² and the surface excavated was larger than 1 m². The size of the projection's surface was dictated by two angles: the angle at which the horizon transected the grid (the strike) and the angle of the horizon's dip. A calculation based on the dip of Layer II-6 shows that its projected surface was in the order of 1.2 m². Clearly, any change in the orientation of the layer's strike would change the grid planar transect and might also change the layer's dip. Elsewhere in this monograph (Chapter 5), we present the absolute excavated surfaces and volumes.

4.1.3.1 Datum and Elevations

Elevations of surfaces and excavated items were read in reference to sea level, based on a fixed excavation datum (benchmark).

Elevations were measured with folding carpenter's rulers and were determined with the use of a string line and line level attached to a nearby datum stake referenced to the site's benchmark (Fig. 4.11).

Elevations were measured on the exposed tilted surface of each excavation unit at the beginning of each excavation session

4.1 Field Methodology

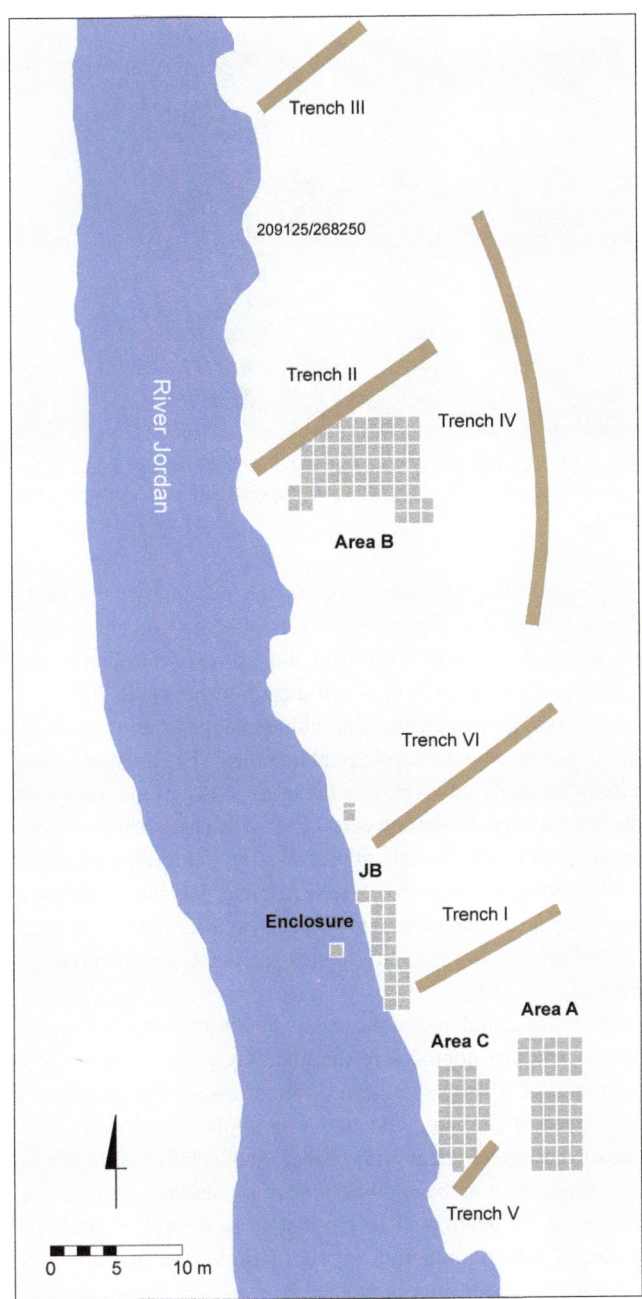

Fig. 4.8 The GBY study area: map of trenches and excavation areas

Fig. 4.9 View of the northern face of Trench II

Fig. 4.10 Excavation of Layer II-1 (the Area B Unconformity)

first on top of the item and the second on the surface below it after its removal, thus providing the item's thickness. Top and bottom elevations of these items were also marked on the maps of each of the archaeological horizons (see below) (Fig. 4.12).

4.1.3.2 Excavation Difficulties and Techniques

A major goal of the GBY excavations was to expose the largest possible surface of each of the archaeological horizons. Because of the tectonically induced structural position of the deposits, this objective was difficult to meet. The layers in the study area dip 25° to 45° southwest (Goren-Inbar et al. 2000), an inclination that poses severe logistical problems. Although it was possible to extend the excavated surface along the strike, its exposure along the dip was limited, making extensive deepening of the excavation unfeasible (unlike at 'Ubeidiya; Bar-Yosef and Goren-Inbar 1993). The exposure was further limited by the fact that the sediments were waterlogged and pumping was necessary, particularly as the excavation deepened.

(i.e. a 0.5 m² sub-square to a depth of 5 cm) and at its end. Initial and final measurements were taken on the tilted exposed surfaces and on the same points of each square: the northeastern (uppermost) point and the southwestern (lowermost) point, each marked on the excavator's sheets. In larger, complex, or very difficult cases (e.g. baulk removal or a combination of many squares with a small surface in each), an additional measuring point was added in the middle of the surface. Thus a 5 cm thick deposit in a sub-square could result in six elevation readings.

Two elevation readings were taken as well for each find (stone, including natural pieces, bone, and wood) larger than 2 cm, the

Fig. 4.11 Measuring elevations during excavation

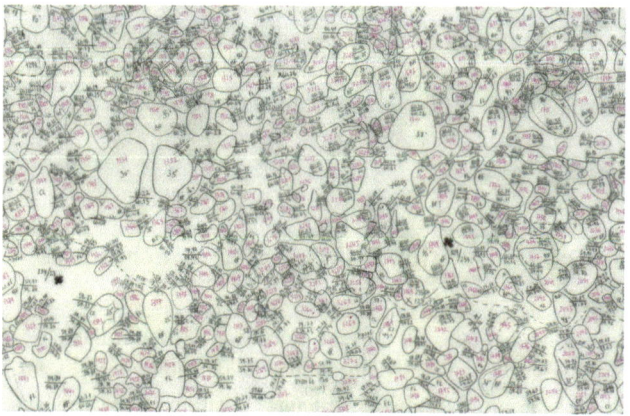

Fig. 4.12 Segment of the field map of Layer II-6 Level 4

It is quite a difficult task to follow the strike and dip while excavating. In accomplishing this with maximal precision, we were aided by the cross-sections of the tilted Pleistocene deposits exposed below the floodplain deposits of the Jordan River (i.e. the Unconformities: Layers V-1, I-1, and II-1). Since the stratigraphy of the archaeological horizons is visible there, they were marked with white strings stretched along the strike

Fig. 4.13 Excavation of Area B; note the white strings laid on the Unconformity (Layer II-1) marking the strike of different levels of Layer II-6

(Fig. 4.13). The excavators began the removal of the tilted deposits from the top (the contact with the Unconformity), advancing downwards. Following the dip was controlled by the removal of a standard unit of sediment thickness (Fig. 4.5).

Excavation was carried out with dental picks and ice picks, and in places with archaeological hammers. The most common excavation tool, however, was the brush. Most of the work was carried out very slowly and carefully, as the basalt artifacts were usually in a very poor state of preservation – soft (clay-like) and easily damaged. In horizons rich in bone, the same prudence was exercised. The slow rate of progress was further dictated by the dark color of the waterlogged sediments and their muddy nature.

It was essential to start the excavation in a manner that would enable an easier approach to the tilted layers. In order to carry out this type of exposure, two different excavation techniques were used at the site. The first was applied to deposits that included archaeological finds (either thick subdivided layers or thin single-item archaeological horizons), and the second was employed for geological layers that were devoid of artifacts, bones, or organic material. Fig. 4.14 shows the two different excavation methods in use in Area B.

Excavation of layers with archaeological finds: The main objective in excavating archaeological horizons was to expose the content of the horizons along their strike and dip. Excavation progressed with the aim of preserving the finds on the surface and enabling the documentation of the original setting of the cultural components. Excavation was carried out in each sub-square to a depth of ca. 5 cm. In instances where a concentration of lithics, bones, and wood was encountered the depth sometimes varied (more or less than 5 cm), though this was always specifically indicated. When the surface (often only one artifact thick) was fully exposed the excavation process was paused and drawing, photography, and general documentation took place.

4.1 Field Methodology

Fig. 4.14 Two excavation methods: a) along the strike and dip, b) from top to bottom

All items larger than 2 cm were drawn and, when identification was possible in the field, the classification (stone artifacts, manuports, bones, or wood, including bark) of the drawn items was noted on the map (Fig. 4.12). Each of these items was assigned a number that appeared both on the maps and on the registration bags and forms (and later on the coding sheets); this was the "map number" of the items. In later stages of analysis, this numbering procedure enabled us to correlate the maps with the database of analyzed material.

When documentation of the exposed surface was complete, the removal of the items took place (a process termed "peeling"). "Peeling" included measurement of four coordinates (x and y provenance and top and bottom elevation) for each item larger than 2 cm. Each such item was packed in a separate bag on which all data were registered (date, area, layer and level, square, sub-square, x, y, and two z coordinates, number on map, and name of excavator). Most basalt items and all the organic material were immersed in water after "peeling" and packed with all the registered information (on mylar paper) both in the water and out of it as a precaution. All other items (smaller than 2 cm) were placed in a "general bag", one for each sub-square, with precise top and bottom elevations. All the sediments excavated from the sub-square were placed in buckets tagged with their provenance data, and were later wet sieved (Section 4.1.5).

Excavation of geological layers: Archaeologically sterile layers were excavated from top to bottom and not according to the strike and dip of the levels. These excavations were of limited extent and served either to create better conditions for exposing an underlying archaeological layer by revealing its contact with the sterile overlying one (e.g. Layer II-5) or simply to go through the sediment at a faster rate (e.g. Layer II-2) (Fig. 4.14).

In contrast with the archaeological layers, sieving of sediments from geological layers was partial. The sampling strategy was changed according to the layers involved – 25% or 50% of the excavated deposit (one/two out of four buckets of sediment). In addition, one sediment bucket from every sub-square was kept unsieved for further analysis. Samples of sediment from archaeological layers were also collected and kept unsieved for future analyses.

4.1.4 Maps and Drafting

In order to facilitate the work along the Jordan River and to provide the precise location of the different excavations in the study area, a detailed map of the river, its banks, and its immediate vicinity was produced by Fantomap at a scale of 1:1000.

Mapping and drawing were carried out continuously throughout the process of excavation. Mylar sheets and pencils were used for all field drawings in order to minimize distortion caused by changes in humidity and the generally waterlogged nature of the site.

4.1.4.1 Maps

The exposed archaeological horizons were mapped in detail at a scale of 1:5. The maps included all visible items (artifacts, bones, wood, and manuports), and their classification (when possible) and elevations on the surface were marked in several locations (Fig. 4.12; Chapter 5). Mapping was conducted as if perpendicularly to the tilted artifact-bearing horizon, a view that is consequently not related to the grid system. Correlation between the grid and each of the maps was accomplished by marking the location of each plumb line (representing the corners of each 1 m² on the grid) on the tilted surface.

Map drawing was facilitated by the use of an aluminum drawing frame (parallel strings spaced 10 cm from one another and forming a grid of 10x10 cm squares), which was placed directly on the tilted surface (Figs. 4.15, 4.16). Great effort was invested in achieving high precision and uniformity between the different drawings of the different layers, as well as between different field seasons.

In cases of superimposed artifact surfaces (levels) such as those of Layer II-6, several lines were marked on the horizontal surface of the Unconformity above the excavated surface (see above). Each of these lines, delineated by parallel white strings, marked the location of an archaeological horizon and its strike. The metal frame was aligned on the newly exposed surface in such a way that the strike's white string was placed on the interior part of the frame (Figs. 4.17, 4.18). Next, the draftsperson marked the location (supplied by the plumb lines) of each 1 m^2 projection on the plan. The precise configuration (alignment) of the strings was also documented by there mapping according to the grid. This facilitated the digitized postioning of the drawn surfaces in the grid.

Whenever possible, a preliminary identification of the item was registered on the map. Among the lithics, this was carried out for handaxes, cleavers, flakes, and cores (usually of large dimensions) with reference to the type of raw material (basalt, limestone, flint). Bone and wood (occasionally bark) were also

Fig. 4.16 Field drawing of an exposed surface; note the aluminum drawing frame laid on the excavated surface and the white strings laid on the Unconformity (Layer II-1) marking the strike of different levels of Layer II-6

Fig. 4.15 Field drawing of an exposed surface; note the aluminum drawing frame laid on the excavated surface

Fig. 4.17 The method of field drafting; note the drawing frame in the foreground, where the interior part of the frame is aligned with the string marking the strike of the level

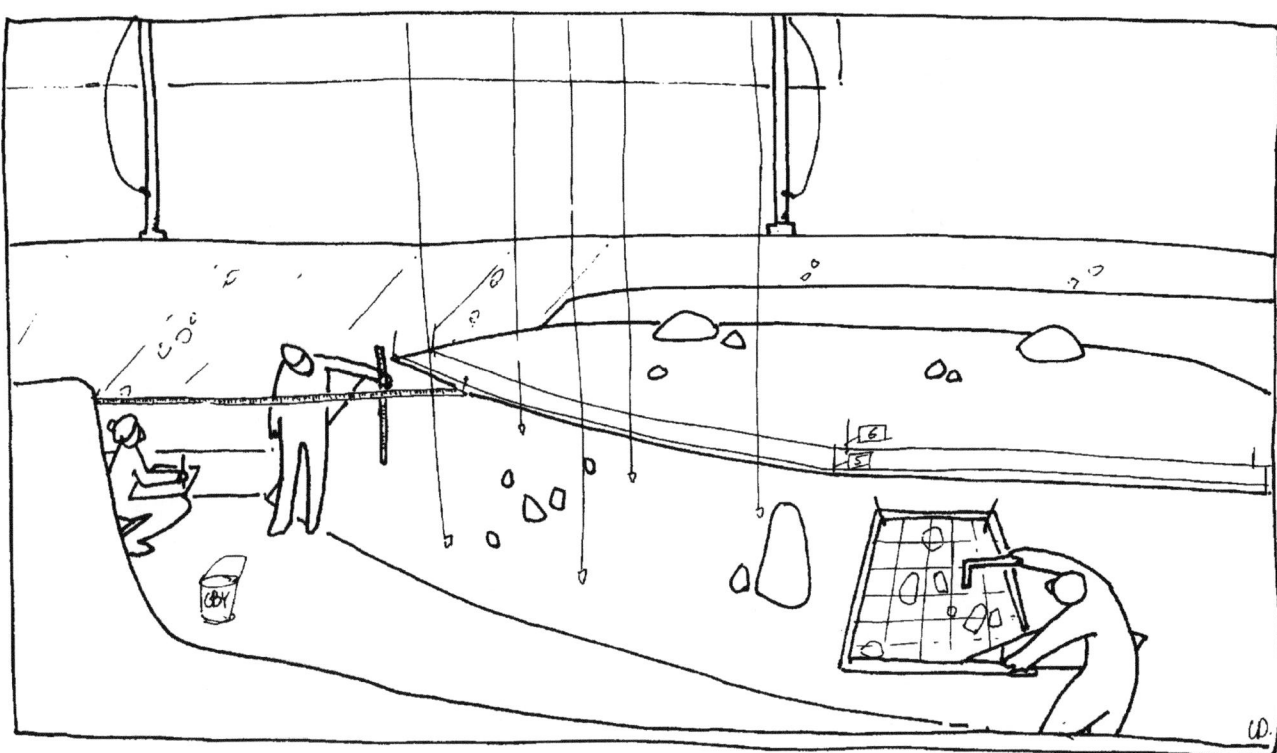

Fig. 4.18 Schematic illustration of different methodologies used during field drafting; in the foreground a drawing of an exposed surface and in the background a drawing of cross-sections perpendicular to the layer; note the freestanding grid and the strings marking the strike of the levels

marked, and when possible the dip of the individual item was noted in addition to its two elevations.

4.1.4.2 Cross-Sections (Profiles)

Cross-sections of two types, both at a scale of 1:10, were part of the routine excavation and registration methodology. The first were drawn perpendicularly to the layer and the second were drawn along the grid lines (Chapter 5).

Perpendicular to the layer: This type of cross-section was used to identify the detailed relationship between the various superimposed levels in Areas B and C. These cross-sections were based on an imaginary horizontal line from which measurements were taken perpendicularly all along the tilted excavated surfaces (Fig. 4.18). Following the "peeling" of the archaeological horizon and the exposure of the underlying one, the same procedure was repeated, thus forming an accumulative cross-section. These are "phantom" cross-sections that do not exist as a standing profile. The method of drawing the cross-sections was as follows:

(1) A folding carpenter's rule or measuring tape was placed horizontally along an imaginary line marking the location of the desired cross-section, with one end placed on the Unconformity. (2) The zero point of the rule was placed on the white string line that marked the particular archaeological level (inclined surface) to be drawn. (3) Two readings (measurements) were taken to locate each point of the cross-section: a horizontal distance from the zero point on the rule and the elevation of the surface at that distance on the inclined surface. The distance was read by using a plumb line from the horizontal rule at particular intervals and marking its location on the tilted surface being drawn. In instances where the cross-section included tools (handaxes, cleavers) or large items of other types (wood fragments, bones), they were integrated into the cross-section and their catalogue number was marked.

In accordance with the grid system: This type of cross-section was required for reconstruction of the configuration of the entire excavated volume. It included the borders of the entire excavated area as well as particular locations within it.

4.1.5 Sieving

All sediments originating in the excavated archaeological horizons were placed in buckets on which their provenance data were registered. The contents of each bucket were wet sieved through mesh. The sieves (plastic boxes with 2 mm plastic mesh sewn into their interior) were tied to a steel cable and submerged in the river (Fig. 4.19). When the sediments were clayey, they had to be soaked for several days before sieving. The wet-

Fig. 4.19 Wet sieving in the Jordan River

Fig. 4.20 Protecting the exposed archaeological horizons with wet geotextile

sieved sediments were then placed on newspapers in cardboard trays for drying and packed in plastic bags when fully dry. The packed sediments were then transported to Jerusalem for the laboratory sorting phase of the analysis (see below).

4.1.6 Conservation

The advantage of the site's waterlogged nature is reflected in the outstandingly good preservation of its organic material. In order to maintain similar conditions for continued preservation and prevent desiccation and hence deformation of these items, it was essential that they remained damp while they were exposed on the archaeological horizon; after their removal they were packed in water. Water was sprayed and poured onto the exposed horizons, which were covered with soaked geotextile fabric or wet foam mattresses (Fig. 4.20) and then by plastic sheets after excavation was halted for the day.

Immediately after their removal all wood samples were placed in plastic bags filled with water in order to prevent desiccation; this process had to take place very quickly on exposure to atmospheric conditions, particularly during the hot summer of the Upper Jordan Valley, which is characterized by high evaporation.

Descriptive data on each retrieved wood or bark fragment (location, size, etc.) were registered on special forms of the wood catalogue and later stored in the computerized database of organic material. Particular methods of conserving and storing the wood have been thoroughly discussed elsewhere (Goren-Inbar et al. 2002b and photographs therein). Several wood fragments were kept sterile and did not undergo conservation. These were selected for DNA analysis during the 1991 field season and were not preserved in water.

Conservation activity in the field also included basalt artifacts. Each basalt artifact that required conservation due to its poor state of preservation (Chapter 3) was placed in water immediately after its removal. It was then transferred to the field laboratory to receive the appropriate conservation treatment, which was based on immersion in different solutions of polyethylene glycol. This was carried out by Orna Cohen, the project's conservator.

The continuous watering of the horizons, and the disturbance inflicted by the excavations on the natural subterranean seepage of water from the Golan Heights (east) towards the Jordan River (west), caused all low areas to fill with water. This, in turn, necessitated pumping. Up to three electric pumps (sucking mud pumps) were used simultaneously in the trenches and areas, while continuous pumping was applied to excavation in the river itself.

4.2 Laboratory Methodology

Over the years of the project, a large amount of lithic items was retrieved from the excavated surfaces, as well as bags of sieved sediment. These were stored at the laboratory of the Institute of Archaeology of the Hebrew University of Jerusalem. The analysis of both categories of finds was carried out continuously from the first excavation season until the completion of this volume. Most of the faunal and botanical remains have been analyzed and published (Goren-Inbar et al. 2002b; Rabinovitch et al. 2012). Wood and bark fragments were conserved by O. Cohen (Goren-Inbar et al. 2002b:17–18) and identified by E. Werker (Hebrew University) (Goren-Inbar et al. 2002b). Seeds and fruits were identified by M.S. Kislev and Y. Melamed (Bar-Ilan University) (Melamed 1997, 2003; Melamed et al. 2016; Goren-Inbar et al. 2000; Goren-Inbar et al. 2002a, 2004). Micromammals were analyzed and identified by the late E. Tchernov (Goren-Inbar et al. 2000), R. Rabinovich, R. Biton, and M. Goldstein (Hebrew University) and crab fragments by S. Ashkenazi (Hebrew University) (Ashkenazi et al. 2005). The malacological assemblage was studied by the late S. Ashkenazi and is currently under study by H.K. Mienis (Hebrew University). Mammal remains were studied by R. Rabinovich (Hebrew University), including a taphonomic study by R. Rabinovich (Hebrew University) and S. Gaudzinski (RGZM, Germany) (Rabinovich and Biton 2011; Rabinovich et al. 2012). Fish remains were analyzed by I. Zohar (Hebrew University) (Zohar and Biton 2011; Zohar et al. 2014). A summary of these faunal and botanical studies is provided in Section 2.3.

4.2.1 Sediment Sorting and Analysis

Sediment sorting was carried out at the Institute of Archaeology on the Mt. Scopus campus of the Hebrew University and lasted twenty years from the end of the first field season. Students of different departments of the Hebrew University examined the contents of the plastic bags containing sieved and dried sediments. The diverse contents of the bags were identified and sorted according to different categories: lithics (both flaked and natural items), organic remains (e.g. wood and bark fragments, seeds, and fruits), and the remains of mammals, birds, crabs, fish, and malacofauna (bivalves and gastropods).

The various sorted components were distributed to their respective specialists for analysis (see above), while the entire lithic component was kept at the Institute of Archaeology for detailed analysis. After the sorting of each bag the residual sediment, unless required for aspects beyond the primary analysis, for example to serve as a matrix in experimental work (Rabinovich at al. 2006) or to be used for museological purposes, was discarded, since sampling during fieldwork had resulted in an extensive archive of unwashed sedimentological samples from different layers of the site.

Lithics

The analysis of the sieved lithic component (not coordinated) comprised several stages. First, the lithics were sorted into natural objects and those that had undergone knapping. The artifacts were then divided into two size categories: larger and smaller than 2 cm. Those measuring 2 cm and above were classified into items bearing ventral surfaces (flakes and flake tools – FFT) and items lacking them (cores and core tools – CCT). The latter underwent detailed attribute analysis in order to enlarge the sample size, which is usually small. The FFT, which are abundantly represented in the excavated assemblages, were all classified by raw material; in addition, 20% of each layer's sieved FFT inventory underwent detailed attribute analysis. Sampling was based on a series of random numbers that were computer generated. For some of the layers (e.g. Layer II-6 Level 1), however, all the sieved material was analyzed. The counts of FFT that originate in the sieved sediments and were not included in the detailed analyses are presented in Chapter 5.

Stone artifacts smaller than 2 cm (microartifacts) formed the bulk of the sorted lithic material. These were subdivided into two groups – items bearing ventral faces (FFT) and those lacking them (CCT). The latter were not subjected to further analysis in the current project but were kept for future studies. Natural items measuring less than 2 cm were also collected and kept for future analysis. These comprise various raw materials: flint, chert, limestone, concretions, weathered basalt (including crystals), quartz, and rarely fossils (e.g. Goren-Inbar et al. 1991b).

The analysis of the FFT microartifacts included identification of raw material and signs of burning on flint (for details, see Alperson-Afil and Goren-Inbar 2010). In addition, it included identification of items classified as *éclats de taille de biface*, products considered to pertain to the final stages of handaxe modification. The detailed analysis of the microartifacts was archived in a special database encompassing the entirety of these finds from all areas and layers.

4.2.2 Attribute Lists and Lithic Analysis

In contrast to the lithic material originating in the sieved sediments described in the previous section, *all* lithic items measuring 2 cm and above that were retrieved from the archaeological horizons during excavation (including both artifacts with three coordinates and the contents of the "general bags") underwent detailed attribute analysis. These were classified into several typo-technological categories and analyzed in detail.

The lithic analysis (both qualitative and quantitative) is based on techno-morpho-typological and stylistic attributes resulting from many years of artifact studies, which encompass the experience gained through previous studies of both Acheulian and Mousterian lithic assemblages (Goren-Inbar 1990; Bar-Yosef and Goren-Inbar 1993). This type of analysis reflects the authors' view that typological analysis is a foundation stone for gaining an understanding of the assemblages, serving as a common denominator and a common "language". While this is definitely not a modern approach, it is a means of communication with the academic community that is engaged with the study of prehistoric cultural entities.

The typological list for the analysis of the GBY lithic assemblages (Appendix 2) is based primarily on that of Bordes (1961), as it takes into account both flakes and flake tools and particularly emphasizes different categories of cores and core tools. The GBY typological definitions also follow Roe (1964, 1968) and Leakey (1971). These underwent slight modifications. The dynamics employed in the formation of the type list are illustrated, for example, by the presence of "pitted stones". This type, absent from Bordes' type list, does exist in that of Leakey (1971) and was integrated in the GBY list to suit the findings on hand. Another example is that of "heavy-duty scrapers" (Leakey 1971), which we integrated in the type list but with a slightly different meaning from that assigned by Leakey. At GBY this type differs from its counterpart at Olduvai Gorge and the term is applied to large robust basalt tools with coarse ("heavy") denticulated edge(s). These edges are characterized not by an angle approaching 90°, as observed in East Africa, but by an acute angle (Goren-Inbar et al. 2008; Chapter 7). Throughout the lithic analysis, the typological classification of chopping tools differentiated between three subtypes (chopper, inverse chopper, and chopping tool). Due to the scarcity of this morphotype in general, and of choppers in particular, all three subtypes were eventually grouped under the single type of chopping tool.

One of the characteristic features of the GBY lithic assemblages is the presence of multiple tools. At GBY, a single artifact could represent one, two, or three defined tools. In order to enable detailed description of all possible combinations, the attribute analysis integrated three typological definitions and their associated attributes (i.e. type, location, and face of retouch). The hierarchical order of the three typologies (Types A, B, and C) was determined individually for each artifact. The hierarchy took into consideration the quality of modification and the extent of the retouched edge. The multiple tools in this volume are presented according to their primary typological classification (Type A), followed by a breakdown of the other typology/ies. For example, a side-scraper with a borer is presented in the section that discusses side-scrapers under the heading of multiple tools, where all side-scrapers that are multiple tools are recorded.

The other facets of the attribute analysis are aimed at gaining information relevant to technological and stylistic features that characterize the realm of stone working. The accumulated experience, and particularly that gained by the long study of the GBY material, was extremely revealing in the sense that it showed time and again how little we really understand the thoughts and mechanisms behind the wealth of artifacts from the excavations. This lesson led to the decision, based on our uncertainties, to accumulate data in the form of detailed attributes, although we could not necessarily explain the value of recording each of them. We even added attributes that as the analysis progressed seemed essential for a more precise description of the unique assemblages. For example, experimental knapping (see below) provided insights into several technological features that result from the type of percussor used and the energy of the inflicted blow (e.g. lipped striking platforms, double cones, *éclat Siret*, and split percussors).

The general aim of the attribute analysis was to characterize the assemblages in as detailed a manner as possible, and through this procedure to achieve a better understanding of the production systems that at present are largely hidden from us.

The analysis used individual lists tailored to the different major typo-technological groups of the assemblages, as they differ substantially in their production and characteristics. Thus, different attribute lists were constructed for the analysis of flakes and flake tools (FFT), cores and core tools (CCT), handaxes, cleavers, and natural nodules. Four attribute lists are presented in detail in Appendices 3–6. All of the different lists have attributes in common that relate to excavation provenance and general information, including excavation date, catalogue number, and stratigraphic and spatial provenance. The different lists also share the following attributes: type of raw material, state of preservation, patination, amount of cortex, and breakage pattern (Appendices 3–6). Each of the lists includes descriptive notes.

Different computer software packages were used over the years, starting with DATABASE 3 and later Microsoft Office Access. Statistical analysis was carried out using different software through the long duration of the project; these included the following applications: SAS (SAS/STAT, 1987), StatView,

4.2 Laboratory Methodology

SPSS 15.0 (2006), and recently Microsoft Office Excel and JMP 7.0 (2007) and JMP 11.0 and 12.0.

4.2.2.1 Flakes and Flake Tools

When analyzing flakes and flake tools, the attributes and attribute states aimed to embrace typological, technological (e.g. direction of blow, striking platform, scar pattern, number of scars), and stylistic (e.g. type, location, face, and form of retouch) characteristics. In addition, a series of size measurements (maximal length, length, width, and thickness) was taken for each of the analyzed FFT (Appendix 3).

The FFT assemblage of Area C, which was made primarily on flint, underwent more detailed analysis (e.g. bulb characteristics, lipping).

4.2.2.2 Cores and Core Tools

In the analysis of CCT we excluded the bifacial component (see below). The same attribute list was used for cores and for core tools and included qualitative and quantitative attributes (Appendix 4). The list of attributes took into consideration the presence of retouch and hence a modified edge (Appendix 4).

During measurement and description of the CCT, the dominant working edge was systematically aligned proximally (Fig. 4.21). The length of the working edge and that of the circumference were measured on the same plane and were limited to the dominant working edge. The dominant debitage face was used for observations of scar pattern and number of scars on cores. Broken CCT were only minimally described and their measurements were restricted to maximal length, width, and thickness.

Cores on flakes were analyzed within the category of CCT and integrated into the description and discussion of the cores component.

4.2.2.3 Bifaces: Handaxes and Cleavers

The bifacial component of GBY (Goren-Inbar and Saragusti 1996; Saragusti et al. 1998, 2005; Saragusti 2002) was analyzed using two separate attribute lists that share, among others, attributes related to the morpho-technological characteristics of these tools (Appendices 5–6). As a working rule it was decided that in cases where bifacial retouch is visible but it is impossible to distinguish whether the item is a cleaver or a handaxe, the biface should be classified and analyzed as a handaxe.

All bifacial tools were measured following Tixier (1957), Bordes (1961), and Roe (1964) according to the tools' symmetry. For both handaxes and cleavers the criterion for distinction between Face 1 and Face 2 was that Face 1 is the more extensively worked face (Goren-Inbar and Saragusti 1996). In the GBY cleaver assemblage, Face 2 is usually a ventral face of the flake on which the cleaver was modified.

For proximally or laterally broken bifaces (substantially fragmented), measurements included only maximal width and thickness. Bifaces that were only minimally broken at their distal end and whose entire form was readable (a common feature in the assemblage; see Goren-Inbar and Saragusti 1996) underwent a full attribute analysis.

When describing the section of bifaces we selected the transversal view, which is generally more informative from a technological point of view than the longitudinal one.

Some of the bifaces bear remnants of their earlier life-histories in the form of scars that pertain to the modification of the core prior to the removal of the blank. In contrast, the flaking activities that are recorded on the biface and are visible as scars that postdate the production of the blank (flakes in most cases) are considered "retouch" in this volume. We distinguish between several types of retouch (Appendices 5-6). The two most common types of retouch on bifacial tools are the "bifacial retouch" and the "thinning". The former, despite its name, refers to an intensive retouch that may appear on one or both faces of the tool and is characteristic of the bifacial component at GBY. Thinning refers to a retouch restricted only to the edges of the tool and consisting on only a few scars, usually used to thin or remove the bulb of percussion area. Other, rarer types of retouch on bifacial tools consist of a "scraper-like" retouch, a flat and limited retouch, and combinations of the above.. Intensive retouch on one face or both is considered "bifacial retouch" (Fig. 4.22). Where only a few scars are observed, and they are restricted to the edge/s and do not extensively cover the surfaces of the biface, the retouch is defined as "thinning" (Fig. 4.23). The

Fig. 4.21 Example of positioning of CCT: a flint core (#9689 Layer V-6) proximally positioned with the dominant working edge and the dominant debitage face on the left

Fig. 4.22 Flint handaxe with bifacial retouch (#6175 Layer II-6 Level 1), Face 1 on the left

latter type of retouch is typical of the removing/flattening of the bulb of percussion area on many of the bifacial tools (Chapter 7).

When a ventral face was observed on a biface, the direction of blow was determined. In regular flakes this was carried out while the biface was placed with its ventral face hidden and the dorsal face facing upward (Fig. 4.24). When items were identified as Kombewa flakes, particularly in the cleavers assemblage, the direction of blow was registered individually for each face.

The typology of bifaces, particularly in the case of handaxes, followed that of Bordes (1961). However, several modifications were necessary. The first is the "amorphous" biface (Type 128), which has enough bifacial retouch on both faces for it to be identified as a biface but is too amorphous to be classified under any of the other handaxe types. What remains unknown is whether any of these items had a cleaver morphology in their earlier life-history. The second modification, Type 129, includes all bifacially worked items that are too broken or rolled to be identified and classified under any of the biface types. The breakage may have resulted from either human or natural agency.

Cleaver identification follows Tixier (1957) and refers to bifaces on which the width of the working edge is larger than half of the maximal width. The large distal scar (not retouched or flaked) that forms the working edge was not measured according to its original flaking orientation but rather followed the positioning of the cleaver, so that the readings of this scar were taken along the length and width of the cleaver (Fig. 4.25). The two angles between the working edge and the lateral edges of the cleaver were measured

Fig. 4.23 Basalt cleaver (#223 Layer II-6 Level 4); note the morphology of the proximal end of the ventral face, where thinning has resulted in a visible hinged scar

Fig. 4.24 Basalt handaxe with visible direction of blow on the ventral face (#5791 Layer II-6 Level 4b), Face 1 on the left

4.3 This Volume

Fig. 4.25 Basalt cleaver with visible direction of blow on the ventral face (#5448 Layer II-6 Level 1), Face 1 on the left

based on a straight line connecting the distal ends of the lateral edges.

An important feature of the study of bifaces at GBY was the insight gained from a series of knapping experiments. The aim of the experiments was to acquire further information on the reduction sequences of both handaxes and cleavers and to compare the results with those of the archaeological material. The experiments, conducted by B. Madsen, were restricted to basalt and involved the acquisition of raw material (basalt slabs) and percussors of different types and sizes, followed by the production of giant cores and large flakes, and finally the modification of bifaces (Madsen and Goren-Inbar 2004; Goren-Inbar 2011; Goren-Inbar et al. 2011; Sharon et al. 2011). The results of the experiments were integrated in the analysis of all of the GBY lithics and threw light on various aspects such as technology (the application of soft antler percussors, particular features of the striking platforms, the number of potential flakes, etc.), mobility, and diverse facets of hominin behavior.

4.2.2.4 Natural Nodules

All natural rocks measuring 2 cm and above were analyzed, including basic granulometric attributes that were registered in a separate database from that of the lithic artifacts. The analysis focused on the size of the rocks and on a series of observations including type of raw material, burning (for flint), length, width, thickness, weight, sphericity, and roundness (Krumbein 1941). This pebble/cobble assemblage forms a separate and detailed archive, one of the largest of the GBY databases. This component is not included in the analysis presented in this volume and will contribute much to future studies of the paleoenvironment and paleoecology of the site.

4.2.3 Digitization of Maps and Cross-Sections

All cross-sections and maps drawn in the field underwent digitization in order to provide a computerized reconstruction of the excavated area and a basis for spatial analysis of the archaeological horizons. The process included scanning of field illustrations and digitization of all features using AutoCAD (Autodesk AutoCAD 2008 17.1), in which the field illustrations were anchored to their original spatial location (Israel Grid and elevations above msl). For the presentation and analysis of the maps and cross-sections we used ArcGIS, a software package that comprises a collection of software and geographic data for capturing, managing, analyzing, and displaying all forms of geographically referenced information. Of the software available, ArcMap (ESRI® ArcMap™ 9.3) was used for the spatial display and analysis of the archaeological data, and ArcScene (ESRI® ArcScene™ 9.3) was used for three-dimensional positioning of the illustrations and the reconstruction of the excavated areas. Some of these computerized reconstructions are presented in Chapter 5.

4.3 This Volume

The lithic analysis of the GBY assemblages lasted from the end of the very first excavation season up to the writing of this volume. Unlike the process of attribute analysis, which treated artifacts of different raw materials in a similar manner, this volume presents individually the results of the analysis for each of the three different raw materials. The extensive differences between the reduction sequences of flint, basalt, and limestone encountered during the analysis necessitated an individual treatment for each. Thus, the internal structure of each chapter is tailored to the particular features of the raw material under discussion. Nevertheless, all three chapters follow a similar general outline of taphonomical, typological, and technological descriptions. The term taphonomy, which originally originally just concerned biological remains, is used here in a broader sense to relate to postdepositional processes that may have influenced all materials from the site, biological as well as lithic. We choose to present the typology before the technology as this provides an understanding of the desired (target) artifacts, and establishes a common framework for the technological discussion.

Table 4.1 Lithic categories

	Flint	Basalt	Limestone	Total
TOOLS[*]				
WASTE				
Debitage				
Cores				
Non-flakes				
Subtotal				
Microartifacts				
FFT not analyzed				
Natural nodules				
Total				

[*] flake tools and core tools

Table 4.2 Typo-technological breakdown of the lithic inventory and the associated typologies and lithic categories

Artifact class	Typological number/s	Lithic category
TOOLS		
Flake tools		
Side-scraper	8–29	Tools
End-scraper	30–31	Tools
Burin	32–33	Tools
Borer	34–35	Tools
Backed knife	36–38	Tools
Notch	42, 54	Tools
Denticulate	43, 51	Tools
Truncation	40	Tools
Retouched flake	106	Tools
Massive scraper	114	Tools
Varia	8, 44, 46–49, 53, 57–58, 62	Tools
Core tools		
Bifaces		
Handaxe	73–83, 86–91, 120	Tools
Cleaver	84–85	Tools
Trihedral	92	Tools
Pick	93	Tools
Biface preform	123	Tools
Amorphous biface	128	Tools
Other		
Chopping tool	59–61	Cores
Polyhedron	94	Cores
Subspheroid	116	Cores
Discoid	115	Tools
Disc	95	Tools
Percussor	96	Non-flakes
Pitted stone	127	Non-flakes
Anvil	112	Non-flakes
WASTE		
Flakes		
Flake	1–3, 100, 108	Debitage
Blade	101	Debitage
Dorsally plain flake	105	Debitage
Éclat de taille de biface	104	Debitage
Biface sharpening flake	124	Debitage
Core tablet	98	Debitage
Core waste	99	Debitage
Non-flakes		
Angular fragment	97	Non-flake
Indeterminate waste	122, 129	Non-flake
Cores		
Levallois core	64–66	Cores
Discoidal core	67	Cores
Globular core	68	Cores
Prismatic core	69	Cores
Pyramidal core	70	Cores
Core on flake	107	Cores
Core on handaxe	121	Cores
Core varia	71	Cores

The disadvantage of presenting each of the raw materials independently (Chapters 6–8) is that it does not provide a coherent picture of each of the assemblages of the different layers and levels. The presentations in Chapters 6–8 consider only assemblages of adequate sample size (the criteria for selecting sample size differ between raw materials and between lithic categories and are specified in each of the chapters). In Chapter 5 we present each of the GBY assemblages within its archaeological context, including its full lithic inventory.

The inventory of each layer is presented in two tables; the first (Table 4.1) provides a general picture of the major lithic categories of the assemblage, while the second (Table 4.2) provides a detailed typo-technological breakdown of the inventory. To clarify the decisions that guided us in forming the different lithic categories, we specify in Table 4.2 which typological types were included in each artifact class and to which lithic category it was ascribed. The tools category consists of all lithic artifacts which have undergone some secondary modification such as retouch. Thus, flake tools include all lithic pieces with an identifiable ventral face which were modified after their removal. The waste category includes all pieces which are not defined as tools. The debitage sub-category consists of all lithic pieces larger than 2 cm with an identifiable ventral face which were not retouched. These categories are subsequently classified into different sub-types as detailed in Table 4.2.

The category of waste includes eclat de taille de biface and biface sharpening flakes which consist of debitage removed during the final stages of bifacial tools production. Both types of flakes are generally characterized by prepared striking platforms, a relatively higher number of dorsal scars, a complex scar pattern and higher surface to thickness ratio (Newcomer 1971; Sharon and Goren-Inbar 1999; Goren-Inbar and Sharon 2006; Andrefsky 2009). These two biface modification flakes differ in their size, as biface sharpening flakes (termed by some as biface trimming flake: Andrefsky 2009: 124, or biface rejuvinantion flakes: Goren-Inbar and Sharon 2006) tend to be larger and more robust than eclat de taille de biface, and often bear remnants of the bifacial working edge.

4.3.1 Illustrating Lithic Artifacts

In the closing stages of the lithic analysis, we selected items for illustration. Items were chosen on the basis of the central role of their type, particular technological features that they display, or the fact that they are characteristic of a particular layer or level. Unique typological or technological features of the assemblages were also selected for illustration.

This volume is the product of long years of research, during which some artifacts were illustrated and photographed. Photographs were taken by Gabi Laron and by members of our excavation team. Line drawings by a number of artists (see the Acknowledgements) have appeared in publications of selected preliminary results. With advances in digital technology, 3D scanning devices (manufactured by Polygon Technology, Darmstadt, Germany) were assembled at the Institute of Archaeology, The Hebrew University of Jerusalem. Together with Dr. Leore Grosman and Prof. Uzy Smilansky, we initiated a continuous and long-term project that involves different aspects of 3D documentation and analyses of the GBY lithics (Grosman et al. 2008, 2014). This project includes, among others, the formation of a digital 3D archive of all the GBY bifaces (handaxes and cleavers), giant cores (Goren-Inbar 2011; Goren-Inbar et al. 2011), and massive basalt scrapers (Goren-Inbar et al. 2008). As the preparation of the present volume progressed, we decided to apply the 3D technology extensively to the production of illustrations. It is particularly useful for the flint component, which includes many very small artifacts that are often extensively retouched and difficult to illustrate. The 3D technology is particularly advantageous for documentation for the following reasons:

1. The complete geometric manifestation of the artifact's surface (3D image) is available for illustrating the artifacts. These 3D images can be appended as a supplement to the digital publication of the artifacts.
2. The 2D documentation in the figures is based on the recently developed software program Artifact3-D (Grosman et al. 2014). The program provides the user with a ready-made template that is based on the conventional layout of stone tools (Inizan et al. 1999).
3. Artifact3-D provides a method for rotating the artifact to a desired position. Consequently, preferred views of the tool or of a particular technological or stylistic feature can be selected for inclusion in the figures.
4. This documentation is free of the subjective interpretation of traditional artistic drawing. The reader is able to observe and assess directly, without intermediary phases, the features observed on the artifact.
5. The object is scanned at high resolution, providing an extremely precise representation of the surface geometry.

Throughout this volume we have utilized a 3D optical scanner (Computerized Archaeology division, Institute of Archaeology, Hebrew University of Jerusalem) to provide precise and complete digitized images of the artifacts (for details see Grosman et al. 2008; Grosman et al. 2011). The digitized scans were processed by the software program Artifact3-D, which allows for both automatic and manual positioning of the artifacts. This program presents the artifacts in five or six views (including sections at desired locations), extracting values such as the location of center of mass, volume, length, width, thickness, and an algorithm that calculates the mean angle between two surfaces on a given artifact. As scanning was introduced to our project at a very advanced stage of the lithic analysis, we did not use the wealth of options for data analysis that the software offers, but made extensive use of the automatic and user-friendly tool for the production of multi-artifact illustrations. The program follows the traditional guidelines for producing plates of the scanned objects. The working mode selected for artifact presentation here was manual positioning, which automatically combines the artifacts to scale. We selected a basically similar presentation for each artifact that includes a presentation of the two faces, a cross-section, and two additional views that are a perpendicular view of each extremity (Figs 4.21–4.25). The perpendicular view of the lower extremity of flakes and flake tools also offers a detailed view of the striking platform. In addition to the above, particularly for the presentation of flakes and flake tools, we sometimes added a view that illustrates a particular retouch or other traits. In such cases, this view is usually located on the right side of the figure. When a series of illustrations accompanies the description of a particular feature, they are usually given in order of the stratigraphic position of the artifacts, from the youngest layer to the oldest.

It is clear that the digitized 3D image provides additional information and precision. The image is free of the interpretation that is conventionally offered by artists, mainly in terms of the direction and orientation of scars. Another advantage is that where conchoidal waves are visible in the flake scars, they are registered on the 3D image. In line drawings no information is available on the artifact's texture (except for location of cortex) or the freshness of flint artifacts. The digital image, on the other hand, is a mesh of points that are generated to form an image by a color code fill (rendering). Artifact3-D uses a gray scale for rendering the 3D models, and consequently the object is of the same color regardless of the texture. Thus, in order to mark the location of cortex on the 3D digitized images, a yellow color was used in most cases.

For the time being, we believe that it is worthwhile to get acquainted with this novel documentation technology and to consider its definite advantages, primarily the high resolution of the scan, which is reflected in the detail of the image, and the option of rotating the artifact to a desired view.

Chapter 5
The Lithic Assemblages in Context

Abstract Chapter 5 presents a detailed account of the excavated archaeological layers at Gesher Benot Ya'aqov. It provides data on the location and stratigraphic position of each excavated unit in the study area, as well as details of the surface exposed, thickness and volume of the layers, and their sedimentological characteristics. When available, the results of faunal and floral analyses are summarized and references are provided. The description of the excavated units is accompanied by a series of illustrations, including photographs, cross-sections, and field maps.

In some cases, information on the strategy applied to sampling and excavation methods in particular settings is given.

For each excavated stratigraphic unit the inventory of the lithic assemblage of all field seasons is presented in detail. This is the only detailed presentation of the complete lithic inventory (regardless of sample size), encompassing all three raw materials and the frequencies of both waste and tools.

Keywords cross-sections · excavated areas · excavated surface · excavated volume · field maps · Gesher Benot Ya'aqov · lithic inventory

This chapter describes the archaeological contexts of the lithic assemblages discussed in this volume. The lithic assemblages originate in different excavation seasons and areas and occur along the entire Early–Middle Pleistocene stratigraphic sequence (Table 5.1). The present chapter discusses different aspects of the excavated layers, including among others their excavated areas and volumes and their lithic inventory, as well as some sedimentological and stratigraphic information. The data are presented in stratigraphic order, from the youngest occurrence to the oldest. The chapter includes a graphic representation of the stratigraphic location of each lithic assemblage and a

Table 5.1 Excavation of areas and trenches by field season

	1989	1990	1991	1995	1996	1997a	1997b
Areas							
Bar*	+						
Area C					+	+	+
Jordan Bank				+	+		
Jordan River		+					
Area A	+						
Area B	+	+	+	+	+	+	
Trenches							
Trench I	+						
Trench II	+						
Trench III				+			
Trench IV				+		+	
Trench V					+		
Trench VI					+		
Trench VII						+	

* material collected from surface; lithic assemblage not included in this volume, see Goren-Inbar et al. 1992

variety of other illustrative materials. For each archaeological horizon the complete inventory (regardless of sample size) is provided, encompassing the different raw materials, the different lithic categories, and the detailed typological classification (clarifications regarding the structure and content of the inventory tables are provided in Chapter 4.3.1). In-depth discussions of the lithic assemblages and their typological and technological traits are presented separately for each of the raw materials in Chapters 6–8.

5.1 The Bar

During the first season of excavations (1989), a bar-like depositional structure was observed near the eastern bank of the Jordan River, its extent of exposure depending on the fluctuations of the water level (Figs. 3.8, 3.11, 5.1). Four lithofacial lake-margin units (primarily gravels and sands) were identified, all tilted by 20–25° and all containing archaeological and paleontological finds. The Bar, whose morphology results from the different lithologies of the sediments as well as tectonic activity, is a segment of the Benot Ya'akov Formation (BYF) and contains viviparids, which are the guide fossil of the formation (Chapter 3.3.4.1). The Bar sediments are the youngest deposits in the study area, overlying the layers of Area C (Figs. 3.5a, 3.13, 3.14a), and are assigned to the youngest paleoclimatic cycle of the site (cycle 6, MIS 18, after Feibel 2001: fig. 5.3, 2004: fig. 11; Goren-Inbar et al. 2000).

At its maximal exposure, the Bar was over 200 m in length and ca. 60 m in width. Although there were clear indications that artifacts and faunal finds were preserved *in situ* in its layers, excavation was not carried out and the archaeological material was collected through survey. It became evident that both the lithic artifacts and the bones had been exposed to taphonomic postdepositional processes, of which the most evident was size sorting (only large items were found).

The survey resulted in the identification of medium-sized and large fossil mammal bones and typical Acheulian lithic artifacts. These assemblages have been published in detail (Goren-Inbar et al. 1992) and are not included in this volume.

5.2 Area C

Area C is located in the southernmost part of the study area, less than one meter east of the Jordan River (Fig. 5.1). Although the excavated area was separated from the water only by a small baulk about half a meter wide, the exposed sediments remained dry (Figs. 5.2–5.3). Excavations in Area C took place during three field seasons (Table 5.1; Fig. 5.4); the field season of September 1997 (1997b) was dedicated entirely to this area.

The excavation of Area C was prompted by the exposure of archaeological layers on the bank of the Jordan River, at the water's edge (Fig. 5.1). The proximity of the river and its fluctuating water level are causing the erosion of these layers. The exposures (Section 5.2.5) revealed exceptionally well-preserved lithic assemblages, primarily on flint, in association with extremely rich and well-preserved faunal assemblages (Rabinovich et al. 2012). Furthermore, one of these exposures is a coquina layer composed primarily of the gastropods *Viviparus apameae galileae* and *Bellamya* sp., the guide fossils of the Middle Pleistocene BYF (Goren-Inbar and Belitzky 1989

Fig. 5.1 General view of the southern sector of the study area

Fig. 5.2 General view of Area C looking north

5.2 Area C

Fig. 5.3 General view of Area C looking south

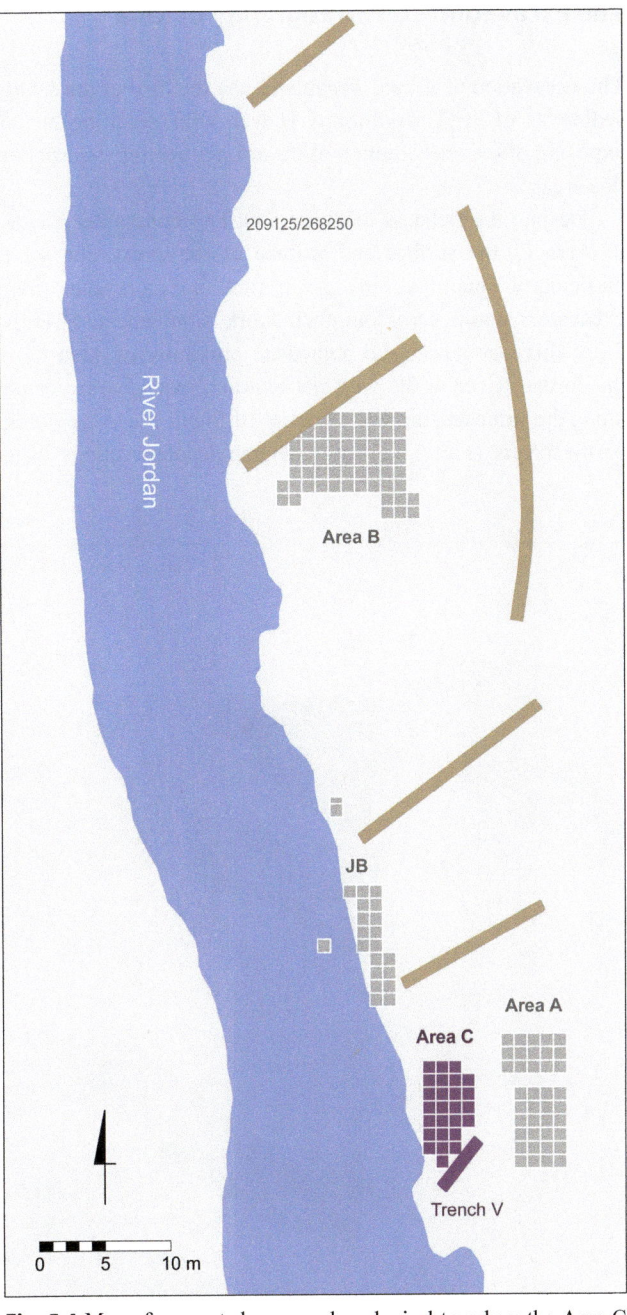

Fig. 5.4 Map of excavated areas and geological trenches; the Area C excavations and the adjacent Trench V are highlighted

and references therein; Ashkenazi et al. 2010), in densities unmatched elsewhere in the Dead Sea Rift.

In places where this coquina (Layer V-5) was eroded to its base, a dark clayey deposit (Layer V-6), extremely rich in lithic and paleontological finds, was exposed. These two layers are in fact the most extensive and longest exposures in the entire study area (Fig. 3.8). A substantial portion of them was destroyed in December 1999 during drainage works (Chapter 2).

A further reason for initiating excavations in Area C was to achieve a better understanding of the BYF by retrieving data on its sedimentary composition and structure. Thus, in 1996 there was good reason to believe that the layers exposed along the Jordan River bank could be detected and excavated on dry land; conditions there would hopefully enable more precise excavation methods and hence better spatial and stratigraphic results. Furthermore, it was expected that the excavations would result in sufficiently large samples for detailed analysis, enabling us to gain an understanding, both geological and archaeological, of the younger part of the depositional record in the study area.

In order to excavate on land the layers that were exposed at the water's edge, we projected their strike eastward.

Two types of field activities took place in Area C: the digging of Trench V (June 30th, 1996) and the excavations (Fig. 5.4). Trench V, located immediately south of the space earmarked for excavation, furnished a detailed stratigraphic and sedimentological archive for this area (Feibel 2001, 2004). The spoils of Trench V yielded lithic material lacking precise stratigraphic provenance (five flint tools, 12 flint flakes, and two basalt flakes). Additional data were provided prior to excavation by observations on the Unconformity of Area C (Layer V-1), which supplied a cross-section in plan view of the stratigraphy of Area C.

The Excavation and Stratigraphy of Area C

The excavation of Area C began with the removal of the sandy sediments of the Unconformity (Layer V-1), resulting in the exposure of the cross-section of the artifact-bearing deposits of this area.

Despite the richness of the artifact and bone assemblages of Area C, the surface and volume of the excavation were particularly small. At any given time no more than four excavators could carry out their work simultaneously (Fig. 5.3). This was due to the immediate proximity of Area C to the Jordan River in the west. In addition, extensive exposure along the strike and dip was extremely difficult due to the tilting of the layers (Fig. 5.3). The stratigraphic observations were further limited by the water level in Trench V, which prevented exposure of the deposit to a further depth.

The stratigraphy of Area C is based on observations carried out in Trench V, and on the exposures and cross-sections in the excavated area. It is composed of a series of layers made up of coquinas and clays that dip by ca. 30° and are tilted northwest to southeast, like other excavated strata in the study area. These exposed layers form the uppermost part of the excavated depositional sequence of GBY and are part of the fifth depositional cycle (Goren-Inbar et al. 2000; Feibel 2001, 2004).

Combining the stratigraphic evidence resulted in the identification of five artifact-bearing layers, which were numbered from the western (younger) end of Trench V to its

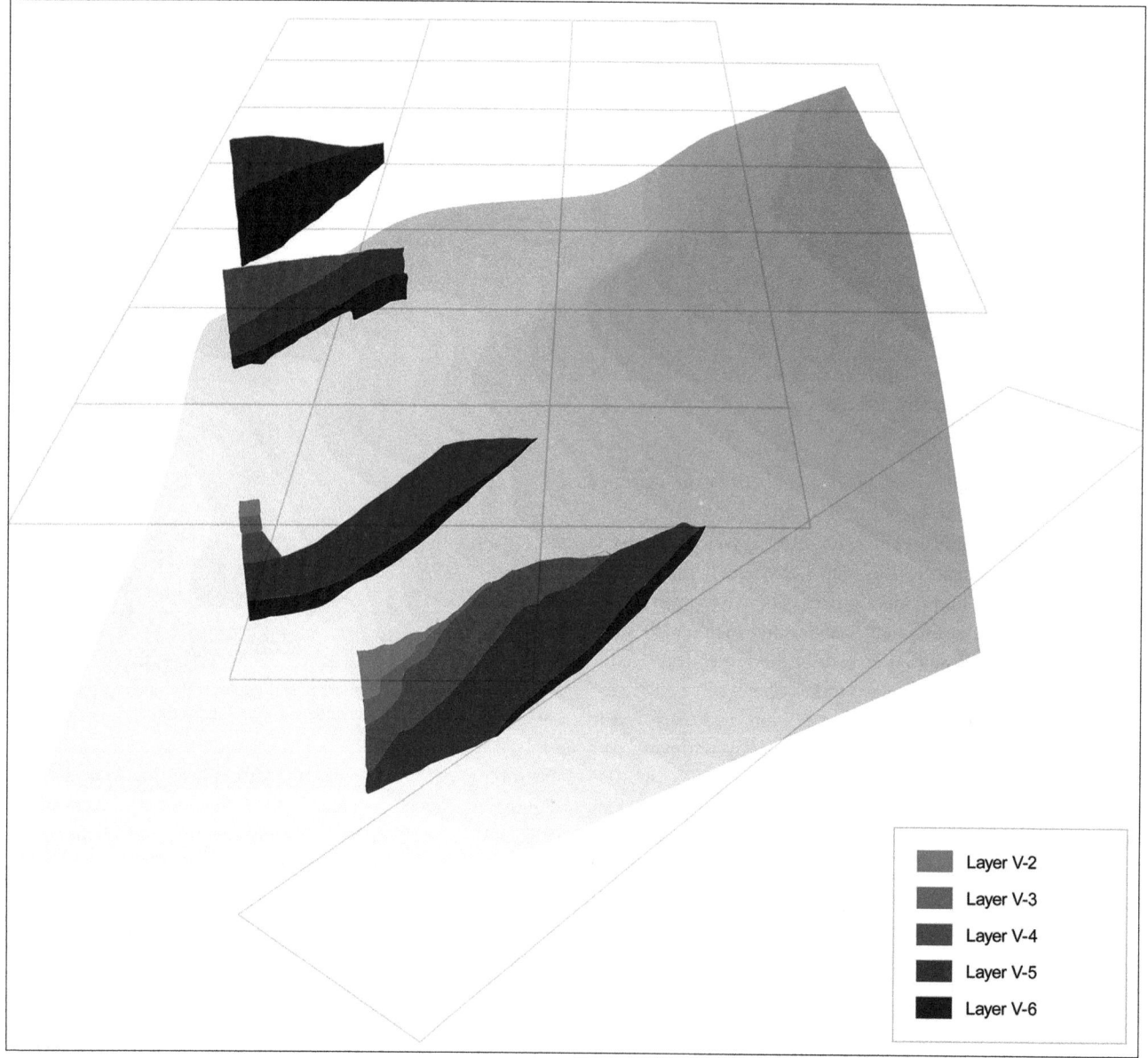

Fig. 5.5 Isometric view of the excavations and cross-sections in Area C, view to north

5.2 Area C

Table 5.2 Data on excavated surface (m²) and volume (m³) of layers exposed in Area C

Layer	Surface	Volume*
V-1 Unconformity	7.90	
V-2	0.20	
V-3	1.10	
V-4	1.88	
V-5	6.39	1.59
V-6	7.04	1.97

* volume is calculated using the thickness of each of the individual layers as recorded in the cross-section of Trench V and in the cross-sections (perpendicular to the layers) of the excavations

Table 5.3 Counts and frequencies (%) of the lithic assemblage of Layer V-1

	Flint	Basalt	Limestone	Total
	% (N)	% (N)	% (N)	N
TOOLS#	39.29 (11)	10.00 (2)	-	13
WASTE				
Debitage	50.00 (14)	60.00 (12)	-	26
Cores	7.14 (2)	5.00 (1)	-	3
Non-flakes	3.57 (1)	25.00 (5)	100.00 (5)	11
Total	28	20	5	53

flake tools and core tools

eastern (older) end: Layers V-2, V-3, V-4, V-5, and V-6 (Fig. 5.5). Lithic artifacts were recovered from all of these layers. However, as a result of the structural geometry of the tilted block, excavation focused mainly on Layers V-5 and V-6. The surface and volume of excavation of each of the layers are presented in Table 5.2.

Among the fauna, the medium-sized and large mammals from Area C are described in detail in GBY Volume III (Rabinovich et al. 2012), while mollusks, crabs, micromammals, amphibians, and reptiles are described by others (Ashkenazi et al. 2005, 2010; Rabinovich and Biton 2011). In addition, wood remains (Goren-Inbar et al. 2002b, 2004) as well as nuts (Goren-Inbar et al. 2002a), seeds and fruits (Melamed et al. 2011), were analyzed.

5.2.1 Layer V-1 (Unconformity C)

Layer V-1 is composed of sandy sediments that unconformably overlies the Middle Pleistocene artifact-bearing deposits of Area C. This layer was found spotted with lithics (Tables 5.3–5.4), bones, and organic material.

5.2.2 Layers V-2, V-3, and V-4

Table 5.4 Counts and frequencies (%) of the typological composition of Layer V-1

	Flint	Basalt	Limestone	Total
	% (N)	% (N)	% (N)	N
TOOLS				
Flake tools				
Side-scrapers	3.57 (1)	-	-	1
Burins	3.57 (1)	-	-	1
Borers	7.14 (2)	-	-	2
Notches	7.14 (2)	-	-	2
Denticulates	7.14 (2)	-	-	2
Truncations	3.57 (1)	-	-	1
Retouched flakes	7.14 (2)	-	-	2
Core tools				
Bifaces				
Handaxes	-	5.00 (1)	-	1
Cleavers	-	5.00 (1)	-	1
Others				
Percussors	-	15.00 (3)	40.00 (2)	5
WASTE				
Flakes				
Flakes	39.29 (11)	55.00 (11)	-	22
Blades	7.14 (2)	-	-	2
Core waste	3.57 (1)	5.00 (1)	-	2
Non-flakes				
Angular fragments	3.57 (1)	-	20 (1)	2
Cores				
Core varia	7.14 (2)	5.00 (1)	-	3
Modified	-	10.00 (2)	40.00 (2)	4
Total	28	20	5	53

Layer V-2 is the youngest artifact-bearing layer exposed in Area C, underlying the Recent sediments of Layer V-1. It is a coquina layer, some 20 cm thick in Trench V (Fig. 3.13), which was only minimally excavated due to the tilting of the layers and the resulting geometry in which the uppermost layers are much more heavily truncated (Fig. 5.5). Thus, Layer V-2, together with the underlying Layers V-3 and V-4, was minimally exposed in comparison with the excavated surfaces of Layers V-5 and V-6. Nevertheless, lithics, bones, and some organic material were recovered from each of these layers and their contacts (Tables 5.5–5.6). In Trench V the thickness of the coquina layer of Layer V-3 is some 10 cm and that of Layer V-4 some 20 cm. Layer V-4 is sandy, with gray clay and broken mollusks.

5.2.3 Layer V-5

Layer V-5 is a coquina layer, some 30 cm thick in Trench V. In the excavated area Layer V-5 was 25 cm thick and was excavated to the extent of 6.39 m² (Table 5.2).

Table 5.5 Counts and frequencies (%)* of the lithic assemblages of Layers V-2–V-4/5

	Flint	Basalt	Limestone	Total
	% (N)	% (N)	% (N)	N
Layer V-2				
TOOLS#	33.33 (1)	62.50 (5)	-	6
Waste				
Debitage	66.67 (2)	37.50 (3)	-	5
Subtotal	3	8	-	11
Microartifacts	7	-	-	7
Natural nodules	-	13	-	13
Total	10	21	-	36
Layer V-2/3				
TOOLS#	100.00 (1)			1
Subtotal	1	-	-	1
Microartifacts	3	-	-	3
Natural nodules	-	-	-	-
Total	4	-	-	4
Layer V-3				
TOOLS#	42.86 (3)	50.00 (1)	-	4
Waste				
Debitage	57.14 (4)	50.00 (1)	-	5
Subtotal	7	2	-	9
Microartifacts	114	2	4	120
Natural nodules	4	2	1	7
Total	125	6	5	136
Layer V-3/4				
TOOLS#	100.00 (1)	92.31 (12)	-	13
Waste				
Debitage	-	7.69 (1)	-	1
Subtotal	1	13	-	14
Microartifacts	-	-	-	-
Natural nodules	-	-	-	-
Total	1	13	-	14
Layer V-4				
TOOLS#	52.38 (11)	16.67 (1)	-	12
Waste				
Debitage	23.81 (5)	66.67 (4)	-	9
Non-flakes	23.81 (5)	16.67 (1)	-	6
Subtotal	21	6	-	27
Microartifacts	470	54	8	532
Natural nodules	6	8	1	15
Total	497	68	9	574
Layer V-4/5				
TOOLS#	44.44 (12)	28.57 (2)	-	14
Waste				
Debitage	40.74 (11)	71.43 (5)	-	16
Cores	3.70 (1)	-	-	1
Non-flakes	11.11 (3)	-	-	3
Subtotal	27	7	-	34
Microartifacts	880	68	11	959
Natural nodules	7	2	2	11
Total	914	77	13	1,004

* % calculated from the subtotal
# flake tools and core tools

Table 5.6 Counts and frequencies (%) of the typological composition of Layers V-2–V-4/5

	Flint	Basalt	Total
	% (N)	% (N)	N
Layer V-2			
TOOLS			
Flake tools			
Retouched flakes	33.33 (1)	-	1
Core tools			
Bifaces			
Handaxes	-	37.50 (3)	3
Cleavers	-	25.00 (2)	2
WASTE			
Flakes			
Flakes	33.33 (1)	37.50 (3)	4
Éclats de taille de biface	33.33 (1)	-	1
Total	3	8	11
Layer V-3			
TOOLS			
Flake tools			
Borers	14.29 (1)	-	1
Notches	28.57 (2)	-	2
Core tools			
Bifaces			
Biface preforms	-	50.00 (1)	1
WASTE			
Flakes			
Flakes	57.14 (4)	50.00 (1)	5
Total	7	2	9
Layer V-3/4			
TOOLS			
Flake tools			
Notches	100.00 (1)	-	1
Core tools			
Bifaces			
Handaxes	-	30.77 (4)	4
Cleavers	-	61.54 (8)	8
WASTE			
Flakes			
Flakes	-	7.69 (1)	1
Total	1	13	14
Layer V-4			
TOOLS			
Flake tools			
Side-scrapers	4.76 (1)	-	1
End-scrapers	4.76 (1)	-	1
Borers	33.33 (7)	16.67 (1)	8
Notches	4.76 (1)	-	1
Retouched flakes	4.76 (1)	-	1
WASTE			
Flakes			
Flakes	23.81 (5)	66.67 (4)	9
Non-flakes			
Angular fragments	4.76 (1)	16.67 (1)	2
Modified	19.05 (4)	-	4
Total	21	6	27

5.2 Area C

Table 5.6 (cont.)

	Flint	Basalt	Total
	% (N)	% (N)	N
	Layer V-4/5		
TOOLS			
Flake tools			
Side-scrapers	7.41 (2)	-	2
End-scrapers	7.41 (2)	-	2
Borers	7.41 (2)	-	2
Notches	14.81 (4)	-	4
Denticulates	3.70 (1)	-	1
Truncations	3.70 (1)	-	1
Core tools	-	-	-
Bifaces	-	-	-
Handaxes	-	28.57 (2)	2
WASTE			
Flakes			
Flakes	25.93 (7)	71.43 (5)	12
Éclats de taille de biface	11.11 (3)	-	3
Non-flakes	-	-	-
Core waste	3.70 (1)	-	1
Cores	-	-	-
Core varia	3.70 (1)	-	1
Modified	11.11 (3)	-	3
Total	27	7	34

Layer V-5 is composed mainly of *Viviparus* and large quantities of deteriorated *Unio* shells, with hardly any sediment matrix (Fig. 5.6). The viviparid specimens of Layer V-5 are the largest encountered at the site and are in a pristine state of preservation (displaying, for instance, the preservation of fossil embryos: Ashkenazi et al. 2010), indicating rapid sealing and minimal postdepositional imprints for this layer. The archaeological horizon of Layer V-5 was rich in faunal and lithic remains found in an excellent state of preservation (Tables 5.7–5.8; Fig. 5.7).

Fig. 5.6 The northern cross-section of the excavations of Area C; visible are Layers V-5 and V-6

Table 5.7 Counts and frequencies (%)* of the lithic assemblage of Layer V-5

	Flint	Basalt	Limestone	Total
	% (N)	% (N)	% (N)	N
TOOLS[#]	31.25 (105)	2.13 (2)	-	107
WASTE				
Debitage	61.61 (207)	88.30 (83)	50.00 (2)	292
Cores	3.27 (11)	-	-	11
Non-flakes	3.87 (13)	9.57 (9)	50.00 (2)	24
Subtotal	336	94	4	434
Microartifacts	30,899	5,702	347	36,948
Natural nodules	72	163	24	259
Total	31,307	5,959	375	37,641

* % calculated from the subtotal
[#] flake tools and core tools

Table 5.8 Counts and frequencies (%) of the typological composition of Layer V-5

	Flint	Basalt	Limestone	Total
	% (N)	% (N)	% (N)	N
TOOLS				
Flake tools				
Side-scrapers	3.57 (12)	-	-	12
End-scrapers	1.19 (4)	-	-	4
Burins	0.30 (1)	-	-	1
Borers	4.76 (16)	-	-	16
Notches	7.74 (26)	-	-	26
Denticulates	6.55 (22)	-	-	22
Truncations	0.60 (2)	-	-	2
Retouched flakes	6.25 (21)	-	-	21
Varia	0.30 (1)	-	-	1
Core tools				
Bifaces	-	-	-	-
Cleavers	-	2.13 (2)	-	2
Others	-	-	-	-
Chopping tools	0.30 (1)	-	-	1
Percussors	0.30 (1)	3.19 (3)	50.00 (2)	6
WASTE				
Flakes				
Flakes	36.01 (121)	80.85 (76)	50.00 (2)	199
Blades	1.49 (5)	1.06 (1)	-	6
Éclats de taille de biface	22.02 (74)	6.38 (6)	-	80
Biface sharpening flakes	0.60 (2)	-	-	2
Core waste	1.49 (5)	-	-	5
Non-flakes				
Angular fragments	1.49 (5)	2.13 (2)	-	7
Cores				
Levallois cores	0.30 (1)	-	-	1
Discoidal cores	0.30 (1)	-	-	1
Cores on flakes	1.19 (4)	-	-	4
Core varia	0.89 (3)	-	-	3
Amorphous cores	0.30 (1)	-	-	1
Modified	2.08 (7)	4.26 (4)	-	11
Total	336	94	4	434

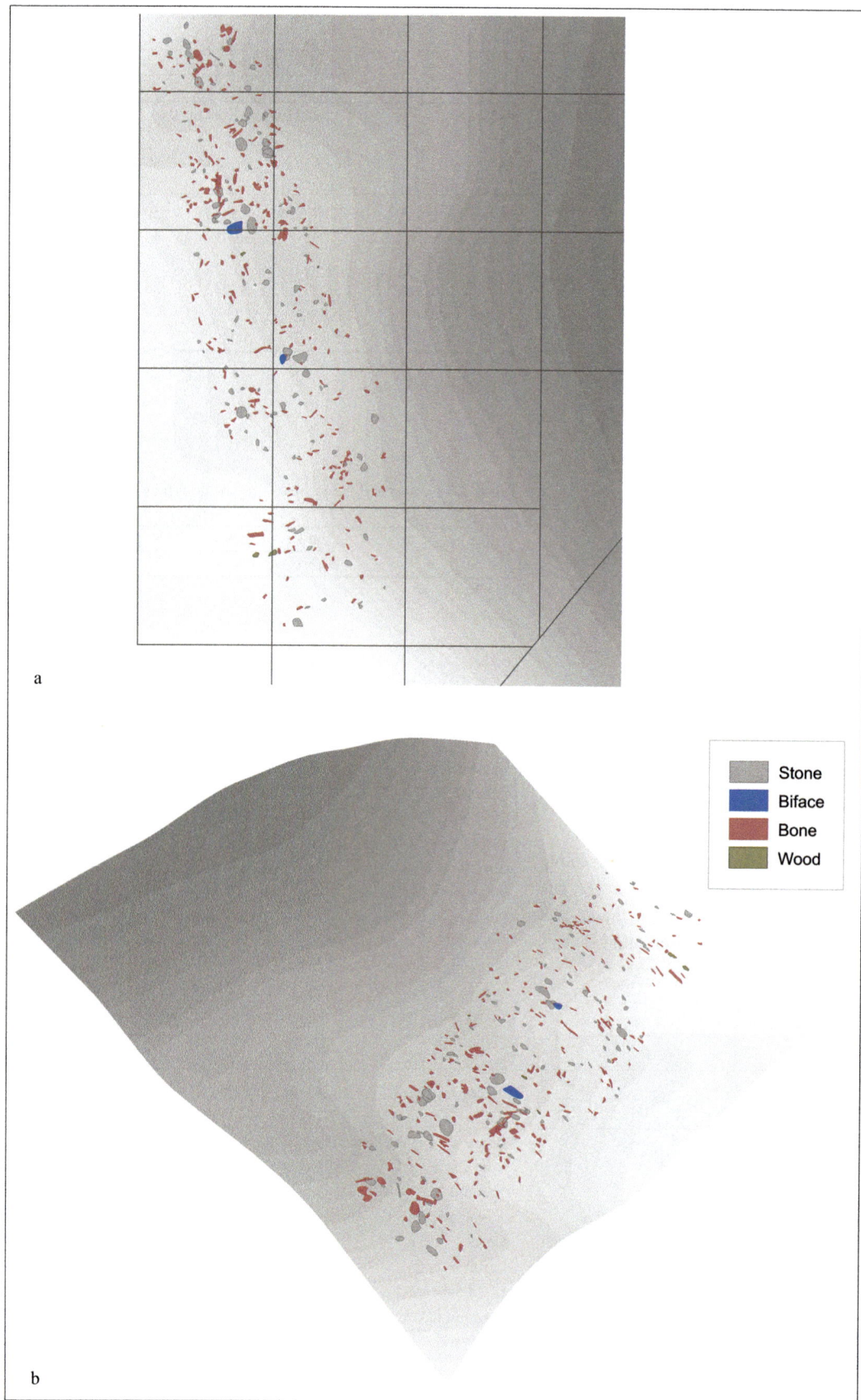

Fig. 5.7 Field map of Layer V-5: a) in plan view, view to north, b) in its original tilted position; view to southeast

5.2 Area C

Fig. 5.8 Layer V-6: a scatter of bones and lithics

5.2.4 Layer V-6

Layer V-6, composed of mud, is about one meter thick in Trench V. There are pockets of mollusks in the contact of Layers V-5/6 and immediately below it. Excavation of Layer V-6 (Figs. 5.3, 5.5–5.6, 5.8) penetrated to a maximal depth of 35 cm (average depth 28 cm) without reaching the base of the layer. During excavation of this layer, we observed an increase in the size of lithic artifacts (Tables 5.9–5.10) and of bones in the lowermost excavated area.

Observation of the archaeological horizon during field work (Fig. 5.9), as well as the analyses of lithic microartifacts and faunal remains, is indicative of minimal postdepositional changes on the surface of Layer V-6. Examples are the anatomical association of crab remains (Ashkenazi et al. 2005), the preservation of a fragile swan skull in anatomical position, and the wealth of information on the spatial organization of basalt and flint micro- and macroartifacts (Alperson-Afil et al. 2009; Alperson-Afil and Goren-Inbar 2010).

Table 5.9 Counts and frequencies (%)* of the lithic assemblage of Layer V-6

	Flint	Basalt	Limestone	Total
	% (N)	% (N)	% (N)	N
TOOLS[#]	8.31 (25)	8.54 (7)	-	32
WASTE				
Debitage	89.04 (268)	67.07 (55)	100.00 (2)	325
Cores	2.33 (7)	3.66 (3)	-	10
Non-flakes	0.33 (1)	20.73 (17)	-	18
Subtotal	301	82	2	385
Microartifacts	4,683	2,020	100	6,803
Natural nodules	5	49	3	57
Total	4,989	2,151	105	6,945

* % calculated from the subtotal
[#] flake tools and core tools

Table 5.10 Counts and frequencies (%) of the typological composition of Layer V-6

	Flint	Basalt	Limestone	Total
	% (N)	% (N)	% (N)	N
TOOLS				
Flake tools				
Side-scrapers	3.65 (11)	-	-	11
End-scrapers	0.66 (2)	-	-	2
Borers	0.33 (1)	-	-	1
Notches	1.33 (4)	-	-	4
Denticulates	1.00 (3)	-	-	3
Retouched flakes	1.00 (3)	-	-	3
Massive scrapers	-	1.22 (1)	-	1
Core tools				
Bifaces				
Handaxes	0.33 (1)	-	-	1
Cleavers	-	7.32 (6)	-	6
Others				
Chopping tools	0.33 (1)	-	-	1
Percussors	-	14.63 (12)	-	12
Pitted anvils	-	1.22 (1)	-	1
WASTE				
Flakes				
Flakes	26.58 (80)	62.20 (51)	50.00 (1)	132
Blades	1.33 (4)	-	-	-
Éclats de taille de biface	58.47 (176)	4.88 (4)	50.00 (1)	181
Biface sharpening flakes	1.66 (5)	-	-	5
Core waste	1.00 (3)	-	-	3
Non-flakes				
Cores				
Levallois cores	1.00 (3)	-	-	3
Cores on flakes	0.33 (1)	-	-	1
Core varia	0.66 (2)	1.22 (1)	-	3
Amorphous cores	-	-	-	1
Giant cores	-	-	-	1
Modified	0.33 (1)	2.44 (2)	-	3
Total	301	82	2	385

5.2.5 Jordan Bank (JB)

The term Jordan Bank (JB) refers to a package of BYF strata that forms a major part of the Jordan River bank and is a

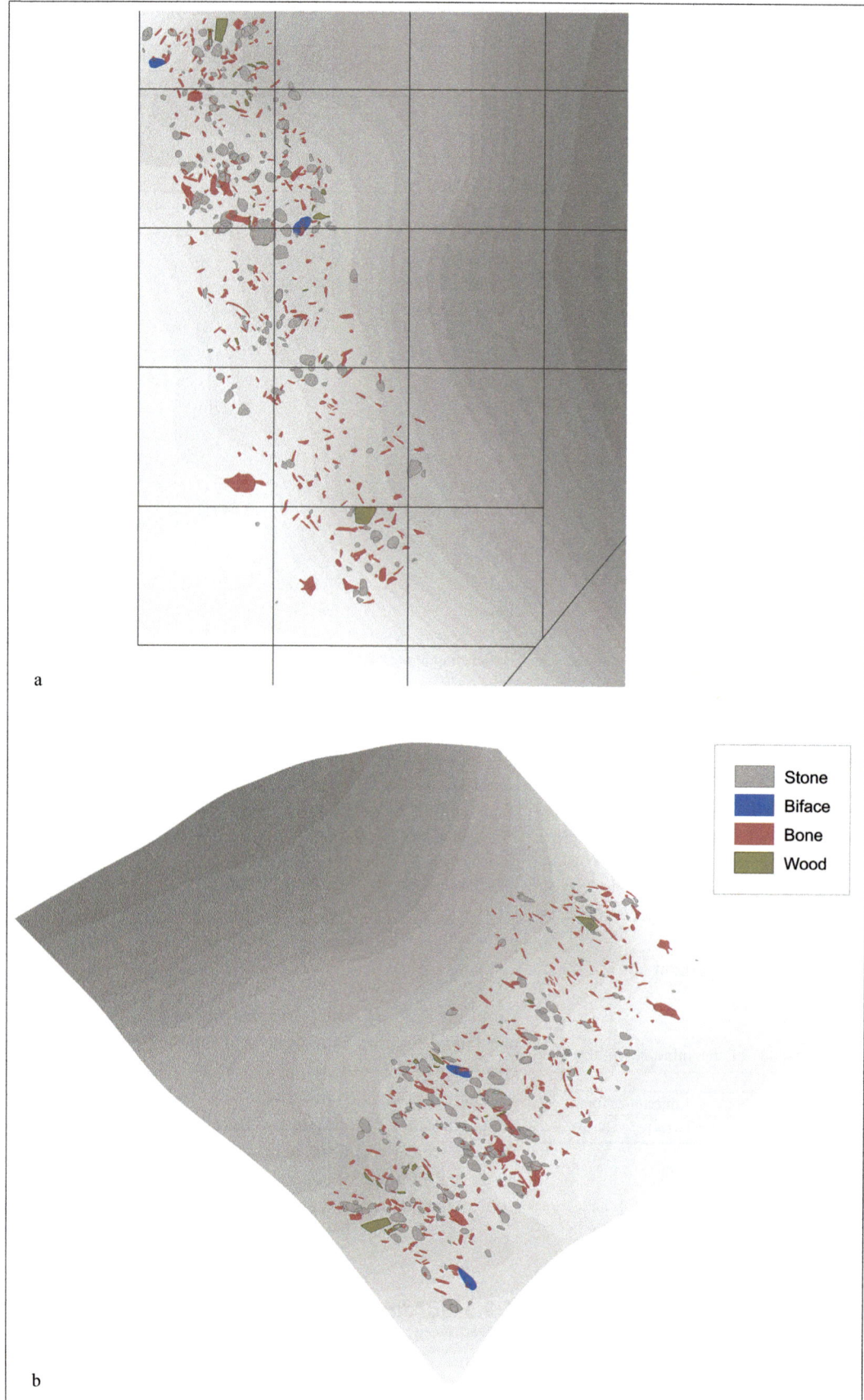

Fig. 5.9 Field map of Layer V-6 a) in plan view, view to north, b) in its original tilted position, view to southeast

5.2 Area C

geologically mappable unit in the study area (Fig. 3.8, Fig. 5.10). This landscape feature is the most extensive known outcrop of the BYF in the study area and in the entire GBY site.

The JB deposit is a tilted cemented coquina composed of a packed viviparid bed, whitish in color (Fig. 5.11). Underlying this coquina (Layer V-5) and protected by it is a black mud layer (Layer V-6). These two layers form the sloping bank of the Jordan River.

The JB is exposed to the erosive effects of the river's fluctuating water level and changing velocity (daily, seasonally,

Fig. 5.11 Close-up view of the Jordan Bank coquina (Layer V-5)

and annually), as well as the destructive activities of the Kinneret Drainage Authority and of the thousands of tourists engaged in water sports, who take a break here and drag their vessels onto dry land over the JB outcrop, exploiting its natural slope and hardness. These activities fracture the hard coquina of the JB and expose the underlying soft and muddy dark clayey layer, which then erodes very quickly, destroying the cultural heritage of the site. The erosive and destructive processes described above result in the exposure of finds in the river and on land, and thus stone artifacts, manuports, mammal bones, wood and bark fragments, and many other archaeological remains become vulnerable to the action of water and to trampling by people, cows, or vehicles.

A remarkable number of identifiable mammal bones and lithic artifacts of great interest have been exposed on the sloping sediments of the JB. These isolated finds, which seem unique in comparison with the content of other layers at the site, have frequently been collected and coordinated in an attempt to gather more information. Apart from the striking number of identifiable bones, some evidence of carcass exploitation (cut marks and clear signatures of marrow exploitation) has been identified.

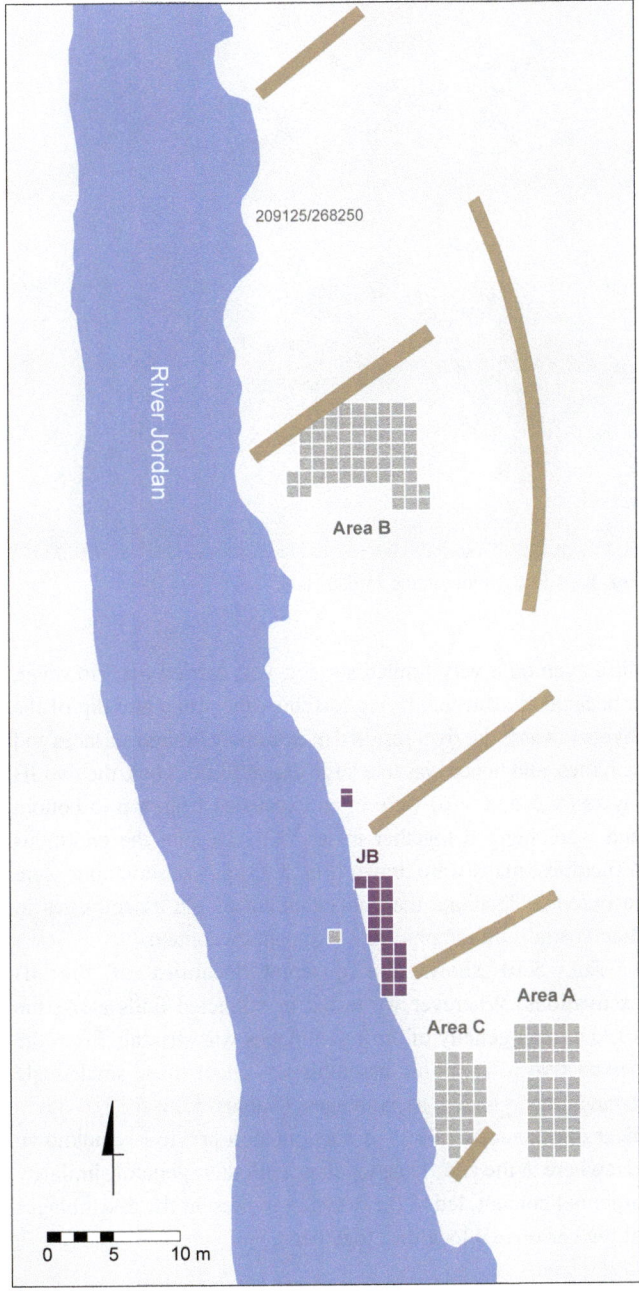

Fig. 5.10 Map of excavated areas and geological trenches; the Jordan Bank excavations are highlighted

The lithic artifacts, in mint condition, were made primarily on flint of extremely homogeneous quality throughout the exposed strata. This contrasts with the generally extensive use of basalt as raw material for artifacts gathered from previously excavated areas.

It was only during the field season of 1995 that a systematic attempt to retrieve *in situ* assemblages from the JB layers took place (Table 5.1; Figs. 5.12–5.14). A segment of the JB was integrated into the grid system and excavation on a very small scale and in very difficult conditions was carried out during two field seasons along a continuous stretch of the riverbank. During the field season of 1996 the excavation was extended southward. In total, some 20 m^2 were excavated in the JB (Fig. 5.13).

In order to enable excavation of the JB layers (V-5 and V-6), an attempt to prevent the river from flooding the excavated surface was essential. Several systems were employed, including building a "wall" made of filled-up plastic bags (Fig. 5.14) and the use of pumps. When conditions permitted, excavation by the general excavation procedure at GBY (along the strike and

Fig. 5.14 Excavation on the Jordan Bank

dip), even on a very limited surface, was carried out. However, it became evident that trying to follow the strike and dip of the layers towards the river resulted in extremely limited surfaces and volumes, and hence was to a large extent futile. Thus, the two JB layers (V-5 and V-6) were often excavated from top to bottom and were lumped together under "JB". Despite the enormous difficulties, maps were drawn (Fig. 5.15) and observations were recorded to facilitate the understanding of the assemblages in their spatial, stratigraphic, and sedimentary context.

Fig. 5.10 shows the different localities of the JB excavations. Wherever we tested or collected finds along the JB, the homogeneity of the assemblages was striking. From the perspective of the lithic assemblages, all of these small-scale excavations yielded assemblages (Tables 5.11–5.12) with a clear dominance of flint – a phenomenon previously unknown elsewhere at the site. This, together with their general similarity in faunal content, led to the decision to present the assemblages of the various JB localities together.

Among the fauna, the medium-sized and large mammals from the JB excavations are described in detail in GBY Volume III

Fig. 5.12 Jordan Bank: excavation of an elephant (*Palaeoloxodon antiquus*) tooth in Layer V-5; note the *Viviparus apameae* in the matrix of the tooth

Fig. 5.13 Excavation on the Jordan Bank

5.2 Area C

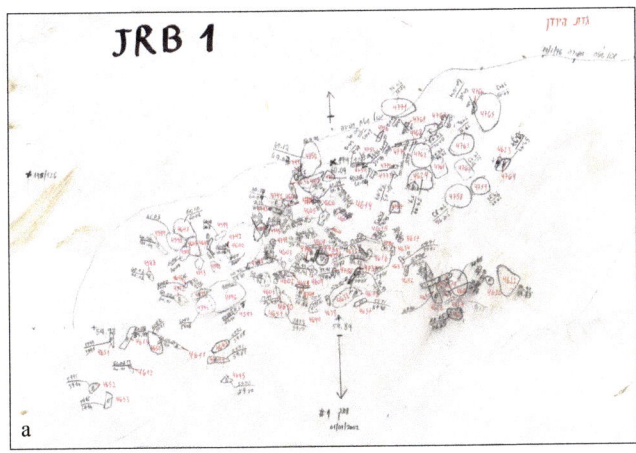

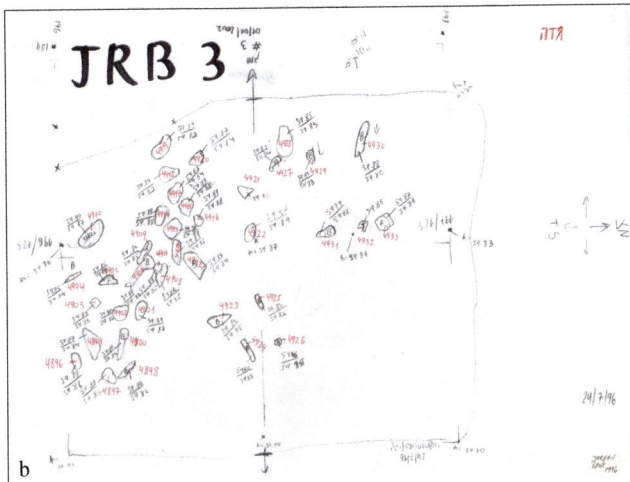

Fig. 5.15 Jordan Bank, southern sector: examples of field maps (1996)

(Rabinovich et al. 2012), while mollusks, crabs, micromammals, amphibians, and reptiles are described by others (Ashkenazi et al. 2005; Rabinovich and Biton 2011). In addition, analyses of the wood remains have been published (Goren-Inbar et al. 2002b).

5.2.6 The Enclosure (Jordan River)

During the field season of 1990, the water of the Jordan River was at its lowest level and a very large bone was observed under water, bedded in the riverbed (Fig. 5.10). An extensive effort was made to expose the bone, which from its size and massiveness clearly belonged to an elephant.

The excavation procedure involved the construction of an "enclosure" composed of plastic bags around the bone. Upon its completion, pumps were introduced to the enclosure and the water was drained from it (Fig. 5.16). The excavation procedure was extremely laborious, as the area had to be pumped continuously in order to enable excavation of the bone. The bone was successfully exposed and the detailed morphology of an elephant vertebra became visible and was drawn. No artifacts were found in the vicinity of the vertebra, but some pebbles

Table 5.11 Counts and frequencies (%)* of the lithic assemblage of JB

	Flint	Basalt	Limestone	Total
	% (N)	% (N)	% (N)	N
TOOLS[#]	20.93 (36)	9.41 (8)	–	44
WASTE				
Debitage	73.84 (127)	49.41 (42)	80.00 (8)	177
Cores	4.07 (7)	2.35 (2)	10.00 (1)	10
Non-flakes	1.16 (2)	38.82 (33)	10.00 (1)	36
Subtotal	172	85	10	267
Microartifacts	906	79	4	989
Natural nodules	7	107	4	118
Total	1,085	271	18	1,374

* % calculated from the subtotal
[#] flake tools and core tools

Table 5.12 Counts and frequencies (%) of the typological composition of JB

	Flint	Basalt	Limestone	Total
	% (N)	% (N)	% (N)	N
TOOLS				
Flake tools				
Side-scrapers	4.07 (7)	–	–	7
Burins	1.16 (2)	–	–	2
Notches	4.65 (8)	–	–	8
Denticulates	3.49 (6)	–	–	6
Retouched flakes	5.81 (10)	–	–	10
Varia	0.58 (1)	–	–	1
Core tools				
Bifaces				
Handaxes	1.16 (2)	–	–	2
Cleavers	–	9.41 (8)	–	8
Others				
Chopping tools	–	2.35 (2)	10.00 (1)	3
Percussors	–	23.53 (20)	10.00 (1)	21
Anvils	–	2.35 (2)	–	2
WASTE				
Flakes				
Flakes	24.42 (42)	45.88 (39)	70.00 (7)	88
Blades	1.74 (3)	–	–	3
Éclats de taille de biface	46.51 (80)	3.53 (3)	10.00 (1)	84
Core waste	1.16 (2)	–	–	2
Non-flake				
Angular fragments	–	9.41 (8)	–	8
Cores				
Levallois cores	0.58 (1)	–	–	1
Core varia	1.74 (3)	–	–	3
Amorphous cores	1.74 (3)	–	–	3
Modified	1.16 (2)	3.53 (3)	–	5
Total	172	85	10	267

Fig. 5.16 Excavation of the elephant vertebra in the Enclosure

(of hammerstone size) were observed in association with the vertebra.

Although detailed stratigraphic observations could not be made, it became evident that while the bone was covered by coquina sediment, most of it was bedded in dark clay. The taxonomic composition of the coquina enabled its identification with Layer V-5 and the underlying layer, the black mud, was identified as Layer V-6 (both layers are described above).

Despite our extensive efforts to extract the entire bone mass with all due precautions and conservation measures (e.g. polyurethane), the bone crumbled into many pieces when it was lifted.

5.3 Area A

The excavations in Area A (Fig. 5.17) were the first attempt to explore the study area at GBY. These were carried out during the field season of 1989 and were never resumed. Since this was our first attempt to gain an understanding of the site, many of our methodological insights were initiated there, although they obviously did not reach the advanced level achieved through our accumulated experience in later field seasons and in other excavation areas (Chapter 4).

The excavations in Area A and the digging of Trench I in the same year provided a key to the stratigraphy of this area. Table 5.13 presents the data of the excavated surface

Table 5.13 Data on excavated surface (m^2) and volume (m^3)* of layers exposed in Area A

Layer	Surface	Volume
I-4	5.25	1.57
I-5	5.0	0.55

* volume is calculated using the thickness of each of the individual layers as recorded in the cross-section of Trench I and in the cross-sections (perpendicular to the layers) of the excavations

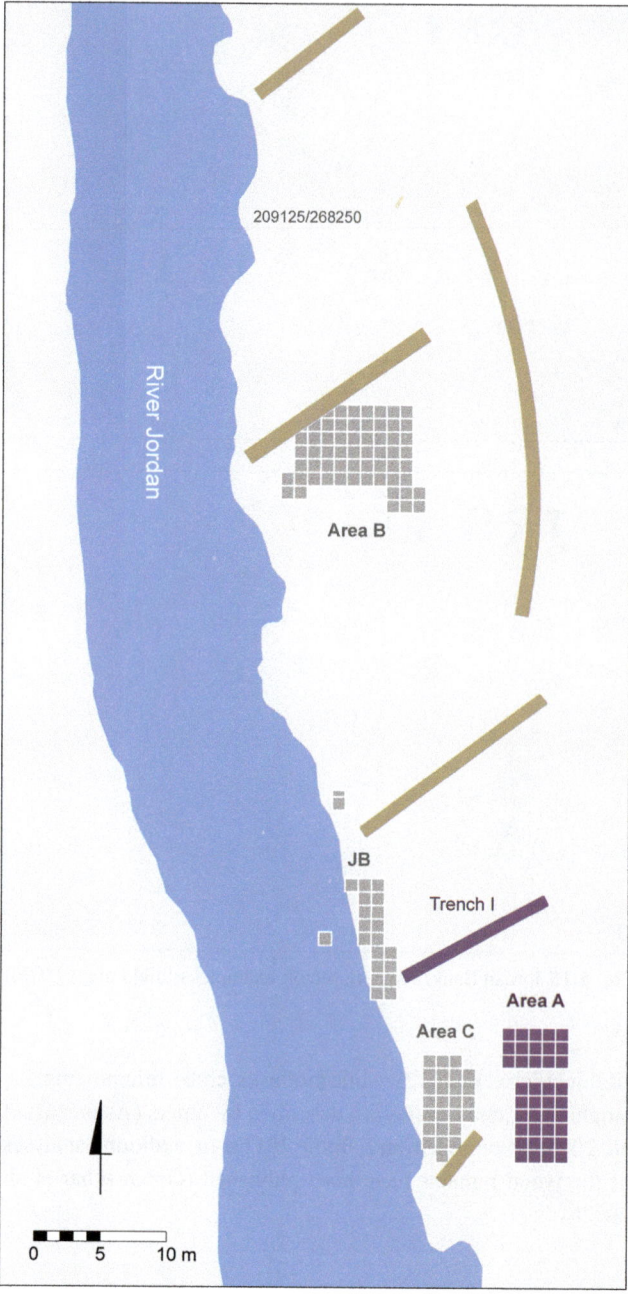

Fig. 5.17 Map of excavated areas and geological trenches; the Area A excavations and the adjacent Trench I are highlighted

and volume and Figs. 5.18–5.19 depict the tilted layers underlying the floodplain deposit of Layer I-1 (Unconformity A). The excavations began with the stripping of Layer I-1 from the surface in order to provide close sedimentological and stratigraphic observations prior to excavation; this type of exposure provided a cross-section in plan view due to the tilt of the layers. The removal of Layer I-1 exposed small pebbles and some larger stones in the contact with the underlying Pleistocene layers.

It was only after the digging of Trench I that each of the observed layers was assigned an individual name (Chapter 4),

5.3 Area A

Fig. 5.18 General view of Area A looking north

Fig. 5.19 General view of Area A looking south

from the youngest (closer to the Jordan River) to the oldest in ascending order; thus, Layer I-4 is younger than the underlying Layer I-5 (Fig. 5.20). It was essential in this first season of excavation at the site to establish a system of nomenclature that would subsequently be used for all other trenches and excavated layers.

Excavation focused on Layers I-4 and I-5, as it was there that most of the artifacts and bones were identified before excavation. The cross-section exposed below Layer I-1 suggested that the archaeological horizon is located at the contact between these two layers (I-4 and I-5). Thus, in order to expose the archaeological finds along their strike and dip (ca. 20°), the deposits of Layer I-4 had to be stripped away to enable the uppermost surface of Layer I-5 to be exposed (Chapter 4). Excavation was first initiated in the five easternmost squares of Area A (enlarged later during the field season) and a row of 1 m^2 was left untouched to serve as a baulk in order to facilitate detailed stratigraphic observations (Fig. 5.17).

In comparison with Area B, which was excavated simultaneously, Area A was poor in lithic artifacts, and therefore excavation there was halted during the first season. In contrast to the scarcity of lithic artifacts, however, Area A was rich in fish and crab remains and contained a wealth of bird bones and organic material.

Wood fragments (Fig. 5.21), bark (Goren-Inbar et al. 2002b), seeds, fruits (Goren-Inbar et al. 2002a; Melamed et al. 2011), pollen (van Zeist and Bottema 2009), mollusks (Ashkenazi et al. 2010), crabs (Ashkenazi et al. 2005), fish (Zohar and Biton 2011, Zohar et al. 2014), amphibians, micrommamals (Goldstein 2011), reptiles, and bird bones (Fig. 5.22) were all found in Area A and are described elsewhere. Fragments of identifiable bones of medium-sized and large mammals include *Palaeoloxodon antiquus* (tooth fragments), *Hippopotamus amphibius*, *Sus scrofa*, Carnivore und., *Bos* sp., *Dama* sp., and a few antler fragments of an unidentified cervid. Other fragments, mainly unidentifiable bone splinters, were also recovered.

5.3.1 Layer I-4

Layer I-4, comprising black mud sediments, is some 29 cm thick in the sections of the excavated area (3 m thick in Trench I). The deposits of Layer I-4 were excavated from top to bottom down to a few centimeters above the contact with Layer I-5 (Fig. 5.23). Where a surface of adequate size was exposed, excavation followed the strike and dip (Chapter 4.1). The excavated deposits of Layer I-4 were sampled for wet sieving (a single full bucket from each sub-square of a given depth unit of excavation). The lithic inventory of Layer I-4 is presented in Tables 5.14–5.15.

5.3.2 Layer I-5

Layer I-5 is a coquina (Figs. 5.21–5.23) with some sandy and clayey lenses. The coquina is predominantly composed of

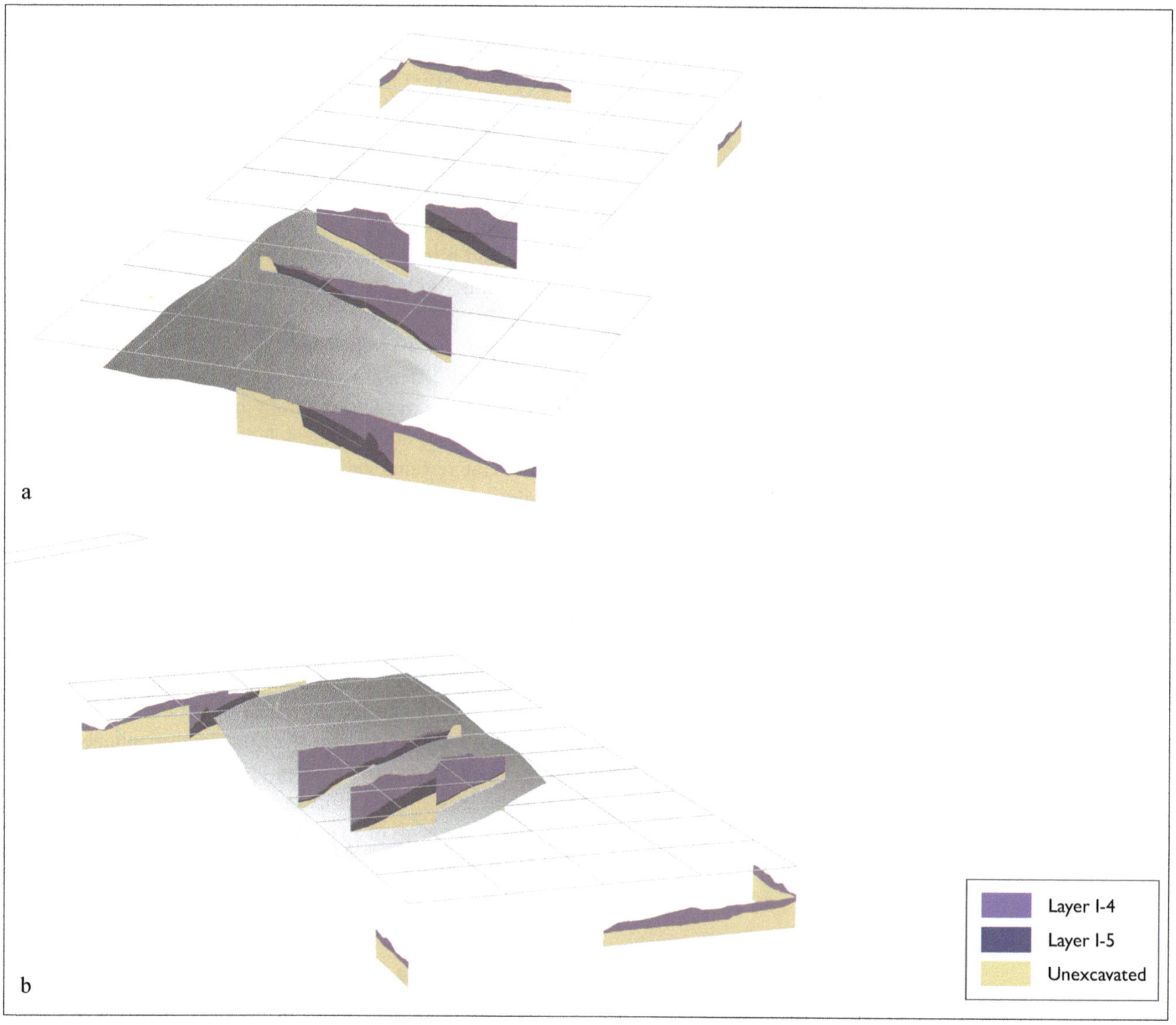

Fig. 5.20 Isometric views of the excavations and cross-sections in Area A: a) view to southeast, b) view to northwest

Fig. 5.21 Wood fragments bedded in the sediments of Layer I-5 (scale 7 cm)

Melanopsis and *Theodoxus* species (with a noted presence of *Viviparus apameae*) and is some 10 cm thick in the sections of the excavated area (2 m thick in Trench I). Layer I-5 was excavated only in its uppermost part and its base was not reached. The top of this layer had a conglomeratic component rich in fragments of weathered basalt in the form of rounded grains, small and rounded flint pebbles, and pieces of red vesicular basalt. The contact between the coquina Layer I-5 and the overlying clayey Layer I-4 was found to be the richest in stone artifacts (Tables 5.16–5.17) in this area. This particular configuration was encountered in other areas at GBY and is described elsewhere by Feibel (2001).

5.3 Area A

Fig. 5.22 A bird bone bedded in the sediments of Layer I-5

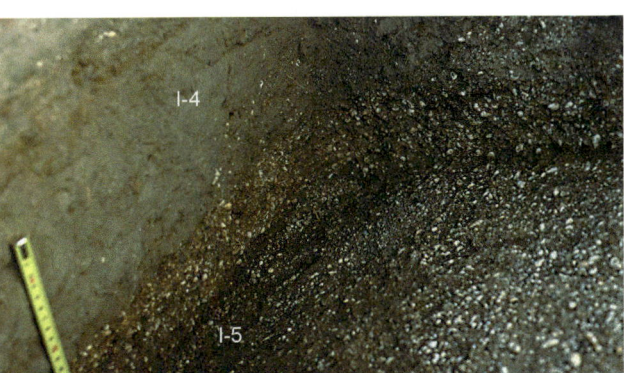

Fig. 5.23 The stratigraphic position of Layers I-4 and I-5 as seen in a corner of an excavated square

Table 5.15 Counts of the typological composition of Layer I-4

	Flint	Basalt	Limestone	Total
TOOLS				
Flake tools				
Side-scrapers	2	-	-	2
End-scrapers	1	-	-	1
Burins	-	-	-	-
Borers	1	-	-	1
Notches	7	-	-	7
Denticulates	3	-	-	3
Truncations	-	-	-	-
Retouched flakes	3	-	-	3
Massive scrapers	-	-	-	-
Varia	1	-	-	1
Core tools				
Bifaces	-	-	-	-
Handaxes	-	-	-	-
Cleavers	-	-	-	-
Others	-	-	-	-
Chopping tools	-	-	-	-
Percussors	-	-	-	-
Pitted anvils	-	-	-	-
Anvils	-	-	-	-
WASTE				
Flakes				
Flakes	26	10	-	36
Blades	-	-	-	-
Éclats de taille de biface	-	-	-	-
Biface sharpening flakes	-	-	-	-
Core waste	1	-	-	1
Non-flakes				
Angular fragments	1	-	-	1
Cores				
Levallois cores	-	-	-	-
Pyramidal cores	-	-	-	-
Cores on flakes	-	-	-	-
Core varia	-	-	-	-
Amorphous cores	-	-	-	-
Giant cores	-	-	-	-
Modified	-	1	-	1
Total	46	11	-	57

Table 5.14 Counts and frequencies (%)* of the lithic assemblage of Layer I-4

	Flint		Basalt		Limestone		Total	
	%	(N)	%	(N)	%	(N)	%	(N)
TOOLS[#]	39.13	(18)	-		-		31.58	(18)
WASTE								
Debitage	58.70	(27)	90.90	(10)	-		64.91	(37)
Cores	-		-		-		-	
Non-flakes	2.17	(1)	9.10	(1)	-		3.51	(2)
Subtotal	46		11		-		57	
Microartifacts	7,449		106		103		7,658	
Total	7,495		117		103		7,715	

* % calculated from the subtotal
[#] flake tools and core tools

Table 5.16 Counts and frequencies (%)* of the lithic assemblage of Layer I-5

	Flint	Basalt	Limestone	Total
	% (N)	% (N)	% (N)	N
TOOLS[#]	33.93 (19)	80.00 (4)	100.00 (1)	24
WASTE				
Debitage	50.00 (28)	-	-	28
Cores	1.78 (1)	-	-	1
Non-flakes	14.29 (8)	20.00 (1)	-	9
Subtotal	56	5	1	62
Microartifacts	15,782	219	135	16,136
Natural nodules				
Total	15,838	224	136	16,198

* % calculated from the subtotal
[#] flake tools and core tools

Table 5.17 Counts of the typological composition of Layer I-5

	Flint	Basalt	Limestone	Total
TOOLS				
Flake tools				
Side-scrapers	3	-	-	3
End-scrapers	2	-	-	2
Burins	-	-	-	-
Borers	-	-	-	-
Notches	6	-	-	6
Denticulates	3	-	-	3
Truncations	-	-	-	-
Retouched flakes	3	-	-	3
Massive scrapers	-	-	-	-
Varia	-	-	-	-
Core tools				
Bifaces				
Handaxes	-	1	-	1
Cleavers	-	-	-	-
Others				
Notches	1	-	-	1
Denticulates	1	-	-	1
Chopping tools	-	-	-	-
Percussors	-	3	1	4
Pitted anvils	-	-	-	-
Anvils	-	-	-	-
WASTE				
Flakes				
Flakes	28	-	-	28
Blades	-	-	-	-
Éclats de taille de biface	-	-	-	-
Biface sharpening flakes	-	-	-	-
Core waste	-	-	-	-
Non-flakes				
Angular fragments	7	-	-	7
Cores				
Levallois cores	1	-	-	1
Pyramidal cores	-	-	-	-
Cores on flakes	-	-	-	-
Core varia	-	-	-	-
Amorphous cores	-	-	-	-
Giant cores	-	-	-	-
Modified	1	1	-	2
Total	56	5	1	62

5.4 Area B

Area B (Figs. 5.24–5.25) was the main focus of the GBY excavations from the first field season. The study of the stratigraphy of Area B has been a key objective in all attempts to understand the depositional history and the cultural sequence embedded in it. The detailed stratigraphy of Area B was compiled from the study of both faces (cross-sections) of Trench II and from observations on the contact between Layer II-1 (the Unconformity) and the underlying Early–Middle Pleistocene strata; the cleaning of this interface revealed a sequence of tilted layers in a horizontal cross-section (Chapter 4). Further information on the stratigraphy of Area B was retrieved from observations carried out during the excavation of each of the layers and levels, documented in a series of cross-sections (Fig. 5.26). Area B furnished the most extensive exposures of artifact-bearing layers and levels (archaeological horizons) at GBY.

Area B was excavated during all field seasons except for 1997b (Table 5.1). Table 5.18 lists the excavation seasons of each of the layers and levels of Area B. This area contains the highest number of archaeological horizons excavated in the study area. If the presence of only a few *in situ* artifacts is considered a site, then Area B contains 19 sites, all Acheulian, and hence represents the longest sequence exposed in the study area and accordingly the longest hominin lake-edge occupation.

Table 5.19 presents the excavated surface and volume of each of the layers and levels of Area B.

Trench II was dug perpendicularly to the strike of the layers (outcropping on the Jordan River bank) and thus cutting through a minimal volume of it. Despite this geometric

Fig. 5.24 General view of Area B looking east at the beginning of excavation

5.4 Area B

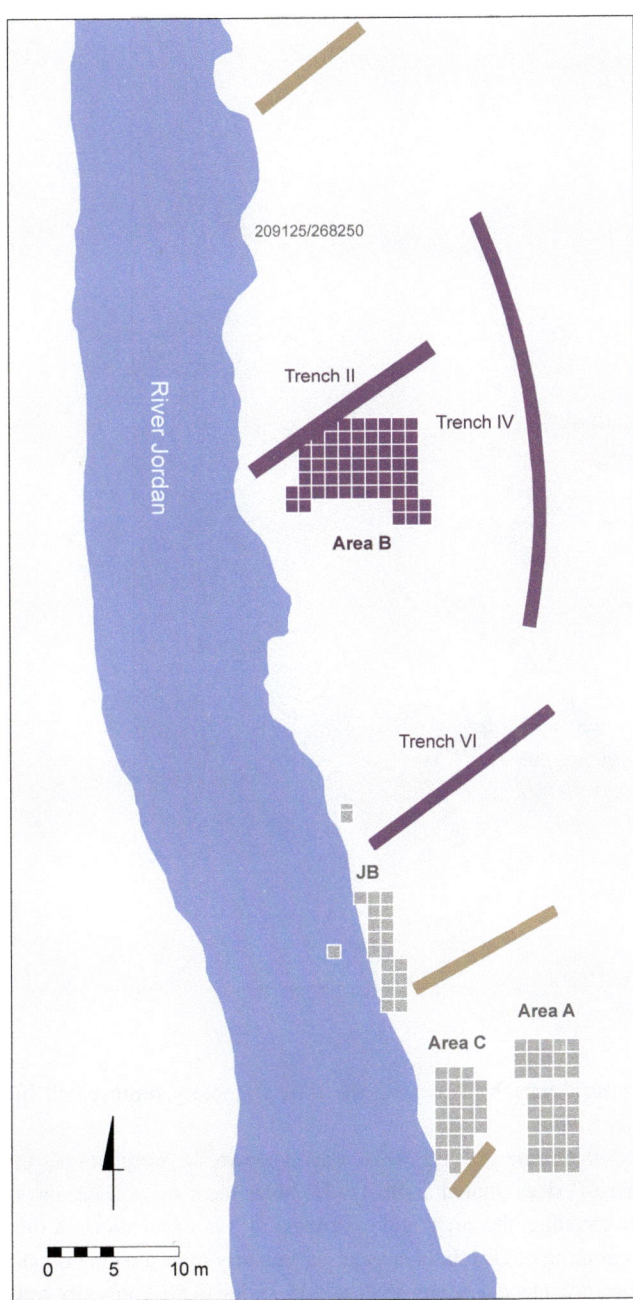

Fig. 5.25 Map of excavated areas and geological trenches; the Area B excavations and the adjacent Trenches II, IV, and VI are highlighted

as limestone percussors, were dug from Trench II of Area B. As the stratigraphy had not yet been established, and as it was evident that the basalt and wood would deteriorate with exposure to the heat and sun, the items were collected and immediately immersed in water. Though they lack a stratigraphic assignment, and despite the facts that the spoils were not sieved, these finds have been analyzed and discussed (e.g. Goren-Inbar et al. 2002b). Among the wood fragments from the spoils of Trench II was the only shaped wooden tool found at GBY, a polished plank made of *Salix* (Belitzky et al. 1991; Goren-Inbar et al. 2002b). The spoils of Trench II included a mass of green clay, which was particularly rich in basalt cleavers and handaxes. Based on stratigraphic considerations it is assigned to Layer II-6. Although the green color disappeared upon exposure due to oxidation, the assignment of the artifacts to a particular type of clay was taken into account during the lithic analysis (items were registered under "Trench II green").

Area B and Trench II yielded an immense archive, not only of Acheulian cultural remains but also of data that are relevant to depositional, sedimentological, taphonomic, paleontological, and chronological aspects. Perhaps most relevant to the archaeological work, Trench II also yielded a better perspective and understanding of the structure of this particular block of the BYF in the study area and enabled detailed measurement of the tilt of the excavated deposits (up to 40° dip). It was because of this dip that, despite a plan to expose extensive surfaces of each layer and level, we eventually had to restrict the excavation to minimal exposures of the strike and dip, all imposed by the tectonic structure. This once again created long, narrow surface exposures, as in all areas of GBY.

The entire depositional sequence in Area B is 10.7 m thick and is composed mainly of deposits of two types: clays and coquinas. This measurement is based on observations carried out in Trench II, which provided thickness measures for other layers and levels in Area B (Fig. 3.14a); unless otherwise specified, all thickness measurements are based on these observations.

Unique to Area B is the series of storm beach deposits (Feibel 2001) that builds Layer II-6. This package of levels, superimposed one on the other, comprises most of the archaeological horizons that were excavated in Area B. The entire sequence is interpreted as representing an oscillating freshwater lake and is assigned to depositional cycles 2–4 of GBY, which represent OIS 19–20 (Goren-Inbar et al. 2000; Feibel 2004).

The richness of finds in Area B is clearly not homogeneous along the excavated sequence. We focused our main excavation effort on the layers that yielded high frequencies of finds (e.g. Layer II-6), but also excavated archaeological horizons that were less rich in finds. This variability yields data on the types and intensity of activities, and particularly

limitation, it is very surprising that on such a small surface exposure there was such a large concentration of stone artifacts as well as other finds. This phenomenon will be discussed in detail in the volume dedicated to sedimentology (Feibel, studies under way).

When Trench II was dug at the beginning of the first field season, it became evident that this part of the site is extremely rich. While the backhoe dumped mainly dark clayey sediments from Trench I of Area A, a few hours later incredible numbers of basalt handaxes and cleavers, as well

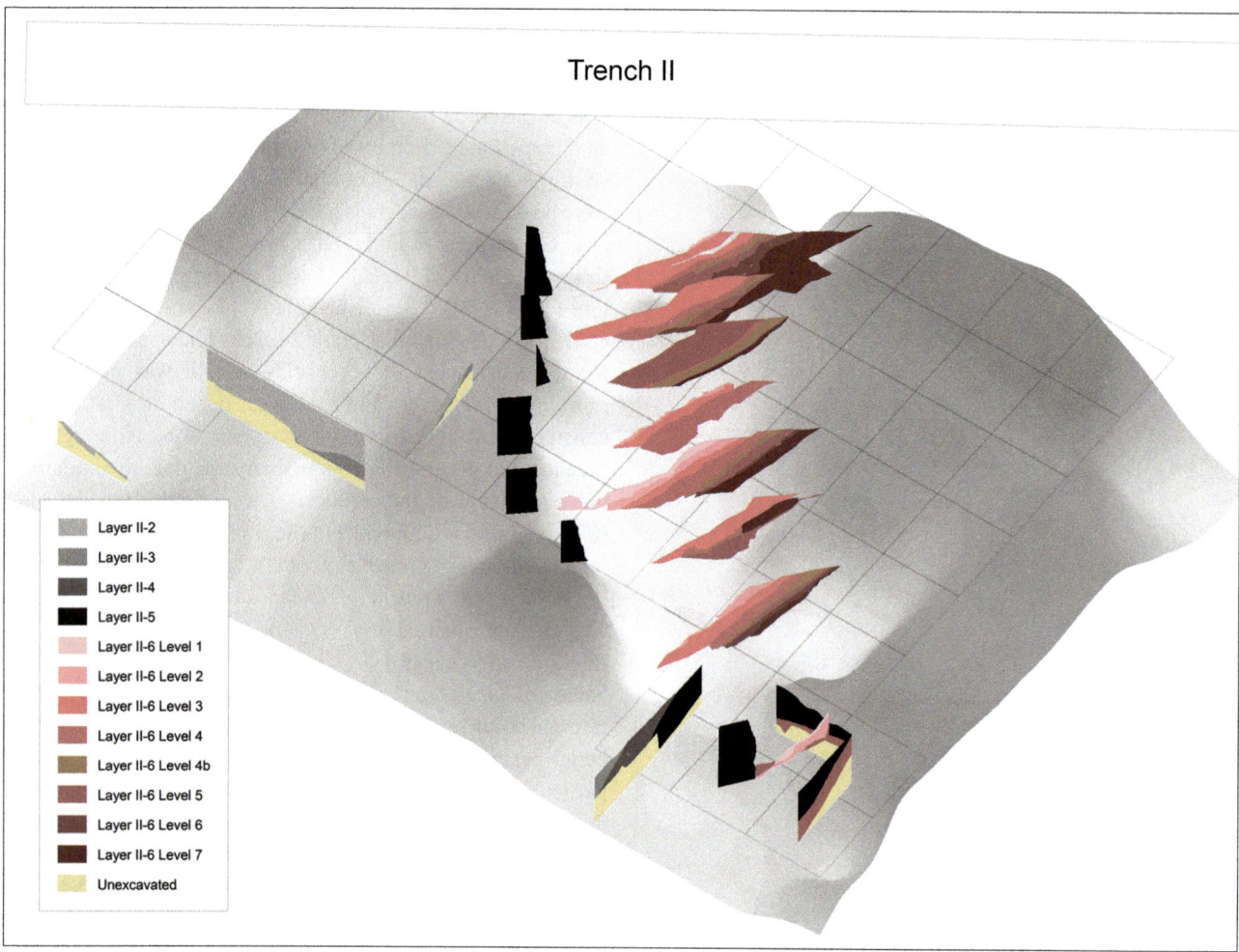

Fig. 5.26 Isometric view of the excavations and cross-sections in Area B, view to northeast

Table 5.18 Excavation of layers, levels, and trenches of Area B by field season

Layer/Level	1989	1990	1991	1995	1996	1997a
II-2	-	+	+	+	-	-
II-2/3	-	+	+	+	-	-
II-3	-	-	+	-	-	-
II-3/4	-	-	+	-	-	-
II-4	-	-	+	-	-	-
II-4/5	-	-	-	-	-	-
II-5	+	+	+	+	-	-
II-5/6	+	+	+	+	-	-
Level 1	+	+	+	+	-	+
Level 2	-	+	+	+	-	+
Level 3	-	+	+	+	-	+
Level 4	-	-	+	+	+	-
Level 4b	-	-	+	+	+	-
Level 5	-	-	-	+	+	-
Level 6	-	-	-	+	+	-
Level 7	-	-	-	-	+	+
II-7/8 (bone)	-	-	-	-	-	+
Trench VII	-	-	-	-	-	+

cultural variability, along the time trajectory represented in Area B.

From the second field season onwards, excavations in Area B necessitated a substantial investment of working days in cleaning the area and preparing it for excavation at the beginning of each field season, particularly after the long break that took place between 1991 and 1995 due to the unusually wet winter of 1991/2.

It was in the excavations of Area B that the most innovative methodological ideas and different excavation modes originated. The methods of handling organic materials with all their difficulties were created and trialed there (Goren-Inbar et al. 2002b). The excavations in Area B, and in particular those of Layer II-6, furnished a substantial insight into some of the postdepositional processes that took place at the site, especially with regard to basalt preservation.

The medium-sized and large mammals from the Area B excavations have been analyzed and published (Goren-Inbar et al. 1994; Rabinovich and Biton 2011), as have the wood (Goren-Inbar et al. 2002b), and fish remains (Alperson-Afil et al. 2009; Zohar and Biton 2011).

5.4 Area B

Table 5.19 Data on excavated surface (m²) and volume (m³)[#] of archaeological horizons exposed in Area B

Layer	Mean thickness[*]	Surface	Volume[#]
II-1 Unconformity	-	-	-
II-2	-	-	-
II-2/3	-	4.67	0.47
II-3	-	-	-
II-3/4	-	-	-
II-4	-	-	-
II-5	-	25.00	13.00
II-5/6	-	19.14	0.38
II-6 Level 1	0.18	23.79	4.28
II-6 Level 2	0.12	25.62	3.07
II-6 Level 3	0.14	17.92	2.50
II-6 Level 4	0.13	16.64	2.16
II-6 Level 4b	0.06	13.69	0.82
II-6 Level 5	0.09	13.39	1.20
II-6 Level 6	0.11	12.62	1.38
II-6 Level 7	0.11; 0.68; 0.55[**]	19.60	5.78
II-6/7	-	-	-
VI-14	-	-	-

[*] mean of excavated thickness based on cross-sections; provided only for the complex of Layer II-6

[**] from left to right: upper horizon, south test pit, north test pit

[#] volume is calculated using the south of each of the individual layers as recorded in the cross-section of Trench II and in the cross-sections (perpendicular to the layers) of the excavations

5.4.1 Layer II-1 (Unconformity B)

The Unconformity of Area B (Figs. 5.24, 5.27), composed of Recent material of the Jordan River floodplain, comprises Layer II-1. Once its sediments were excavated, the underlying truncated and tilted BYF deposits were revealed in plan view, serving as guidelines for the relative chronology of the different layers and for all excavation activities that took place in Area B. After the removal of Layer II-1, the horizontal surface displayed distinct lines of stone artifacts, bones, and sometimes wood, marking the location of the archaeological horizons and their strike. These lines of finds were separated by sterile clayey deposits.

The surface at the base of Layer II-1 not only served as a stratigraphic reference but also yielded rich archaeological (Tables 5.20–5.21) and paleontological information. It was physically used for methodological purposes by furnishing a horizontal surface to which the tilted surfaces related and on which strikes were marked (Chapter 4).

The artifacts scattered at the base of Layer II-1 had been truncated/eroded from their original depositional context by the Jordan River. The date of this event or events, clearly predating erosional events of the Jordan River in Recent times, remains unknown.

Fig. 5.27 General view of Area B looking east

Excavating the Unconformity included the following procedures:

1) Excavation and removal of Layer II-1, excavated from top to bottom in accordance with its semi-horizontal to horizontal deposits.
2) Detailed drawing of cross-sections of Layer II-1 in a variety of locations to illustrate its components, thickness, variability, and other features.
3) Exposure, and detailed mapping and recording, of the coordinates of finds on the uppermost truncated surface of the Pleistocene deposits underlying Layer II-1, including all items larger than 2 cm (natural boulders and cobbles, stone artifacts, bones, and wood).
4) Immersion in water (only for large wood segments and the most indicative basalt artifacts).
5) The deposits from which the artifacts were removed were not sieved, as the finds were not *in situ*.
6) Basalt boulders and cobbles were not collected but their sizes were registered (by either maximal length axis or [granolumetric] b axis). The largest measured 25, 22, 19, and 17 cm along their b axis and were identified as basalt slabs, modified artifacts, and percussors. Items smaller than 2 cm were not observed, most probably as a result of winnowing by the flows, floods, and general erosive forces that affected the Unconformity.
7) Some modern items were exposed during excavation,

Table 5.20 Counts and frequencies (%) of the lithic assemblage of Layer II-1

	Flint	Basalt	Limestone	Total
	% (N)	% (N)	% (N)	N
TOOLS[#]	31.46 (67)	21.79 (61)	14.29 (3)	131
WASTE				
Debitage	52.11 (111)	70.00 (196)	47.62 (10)	317
Cores	7.51 (16)	1.43 (4)	19.05 (4)	24
Non-flakes	8.92 (19)	6.79 (19)	19.05 (4)	42
Total	213	280	21	514

[#] flake tools and core tools

Table 5.21 Counts and frequencies (%) of the typological composition of Layer II-1

	Flint	Basalt	Limestone	Total
	% (N)	% (N)	% (N)	N
TOOLS				
Flake tools				
Side-scrapers	4.23 (9)	0.71 (2)	4.76 (1)	12
End-scrapers	1.41 (3)	-	-	3
Borers	0.47 (1)	-	-	1
Notches	9.86 (21)	-	-	21
Denticulates	4.23 (9)	0.71 (2)	-	11
Truncations	0.47 (1)	-	-	1
Retouched flakes	7.98 (17)	0.71 (2)	4.76 (1)	20
Varia	-	0.71 (2)	-	2
Core tools				
Bifaces				
Handaxes	2.82 (6)	15.36 (43)	-	49
Cleavers	-	2.14 (6)	-	6
Biface preforms	-	1.43 (4)	-	4
Others				
Chopping tools	0.94 (2)	-	14.29 (3)	5
Subspheroid	-	-	4.76 (1)	1
Discoid	-	-	4.76 (1)	1
Percussors	-	2.14 (6)	9.52 (2)	8
Pitted anvils	-	1.07 (3)	-	3
WASTE				
Flakes				
Flakes	42.25 (90)	66.79 (187)	42.86 (9)	286
Blades	1.41 (3)	1.07 (3)	-	6
Éclats de taille de biface	6.57 (14)	1.79 (5)	-	19
Biface sharpening flakes	1.41 (3)	-	-	3
Core waste	0.47 (1)	0.36 (1)	4.76 (1)	3
Non-flakes				
Angular fragments	3.76 (8)	0.71 (2)	-	10
Cores				
Discoidal cores	0.94 (2)	-	-	2
Globular cores	0.47 (1)	-	-	1
Pyramidal cores	-	0.36 (1)	-	1
Cores on flakes	0.47 (1)	-	-	1
Core varia	2.82 (6)	0.36 (1)	-	7
Amorphous cores	1.88 (4)	0.36 (1)	-	5
Giant cores	-	0.36 (1)	-	1
Modified	5.16 (11)	2.86 (8)	9.52 (2)	21
Total	213	280	21	514

Fig. 5.28 General view of Area B looking southeast

truncated tilted deposits of Layer II-6 in order to facilitate excavation of that layer (the strike, the thickness of the levels, the nature of each of the horizons, etc.). It was also excavated for the same purposes above the archaeological horizon of Layer II-2/3, as well as in some other areas (Fig. 5.28).

5.4.2 Layer II-2

Layer II-2 is a thick (1.7 m) coquina with brown clay bands between reddish pockets of mollusks. Layer II-2 was excavated primarily in order to reach the archaeological horizon of Layer II-2/3; this was carried out according to the grid from top to bottom and not along the strike and dip (Figs. 5.28–5.29). The excavated material was sieved in its entirety until July 17th, 1991, when we initiated a sampling procedure (one in four buckets of sediment) that lasted until the end of the excavation. The excavated sediments contained wood segments (Goren-Inbar et al. 2002b) and remains of medium-sized and large mammals, including *Hippopotamus amphibius*, *Dama* sp., *Gazella* cf. *gazella*, and *Palaeoloxodon antiquus* (Goren-Inbar

confirming the assignment of the floodplain material to the Recent activity of the Jordan River. The recovered items included aluminum foil, cigarette butts, shells of modern pistachio nuts, and a Syrian army communications set.

8) The deposits of Layer II-1 were sampled for radiocarbon dating in order to verify their assignment to the Recent floodplain material. Only two dates are available, both within the Holocene (Chapter 3).

Layer II-1 was excavated to the greatest extent above the

Fig. 5.29 General view of Area B looking northeast

5.4 Area B

Table 5.22 Counts and frequencies (%)* of the lithic assemblage of Layer II-2

	Flint	Basalt	Limestone	Total
	% (N)	% (N)	% (N)	N
TOOLS[#]	48.28 (56)	-	-	56
WASTE				
Debitage	44.83 (52)	94.12 (16)	83.33 (10)	78
Cores	3.45 (4)	-	-	4
Non-flakes	3.45 (4)	5.88 (1)	16.67 (2)	7
Subtotal	116	17	12	145
Microartifacts	3,745	76	170	3,991
Not analyzed FFT	67	9	-	76
Natural nodules	178	112	67	357
Total	4,106	214	249	4,569

* % calculated from the subtotal
[#] flake tools and core tools

Table 5.23 Counts and frequencies (%) of the typological composition of Layer II-2

	Flint	Basalt	Limestone	Total
	% (N)	% (N)	% (N)	N
TOOLS				
Flake tools				
Side-scrapers	6.03 (7)	-	-	7
End-scrapers	2.59 (3)	-	-	3
Borers	8.62 (10)	-	-	10
Notches	16.38 (19)	-	-	19
Denticulates	5.17 (6)	-	-	6
Retouched flakes	8.62 (10)	-	-	10
Core tools				
Bifaces				
Handaxes	0.86 (1)	-	-	1
Others				
Chopping tools	0.86 (1)	-	-	1
Percussors	-	5.88 (1)	16.67 (2)	3
WASTE				
Flakes				
Flakes	43.10 (50)	88.24 (15)	83.33 (10)	75
Éclats de taille de biface	1.72 (2)	5.88 (1)	-	3
Non-flakes				
Angular fragments	2.59 (3)	-	-	3
Cores				
Levallois cores	0.86 (1)	-	-	1
Cores on flakes	0.86 (1)	-	-	1
Core varia	0.86 (1)	-	-	1
Modified	0.86 (1)	-	-	1
Total	116	17	12	145

et al. 2009; for micromammals, see Rabinovich and Biton 2011). Unmodified large basalt slabs were observed within Layer II-2 in the section of Trench II, the largest measuring 35 cm along the longest axis. At the time of observation the presence of extremely large basalt slabs was noted, although their meaning was understood only several years later (Chapter 7). At the same time this component was also identified in other parts of Layer II-2, and it is suggested (Goren-Inbar 2011b; Goren-Inbar et al. 2011) that hominins brought these slabs to this locality, a conclusion supported by sedimentological and geological considerations. Although Layer II-2 included a variety of lithic artifacts (Tables 5.22–5.23), no archaeological horizon was identified. This is only partially due to the fact that the excavation did not follow the strike and dip of this layer.

5.4.3 Layer II-2/3

The truncated deposits of Layer II-2/3 were observed at the base of Layer II-1. It appeared as a row of stones arranged along the strike of the contact between Layers II-2 and II-3. Excavation (Table 5.19) aimed to expose the surface of the contact along its strike and dip and indeed revealed an archaeological horizon (Figs. 5.28, 5.30–5.31). This horizon is composed of wood segments (Goren-Inbar et al. 2002b), bones, unmodified stones, and lithic artifacts, which underlie the coquina of Layer II-2 and overlie the black mud of Layer II-3.

Fig. 5.30 General view of Area B looking east

Fig. 5.31 Close-up view of Layer II-2/3

Table 5.24 Counts and frequencies (%) of the lithic assemblage of Layer II-2/3

	Flint	Basalt	Limestone	Total
	% (N)	% (N)	% (N)	N
TOOLS[#]	27.35 (32)	12.73 (7)	-	39
WASTE				
Debitage	57.26 (67)	56.36 (31)	100.00 (3)	101
Cores	6.84 (8)	1.82 (1)	-	9
Non-flakes	8.55 (10)	29.09 (16)	-	26
Subtotal	117	55	3	175
Microartifacts	394	27	6	427
Not analyzed FFT	6	1	-	7
Natural nodules	383	7	55	445
Total	900	90	64	1,054

[*] % calculated from the subtotal
[#] flake tools and core tools

Table 5.25 Counts and frequencies (%) of the typological composition of Layer II-2/3

	Flint	Basalt	Limestone	Total
	% (N)	% (N)	% (N)	N
TOOLS				
Flake tools				
Side-scrapers	1.71 (2)	-	-	2
End-scrapers	0.85 (1)	-	-	1
Burins	0.85 (1)	-	-	1
Borers	1.71 (2)	-	-	2
Notches	12.82 (15)	-	-	15
Denticulates	2.56 (3)	-	-	3
Truncations	1.71 (2)	-	-	2
Retouched flakes	5.13 (6)	-	-	6
Core tools				
Bifaces				
Handaxes	-	10.91 (6)	-	6
Biface preforms	-	1.82 (1)	-	1
Others				
Percussors	-	9.09 (5)	-	5
Pitted anvils	-	1.82 (1)	-	1
WASTE				
Flakes				
Flakes	55.56 (65)	56.36 (31)	100.00 (3)	99
Éclats de taille de biface	0.85 (1)	-	-	1
Core waste	0.85 (1)	-	-	1
Non-flakes				
Angular fragments	5.13 (6)	14.55 (8)	-	14
Cores				
Levallois cores	1.71 (2)	-	-	2
Core varia	3.42 (4)	-	-	4
Amorphous cores	1.71 (2)	-	-	2
Giant cores	-	1.82 (1)	-	1
Modified	3.42 (4)	3.64 (2)	-	6
Total	117	55	3	175

The morphology of this contact layer is not entirely flat but is that of sculptured steps within the clays of Layer II-3, caused by the deposition of the coquina of Layer II-2 on the wet surface of the black mud. The horizon of Layer II-2/3 is classified as an archaeological entity bedded on top of fine-grained sediments and capped by coarse-grained sediments (Feibel 2001). Despite the small size of the excavated surface, this horizon contributes to our understanding of the cultural affinities and formation processes of the site. Layer II-2/3 is a typical Acheulian entity (Tables 5.24–5.25) associated with faunal remains of *Hippopotamus amphibius*, *Dama* sp., Bovidae gen. et sp. indet., and *Palaeoloxodon antiquus* (Goren-Inbar et al. 2009), as well as micromammals (Rabinovich and Biton 2011).

5.4.4 Layer II-3

Layer II-3 is composed of black mud some 50 cm thick. The sediments are spotted with reddish-yellow sparkling

5.4 Area B

concretions of limonite and gypsum. These concretions are a result of iron oxidation (identified in the field using FTIR by S. Weiner of the Weizmann Institute of Science). The concretions directly underlie the archaeological horizon of Layer II-2/3 and continue through the entire thickness of Layer II-3. At the base of Layer II-3, more small pebbles were observed than elsewhere in this layer. This layer, like others, is characterized by a high calcite content. Stone artifacts observed in the Trench II cross-sections suggested that there is an archaeological horizon in the lowermost part of this layer, close to the contact with Layer II-4. A very small surface was excavated in order to reach the artifact-bearing contact horizon of Layer II-3/4. The excavated sediments were sampled (one in four buckets) and, although no distinct archaeological horizon was observed within Layer II-3, lithic artifacts were recorded (Tables 5.26–5.27), as well as faunal remains of *Cervus* cf. *elaphus* (Goren-Inbar et al. 2009).

Table 5.26 Counts and frequencies (%)* of the lithic assemblage of Layer II-3

	Flint	Basalt	Total
	% (N)	% (N)	N
TOOLS#	60.00 (3)	-	3
WASTE			
Debitage	20.00 (1)	-	1
Cores	20.00 (1)	-	1
Non-flakes	-	100.00 (1)	1
Subtotal	5	1	6
Microartifacts	-	-	-
Not analyzed FFT	7	-	7
Natural nodules	-	-	-
Total	12	1	13

* % calculated from the subtotal
# flake tools and core tools

Table 5.27 Counts and frequencies (%) of the typological composition of Layer II-3

	Flint	Basalt	Total
	% (N)	% (N)	N
TOOLS			
Flake tools			
Side-scrapers	20.00 (1)	-	1
Notches	20.00 (1)	-	1
Denticulates	20.00 (1)	-	1
Core tools			
Others			
Chopping tools	20.00 (1)	-	1
WASTE			
Flakes			
Flakes	20.00 (1)	-	1
Non-flakes			
Angular fragments	-	100.00 (1)	1
Total	5	1	6

5.4.5 Layer II-3/4

The color of the contact Layer II-3/4 was reddish on excavation as a result of oxidation, as seen elsewhere in GBY. Although the excavation of this surface yielded a disappointingly low density of finds, it was nevertheless rewarding since it provided a group of very large (giant) basalt slabs (Figs. 5.30, 5.32). The maximal length axis of these blocks measured 40 cm or more. Although measurements and notes were taken at the time of exposure, the full meaning of these and other similar objects was not fully understood; only later did it become apparent that these unflaked basalt slabs (see discussion in Chapter 7) were brought to the site by hominins.

An additional component of this layer was four wood segments (Goren-Inbar et al. 2002b). Thirteen flint flakes eroded from this contact in its exposure in Trench II; although collected, they were not analyzed, as their spatial provenance is unknown.

5.4.6 Layer II-4

This layer is a 30 cm thick coquina with a few wood segments (Goren-Inbar et al. 2002b) and scarce remains of *Palaeoloxodon antiquus* (Rabinovich et al. 2006; Goren-Inbar et al. 2009). The excavated sediments were sampled (one in four buckets), resulting in assemblages of micrommamals, fish, and herpetofauna (Goren-Inbar et al. 2009; Zohar personal communication). The lithic inventory of Layer II-4 is presented in Tables 5.28–5.29.

Fig. 5.32 Close-up view of the archaeological horizons of Layers II-2 and II-3/4 (W=wood); in the background Layer II-6 Level 4

Table 5.28 Counts and frequencies (%)* of the lithic assemblage of Layer II-4

	Flint	Basalt	Limestone	Total
	% (N)	% (N)	% (N)	N
TOOLS[#]	11.11 (1)	-	-	1
WASTE				
Debitage	44.44 (4)	80.00 (4)	100.00 (1)	9
Cores	22.22 (2)	-	-	2
Non-flakes	22.22 (2)	20.00 (1)	-	3
Subtotal	9	5	1	15
Microartifacts	15	-	-	15
Not analyzed FFT	30	3	-	33
Natural nodules	-	-	-	-
Total	54	8	1	63

* % calculated from the subtotal
[#] flake tools and core tools

Table 5.29 Counts and frequencies (%) of the typological composition of Layer II-4

	Flint	Basalt	Limestone	Total
	% (N)	% (N)	% (N)	N
TOOLS				
Flake tools				
Retouched flakes	11.11 (1)	-	-	1
Flake tools				
Others				
Percussors	-	20.00 (1)	-	1
WASTE				
Flakes				
Flakes	33.33 (3)	80.00 (4)	100.00 (1)	8
Éclats de taille de biface	11.11 (1)	-	-	1
Non-flakes				
Angular fragments	11.11 (1)	-	-	1
Cores				
Core varia	22.22 (2)	-	-	2
Modified	11.11 (1)	-	-	1
Total	9	5	1	15

5.4.7 Layer II-5

Layer II-5 is a layer of black mud 70 cm thick and was excavated from top to bottom in order to expose the underlying archaeological horizon (Layer II-6) along its strike and dip. During the field season of 1990 it became evident that finds organized in a line oriented NW-SE were bedded in the dark mud. Due to the nature and color of the sediment, it was impossible to identify the artifacts, particularly those of basalt. Thus, while no exposure along the strike and dip was feasible, each artifact 2 cm and larger was plotted in three coordinates. Later on in this season, the discovery of handaxes verified the presence of an archaeological horizon one item thick. The lithic assemblage is presented in Tables 5.30–5.31. Until the middle of the 1991

Table 5.30 Counts and frequencies (%) of the lithic assemblage of Layer II-5

	Flint	Basalt	Limestone	Total
	% (N)	% (N)	% (N)	N
TOOLS[#]	37.50 (90)	11.50 (13)	21.43 (3)	106
WASTE				
Debitage	43.75 (105)	81.42 (92)	42.86 (6)	203
Cores	10.00 (24)	0.88 (1)	-	25
Non-flakes	8.75 (21)	6.19 (7)	35.71 (5)	33
Subtotal	240	113	14	367
Microartifacts	531	24	18	573
Not analyzed FFT	136	83	3	222
Natural nodules	72	135	15	222
Total	979	355	50	2,324

* % calculated from the subtotal
[#] flake tools and core tools

Table 5.31 Counts and frequencies (%) of the typological composition of Layer II-5

	Flint	Basalt	Limestone	Total
	% (N)	% (N)	% (N)	N
TOOLS				
Flake tools				
Side-scrapers	7.08 (17)	0.88 (1)	-	18
End-scrapers	2.50 (6)	-	-	6
Borers	6.67 (16)	-	-	16
Notches	10.00 (24)	0.88 (1)	21.43 (3)	28
Denticulates	3.33 (8)	-	-	8
Truncations	0.83 (2)	-	-	2
Retouched flakes	7.08 (17)	3.54 (4)	-	21
Core tools				
Bifaces				
Handaxes	-	2.65 (3)	-	3
Cleavers	-	1.77 (2)	-	2
Biface preforms	-	1.77 (2)	-	2
Others				
Chopping tools	0.42 (1)	-	-	1
Percussors	0.42 (1)	4.42 (5)	7.14 (1)	7
WASTE				
Flakes				
Flakes	40.42 (97)	77.88 (88)	42.86 (6)	191
Blades	0.83 (2)	0.88 (1)	-	3
Éclats de taille de biface	1.67 (4)	2.65 (3)	-	7
Core waste	0.83 (2)	-	-	2
Non-flakes				
Angular fragments	5.00 (12)	1.77 (2)	-	14
Cores				
Levallois cores	0.83 (2)	-	-	2
Globular cores	0.42 (1)	-	-	1
Pyramidal cores	0.42 (1)	-	-	1
Cores on flakes	1.25 (3)	0.88 (1)	-	4
Core varia	5.00 (12)	-	-	12
Amorphous cores	1.67 (4)	-	-	4
Modified	3.33 (8)	-	28.57 (4)	12
Total	240	113	14	367

5.4 Area B

season, only one bucket was sampled from every excavated sub-square. From July 15th, 1991, the sampled sediments included one in four excavated buckets. Remains of medium-sized and large mammal include *Hippopotamus amphibius*, *Dama* sp., and *Palaeoloxodon antiquus* (Goren-Inbar et al. 2009). In addition, wood segments (Goren-Inbar et al. 2002b), micromammals, fish, and herpetofauna were found (Goren-Inbar et al. 2009; Rabinovich and Biton 2011; Zohar pers. comm.).

5.4.8 Layer II-5/6

Between the black mud of Layer II-5 and the sediments of Layer II-6 Level 1 was a sharp contact characterized by the discontinuous presence of small pebbles and by the presence of several very long pieces of wood (Goren-Inbar et al. 2002b). It contained lithics (Tables 5.32–5.33) and fossil fauna and flora, which are similar in character to the finds of Layer II-6. Paleontological remains include carnivores, *Hippopotamus amphibius*, *Dama* sp., and *Palaeoloxodon antiquus* (Goren-Inbar et al. 2009). In addition, micromammals, fish, and herpetofauna were found (Goren-Inbar et al. 2009; Rabinovich and Biton 2011; Zohar personal communication).

5.4.9 Layer II-6

Layer II-6 was identified and defined in Trench II, where it reaches a thickness of 2.2 m. It attains an even greater thickness along the strike towards the southeast (as in Fig. 3.13). The southeasternmost exposure of Layer II-6 was identified in the easternmost end of Trench VI, Layer VI-14 (see below).

Layer II-6 is the only case of stratigraphy within a layer at GBY. The stratigraphic complexity of Layer II-6 was initially observed in Trench II and later in the cross-sections in plan view of the truncated deposits below Layer II-1. In the cross-sections (Fig. 5.26) we could discern stratigraphic units one artifact thick, later referred to as levels (1, 2, 3, 4, 5, 6, and 7). This observation was confirmed during the excavation, with the addition of Level 4b, which was not visible in any of the grid cross-sections. Each of the levels was considered a separate archaeological entity. This was confirmed following excavation by the analyses of Layer II-6, which resulted in the identification of markedly different compositions and frequencies between levels, though all are assigned to the Acheulian culture.

The sediments of Layer II-6 consist of gravelly or molluskan sands (rarely as blocks of cemented coquina) deriving from storm events that deposited coarse sediments from the beach and near-shore environment onto the beach face. Hominin

Table 5.32 Counts and frequencies (%)* of the lithic assemblage of Layer II-5/6

	Flint	Basalt	Limestone	Total
	% (N)	% (N)	% (N)	N
TOOLS#	31.82 (49)	14.29 (7)	-	56
WASTE				
Debitage	44.16 (68)	77.55 (38)	50.00 (2)	108
Cores	11.69 (18)	-	50.00 (2)	20
Non-flakes	12.34 (19)	8.16 (4)	-	23
Subtotal	154	49	4	207
Microartifacts	266	2	1	269
Not analyzed FFT	88	70	2	160
Natural nodules	115	139	19	273
Total	623	260	26	909

* % calculated from the subtotal
# flake tools and core tools

Table 5.33 Counts and frequencies (%) of the typological composition of Layer II-5/6

	Flint	Basalt	Limestone	Total
	% (N)	% (N)	% (N)	N
TOOLS				
Flake tools				
Side-scrapers	4.55 (7)	2.04 (1)	-	8
End-scrapers	1.95 (3)	-	-	3
Burins	0.65 (1)	-	-	1
Borers	1.95 (3)	-	-	3
Notches	13.64 (21)	2.04 (1)	-	22
Denticulates	1.95 (3)	-	-	3
Truncations	0.65 (1)	-	-	1
Retouched flakes	6.49 (10)	4.08 (2)	-	12
Massive scrapers	-	2.04 (1)	-	1
Core tools				
Bifaces				
Handaxes	-	2.04 (1)	-	1
Cleavers	-	2.04 (1)	-	1
Others				
Chopping tools	0.65 (1)	-	-	1
Percussors	-	2.04 (1)	-	1
WASTE				
Flakes				
Flakes	41.56 (64)	77.55 (38)	50.00 (2)	104
Éclats de taille de biface	0.65 (1)	-	-	1
Core waste	1.95 (3)	-	-	3
Non-flakes				
Angular fragments	6.49 (10)	2.04 (1)	-	11
Cores				
Levallois cores	1.30 (2)	-	-	2
Cores on flakes	1.30 (2)	-	25.00 (1)	3
Core varia	5.84 (9)	-	25.00 (1)	10
Amorphous cores	2.60 (4)	-	-	4
Modified	5.84 (9)	4.08 (2)	-	11
Total	154	49	4	207

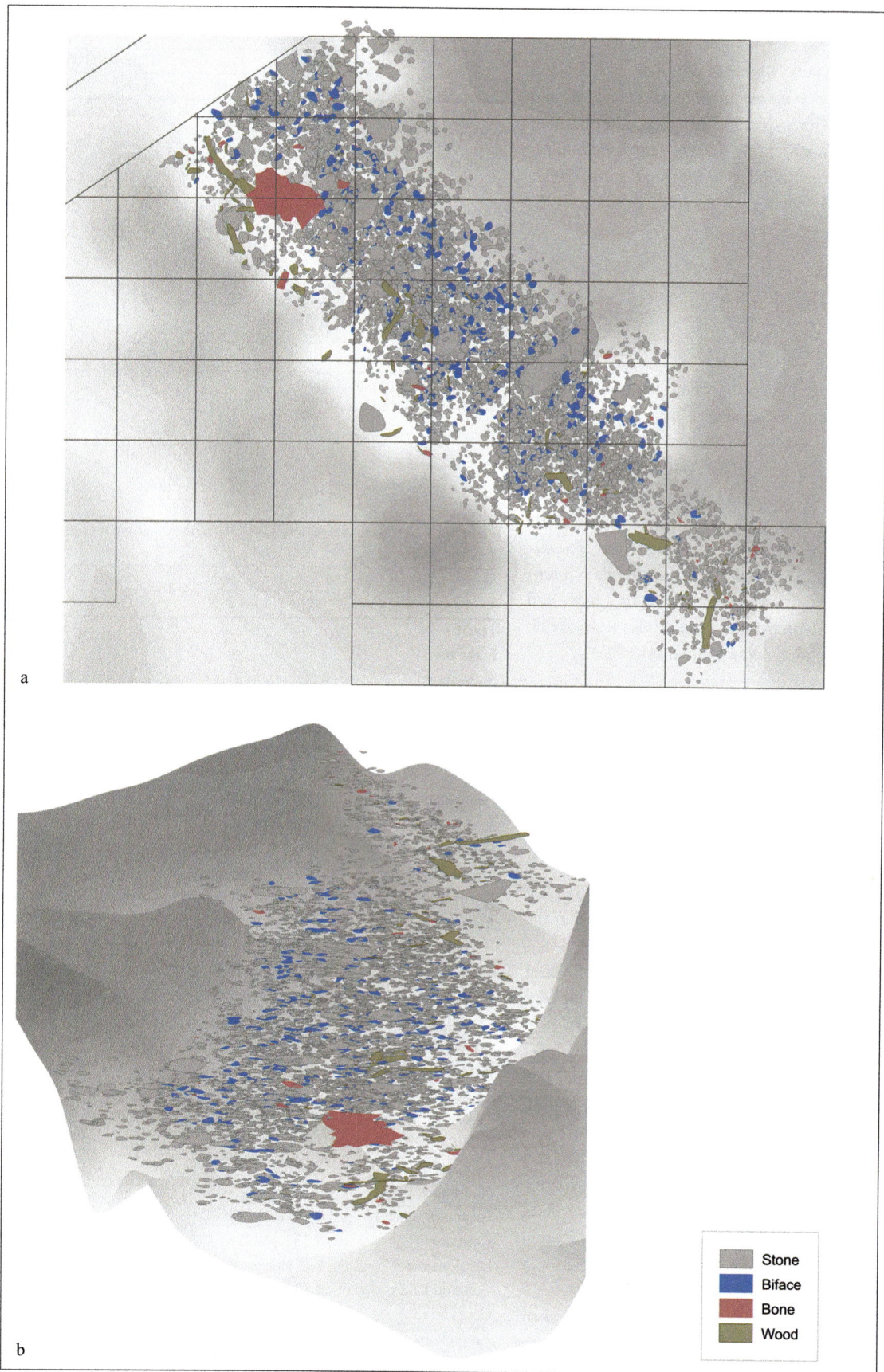

Fig. 5.33 Field maps of Layer II-6 Levels 1–7: a) in plan view, b) in its original tilted position; view to southeast

activity within the eight sequential levels of Layer II-6 was carried out following the formation of a stable surface on the beach, evidenced by the high degree of abrasion and polish on fragmented mollusks and by the sorting of the detrital clastics. Subsequent storm events covered each of the archaeological levels and "the freshness of most of the artifact edges implies that they accumulated well above the strand line, on the upper beach face" (Feibel 2001:137).

The stratigraphic and cultural sequence of Layer II-6 (Fig. 5.33) provides a unique sequence of Acheulian occupations, superimposed one on the other and ranging in thickness between 10 and 20 cm (Fig. 3.13), with the exception of the thicker Level 7 (discussed below). All levels yielded rich assemblages of organic material (Goren-Inbar et al. 2002a; Goren-Inbar et al. 2002b; Melamed et al. 2011) and paleontological remains (fish: Zohar and Biton 2011; crabs: Ashkenazi et al. 2005; micromammals, herpetofauna: Rabinovich and Biton 2011; birds, medium-sized and large mammals: Goren-Inbar et al. 2009). Evidence for continuous use and control of fire was identified within each of the Layer II-6 levels (Goren-Inbar et al. 2004; Alperson-Afil 2008; Alperson-Afil and Goren-Inbar 2010).

Layer II-6 Level 1

Excavation of Layer II-6 Level 1 began after the digging of Trench II and the discovery of an extremely rich lithic assemblage, composed mainly of basalt handaxes and cleavers, in the spoils of this trench. Level 1 was exposed on removal of the dark clay that constitutes Layer II-5 and careful exposure of the contact of Layers II-5/6. The precise thickness of Level 1 was unknown at the beginning of its excavation and the aim was to expose as much of the surface as possible along the strike and dip.

The first field season was dedicated in its entirety to the excavation of an elephant skull (Fig. 5.34) and its associated bone fragments (Goren-Inbar et al. 1994). Level 1 was further excavated in its southernmost part until the 1997a season (Fig. 5.35).

Level 1 is the richest in organic material of GBY's horizons (Goren-Inbar et al. 2002b). In addition to the largest wood assemblage (Fig. 5.35), this level yielded the second largest assemblage of fruits and seeds (Melamed 1997, 2003). From a stratigraphic perspective there is no doubt that the wood, fruits, and seeds, the mammal bones (mainly those of the elephant), and the entire stone tool assemblage all belong to a single archaeological entity. Level 1 also turned out to be the second thickest sedimentological unit in the complex of Layer II-6 (Table 5.19).

Taphonomic study of Level 1 indicates that very little movement occurred during postdepositional times. Several lines of evidence support this conclusion:

1) There is no evidence of winnowing by the water of the lake, as Level 1 is extremely rich in small lithic debris of all raw materials (Alperson-Afil and Goren-Inbar 2010).

2) There is no evidence for clast-size sorting in the lithic assemblages of Level 1. Items such as the giant cores (over 16 kg) are associated with the debris; furthermore, all size clasts are represented.

3) Elongated artifacts, and particularly handaxes and cleavers, do not reflect a single or double dominant orientation mode that could indicate postdepositional re-orientation resulting from possible shore currents. Rather, they are oriented in all directions.

4) Several bone fragments that were found on the surface of Level 1 were conjoined with the main bone mass of the elephant skull (Goren-Inbar et al. 1994). These conjoinable pieces clearly show that minimal movement took place on the horizon before it was sealed by the dark clays of Layer II-5. Both small and large pieces are fragments of the skull and were found in its proximity. In addition: "Thousands of tiny bone fragments were found embedded in the sediment adhering to and in direct proximity to the skull. About half of these fragments have the thin-walled appearance typical of the inner lining of the braincase and sinuses of an elephant skull, and are very likely to belong to it" (Goren-Inbar et al. 1994:100).

5) Excavation in Level 1 furnished evidence of a minimal role of postdepositional processes in the form of small, sand-size, broken, and crushed mollusks found around the base of the elephant skull. While these are indicative of the motion of low-energy water (subsequent to the discard of the skull), this low energy was not strong enough to cause winnowing of the microartifacts and of the small skull fragments.

The detailed stratigraphy and the above observations rule out a taphonomic interpretation that views Level 1 as resulting from a series of palimpsest events (see below).

Remains of medium-sized and large mammals include *Vulpes* sp., Carnivore und., *Hippopotamus amphibius*, *Cervus* cf.

Fig. 5.34 Close-up view of Layer II-6 Level 1

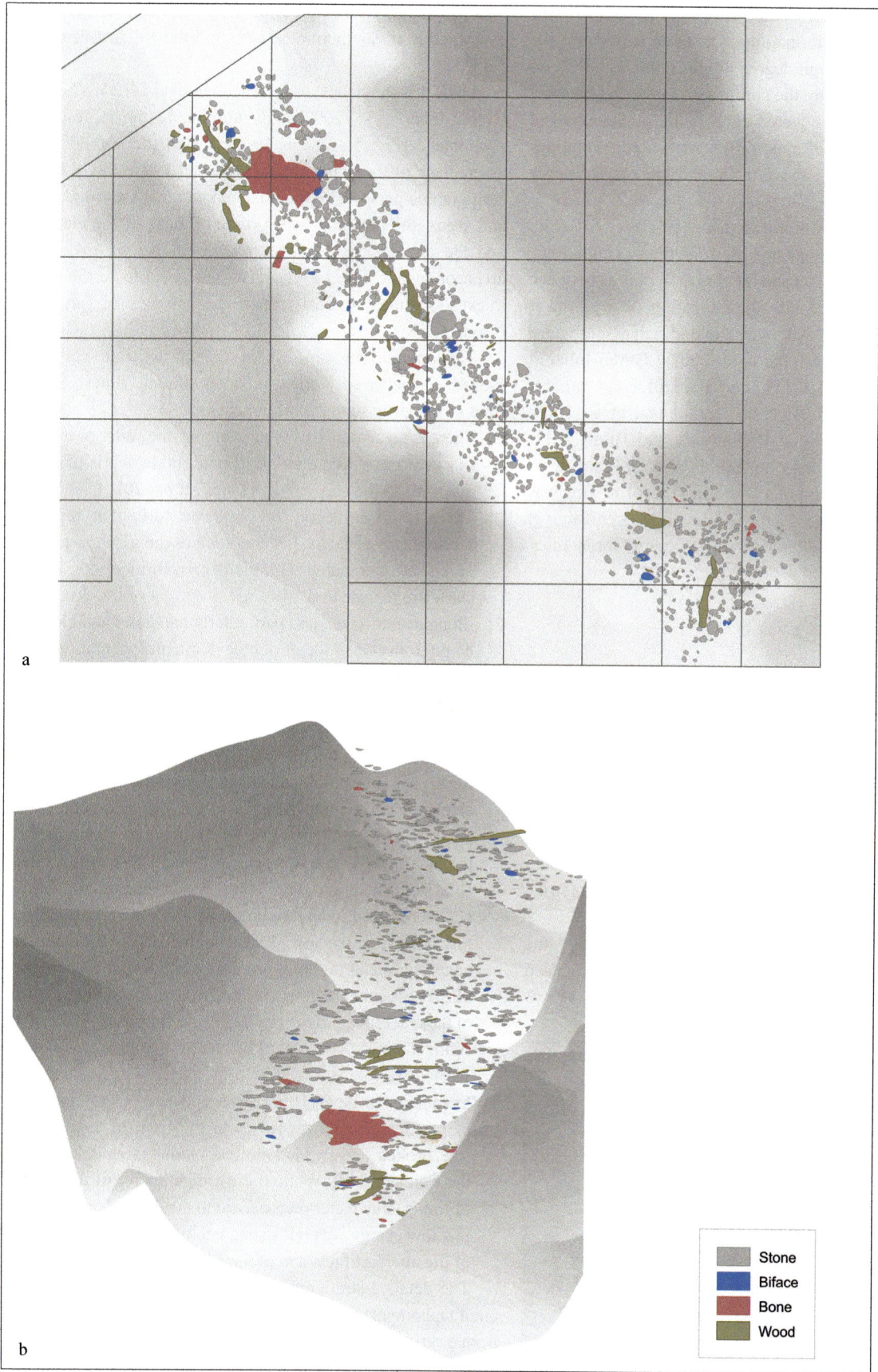

Fig. 5.35 The field map of Layer II-6 Level 1: a) in plan view, b) in its original tilted position; view to southeast

5.4 Area B

elaphus, *Dama* sp., *Gazella* cf. *G. gazella*, and *Palaeoloxodon antiquus* (Goren-Inbar et al. 2009) (for micromammals, see Rabinovich and Biton 2011).

Level 1 was extremely rich in finds and primarily in lithic artifacts (Tables 5.34–5.35). Interestingly, this extensive occupation has yielded insight into some of the most important issues studied at the site, including the production of bifacial tools (Goren-Inbar et al. 1992, 2000; Brande and Saragusti 1996; Goren-Inbar and Saragusti 1996; Madsen and Goren-Inbar 2004), the presence of edible nuts (Goren-Inbar et al. 2002a), and the elephant skull and its associated wood and lithic artifacts (Goren-Inbar et al. 1994). While most of these topics will be dealt with elsewhere in this volume (Chapters 6–9), the elephant skull and its associated finds merit some additional description here.

The elephant skull, that of a female or a young male, was found in association with a number of items, an arrangement that is considered to document the arrangement of these objects at the time of discard. The items in this association included the following (for further details see Goren-Inbar et al. 1994):

a) A skull (89 x 52 cm) (Fig. 5.36) of a straight-tusked elephant, *Palaeoloxodon antiquus* (Falc. and Caut.), and isolated bone fragments belonging to it, including skull fragments (unidentified parts of the cranium), two fragments that may represent parts of the zygomatic arch, two portions of premaxillary bone (conjoined with the skull), thousands of small fragments of which about half belong to the inner lining of the braincase and the other half cannot be assigned to a particular anatomical element, the left occipital condyle, tusk fragments, molar fragments, and a fragment of a limb bone. An elephant stylohyoid of a superior-posterior length of 135 mm (Shoshani et al. 2001) that was discovered in Layer II-1 in the proximity of Layer II-6 may belong to the same animal.

b) A wooden log (common oak, *Quercus calliprinos* Webb.), 106 cm long and 13 cm in diameter (Fig. 5.36).

c) A giant basalt Levallois core (32 x 26 x 17 cm; working edge 89 cm long; weight 16.5 kg) with an entirely flaked debitage surface and a limited flaked surface on the opposite face (Goren-Inbar et al. 1994; Madsen and Goren-Inbar 2004 and detailed description therein; Goren-Inbar 2011; Goren-Inbar et al. 2011) (Fig. 5.36).

d) A giant basalt boulder of pebble morphology interpreted as a percussor.

e) A variety of stone artifacts, mainly basalt but some flint items as well, were found in association with the skull.

Table 5.34 Counts and frequencies (%)* of the lithic assemblage of Layer II-6 Level 1

	Flint	Basalt	Limestone	Total
	%(N)	%(N)	%(N)	N
TOOLS#	24.22 (264)	13.40 (147)	24.39 (10)	421
WASTE				
Debitage	49.72 (542)	74.02 (812)	43.90 (18)	1372
Cores	8.99 (98)	1.00 (11)	7.32 (3)	112
Non-flakes	17.06 (186)	11.58 (127)	24.39 (10)	323
Subtotal	1,090	1,097	41	2,228
Microartifacts	56,066	2,896	1,674	60,636
Not analyzed FFT	33	33	2	68
Natural nodules	2,463	3,957	380	6,800
Total	59,652	7,983	2,097	69,732

* % calculated from the subtotal
# flake tools and core tools

Table 5.35 Counts and frequencies (%) of the typological composition of Layer II-6 Level 1

	Flint	Basalt	Limestone	Total
	%(N)	%(N)	%(N)	N
TOOLS				
Flake tools				
Side-scrapers	2.20 (24)	1.28 (14)	2.44 (1)	39
End-scrapers	2.29 (25)	0.18 (2)	-	27
Burins	0.28 (3)	0.09 (1)	-	4
Borers	3.30 (36)	0.09 (1)	-	37
Notches	7.71 (84)	1.73 (19)	2.44 (1)	104
Denticulates	2.29 (25)	1.37 (15)	-	40
Truncations	0.83 (9)	-	-	9
Retouched flakes	4.95 (54)	1.00 (11)	4.88 (2)	67
Massive scrapers	-	1.00 (11)	2.44 (1)	12
Varia	-	0.09 (1)	-	1
Core tools				
Bifaces				
Handaxes	0.37 (4)	4.92 (54)	9.76 (4)	62
Cleavers	-	1.55 (17)	-	17
Biface preforms	-	0.09 (1)	2.44 (1)	2
Others				
Chopping tools	0.92 (10)	-	4.88 (2)	12
Percussors	0.09 (1)	3.37 (37)	9.76 (4)	42
Anvils	-	0.73 (8)	-	8
WASTE				
Flakes				
Flakes	45.05 (491)	72.84 (799)	43.90 (18)	1,308
Blades	0.46 (5)	-	-	5
Éclats de taille de biface	1.83 (20)	0.55 (6)	-	26
Biface sharpening flakes	0.09 (1)	0.09 (1)	-	2
Core waste	2.29 (25)	0.55 (6)	-	31
Non-flakes				
Angular fragments	11.38 (124)	4.28 (47)	-	171
Indeterminate waste	-	0.09 (1)	-	1
Cores				
Levallois cores	0.37 (4)	0.09 (1)	-	5
Discoidal cores	0.18 (2)	-	-	2
Cores on flakes	0.83 (9)	0.09 (1)	-	10
Core varia	4.95 (54)	0.27 (3)	2.44 (1)	58
Amorphous cores	1.74 (19)	0.09 (1)	-	20
Giant cores	-	0.46 (5)	-	5
Modified	5.60 (61)	3.10 (34)	14.63 (6)	101
Total	1,090	1,097	41	2,228

Fig. 5.36 The elephant skull and its associated artifacts from Layer II-6 Level 1: a) detailed view of the premaxillary part (ventral face) of the elephant skull (scale 1 m), b) a fragment of the premaxillary bone (conjoined with the skull), c) the elephant skull and associated finds; note the heavily fractured bone mass (scale 1 m), d) the wooden log and associated lithic artifacts, e) the wooden log overlying flint artifacts, f) the elephant skull entirely exposed and ready for plaster of Paris casting; note that a small part of a giant core is visible, the rest underlies the skull (scale 0.5 m), g) the log after final exposure and cleaning, h) the elephant skull after total exposure with close-up view of the trunk foramen (scale 20 cm)

5.4 Area B

Apart from the two large basalt items described above, there were bifacial tools (both handaxes and cleavers) and other flakes and flake tools. Of special importance is the presence of giant cores in close vicinity to the skull, visible in all photographs and drawings (Figs 5.34–5.36) and more frequent in this level than anywhere else at the site (Chapter 7). Lithic artifacts were found under, on, and around the elephant skull. There is no doubt that this association reflects the original composition and that taphonomic processes were extremely limited in amplitude.

The spatial configuration of the items has been interpreted as documenting the driving, hunting, and carcass processing of the elephant. This interpretation is based on the following components: the presence of the wooden log under the skull, the presence of the two large basalt objects (percussor and giant core) under the complete premaxillaries, and the intentional removal of the "… entire palatal and basicranial region of the skull, including the molars and their alveoli… and the remaining surface is totally crushed" (Goren-Inbar et al. 1994:100).

In addition: "An unusual feature is a damaged area approximately 30 cm long by 6 cm across, just below the nasal opening (Pl. II: 3). In this area, the surface of the bone has been removed to a depth of between one and three centimetres. The feature is very localized and has clear edges unlike normal pre- or postdepositional damage. It is as though bone had been removed by percussion along a narrow strip of the cranium" (Goren-Inbar et al. 1994:100–101) (Fig. 5.36).

Layer VI-14

Trench VI (Fig. 5.37) was dug in 1996 to provide further clarification of various structural and sedimentological issues (Feibel, additional studies under way). The trench sediments revealed a tilted sequence of layers, among which the majority consisted of clays and only a few were rich in malacofauna. Within the spoils of Trench VI, several handaxes were found (eight on basalt and one on flint) as well as a single basalt waste artifact. It was only during the field season of 1997, during cleaning of the easternmost end of the trench that additional layers were exposed (Layers VI-13 and 14), and an archaeological horizon (Layer VI-14) was identified. Following this discovery, excavations were initiated, limited to the ca. 1 m width of Trench VI (Fig. 5.25, Figs. 5.37–5.38) and following the strike and dip (35°). Like Layer II-6 Level 1, this layer turned out to be very rich in lithic artifacts (Tables 5.36–5.37) and fragments of wood, particularly long ones. Layer VI-14 is the easternmost lateral continuation of Layer II-6 Level 1 in the study area, located some 17.03 m from Trench II where it was first identified.

Layer VI-14 is 60 cm thick (Fig. 3.13), although only its

Fig. 5.37 General view of Trench VI: a) looking west, b) looking east with Layer VI-14 in view

5.4 Area B

Fig. 5.38 Close-up view of Layer VI-14

uppermost surface was exposed and excavated (Fig. 5.38). It is also characterized by the presence of large basalt stones. A large concretion was exposed, similar in composition and characteristics to others found in Layer II-6.

Layer II-6 Level 2

This level was excavated during several field seasons. It is 12 cm thick and stretches along the strike similarly to Level 1 (Fig. 5.39–5.40), exhibiting the largest excavated surface exposure at the site (Fig. 5.41). This level was selected for a detailed study that examined the association between a hearth and different assemblages (lithic, faunal, and botanical). The results of this analysis suggest that Level 2 preserves discrete patterning of various categories of finds, providing evidence of spatial organization of hominin activities (Alperson-Afil et al. 2009). These patterns, which were not observed during excavation, occur in other archaeological horizons at the site as well (Alperson-Afil additional studies under way). The spatial patterns of Level 2 provide important information on site formation processes at GBY. They show that minimal involvement of taphonomic processes is recorded in Layer II-6.

Tables 5.38–5.39 present the lithic assemblages of Layer II-6 Level 2. Remains of medium-sized and large mammals include Carnivore und., *Sus scrofa*, *Dama* sp., and *Palaeoloxodon antiquus* (Goren-Inbar et al. 2009) (for micromammals, see Rabinovich and Biton 2011).

Table 5.37 Counts and frequencies (%) of the typological composition of Layer VI-14

	Flint	Basalt	Limestone	Total
	%(N)	%(N)	%(N)	N
TOOLS				
Flake tools				
Side-scrapers	3.33 (1)	-	-	1
End-scrapers	3.33 (1)	-	-	1
Borers	3.33 (1)	-	-	1
Notches	13.35 (4)	1.01 (1)	-	5
Denticulates	6.67 (2)	-	-	2
Retouched flakes	6.67 (2)	1.01 (1)	-	3
Massive scrapers	-	2.02 (2)	-	2
Core tools				
Bifaces				
Handaxes	3.33 (1)	9.09 (9)	-	10
Others				
Percussors	-	3.03 (3)	-	3
WASTE				
Flakes				
Flakes	50.00 (15)	80.80 (80)	100.00 (2)	97
Blades	-	2.02 (2)	-	2
Éclats de taille de biface	3.33 (1)	-	-	1
Biface sharpening flakes	-	1.01 (1)	-	1
Cores				
Levallois cores	3.33 (1)	-	-	1
Cores on flakes	3.33 (1)	-	-	1
Total	30	99	2	131

Table 5.36 Counts and frequencies (%)* of the lithic assemblage of Layer VI-14

	Flint	Basalt	Limestone	Total
	%(N)	%(N)	%(N)	N
TOOLS#	37.93 (11)	4.44 (4)	-	15
WASTE				
Debitage	55.17 (16)	92.22 (83)	100.00 (2)	101
Cores	6.90 (2)	-	-	2
Non-flakes	-	3.33 (3)	-	3
Subtotal	29	90	2	121
Microartifacts	9	1	-	10
Natural nodules	-	-	-	-
Total	38	91	2	131

* % calculated from the subtotal
# flake tools and core tools

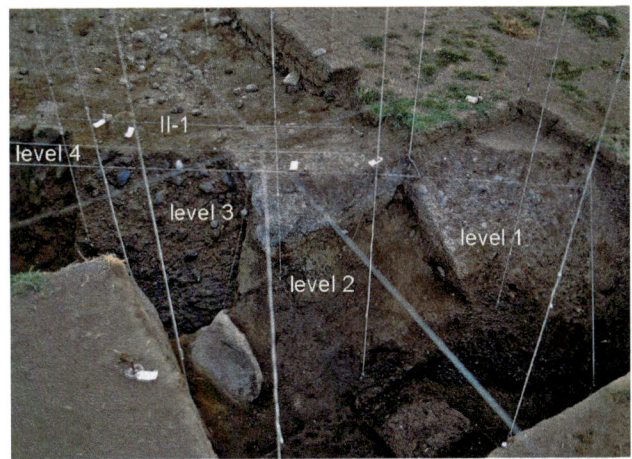

Fig. 5.39 Close-up view of Layer II-6 Levels 1–4

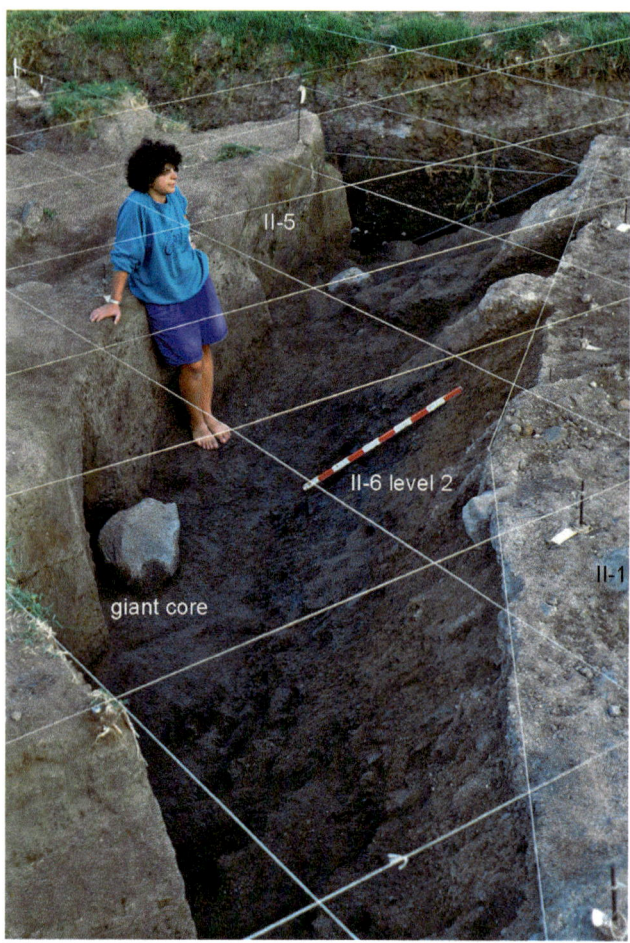

Fig. 5.40 Close-up view of Layer II-6 Level 2 looking northeast

Layer II-6 Level 3

This level is some 14 cm thick and resembles the overlying Level 2 in its sedimentological character. It is, however, very

Table 5.38 Counts and frequencies (%)* of the lithic assemblage of Layer II-6 Level 2

	Flint	Basalt	Limestone	Total
	%(N)	%(N)	%(N)	N
TOOLS[#]	26.99 (122)	9.00 (89)	-	211
WASTE				
Debitage	43.36 (196)	76.74 (759)	56.52 (13)	968
Cores	9.96 (45)	1.52 (15)	8.70 (2)	62
Non-flakes	19.69 (89)	12.74 (126)	34.78 (8)	223
Subtotal	452	989	23	1,464
Microartifacts	73,655	3,930	2,184	79,769
Not analyzed FFT	70	125	3	198
Natural nodules	955	880	107	1,942
Total	75,132	5,924	2,317	83,373

* % calculated from the subtotal
[#] flake tools and core tools

Table 5.39 Counts and frequencies (%) of the typological composition of Layer II-6 Level 2

	Flint	Basalt	Limestone	Total
	%(N)	%(N)	%(N)	N
TOOLS				
Flake tools				
Side-scrapers	2.88 (13)	0.20 (2)	-	15
End-scrapers	2.88 (13)	-	-	13
Burins	-	0.30 (3)	-	3
Borers	3.54 (16)	-	-	16
Notches	7.96 (36)	1.31 (13)	-	49
Denticulates	1.55 (7)	0.51 (5)	-	12
Truncations	0.88 (4)	-	-	4
Retouched flakes	5.75 (26)	1.82 (18)	-	44
Massive scrapers	-	1.11 (11)	-	11
Varia	0.88 (4)	0.40 (4)	-	8
Core tools				
Bifaces				
Handaxes	0.66 (3)	2.02 (20)	-	23
Cleavers	-	1.01 (10)	-	10
Biface preforms	-	0.30 (3)	-	3
Others				
Chopping tools	2.21 (10)	0.20 (2)	8.70 (2)	14
Percussors	-	4.55 (45)	26.09 (6)	51
Pitted anvils	-	0.10 (1)	-	1
Anvils	-	0.71 (7)	-	7
WASTE				
Flakes				
Flakes	40.04 (181)	75.33 (745)	56.52 (13)	939
Blades	-	0.30 (3)	-	3
Éclats de taille de biface	0.88 (4)	0.71 (7)	-	11
Biface sharpening flakes	0.44 (2)	0.10 (1)	-	3
Non-flakes				
Core waste	1.99 (9)	0.30 (3)	-	12
Angular fragments	10.84 (49)	3.03 (30)	-	79
Indeterminate waste	-	0.10 (1)	-	1
Cores				
Levallois cores	0.88 (4)	-	-	4
Discoidal cores	0.44 (2)	-	-	2
Globular cores	0.22 (1)	-	-	1
Cores on flakes	1.33 (6)	0.10 (1)	-	7
Core varia	3.32 (15)	-	-	15
Amorphous cores	1.55 (7)	0.40 (4)	-	11
Giant cores	-	0.81 (8)	-	8
Modified	8.85 (40)	4.25 (42)	8.70 (2)	84
Total	452	989	23	1,464

different from the underlying Level 4, which is considerably denser and exhibits clear dominance of large basalt artifacts (Figs. 5.39, 5.42–5.43).

Remains of medium-sized and large mammals include *Hippopotamus amphibius*, *Dama* sp., and *Palaeoloxodon antiquus* (Goren-Inbar et al. 2009) (for micromammals, see Rabinovich and Biton 2011). The lithic inventory of Layer II-6 Level 3 is presented in Tables 5.40–5.41.

5.4 Area B

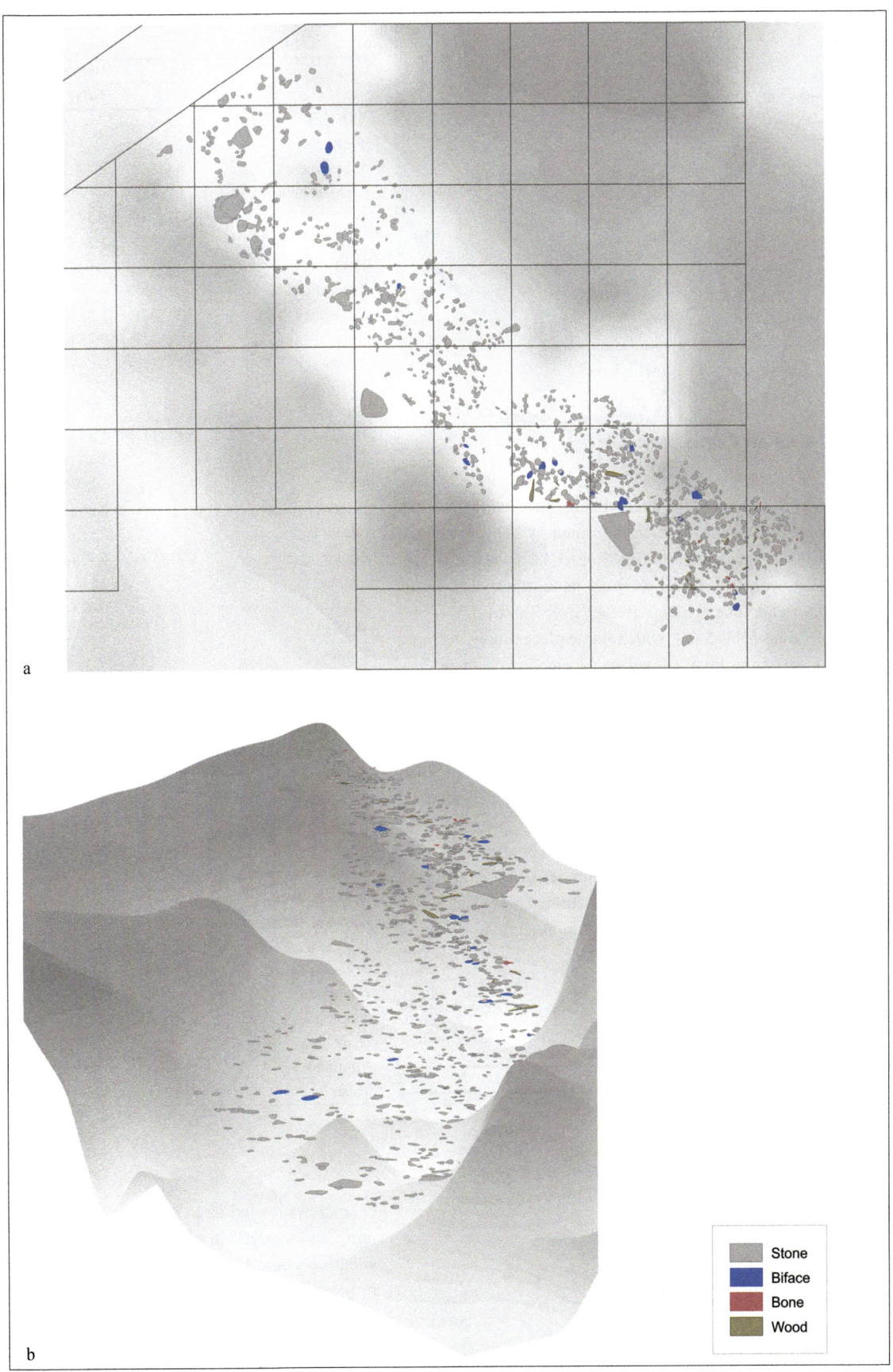

Fig. 5.41 Field map of Layer II-6 Level 2: a) in plan view, b) in its original tilted position; view to southeast

Fig. 5.42 Close-up view of Layer II-6 Level 3

Layer II-6 Level 4 and Level 4b

Levels 4 (Figs. 5.44–5.48) and 4b are particularly distinct archaeological horizons, as each is formed by densely aligned basalt artifacts ("pavement"). Figs 5.45–5.48 present different sectors of Layer II-6 Level 4 from its northwestern to its southeastern end. The artifacts (Level L-4: Tables 5.42–5.43; Level 4b: Tables 5.44–5.45), with a single piece usually forming the thickness of the level (Level 4: 13 cm; Level 4b: 6 cm), uniformly cover the entire exposed surface without sediment patches between them (Figs. 5.49–5.50), and are therefore highly visible from a sedimentological point of view. These "biface-paved" surfaces of Levels 4 (Goren-Inbar and Saragusti 1996) and 4b are a typical feature of the African Acheulian Complex.

Level 4b is much smaller in excavated volume than Level 4, and the same is true for the total number of lithic artifacts. Level 4b is limited in extent and immediately underlies Level 4, while resembling it in its lithic assemblages. These

Table 5.40 Counts and frequencies (%)* of the lithic assemblage of Layer II-6 Level 3

	Flint %(N)	Basalt %(N)	Limestone %(N)	Total N
TOOLS[#]	28.33 (166)	6.96 (56)	6.06 (2)	224
WASTE				
Debitage	40.44 (237)	83.98 (676)	66.67 (22)	935
Cores	14.68 (86)	1.12 (9)	6.06 (2)	97
Non-flakes	16.55 (97)	7.95 (64)	21.21 (7)	168
Subtotal	586	805	33	1,424
Microartifacts	99,434	2,771	1,636	103,841
Not analyzed FFT	418	513	7	938
Natural nodules	1,420	854	236	2,510
Total	101,858	4,943	1,912	108,713

* % calculated from the subtotal
[#] flake tools and core tools

Table 5.41 Counts and frequencies (%) of the typological composition of Layer II-6 Level 3

	Flint %(N)	Basalt %(N)	Limestone %(N)	Total N
TOOLS				
Flake tools				
Side-scrapers	4.44 (26)	0.37 (3)	-	29
End-scrapers	3.07 (18)	-	3.03 (1)	19
Borers	4.27 (25)	0.37 (3)	-	28
Notches	5.46 (32)	1.24 (10)	3.03 (1)	43
Denticulates	2.39 (14)	0.62 (5)	-	19
Truncations	1.02 (6)	-	-	6
Retouched flakes	7.34 (43)	0.87 (7)	-	50
Massive scrapers	-	0.25 (2)	-	2
Varia	0.17 (1)	-	-	1
Core tools				
Bifaces				
Handaxes	0.17 (1)	1.37 (11)	-	12
Cleavers	-	1.49 (12)	-	12
Picks	-	0.12 (1)	-	1
Biface preforms	-	0.25 (2)	-	2
Other				
Chopping tools	1.71 (10)	0.12 (1)	3.03 (1)	12
Percussors	-	2.11 (17)	-	17
Anvils	-	0.62 (5)	-	5
WASTE				
Flakes				
Flakes	36.86 (216)	82.61 (665)	63.64 (21)	902
Blades	0.17 (1)	0.12 (1)	3.03 (1)	3
Éclats de taille de biface	0.85 (5)	0.75 (6)	-	11
Biface sharpening flakes	0.68 (4)	0.12 (1)	-	5
Core waste	1.88 (11)	0.37 (3)	-	14
Non-flakes				
Angular fragments	6.14 (36)	1.24 (10)	3.03 (1)	47
Cores				
Levallois cores	0.85 (5)	-	-	5
Discoidal cores	0.17 (1)	-	-	1
Globular cores	0.34 (2)	-	-	2
Prismatic cores	0.17 (1)	-	-	1
Cores on flakes	2.05 (12)	0.12 (1)	-	13
Core varia	5.29 (31)	0.25 (2)	-	33
Amorphous cores	4.10 (24)	-	3.03 (1)	25
Giant cores	-	0.62 (5)	-	5
Modified	10.41 (61)	3.98 (32)	18.18 (6)	99
Total	586	805	33	1,424

characteristics led us to consider the two levels to be a single unit, subdivided stratigraphically into two horizons. This situation could be interpreted either as a single level, thicker in places, or as two discrete horizons of different occupation events, an issue that will be resolved only by additional analyses (Feibel, additional studies under way).

At first glance, the dominance of basalt artifacts in this levels is striking (Figs. 5.44–5.48). However, analysis of

5.4 Area B

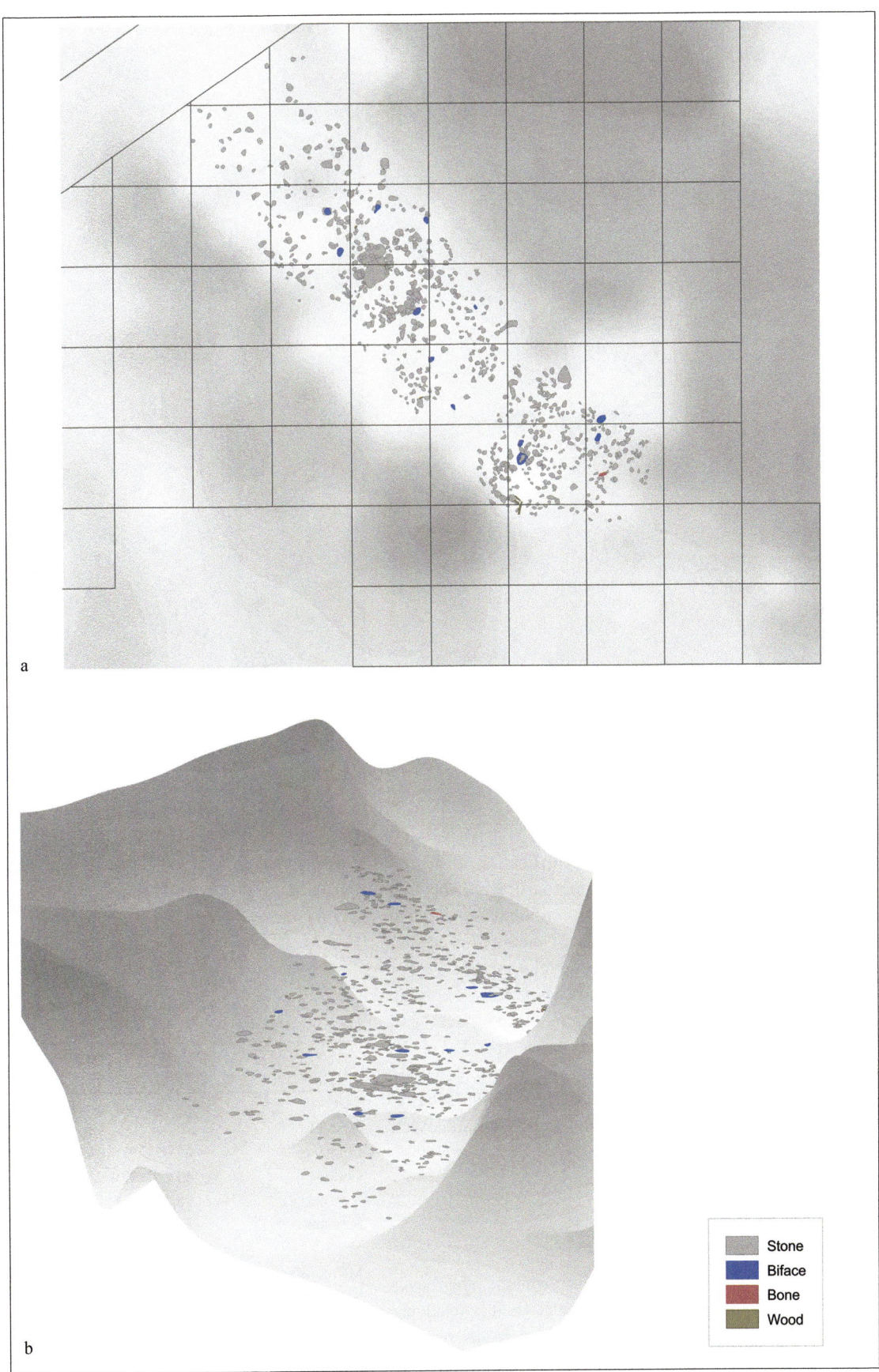

Fig. 5.43 Field map of Layer II-6 Level 3: a) in plan view, b) in its original tilted position; view to southeast

Fig. 5.44 View of Layer II-6 Level 4

Fig. 5.45 Close-up view of Layer II-6 Level 4

the small fraction resulted in the identification of a high frequency of flint and limestone microartifacts. The two levels are also characterized by a minimal appearance of wood and fossil bones. Remains of medium-sized and large mammals found in Layer II-6 Level 4 include *Vulpes* sp., Carnivore und., *Hippopotamus amphibius*, Cervidae sp., *Dama* sp., *Bos* sp., Bovidae gen. et sp. indet., and *Palaeoloxodon antiquus*. Those of Layer II-6 Level 4b include *Cervus* cf. *elaphus*, Cervidae sp., *Bos* sp., Bovidae gen. et indet., and *Palaeoloxodon antiquus* (Goren-Inbar et al. 2009) (for micromammals of both levels, see Rabinovich and Biton 2011).

Layer II-6 Level 5

Level 5 is only 9 cm thick. In comparison with the other Layer II-6 levels, particularly the overlying Level 4b, it is relatively sparse in lithic finds (Tables 5.46–5.47) and faunal remains (Figs. 5.51–5.52). The latter include *Sus scrofa*, *Hippopotamus amphibius*, *Cervus* cf. *elaphus*, *Dama* sp., and Bovini indet. (Goren-Inbar et al. 2009 [ISF]) (for micromammals, see Rabinovich and Biton 2011).

Table 5.42 Counts and frequencies (%)* of the lithic assemblage of Layer II-6 Level 4

	Flint	Basalt	Limestone	Total
	%(N)	%(N)	%(N)	N
TOOLS#	29.36 (202)	25.29 (282)	4.76 (2)	486
WASTE				
Debitage	37.65 (259)	63.14 (704)	40.48 (17)	980
Cores	14.83 (102)	0.99 (11)	2.38 (1)	114
Non-flakes	18.17 (125)	10.58 (118)	52.38 (22)	265
Subtotal	688	1,115	42	1,845
Microartifacts	109,868	7,895	4,906	122,669
Not analyzed FFT	913	478	20	1,411
Natural nodules	768	929	398	2,095
Total	112,237	10,417	5,366	128,020

* % calculated from the subtotal
# flake tools and core tools

5.4 Area B

Fig. 5.46 Close-up view of Layer II-6 Level 4

Fig. 5.47 Close-up view of Layer II-6 Level 4

Layer II-6 Level 6

Level 6, 11 cm thick, is sedimentologically similar to the overlying Level 5 and is also relatively sparse in lithic finds (Tables 5.48–5.49) and faunal remains (Figs. 5.53–5.54). The remains of medium-sized and large mammals include Carnivore und., *Sus scrofa*, *Hippopotamus amphibius*, *Cervus* cf. *elaphus*, Cervidae sp., *Dama* sp., *Gazella* cf. *G. gazella*, *Equus* sp., and *Palaeoloxodon antiquus* (Goren-Inbar et al. 2009) (for micromammals, see Rabinovich and Biton 2011).

Layer II-6 Level 7

This level is a sedimentological unit that differs quite distinctly from the other levels of Layer II-6. It attains a thickness of ca. 40 cm in the section of Trench II and in the cross-section in plan view under Layer II-1. This stratigraphic unit seems to be sedimentologically sorted upwards and is the oldest part of the complex of Layer II-6. It is composed of sediments that range from sand (in the upper part) to gravel of cobble and boulder size (in the lower part). The sand deposit is primarily comprised of broken mollusks, though in some instances they are complete.

As Level 7 was excavated during the very last seasons, it was not exposed to its full depth all along the strike, but some control over its thickness was gained through selection of two different areas. In these two "windows", situated far apart along the strike, excavation was aimed at revealing the full extent of the level's thickness (Figs. 5.55–5.58) and did indeed reach the base of this level – the contact between Layer II-6 (Level 7) and the underlying Layer II-7.

The excavation of these two "windows" (the north and south test pits) revealed a sorted depositional sequence that becomes finer towards the top. At the base, very coarse gravel overlies the clayey deposit of Layer II-7. The largest boulder measured 38x30 cm (not fully exposed) (Figs. 5.57–5.58). Above the gravel is a thin clayey layer, which is overlain by a thick series of sands (Figs. 5.57–5.58).

Methodologically, the excavations of Level 7 were divided into three parts, each excavated and mapped individually. These include the two test pits and the upper surface (excavated along the strike) between them (Figs. 5.56, 5.59).

Table 5.43 Counts and frequencies (%) of the typological composition of Layer II-6 Level 4

	Flint %(N)	Basalt %(N)	Limestone %(N)	Total N
TOOLS				
Flake tools				
Side-scrapers	4.51 (31)	0.72 (8)	2.38 (1)	40
End-scrapers	1.31 (9)	-	-	9
Burins	0.87 (6)	-	-	6
Borers	4.22 (29)	-	-	29
Notches	7.41 (51)	1.61 (18)	2.38 (1)	70
Denticulates	2.47 (17)	0.36 (4)	-	21
Truncations	1.16 (8)	-	-	8
Retouched flakes	6.69 (46)	1.52 (17)	-	63
Massive scrapers	-	0.72 (8)	-	8
Varia	0.29 (2)	0.27 (3)	-	5
Core tools				
Bifaces				
Handaxes	0.44 (3)	14.44 (161)	-	164
Cleavers	-	5.38 (60)	-	60
Trihedrals	-	0.09 (1)	-	1
Picks	-	0.09 (1)	-	1
Biface preforms	-	0.09 (1)	-	1
Others				
Chopping tools	1.60 (11)	0.18 (2)	-	13
Percussors	-	4.13 (46)	26.19 (11)	57
Pitted anvils	-	0.09 (1)	-	1
Anvils	-	0.63 (7)	-	7
WASTE				
Flakes				
Flakes	33.28 (229)	60.99 (680)	35.71 (15)	924
Blades	0.44 (3)	0.27 (3)	2.38 (1)	7
Éclats de taille de biface	1.31 (9)	1.43 (16)	-	25
Biface sharpening flakes	0.44 (3)	0.09 (1)	-	4
Core waste	2.18 (15)	0.36 (4)	2.38 (1)	20
Non-flakes				
Angular fragments	10.76 (74)	2.51 (28)	4.76 (2)	104
Indeterminate waste	-	0.27 (3)	-	3
Cores				
Levallois cores	1.74 (12)	-	2.38 (1)	13
Discoidal cores	0.29 (2)	-	-	2
Globular cores	0.15 (1)	-	-	1
Prismatic cores	0.15 (1)	-	-	1
Cores on flakes	2.33 (16)	0.09 (1)	-	17
Cores on handaxes	0.15 (1)	-	-	1
Core varia	5.09 (35)	0.09 (1)	-	36
Amorphous cores	3.34 (23)	0.18 (2)	-	25
Giant cores	-	0.45 (5)	-	5
Modified	7.41 (51)	2.96 (33)	21.43 (9)	93
Total	688	1,115	42	1,845

Fig. 5.48 Close-up view of Layer II-6 Level 4

The entire Level 7 sequence is interpreted on the basis of paleontological and paleobotanical remains and diverse taphonomic observations as a buried channel, differing from the overlying sequence of Layer II-6 Levels 1–6, which are considered to be a series of storm beaches (Goren-Inbar et al. 2000; Feibel 2001, 2004). The coarse clastic component of Level 7 is the only record of natural gravel in its primary depositional environment, unlike Levels 1–6, where coarse sediments originate from the transport of cobbles/boulders by hominins.

Level 7 also differs from Levels 1–6 in respect to other components. These include mammal bones and large quantities of organic material (fruits, seeds, and in particular wood fragments). Most of the organic material was found in the sandy component of Level 7, together with thin layers of organic material, dark in color, which were stratigraphically separated by layers of sandy deposit. The lithic inventory of Level 7 presented in Tables 5.50–5.51 includes the entire excavated volume of this level.

Remains of medium-sized and large mammals include

5.4 Area B

Table 5.44 Counts and frequencies (%)* of the lithic assemblage of Layer II-6 Level 4b

	Flint	Basalt	Limestone	Total
	%(N)	%(N)	%(N)	N
TOOLS#	38.00 (57)	16.72 (103)	25.00 (7)	167
WASTE				
Debitage	46.67 (70)	74.51 (459)	50.00 (14)	543
Cores	6.67 (10)	-	7.14 (2)	12
Non-flakes	8.67 (13)	8.77 (54)	17.86 (5)	72
Subtotal	150	616	28	794
Microartifacts	7,521	1,083	159	8,763
Not analyzed FFT	140	126	4	270
Natural nodules	219	498	42	759
Total	8,030	2,323	233	10,586

* % calculated from the subtotal
# flake tools and core tools

Table 5.45 Counts and frequencies (%) of the typological composition of Layer II-6 Level 4b

	Flint	Basalt	Limestone	Total
	%(N)	%(N)	%(N)	N
TOOLS				
Flake tools				
Side-scrapers	3.33 (5)	0.49 (3)	3.57 (1)	9
End-scrapers	3.33 (5)	-	-	5
Burins	0.67 (1)	0.16 (1)	-	2
Borers	8.67 (13)	-	-	13
Notches	8.00 (12)	0.97 (6)	3.57 (1)	19
Denticulates	5.33 (8)	0.32 (2)	3.57 (1)	11
Truncations	0.67 (1)	-	3.57 (1)	2
Retouched flakes	7.33 (11)	0.97 (6)	3.57 (1)	18
Massive scrapers	-	1.14 (7)	-	7
Varia	-	0.16 (1)	-	1
Core tools				
Bifaces				
Handaxes	0.67 (1)	8.77 (54)	7.14 (2)	57
Cleavers	-	2.92 (18)	-	18
Biface preforms	-	0.81 (5)	-	5
Others				
Chopping tools	-	-	7.14 (2)	2
Percussors	-	4.87 (30)	14.29 (4)	34
Anvils	-	0.49 (3)	-	3
WASTE				
Flakes				
Flakes	45.33 (68)	73.38 (452)	50.00 (14)	534
Blades	-	0.16 (1)	-	1
Éclats de taille de biface	0.67 (1)	0.81 (5)	-	6
Biface sharpening flakes	0.67 (1)	0.16 (1)	-	2
Non-flakes				
Angular fragments	2.67 (4)	2.76 (17)	3.57 (1)	22
Indeterminate waste	-	0.32 (2)	-	2
Cores				
Levallois cores	0.67 (1)	-	-	1
Core varia	5.33 (8)	-	-	8
Amorphous cores	0.67 (1)	-	-	1
Modified	6.00 (9)	0.32 (2)	-	11
Total	150	616	28	794

Vulpes sp., Carnivore und., *Sus scrofa*, *Hippopotamus amphibius*, *Megalocerus* sp., *Cervidae* sp., *Dama* sp., Bovini sp., *Gazella* cf. *G. gazella*, Bovidae gen. et sp. indet., *Dicerorhinus hemitoechos*, *Equus* sp., and *Palaeoloxodon antiquus* (Goren-Inbar et al. 2009) (for micrommamals, see Rabinovich and Biton 2011).

Lithic artifacts occur through the entire thickness of Level 7, even in the coarsest component of this level (the gravel), down to the contact with the clayey Layer II-7; they are more abundant in the north test pit.

The excavation of the gravel component of Level 7 is an important contribution to the understanding of GBY's depositional and archaeological record. It is the only coarse gravel at GBY that was excavated methodically and has yielded biological and cultural assemblages. In view of its geomorphologic interpretation as a large channel, the record of Level 7 makes a most valuable contribution to the understanding of the diversity and variability of these assemblages and their lake margin environment. The presence of hominin occupation in a different type of environment (a river) provides additional insight into the occupation and exploitation of different habitats, all components of the lake margin paleolandscape.

5.4.10 Layer II-6/7

Excavations of the contact Layer II-6/7 were extremely limited and focused on the north test pit of Layer II-6 Level 7. The artifacts (Tables 5.52–5.53) were found under the large clasts of the base of Layer II-6 Level 7, bedded on a black mud sediment (Layer II-7). As in other instances where archaeological remains were found in the contact of two sedimentological entities, the remains included wood segments (Goren-Inbar et al. 2002b).

5.4.11 Layer II-7 (Layer IV-7)

Layer II-7 underlies the gravel of Layer II-6 Level 7. Sedimentologically, it is composed of black mud, and it is exposed in both Trench II (the surface underlying Layer II-6/7) and Trench IV (in section). It is 60 cm thick in Trench II

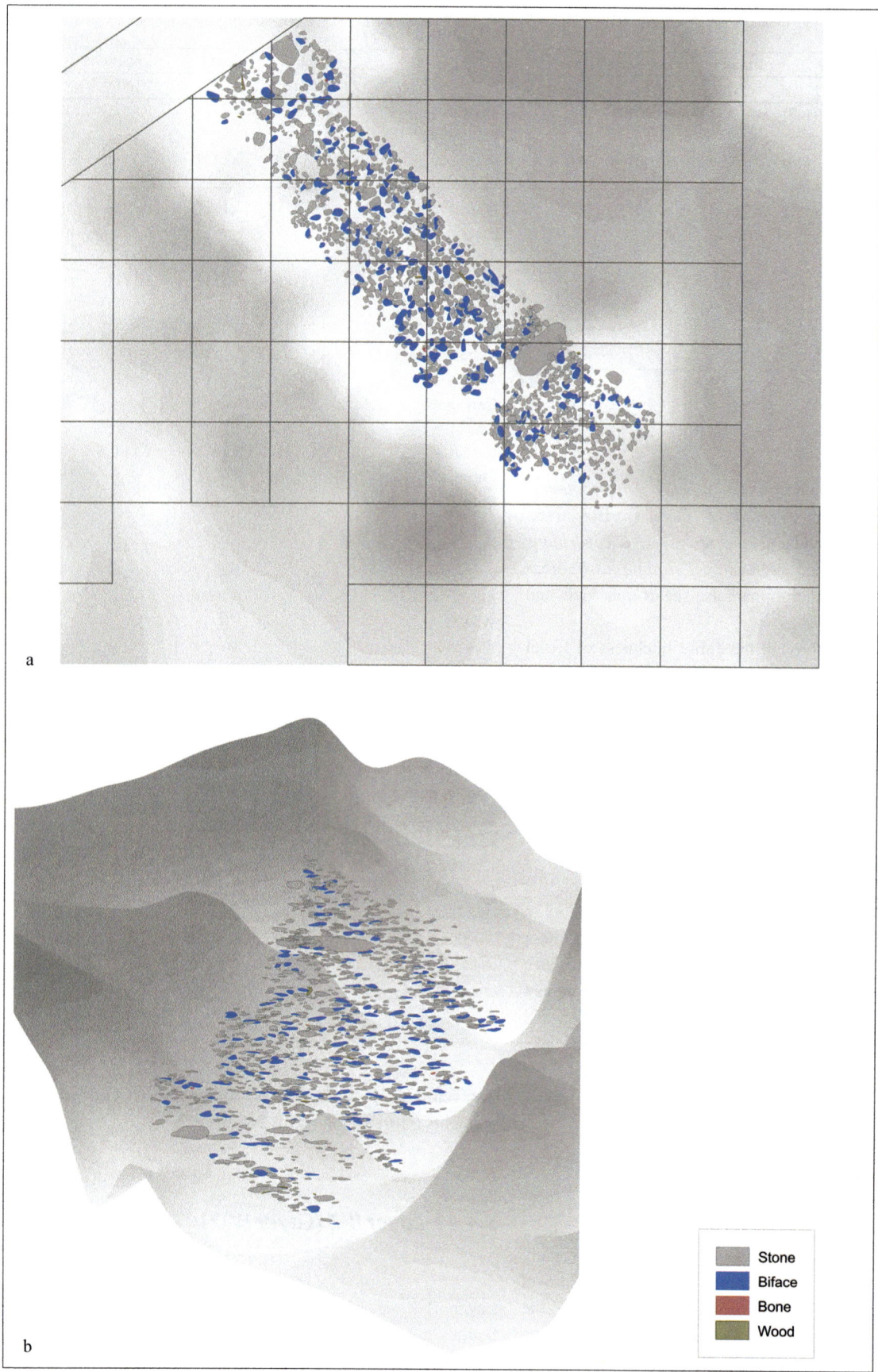

Fig. 5.49 Field map of Layer II-6 Level 4: a) in plan view, b) in its original tilted position; view to southeast

5.4 Area B

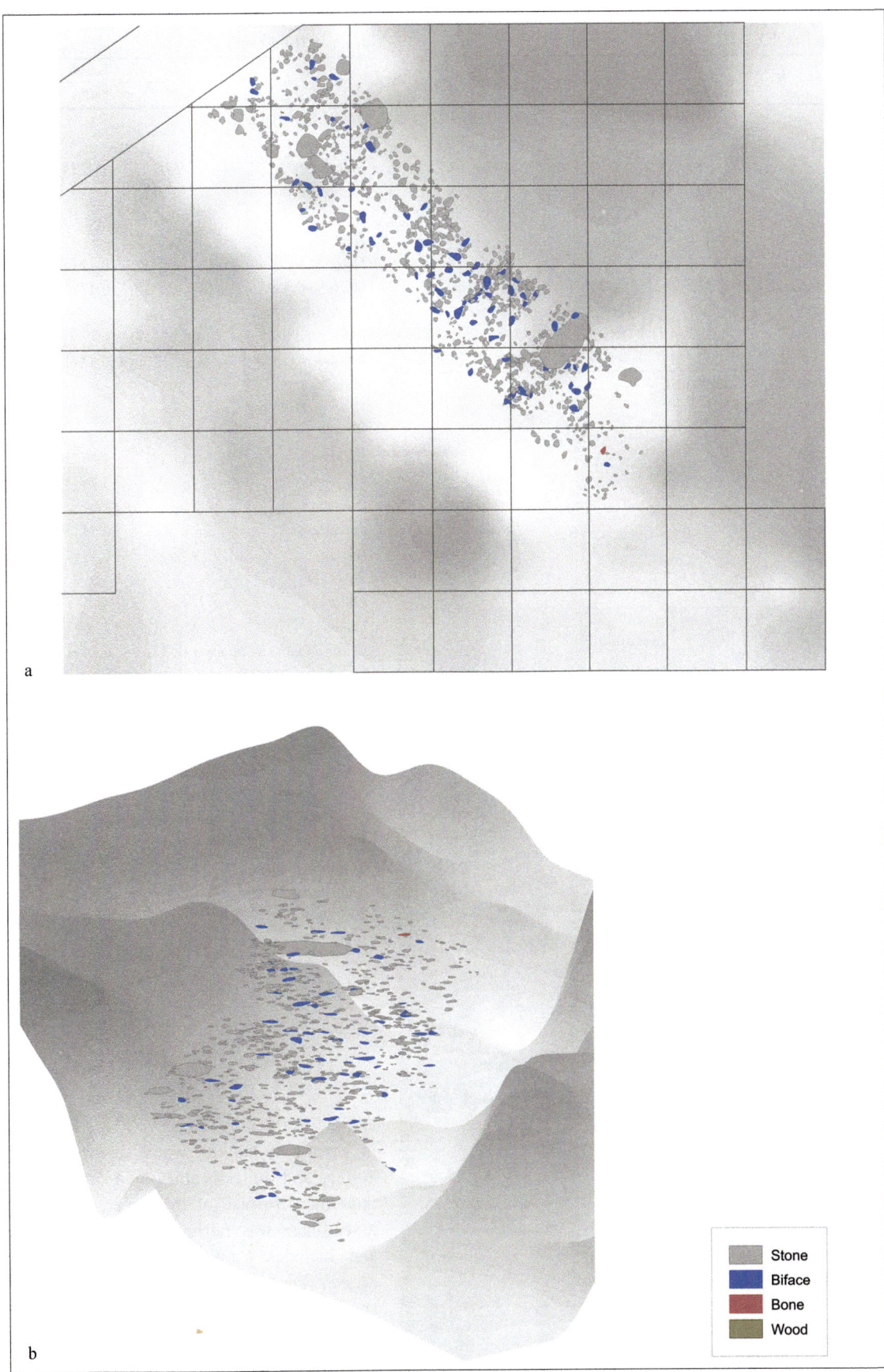

Fig. 5.50 Field map of Layer II-6 Level 4b: a) in plan view, b) in its original tilted position; view to southeast

Table 5.46 Counts and frequencies (%)* of the lithic assemblage of Layer II-6 Level 5

	Flint %(N)	Basalt %(N)	Limestone %(N)	Total N
TOOLS#	40.30 (81)	11.68 (25)	6.90 (2)	108
WASTE				
Debitage	38.31 (77)	76.64 (164)	79.31 (23)	264
Cores	10.95 (22)	1.87 (4)	-	26
Non-flakes	10.45 (21)	9.81 (21)	13.79 (4)	46
Subtotal	201	214	29	444
Microartifacts	32,565	2,570	2,047	37,182
Not analyzed FFT	125	140	10	275
Natural nodules	420	882	155	1,457
Total	33,311	3,806	2,241	39,358

* % calculated from the subtotal
# flake tools and core tools

Table 5.47 Counts and frequencies (%) of the typological composition of Layer II-6 Level 5

	Flint %(N)	Basalt %(N)	Limestone %(N)	Total N
TOOLS				
Flake tools				
Side-scrapers	5.97 (12)	1.87 (4)	-	16
End-scrapers	2.49 (5)	-	-	5
Burins	1.00 (2)	-	-	2
Borers	5.47 (11)	-	-	11
Notches	7.46 (15)	0.47 (1)	3.45 (1)	17
Denticulates	2.49 (5)	0.47 (1)	-	6
Truncations	1.00 (2)	-	-	2
Retouched flakes	11.94 (24)	-	-	24
Massive scrapers	-	0.47 (1)	-	1
Varia	1.49 (3)	-	3.45 (1)	4
Core tools				
Bifaces				
Handaxes	1.00 (2)	4.67 (10)	-	12
Cleavers	-	3.74 (8)	-	8
Others				
Percussors	-	6.54 (14)	6.90 (2)	16
Anvils	-	0.47 (1)	-	1
WASTE				
Flakes				
Flakes	37.31 (75)	70.09 (150)	79.31 (23)	248
Éclats de taille de biface	1.00 (2)	6.54 (14)	-	16
Non-flakes				
Angular fragments	6.47 (13)	1.40 (3)	3.45 (1)	17
Indeterminate waste	1.00 (2)	-	-	2
Cores				
Levallois cores	1.00 (2)	-	-	2
Globular cores	0.50 (1)	0.93 (2)	-	3
Cores on flakes	1.00 (2)	-	-	2
Core varia	6.47 (13)	-	-	13
Amorphous cores	1.99 (4)	-	-	4
Giant cores	-	0.93 (2)	-	2
Modified	2.99 (6)	1.40 (3)	3.45 (1)	10
Total	201	214	29	444

Fig. 5.51 Close-up view of Layer II-6 Level 5

and 2.4 m thick in Trench IV. This layer was not excavated in Trench II and its wood segments (Goren-Inbar et al. 2002b) and lithics (Tables 5.54–5.55) derive only from the uppermost surface of this layer. In Trench IV the faunal assemblage was retrieved from the section of the trench and includes Carnivore und., *Hippopotamus amphibius*, *Megalocerus* sp., *Cervus* cf. *elaphus*, Cervidae sp., *Dama* sp., Bovini indet., Bovidae gen. et sp. indet., *Gazella* cf. *G. gazella*, Caprini, *Dicerorhinus* cf. *hemitoechos*, *Equus* sp., *Equus* cf. *africanus*, and *Palaeoloxodon antiquus* (Goren-Inbar et al. 2009).

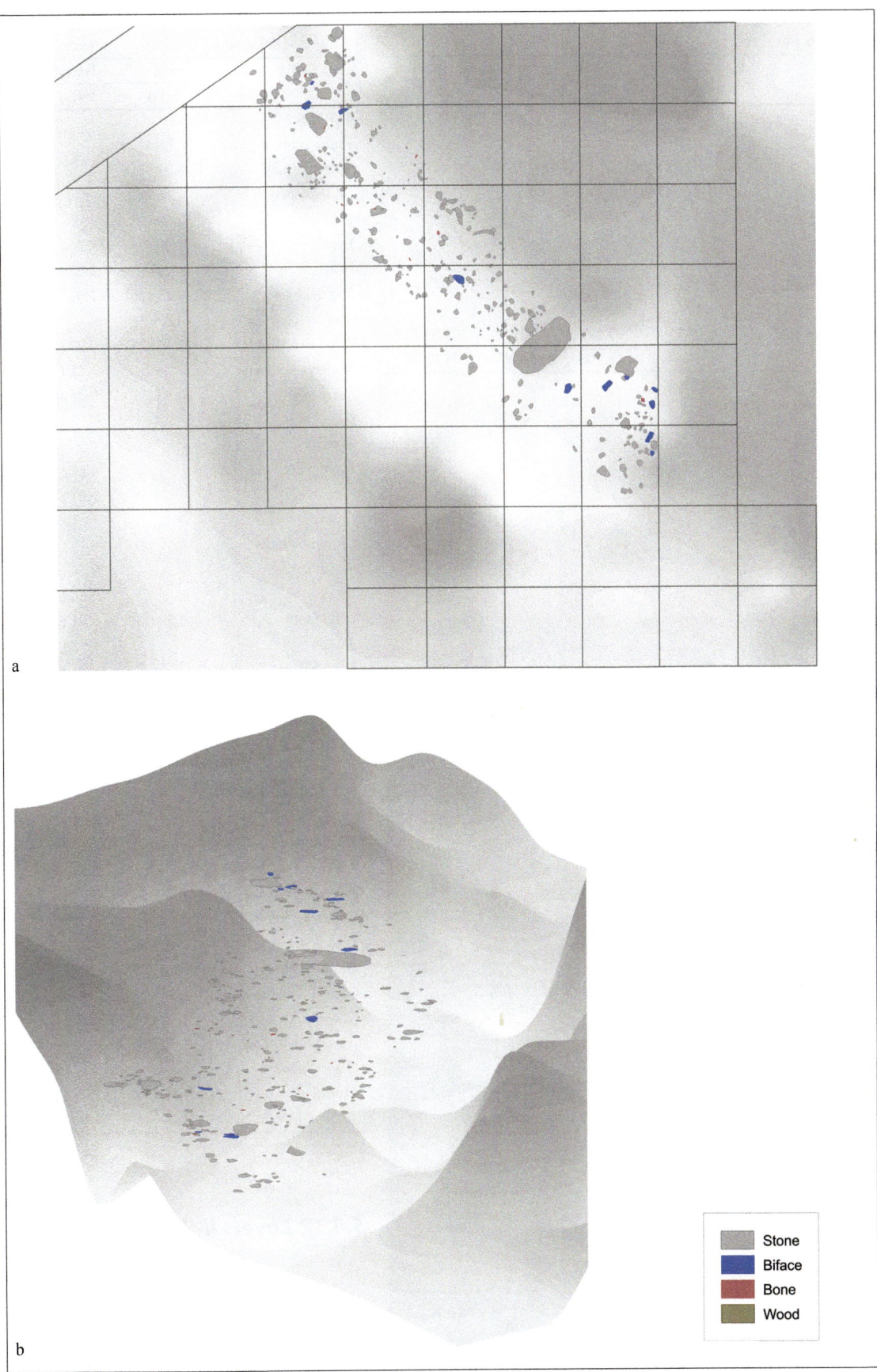

Fig. 5.52 Field map of Layer II-6 Level 5: a) in plan view, b) in its original tilted position; view to southeast

Table 5.48 Counts and frequencies (%)* of the lithic assemblage of Layer II-6 Level 6

	Flint %(N)	Basalt %(N)	Limestone %(N)	Total N
TOOLS[#]	41.72 (141)	12.53 (49)	10.34 (3)	193
WASTE				
Debitage	41.72 (141)	74.17 (290)	75.86 (22)	453
Cores	12.43 (42)	1.53 (6)	-	48
Non-flake	4.14 (14)	11.76 (46)	13.79 (4)	64
Subtotal	338	391	29	758
Microartifacts	12,619	522	384	13,525
Not analyzed FFT	187	139	8	334
Natural nodules	497	2,238	217	2,952
Total	13,641	3,290	638	17,569

* % calculated from the subtotal
[#] flake tools and core tools

Table 5.49 Counts and frequencies (%) of the typological composition of Layer II-6 Level 6

	Flint %(N)	Basalt %(N)	Limestone %(N)	Total N
TOOLS				
Flake tools				
Side-scrapers	6.21 (21)	-	-	21
End-scrapers	2.66 (9)	-	-	9
Burins	0.59 (2)	-	-	2
Borers	7.40 (25)	0.26 (1)	-	26
Notches	11.54 (39)	1.28 (5)	-	44
Denticulates	4.73 (16)	1.02 (4)	-	20
Truncations	0.89 (3)	-	-	3
Retouched flakes	7.40 (25)	0.77 (3)	6.90 (2)	30
Massive scrapers	-	1.28 (5)	-	5
Varia	0.30 (1)	0.26 (1)	-	2
Core tools				
Bifaces				
Handaxes	-	4.60 (18)	-	18
Cleavers	-	2.81 (11)	3.45 (1)	12
Others				
Chopping tools	0.59 (2)	-	-	2
Subspheroids	-	0.26 (1)	-	1
Discs	-	0.26 (1)	-	1
Percussors	0.59 (2)	6.91 (27)	13.79 (4)	33
Anvils	-	1.53 (6)	-	6
WASTE				
Flakes				
Flakes	37.87 (128)	71.61 (280)	75.86 (22)	430
Blades	0.59 (2)	0.26 (1)	-	3
Éclats de taille de biface	1.78 (6)	1.79 (7)	-	13
Biface sharpening flakes	0.89 (3)	-	-	3
Core waste	0.59 (2)	0.51 (2)	-	4
Non-flakes				
Angular fragments	1.78 (6)	1.53 (6)	-	12
Cores				
Levallois cores	0.89 (3)	-	-	3
Discoidal cores	0.30 (1)	-	-	1
Globular cores	0.89 (3)	-	-	3
Cores on flakes	3.85 (13)	-	-	13
Core varia	4.14 (14)	0.26 (1)	-	15
Amorphous cores	1.78 (6)	0.26 (1)	-	7
Giant cores	-	0.77 (3)	-	3
Modified	1.78 (6)	1.79 (7)	-	13
Total	338	391	29	758

Fig. 5.53 Close-up view of Layer II-6 Level 6

5.4.12 Layers II-7/8 and II-8

A large bone was excavated in Trench II during the field season of 1997. It was found on the northern face of the trench and initially seemed to be stratigraphically bedded in the contact of Layer II-7/8. Detailed stratigraphic observations revealed

5.4 Area B

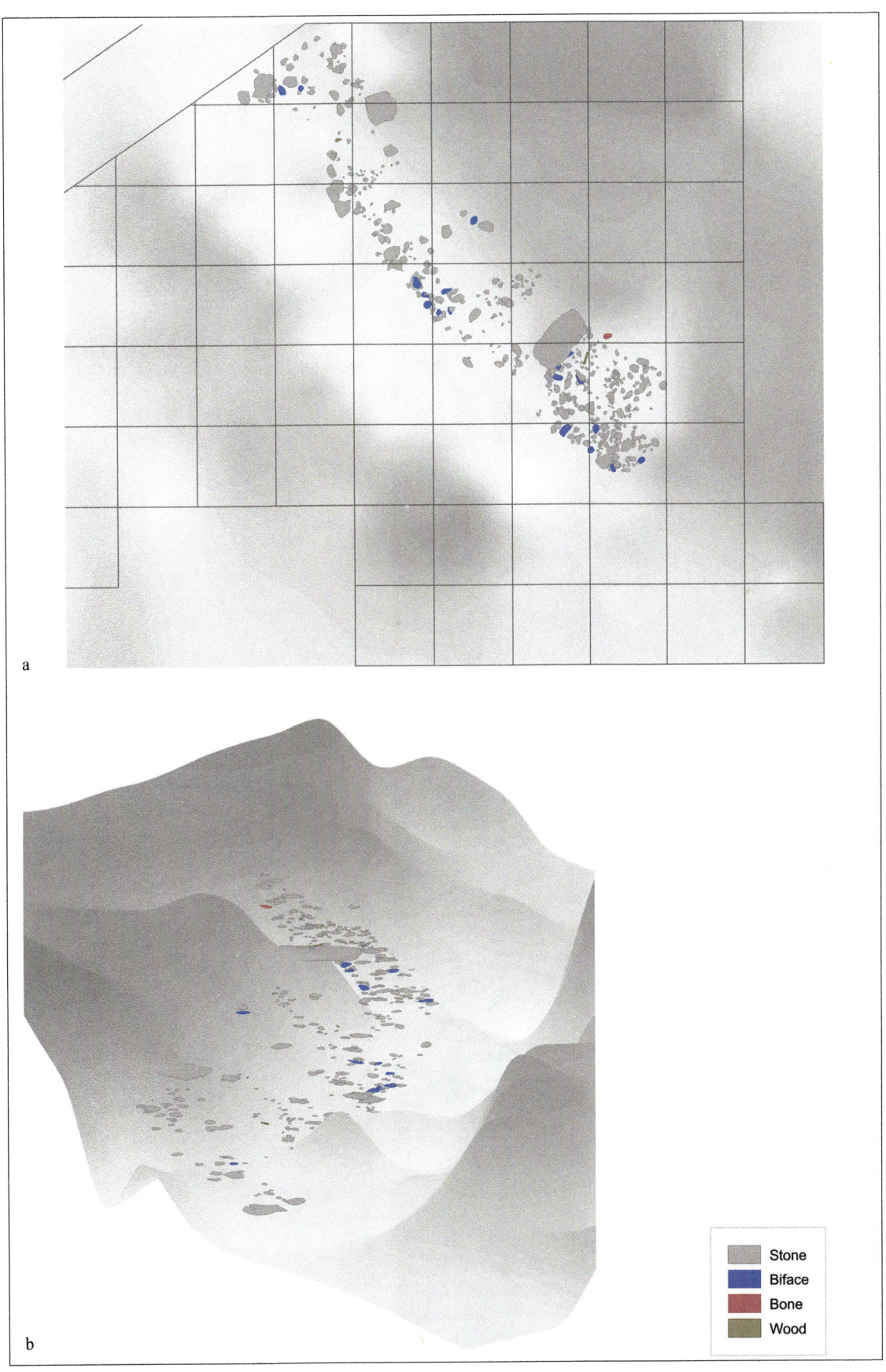

Fig. 5.54 Field map of Layer II-6 Level 6: a) in plan view, b) in its original tilted position; view to southeast

Fig. 5.55 Close-up view of Layer II-6 Level 7; S=south test pit, N=north test pit

Fig. 5.56 Close-up view of Layer II-6 Level 7; S=south test pit, N=north test pit

Fig. 5.57 Close-up view of the north test pit of Layer II-6 Level 7: a) lower part (gravel), b) upper part (sand)

5.4 Area B

Fig. 5.58 Close-up view of the south test pit of Layer II-6 Level 7: a) lower part (gravel), b) upper part in section (sand)

Table 5.50 Counts and frequencies (%)* of the lithic assemblage of Layer II-6 Level 7

	Flint	Basalt	Limestone	Total
	%(N)	%(N)	%(N)	N
TOOLS[#]	38.41 (338)	7.75 (48)	3.30 (3)	389
WASTE				
Debitage	47.61 (419)	79.32 (491)	80.22 (73)	983
Cores	7.27 (64)	1.29 (8)	1.10 (1)	73
Non-flakes	6.70 (59)	11.63 (72)	15.38 (14)	145
Subtotal	880	619	91	1,590
Microartifacts	41,843	1,969	1,623	45,435
Not analyzed FFT	216	182	6	404
Natural nodules	992	2,945	292	4,229
Total	43,931	5,715	2,012	53,248

* % calculated from the subtotal
[#] flake tools and core tools

Table 5.51 Counts and frequencies (%) of the typological composition of Layer II-6 Level 7

	Flint	Basalt	Limestone	Total
	%(N)	%(N)	%(N)	N
TOOLS				
Flake tools				
Side-scrapers	7.16 (63)	0.16 (1)	-	64
End-scrapers	2.16 (19)	-	-	19
Burins	0.23 (2)	-	-	2
Borers	4.89 (43)	0.16 (1)	1.10 (1)	45
Notches	11.48 (101)	0.65 (4)	-	105
Denticulates	2.39 (21)	0.32 (2)	-	23
Truncations	1.36 (12)	-	-	12
Retouched flakes	7.95 (70)	1.13 (7)	1.10 (1)	78
Massive scrapers	-	0.16 (1)	-	1
Varia	0.34 (3)	-	-	3
Core tools				
Bifaces				
Handaxes	0.45 (4)	3.23 (20)	1.10 (1)	25
Cleavers	-	1.94 (12)	-	12
Others				
Chopping tools	1.02 (9)	-	-	9
Percussors	-	6.62 (41)	9.89 (9)	50
Pitted anvils	-	0.32 (2)	-	2
Anvils	-	0.97 (6)	-	6
WASTE				
Flakes				
Flakes	41.93 (369)	77.22 (478)	78.02 (71)	918
Blades	0.57 (5)	0.32 (2)	1.10 (1)	8
Éclats de taille de biface	2.84 (25)	1.62 (10)	1.10 (1)	36
Biface sharpening flakes	0.91 (8)	-	-	8
Core waste	1.36 (12)	0.16 (1)	-	13
Non-flakes				
Angular fragments	2.61 (23)	1.45 (9)	1.10 (1)	33
Cores				
Levallois cores	0.80 (7)	-	-	7
Pyramidal cores	0.23 (2)	-	-	2
Cores on flakes	1.36 (12)	-	-	12
Core varia	2.73 (24)	0.16 (1)	1.10 (1)	26
Amorphous cores	1.14 (10)	-	-	10
Giant cores	-	1.13 (7)	-	7
Modified	4.09 (36)	2.26 (14)	4.40 (4)	54
Total	880	619	91	1,590

that most of the bone was buried in the gray mud sediments of Layer II-8 and that it was partially covered by the black mud sediments of Layer II-7 (Figs. 5.60–5.61). The top of the bone was covered by the sediments of Layer II-1 due to the erosion of the sediments and the proximity of the bone to the truncated surface. Several basalt pebbles were found near the bone, mainly on top of it, and are suspected to be a component of Layer II-1.

As the bone was located so close to the surface, it was heavily fissured and rootlets were growing through parts of it. This type of taphonomic situation is extremely

Fig. 5.59 Field map of Layer II-6 Level 7 (excluding the test pits): a) in plan view, b) in its original tilted position; view to southeast

5.4 Area B

Table 5.52 Counts and frequencies (%)* of the lithic assemblage of Layer II-6/7

	Flint %(N)	Basalt %(N)	Limestone %(N)	Total N
TOOLS#	40.00 (12)	-	-	12
WASTE				
Debitage	53.33 (16)	100.00 (24)	100.00 (5)	45
Cores	6.67 (2)	-	-	2
Subtotal	30	24	5	59
Microartifacts	412	53	37	502
Not analyzed FFT	41	49	3	93
Natural nodules	3	37	-	40
Total	486	163	45	694

* % calculated from the subtotal
# flake tools and core tools

Table 5.53 Counts and frequencies (%) of the typological composition of Layer II-6/7

	Flint %(N)	Basalt %(N)	Limestone %(N)	Total N
TOOLS				
Flake tools				
Side-scrapers	10.00 (3)	-	-	3
Borers	3.33 (1)	-	-	1
Notches	13.33 (4)	-	-	4
Denticulates	3.33 (1)	-	-	1
Truncations	6.67 (2)	-	-	2
Retouched flakes	3.33 (1)	-	-	1
WASTE				
Flakes				
Flakes	40.00 (12)	95.83 (23)	100.00 (5)	40
Éclats de taille de biface	13.33 (4)	4.17 (1)	-	5
Cores				
Cores on flakes	6.67 (2)	-	-	2
Total	30	24	5	59

Table 5.54 Counts and frequencies (%) of the lithic assemblage of Layer II-7

	Flint %(N)	Basalt %(N)	Total N
TOOLS#	44.12 (15)	-	15
WASTE			
Debitage	50.00 (17)	100.00 (3)	20
Cores	5.88 (2)	-	2
Total	34	3	37

flake tools and core tools

Table 5.55 Counts and frequencies (%) of the typological composition of Layer II-7

	Flint %(N)	Basalt %(N)	Total N
TOOLS			
Flake tools			
Side-scrapers	2.94 (1)	-	1
Borers	2.94 (1)	-	1
Notches	11.76 (4)	-	4
Denticulates	8.82 (3)	-	3
Truncations	2.94 (1)	-	1
Retouched flakes	11.76 (4)	-	4
Varia	2.94 (1)	-	1
WASTE			
Flakes			
Flakes	38.24 (13)	100.00 (3)	16
Éclats de taille de biface	8.82 (3)	-	3
Biface sharpening flakes	2.94 (1)	-	1
Cores			
Globular cores	2.94 (1)	-	1
Core varia	2.94 (1)	-	1
Total	34	3	37

rare at the site and can be associated with its particular position. Worth noting also are the multitude of burrows (of crabs?) that surrounded the upper parts of the bone and made it impossible to carry out detailed stratigraphic and taphonomic observations on the contact of Layers II-1/II-7, II-8.

The bone was identified as the left part of a hippopotamus pelvis (R. Rabinovich, personal communication).

5.4.13 Layer II-12

The longest piece of wood in the botanical assemblages of GBY was identified in Layer II-12, which is visible only in the northern face of Trench II (Goren-Inbar et al. 2002b: 82, fig. 20). It was cleaned, excavated, and preserved. This piece (*Quercus ithaburensis*) retained its bark in its anatomical position on the underside of the log. A single flint microartifact was found in the vicinity of the wood.

This artifact is extremely important as it illustrates the possibility that II-12 is actually an archaeological layer exposed only at its margins. The location of the trenches considered the strike and dip of the layers but was not guided by any landmarks. Thus, the fact that so many rich archaeological horizons were identified was unexpected. Clearly, some of the layers that are considered geological (or devoid of archaeological material) may actually be archaeological horizons. In some of these, edible macrobotanical remains were found in relatively high densities. It is possible that these not only represent the natural flora but also reflect hominin activities that did not require many lithic artifacts (Goren-Inbar et al. 2014; Melamed et al. 2016).

Fig. 5.60 General view of Trench II; a large bone (not visible in the photograph) is located where the person's hands are placed

5.4.14 Trench II

Lithic artifacts were collected from the digging of Trench II. These are divided into two assemblages, one generally originating in the spoils of Trench II (Tables 5.56–5.57) and the other restricted to the spoils of green sediments (Tables 5.58–5.59) discussed above.

Table 5.56 Counts and frequencies (%) of the lithic assemblage of Trench II

	Flint	Basalt	Limestone	Total
	%(N)	%(N)	%(N)	N
TOOLS[#]	37.50 (3)	50.00 (19)	-	22
WASTE				
Debitage	37.50 (3)	18.42 (7)	-	10
Cores	-	-	25.00 (1)	1
Non-flakes	25.00 (2)	31.58 (12)	75.00 (3)	17
Total	8	38	4	50

[#] flake tools and core tools

Table 5.57 Counts and frequencies (%) of the typological composition of Trench II

	Flint	Basalt	Limestone	Total
	%(N)	%(N)	%(N)	N
TOOLS				
Flake tools				
Side-scrapers	-	2.63 (1)	-	1
Notches	-	2.63 (1)	-	1
Truncations	25.00 (2)	-	-	2
Core tools				
Bifaces				
Handaxes	12.50 (1)	23.68 (9)	-	10
Cleavers	-	21.05 (8)	-	8
Others				
Chopping tools	-	-	25.00 (1)	1
Percussors	-	7.89 (3)	-	3
Anvils	-	5.26 (2)	-	2
WASTE				
Flakes				
Flakes	37.50 (3)	18.42 (7)	-	10
Non-flakes				
Angular fragments	25.00 (2)	2.63 (1)	-	3
Indeterminate waste	-	2.63 (1)	-	1
Modified	-	13.16 (5)	75.00 (3)	8
Total	8	38	4	50

Fig. 5.61 Views of the hippopotamus pelvis in Trench II: a) general, b) close-up

5.4 Area B

Table 5.58 Counts and frequencies (%) of the lithic assemblage of Trench II green

	Flint %(N)	Basalt %(N)	Limestone %(N)	Total N
TOOLS#	46.43 (26)	34.65 (35)	28.57 (2)	63
WASTE				
Debitage	39.29 (22)	52.48 (53)	14.29 (1)	76
Cores	12.50 (7)	0.99 (1)	57.14 (4)	12
Non-flakes	1.79 (1)	11.88 (12)	-	13
Total	56	101	7	164

flake tools and core tools

Table 5.59 Counts and frequencies (%) of the typological composition of Trench II green

	Flint %(N)	Basalt %(N)	Limestone %(N)	Total N
TOOLS				
Flake tools				
Side-scrapers	7.14 (4)	-	-	4
End-scrapers	3.57 (2)	-	-	2
Burins	1.79 (1)	-	-	1
Borers	5.36 (3)	-	-	3
Notches	8.93 (5)	0.99 (1)	-	6
Denticulates	5.36 (3)	-	-	3
Truncations	1.79 (1)	-	-	1
Retouched flakes	7.14 (4)	3.96 (4)	-	8
Massive scrapers	-	1.98 (2)	-	2
Core tools				
Bifaces				
Handaxes	5.36 (3)	17.82 (18)	14.29 (1)	22
Cleavers	-	8.91 (9)	-	9
Biface preforms	-	0.99 (1)	14.29 (1)	2
Others				
Chopping tools	1.79 (1)	-	42.86 (3)	4
Percussors	-	2.97 (3)	-	3
Pitted anvils	-	1.98 (2)	-	2
WASTE				
Flakes				
Flakes	35.71 (20)	52.48 (53)	14.29 (1)	74
Éclats de taille de biface	3.57 (2)	-	-	2
Non-flakes				
Angular fragments	1.79 (1)	-	-	1
Indeterminate waste	-	3.96 (4)	-	4
Cores				
Levallois cores	7.14 (4)	-	-	4
Core varia	3.57 (2)	-	14.29 (1)	3
Amorphous cores	-	0.99 (1)	-	1
Modified	-	2.97 (3)	-	3
Total	56	101	7	164

5.4.15 Layer IV-25

Trench IV (Fig. 5.25) was dug in order to drain the site during excavation. It is the only one of the six trenches that was not dug perpendicularly to the strike of the layers. Despite this, the trench furnished important information on each of the layers as well as on their lateral appearance. Of the entire exposure of Trench IV, lithic artifacts were found in Layers IV-6, IV-7, and IV-25. In addition, two basalt handaxes were found in the spoils of Trench IV.

Layer IV-25 is the deepest stratum of the stratigraphic sequence, and hence the oldest in the study area (Fig. 3.13). It is composed of very rounded basalt gravel without clasts of other types of raw materials or bonding matrix. This composition testifies to the existence of a major fluvial system that probably drained a basaltic relief situated somewhere east of the Jordan River. The extent of the channel and its orientation are unknown, as the exposure of the layer was minimal.

Piles (spoils) of basalt boulders resulting from the mechanical excavation of the trench were placed on the edges of the trench and flint artifacts were retrieved from them. Figs. 5.62–5.71 illustrate some of these finds. Most of the

Fig. 5.62 Flint double convex-concave side-scraper (#17356 Layer IV-25)

Fig. 5.63 Flint side-scraper on ventral face (#17357 Layer IV-25)

Fig. 5.64 Flint offset scraper (#17354 Layer IV-25)

Fig. 5.67 Flint typical borer **1** multiple views **2** detailed view of borer (#17360 Layer IV-25)

Fig. 5.65 Flint atypical end-scraper (#17359 Layer IV-25)

Fig. 5.68 Flint notch (#17362 Layer IV-25)

Fig. 5.66 Flint typical borer **1** multiple views **2** detailed view of borer (#17358 Layer IV-25)

Fig. 5.69 Flint *éclat de taille de biface* (#17361 Layer IV-25)

artifacts are not fresh, being abraded to different degrees. It is evident that this small assemblage is a redeposited one and that it was transported with the boulders to its present location. This assemblage indicates that somewhere in the GBY paleolandscape, and beyond the immediate surroundings of the study area, there was another site. The abraded character of the lithics indicates that their taphonomic life-histories are long. Fluvial transportation of Acheulian artifacts (primarily handaxes and cleavers) is documented in a conglomerate overlying the BYF north of the bridges that is much younger than the Acheulian site (Sharon et al. 2010; Sharon 2011), but such an Early Pleistocene event (MIS 20) is unique and is documented only by the few artifacts collected from the spoils of Layer IV-25.

The small lithic assemblage of this layer comprises only a few flint artifacts and is not included in the detailed analyses of Chapters 6–8. However, considering its antiquity, we provide the results of the analyses of the lithic component here.

5.4 Area B

Fig. 5.70 Flint *éclat de taille de biface* (#17365 Layer IV-25)

Fig. 5.71 Flint burin spall (#17364 Layer IV-25)

Observations on both patination and preservation illustrate the depositional history of the artifacts. Heavy weathering was encountered on 54.54% of the cases and some 27% were fresh.

Table 5.60 Counts and frequencies (%) of the typological composition of Layer IV-25

	Flint
	%(N)
Double convex side-scrapers	9.09 (1)
Offset side-scrapers	9.09 (1)
Side-scrapers on ventral face	9.09 (1)
Atypical end-scrapers	9.09 (1)
Typical borers	18.18 (2)
Notches	9.09 (1)
Éclats de taille de biface	27.27 (3)
Burin spalls	9.09 (1)
Total	11

Table 5.61 Descriptive statistics of flint artifacts of Layer IV-25 (mm)

	Length	Width	Thickness
Mean	34.18	34.45	9.72
Std dev	18.69	16.88	5.17
Std err mean	5.63	5.09	1.56
N	11	11	11

This double modality shows that the life-histories of the artifacts were not identical and perhaps represent more than one source of the artifacts. Patination and double patination occur in similar frequencies, that of patinated artifacts being slightly higher (ca. 54%). Another taphonomic observation, that of breakage, shows that over 60% of the artifacts are complete.

The lithic inventory of Layer IV-25 includes 11 flint artifacts, all of which are FFT. Of these, most are retouched. Each type is represented by a single tool, except for borers and *éclats de taille de biface* (Table 5.60). In addition, and similarly to the large lithic assemblages of other layers at the site, on three artifacts there are additional types (multiple tools), including a single convex side-scraper, a denticulate, and a retouched flake. Although the sample size is small, the retouched artifacts are similar typologically to those of the other assemblages of the site, as is the small size of the artifacts (Table 5.61). It is of interest to note that biface production is represented by three *éclats de taille de biface*. The presence of these three flakes is noteworthy, as it provides a clue to the Acheulian origin of this small flint assemblage. It testifies to the fact that there was an Acheulian entity in the vicinity of the site during Early Pleistocene times, the earliest of the GBY record, as all the other assemblages are located above the Matuyama/Brunhes chron boundary.

Chapter 6
The Flint Component

Abstract Chapter 6 aims to provide a comprehensive description of the flint assemblages, a major component of the archaeological horizons of Gesher Benot Ya'aqov. Analyses consist of taphonomic, morphological, technological, and typological observations, which enable characterization and reconstruction of the operational sequences of flint.

Both cores and core tools and flakes and flake tools are small, probably a result of the small size of the flint nodules chosen as blanks. In contrast, flint bifaces are modified on larger nodules. The bifacial component is represented in this chapter only through the analysis of its shaping and finishing products. These resemble products of the Levallois method, which is discussed in detail, particularly in the analyses of Levallois cores.

Flake tools exhibit a variety of types and occur in high frequencies through the entire cultural sequence. Special attention is paid here to the modification of the proximal ends of flakes, interpreted as a preparatory stage for hafting.

Keywords cores and core tools • flakes and flake tools • flint • Gesher Benot Ya'aqov • hafting • Levallois • reduction sequence

Flint artifacts are recorded for each of the archaeological horizons at Gesher Benot Ya'aqov (GBY) and commonly include cores and core tools (CCT), flakes and flake tools (FFT), and microartifacts. Flint bifaces (BF) either occur in very small frequencies or are absent from the archaeological horizons. However, other components of the flint assemblage (thinning and finishing flakes, e.g. *éclats de taille de biface*) provide evidence for modification of bifaces, even when the tools themselves are not present. Flint is additionally present in the form of small, natural, unmodified nodules. This component was naturally introduced into the lake margin environment and is not discussed in the framework of this volume (Feibel, addional studies under way). Counts of this component are presented in Chapter 5 for each of the archaeological horizons.

This chapter presents and discusses the taphonomy, typology, and technology of the flint CCT and FFT of GBY. The flint bifaces are discussed separately below (Chapter 7). Unlike other components of the lithic assemblages, bifaces of all three raw materials are presented and discussed together. This resolution derives from the relatively small number of the flint bifaces and in order to avoid extensive repetition of identical typological classifications, description of measurements, and other attributes.

6.1 Inventory and Taphonomy of the Flint Component

6.1.1 Inventory of the Flint Component

The complete flint inventory of GBY throughout the stratigraphic sequence is presented in Table 6.1. In all tables the data are given in chronological order from the youngest (top) to the oldest (bottom) layer. As is evident from Table 6.1, the frequencies of flint in general, as well as those of particular technological classes, vary among the different layers.

In some of the following presentations, we include only assemblages that are composed of a minimum number of 30 artifacts and are hence suitable samples for analyses.

6.1.2 Taphonomy of the Flint Component

State of preservation, patination, and breakage patterns can be used to determine the taphonomic history of the flint component. Table 6.2 details the preservation state of flint throughout the GBY sequence. The differences in preservation reflect different depositional environments (Feibel 2001, 2004), which include low-energy lake margin environments (e.g. less abraded artifacts in Layer II-6 Levels 1–4b) alongside

Table 6.1 Counts of flint artifacts and microartifacts throughout the GBY stratigraphic sequence

Area	Layer	Artifacts				Microartifacts
		CCT	FFT	BF	Total	
C	V	-	17	-	17	282
	V-2	-	3	-	3	7
	V-2/3	-	1	-	1	3
	V-3	-	7	-	7	104
	V-3/4	-	1	-	1	-
	V-4	6	15	-	21	470
	V-4/5	6	21	-	27	20
	V-5	25	311	-	336	30,875
	V-6	9	291	1	331	4,507
	JB	12	158	2	172	928
A	I-4	-	46	-	46	7,449
	I-5	11	45	-	56	15,782
B	II-2	7	108	1	116	3,745
	II-2/3	18	99	-	117	7,374
	II-3	1	4	-	5	-
	II-4	4	5	-	9	15
	II-5	50	190	-	240	3,615
	II-5/6	39	115	-	154	8,887
	Trench II	3	5	1	9	-
	Trench II green	8	45	3	56	2
	II-6/L1	322	761	4	1,087	53,835
	II-6/L2	148	305	4	457	73,627
	II-6/L3	200	385	1	586	91,927
	II-6/L4	252	433	3	688	106,428
	II-6/L4b	25	124	1	150	7,625
	II-6/L5	42	156	4	202	33,073
	II-6/L6	64	274	3	338	12,533
	II-6/L7	137	740	-	880	24,132
	II-6/7	2	28	-	30	412
	II-7	2	32	-	34	-
	VI-14	2	18	1	21	9
	IV-25	-	12	-	12	-
	Unconformity B	36	173	7	-	-
	Unconformity C	4	24	-	-	-
	Total	1,435	4,942	36	6,238	487,664

fluvial systems of higher energy (e.g. more abraded artifacts in Layer II-6 Levels 5–7).

These differences are further expressed in the preservation of FFT vs. CCT. In low-energy environments CCT are significantly more abraded than FFT. This contrasts with higher-energy environments, where the preservation of CCT and of FFT is more similar.

The youngest assemblages (Layers V-5, V-6, and JB), which are associated with environments of lower energy than in Layer II-6, display the highest values of fresh items. Even there, however, CCT are more abraded than FFT (Table 6.2).

We suspect that the observed differences in preservation between CCT and FFT may be rooted in subsurface weathering processes. More particularly, it is possible that CCT and FFT, which are of different sizes and hence surface areas, act differently in response to varying depositional environments. Analysis of the relationship between the size and preservation of CCT and FFT demonstrates a possible relationship (Fig. 6.1).

Patination of flint artifacts at GBY has resulted from different agents, such as exposure to atmospheric conditions and burial in anaerobic waterlogged conditions. In our analyses we did not differentiate between different colors of patination, which at GBY vary between white-yellow and black. Table 6.3 presents the frequencies of patination and breakage of CCT and FFT, suggesting that CCT are more patinated. These patterns seem to characterize the entire stratigraphic sequence, regardless of depositional environment. The only exception is Layer II-6 Level 7, which exhibits the opposite tendency, with FFT being more patinated (Table 6.3). Despite the good preservation of the flint artifacts from Area C, they are mostly patinated, even more so than those of Area B (Table 6.3). As documented by taphonomic analysis of bone (e.g. Area C in Rabinovich et al. 2012), extensive patination is not necessarily a result of longer exposure or taphonomic life-history but rather may reflect the particular type of waterlogged depositional environment.

The general tendency of CCT to exhibit higher frequencies of patinated artifacts may result from processes similar to those that resulted in higher frequencies of abraded artifacts among CCT. Nevertheless, observations of patination and preservation are independent and do not record complementary data that could be associated with the same processes. Thus, fresh flint artifacts (e.g. from Layers V-5 and V-6) may exhibit high frequencies of patination.

The analyses of preservation and patination demonstrate that CCT and FFT exhibit different patterns of abrasion and patination. Similarly, the breakage state of these two lithic categories varies, and CCT are less broken than FFT (Table 6.3). This discrepancy is not necessarily a product of taphonomy and could rather be assigned to fracture mechanics of flaking in which spontaneous breaks occur during knapping (see Section 6.2.1).

The frequencies of complete FFT are quite homogeneous throughout the sequence, even when the conglomeratic base of Layer II-6 (II-6/7) is included. Nevertheless, when we examine the different types of breakages, some vary among layers. In the layers of Layer II-6, for example, distal breaks are less frequent than in Area C (Table 6.4). The higher frequencies of distally broken flint flakes are possibly associated with the fact that most of the flint flakes in Area C are products of biface modification, which are often thin and tend to break spontaneously during the knapping process, particularly in the final stages of thinning and finishing (Herzlinger et al. 2015). The assumption that the breakage of flint FFT is mostly spontaneous rather than taphonomic and is not associated with tool use is further supported by the fact that flint flake tools are more often complete than unretouched flint flakes (Section 6.2.1).

6.2 Typology of Flint Flake Tools and Core Tools

Table 6.2 Frequencies (%) of state of preservation of flint CCT and FFT

Layer		Fresh	Slightly abraded	Abraded	Heavily abraded	N
V-5	FFT	73.31	16.40	9.00	1.29	311
	CCT	48.00	40.00	12.00	-	25
V-6	FFT	86.94	11.00	2.06	-	291
	CCT	77.78	22.22	-	-	9
JB	FFT	73.42	20.89	4.43	1.27	158
	CCT	75.00	16.67	8.33	-	12
I-4	FFT	21.74	50.00	26.09	2.17	46
	CCT	-	-	-	-	-
I-5	FFT	31.11	48.89	20.00	-	45
	CCT	18.18	45.45	36.36	-	11
II-2	FFT	14.81	46.30	33.33	5.56	108
	CCT	14.29	42.86	42.86	-	7
II-2/3	FFT	36.36	41.41	19.19	3.03	99
	CCT	27.78	50.00	22.22	-	18
II-5	FFT	39.47	31.58	22.11	6.84	190
	CCT	6.00	54.00	36.00	4.00	50
II-5/6	FFT	32.17	33.91	31.30	2.61	115
	CCT	10.26	48.72	30.77	10.26	39
Trench II green	FFT	31.11	40.00	24.44	4.44	45
	CCT	-	62.50	37.50	-	8
II-6/L1	FFT	33.25	37.98	27.73	1.05	761
	CCT	9.63	55.90	33.85	0.62	322
II-6/L2	FFT	26.56	36.72	27.21	9.51	305
	CCT	11.49	53.38	35.14	-	148
II-6/L3	FFT	34.29	37.14	25.45	3.12	385
	CCT	9.00	65.50	23.50	2.00	200
II-6/L4	FFT	37.18	36.72	20.09	6.00	433
	CCT	11.90	70.24	17.06	0.79	252
II-6/L4b	FFT	23.39	35.48	26.61	14.52	124
	CCT	8.00	60.00	32.00	-	25
II-6/L5	FFT	5.13	53.21	32.69	8.97	156
	CCT	9.52	57.14	33.33	-	42
II-6/L6	FFT	10.58	39.42	37.23	12.77	274
	CCT	12.50	53.13	34.38	-	64
II-6/L7	FFT	8.11	29.19	43.51	19.19	740
	CCT	10.95	56.93	30.66	1.46	137
II-6/7	FFT	42.86	21.43	28.57	7.14	28
	CCT	-	100.00	-	-	2
II-7	FFT	34.38	53.13	12.50	-	32
	CCT	-	50.00	50.00	-	2

6.2 Typology of Flint Flake Tools and Core Tools

This section is concerned with the typological aspects of both flake tools and core tools. We begin the presentation with the typological classification of flint flake tools, which are more abundant than core tools. This presentation is accompanied only by those technological observations that are relevant to the typological modification of the tools. The detailed technological analyses of the flint component are presented in the following Sections 6.3–6.4.

6.2.1 Typology of Flake Tools

The detailed typological inventory of each of the GBY layers is presented in Chapter 5. In the following typological discussion

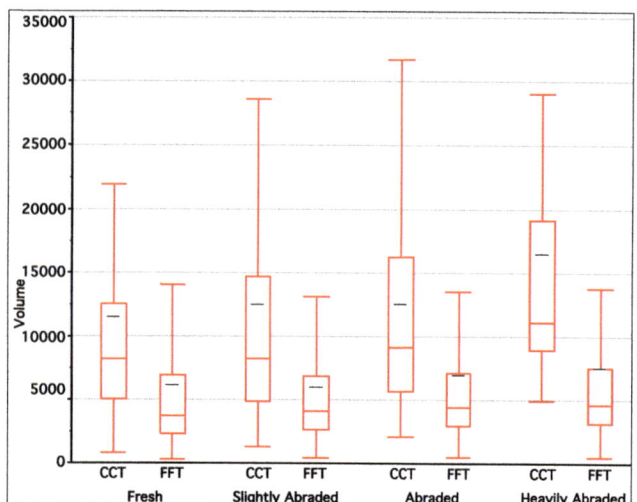

Fig. 6.1 Length and preservation state of flint CCT and FFT

we include assemblages with a minimum number of 30 FFT. Table 6.5 details the counts and frequencies of flint flake tools in those assemblages. Unretouched flint flakes are excluded from the following typological presentation and are discussed elsewhere (Sections 6.2.1.1, 6.3.1).

At GBY, small-sized flake tools form the bulk of the tool assemblages and they are modified primarily on flint. Overall, the typological composition of the different Acheulian assemblages at GBY is similar throughout the stratigraphic sequence (Table 6.5). Where the sample is large enough, the same tool types are represented in each of the lithic assemblages.

The inventory of tools in each of the GBY layers illustrates the typological diversity (Table 6.5). Interestingly, there is a positive relationship between the lithic sample size of the archaeological horizons and the typological diversity. Clearly, when sample sizes are of sufficient size there is also a greater diversity in the tool types. Thus, when this condition is fulfilled certain tool types occur in higher frequencies, while in other, smaller samples they are minimally represented. This is clearly the case for the frequency and diversity of side-scraper types that will be detailed below, but it is true for other tool types as well, such as tools of small pointed-edge morphology (borers) and typical and atypical end-scrapers.

As can be seen in Table 6.5, the most common tool types are those that have notched or denticulated edges. Following in descending order of frequency are retouched flakes (a very general category) and the group of borers. The results of the typological analysis provide an extremely homogeneous picture in terms of the representation and frequencies of flint tool types along the cultural sequence.

The following presentation describes the results of the typological analyses in descending order of frequency, from the dominant tool type to the least common. The single exception are retouched flakes, which are the second most frequent group of retouched tools; because of their general character, however, they are described at the end of the chapter.

In the following data tables, which describe the characteristics of different tool types, we include only those layers in which the typological category consists of a minimum number of 10 tools (with the exception of truncations; see below). In certain layers the samples are very small and hence our discussion is drawn mostly from assemblages that have the highest frequencies of items. In the presentation of Layer II-6 and its sub-levels (1–7) we present the full inventory in order to emphasize the inter-level variability of typological occurrence.

Notches

The definition of a notch at GBY refers to artifacts that exhibit a small retouched concavity on one lateral edge (Figs. 6.2–6.4) or on two lateral edges (Figs. 6.3–6.4); only one Clactonian notch, i.e. non-retouched concavity, was identified in the entire GBY lithic assemblage. The vast majority of the concave morphologies of notches were modified by retouch (95.88%; N=443 in the entire GBY flint notch assemblage), predominantly notch-like (88.5%), and only very infrequently shaped by irregular, abrupt, or semi-abrupt retouch. However, there are some cases in which the morphology of the notch resulted from use (4.11%). Among lithic tools at GBY, signs of use were not considered retouch except for cases identified as notches. In these instances a clear notch morphology was identified, but the retouch was not homogeneous and continuous. In these particular cases it is possible that the notches are the result of use rather than an intentional modification of the edge prior to its use.

Notches are the most common tool type in the flint flake tool assemblage and their frequencies increase in accordance with sample size. The high typological frequencies of the notches "family" are consistent throughout the stratigraphic sequence (Table 6.5).

While the location of a notch on a lateral edge is not considered a typological marker, its occurrence on the distal end has historically been identified and viewed as an independent class, the end-notch (Bordes 1961). It is evident that end-notches are a typological component of the GBY assemblages and they were classified independently during the typological analysis (Fig. 6.5; Table 6.5). However, since notches and end-notches share the same form and stylistic characteristics with retouch, we consider them as belonging to the same morphotype and are therefore analyzed jointly in the following discussion.

The frequencies of notches would have been even higher if we had included the notches used for the production of denticulates and borers. However, we have preferred to limit the presentation to notched tools *per se*.

6.2 Typology of Flint Flake Tools and Core Tools

Table 6.3 Frequencies (%) of breakage and patination of flint CCT and FFT

Layer		Breakage		Patination			N*
		Complete	Broken	No patina	Patina	Double patina	
V-5	FFT	56.27	43.73	1.29	80.71	18.01	311
	CCT	96.00	4.00	-	92.00	8.00	25
V-6	FFT	55.33	44.67	6.19	82.82	11.00	291
	CCT	88.89	11.11	-	66.67	33.33	9
JB	FFT	53.80	46.20	0.63	87.97	11.39	158
	CCT	90.91	9.09	8.33	66.67	25.00	11;12
I-4	FFT	41.3	58.7	21.73	67.40	10.87	46
	CCT	-	-	-	-	-	-
I-5	FFT	63.64	36.36	44.44	48.89	6.67	44
	CCT	90.90	9.10	-	63.64	36.36	11
II-2	FFT	41.67	58.33	0.93	60.19	38.89	108
	CCT	85.71	14.29	-	57.14	42.86	7
II-2/3	FFT	66.67	33.33	6.06	79.80	14.14	99
	CCT	94.44	5.56	-	77.78	22.22	18
II-5	FFT	56.32	43.68	10.53	64.21	25.26	190
	CCT	86.00	14.00	-	78.00	22.00	50
II-5/6	FFT	50.43	49.57	7.83	73.04	19.13	115
	CCT	84.62	15.38	-	92.31	7.69	39
Trench II green	FFT	60.00	40.00	15.56	71.11	13.33	45
	CCT	87.50	12.50	-	87.50	12.50	8
II-6/L1	FFT	57.29	42.71	17.21	64.39	18.40	761
	CCT	77.33	22.67	1.24	68.63	30.12	322
II-6/L2	FFT	53.03	41.97	4.92	72.79	22.30	305
	CCT	76.35	23.65	-	76.71	23.29	148; 146
II-6/L3	FFT	55.32	44.68	16.36	64.68	18.96	385
	CCT	86.22	13.78	1.00	82.50	16.50	196; 200
II-6/L4	FFT	56.12	43.88	11.55	67.21	21.25	433
	CCT	88.80	11.20	0.40	78.17	21.43	250; 252
II-6/L4b	FFT	50.81	49.19	4.84	66.94	28.23	124
	CCT	76.00	24.00	-	76.00	24.00	25
II-6/L5	FFT	61.54	38.46	0.64	64.10	35.26	156
	CCT	47.62	52.38	-	66.67	33.33	42
II-6/L6	FFT	58.03	41.97	2.92	68.61	28.47	274
	CCT	71.88	28.13	-	60.32	39.68	64; 63
II-6/L7	FFT	48.92	51.08	1.08	74.86	24.05	740
	CCT	78.83	21.17	0.73	65.69	33.58	137
II-6/7	FFT	53.57	46.43	-	89.29	10.71	28
	CCT	100.00	100.00	-	50.00	50.00	2
II-7	FFT	43.75	56.25	-	78.13	21.88	32
	CCT	100.00	-	-	100.00	-	2

* in cases where N differs between the two attributes, the first value refers to counts of breakage and the second to those of patination

The preservation state of notches (Table 6.6) exhibits the general similarity of weathering patterns observed for retouched and unretouched flint flakes, among which slightly abraded artifacts are the common mode. Notches are characterized by high frequencies of double patinated artifacts (Table 6.6), significantly higher than those observed for non-retouched flakes ($\chi^2=11.74$; $df=2$; $p=0.0028$). Another difference between notches and unretouched flakes is that among notches there are higher frequencies of complete items. Examination of the breakage patterns of notches shows that the two most common locations of breakage are the proximal and distal edges, with a minimal occurrence of notches on fragments (Table 6.6).

For many of the notches the direction of blow is indeterminable, but where it can be determined the most common direction is end-struck (Table 6.6). In terms of cortical coverage (Table 6.6) and size (Table 6.7), notches and unretouched flakes exhibit similar patterns.

The most common locations of notches are on the distal, left, and right edges (Table 6.8). The right edge is often more

Table 6.4 Frequencies (%) of breakage patterns of flint FFT

Layer	Complete	Distal	Proximal	Lateral	Distal & lateral	Distal & proximal	Proximal & lateral	Fragment	Indet.	N*
V-5	56.27	13.18	7.40	4.50	3.22	3.86	1.29	10.29	-	311
V-6	55.33	16.84	8.59	5.84	4.12	2.06	1.72	5.50	-	291
JB	53.80	15.82	10.13	8.23	-	1.90	1.90	8.23	-	158
I-4	41.30	8.70	21.74	10.87	2.17	6.52	4.35	4.35	-	46
I-5	63.64	11.36	11.36	2.27	-	2.27	4.55	4.55	-	44
II-2	41.67	12.96	8.33	13.89	0.93	0.93	5.56	15.74	-	108
II-2/3	66.67	9.09	11.11	2.02	-	4.04	2.02	4.04	1.01	99
II-5	56.32	13.16	8.95	4.74	2.63	3.68	4.21	5.79	0.53	190
II-5/6	50.43	14.78	10.43	6.09	0.87	5.22	6.96	5.22	-	115
Trench II green	60.00	11.11	20.00	2.22	-	2.22	2.22	2.22	-	45
II-6/L1	57.29	10.38	12.88	5.65	1.18	3.68	1.71	7.10	0.13	761
II-6/L2	58.03	9.84	8.52	7.54	2.95	4.95	1.64	6.89	-	305
II-6/L3	55.32	10.91	13.51	10.65	0.78	2.08	1.30	5.45	-	385
II-6/L4	56.12	12.01	11.55	4.85	3.93	3.23	2.08	6.24	-	433
II-6/L4b	50.81	12.10	12.10	12.10	2.42	4.84	0.81	4.84	-	124
II-6/L5	61.54	12.18	7.69	7.05	1.92	1.28	1.28	7.05	-	156
II-6/L6	58.03	13.87	8.03	7.66	1.46	2.19	1.09	7.66	-	274
II-6/L7	48.92	12.70	12.16	7.43	2.43	3.24	3.11	10.00	-	740
II-6/7	53.57	17.86	-	17.86	3.57	-	-	7.14	-	28
II-7	43.75	12.50	9.38	9.38	9.38	-	3.13	12.50	-	32

* only layers with a minimum number of 30 FFT are included

Table 6.5 Typological frequencies (%) of flint flake tools by layer and level

Type/Layer	V-5	V-6	JB	I-4	I-5	II-2	II-2/3	II-5	II-5/6	Trench II green
N	104	25	34	18	17	54	32	86	49	23
Single straight side-scraper	1.92	12.00	8.82	-	5.88	1.85	-	4.65	2.04	-
Single convex side-scraper	0.96	12.00	8.82	5.56	11.76	5.56	-	3.49	4.08	13.04
Single concave side-scraper	0.96	8.00	-	-	-	1.85	-	1.16	2.04	-
Double straight side-scraper	-	-	-	-	-	-	-	-	-	-
Double straight-convex side-scraper	-	-	-	-	-	-	-	-	-	-
Double convex side-scraper	-	-	-	-	-	-	-	-	-	-
Double convex-concave side-scraper	-	-	-	-	-	-	-	-	-	-
Convergent straight side-scraper	0.96	-	-	-	-	-	-	-	-	-
Convergent convex side-scraper	-	4.00	-	-	-	-	-	-	-	-
Offset (*déjeté*) side-scraper	-	-	-	-	-	-	-	-	2.04	4.35
Straight transverse side-scraper	0.96	4.00	-	-	-	-	-	-	2.04	-
Convex transverse side-scraper	0.96	-	-	-	-	1.85	3.13	-	-	-
Concave transverse side-scraper	-	-	-	-	-	-	-	1.16	-	-
Side-scraper on ventral face	0.96	4.00	2.94	-	-	1.85	-	3.49	-	-
Abruptly retouched side-scraper	3.85	-	-	5.56	-	-	3.13	4.65	-	-
Alternately retouched side-scraper	-	-	-	-	-	-	-	-	2.04	-
Typical end-scraper	-	4.00	-	-	-	1.85	-	2.33	2.04	-
Atypical end-scraper	3.85	4.00	-	5.56	11.76	3.70	3.13	4.65	4.08	8.70
Typical burin	-	-	2.94	-	-	-	-	-	-	-
Atypical burin	0.96	-	2.94	-	-	-	3.13	-	2.04	4.35
Typical borer	3.85	-	-	-	-	-	3.13	3.49	-	-
Atypical borer	11.54	4.00	-	5.56	-	16.67	3.13	15.12	6.12	13.04
Atypical backed knife	1.92	4.00	5.88	-	-	-	-	-	-	-
Truncation	1.92	-	-	-	-	-	6.25	2.33	2.04	4.35
Notch	20.19	16.00	11.76	33.33	23.53	27.78	40.63	19.77	38.78	21.74
Denticulate	21.15	12.00	17.65	16.67	17.65	11.11	6.25	8.14	6.12	13.04
Denticulated point	-	-	-	-	-	3.13	-	-	-	-
End-notched piece	1.92	-	5.88	5.56	11.76	7.41	6.25	5.81	4.08	-
Varia	0.96	-	2.94	5.56	-	-	-	-	-	-
Retouched flake	20.19	12.00	29.41	16,67	17.65	18.52	18.75	19.77	20.41	17.39

6.2 Typology of Flint Flake Tools and Core Tools

Table 6.5 (cont.)

Type/Layer	II-6/L1	II-6/L2	II-6/L3	II-6/L4	II-6/L4b	II-6/L5	II-6/L6	II-6/L7	II-6/7	II-7
N	236	113	152	188	54	79	133	320	12	15
Single straight side-scraper	0.42	-	0.66	2.66	-	2.53	3.01	3.13	-	-
Single convex side-scraper	2.97	-	3.29	3.72	1.85	3.80	-	2.50	-	-
Single concave side-scraper	0.85	1.77	1.32	1.06	1.85	2.53	1.50	1.56	-	-
Double straight side-scraper	-	-	-	-	-	-	0.75	0.63	-	-
Double straight-convex side-scraper	0.42	-	-	-	-	-	-	-	-	-
Double convex side-scraper	-	-	0.66	-	-	-	-	-	-	-
Double convex-concave side-scraper	-	0.88	-	-	-	-	0.75	-	-	-
Convergent convex side-scraper	-	0.88	0.66	-	-	-	-	0.94	8.33	-
Offset (*déjeté*) side-scraper	-	-	-	-	-	-	-	1.25	-	-
Straight transverse side-scraper	0.85	-	-	-	-	1.27	-	-	-	-
Convex transverse side-scraper	-	-	-	1.06	-	1.27	1.50	1.25	8.33	-
Concave transverse side-scraper	-	-	-	-	-	-	-	-	-	-
Side-scraper on ventral face	1.69	3.54	-	1.06	1.85	-	0.75	3.44	-	-
Abruptly retouched side-scraper	2.12	2.65	0.66	3.72	1.85	2.53	4.51	3.44	8.33	6.67
Thinned back side-scraper	-	-	-	-	-	1.27	-	-	-	-
Bifacially retouched side-scraper	-	-	0.66	-	-	-	-	-	-	-
Alternately retouched side-scraper	-	-	-	-	-	-	0.75	0.94	-	-
Typical end-scraper	2.97	2.65	2.63	2.13	5.56	2.53	1.50	1.88	-	-
Atypical end-scraper	5.08	8.85	9.21	2.66	3.70	3.80	5.26	4.06	-	-
Typical burin	0.42	-	-	-	-	-	0.75	-	-	-
Atypical burin	-	-	-	2.13	1.85	2.53	0.75	0.63	-	-
Typical borer	2.12	4.42	3.29	2.13	7.41	-	0.75	1.25	-	-
Atypical borer	11.02	9.73	12.50	12.77	16.67	13.92	17.29	11.88	8.33	6.67
Atypical backed knife	0.85	-	-	0.53	-	1.27	-	0.31	-	-
Truncation	3.81	3.54	3.95	4.26	1.85	2.53	2.26	3.75	16.67	6.67
Notch	24.58	20.35	13.82	21.28	20.37	15.19	22.56	22.81	33.33	6.67
Denticulate	9.32	5.31	8.55	9.04	11.11	5.06	11.28	5.63	8.33	13.33
Denticulated point	0.42	-	-	-	1.85	1.27	-	-	-	6.67
End-notched piece	6.78	8.85	5.26	4.26	1.85	2.53	3.76	5.94	-	20.00
Varia	0.42	3.54	0.66	1.06	-	3.80	1.50	0.94	-	6.67
Retouched flake	22.88	23.01	28.29	24.47	20.37	30.38	18.80	21.88	8.33	26.67

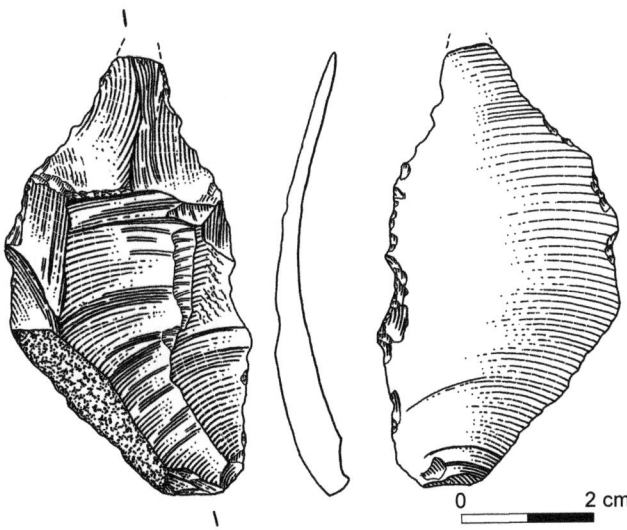

Fig. 6.2 Flint notch (#4716 Layer V-5)

frequent than the left. There are instances in which more than one notch was modified, on different edges and in various combinations. An interesting group of notches is that modified on the proximal end (Table 6.8). Additional cases in which the notch was produced by the removal of the entire striking platform are listed in the discussion of removed striking platforms (Section 6.2.1.1). When the face of retouch is examined, the dorsal face is significantly more frequent than the other alternatives (Table 6.8). The angle of retouch is extremely consistent (Table 6.7).

On 73 of the notched tools (15.80% of the entire notched tool assemblage) we observed an additional tool type. Table 6.9 presents the detailed breakdown of the primary notched tools (Typology A) and their associated secondary tool type (Typology B). The table shows that notches and end-notched pieces may occur on the same artifact. Notched tools are associated primarily with retouched flakes, followed by

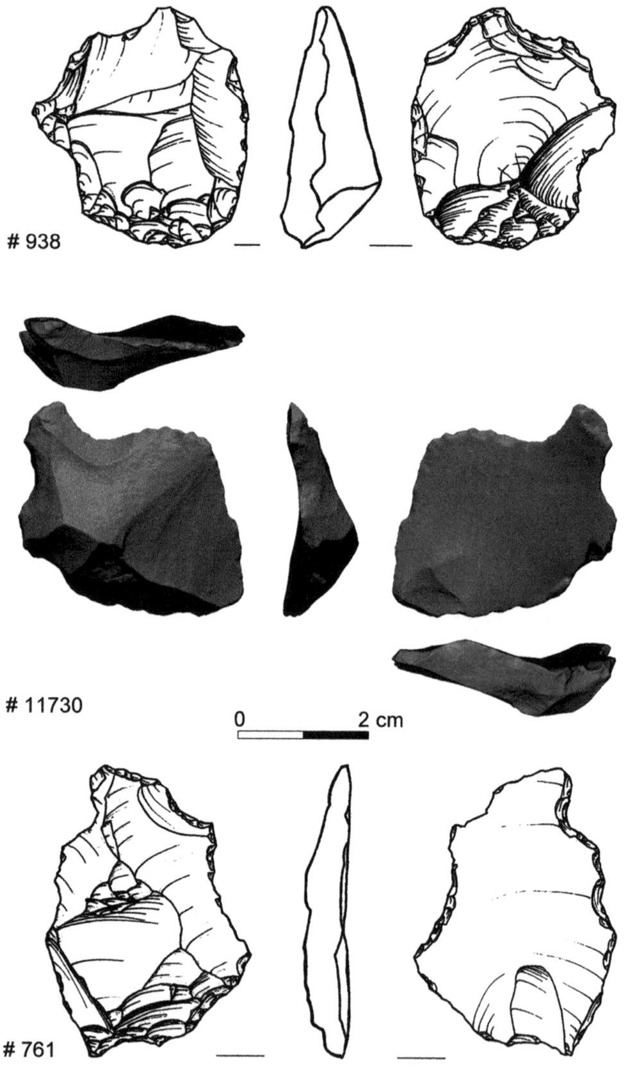

Fig. 6.3 Flint notches (#938 Layer II-6 Level 1, #11730 Layer II-6 Level 7, #761 Layer II-5)

Fig. 6.4 Flint notches (#17235 Layer II-6 Level 1, #13866 Layer II-6 Level 7, **1** note bulb of percussion on dorsal face **2** detailed view of notch)

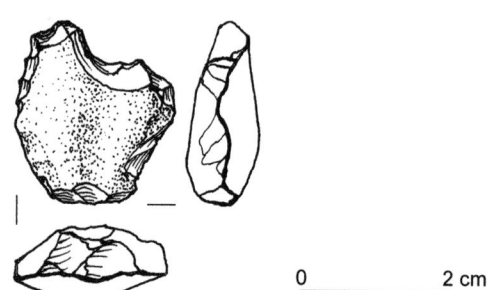

Fig. 6.5 Flint notch (#6136 Layer II-6 Level 1)

denticulates (Fig. 6.6). Notches were also recorded as a second or third tool on other tool types. These cases are described below according to the dominant typological classification (Typology A).

In the entire assemblage of notched tools at GBY, only four notches are associated with two additional tool types. The recorded combinations (Typology A/B/C) are: notch/notch/end-notch, notch/end-notch/retouched flake, notch/denticulate/retouched flake (Fig. 6.7), and notch/retouched flake/sidescraper on ventral face.

To sum up, our study of notches does not assume any mode of use or function for these tools. Clearly, in order to mimic the formation of notched edges, experimental work is needed. The fact that there are high frequencies of notches on complete blanks rules out the involvement of a major taphonomic

6.2 Typology of Flint Flake Tools and Core Tools

Table 6.6 Counts and frequencies (%) of characteristics of flint notches

	Layer												
						II-6 Levels							
	V-5	II-2	II-2/3	II-5	II-5/6	L1	L2	L3	L4	L4b	L5	L6	L7
N	23	19	15	22	21	74	33	29	48	12	14	35	92
Preservation													
Fresh	16	1	3	9	10	24	9	11	18	1	1	5	8
	69.57	5.26	20.00	40.91	47.62	32.43	27.27	37.93	37.50	8.33	7.14	14.29	8.70
Slightly abraded	3	10	8	4	6	30	13	6	18	8	9	11	27
	13.04	52.63	53.33	18.18	28.57	40.54	39.39	20.69	37.50	66.67	64.29	31.43	29.35
Abraded	4	8	4	6	5	20	8	11	10	1	4	15	42
	17.39	42.11	26.67	27.27	23.81	27.03	24.24	37.93	20.83	8.33	28.57	42.86	45.65
Heavily abraded	-	-	-	3	-	-	3	1	2	2	-	4	15
				13.64			9.09	3.45	4.17	16.67		11.43	16.30
Patination													
No patina	-	-	-	2	2	9	-	3	4	1	-	-	1
				9.09	9.52	12.16		10.34	8.33	8.33			1.09
Patinated	20	10	12	12	14	46	26	19	32	7	10	21	59
	86.96	52.63	80.00	54.55	66.67	62.16	78.79	65.52	66.67	58.33	71.43	60.00	64.13
Double patinated	3	9	3	8	5	19	7	7	12	4	4	14	32
	13.04	47.37	20.00	36.36	23.81	25.68	21.21	24.14	25.00	33.33	28.57	40.00	34.78
Breakage													
Complete	15	6	12	13	10	52	19	22	30	6	9	17	51
	65.22	31.58	80.00	59.09	47.62	70.27	57.58	75.86	62.50	50.00	64.29	48.57	55.43
Distal	4	3	1	4	5	9	4	3	5	3	-	3	10
	17.39	15.79	6.67	18.18	23.81	12.16	12.12	10.34	10.42	25.00		8.57	10.87
Lateral	-	3	-	-	1	1	1	3	1	1	2	2	6
		15.79			4.76	1.35	3.03	10.34	2.08	8.33	14.29	5.71	6.52
Proximal	1	4	1	2	4	8	4	-	7	2	3	7	13
	4.35	21.05	6.67	9.09	19.05	10.81	12.12		14.58	16.67	21.43	20.00	14.13
Distal & lateral	1	-	-	-	-	-	1	-	1	-	-	1	1
	4.35						3.03		2.08			2.86	1.09
Distal & proximal	-	-	1	1	1	3	2	-	3	-	-	1	4
			6.67	4.55	4.76	4.05	6.06		6.25			2.86	4.35
Fragment	2	3	-	2	-	1	2	1	-	-	-	3	4
	8.70	15.79		9.09		1.35	6.06	3.45				8.57	4.35
Proximal & lateral	-	-	-	-	-	-	-	-	1	-	-	1	3
									2.08			2.86	3.26
Direction of blow													(N=91)
Indeterminate	8	14	1	7	4	11	8	4	10	3	3	14	42
	34.78	73.68	6.67	31.82	19.05	14.86	24.24	13.79	20.83	25.00	21.43	40.00	46.15
End-struck	7	-	9	12	11	43	19	18	29	7	10	11	21
	30.43		60.00	54.55	52.38	58.11	57.58	62.07	60.42	58.33	71.43	31.43	23.08
Side-struck	6	4	5	3	5	16	5	6	4	2	1	9	18
	26.09	21.05	33.33	13.64	23.81	21.62	15.15	20.69	8.33	16.67	7.14	25.71	19.78
Special side-struck	2	1	-	-	1	4	1	1	5	-	-	1	10
	8.70	5.26			4.76	5.41	3.03	3.45	10.42			2.86	10.99
Cortex													
No cortex	14	11	8	12	11	45	18	20	27	10	11	23	58
	60.87	57.89	53.33	54.55	52.38	60.81	54.55	68.97	56.25	83.33	78.57	65.71	63.04
1–25%	2	2	4	4	2	17	1	5	6	1	2	3	8
	8.70	10.53	26.67	18.18	9.52	22.97	3.03	17.24	12.50	8.33	14.29	8.57	8.70
26–50%	2	1	1	2	5	3	7	-	4	-	-	5	8
	8.70	5.26	6.67	9.09	23.81	4.05	21.21		8.33			14.29	8.70
51–75%	-	2	-	1	2	5	1	1	4	-	-	2	5
		10.53		4.55	9.52	6.76	3.03	3.45	8.33			5.71	5.43
76–100%	5	3	2	2	1	4	5	3	7	1	1	2	10
	21.74	15.79	13.33	9.09	4.76	5.41	15.15	10.34	14.58	8.33	7.14	5.71	10.87
Indeterminate	-	-	-	1	-	-	1	-	-	-	-	-	3
				4.55			3.03						3.26

Table 6.7 Descriptive statistics of flint notches (size in mm; angle in degrees)

	Layer												
						II-6 Levels							
	V-5	II-2	II-2/3	II-5	II-5/6	L1	L2	L3	L4	L4b	L5	L6	L7
N	23	19	15	22	21	74	33	29	48	12	14	35	92
Maximum length													
Maximum	68.00	45.00	45.00	39.00	42.00	67.00	52.00	49.00	44.00	39.00	54.00	44.00	49.00
Median	27.00	25.00	27.00	25.00	26.00	28.00	26.00	27.00	28.00	26.00	25.50	28.00	27.00
Minimum	21.00	21.00	21.00	16.00	19.00	20.00	20.00	21.00	18.00	22.00	22.00	21.00	21.00
Mean	30.87	25.32	29.20	27.09	27.10	29.18	29.09	30.24	28.98	26.92	28.50	29.23	27.73
Std dev	11.52	5.22	6.42	6.33	5.53	7.70	8.58	7.06	6.05	4.64	8.57	6.19	5.45
Std err mean	2.40	1.20	1.66	1.35	1.21	0.90	1.49	1.31	0.87	1.34	2.29	1.05	0.57
Length													
Maximum	66.00	34.00	45.00	38.00	34.00	67.00	52.00	49.00	44.00	37.00	49.00	42.00	44.00
Median	23.00	19.00	24.00	24.00	26.00	24.50	24.00	25.00	26.50	24.50	24.50	23.00	23.00
Minimum	12.00	15.00	16.00	11.00	16.00	10.00	13.00	11.00	17.00	16.00	13.00	14.00	12.00
Mean	26.78	20.79	25.13	25.50	24.43	25.86	26.45	27.62	27.35	24.58	26.64	25.17	23.96
Std dev	11.73	4.74	7.74	7.03	5.25	8.87	9.52	8.26	6.46	5.85	8.65	6.34	6.43
Std err mean	2.45	1.09	2.00	1.50	1.15	1.03	1.66	1.53	0.93	1.69	2.31	1.07	0.67
Width													
Maximum	50.00	41.00	31.00	34.00	42.00	43.00	51.00	39.00	41.00	31.00	36.00	43.00	37.00
Median	22.00	22.00	22.00	19.00	20.00	24.50	22.00	23.00	21.00	19.50	19.00	22.00	23.00
Minimum	11.00	13.00	13.00	14.00	14.00	12.00	7.00	13.00	13.00	13.00	11.00	14.00	9.00
Mean	22.74	21.89	22.67	20.59	22.00	23.77	23.45	22.76	23.00	20.83	20.36	23.43	22.62
Std dev	8.90	6.05	4.88	5.94	6.32	6.00	9.10	5.51	6.96	5.56	6.63	6.59	5.86
Std err mean	1.86	1.39	1.26	1.27	1.38	0.70	1.58	1.02	1.00	1.60	1.77	1.11	0.61
Thickness													
Maximum	13.00	12.00	15.00	30.00	12.00	21.00	17.00	23.00	18.00	12.00	13.00	15.00	23.00
Median	7.00	8.00	9.00	9.50	8.00	10.00	10.00	10.00	9.00	8.00	9.50	8.00	8.00
Minimum	4.00	5.00	5.00	4.00	5.00	4.00	4.00	4.00	5.00	6.00	4.00	6.00	4.00
Mean	7.70	8.26	9.47	10.09	8.10	9.84	9.79	10.59	9.52	8.83	8.64	9.00	8.66
Std dev	2.36	1.63	3.23	5.41	1.81	3.55	3.63	4.02	3.05	2.12	2.84	2.96	2.75
Std err mean	0.49	0.37	0.83	1.15	0.40	0.41	0.63	0.75	0.44	0.61	0.76	0.50	0.29
Angle of retouch	(N=16)	(N=18)	(N=15)	(N=18)	(N=9)	(N=61)	(N=27)	(N=23)	(N=40)	(N=8)	(N=10)	(N=28)	(N=81)
Maximum	96.00	103.00	89.00	113.00	111.00	118.00	93.00	113.00	92.00	88.00	105.00	98.00	106.00
Median	79.50	84.00	78.00	82.50	82.00	81.00	79.00	83.00	81.50	83.50	80.00	80.50	81.00
Minimum	67.00	66.00	61.00	50.00	66.00	67.00	67.00	56.00	72.00	74.00	70.00	80.00	67.00
Mean	80.44	82.50	77.33	81.22	82.78	82.87	80.48	81.65	81.7	82.13	81.6	81.50	81.20
Std dev	7.64	8.72	6.58	12.56	14.13	8.42	5.56	10.63	5.65	4.55	9.55	7.42	7.94
Std err mean	1.91	2.06	1.70	2.96	4.71	1.08	1.07	2.21	0.89	1.61	3.02	1.40	0.88

agent in the life-histories of these tools. This assumption is corroborated by the fact that only one Clactonian notch was recorded in the entire GBY lithic assemblage. Additional support is provided by the minimal frequencies of notches on fragments.

Denticulates

Denticulates are defined as artifacts that exhibit a series of more than two notches on the same edge (Fig. 6.8). They are conceptually related to the notched tools, although they were not included in the above analyses of notches and are analyzed and discussed separately here.

Since denticulates occur in relatively low frequencies, the analyzed samples of each layer are fairly small. However, the common preservation state seems to be slightly abraded, with the exception of Layer V-5, where fresh denticulates occur in the highest frequencies (Table 6.10). This pattern is similar to the preservation state observed on other tool types, and again the exception is Layer II-6 Level 7, which exhibits higher frequencies of abraded tools. Similarly, as observed for other tool types, denticulates are mostly patinated, with some double patinated tools (Table 6.10).

6.2 Typology of Flint Flake Tools and Core Tools

Table 6.8 Counts and frequencies (%) of retouch characteristics of flint notches

	Layer												
						II-6 Levels							
	V-5	II-2	II-2/3	II-5	II-5/6	L1	L2	L3	L4	L4b	L5	L6	L7
N	23	19	15	22	21	73	33	28	46	12	14	35	91
Location of retouch													
Distal	3	6	3	6	4	23	14	8	9	3	1	10	20
	13.04	31.58	20.00	27.27	19.05	31.51	42.42	28.57	19.57	25.00	7.14	28.57	21.98
Proximal	3	1	-	2	-	9	2	4	2	1	-	3	8
	13.04	5.26		9.09		12.33	6.06	14.29	4.35	8.33		8.57	8.79
Left	4	7	5	5	6	17	8	6	13	4	3	5	27
	17.39	36.84	33.33	22.73	28.57	23.29	24.24	21.43	28.26	33.33	21.43	14.29	29.67
Right	12	4	6	6	10	21	8	9	19	3	6	15	26
	52.17	21.05	40.00	27.27	47.62	28.77	24.24	32.14	41.30	25.00	42.86	42.86	28.57
Both edges	-	-	-	2	1	1	-	1	2	1	3	2	6
				9.09	4.76	1.37		3.57	4.35	8.33	21.43	5.71	6.59
Distal & both edges	1	-	-	-	-	-	-	-	-	-	-	-	-
	4.35												
Distal & right	-	-	1	1	-	1	-	-	-	-	-	-	2
			6.67	4.55		1.37							2.20
Distal & left	-	1	-	-	-	-	-	-	-	-	-	-	-
		5.26											
Convergent	-	-	-	-	-	-	-	-	-	-	-	-	1
													1.10
Circumference	-	-	-	-	-	1	-	-	1	-	1	-	1
						1.37			2.17		7.14		1.10
Indeterminate	-	-	-	-	-	-	1	-	-	-	-	-	-
							3.03						
Face of retouch	(N=22)		(N=13)		(N=14)		(N=32)	(N=27)		(N=11)	(N=12)		(N=90)
Dorsal	8	14	10	11	8	46	27	15	29	4	7	21	60
	36.36	73.68	76.92	50.00	57.14	63.01	84.38	55.56	63.04	36.36	58.33	60.00	66.67
Ventral	9	2	3	4	3	16	2	9	13	5	3	11	18
	40.91	10.53	23.08	18.18	21.43	21.92	6.25	33.33	28.26	45.45	25.00	31.43	20.00
Dorsal & ventral	4	1	-	3	2	5	2	-	3	1	2	1	3
	18.18	5.26		13.64	14.29	6.85	6.25		6.52	9.09	16.67	2.86	3.33
Side	1	2	-	4	1	6	1	3	1	1	-	2	9
	4.55	10.53		18.18	7.14	8.22	3.13	11.11	2.17	9.09		5.71	10.00

Fig. 6.6 Flint multiple tool: notch and denticulate (#9186 Layer II-6 Level 3)

Table 6.9 Counts of multiple tools on flint notches

Typology B	Typology A		Total
	Notch	End-notch	
Retouched flake	35	5	40
Notch	-	1	1
End-notch	7	-	7
Denticulate	11	-	11
Truncation	-	1	1
Varia	2	-	2
Total	55	7	62

Fig. 6.7 Flint multiple tool: notch, denticulate and retouched flake with removed striking platform **1** multiple views **2** detailed view of notch (#5942 Layer II-6 Level 1)

The denticulates are typically complete, and when broken they exhibit mostly distal breaks (with the exception of Layer II-6 Level 4, in which proximal breaks are more common).

The blanks on which denticulates are modified are mostly end-struck, followed by side-struck. As with other tool types, there is an apparent selection of non-cortical blanks for the production of denticulates.

When the location of the denticulated edge is examined, the most common mode is that of the left edge, followed by the circumference. Layer V-5 deviates from this pattern, with equal proportions of distal and right edges (Table 6.11). Denticulates

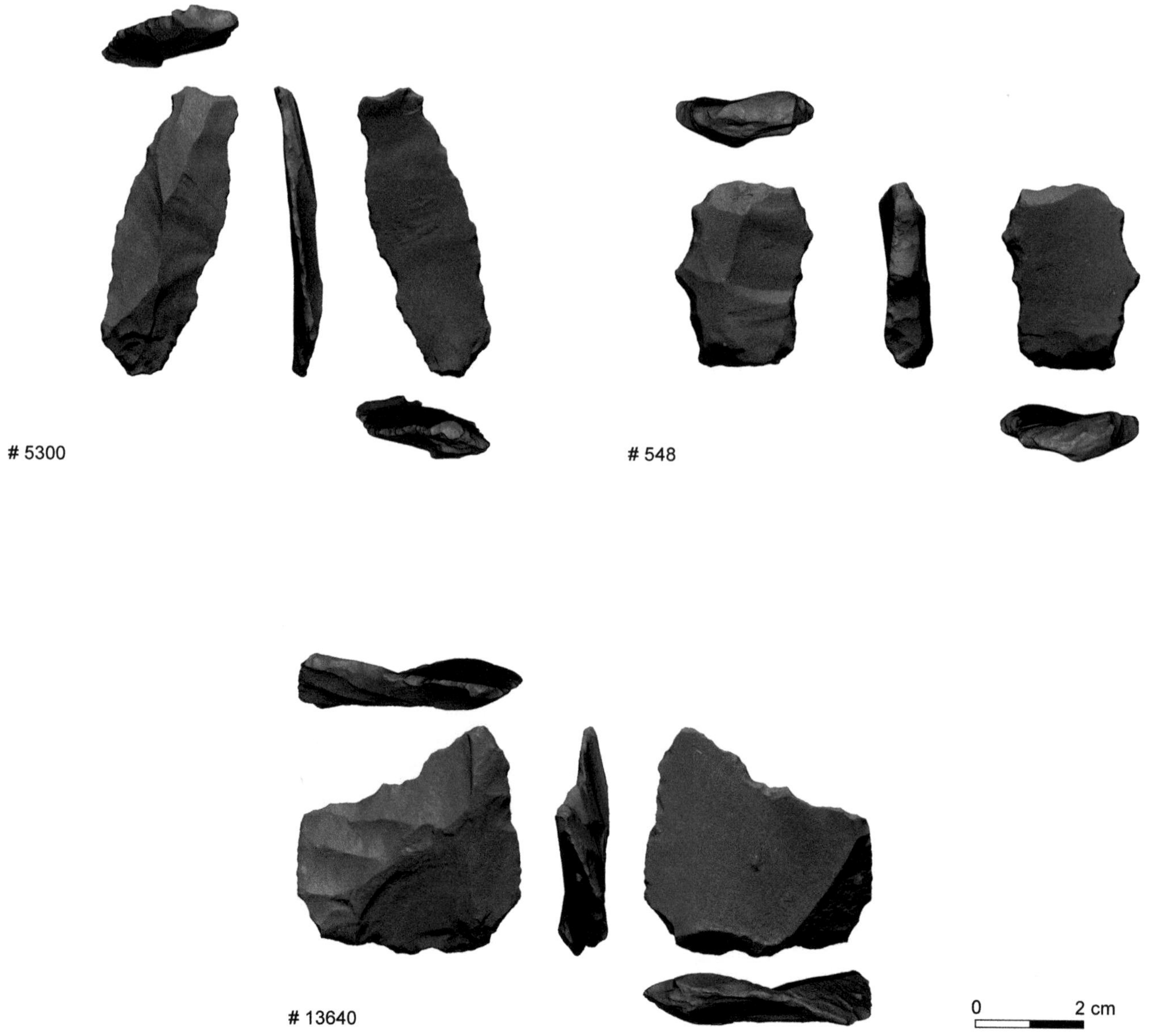

Fig. 6.8 Flint denticulates (#5300 Layer V-5 with signs of utilization, #548 Layer II-3 with removed striking platform, #13640 Layer II-6/7 on biface sharpening flake)

6.2 Typology of Flint Flake Tools and Core Tools

Table 6.10 Counts and frequencies (%) of characteristics of flint denticulates

	Layer								
					II-6 Levels				
	V-5	L1	L2	L3	L4	L4b	L5	L6	L7
N	22	22	6	13	17	6	4	15	18
Preservation									
Fresh	15	7	2	3	8	1	-	3	5
	68.18	31.82	33.33	23.08	47.06	16.67		20.00	27.78
Slightly abraded	7	13	1	6	4	3	2	8	5
	31.82	59.09	16.67	46.15	23.53	50.00	50.00	53.33	27.78
Abraded	-	2	2	3	5	1	2	4	7
		9.09	33.33	23.08	29.41	16.67	50.00	26.67	38.89
Heavily abraded	-	-	1	1	-	1	-	-	1
			16.67	7.69		16.67			5.56
Patination									
No patina	-	1	-	1	1	1	-	-	-
		4.55		7.69	5.88	16.67			
Patinated	16	17	5	9	10	4	2	10	12
	72.73	77.27	83.33	69.23	58.82	66.67	50.00	66.67	66.67
Double patinated	6	4	1	3	6	1	2	5	6
	27.27	18.18	16.67	23.08	35.29	16.67	50.00	33.33	33.33
Breakage									
Complete	13	16	5	9	10	5	3	10	14
	59.09	72.73	83.33	69.23	58.82	83.33	75.00	66.67	77.78
Distal	4	3	-	1	1	-	-	2	-
	18.18	13.64		7.69	5.88			13.33	
Lateral	3	1	-	-	2	1	-	2	2
	13.64	4.55			11.76	16.67		13.33	11.11
Proximal	-	1	-	2	4	-	1	-	-
		4.55		15.38	23.53		25.00		
Distal & proximal	1	1	1	-	-	-	-	-	1
	4.55	4.55	16.67						5.56
Fragment	1	-	-	1	-	-	-	1	1
	4.55			7.69				6.67	5.56
Direction of blow									(N=17)
End-struck	9	15	5	5	7	3	1	7	4
	40.91	68.18	83.33	38.46	41.18	50.00	25.00	46.67	23.53
Side-struck	4	5	-	6	5	2	-	1	6
	18.18	22.73		46.15	29.41	33.33		6.67	35.29
Special side-struck	2	-	-	-	1	-	1	2	2
	9.09				5.88		25.00	13.33	11.76
Indeterminate	7	2	1	2	4	1	2	5	5
	31.82	9.09	16.67	15.38	23.53	16.67	50.00	33.33	29.41
Cortex									
No cortex	16	14	4	9	11	2	1	9	12
	72.73	63.64	66.67	69.23	64.71	33.33	25.00	60.00	66.67
1–25%	4	2	1	2	3	1	1	5	-
	18.18	9.09	16.67	15.38	17.65	16.67	25.00	33.33	
26–50%	1	2	-	1	-	2	1	1	-
	4.55	9.09		7.69		33.33	25.00	6.67	
51–75%	-	2	1	-	-	1	-	-	2
		9.09	16.67			16.67			11.11
76–100%	1	2	-	1	3	-	1	-	4
	4.55	9.09		7.69	17.65		25.00		22.22

Table 6.11 Counts and frequencies (%) of retouch characteristics of flint denticulates

					Layer				
					II-6 Levels				
	V-5	L1	L2	L3	L4	L4b	L-5	L6	L7
N	21	22	6	12	17	6	4	15	18
Location of retouch									
Distal	3	5	-	3	2	1	1	1	1
	14.29	22.73		25.00	11.76	16.67	25.00	6.67	5.56
Proximal	-	1	-	-	1	-	-	-	-
		4.55			5.88				
Left	7	5	2	5	3	1	-	6	4
	33.33	22.73	33.33	41.67	17.65	16.67		40.00	22.22
Right	3	2	2	1	4	2	-	2	4
	14.29	9.09	33.33	8.33	23.53	33.33		13.33	22.22
Both edges	1	-	-	-	1	-	1	1	2
	4.76				5.88		25.00	6.67	11.11
Distal & right	2	-	-	1	-	-	-	-	1
	9.52			8.33					5.56
Distal & left	-	2	-	-	-	-	-	1	2
		9.09						6.67	11.11
Distal & both edges	-	-	-	-	-	1	-	2	-
						16.67		13.33	
Proximal & both edges	1	-	-	-	-	-	-	-	-
	4.76								
Convergent	2	2	1	-	2	-	-	1	-
	9.52	9.09	16.67		11.76			6.67	
Circumference	2	5	1	2	4	-	2	1	4
	9.52	22.74	16.67	16.67	23.53		50.00	6.67	22.22
Indeterminate	-	-	-	-	-	1	-	-	-
						16.67			
Face of retouch				(N=10)		(N=4)	(N=3)	(N=14)	
Dorsal	14	13	5	5	7	3	2	9	8
	66.67	59.09	83.33	50.00	41.18	75.00	66.67	64.29	44.44
Ventral	2	2	-	3	3	1	-	1	3
	9.52	9.09		30.00	17.65	25.00		7.14	16.67
Dorsal & ventral	5	7	1	2	6	-	1	3	7
	23.81	31.82	16.67	20.00	35.29		33.33	21.43	38.89
Side	-	-	-	-	1	-	-	1	-
					5.88			7.14	

are modified primarily on the dorsal face, followed by the combination of dorsal and ventral faces (Table 6.11). Despite the typological similarity of the edge forms of denticulates and notches, different preferences are evident in denticulates with regard to retouch location and face of retouch.

As observed for the entire flint component, both retouched and unretouched, denticulates are generally small, with quite similar dimensions of length and width (Table 6.12). A general similarity in size between denticulates and notches is observed, with the exception of Layer II-6 Level 1, where denticulates appear to be slightly larger in all dimensions.

On 15 denticulates (8.47% of all denticulated tools at GBY) we observed an additional tool type. Of these 11 are retouched flakes, one is an atypical borer, and one is an end-notch. Denticulates were also recorded as a secondary tool on other tool types. These cases are presented elsewhere in accordance with the dominant typological classification (Typology A).

Denticulated Points

This tool category, a Levantine version of the Tayac point, was included in the category of denticulates because of the

6.2 Typology of Flint Flake Tools and Core Tools

Table 6.12 Descriptive statistics of flint denticulates (size in mm; angle in degrees)

	Layer								
		II-6 Levels							
	V-5	L1	L2	L3	L4	L4b	L5	L6	L7
N	22	22	6	13	17	6	4	15	18
Maximum length									
Maximum	50.00	53.00	51.00	41.00	51.00	40.00	43.00	45.00	39.00
Median	30.00	31.00	28.50	27.00	30.00	35.00	36.00	28.00	25.50
Minimum	21.00	21.00	21.00	21.00	22.00	21.00	22.00	23.00	22.00
Mean	31.86	32.23	33.00	28.46	30.82	32.17	34.25	30.73	26.78
Std dev	8.96	7.98	12.95	6.50	7.05	7.55	8.85	6.70	4.98
Std err mean	1.91	1.70	5.29	1.80	1.71	3.08	4.42	1.73	1.17
Length									
Maximum	50.00	42.00	51.00	38.00	51.00	39.00	43.00	44.00	28.00
Median	29.50	29.50	27.50	23.00	24.00	31.50	30.50	27.00	22.00
Minimum	18.00	19.00	17.00	13.00	20.00	21.00	22.00	15.00	18.00
Mean	29.77	29.41	31.83	23.92	27.53	30.83	31.50	27.93	22.22
Std dev	8.35	6.75	14.12	7.65	7.79	6.88	8.66	6.95	2.96
Std err mean	1.78	1.44	5.76	2.12	1.89	2.81	4.33	1.80	0.70
Width									
Maximum	43.00	51.00	39.00	36.00	42.00	37.00	35.00	37.00	39.00
Median	19.50	25.00	21.50	23.00	26.00	29.50	27.00	23.00	22.00
Minimum	13.00	13.00	10.00	18.00	15.00	15.00	18.00	14.00	12.00
Mean	22.36	25.45	22.67	25.00	26.06	26.33	26.75	24.53	22.94
Std dev	7.57	8.49	9.37	5.49	6.80	9.33	6.99	6.93	6.70
Std err mean	1.61	1.81	3.83	1.52	1.65	3.81	3.50	1.79	1.58
Thickness									
Maximum	13.00	16.00	14.00	14.00	16.00	19.00	14.00	15.00	14.00
Median	8.00	10.00	9.50	8.00	9.00	10.50	10.00	11.00	9.00
Minimum	5.00	3.00	5.00	6.00	4.00	6.00	8.00	5.00	4.00
Mean	8.27	9.59	9.50	9.00	9.47	11.50	10.50	9.93	8.44
Std dev	2.47	3.47	3.02	2.80	3.91	5.09	2.65	2.87	3.01
Std err mean	0.53	1.74	1.23	0.78	0.95	2.08	1.32	0.74	0.71
Angle of retouch	(N=13)	(N=21)	(N=6)	(N=8)	(N=16)	(N=6)	(N=2)	(N=11)	(N=15)
Maximum	82.00	105.00	89.00	89.00	88.00	91.00	87.00	108.00	88.00
Median	78.00	82.00	85.00	81.00	82.50	83.00	77.00	79.00	78.00
Minimum	69.00	63.00	75.00	67.00	72.00	76.00	67.00	66.00	69.00
Mean	77.31	82.19	83.17	80.50	81.63	83.50	77.00	81.91	79.27
Std dev	3.84	7.78	5.60	7.69	5.88	5.54	14.14	11.05	5.30
Std err mean	1.06	1.70	2.29	2.72	1.47	2.26	10.00	3.33	1.37

similarity of their edges. As these tools have denticulated convergent edges (Fig. 6.9), they are considered here to be a sub-group of the denticulate category, and are integrated into the above analysis.

Four tools are classified under the typological category of denticulated points. Of these, three are cortical flakes and one is a broken tip. All share the same characteristic of converging lateral edges, retouched in a denticulated fashion, resulting in a pointed plan form. One of the tools (#14962, Fig. 6.9) is additionally retouched in the same denticulated manner on its proximal end. On three items the retouch is limited to the dorsal face, while on the fourth, a heavily abraded tool, a few retouched removals can be discerned on the ventral face as well.

Borers

This category includes a variety of tools that are all characterized by a small-scale pointed morphology (Figs. 6.10–6.13). The point is often modified by two notches and varies in thickness, corresponding to the thickness of the tool. Included in this typological category are tools such as perçoirs, becs (beaks), and awls. They perhaps lack the

Fig. 6.9 Flint denticulated points (#621 Layer II-2/3, #7343 Layer II-6 Level 4, #14962 Layer II-7)

standardization that characterizes this tool type in later periods, when the pointed edge is systematically located in the same place, but at GBY these tools are nevertheless distinct and recur throughout the sequence. The frequencies of typical and atypical borers are presented in detail in Table 6.5. In the following description we have combined them into a single group.

Borers are clearly an important component of early tool inventories and appear in many other Lower Paleolithic sites, whether Acheulian or not (e.g. 'Ubeidiya: Bar-Yosef and Goren-Inbar 1993; Bizat Ruhama: Zaidner 2010, Zaidner et al. 2010; Revadim: Malinsky-Buller et al. 2011; Solodenko 2010).

At GBY an additional aspect of the typology and technology of borers is seen in artifacts with a removed striking platform that was modified in the form of a borer. These are not included in the following discussion and are described in detail below (Section 6.2.1.1). However, in some of these cases the distinctive borer characteristics were sufficiently prominent to allow the artifacts to be classified as borers (e.g. #1675, Fig. 6.13)

The following discussion of borers refers to Layer II-6 Levels 1–7 and Layer V-5, for each of which a minimum number of 10 borers is recorded.

Table 6.13 presents data on the physical properties of the borers. It is clear that the preservation of most of the borers varies between fresh and slightly abraded and that they are distributed evenly (when sample size is large enough) between these categories. An exception is Layer V-5, in which most of the artifacts are fresher, a phenomenon encountered in other tool types and waste product categories and relates to the general depositional nature of the assemblages excavated in Area C.

When patination is examined, the patinated category is the most frequent in all layers. The only deviation from this pattern is encountered in Layer II-6 Level 5, where double patinated artifacts are more frequent (Table 6.13).

Most of the borers are complete in all layers and levels, even where the frequencies are extremely small. Similarly to other tool types, it seems that the frequencies of complete borers are higher than those of unretouched flakes. Technologically, the borers are mostly made on end-struck flakes, followed by side-struck flakes (Table 6.13). The majority of borers are modified on non-cortical flakes or on flakes with minimal cortical coverage (Table 6.13).

The dominant mode of production of the borers' working edge is by notches, although other retouch types occur as well in smaller frequencies: regular and abrupt retouch are the next most frequent retouch categories in the GBY borer group (Table 6.14). The pointed morphology of the borers is mostly that of a single tooth, followed by much lower frequencies of borers shaped by convergent retouch.

A great variety is observed in the location of retouch of

6.2 Typology of Flint Flake Tools and Core Tools

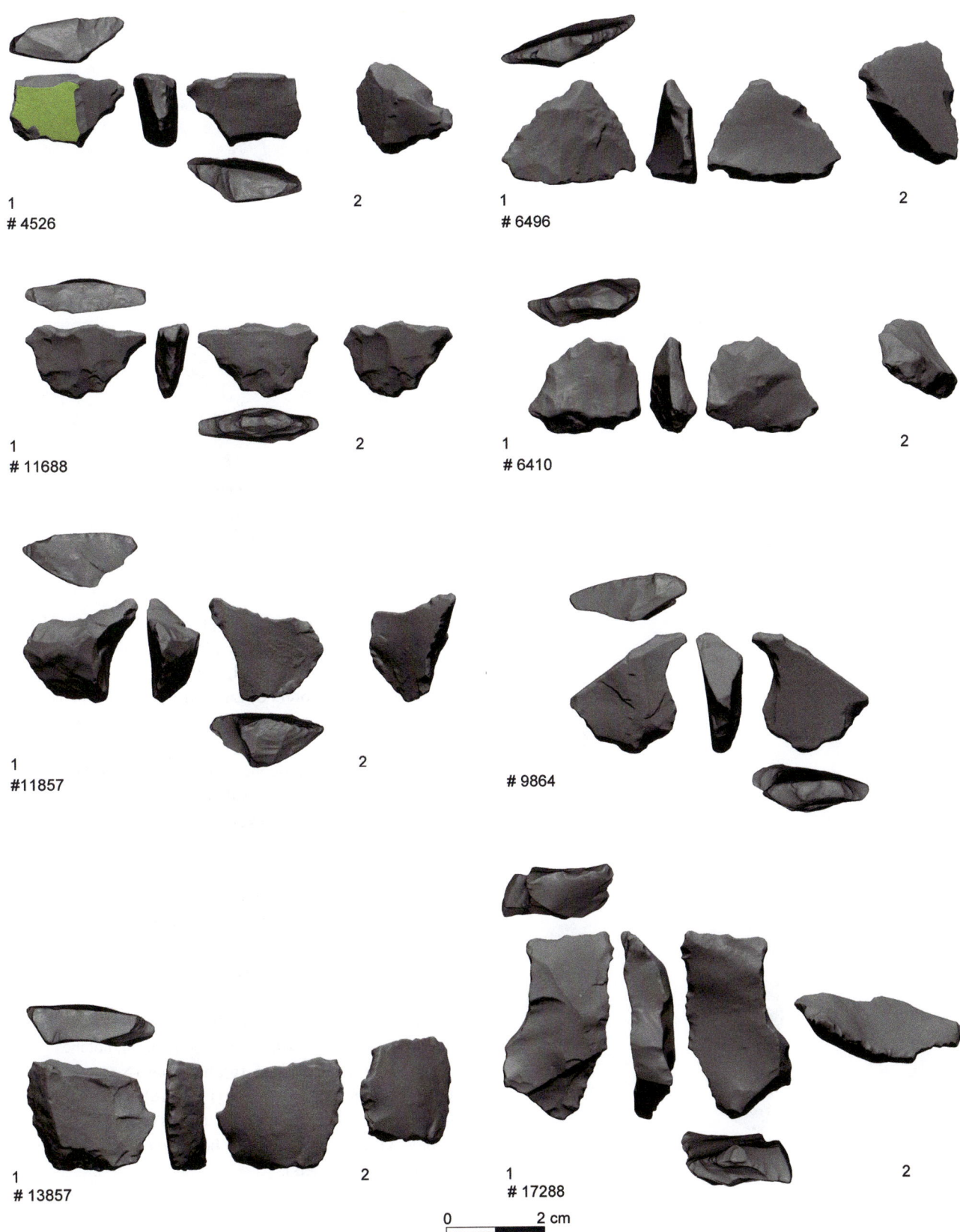

Fig. 6.10 Flint borers **1** multiple views **2** detailed view of borer (#4526 Layer V-5, #11688 Layer II-6 Level 1 with signs of utilization, #11857 Layer II-6 Level 1, #13857 Layer II-6 Level 1 with signs of utilization, #6496 Layer II-6 Level 1 with signs of utilization, #6410 Layer II-6 Level 1, #9864 Layer II-6 Level 7, #17288 Layer II-6 Level 6 on *débordant* flake)

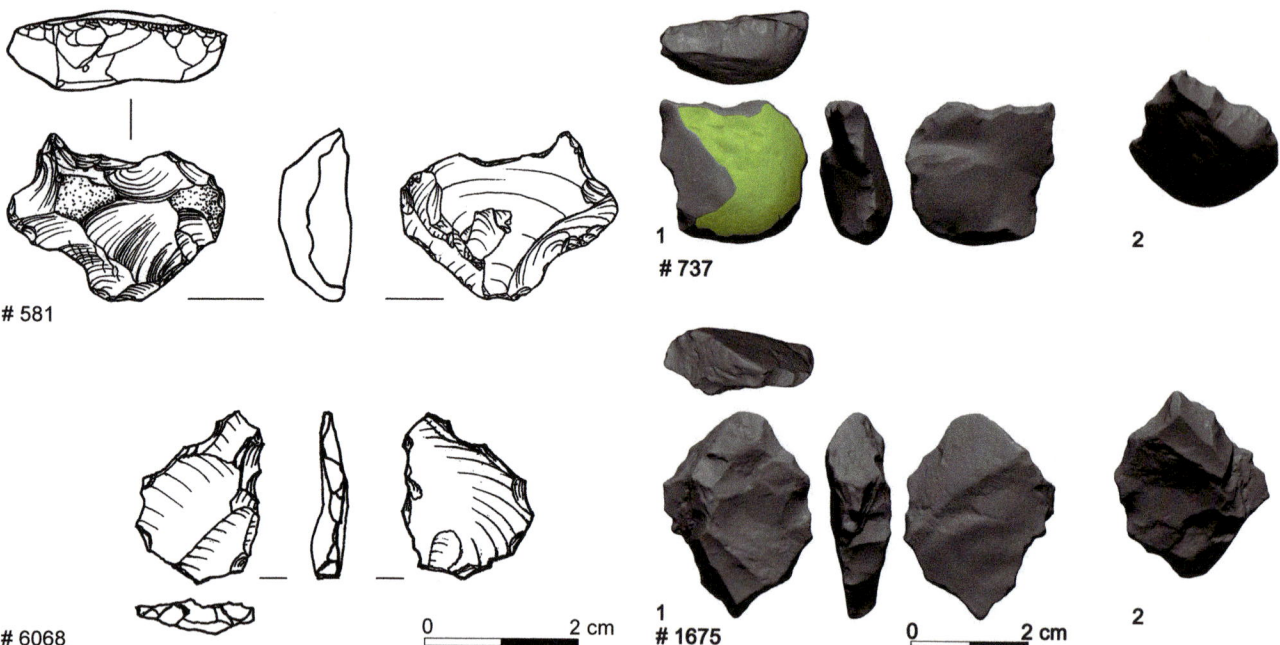

Fig. 6.11 Flint borers (#581 Layer II-2/3 with removed striking platform, #6068 Layer II-6 Level 1 with signs of utilization)

Fig. 6.13 Flint borers **1** multiple views **2** detailed view of borer (#737 Layer II-5, #1675 Layer II-6 Level 4b)

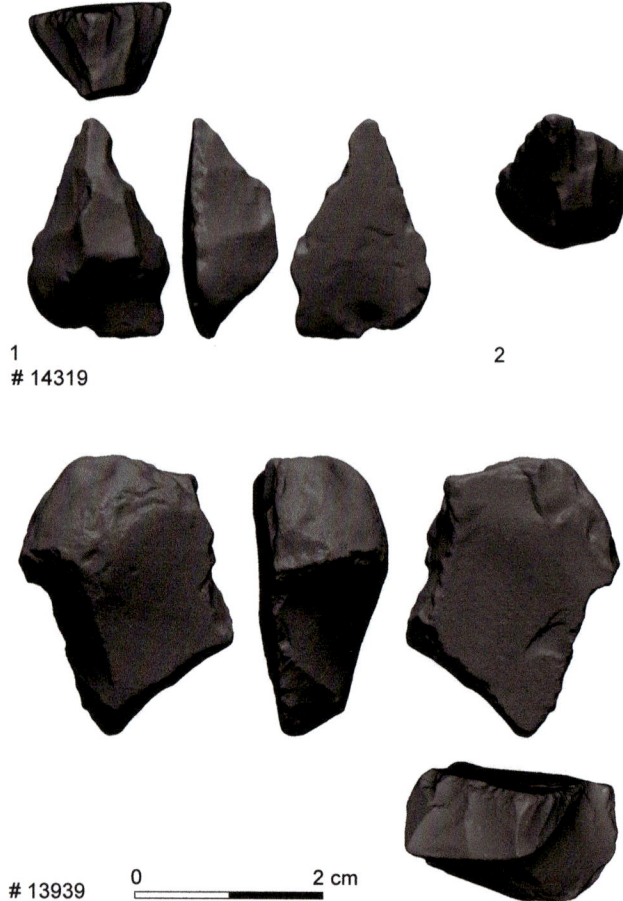

Fig. 6.12 Flint borers (#14319 Layer II-6 Level 7 **1** multiple views **2** detailed view of borer, #13939 Layer II-6 Level 7 with removed striking platform)

borers (Table 6.14). There is no preferred location and the tools are modified on the entire plan form of the artifacts. However, the locations of distal, left, and right are the most common. Within the borers of Layer II-6 Level 7 (the richest borer assemblage), there are high frequencies of all locations of retouch.

Borers are often retouched on their dorsal faces and less commonly on both dorsal and ventral faces; when the sample size is large, however, ventral frequencies rise substantially.

There is a great similarity in the dimensions of borers throughout the cultural sequence. This holds true for all size attributes, as well as for the angle of retouch of the borers. The homogeneity of size (Table 6.15) is not restricted to borers and similar sizes are recorded for other tool types (Section 6.2.1.1).

Some 25% of the borers display an additional tool type (Table 6.16). These include retouched flakes (Fig. 6.14), notches (Fig. 6.15), and side-scrapers (Fig. 6.16), but a few (N=4) exhibit an additional borer on a different edge (Fig. 6.16). It should also be noted that among the multiple tools of GBY, 14 side-scrapers (Typology A) are associated with borers (see detailed counts below under side-scrapers).

Side-scrapers

Side-scrapers, classified following Bordes (1961; see Chapter 4), occur in relatively high frequencies. Their detailed typological

6.2 Typology of Flint Flake Tools and Core Tools

Table 6.13 Counts and frequencies (%) of characteristics of flint borers

	Layer									
			\multicolumn{8}{c}{II-6 Levels}
	V-5	II-5	L1	L2	L3	L4	L4b	L5	L6	L7
N	16	16	31	16	24	28	13	11	24	42
Preservation										
Fresh	15	7	16	4	12	13	6	1	2	-
	93.75	43.75	51.61	25.00	50.00	46.43	46.15	9.09	8.33	
Slightly abraded	1	6	13	10	11	10	6	4	14	15
	6.25	37.50	41.94	62.50	45.83	35.71	46.15	36.36	58.33	35.71
Abraded	-	3	2	2	1	5	1	6	6	21
		18.75	6.45	12.50	4.17	17.86	7.69	54.55	25.00	50.00
Heavily abraded	-	-	-	-	-	-	-	-	2	6
									8.33	14.29
Patination										
No patina	-	-	5	1	7	1	1	-	-	-
			16.13	6.25	29.17	3.57	7.69			
Patinated	11	10	19	9	13	21	11	4	18	29
	68.75	62.50	61.29	56.25	54.17	75.00	84.62	36.36	75.00	69.05
Double patinated	5	6	7	6	4	6	1	7	6	13
	31.25	37.50	22.58	37.50	16.67	21.43	7.69	63.64	25.00	30.95
Breakage										
Complete	14	10	20	11	15	24	8	6	20	32
	87.50	62.50	64.52	68.75	62.50	85.71	61.54	54.55	83.33	76.19
Distal	-	-	3	1	3	1	1	1	2	3
			9.68	6.25	12.50	3.57	7.69	9.09	8.33	7.14
Lateral	-	3	2	-	4	1	3	1	2	2
		18.75	6.45		16.67	3.57	23.08	9.09	8.33	4.76
Proximal	1	1	5	3	2	1	-	3	-	2
	6.25	6.25	16.13	18.75	8.33	3.57		27.27		4.76
Distal & proximal	-	-	-	1	-	1	-	-	-	1
				6.25		3.57				2.38
Fragment	1	1	1	-	-	-	1	-	-	1
	6.25	6.25	3.23				7.69			2.38
Proximal & lateral	-	1	-	-	-	-	-	-	-	1
		6.25								2.38
Direction of blow										
End-struck	5	7	10	9	13	15	8	4	10	15
	31.25	43.75	32.26	56.25	54.17	53.57	61.54	36.36	41.67	35.71
Side-struck	4	4	14	4	8	10	4	2	8	6
	25.00	25.00	45.16	25.00	33.33	35.71	30.77	18.18	33.33	14.29
Special side-struck	2	-	-	-	1	1	1	1	3	8
	12.50				4.17	3.57	7.69	9.09	12.50	19.05
Indeterminate	5	5	7	3	2	2	-	4	3	13
	31.25	31.25	22.58	18.75	8.33	7.14		36.36	12.50	30.95
Cortex										
No cortex	13	6	21	12	17	17	7	9	16	31
	81.25	37.50	67.74	75.00	70.83	60.71	53.85	81.82	66.67	73.81
1–25%	1	2	3	2	4	4	3	1	2	6
	6.25	12.50	9.68	12.50	16.67	14.29	23.08	9.09	8.33	14.29
26–50%	1	4	4	-	1	4	-	1	2	1
	6.25	25.00	12.90		4.17	14.29		9.09	8.33	2.38
51–75%	-	1	1	-	-	2	1	-	-	-
		6.25	3.23			7.14	7.69			
76–100%	1	3	2	2	2	1	2	-	4	3
	6.25	18.75	6.45	12.50	8.33	3.57	15.38		16.67	7.14
Indeterminate	-	-	-	-	-	-	-	-	-	1
										2.38

Table 6.14 Counts and frequencies (%) of retouch characteristics of flint borers

			Layer							
						II-6 Levels				
	V-5	II-5	L1	L2	L3	L4	L4b	L5	L6	L7
N	16	15	31	16	24	28	13	11	24	42
Type of retouch	(N=14)			(N=15)	(N=23)					(N=41)
Regular	2 / 14.29	2 / 13.33	3 / 9.68	1 / 6.67	2 / 8.70	1 / 3.57	1 / 7.69	-	-	6 / 14.63
End-scraper	1 / 7.14	-	-	-	-	-	-	-	1 / 4.17	-
Side-scraper	-	-	1 / 3.23	-	-	-	-	-	-	3 / 7.32
Notch	7 / 50.00	10 / 66.67	21 / 67.74	10 / 66.67	18 / 78.26	21 / 75.00	8 / 61.54	6 / 54.55	14 / 58.33	27 / 65.85
Invasive	-	-	-	-	1 / 4.35	-	-	1 / 9.09	1 / 4.17	-
Quina	-	-	-	-	-	1 / 3.57	-	-	-	-
Semi-abrupt	-	-	-	-	-	-	-	-	1 / 4.17	-
Abrupt	3 / 21.43	2 / 13.33	2 / 6.45	2 / 13.33	2 / 8.70	2 / 7.14	1 / 7.69	1 / 9.09	-	2 / 4.88
Mixed	-	-	-	-	-	1 / 3.57	-	1 / 9.09	1 / 4.17	-
Irregular	-	-	2 / 6.45	2 / 13.33	-	2 / 7.14	-	-	2 / 8.33	2 / 4.88
Signs of use	1 / 7.14	-	1 / 3.23	-	-	-	2 / 15.38	-	3 / 12.50	-
Indeterminate	-	1 / 6.67	1 / 3.23	-	-	-	1 / 7.69	2 / 18.18	1 / 4.17	1 / 2.44
Location of Retouch										
Distal	8 / 50.00	3 / 20.00	8 / 25.81	4 / 25.00	7 / 29.17	11 / 39.29	5 / 38.46	2 / 18.18	7 / 29.17	5 / 11.90
Proximal	-	2 / 13.33	3 / 9.68	-	2 / 8.33	2 / 7.14	1 / 7.69	-	3 / 12.50	2 / 4.76
Left	2 / 12.50	3 / 20.00	7 / 22.58	3 / 18.75	4 / 16.67	7 / 25.00	1 / 7.69	-	4 / 16.67	5 / 11.90
Right	2 / 12.50	1 / 6.67	2 / 6.45	3 / 18.75	3 / 12.50	1 / 3.57	2 / 15.38	2 / 18.18	2 / 8.33	5 / 11.90
Both edges	-	-	2 / 6.45	1 / 6.25	2 / 8.33	-	1 / 7.69	-	1 / 4.17	7 / 16.67
Distal & both edges	-	-	1 / 3.23	-	1 / 4.17	-	-	-	-	-
Distal & right	-	2 / 13.33	1 / 3.23	-	-	3 / 10.71	3 / 23.08	2 / 18.18	3 / 12.50	6 / 14.29
Distal & left	1 / 6.25	1 / 6.67	1 / 3.23	1 / 6.25	2 / 8.33	1 / 3.57	-	-	2 / 8.33	5 / 11.90
Proximal & left	-	-	1 / 3.23	1 / 6.25	1 / 4.17	1 / 3.57	-	-	-	-
Convergent	3 / 18.75	2 / 13.33	5 / 16.13	3 / 18.75	2 / 8.33	2 / 7.14	-	5 / 45.45	2 / 8.33	7 / 16.67
Circumference	-	1 / 6.67	-	-	-	-	-	-	-	-
Face of retouch				(N=15)	(N=22)	(N=27)	(N=12)			
Dorsal	11 / 68.75	10 / 66.67	24 / 77.42	6 / 40.00	15 / 68.18	18 / 66.67	9 / 75.00	4 / 36.36	14 / 58.33	23 / 54.76
Ventral	1 / 6.25	4 / 26.67	-	3 / 20.00	2 / 9.09	3 / 11.11	-	2 / 18.18	3 / 12.50	8 / 19.05
Dorsal & ventral	4 / 25.00	1 / 6.67	7 / 22.58	4 / 26.67	4 / 18.18	5 / 18.52	2 / 16.67	4 / 36.36	5 / 20.83	10 / 23.81
Side	-	-	-	2 / 13.33	1 / 4.55	1 / 3.70	1 / 8.33	1 / 9.09	2 / 8.33	1 / 2.38

6.2 Typology of Flint Flake Tools and Core Tools

Table 6.15 Descriptive statistics of flint borers (size in mm; angle in degrees)

	Layer									
					II-6 Levels					
	V-5	II-5	L1	L2	L3	L4	L4b	L5	L6	L7
N	16	16	31	16	24	28	13	11	24	42
Maximum length										
Maximum	68.00	49.00	67.00	41.00	43.00	42.00	41.00	31.00	44.00	52.00
Median	27.00	27.00	28.00	28.00	27.50	29.50	28.00	27.00	27.00	27.50
Minimum	21.00	21.00	21.00	20.00	20.00	21.00	21.00	21.00	21.00	21.00
Mean	28.63	28.44	28.81	28.31	28.42	30.46	28.38	26.36	29.29	28.57
Std dev	11.17	7.28	8.05	6.41	5.68	5.76	5.09	3.14	6.85	7.78
Std err mean	2.79	1.82	1.45	1.60	1.16	1.09	1.41	0.95	1.40	1.20
Length										
Maximum	66.00	49.00	67.00	41.00	38.00	42.00	38.00	31.00	40.00	49.00
Median	23.50	24.50	23.00	21.50	25.00	26.50	25.00	23.00	23.00	23.00
Minimum	16.00	14.00	15.00	13.00	13.00	16.00	17.00	20.00	15.00	14.00
Mean	25.88	25.81	24.74	24.63	24.75	27.25	25.69	23.73	24.42	24.81
Std dev	11.82	8.28	9.11	8.44	5.99	6.48	6.21	3.20	6.93	7.37
Std err mean	2.96	2.07	1.64	2.11	1.22	1.23	1.72	0.96	1.41	1.14
Width										
Maximum	34.00	31.00	43.00	38.00	43.00	40.00	29.00	29.00	43.00	52.00
Median	19.50	23.50	24.00	22.00	23.00	23.50	24.00	20.00	23.00	22.00
Minimum	9.00	14.00	14.00	12.00	13.00	15.00	20.00	14.00	12.00	14.00
Mean	19.56	21.50	23.90	22.88	23.92	24.54	24.15	20.82	23.83	23.31
Std dev	5.62	5.75	5.56	7.24	6.65	7.32	2.70	5.38	6.82	7.00
Std err mean	1.41	1.44	1.00	1.81	1.36	1.38	0.75	1.62	1.39	1.08
Thickness										
Maximum	10.00	13.00	14.00	15.00	18.00	16.00	16.00	14.00	19.00	18.00
Median	7.50	9.00	8.00	8.00	8.50	9.00	8.00	9.00	9.50	9.50
Minimum	4.00	5.00	5.00	4.00	4.00	6.00	5.00	5.00	4.00	4.00
Mean	7.25	9.00	8.29	8.63	8.63	9.64	8.62	9.09	9.71	9.36
Std dev	1.91	2.61	2.48	3.16	2.92	2.71	3.10	2.43	3.04	3.11
Std err mean	0.48	0.65	0.45	0.79	0.60	0.51	0.86	0.73	0.62	0.48
Angle of retouch	(N=10)	(N=14)	(N=29)	(N=13)	(N=22)	(N=26)	(N=11)	(N=9)	(N=16)	(N=35)
Maximum	88.00	92.00	95.00	85.00	120.00	91.00	119.00	84.00	86.00	98.00
Median	83.50	77.50	83.00	82.00	84.00	84.00	86.00	80.00	81.50	83.00
Minimum	71.00	50.00	67.00	72.00	75.00	67.00	77.00	71.00	71.00	71.00
Mean	82.10	74.36	81.59	79.85	84.41	82.62	88.55	79.22	80.13	84.06
Std dev	5.04	12.71	5.89	4.79	9.25	5.97	11.36	4.87	4.32	7.17
Std err mean	1.59	3.40	1.09	1.33	1.97	1.17	3.43	1.62	1.08	1.21

Table 6.16 Counts of multiple tools on flint borers

Additional tool type	N
Single straight side-scraper	1
Single convex side-scraper	2
Abruptly retouched side-scraper	2
Borer	4
Notch	21
Denticulate	4
Varia	2
Retouched flake	22

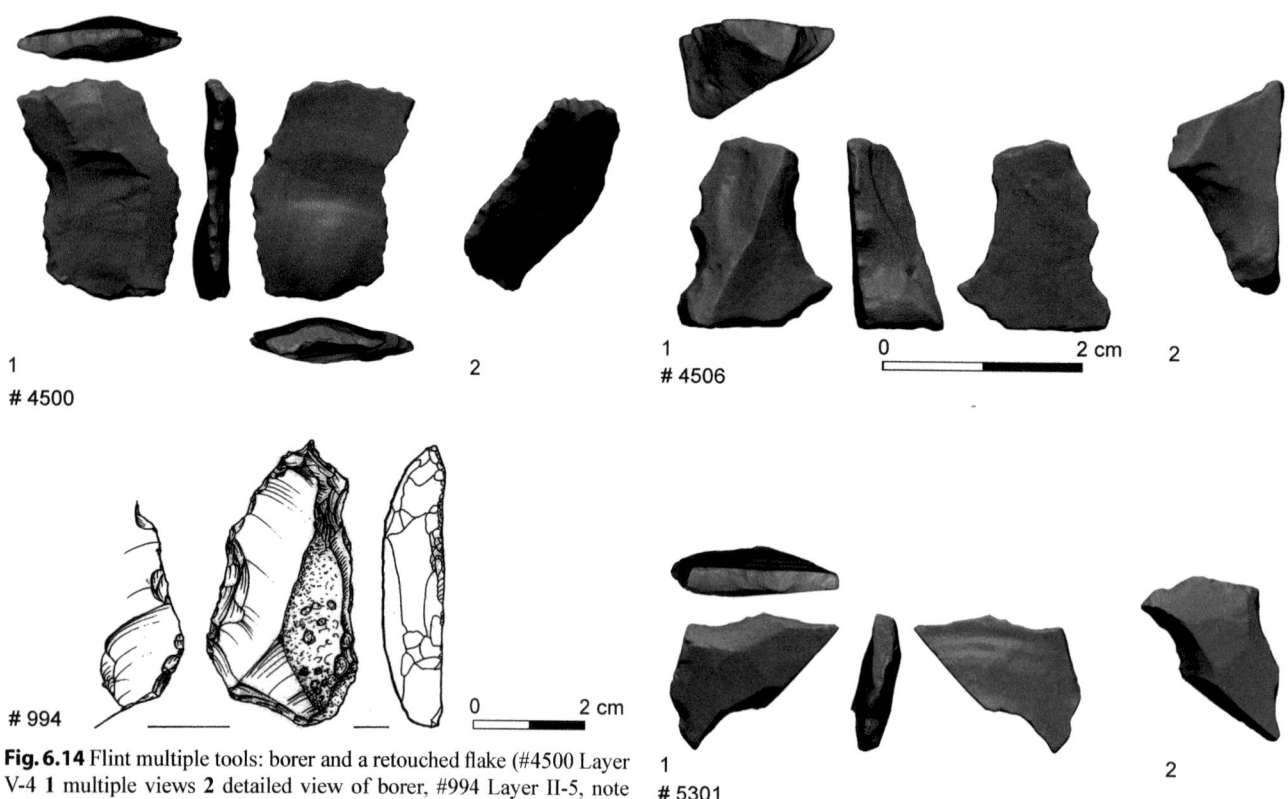

Fig. 6.14 Flint multiple tools: borer and a retouched flake (#4500 Layer V-4 **1** multiple views **2** detailed view of borer, #994 Layer II-5, note ventral removal)

Fig. 6.15 Flint multiple tools **1** multiple views **2** detailed view of borer (#4506 Layer V-4 borer and notch, #5301 Layer V-5 borer and notch, #11064 Layer II-6 Level 1 borer and end-notch)

classification is presented in Table 6.5 according to the different subtypes.

For the present description of the physical properties of this tool type, all subtypes of side-scrapers in each layer were grouped together. With respect to the preservation of side-scrapers, the most common mode is that of slightly abraded. The only exception is Layer II-6 Level 7, where abraded artifacts are more common, similar to the preservation states observed for notches, borers, and retouched flakes of that layer. However, when compared to unretouched flint flakes, the side-scrapers show higher frequencies of fresh artifacts (Table 6.17). This observation was also made for the preservation states of other major tool types. When patination is examined, the most common mode is that of patinated tools, similarly to both retouched and unretouched flakes.

The majority of side-scrapers, in every layer, are complete; when breakage occurs it is mostly at the distal end of the tool, followed by proximal and lateral breakages in decreasing frequencies (Table 6.17). Another characteristic in which side-scrapers resemble other tool types is the dominance of the end-struck production mode (Table 6.17). The high frequency of indeterminate direction of blow in Layer II-6 Level 7 can be attributed to the fact that this assemblage is generally more abraded. Side-scrapers are modified mostly on non-cortical flakes or on flakes with minimal cortical coverage (Table 6.17), and are significantly less cortical than unretouched flakes ($\chi^2=12.07$; $df=4$; $p=0.0168$).

The descriptive statistics of side-scrapers suggest a general homogeneity among layers in the selection of blanks (Table 6.18). The small size of the side-scrapers, which rarely exceed 7 cm in length, reflects the overall diminutive dimensions of

Fig. 6.16 Flint borers **1** multiple views **2** detailed view of borer (#9803 Layer II-6 Level 7 multiple tool borer and single straight side-scraper, #4346 Layer II-6 Level 5)

the entire flint assemblage, both retouched and unretouched. In Layer II-6 Levels 4–5 and 7 the average length seems to be smaller than in other layers. This may stem from the fact that broken tools are included in the statistics. In order to gain a better view of this issue, the sizes of complete side-scrapers were statistically examined for the eight levels of Layer II-6 and no significant different was found. However, when only complete artifacts are examined, side-scrapers in general are significantly larger than unretouched flakes (length: t ratio=-5.97; df=1644; $p(tt)$<0.0001; width: t ratio=-6.12; df=1644; $p(tt)$<0.0001).

Despite their homogeneity of size, the side-scrapers vary widely in blank morphology, suggesting that no particular morphology was preferred in the selection of blanks for modification of side-scrapers.

As seen in Table 6.19, the single side-scraper (N=93, 44.28% of the flint side-scraper assemblage) is the dominant type when the three different edge morphologies (straight, Fig. 6.17, convex, Fig. 6.17, and concave) are combined. Abruptly retouched side-scrapers (Figs. 6.18–6.19) are the second most frequent group (N=45, 21.42% of the flint side-scraper assemblage), followed by side-scrapers on the ventral face (N=30, 14.28% of the flint side-scraper assemblage) (Fig. 6.20). All other types of side-scrapers occur in smaller frequencies (e.g. offset, convergent, Fig. 6.21, transversal, and alternately retouched, Fig. 6.21). The lowest frequency is that of double side-scrapers (Fig. 6.22), which are also the least regular type regarding the varied combinations of edge morphologies (Table 6.20). It seems that the dominant form of edge among side-scrapers is the convex one, whether on single, convergent, or transverse side-scrapers. The typological variability of the side-scrapers is extensive, with on the one hand certain types that are very common throughout the sequence, and on the other hand individual cases of types that occur in minimal frequencies and in a single layer (e.g. bifacially retouched side-scraper) (Table 6.19).

The variability of side-scrapers is further reflected in the edge morphology, type of retouch, and face of retouch (Table 6.20). It is evident that the larger the sample, the more diverse the side-scrapers, as is clearly seen for Layer II-6 Level 7. Two types of retouch occur in higher frequencies – typical side-scraper retouch, followed by abrupt retouch (Table 6.20). While the above observations are derived from the typological definition, the location of retouch does not. This shows a clear preference for the right edge, with the distal and left edges occurring in lower frequencies (Table 6.20). Another observation that is worth noting is the meager presence of multi-edge modifications (e.g. left/right retouch combined with distal/proximal retouch). Thus, the modification of side-scrapers seems to be restricted almost entirely to a single edge (Table 6.20). The dorsal face is the dominant face of retouch among side-scrapers (Table 6.20).

Additional Tool Types (Typologies B and C)

On 102 side-scrapers (38.49% of the entire side-scraper assemblage of GBY) we observed an additional tool type or a classified unretouched artifact (e.g. biface sharpening flake). The detailed typological combinations are listed in Table 6.21. Of the combinations present, the most common is that with lateral notches (Fig. 6.23), followed in descending order by retouched flakes (Fig. 6.23), borers, other side-scrapers (Fig. 6.24), burins (Fig. 6.17), and varia (Table 6.21).

The multiple tools also include artifacts that were modified

Table 6.17 Counts and frequencies (%) of characteristics of flint side-scrapers

				Layer							
							II-6 Levels				
	V-5	V-6	II-5	L1	L2	L3	L4	L4b	L5	L6	L7
N	12	11	16	22	11	18	25	4	12	18	61
Preservation											
Fresh	11	10	11	9	2	7	11	-	-	-	6
	91.67	90.91	68.75	40.91	18.18	38.89	44.00				9.84
Slightly abraded	-	1	4	12	5	9	11	3	8	13	19
		9.09	25.00	54.55	45.45	50.00	44.00	75.00	66.67	72.22	31.15
Abraded	-	-	1	1	2	2	1	1	3	5	33
			6.25	4.55	18.18	11.11	4.00	25.00	25.00	27.78	54.10
Heavily abraded	1	-	-	-	2	-	2	-	1	-	3
	8.33				18.18		8.00		8.33		4.92
Patination											
No patina	-	-	4	3	-	4	4	-	-	-	-
			25.00	13.64		22.22	16.00				
Patinated	9	11	9	12	7	8	17	2	10	10	50
	75.00	100.00	56.25	54.55	63.64	44.44	68.00	50.00	83.33	55.56	81.97
Double patinated	3	-	3	7	4	6	4	2	2	8	11
	25.00		18.75	31.82	36.36	33.33	16.00	50.00	16.67	44.44	18.03
Breakage											
Complete	7	6	8	13	5	10	18	1	7	13	29
	58.33	54.55	50.00	59.09	45.45	55.56	72.00	25.00	58.33	72.22	47.54
Distal	1	1	3	3	-	5	2	-	1	3	12
	8.33	9.09	18.75	13.64		27.78	8.00		8.33	16.67	19.67
Lateral	-	-	-	3	1	1	1	2	2	1	4
				13.64	9.09	5.56	4.00	50.00	16.67	5.56	6.56
Proximal	3	1	2	3	3	2	4	1	2	-	11
	25.00	9.09	12.50	13.64	27.27	11.11	16.00	25.00	16.67		18.03
Distal & lateral	-	2	1	-	-	-	-	-	-	-	1
		18.18	6.25								1.64
Distal & proximal	1	-	1	-	1	-	-	-	-	1	1
	8.33		6.25		9.09					5.56	1.64
Fragment	-	1	1	-	1	-	-	-	-	-	3
		9.09	6.25		9.09						4.92
Direction of blow											
End-struck	5	6	11	10	9	11	7	2	5	7	15
	41.67	54.55	68.75	45.45	81.82	61.11	28.00	50.00	41.67	38.89	24.59
Side-struck	2	2	2	5	1	3	10	-	2	4	10
	16.67	18.18	12.50	22.73	9.09	16.67	40.00		16.67	22.22	16.39
Special side-struck	-	1	-	2	-	2	4	-	2	2	10
		9.09		9.09		11.11	16.00		16.67	11.11	16.39
Indeterminate	5	2	3	5	1	2	4	2	3	5	26
	41.67	18.18	18.75	22.73	9.09	11.11	16.00	50.00	25.00	27.78	42.62
Cortex											
No cortex	8	6	9	11	5	12	18	2	10	14	40
	66.67	54.55	56.25	50.00	45.45	66.67	72.00	50.00	83.33	77.78	65.57
1–25%	2	3	6	7	4	4	2	-	1	1	6
	16.67	27.27	37.50	31.82	36.36	22.22	8.00		8.33	5.56	9.84
26–50%	-	1	1	-	-	1	2	-	-	1	8
		9.09	6.25			5.56	8.00			5.56	13.11
51–75%	1	-	-	3	1	-	1	-	-	1	1
	8.33			13.64	9.09		4.00			5.56	1.64
76–100%	1	1	-	1	1	1	2	1	1	1	6
	8.33	9.09		4.55	9.09	5.56	8.00	25.00	8.33	5.56	9.84
Indeterminate	-	-	-	-	-	-	-	1	-	-	-
								25.00			

6.2 Typology of Flint Flake Tools and Core Tools

Table 6.18 Descriptive statistics of flint side-scrapers (size in mm; angle in degrees)

	Layer										
				II-6 Levels							
	V-5	V-6	II-5	L1	L2	L3	L4	L4b	L5	L6	L7
N	12	11	16	22	11	18	25	4	12	18	61
Maximum length											
Maximum	50.00	71.00	47.00	53.00	46.00	55.00	44.00	49.00	48.00	42.00	55.00
Median	35.50	50.00	31.00	30.50	30.00	35.00	30.00	31.00	26.50	31.50	28.00
Minimum	21.00	24.00	22.00	22.00	22.00	22.00	23.00	25.00	21.00	24.00	21.00
Mean	35.58	45.82	31.31	30.82	31.36	35.39	30.00	34.00	28.17	32.83	30.84
Std dev	9.48	16.42	6.63	7.42	6.96	8.47	6.43	10.42	6.84	5.71	8.37
Std err mean	2.73	4.95	1.66	1.58	2.09	1.99	1.29	5.212	1.97	1.35	1.07
Length											
Maximum	45.00	68.00	47.00	53.00	43.00	55.00	42.00	49.00	45.00	42.00	50.00
Median	32.00	48.00	28.50	25.00	28.00	32.00	23.00	27.50	24.50	29.00	25.00
Minimum	21.00	16.00	22.00	18.00	22.00	16.00	13.00	23.00	15.00	20.00	11.00
Mean	31.17	40.73	29.69	27.09	29.64	32.39	24.64	31.75	25.00	29.39	26.57
Std dev	7.95	19.08	6.45	8.19	7.131	9.666	7.24	11.87	7.711	6.71	7.511
Std err mean	2.29	5.75	1.61	1.75	2.15	2.278	1.45	5.935	2.226	1.58	0.962
Width											
Maximum	41.00	53.00	36.00	37.00	28.00	45.00	41.00	30.00	30.00	42.00	55.00
Median	28.50	36.00	23.50	23.00	24.00	27.00	25.00	24.50	22.00	26.00	25.00
Minimum	12.00	13.00	14.00	12.00	14.00	17.00	14.00	17.00	16.00	20.00	11.00
Mean	26.92	34.45	24.19	24.18	23.09	27.39	26.48	24.00	22.00	27.39	25.75
Std dev	10.81	11.33	6.84	6.82	4.571	6.997	7.85	6.055	4.348	6.18	9.075
Std err mean	3.12	3.415	1.71	1.45	1.378	1.649	1.57	3.028	1.255	1.46	1.162
Thickness											
Maximum	12.00	19.00	18.00	20.00	12.00	16.00	17.00	13.00	12.00	17.00	21.00
Median	9.50	10.00	9.50	9.00	10.00	10.50	9.00	10.00	8.50	11.50	9.00
Minimum	5.00	1.00	6.00	5.00	5.00	6.00	5.00	6.00	5.00	7.00	5.00
Mean	9.00	9.72	10.06	10.64	8.90	10.44	9.36	9.75	8.333	11.78	9.88
Std dev	2.29	4.31	3.42	4.12	2.25	2.89	2.75	2.98	2.27	3.14	3.16
Std err mean	0.66	1.30	0.85	0.88	0.68	0.68	0.55	1.49	0.65	0.74	0.40
Angle of Retouch	(N=7)			(N=22)	(N=10)	(N=12)	(N=24)		(N=10)	(N=15)	(N=58)
Maximum	92.00	87.00	95.00	103.00	110.00	87.00	93.00	92.00	90.00	92.00	94.00
Median	82.00	70.00	77.00	85.50	85.00	78.50	83.00	86.50	80.00	80.00	78.00
Minimum	65.00	40.00	58.00	72.00	70.00	70.00	63.00	77.00	67.00	66.00	68.00
Mean	80.29	65.45	77.00	86.23	85.90	78.42	80.92	85.50	77.90	81.20	79.14
Std dev	8.34	16.43	9.11	9.26	11.26	5.99	7.74	6.86	7.17	6.39	6.38
Std err mean	3.15	4.95	2.44	1.97	3.56	1.73	1.58	3.43	2.27	1.65	0.84

to form three different tool types, though their frequencies are extremely low. In the assemblage of side-scrapers, only seven were classified as exhibiting three tool types (e.g. Fig. 6.25 Table 6.22).

End-scrapers

The category of end-scrapers includes two subtypes (typical and atypical; Table 6.5) that were combined in the following discussion due to the generally low frequencies of the entire category of end-scrapers. Despite their relatively low frequencies, end-scrapers are yet another group that is represented in some layers by more than 10 artifacts.

End-scrapers are artifacts on which the working edge is typically distal (whenever a typical end-scraper retouch was identified, the tool was classified as an end-scraper even if its retouch was not located at the distal end of the tool) and usually exhibits a convex morphology achieved by a typical end-scraper retouch. This retouch is clearly distinguished from that of side-scrapers and, in contrast to the variability observed

Table 6.19 Counts and typological frequencies (%) of flint side-scrapers

Type	Layer										
	V-5	V-6	II-5	II-6 Levels							
				L1	L2	L3	L4	L4b	L5	L6	L7
Single straight	2 16.67	3 27.27	4 25.00	1 4.55	-	1 5.56	5 20.00	-	2 16.67	4 22.22	10 16.39
Single convex	1 8.33	3 27.27	3 18.75	7 31.82	-	5 27.78	7 28.00	1 25.00	3 25.00	-	9 14.75
Single concave	1 8.33	2 18.18	1 6.25	2 9.09	2 18.18	2 11.11	2 8.00	1 25.00	2 16.67	2 11.11	5 8.20
Double straight	-	-	-	-	-	-	-	-	-	1 5.56	2 3.28
Double straight-convex	-	-	-	1 4.55	-	-	-	-	-	-	-
Double convex	-	-	-	-	-	1 5.56	-	-	-	-	-
Double convex-concave	-	-	-	-	1 9.09	-	-	-	-	1 5.56	-
Convergent straight	1 8.33	-	-	-	-	-	-	-	-	-	-
Convergent convex	-	1 9.09	-	-	1 9.09	1 5.56	-	-	-	-	3 4.92
Offset	-	-	-	-	-	-	-	-	-	-	3 4.92
Straight transverse	1 8.33	1 9.09	-	2 9.09	-	-	-	-	1 8.33	-	-
Convex transverse	1 8.33	-	-	-	-	2 11.11	2 8.00	-	1 8.33	2 11.11	4 6.56
Concave transverse	-	-	1 6.25	-	-	1 5.56	-	-	-	-	-
On ventral face	1 8.33	1 9.09	3 18.75	4 18.18	3 27.27	3 16.67	2 8.00	1 25.00	-	1 5.56	11 18.03
Abruptly retouched	4 33.33	-	4 25.00	5 22.73	4 36.36	1 5.56	7 28.00	1 25.00	2 16.67	6 33.33	11 18.03
With thinned back	-	-	-	-	-	-	-	-	1 8.33	-	-
With bifacial retouch	-	-	-	-	-	1 5.56	-	-	-	-	-
Alternately retouched	-	-	-	-	-	-	-	-	-	1 5.56	3 4.92
Total	12	11	16	22	11	18	25	4	12	18	61

in side-scraper types and their retouch, end-scrapers are more homogeneous in the morphology of their working edge. However, the morphology of the working edge of end-scrapers varies in form (Fig. 6.26) and can resemble classic end-scrapers, nosed end-scrapers (associated with a distal notch or notches), and carinated varieties (often modified on thicker flakes).

The preservation of end-scrapers (Table 6.23) resembles that of other tool types and they are usually slightly abraded with the exception of more abraded artifacts in Layer II-6 Level 7, as noted above for other tool types. Most of the end-scrapers are patinated, with a slightly lower frequency of double patinated artifacts than in previously discussed tool types (Table 6.23). Analysis of breakage patterns shows that most end-scrapers are complete, followed by proximal breaks in lower frequencies (Table 6.23). This pattern differs from that observed on other tool types, in which distal breaks are more common. Clearly, where distal breaks are recorded they occur on atypical end-scrapers, since these are sometimes modified on lateral edges.

Generally, end-scrapers are modified on non-cortical blanks. In Layer II-6 Level 7 the frequencies of cortical flakes (76–100%) are significantly higher than elsewhere (Table 6.23).

6.2 Typology of Flint Flake Tools and Core Tools

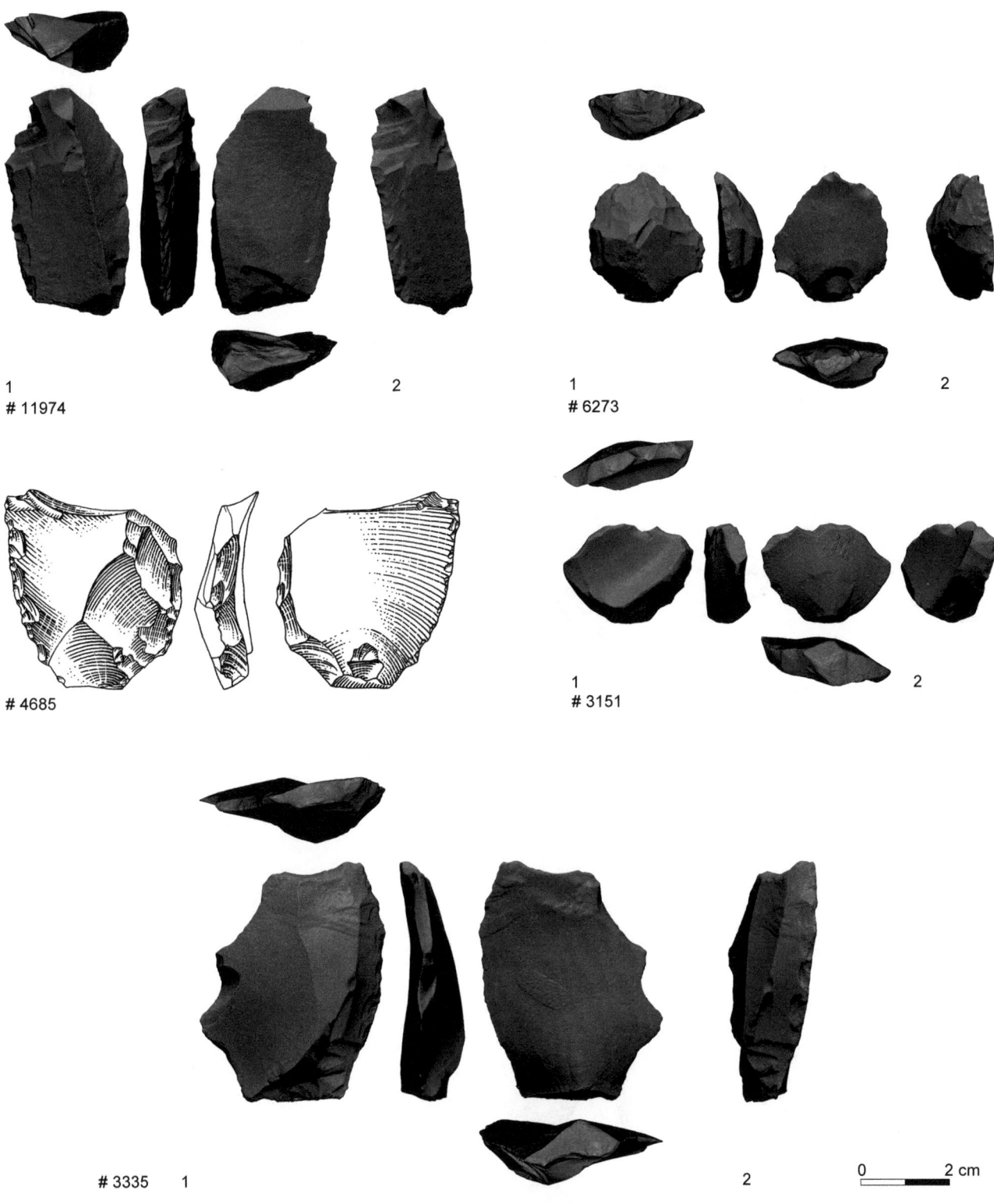

Fig. 6.17 Flint side-scrapers **1** multiple views **2** detailed view of working edge (#11974 JB single straight side-scraper and burin, note removed striking platform, #4685 Layer V-5 single convex side-scraper on *débordant* flake, #6273 Layer II-6 Level 1 single convex side-scraper with signs of utilization, #3151 Layer II-6 Level 4 single convex side-scraper with signs of utilization, #3335 JB single convex side-scraper with signs of utilization)

The common technological production mode of end-scraper blanks is that of end-struck blank, followed by side-struck. Layer II-6 Level 7 deviates from this pattern, with relatively small frequencies of end-struck blanks, most likely influenced by the high frequencies of indeterminate direction of blow (Table 6.23).

The overall size of end-scrapers is homogeneous throughout the sequence. Their dimensions are similar to those of the generally small-sized flint flakes, both retouched and unretouched (Table 6.24).

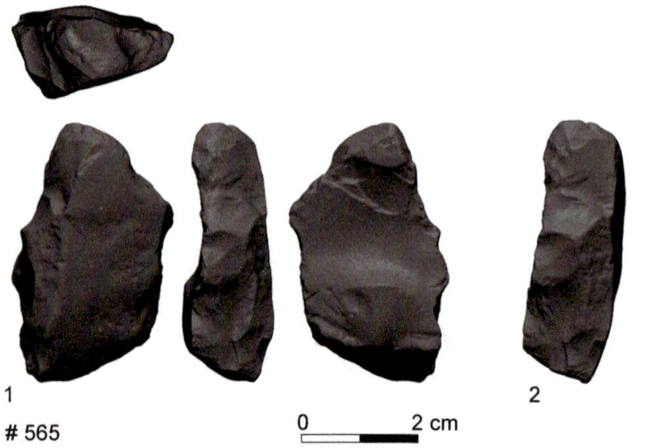

The results of the detailed analysis illustrate a lack of standardization in the type and morphology of end-scraper retouch, which is expressed in the variety of the types and location of the retouch (Table 6.25). The single standardized characteristic is that end-scrapers are nearly always modified on the dorsal face.

End-scrapers also occur on multiple tools (Table 6.26) and are particularly associated with notches and retouched flakes (Fig. 6.27). Similar configurations occur with triple tools (Fig. 6.27).

Other Flake Tools

In the following description we refer to typological groups of minimal occurrence (truncations, knives, burins, and varia tools) that do not meet the constraints noted at the beginning of this chapter (a minimum number of 10 tools per layer/level). Consequently, descriptive statistics of size are not provided here but are given in Section 6.2.1.1.

Truncations

Truncations are tools that are predominantly abruptly retouched, generally all along their distal edge. In the following description we have deviated from the norm of presenting only horizons with a minimum number of 10 tools and included

Fig. 6.18 Flint abruptly retouched side-scraper **1** multiple views **2** detailed view of scraper edge (#565 Layer II-6 Level 1)

Fig. 6.19 Flint abruptly retouched side-scraper **1** multiple views **2** detailed view of scraper edge (#13007 Layer II-6 Level 1 with removed striking platform)

6.2 Typology of Flint Flake Tools and Core Tools

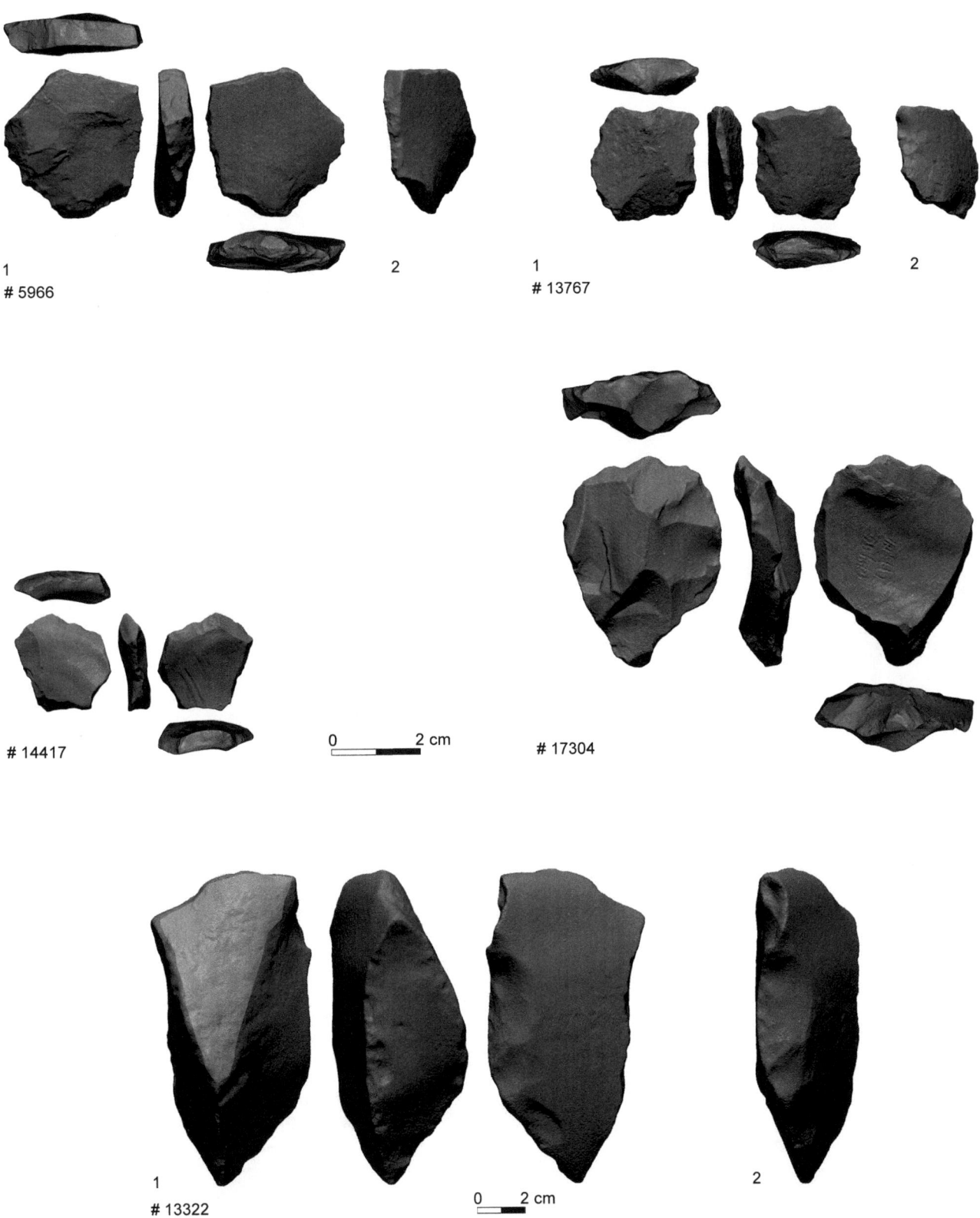

Fig. 6.20 Flint side-scrapers on ventral face **1** multiple views **2** detailed view of scraper edge (#5966 Layer II-6 Level 1 with signs of utilization, #14417 Layer II-6 Level 7, #13767 Layer II-6 Level 7, #17304 Layer II-6 Level 7 on biface sharpening flake, #13322 Layer II-6 Level 7 on hinged flake with signs of utilization)

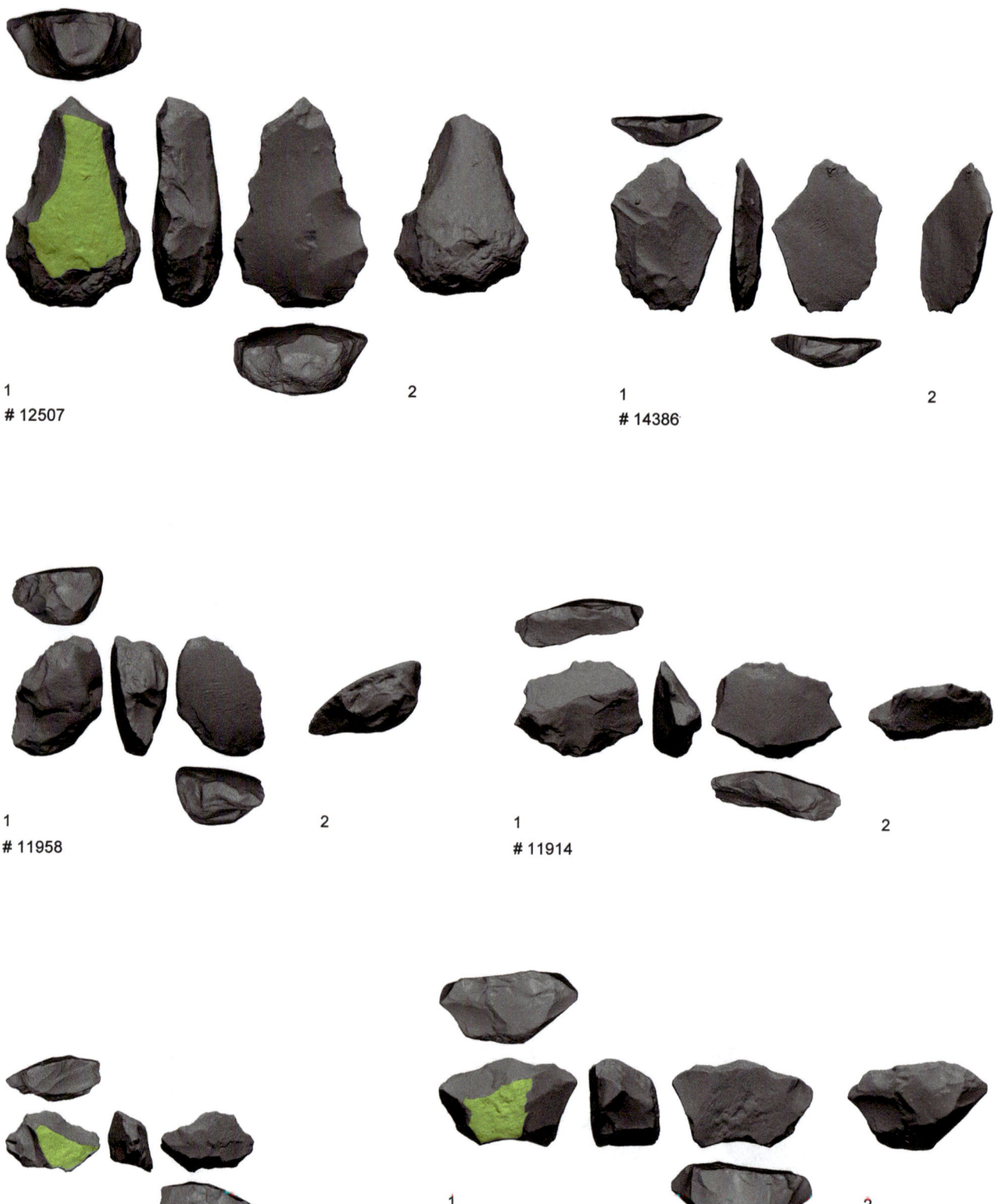

Fig. 6.21 Flint side-scrapers **1** multiple views **2** detailed view of scraper edge (#12507 Layer II-6/7 convergent-convex side-scraper with ventral removal, note battering marks on cortex, #11958 Layer II-6 Level 6 convex transversal scraper, #9393 Layer II-6 Level 7 convex transversal scraper, #14386 Layer II-6 Level 7 alternately retouched side-scraper, #11914 Layer II-6 Level 7 convex transversal scraper, #9568 Layer II-6 Level 7 convex transversal scraper with removed striking platform)

6.2 Typology of Flint Flake Tools and Core Tools

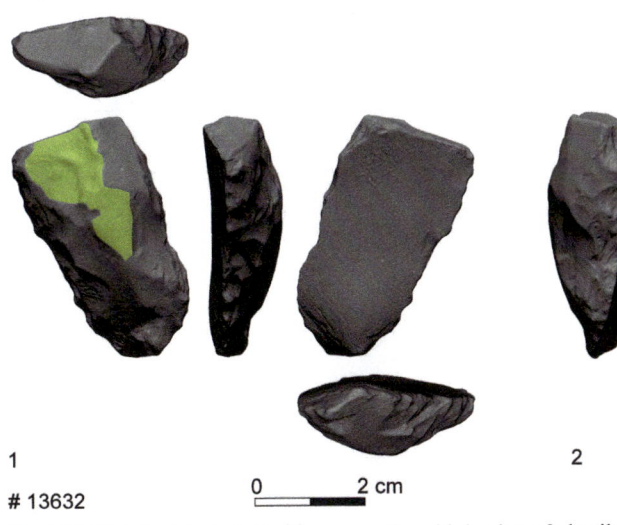

Fig. 6.22 Flint double straight side-scraper **1** multiple views **2** detailed view of scraper edge (#13632 Layer II-6 Level 7)

the entire sequence of Layer II-6. Despite their relatively low frequencies, truncations are a consistent and easily identified morphotype.

The physical properties of truncations, as well as some of their technological features, are presented in Table 6.27. The low frequencies of truncations do not permit thorough discussion for each of the levels of Layer II-6. Generally, truncations appear to be frequently fresh, patinated, and complete. The blank of truncations is often end-struck and non-cortical (Table 6.27).

Analysis of the retouch suggests that the modification of the truncated edge is predominantly achieved by an abrupt retouch located distally and on the dorsal face (Table 6.28). The morphology of the retouched edge is diversified and includes a variety of forms (Table 6.28).

About a third of the truncations are multiple tools, associated with one additional type. These include retouched

Table 6.20 Counts and frequencies (%) of retouch characteristics of flint side-scrapers

	Layer										
				II-6 Levels							
	V-5	V-6	II-5	L1	L2	L3	L4	L4b	L5	L6	L7
N	11	11	16	22	11	18	25	4	12	18	61
Type of retouch											
Regular	1 9.09	-	-	-	-	-	-	-	1 8.33	-	1 1.64
End-scraper	-	-	2 12.50	1 4.55	-	1 5.56	-	-	-	-	1 1.64
Side-scraper	5 45.45	10 90.91	8 50.00	14 63.64	3 27.27	13 72.22	18 72.00	3 75.00	8 66.67	8 44.44	45 73.77
Notch	1 9.09	-	1 6.25	1 4.55	1 9.09	1 5.56	-	-	-	2 11.11	3 4.92
Invasive	-	-	2 12.50	-	1 9.09	-	-	-	1 8.33	2 11.11	3 4.92
Quina	-	-	1 6.25	-	-	-	-	-	-	1 5.56	2 3.28
Semi-abrupt	1 9.09	1 9.09	-	-	-	-	2 8.00	-	-	1 5.56	-
Abrupt	3 27.27	-	1 6.25	6 27.27	4 36.36	1 5.56	4 16.00	1 25.00	2 16.67	3 16.67	5 8.20
Bifacial	-	-	-	-	-	1 5.56	-	-	-	-	-
Mixed	-	-	-	-	-	1 5.56	-	-	-	-	-
Nahr Ibrahim	-	-	-	-	1 9.09	-	1 4.00	-	-	-	-
Irregular	-	-	-	-	-	-	-	-	-	-	1 1.64
Indeterminate	-	-	1 6.25	-	1 9.09	-	-	-	-	1 5.56	-
Location of retouch											
Distal	2 18.18	1 9.09	2 12.50	2 9.09	2 18.18	6 33.33	6 24.00	-	2 16.67	3 16.67	7 11.48
Proximal	-	-	-	-	2 18.18	2 11.11	3 12.00	-	1 8.33	1 5.56	1 1.64
Left	5 45.45	6 54.55	9 56.25	8 36.36	3 27.27	4 22.22	5 20.00	2 50.00	4 33.33	5 27.78	19 31.15

Table 6.20 (cont.)

	Layer										
				II-6 Levels							
	V-5	V-6	II-5	L1	L2	L3	L4	L4b	L5	L6	L7
N	11	11	16	22	11	18	25	4	12	18	61
Right	3	3	5	11	3	5	10	2	5	5	23
	27.27	27.27	31.25	50.00	27.27	27.78	40.00	50.00	41.67	27.78	37.70
Both edges	1	-	-	-	-	-	-	-	-	-	3
	9.09										4.92
Distal & right	-	-	-	-	-	-	-	-	-	-	3
											4.92
Distal & left	-	-	-	-	-	-	-	-	-	1	3
										5.56	4.92
Proximal & left	-	-	-	-	-	-	-	-	-	1	-
										5.56	
Convergent	-	1	-	1	1	-	1	-	-	2	2
		9.09		4.55	9.09		4.00			11.11	3.28
Circumference	-	-	-	-	-	1	-	-	-	-	-
						5.56					
Face of retouch		(N=10)		(N=21)		(N=16)	(N=24)		(N=16)		
Dorsal	8	8	13	17	4	10	16	2	10	11	40
	72.73	80.00	81.25	80.95	36.36	62.50	66.67	50.00	83.33	68.75	65.57
Ventral	1	1	2	1	4	5	3	1	1	1	13
	9.09	10.00	12.50	4.76	36.36	31.25	12.50	25.00	8.33	6.25	21.31
Dorsal & ventral	-	1	1	2	1	1	2	-	-	2	7
		10.00	6.25	9.52	9.09	6.25	8.33			12.50	11.48
Side	2	-	-	1	2	-	3	1	1	2	1
	18.18			4.76	18.18		12.50	25.00	8.33	12.50	1.64
Form of retouched edge				(N=21)		(N=15)	(N=23)				(N=60)
Straight	4	3	5	5	2	2	6	2	3	6	11
	36.36	27.27	33.33	23.81	18.18	13.33	26.09	50.00	25.00	33.33	18.33
Convex	4	6	7	9	2	9	11	1	4	6	27
	36.36	54.55	46.67	42.86	18.18	60.00	47.83	25.00	33.33	33.33	45.00
Concave	1	2	1	4	4	3	3	1	3	3	12
	9.09	18.18	6.67	19.05	36.36	20.00	13.04	25.00	25.00	16.67	20.00
Convergent	1	-	-	1	1	-	1	-	-	1	3
	9.09			4.76	9.09		4.35			5.56	5.00
Wavy	-	-	-	-	1	-	2	-	2	-	5
					9.09		8.70		16.67		8.33
Denticulate	-	-	-	-	-	-	-	-	-	-	2
											3.33
One tooth	-	-	-	1	-	-	-	-	-	2	-
				4.76						11.11	
Oblique	-	-	1	1	1	1	-	-	-	-	-
			6.67	4.76	9.09	6.67					
Indeterminate	1	-	1	-	-	-	-	-	-	-	-
	9.09		6.67								

6.2 Typology of Flint Flake Tools and Core Tools

Table 6.21 Counts and frequencies (%) of multiple (double) artifacts on flint side-scrapers

Typology B \ Typology A	Single straight side-scraper	Single convex side-scraper	Single concave side-scraper	Double straight side-scraper	Double straight-convex side-scraper	Double convex side-scraper	Double convex-concave side-scraper	Convergent convex side-scraper	Offset side-scraper	Straight transverse side-scraper	Convex transverse side-scraper	Concave transverse side-scraper	Side-scraper on ventral face	Abruptly retouched side-scraper	Thinned back side-scraper	Alternately retouched side-scraper	Total
Side-scraper on ventral face	-	2 9.52	-	-	-	-	-	-	2 100.00	-	-	-	-	-	-	-	4
Abruptly retouched side-scraper	2 8.33	1 4.76	1 6.67	-	-	-	-	-	-	1 50.00	-	-	1 11.11	2 14.29	-	-	8
Atypical end-scraper	1 4.17	-	-	-	-	-	-	-	-	-	1 20.00	1 100.00	1 11.11	-	-	-	4
Atypical burin	3 12.50	-	2 13.33	-	-	-	-	-	-	-	-	-	-	-	-	-	5
Typical borer	-	1 4.76	-	-	-	-	-	-	-	-	-	-	1 11.11	1 7.14	-	-	3
Atypical borer	2 8.33	3 14.29	1 6.67	-	2 100.00	-	-	-	-	1 50.00	2 40.00	-	-	-	-	-	11
Typical backed knife	-	1 4.76	-	-	-	-	-	-	-	-	-	-	-	-	-	-	1
Truncation	1 4.17	2 9.52	1 6.67	-	-	-	-	-	-	-	-	-	-	1 7.14	-	-	5
Notch	8 33.33	5 23.81	4 26.67	1 100.00	-	-	-	1 100.00	-	-	1 20.00	-	3 33.33	4 28.57	-	1 100.00	28
Denticulate	-	2 9.52	-	-	-	-	-	-	-	-	-	-	-	2 14.29	-	-	4
Retouch on ventral face	-	1 4.76	-	-	-	-	-	-	-	-	-	-	-	-	-	-	1
End-notched piece	-	1 4.76	-	-	-	-	1 50.00	-	-	-	-	-	-	-	-	-	2
Varia	1 4.17	-	1 6.67	-	-	1 100.00	1 50.00	-	-	-	-	-	-	-	-	-	4
Retouched flake	6 25.00	2 9.52	4 26.67	-	-	-	-	-	-	-	1 20.00	-	2 22.22	4 28.57	1 100.00	-	20
Biface sharpening flake	-	-	-	-	-	-	-	-	-	-	-	-	1 11.11	-	-	-	1
Core on flake	-	-	1 6.67	-	-	-	-	-	-	-	-	-	-	-	-	-	1
Total	24	21	15	1	2	1	2	1	2	2	5	1	9	14	1	1	102

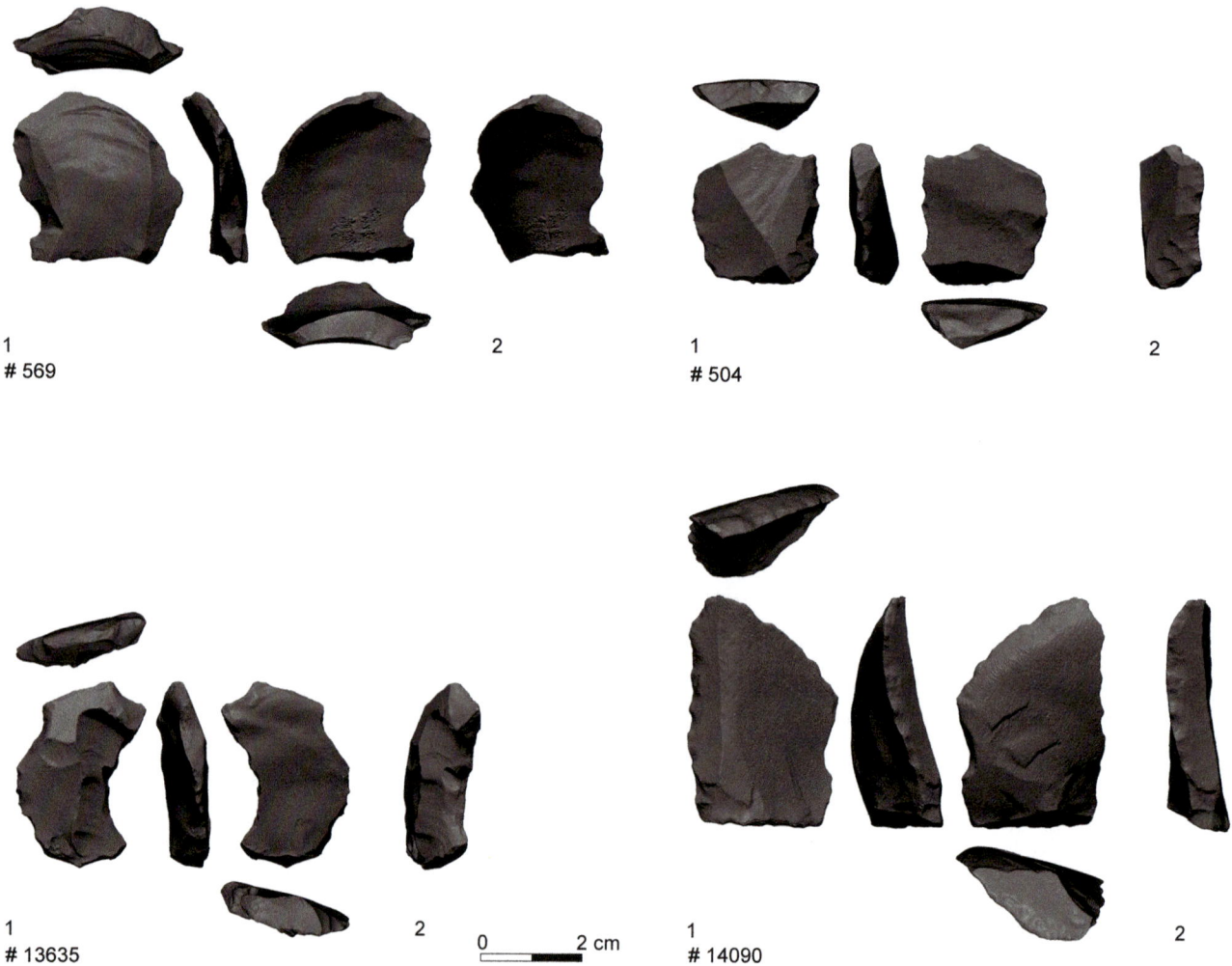

Fig. 6.23 Flint multiple tools **1** multiple views **2** detailed view of scraper edge (#569 Layer II-5 side-scraper on ventral face and notch, #13635 Layer II-6 Level 7 single concave side-scraper and notch on a hinged flake, #504 Layer II-5 single concave side-scraper and retouched flake with signs of utilization, #14090 Layer II-6 Level 7 single straight side-scraper and retouched flake on hinged flake)

flakes (N=10), notches (N=8), denticulates (N=5; Fig. 6.28), a single side-scraper, and a varia tool.

Backed Knives

The category of backed knives includes two subtypes, backed knives on which the back is modified by retouch, either abrupt (Fig. 6.29) or bipolar (Fig. 6.30), and naturally backed knives on which the back is cortical (Fig. 6.29). Both subtypes exhibit either retouch or signs of use wear on the opposite lateral edge, suggesting that they derive from a particular use pattern.

These tools occur rarely throughout the GBY sequence and are thus discussed as a group that includes all the artifacts represented at GBY. As a group, the general aspect of these tools (when complete), both naturally backed and retouched, is elongated.

The abruptly retouched backed knives (N=9) display a standardized continuous retouch that is occasionally (in four cases) associated with residual cortex. On some artifacts (N=4) the abrupt retouch is bifacial (i.e. on both ventral and dorsal faces). Other characteristics of this group are presented below (Section 6.2.1.1).

Very few knives (N=4) are associated with an additional tool type; they include a notch, a denticulate (Fig. 6.29), a retouched flake, and a tool classified as varia.

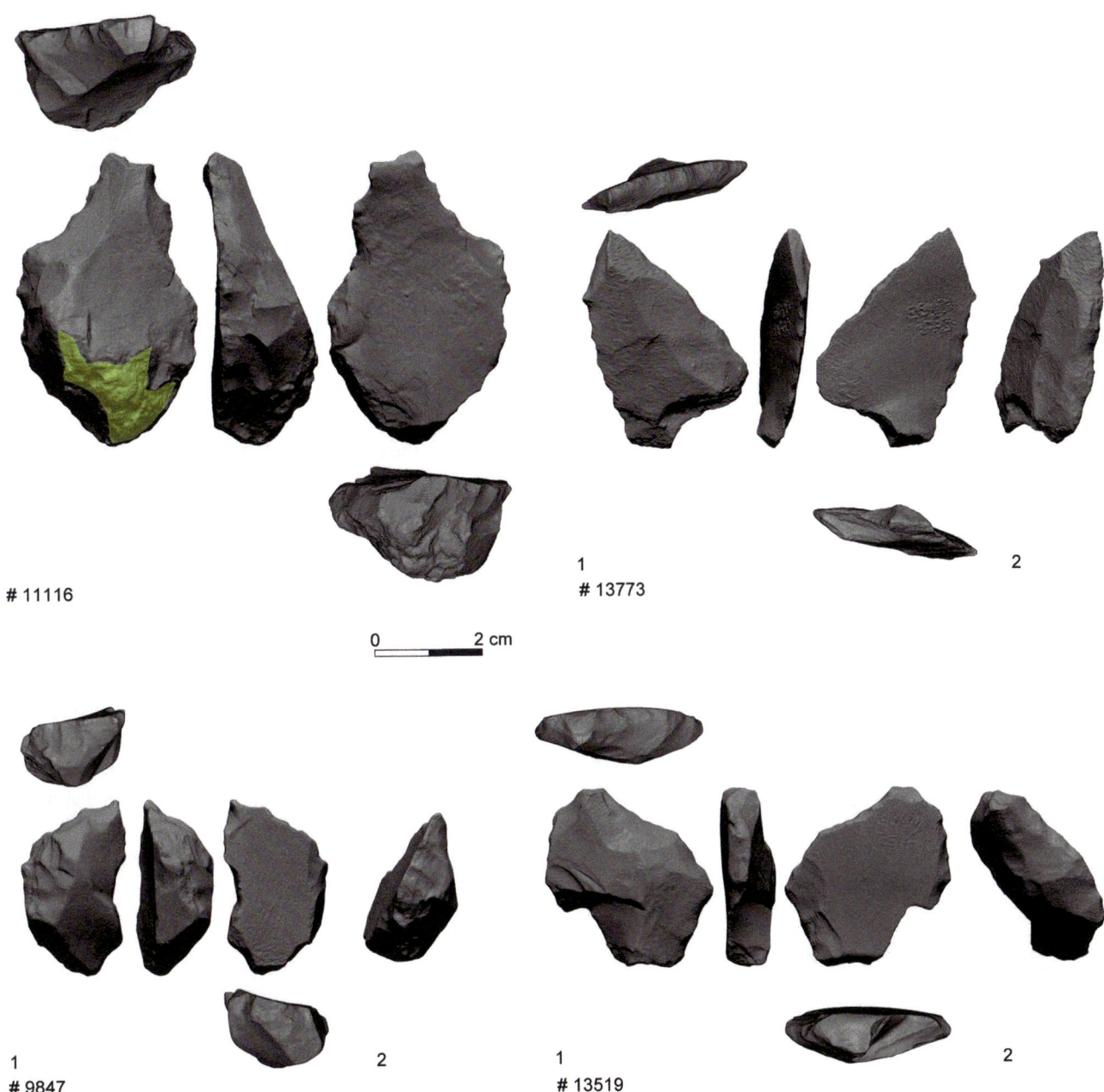

Fig. 6.24 Flint multiple tools **1** multiple views **2** detailed view of scraper edge (#11116 Layer II-6 Level 1 single convex side-scraper and borer, #9847 Layer II-6 Level 7 abruptly retouched side-scraper and borer with removed striking platform, #13773 Layer II-6 Level 7 offset side-scraper and side-scraper on ventral face, #13519 Layer II-6 Level 7 offset side-scraper and side-scraper on ventral face)

Burins

The rarest tool type on the typological spectrum is the burin. In the small assemblages it does not occur at all, and when present it is often represented by a single item. Exceptions to this are the cases of Layer II-6 Level 4, where burins are represented by four items, and Levels 5 and 7, in which two items occur. Although burins do appear in earlier assemblages (such as that of 'Ubeidiya), they are always scarce.

Only one burin spall is recorded (from Layer V-5, Fig. 6.31). The burins are sometimes well formed and show a high degree of variation. The entire collection of burins comprises 20 tools. Seven are formed on the ventral face of a flake (Fig. 6.31); of these, two are multiple burins, one is a double burin on a truncation, one is a dihedral burin on a break, and another is a burin on small facets. The burin removals of this group are clearly distinguishable flat elongated scars; of these, four are made by multiple removals and three by a single one.

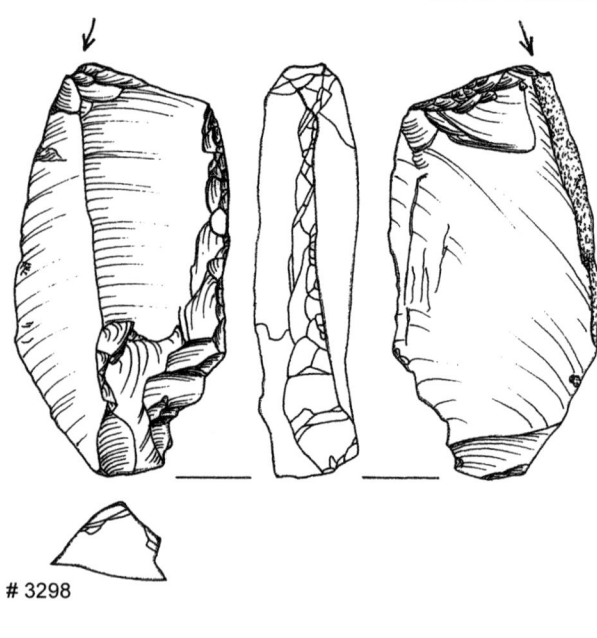

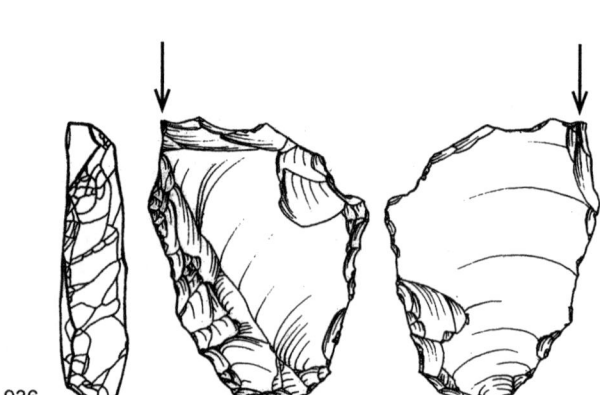

Fig. 6.25 Flint multiple (triple) tools (#3298 JB single straight sidescraper, burin, and truncation, #936 Layer II-5 single straight sidescraper, burin, and denticulate)

Technologically, three burins were made by a series of miniature flake removals from opposite directions (dihedral) and two were made on a flat surface formed by a scar located on the dorsal face of the tool. Other burins are transversal, with a single blow (Figs. 6.31–6.32) or multiple blows. Two burins are modified on truncations (Fig. 6.32). Five items are classified as dihedral burins (Fig. 6.32); of these, four are offset and one is more of an angle type.

Other characteristics of this group are presented below (Section 6.2.1.1).

Varia Tools

The category of varia assembles all the retouched tools that do not fit into the classifications of Bordes' typological list. In addition, several morphotypes that stand alone and occur in very low frequencies were integrated into this category: alternately retouched bec (beak) (N=6) (Figs. 6.33–6.34), limace (N=2) (Fig. 6.33), abrupt and alternate retouch (thick) (N=1), bifacial retouch (N=1), and tanged tools (N=1).

Interestingly, in the entire GBY sequence very few flint tools were classified as varia tools. This may be influenced by the fact that the category of retouched flakes (described below) comprises a large variety of tools that by definition do not fit into the classic Bordesian typology. Nevertheless, the difference between the retouched flakes and the varia group is that in the latter each tool exhibits more than one type of retouch (Fig. 6.33) and may also display various combinations of ventral removals/signs of use, which often occur on different edges (Figs. 6.33–6.34).

Table 6.22 List of multiple (triple) artifacts on flint sidescrapers (3 typological classifications)

Typology A (side-scrapers)	Typology B	Typology C
Simple straight	Atypical end-scraper	Atypical borer
Simple straight	Atypical burin	Denticulate
Simple straight	Atypical burin	Truncation
Simple convex	Side-scraper on ventral face	Notch
Transversal convex	Notch	Retouched flake
On ventral face	Abruptly retouched side-scraper	Retouched flake
Thinned back	Retouched flake	Éclat de taille de biface

6.2 Typology of Flint Flake Tools and Core Tools

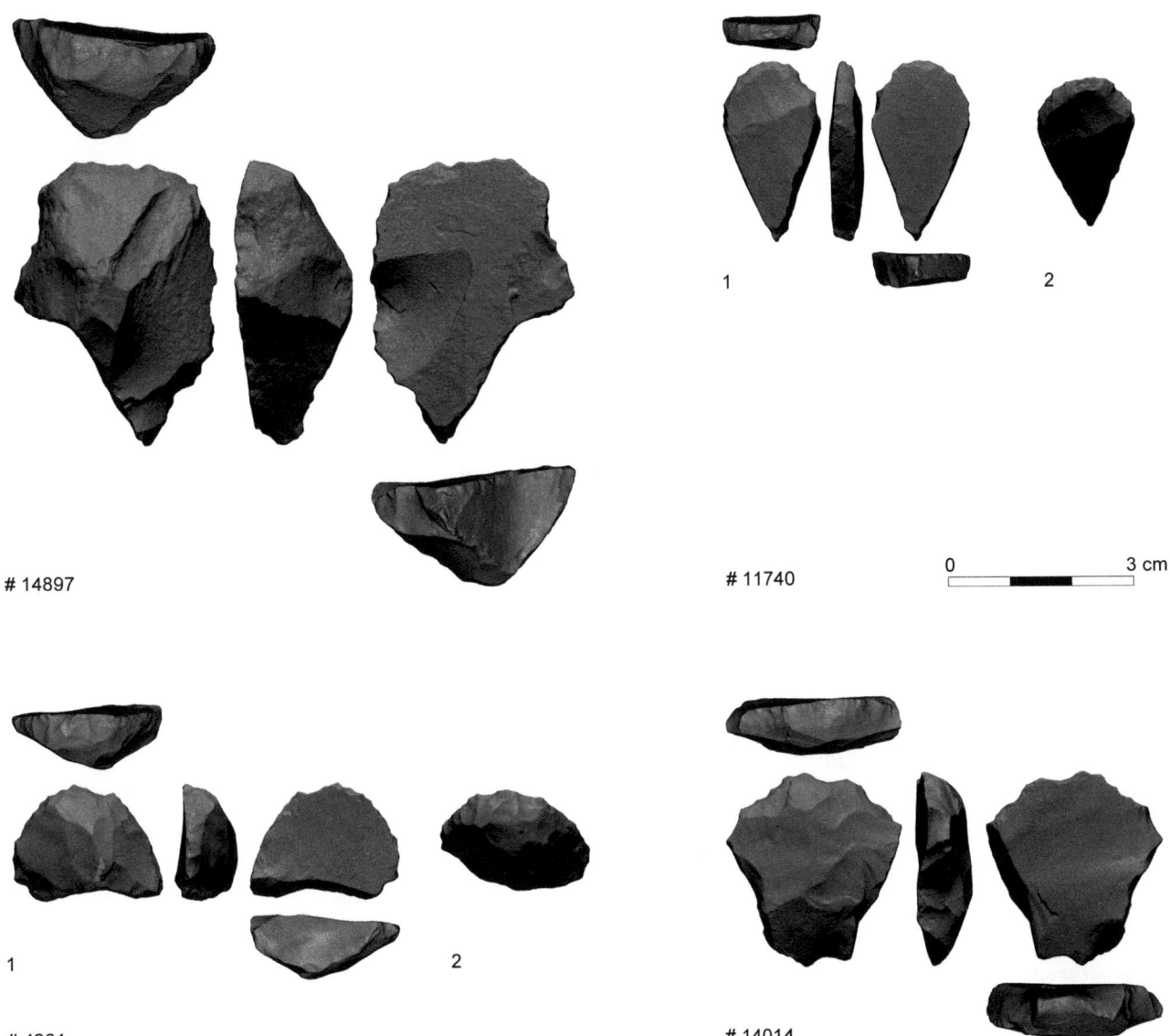

Fig. 6.26 Flint end-scrapers **1** multiple views **2** detailed view of scraper edge (#14897 Layer II-6 Level 3, #4261 Layer II-6 Level 6, #11740 Layer II-6 Level 7, #14014 Layer II-6 Level 7)

Retouched Flakes

The second most frequent tool type at GBY is retouched flakes. This category encompasses a variety of tools with the common denominator of retouch extending over a minimum length of 1 cm that cannot be classified under any other type in the typological list (Fig. 6.35). Consequently, this category assembles a variety of styles of retouch – type, location, face, etc. (see further discussion below).

The physical properties of retouched flakes are detailed in Table 6.29, which shows that in most assemblages retouched flakes are either fresh or slightly abraded, in similar frequencies. One exception is Layer II-6 Level 7, in which there is a higher degree of weathering. As for patination, most of the retouched flakes are patinated, with a substantial component of double patinated tools.

Retouched flakes are predominantly complete and when broken exhibit mostly distal breaks (Table 6.29). As observed for other tool types, when direction of blow is discernible the dominant blank among retouched flakes is end-struck, followed by side-struck.

When retouch is examined, the dominant retouch type is irregular, followed by regular (Table 6.30). In several assemblages that exhibit the highest frequencies of retouched flakes, however, the frequencies of regular and irregular retouch are very similar (e.g. Layer II-6 Levels 4, 5, and 7).

Table 6.23 Counts and frequencies (%) of characteristics of flint end-scrapers

	Layer II-6 Levels							
	L1	L2	L3	L4	L4b	L5	L6	L7
N	19	13	18	9	5	5	9	19
Preservation								
Fresh	6	6	2	5	2	-	1	-
	31.58	46.15	11.11	55.56	40.00		11.11	
Slightly abraded	9	3	11	3	2	4	6	5
	47.37	23.08	61.11	33.33	40.00	80.00	66.67	26.32
Abraded	4	4	5	1	1	1	2	14
	21.05	30.77	27.78	11.11	20.00	20.00	22.22	73.68
Patination								
No patina	3	2	1	2	-	-	1	-
	15.79	15.38	5.56	22.22			11.11	
Patinated	13	7	14	6	3	1	6	15
	68.42	53.85	77.78	66.67	60.00	20.00	66.67	78.95
Double patinated	3	4	3	1	2	4	2	4
	15.79	30.77	16.67	11.11	40.00	80.00	22.22	21.05
Breakage								
Complete	14	5	10	7	4	3	5	9
	73.68	38.46	55.56	77.78	80.00	60.00	55.56	47.37
Distal	-	-	1	-	-	2	2	1
			5.56			40.00	22.22	5.26
Lateral	1	4	1	1	-	-	1	1
	5.26	30.77	5.56	11.11			11.11	5.26
Proximal	4	3	6	1	1	-	1	6
	21.05	23.08	33.33	11.11	20.00		11.11	31.58
Fragment	-	1	-	-	-	-	-	1
		7.69						5.26
Proximal & lateral	-	-	-	-	-	-	-	1
								5.26
Direction of blow								
End-struck	9	7	8	5	3	3	3	4
	47.37	53.85	44.44	55.56	60.00	60.00	33.33	21.05
Side-struck	7	2	5	4	1	-	2	3
	36.84	15.38	27.78	44.44	20.00		22.22	15.79
Special side-struck	-	1	1	-	-	-	-	1
		7.69	5.56					5.26
Indeterminate	3	3	4	-	1	2	4	11
	15.79	23.08	22.22		20.00	40.00	44.44	57.89
Cortex								
No cortex	11	6	11	4	3	4	5	14
	57.89	46.15	61.11	44.44	60.00	80.00	55.56	73.68
1–25%	5	4	3	2	-	-	2	-
	26.32	30.77	16.67	22.22			22.22	
26–50%	1	1	1	1	-	-	1	1
	5.26	7.69	5.56	11.11			11.11	5.26
51–75%	-	-	1	1	1	1	-	-
			5.56	11.11	20.00	20.00		
76–100%	2	2	2	1	1	-	1	4
	10.53	15.38	11.11	11.11	20.00		11.11	21.05

6.2 Typology of Flint Flake Tools and Core Tools

Table 6.24 Descriptive statistics of flint end-scrapers (size in mm; angle in degrees)

	Layer II-6 Levels							
	L1	L2	L3	L4	L4b	L5	L6	L7
N	19	13	18	9	5	5	9	19
Maximum length								
Maximum	35.00	38.00	47.00	37.00	66.00	40.00	45.00	35.00
Median	28.00	28.00	26.50	29.00	36.00	29.00	25.00	25.00
Minimum	21.00	23.00	20.00	20.00	24.00	23.00	22.00	21.00
Mean	27.58	29.23	28.50	28.56	39.20	29.60	28.44	26.42
Std dev	4.48	4.59	5.93	5.55	16.33	6.31	8.41	4.30
Std err mean	1.03	1.27	1.40	1.85	7.30	2.82	2.80	0.99
Length								
Maximum	35.00	38.00	33.00	35.00	64.00	40.00	42.00	34.00
Median	25.00	25.00	25.00	28.00	32.00	27.00	22.00	22.00
Minimum	12.00	20.00	17.00	19.00	13.00	21.00	18.00	13.00
Mean	24.16	26.54	25.50	26.89	35.00	28.40	26.33	22.84
Std dev	6.32	5.62	4.34	5.64	18.53	7.13	9.00	5.64
Std err mean	1.45	1.56	1.02	1.88	8.29	3.19	3.00	1.29
Width								
Maximum	33.00	32.00	45.00	37.00	39.00	28.00	34.00	32.00
Median	22.00	22.00	22.00	24.00	26.00	22.00	20.00	22.00
Minimum	14.00	15.00	12.00	16.00	20.00	20.00	14.00	13.00
Mean	22.21	22.62	23.50	24.44	29.00	23.00	21.56	22.47
Std dev	5.20	5.49	8.04	6.52	8.51	3.00	6.75	5.39
Std err mean	1.19	1.52	1.89	2.17	3.81	1.34	2.25	1.24
Thickness								
Maximum	12.00	18.00	18.00	13.00	13.00	15.00	12.00	20.00
Median	9.00	9.00	10.00	8.00	10.00	10.00	9.00	10.00
Minimum	6.00	5.00	5.00	5.00	10.00	6.00	7.00	5.00
Mean	8.42	9.69	10.39	8.56	11.20	10.60	9.00	10.53
Std dev	1.68	3.54	3.36	2.40	1.64	3.85	1.58	3.81
Std err mean	0.38	0.98	0.79	0.80	0.73	1.72	0.53	0.87
Angle of retouch		(N=12)	(N=17)	(N=8)		(N=3)		
Maximum	100.00	90.00	102.00	92.00	90.00	74.00	90.00	102.00
Median	86.00	82.50	82.00	86.50	80.00	67.00	78.00	84.00
Minimum	53.00	68.00	74.00	74.00	76.00	67.00	62.00	76.00
Mean	83.21	82.00	84.06	84.00	81.40	69.33	77.89	86.32
Std dev	10.21	7.15	6.52	6.52	5.46	4.04	7.80	7.36
Std err mean	2.34	2.06	1.58	2.31	2.44	2.33	2.60	1.69

Table 6.25 Counts and frequencies (%) of retouch characteristics of flint end-scrapers

	Layer II-6 Levels							
Layer	L1	L2	L3	L4	L4b	L5	L6	L7
N	19	13	18	9	5	5	9	19
Type of retouch								
Regular	-	1 / 7.69	1 / 5.56	-	-	-	-	-
End-scraper	16 / 84.21	8 / 61.54	13 / 72.22	4 / 44.44	4 / 80.00	3 / 60.00	4 / 44.44	16 / 84.21
Side-scraper	-	1 / 7.69	1 / 5.56	1 / 11.11	1 / 20.00	-	-	-
Notch	2 / 10.53	2 / 15.38	-	3 / 33.33	-	1 / 20.00	4 / 44.44	1 / 5.26
Invasive	-	-	1 / 5.56	-	-	-	-	-

Table 6.25 (Cont.)

Layer	Layer II-6 Levels							
	L1	L2	L3	L4	L4b	L5	L6	L7
N	19	13	18	9	5	5	9	19
Abrupt	-	-	-	1 11.11	-	-	-	1 5.26
Mixed	1 5.26	-	-	-	-	-	-	1 5.26
Signs of use	-	1 7.69	2 11.11	-	-	-	1 11.11	-
Indeterminate	-	-	-	-	-	1 20.00	-	-
Location of retouch								
Distal	15 78.95	9 69.23	13 72.22	7 77.78	4 80.00	1 20.00	5 55.56	13 68.42
Truncation	-	1 7.69	-	-	-	-	-	-
Proximal	-	-	1 5.56	1 11.11	1 20.00	1 20.00	1 11.11	-
Left	1 5.26	1 7.69	1 5.56	-	-	1 20.00	1 11.11	1 5.26
Right	2 10.53	2 15.38	-	1 11.11	-	1 20.00	-	1 5.26
Both edges	-	-	1 5.56	-	-	-	1 11.11	-
Distal & right	-	-	1 5.56	-	-	1 20.00	-	1 5.26
Distal & left	-	-	1 5.56	-	-	-	-	2 10.53
Proximal & both edges	-	-	-	-	-	-	1 11.11	-
Circumference	1 5.26	-	-	-	-	-	-	-
Indeterminate	-	-	-	-	-	-	-	1 5.26
Face of retouch	(N=18)	(N=12)	(N=17)					
Dorsal	15 83.33	8 66.67	16 94.12	7 77.78	5 100.00	4 80.00	7 77.78	17 89.47
Ventral	1 5.56	2 16.67	-	1 11.11	-	1 20.00	1 11.11	1 5.26
Dorsal & ventral	2 11.11	-	1 5.88	-	-	-	1 11.11	-
Side	-	2 16.67	-	1 11.11	-	-	-	1 5.26
Form of retouched edge		(N=12)	(N=14)					
Straight	1 5.26	-	1 7.14	-	1 20.00	-	1 11.11	-
Convex	11 57.89	10 83.33	13 92.86	5 55.56	4 80.00	3 60.00	5 55.56	17 89.47
Concave	1 5.26	1 8.33	-	2 22.22	-	1 20.00	2 22.22	-
Wavy	3 15.79	-	-	-	-	-	-	2 10.53
Denticulate	2 10.53	1 8.33	-	1 11.11	-	1 20.00	1 11.11	-
One tooth	-	-	-	1 11.11	-	-	-	-
Indeterminate	1 5.26	-	-	-	-	-	-	-

6.2 Typology of Flint Flake Tools and Core Tools

Table 6.26 Counts of multiple tools on flint end-scrapers

Type	Side-scraper	Borer	Truncation	Notch	Denticulate	Varia	Retouched flake	Pitted stone
Double tools								
Typical end-scraper	-	2	1	6	2	-	4	1
Atypical end-scraper	-	1	-	16	4	1	6	-
Total	-	3	1	22	6	1	10	1
Triple tools								
Typical end-scraper & notch	-	-	-	-	-	-	2	-
Atypical end-scraper & notch	1	-	-	1	-	-	1	-
Atypical end-scraper & borer	-	-	-	-	-	-	1	-
Total	-	-	-	1	-	-	4	-

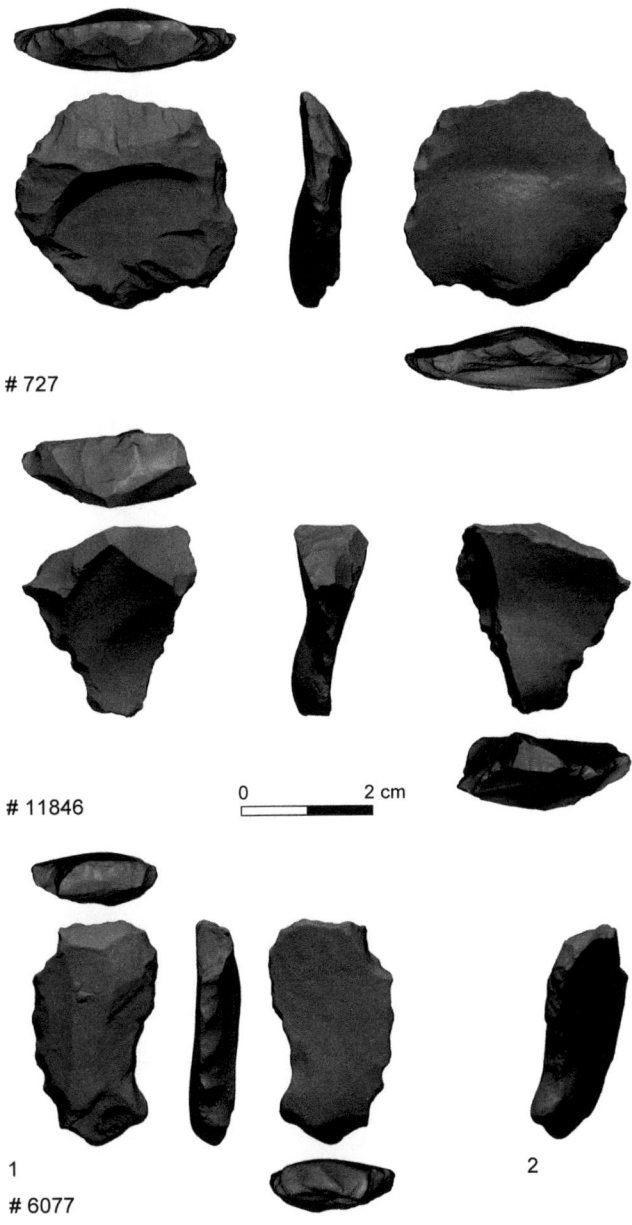

Fig. 6.27 Flint multiple tools (#727 Layer II-6 Level 4 end-scraper and retouched flake, #11846 Layer II-5/6 end-scraper and borer on *débordant* flake, #6077 Layer II-6 Level 1 end-scraper, notch, and retouched flake **1** multiple views **2** detailed view of retouched edge)

Outstanding among these assemblages is Layer II-6 Level 1, in which regular retouch is represented by only a few tools (Table 6.30).

The generalized character of this tool type is further reflected in the presence (in varying frequencies) of signs of retouch that we classified as signs of use. These are minimally represented in most assemblages, but occur in higher frequencies in Layer II-6 Levels 4, 5, and 6. Retouch is defined as extending over a minimum of 1 cm; hence these retouched flakes are artifacts in which the retouch is clear but too limited (less than 1 cm) to be classified as retouch, and is thus classified as signs of use.

When the location of retouch is examined, the dominant location is a lateral one, predominantly on the right edge and with slightly smaller frequencies of the left edge (Table 6.30). Layer II-6 Level 1 shows the reverse, in which the left edge is more commonly retouched. Other locations include the distal edge, both edges, and the circumference.

Retouch is observed mostly on the dorsal face, with the combination of dorsal and ventral faces in significantly lower frequencies (Table 6.30).

The generalized character of this category is further reflected in the diversity of forms of the retouched edges (Table 6.30), which show similar frequencies of straight and convex edges and lower frequencies of concave ones. Layer II-6 Level 7 deviates from this in having high frequencies of convex edges, followed by wavy edges, and then straight and concave edges in similar frequencies. Wavy edges seem fairly common in other assemblages as well.

The different size measurements of retouched flakes (Table 6.31) do not show any deviations among the various layers and in general reflect the overall small size of the flint tools.

On 3.35% of the entire retouched flake assemblage we observed an additional tool type. Of these, 10 tools are classified twice as retouched flakes. In these cases the type, location, face, and form of the retouch are different for each of the individually classified retouched flakes. The other additional tool types

Table 6.27 Counts and frequencies (%) of characteristics of flint truncations

	Layer II-6 Levels							
	L1	L2	L3	L4	L4b	L5	L6	L7
N	9	4	6	8	1	2	3	12
Preservation								
Fresh	7 77.78	-	3 50.00	3 37.50	1 100.00	-	-	1 8.33
Slightly abraded	2 22.22	1 25.00	3 50.00	4 50.00	-	2 100.00	1 33.33	3 25.00
Abraded	-	-	-	1 12.50	-	-	2 66.67	6 50.00
Heavily abraded	-	3 75.00	-	-	-	-	-	2 16.67
Patination								
No patina	1 11.11	-	3 50.00	1 12.50	-	-	-	1 8.33
Patinated	7 77.78	1 25.00	3 50.00	6 75.00	1 100.00	1 50.00	2 66.67	6 50.00
Double patinated	1 11.11	3 75.00	-	1 12.50	-	1 50.00	1 33.33	5 41.67
Breakage								
Complete	5 55.56	4 100.00	4 66.67	7 87.50	-	2 100.00	2 66.67	7 58.33
Lateral	-	-	-	-	-	-	-	1 8.33
Proximal	3 33.33	-	2 33.33	1 12.50	1 100.00	-	-	3 25.00
Distal & proximal	1 11.11	-	-	-	-	-	1 33.33	-
Fragment	-	-	-	-	-	-	-	1 8.33
Direction of blow								(N=11)
End-struck	7 77.78	1 25.00	3 50.00	5 62.50	1 100.00	2 100.00	3 100.00	5 45.45
Side-struck	-	1 25.00	3 50.00	2 25.00	-	-	-	2 18.18
Special side-struck	-	1 25.00	-	-	-	-	-	-
Indeterminate	2 22.22	1 25.00	-	1 12.50	-	-	-	4 36.36
Cortex								
No cortex	5 55.56	2 50.00	1 16.67	5 62.50	1 100.00	2 100.00	2 66.67	9 75.00
1–25%	2 22.22	-	-	1 12.50	-	-	1 33.33	2 16.67
26–50%	2 22.22	-	1 16.67	2 25.00	-	-	-	-
51–75%	-	1 25.00	1 16.67	-	-	-	-	-
76–100%	-	1 25.00	3 50.00	-	-	-	-	1 8.33

6.2 Typology of Flint Flake Tools and Core Tools

Table 6.28 Counts and frequencies (%) of retouch characteristics of flint truncations

	Layer II-6 Levels							
	L1	L2	L3	L4	L4b	L5	L6	L7
N	9	4	6	7	1	2	3	12
Type of retouch			(N=5)	(N=6)				
Regular	2 22.22	-	2 40.00	1 16.67	-	1 50.00	-	-
End-scraper	-	-	1 20.00	1 16.67	-	-	-	-
Notch	1 11.11	1 25.00	-	-	-	-	1 33.33	-
Invasive	1 11.11	-	-	-	-	-	1 33.33	-
Semi-abrupt	1 11.11	-	-	2 33.33	-	-	-	1 8.33
Abrupt	3 33.33	2 50.00	2 40.00	2 33.33	-	1 50.00	1 33.33	9 75.00
Bipolar	-	-	-	-	1 100.00	-	-	-
Mixed	-	-	-	-	-	-	-	1 8.33
Irregular	1 11.11	-	-	-	-	-	-	1 8.33
Indeterminate	-	1 25.00	-	-	-	-	-	-
Location of retouch								
Distal	6 66.67	4 100.00	5 83.33	6 85.71	1 100.00	2 100.00	3 100.00	10 83.33
Proximal	1 11.11	-	-	1 14.29	-	-	-	1 8.33
Left	1 11.11	-	-	-	-	-	-	1 8.33
Right	-	-	1 16.67	-	-	-	-	-
Both edges	1 11.11	-	-	-	-	-	-	-
Face of retouch								
Dorsal	8 88.89	1 25.00	6 100.00	4 57.14	1 100.00	1 50.00	1 33.33	6 50.00
Ventral	1 11.11	-	-	2 28.57	-	1 50.00	-	3 25.00
Dorsal & ventral	-	-	-	-	-	-	1 33.33	-
Side	-	3 75.00	-	1 14.29	-	-	1 33.33	3 25.00
Form of retouched edge								(N=11)
Straight	3 33.33	1 25.00	-	1 14.29	-	-	1 33.33	5 45.45
Convex	3 33.33	-	3 50.00	3 42.86	-	1 50.00	-	4 36.36
Concave	2 22.22	2 50.00	1 16.67	-	1 100.00	-	1 33.33	1 9.09
Wavy	-	-	1 16.67	-	-	-	-	1 9.09
Denticulate	-	-	-	-	-	-	1 33.33	-
One tooth	-	-	-	1 14.29	-	-	-	-
Oblique	1 11.11	1 25.00	1 16.67	2 28.57	-	1 50.00	-	-

Fig. 6.28 Flint multiple tools (#2221 Trench II green truncation and denticulate with removed striking platform, #1317 Layer II-6 Level 4 truncation and denticulate)

observed on the retouched flakes include two notches, an atypical borer, and an atypical end-scraper.

6.2.1.1 Discussion: Typology and Characteristics of Flake Tools

In the above description of the flint tools, the results of the analyses were presented according to the different layers and levels. This presentation demonstrated that the frequencies of different tool types vary along the archaeological sequence.

In the following discussion we combine particular tool types from different layers in order to attain sufficiently large samples for each of the typological groups. Such a procedure ignores the temporal factor but provides us with adequate samples for each of the tool types. This enables better comparison of particular attributes between different types, enabling us to determine whether we can discern patterns of intentional selection of certain traits. We also examine the group of unretouched flint flakes (flakes, blades, and *éclats de taille de biface*) and integrate them into the following discussion in order to facilitate the comparison. In the framework of this typological presentation, and throughout the comparison between the different tool types and the unretouched flakes, we have included in the analysis several technological attributes that seem to further characterize each of the examined categories.

Analysis of the physical properties of the tools in comparison with the unretouched flakes reveals several interesting patterns (Table 6.32). When preservation is examined, for retouched tools the dominant mode seems to be that of slightly abraded, while for unretouched flakes the prevailing mode is that of fresh (Table 6.32). However, the unretouched flakes also display the highest frequency of heavily abraded artifacts. This may be explained by the contribution of the flint assemblage of Layer II-6 Level 7, previously demonstrated to be more heavily abraded.

There are variations in preservation between the different tool categories as well; for example, notches and end-scrapers are less fresh than other tool types. This observation is not correlated with their patination, since the dominant mode for both retouched and unretouched flakes is that of patinated artifacts. However, unretouched flakes exhibit far fewer double patinated artifacts than do retouched tools (Table 6.32). In particular, notches and denticulates, borers, and retouched flakes display relatively high frequencies of double patinated artifacts. This pattern shows that double patination is not a taphonomically induced phenomenon but is related to the life-history of the tool. The fact that the flint assemblage is minimally affected by taphonomic processes is further supported by the breakage patterns: the different categories of tools exhibit higher frequencies of complete blanks in comparison to the unretouched flakes (Table 6.32). Two tool categories (denticulates and borers) exhibit the highest

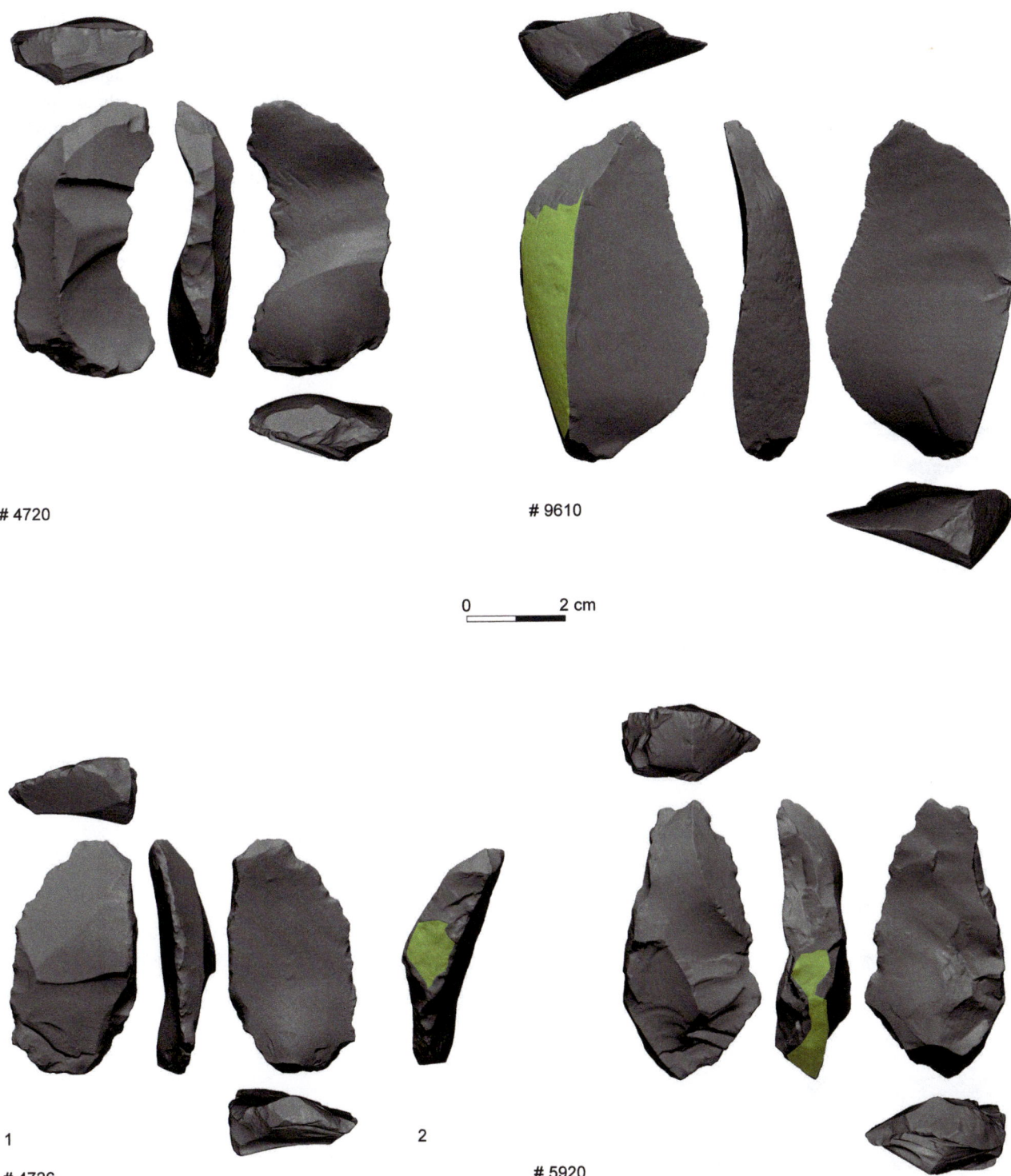

Fig. 6.29 Flint backed knives (#4720 Layer V-5 backed knife with signs of utilization, #4726 Layer V-5 backed knife with signs of utilization **1** multiple views **2** detailed view of backed edge, #9610 JB naturally backed knife, #5920 V-5 naturally backed knife and denticulate with signs of utilization, note steps on proximal end)

occurrences of complete blanks, while the lowest value of complete blanks is recorded for the unretouched flakes. Among the broken artifacts, there is no distinct breakage pattern that is associated with a particular tool type. The only exception is the end-scrapers, which display higher frequencies of proximal breakages.

Most tool types are modified on end-struck blanks, exhibiting frequencies similar to those observed for the unretouched flakes (Table 6.32). Side-struck blanks occur in higher frequencies for borers and end-scrapers (Table 6.32). Another indication of a possible selection of blanks for borers can be found in the cortical coverage. Unlike other tool types, which display cortical coverage similar to that of unretouched flakes, borers

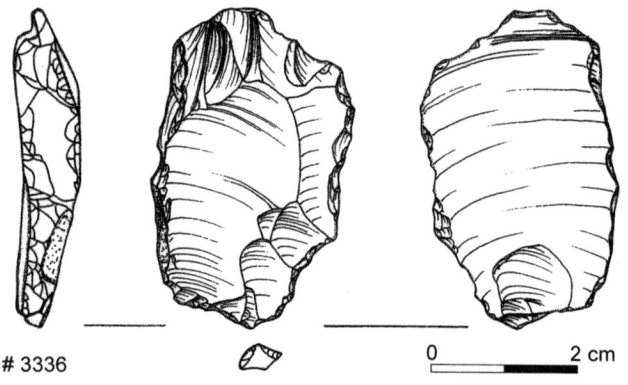

Fig. 6.30 Flint multiple tool (#3336 JB backed knife with bipolar retouch and retouched flake with removed striking platform)

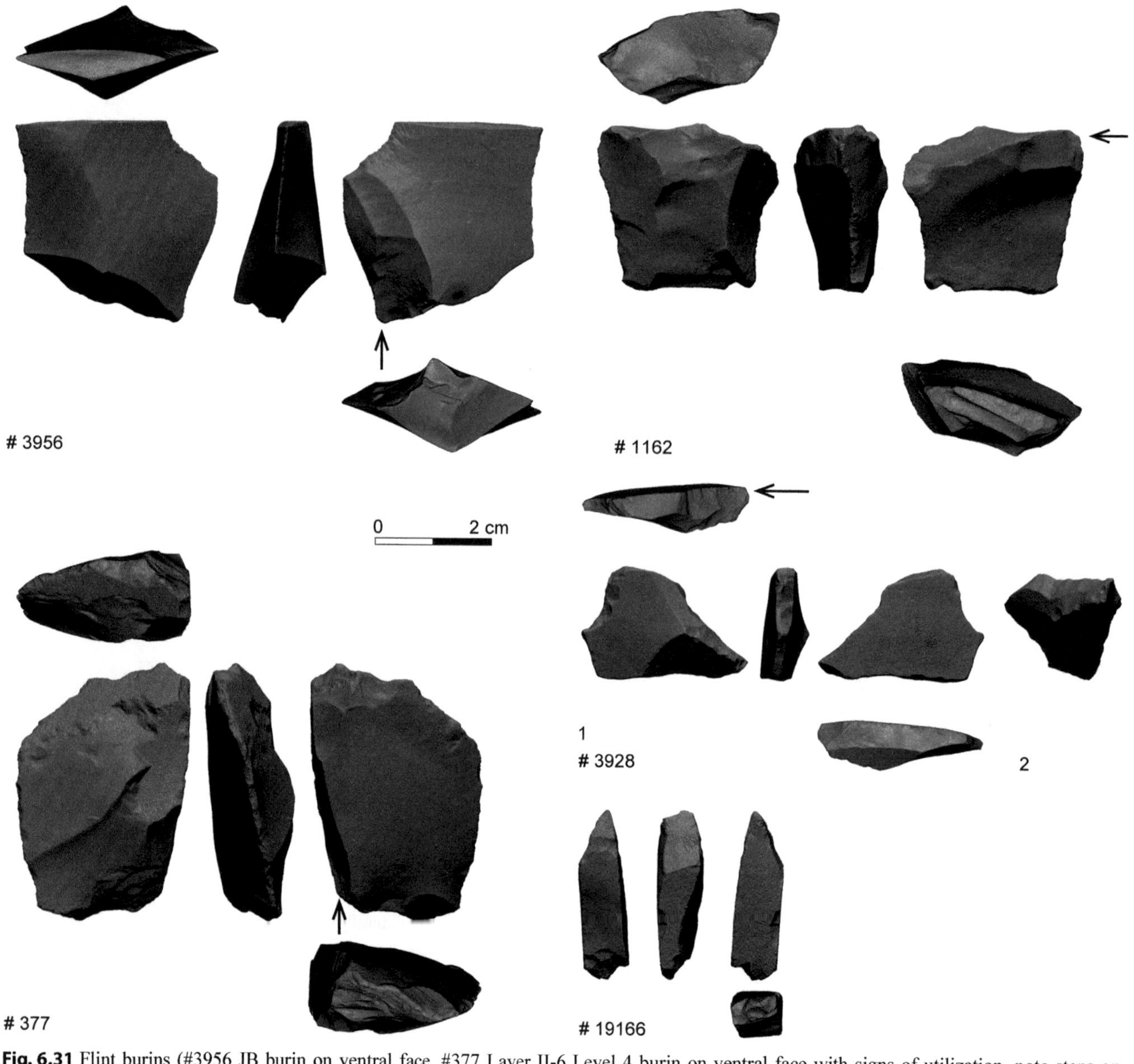

Fig. 6.31 Flint burins (#3956 JB burin on ventral face, #377 Layer II-6 Level 4 burin on ventral face with signs of utilization, note steps on proximal end, #1162 Layer II-6 Level 4b flint multiple tool: burin on ventral face and truncation, #3928 Layer II-6 Level 4 transversal burin, **1** multiple views **2** detailed view of burin, #19166 Layer V-5 burin spall)

6.2 Typology of Flint Flake Tools and Core Tools

Fig. 6.32 Flint burins (#9174 Layer II-6 Level 6 multiple tool: burin and retouched flake, #16142 Trench II burin on truncation, #13760 Layer II-6 Level 7 burin with signs of utilization **1** multiple views **2** detailed view of burin)

appear to be less cortical (Table 6.32). This assumption may apply to side-scrapers as well.

Comparison between the different tool categories and that of unretouched flakes prompts additional observations (Table 6.33). The scar pattern reveals extensive variability, reflecting previous stages of the flint reduction processes that result from different patterns of core preparation and the life-history of cores (Table 6.33). The identification and classification of the scar pattern on flakes is often impossible, resulting in indeterminate patterns that can form as much as a third of the studied sample. Overall, there is a general similarity among the different tool categories and the unretouched flakes, with the two dominant patterns being the along axis and the radial flaking modes. However, radial flaking is significantly more frequent on flake tools than on unretouched flakes ($\chi^2=24.53$; $df=12$; $p<0.0172$), along axis flaking is more frequent on unretouched flakes, but the only scar pattern that is significantly more frequent on unretouched flakes is the cortical pattern ($\chi^2=38.57$; $df=12$; $p=0.0001$). These results demonstrate that cortical flakes were not selected as blanks for tools, but flakes of radial scar pattern were preferred. This preference may be related to the fact that flakes of radial scar pattern are significantly longer and thicker than all other flakes (e.g. in comparison with the along axis scar pattern: length: t ratio=4.02; $df=2988$; $p(tt)<0.0001$; thickness: t ratio=6.89; $df=2988$; $p(tt)<0.0001$). Although these differences are expressed in minimal values (length=1.91 mm; thickness=2.79 mm), a greater thickness is considered important because of the average low values of thickness.

Removed Striking Platforms

As with other technological attributes, the identification of striking platforms is not always possible, resulting in relatively high frequencies of indeterminate platforms. This is the case, and in even higher frequencies, for the unretouched flakes (Table 6.33). Another substantial portion of the striking platforms is broken, so that we are left with only two thirds of the striking platforms for analysis. When identification is feasible, a variety of striking platforms is recorded for the different tool categories. The dominant patterns are plain and removed striking platforms, followed by facetted ones (Table 6.33). The distribution of striking platform types shows different stages of the reduction process, beginning with cortical platforms at early stages of knapping and proceeding to platform modification, which occurred after the production of the artifacts. The latter is seen in significantly lower frequencies in unretouched flakes than in flake tools ($\chi^2=41.41$; $df=6$; $p<0.0001$; categories of broken and indeterminate striking platforms are excluded). Although the striking platform is a technological attribute, the removal of striking platforms is clearly not a component of the reduction sequence

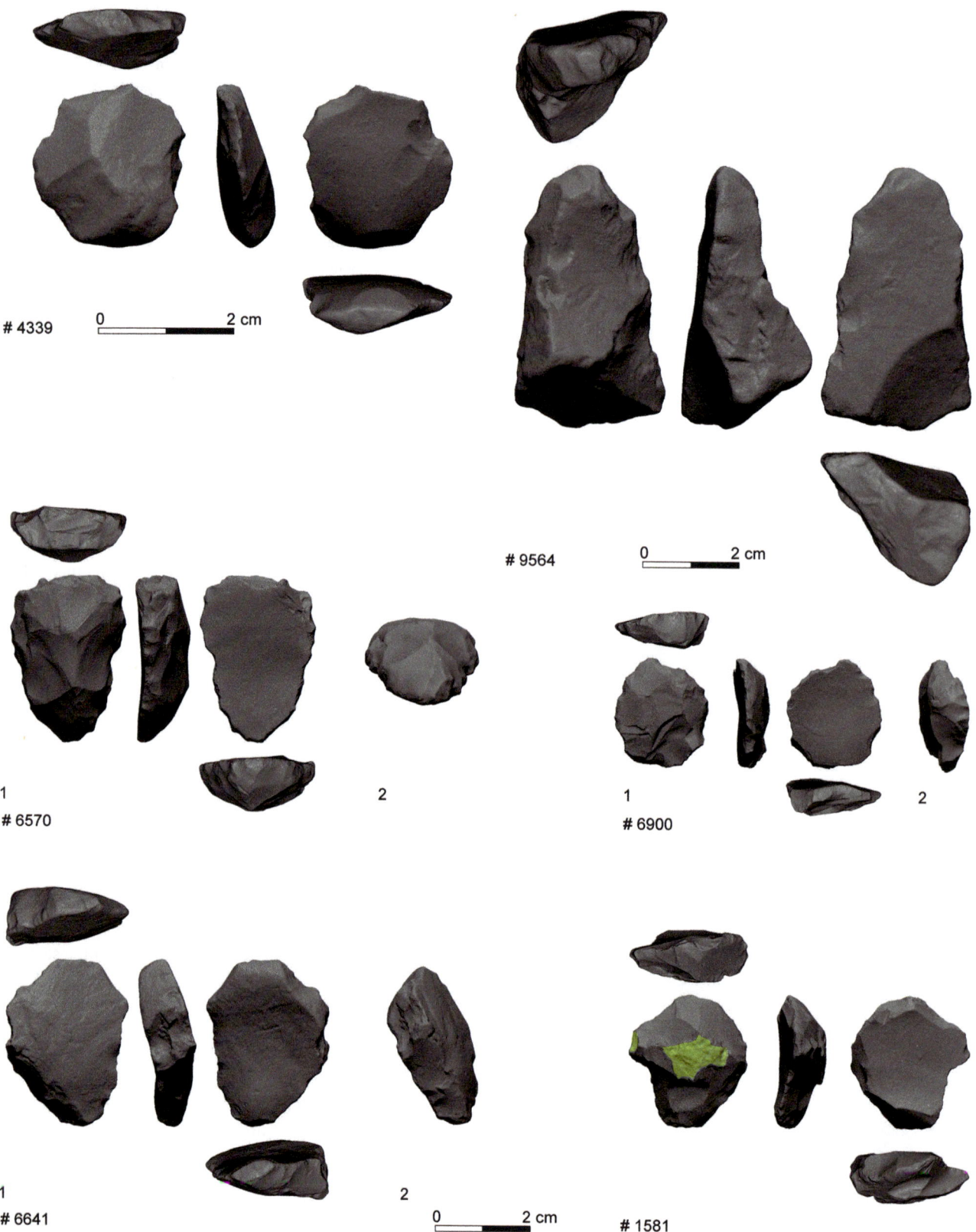

Fig. 6.33 Flint varia tools (#4339 Layer II-6 Level 6 alternately retouched bec (beak) with signs of utilization and ventral removals, #6570 Layer II-6 Level 1 limace with removed striking platform, #6641 Layer II-6 Level 2 varia tool, #9564 Layer II-6 Level 2 varia tool with ventral removals and removed striking platform, #6900 Layer II-6 Level 2 varia tool with signs of utilization, #1581 Layer II-6 Level 3 varia tool with Nahr Ibrahim technique, signs of utilization and removed striking platform)

6.2 Typology of Flint Flake Tools and Core Tools

Fig. 6.34 Flint varia tools (#14092 Layer II-6 Level 7 alternately retouched bec (beak) with signs of utilization **1** multiple views **2** detailed view of retouched edge, #14961 Layer IV-7 varia tool with ventral removals and removed striking platform)

but rather should be linked with the realm of modification of flakes and flake tools.

The high frequencies of removed striking platforms necessitated additional examination, resulting in the identification of various types of platform removals that were classified into five major groups, as follows:

1) A single tooth-like edge made by a combination of two small concavities. These artifacts seem to form a group within the general sphere of tools like becs (beaks), borers, awls, and perçoirs. This group includes some specimens that morphologically resemble tanged artifacts (Fig. 6.36). Despite its morphology, the tooth-like edge was not classified as a borer except in the rare cases where it protruded conspicuously enough to justify such a classification (Fig. 6.37). The tooth-like edge resulting from removal of the proximal end was modified by removals that range between

Fig. 6.35 Flint retouched flakes (#14847 Layer II-6 Level 3 on *outrepassé* flake with removed striking platform, #3924 Layer II-6 Level 4 with Nahr Ibrahim technique, #4231 Layer II-6 Level 6)

small scars and true retouch (notch). In some cases the tooth-like edge was modified by two notches, either on one face or alternately (Fig. 6.37). The length of the tooth-like edge is variable: small, large, elongated, or tanged-like. It is worth noting that some of these artifacts have more than one retouched edge (multiple tools) (Figs. 6.36, 6.38).

2) Individual flake removals that resulted in one or several scars on the proximal end of the artifact. When the scars have been alternately removed, the proximal end resembles that of the previous tooth-like category, but differs in the fact that the scar removals do not have the retouch intensity typical of that category (Fig. 6.39).

3) Scraper-like modifications located on the ventral face. Artifacts of this group resemble side-scrapers and their retouched edge is oriented obliquely to the knapping axis. This retouch removed the striking platform and hence thinned the flake. The scraper-like removal of the striking platform often did not result in a classic side-scraper retouch (Fig. 6.39). In other, rare cases, the scraper retouch was sufficiently dominant to justify a typological classification as a side-scraper on the ventral face (Figs. 6.39–6.40).

4) Concave truncations, intensively retouched so that the general morphology of the striking platform area is eliminated. The concavity of these items extends along the entire width of the proximal edge. As in the other classes described above, the retouch is found on both ventral and dorsal faces and on the side (thickness) of the flake as well. In some cases, when the concave proximal morphology is very dominant, the proximal end is considered a notch or an abruptly retouched side-scraper (Fig. 6.40) in the typological classification (Fig. 6.7). When the proximal modification did not resemble typical tool types, these artifacts were classified either as unretouched flakes (when all edges apart from the proximal end are unretouched) or as tools (when edges in addition to the proximal end are retouched) (Figs. 6.40–6.41).

5) Proximal retouch that resulted in the removal of the striking platform. Included in this group are different modes of retouch that produced varied morphologies of the proximal end (Figs. 6.8, 6.33, 6.41). Some of these artifacts are unretouched except for their proximal modification and others are retouched tools, including multiple tools. This group also demonstrates variability in the extent and intensity of the retouch. Due to this variability, it is the least uniform group of the five.

To conclude, the removal of striking platforms resulted in reduction of the thickness of flakes, and possibly their length. Another means of reducing the thickness of flakes is through thinning of the bulb of percussion. In the GBY flint assemblage, this feature is recorded on only three flakes and hence can be considered negligible.

The fact that the frequencies of removed striking platforms

6.2 Typology of Flint Flake Tools and Core Tools

Table 6.29 Counts and frequencies (%) of characteristics of retouched flint flakes

	Layer												
						II-6 Levels							
	V-5	JB	II-2	II-5	II-5/6	L1	L2	L3	L4	L4b	L5	L6	L7
N	21	10	10	17	10	54	26	43	46	11	24	25	70
Preservation													
Fresh	10	6	2	8	4	21	11	19	14	5	-	1	5
	47.62	60.00	20.00	47.06	40.00	38.89	42.31	44.19	30.43	45.45		4.00	7.14
Slightly abraded	7	4	4	6	2	22	11	17	20	3	13	9	21
	33.33	40.00	40.00	35.29	20.00	40.74	42.31	39.53	43.48	27.27	54.17	36.00	30.00
Abraded	3	-	3	2	3	10	3	7	10	2	9	10	37
	14.29		30.00	11.76	30.00	18.52	11.54	16.28	21.74	18.18	37.50	40.00	52.86
Heavily abraded	1	-	1	1	1	1	1	-	2	1	2	5	7
	4.76		10.00	5.88	10.00	1.85	3.85		4.35	9.09	8.33	20.00	10.00
Patination													
No patina	-	-	-	1	-	8	2	9	3	1	-	1	2
				5.88		14.81	7.69	20.93	6.52	9.09		4.00	2.86
Patinated	18	8	6	11	7	33	16	26	31	8	12	15	47
	85.71	80.00	60.00	64.71	70.00	61.11	61.54	60.47	67.39	72.73	50.00	60.00	67.14
Double patinated	3	2	4	5	3	13	8	8	12	2	12	9	21
	14.29	20.00	40.00	29.41	30.00	24.07	30.77	18.60	26.09	18.18	50.00	36.00	30.00
Breakage													
Complete	13	4	3	7	5	34	13	28	21	3	21	15	41
	61.90	40.00	30.00	41.18	50.00	62.96	50.00	65.12	45.65	27.27	87.50	60.00	58.57
Distal	3	1	3	3	3	6	6	5	9	2	3	2	9
	14.29	10.00	30.00	17.65	30.00	11.11	23.08	11.63	19.57	18.18	12.50	8.00	12.86
Lateral	1	3	1	1	-	2	-	4	3	2	-	3	3
	4.76	30.00	10.00	5.88		3.70		9.30	6.52	18.18		12.00	4.29
Proximal	1	2	-	1	-	4	2	3	4	3	-	3	7
	4.76	20.00		5.88		7.41	7.69	6.98	8.70	27.27		12.00	10.00
Distal & lateral	1	-	-	-	-	-	1	-	2	-	-	-	2
	4.76						3.85		4.35				2.86
Distal & proximal	1	-	-	1	-	4	3	1	1	1	-	-	2
	4.76			5.88		7.41	11.54	2.33	2.17	9.09			2.86
Fragment	1	-	2	2	-	2	-	2	4	-	-	2	4
	4.76		20.00	11.76		3.70		4.65	8.70			8.00	5.71
Proximal & lateral	-	-	1	2	2	2	1	-	2	-	-	-	2
			10.00	11.76	20.00	3.70	3.85		4.35				2.86
Direction of blow													(N=69)
End-struck	6	5	2	9	4	28	10	26	22	6	9	11	17
	28.57	50.00	20.00	52.94	40.00	51.85	38.46	60.47	47.83	54.55	37.50	44.00	24.64
Side-struck	1	-	-	3	3	12	7	8	9	-	5	4	14
	4.76			17.65	30.00	22.22	26.92	18.60	19.57		20.83	16.00	20.29
Special side-struck	2	1	-	-	-	2	1	2	1	-	2	2	9
	9.52	10.00				3.70	3.85	4.65	2.17		8.33	8.00	13.04
Indeterminate	12	4	8	5	3	12	8	7	14	5	8	8	29
	57.14	40.00	80.00	29.41	30.00	22.22	30.77	16.28	30.43	45.45	33.33	32.00	42.03
Cortex													
No cortex	15	6	10	8	6	37	19	25	28	6	16	10	45
	71.43	60.00	100.00	47.06	60.00	68.52	73.08	58.14	60.87	54.55	66.67	40.00	64.29
1–25%	4	3	-	2	3	4	3	10	5	2	2	6	9
	19.05	30.00		11.76	30.00	7.41	11.54	23.26	10.87	18.18	8.33	24.00	12.86
26–50%	-	1	-	4	1	4	1	3	3	1	3	3	5
		10.00		23.53	10.00	7.41	3.85	6.98	6.52	9.09	12.50	12.00	7.14
51–75%	-	-	-	2	-	1	-	3	4	1	3	2	1
				11.76		1.85		6.98	8.70	9.09	12.50	8.00	1.43
76–100%	2	-	-	1	-	8	3	2	6	1	-	4	7
	9.52			5.88		14.81	11.54	4.65	13.04	9.09		16.00	10.00
Indeterminate	-	-	-	-	-	-	-	-	-	-	-	-	3
													4.29

Table 6.30 Counts and frequencies (%) of retouch characteristics of retouched flint flakes

						Layer							
									II-6 Levels				
	V-5	JB	II-2	II-5	II-5/6	L1	L2	L3	L4	L4b	L5	L6	L7
N	21	10	10	17	9	54	26	42	45	11	24	25	64
Type of retouch	(N=20)												(N=63)
Regular	4 20.00	3 30.00	-	1 5.88	4 44.44	6 11.11	3 11.54	14 33.33	9 20.00	3 27.27	4 16.67	4 16.00	12 19.05
End-scraper	1 5.00	-	-	-	-	-	-	-	-	-	1 4.17	1 4.00	-
Side-scraper	-	-	1 10.00	6 35.29	1 11.11	3 5.56	3 11.54	1 2.38	1 2.22	-	3 12.50	-	6 9.52
Notch	-	1 10.00	1 10.00	1 5.88	-	5 9.26	-	1 2.38	-	-	3 12.50	3 12.00	4 6.35
Invasive	2 10.00	-	1 10.00	-	1 11.11	1 1.85	2 7.69	-	2 4.44	-	-	-	1 1.59
Semi-Quina	1 5.00	-	-	-	-	-	-	-	-	-	-	-	-
Quina	-	-	-	-	-	-	-	-	1 2.22	-	-	-	-
Raclette	-	-	-	-	-	-	1 3.85	-	-	-	2 8.33	-	-
Semi-abrupt	2 10.00	-	-	1 5.88	-	-	-	2 4.76	2 4.44	-	-	-	2 3.17
Abrupt	-	1 10.00	-	1 5.88	-	12 22.22	2 7.69	3 7.14	6 13.33	1 9.09	1 4.17	4 16.00	5 7.94
Bifacial	-	-	-	-	-	-	-	-	-	-	-	-	2 3.17
Thinning	-	-	-	-	-	-	1 3.85	-	-	-	-	-	-
Mixed	2 10.00	-	-	1 5.88	-	-	-	-	3 6.67	2 18.18	1 4.17	-	3 4.76
Nahr Ibrahim	-	1 10.00	-	-	-	-	-	1 2.38	2 4.44	-	-	-	-
Irregular	1 5.00	2 20.00	4 40.00	1 5.88	1 11.11	19 35.19	10 38.46	10 23.81	9 20.00	2 18.18	5 20.83	3 12.00	16 25.40
Signs of use	2 10.00	2 20.00	1 10.00	4 23.53	1 11.11	4 7.41	2 7.69	6 14.29	9 20.00	2 18.18	2 8.33	5 20.00	4 6.35
Indeterminate	5 25.00	-	2 20.00	1 5.88	1 11.11	4 7.41	2 7.69	4 9.52	1 2.22	1 9.09	2 8.33	5 20.00	8 12.70
Location of retouch													
Distal	2 9.52	2 20.00	2 20.00	5 29.41	2 22.22	8 14.81	4 15.38	5 11.90	6 13.33	1 9.09	1 4.17	4 16.00	17 26.56
Proximal	-	1 10.00	-	1 5.88	-	3 5.56	1 3.85	3 7.14	5 11.11	-	-	-	-
Left	4 19.05	1 10.00	3 30.00	3 17.65	1 11.11	16 29.63	5 19.23	12 28.57	12 26.67	4 36.36	9 37.50	9 36.00	9 14.06
Right	6 28.57	2 20.00	3 30.00	6 35.29	4 44.44	16 29.63	12 46.15	16 38.10	15 33.33	4 36.36	7 29.17	7 28.00	15 23.44
Both edges	5 23.81	1 10.00	-	1 5.88	-	5 9.26	1 3.85	1 2.38	3 6.67	1 9.09	1 4.17	2 8.00	7 10.94
Distal & both edges	2 9.52	-	-	-	-	-	-	-	-	-	1 4.17	-	2 3.13
Distal & right	-	-	1 10.00	-	-	-	-	-	-	-	2 8.33	-	3 4.69
Distal & left	-	1 10.00	-	-	-	1 1.85	-	1 2.38	1 2.22	-	-	-	3 4.69
Proximal & both edges	-	-	-	-	1 11.11	-	-	-	-	-	-	-	-
Proximal & left	-	-	-	-	-	-	-	-	1 2.22	-	-	-	-

6.2 Typology of Flint Flake Tools and Core Tools

Table 6.30 (cont.)

						Layer							
								II-6 Levels					
	V-5	JB	II-2	II-5	II-5/6	L1	L2	L3	L4	L4b	L5	L6	L7
N	21	10	10	17	9	54	26	42	45	11	24	25	64
Proximal & right	-	1 10.00	-	-	-	-	-	-	-	-	-	-	-
Convergent	-	1 10.00	-	-	-	1 1.85	-	1 2.38	-	-	-	-	1 1.56
Circumference	2 9.52	-	1 10.00	1 5.88	1 11.11	4 7.41	2 7.69	2 4.76	2 4.44	1 9.09	3 12.50	2 8.00	5 7.81
Indeterminate	-	-	-	-	-	-	1 3.85	1 2.38	-	-	-	1 4.00	2 3.13
Face of retouch	(N=19)		(N=8)	(N=16)	(N=8)		(N=25)	(N=37)	(N=44)		(N=23)		(N=63)
Dorsal	10 52.63	2 20.00	8 100.00	13 81.25	5 62.50	34 62.96	14 56.00	22 59.46	30 68.18	5 45.45	16 66.67	11 47.83	36 57.14
Ventral	3 15.79	4 40.00	-	-	2 25.00	5 9.26	4 16.00	8 21.62	7 15.91	2 18.18	1 4.17	4 17.39	10 15.87
Dorsal & ventral	6 31.58	3 30.00	-	2 12.50	-	12 22.22	6 24.00	5 13.51	6 13.64	4 36.36	5 20.83	3 13.04	13 20.63
Side	-	1 10.00	-	1 6.25	1 12.50	3 5.56	1 4.00	2 5.41	1 2.27	-	2 8.33	5 21.74	4 6.35
Form of retouched edge			(N=8)	(N=15)	(N=8)	(N=51)	(N=21)	(N=36)	(N=35)	(N=10)	(N=21)	(N=24)	(N=56)
Straight	4 19.05	3 30.00	1 12.50	5 33.33	-	12 23.53	4 19.05	11 30.56	9 25.71	2 20.00	8 38.10	4 16.67	10 17.86
Convex	3 14.29	2 20.00	3 37.50	4 26.67	4 50.00	14 27.45	7 33.33	10 27.78	11 31.43	2 20.00	6 28.57	11 45.83	21 37.50
Concave	5 23.81	-	1 12.50	5 33.33	2 25.00	10 19.61	2 9.52	8 22.22	4 11.43	2 20.00	3 14.29	5 20.83	8 14.29
Convergent	-	2 20.00	-	-	-	2 3.92	-	1 2.78	1 2.86	-	-	-	1 1.79
Wavy	4 19.05	2 20.00	-	1 6.67	2 25.00	9 17.65	7 33.33	1 2.78	8 22.86	3 30.00	1 4.76	3 12.50	11 19.64
Denticulate	1 4.76	-	-	-	-	2 3.92	-	1 2.78	-	-	2 9.52	-	2 3.57
One tooth	-	1 10.00	-	-	-	-	-	2 5.56	-	-	1 4.76	-	-
Oblique	-	-	-	-	-	-	-	-	2 5.71	-	-	-	1 1.79
Indeterminate	4 19.05	-	3 37.50	-	-	2 3.92	1 4.76	2 5.56	-	1 10.00	-	1 4.17	2 3.57

Table 6.31 Descriptive statistics of retouched flint flakes (size in mm; angle in degrees)

						Layer							
									II-6 Levels				
	V-5	JB	II-2	II-5	II-5/6	L1	L2	L3	L4	L4b	L5	L6	L7
N	21	10	10	17	10	54	26	43	46	11	24	25	70
Maximum length													
Maximum	58.00	66.00	36.00	38.00	42.00	39.00	40.00	49.00	36.00	41.00	49.00	38.00	57.00
Median	24.00	38.00	27.00	26.00	25.00	27.00	28.00	29.00	27.00	31.00	25.50	27.00	26.50
Minimum	20.00	21.00	21.00	21.00	21.00	21.00	20.00	20.00	21.00	22.00	20.00	21.00	21.00
Mean	26.10	38.40	27.00	27.29	26.70	27.59	26.92	29.67	27.70	29.73	26.13	27.52	28.40
Std dev	9.15	13.45	5.01	5.12	6.24	4.17	4.72	7.29	4.18	5.87	5.71	5.08	6.83
Std err mean	2.00	4.25	1.58	1.24	1.97	0.57	0.92	1.11	0.62	1.77	1.16	1.02	0.82
Length													
Maximum	50.00	59.00	33.00	33.00	41.00	39.00	34.00	49.00	36.00	39.00	43.00	38.00	57.00
Median	22.00	35.50	24.00	25.00	21.50	25.00	23.50	26.00	25.00	24.00	22.00	24.00	24.00
Minimum	16.00	17.00	13.00	17.00	16.00	14.00	15.00	14.00	15.00	16.00	17.00	14.00	13.00
Mean	24.19	33.80	23.70	24.53	22.80	24.83	24.23	27.65	24.59	24.91	23.46	24.48	25.27
Std dev	7.99	13.02	7.04	3.97	7.10	5.26	5.35	8.10	4.87	7.44	5.18	6.28	7.68
Std err mean	1.74	4.12	2.23	0.96	2.24	0.72	1.05	1.24	0.72	2.24	1.06	1.26	0.92
Width													
Maximum	43.00	38.00	25.00	38.00	32.00	38.00	35.00	36.00	36.00	34.00	49.00	34.00	38.00
Median	15.00	28.50	20.00	20.00	20.50	21.00	23.00	21.00	21.50	22.00	21.00	20.00	23.00
Minimum	12.00	17.00	13.00	14.00	10.00	11.00	13.00	13.00	12.00	8.00	14.00	15.00	13.00
Mean	17.95	28.10	19.90	22.24	20.00	21.07	22.38	22.65	22.41	23.27	21.83	21.20	23.06
Std dev	8.33	7.61	3.67	6.42	7.09	5.77	5.55	6.24	5.63	7.81	7.67	4.94	5.99
Std err mean	1.82	2.41	1.16	1.56	2.24	0.79	1.09	0.95	0.83	2.36	1.57	0.99	0.72
Thickness													
Maximum	13.00	12.00	14.00	15.00	13.00	21.00	14.00	15.00	21.00	13.00	15.00	16.00	20.00
Median	7.00	8.00	7.50	8.00	7.50	8.50	8.00	9.00	10.50	7.00	8.00	9.00	9.00
Minimum	1.00	4.00	5.00	5.00	6.00	4.00	5.00	4.00	4.00	6.00	5.00	3.00	4.00
Mean	7.48	8.30	7.90	8.88	8.60	9.15	8.23	9.16	10.39	8.45	8.46	9.24	9.10
Std dev	2.75	2.67	2.96	3.59	2.80	3.58	2.64	2.72	4.06	2.38	2.55	2.92	3.04
Std err mean	0.60	0.84	0.94	0.87	0.88	0.49	0.52	0.42	0.60	0.72	0.52	0.58	0.36
Angle of retouch	(N=11)	(N=3)	(N=7)	(N=15)	(N=3)	(N=44)	(N=19)	(N=32)	(N=32)	(N=9)	(N=14)	(N=16)	(N=53)
Maximum	108.00	82.00	113.00	88.00	81.00	100.00	90.00	90.00	102.00	90.00	93.00	98.00	102.00
Median	88.00	73.00	76.00	78.00	70.00	82.50	80.00	81.50	82.50	88.00	74.00	83.00	78.00
Minimum	74.00	53.00	56.00	58.00	60.00	51.00	69.00	68.00	67.00	73.00	64.00	1.00	58.00
Mean	88.91	69.33	78.00	76.80	70.33	82.07	80.26	81.25	82.34	83.78	75.43	68.94	80.75
Std dev	11.54	14.84	17.25	9.14	10.50	9.27	5.98	5.67	7.64	6.63	8.06	33.38	9.94
Std err mean	3.48	8.57	6.52	2.36	6.06	1.40	1.37	1.00	1.35	2.21	2.16	8.34	1.37

are much higher among retouched tools than among unretouched flakes (Table 6.33) suggests that this phenomenon is associated with the tools and hence that it is a typological feature.

The removal of the striking platform produced new morphologies of the proximal end that bear similarities to existing types such as notches, scrapers, truncations, retouched flakes, and borers. Worth noting is the extensive use of removal of striking platforms and of bulbs of percussions in the modification of basalt bifacial tools (handaxes and cleavers; Chapter 7). This focus of attention on the proximal end of basalt bifaces is a major component of the technological, stylistic, and typological conception of bifacial tools made on basalt, and it is very likely that this attention extended to the reduction sequence of flint artifacts as well. However, the particulars of the implementation of the removals differ radically between basalt and flint.

Table 6.34 provides five measures of descriptive statistics for the size of major typological categories and of unretouched flakes (only complete artifacts are included). The most striking feature that emerges from the data is the homogeneity observed within the different tool categories. This homogeneity is expressed in each of the five measures of size and in their statistics, particularly in the width and thickness (Table 6.34). Moreover, a general similarity is

6.2 Typology of Flint Flake Tools and Core Tools

Table 6.32 Counts and frequencies (%) of characteristics of different flint flake tool categories and of unretouched flint flakes

Type	Notch	Denticulate	Side-scraper	Borer	End-scraper	Knife	Truncation	Burin	Retouched flake	Unretouched flake
N	469	161	243	241	121	19	54	18	390	2,791
Preservation										
Fresh	127	57	77	83	32	12	19	9	114	996
	27.08	35.40	31.69	34.44	26.45	63.16	35.19	50.00	29.23	35.69
Slightly abraded	168	64	101	99	52	-	19	7	149	836
	35.82	39.75	41.56	41.08	42.98		35.19	38.89	38.21	29.95
Abraded	143	35	56	50	37	5	11	2	104	729
	30.49	21.74	23.05	20.75	30.58	26.32	20.37	11.11	26.67	26.12
Heavily abraded	31	5	9	9	-	2	5	-	23	230
	6.61	3.11	3.70	3.73		10.53	9.26		5.90	8.24
Patination										
No patina	27	8	19	16	11	2	6	2	30	250
	5.76	4.97	7.82	6.64	9.09	10.53	11.11	11.11	7.69	8.96
Patinated	311	111	172	155	79	13	35	13	251	2051
	66.31	68.94	70.78	64.32	65.29	68.42	64.81	72.22	64.36	73.49
Double patinated	131	42	52	70	31	4	13	3	109	490
	27.93	26.09	21.40	29.05	25.62	21.05	24.07	16.67	27.95	17.56
										(N=2,790)
Breakage										
Complete	280	112	135	171	70	14	38	6	223	1411
	59.70	69.57	55.56	70.95	57.85	73.68	70.37	33.33	57.18	50.57
Distal	58	11	35	15	7	1	-	3	56	369
	12.37	6.83	14.40	6.22	5.79	5.26		16.67	14.36	13.23
Lateral	22	13	19	22	13	1	2	2	25	196
	4.69	8.07	7.82	9.13	10.74	5.26	3.70	11.11	6.41	7.03
Proximal	60	14	37	21	27	3	11	3	33	287
	12.79	8.70	15.23	8.71	22.31	15.79	20.37	16.67	8.46	10.29
Distal & lateral	6	1	4	-	-	-	-	-	7	80
	1.28	0.62	1.65						1.79	2.87
Distal & proximal	18	4	7	4	-	-	2	-	14	95
	3.84	2.48	2.88	1.66			3.70		3.59	3.41
Fragment	18	6	6	6	3	-	1	2	19	273
	3.84	3.73	2.47	2.49	2.48		1.85	11.11	4.87	9.78
Proximal & lateral	7	-	-	2	1	-	-	2	13	76
	1.49			0.83	0.83			11.11	3.33	2.72
Indeterminate	-	-	-	-	-	-	-	-	-	3
										0.11
Direction of blow	(N=468)	(N=159)	(N=242)			(N=18)	(N=53)	(N=17)	(N=387)	(N=2,694)
End-struck	213	66	104	105	49	13	31	6	166	1091
	45.51	41.51	42.98	43.57	40.50	72.22	58.49	35.29	42.89	40.50
Side-struck	93	38	46	67	32	-	8	1	71	533
	19.87	23.90	19.01	27.80	26.45		15.09	5.88	18.35	19.78
Special side-struck	27	15	25	17	5	3	3	1	22	196
	5.77	9.43	10.33	7.05	4.13	16.67	5.66	5.88	5.68	7.28
Indeterminate	135	40	67	52	35	2	11	9	128	874
	28.85	25.16	27.69	21.58	28.93	11.11	20.75	52.94	33.07	32.44
Cortex						(N=18)				(N=2,787)
No cortex	287	101	157	165	75	6	31	14	250	1698
	61.19	62.73	64.61	68.46	61.98	33.33	57.41	77.78	64.10	60.93
1–25%	61	28	41	30	22	5	9	2	56	378
	13.01	17.39	16.87	12.45	18.18	27.78	16.67	11.11	14.36	13.56
26–50%	44	11	16	18	6	4	6	-	29	186
	9.38	6.83	6.58	7.47	4.96	22.22	11.11		7.44	6.67
51–75%	24	9	11	5	4	1	3	2	18	136
	5.12	5.59	4.53	2.07	3.31	5.56	5.56	11.11	4.62	4.88
76–100%	48	12	17	22	14	2	5	-	34	314
	10.23	7.45	7.00	9.13	11.57	11.11	9.26		8.72	11.27
Indeterminate	5	-	1	1	-	-	-	-	3	75
	1.07		0.41	0.41					0.77	2.69

Table 6.33 Counts and frequencies (%) of technological characteristics of different flint flake tool categories and of unretouched flint flakes

Type	Notch	Denticulate	Side-scraper	Borer	End-scraper	Knife	Truncation	Burin	Retouched flake	Unretouched flake
N	461	159	243	239	121	19	54	18	385	2,684
Scar pattern				(N=238)	(N=120)			(N=17)		
Cortical	57	21	27	20	16	1	6	-	37	347
	12.36	13.21	11.11	8.40	13.33	5.26	11.11		9.61	12.93
Plain	16	3	8	5	6	3	-	-	10	76
	3.47	1.89	3.29	2.10	5.00	15.79			2.60	2.83
Along axis	65	19	37	41	24	8	12	5	70	451
	14.10	11.95	15.23	17.23	20.00	42.11	22.22	29.41	18.18	16.80
Parallel	2	1	1	3	-	-	1	-	5	12
	0.43	0.63	0.41	1.26			1.85		1.30	0.45
Convergent	5	4	10	4	3	-	1	-	9	65
	1.08	2.52	4.12	1.68	2.50		1.85		2.34	2.42
Opposed	14	2	4	4	-	-	-	1	5	28
	3.04	1.26	1.65	1.68				5.88	1.30	1.04
Radial	75	36	39	36	20	-	10	5	66	275
	16.27	22.64	16.05	15.13	16.67		18.52	29.41	17.14	10.25
Ridged	13	5	2	11	5	-	-	-	10	61
	2.82	3.14	0.82	4.62	4.17				2.60	2.27
Side	18	5	14	7	3	1	3	-	13	89
	3.90	3.14	5.76	2.94	2.50	5.26	5.56		3.38	3.32
Along axis & side	21	11	15	18	9	1	9	1	23	153
	4.56	6.92	6.17	7.56	7.50	5.26	16.67	5.88	5.97	5.70
Along axis & opposed	16	7	9	10	4	2	3	1	12	86
	3.47	4.40	3.70	4.20	3.33	10.53	5.56	5.88	3.12	3.20
Opposed & side	4	1	2	4	-	-	2	1	6	30
	0.87	0.63	0.82	1.68			3.70	5.88	1.56	1.12
Along axis & radial	-	-	4	-	-	1	-	-	1	3
			1.65			5.26			0.26	0.11
Indeterminate	155	44	71	75	30	2	7	3	118	1,008
	33.62	27.67	29.22	31.51	25.00	10.53	12.96	17.65	30.65	37.56
Striking platform			(N=242)							(N=2,683)
Cortical	28	10	12	15	6	1	-	-	15	159
	6.07	6.29	4.96	6.28	4.96	5.26			3.90	5.93
Punctiform	24	7	7	11	4	-	2	2	16	115
	5.21	4.40	2.89	4.60	3.31		3.70	11.11	4.16	4.29
Plain	121	33	56	46	22	5	9	3	93	735
	26.25	20.75	23.14	19.25	18.18	26.32	16.67	16.67	24.16	27.39
Dihedral	8	2	9	1	2	-	3	-	11	81
	1.74	1.26	3.72	0.42	1.65		5.56		2.86	3.02
Facetted	37	21	31	29	12	3	9	-	47	268
	8.03	13.21	12.81	12.13	9.92	15.79	16.67		12.21	9.99
Crushed	11	2	5	10	2	-	1	-	4	84
	2.39	1.26	2.07	4.18	1.65		1.85		1.04	3.13
Removed	51	36	48	54	27	4	13	3	52	112
	11.06	22.64	19.83	22.59	22.31	21.05	24.07	16.67	13.51	4.17
Broken	128	34	56	52	36	4	14	9	111	764
	27.77	21.38	23.14	21.76	29.75	21.05	25.93	50.00	28.83	28.48
Indeterminate	53	14	18	21	10	2	3	1	36	365
	11.50	8.81	7.44	8.79	8.26	10.53	5.56	5.56	9.35	13.60

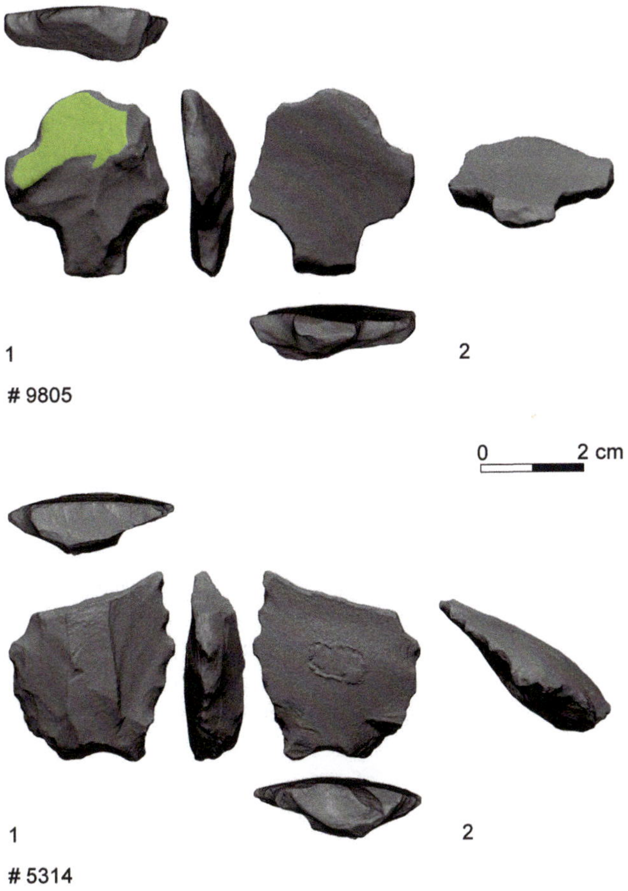

6.2 Typology of Flint Flake Tools and Core Tools

Fig. 6.36 Flint flakes with removed striking platforms **1** multiple views **2** detailed view of proximal end (#9805 Layer II-6 Level 7 tanged tool and retouched flake, #5314 Layer V-5 denticulate)

observed among the different tool categories and between them and unretouched flakes, though within this overall similarity there are some minor deviations in the case of several tool types (knives, side-scrapers, and truncations). The category of knives (naturally, typically, and atypically backed) clearly deviates from the norm. Although the sample size is much smaller than for other tool types, this category is significantly longer than all other tool types (Appendix 6.7) and is also characterized by larger standard deviations (Table 6.34). The elongated morphology of knives is also expressed in their L/W value (Table 6.34). Another such category is that of side-scrapers, which are significantly larger than other tool types as well as than unretouched flakes (Section 6.2 and Appendix 6.7). A tendency toward somewhat longer artifacts is observed in the category of truncations, which may be significant considering the fact that truncations are modified mostly on the distal end of the artifact and the original length of the blank is consequently reduced. Truncations are significantly longer than unretouched flakes as well (Appendix 6.7). Generally, the L/W ratio (Table 6.34)

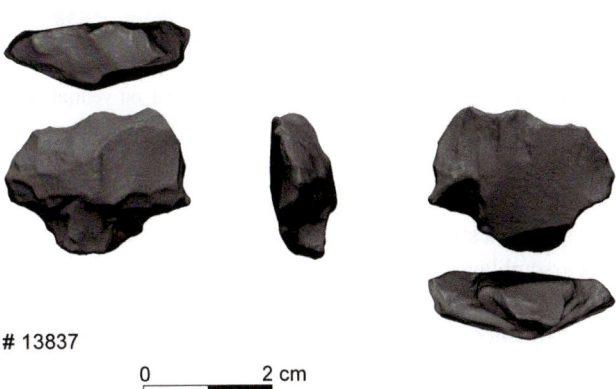

Fig. 6.37 Flint flakes with removed striking platforms (#4343 Layer II-6 Level 6 single straight side-scraper, end-scraper, and borer, #13933 Layer II-6 Level 7 borer, single convex side-scraper, and notch, #13837 Layer II-6 Level 7 varia tool)

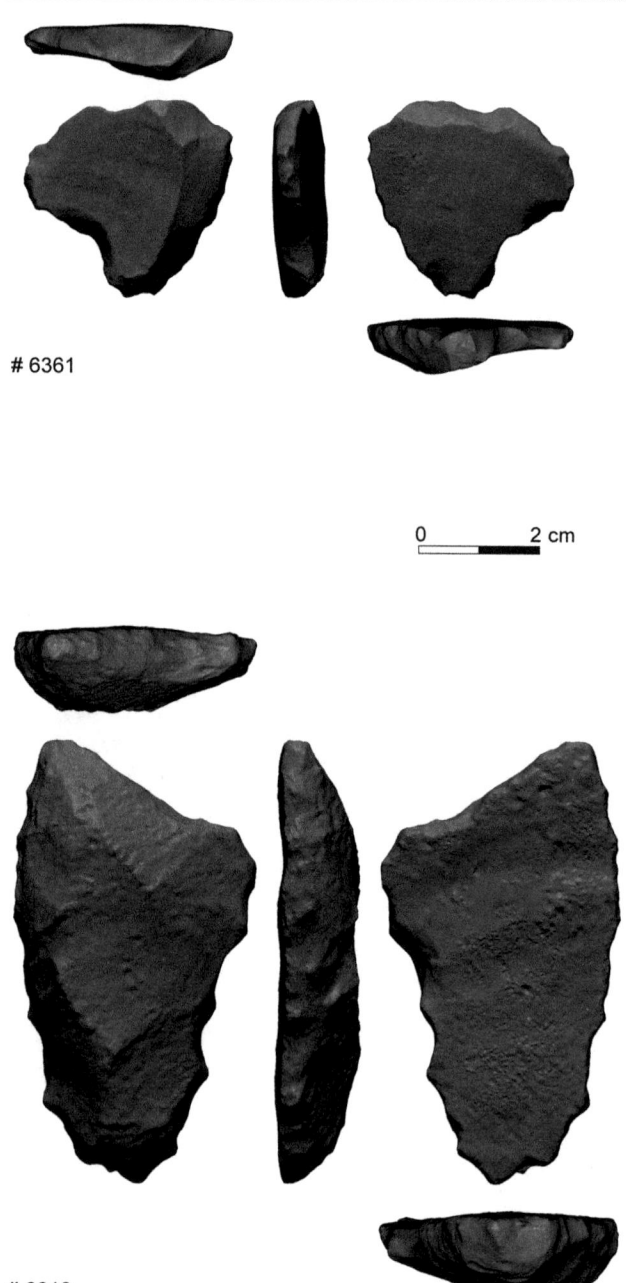

Fig. 6.38 Flint flakes with removed striking platforms (#6361 Layer II-6 Level 1 single convex side-scraper, side-scraper on ventral face, and notch; #6610 Layer II-6 Level 6 varia tool limace)

provides data showing that, with the exception of knives, the flint assemblages are non-laminar.

The size analyses presented above demonstrate the general similarity among the tools and between them and the unretouched flakes. We note, however, that the retouched tools are slightly larger than the unretouched blanks (Appendix 7). Taking into consideration the fact that the original size of the retouched tools prior to retouch and removal of the striking platform was larger than the observed size of the artifacts, we may suggest that

there was a selection of somewhat larger items to be retouched. The GBY hominins selected flint blanks not as particular target objects for tool modification but rather as relatively large items among the available small-sized flint flakes. While there is an evident consistency in the production of the different tool types, the behavioral pattern differs drastically from that observed in assemblages of later periods, where selection of blanks was targeted toward the modification of particular tool types (e.g. Mousterian side-scrapers).

Other characteristics of the retouched tools were examined to explore the extent of variability among the different tool types. With regard to the retouch of different tool categories, it seems that each was modified by a distinct type of retouch. For example, when the entire tool assemblage is examined, notch was the primary retouch type used in the modification of notches (89.30% of a total of 458), denticulates (87.20% of a total of 164), and borers (67.52% of a total of 234). Similarly, side-scrapers were modified primarily by typical side-scraper retouch (64.44% of a total of 239), followed by abrupt retouch (14.23% of a total of 239). As described in detail above, retouched flakes exhibit the highest extent of variability in retouch types, reflecting the unstandardized nature of this tool category.

When the location of retouch is examined, several observations emerge (Table 6.35). For certain tool types, the location of retouch is embedded in their typological classification (e.g. distal modification for end-scrapers and truncations). For other tool types, the location of the retouch may vary within the typological classification, so that possible patterns of preference can be detected. Depending on the tool type, the common locations are observed on the right, left, and distal edges in varying frequencies (Table 6.35). The location of retouch is a distinct characteristic that emphasizes the difference between denticulates and notches. Despite the general similarity in the mode of modification (i.e. notches), the two categories differ in the edge selected for modification. While notches are modified predominantly on the right edge, denticulates occur primarily on the left edge and on the circumference of the tool (Table 6.35). Notches, side-scrapers, and retouched flakes, which form the largest categories of retouched tools, show a selection of the right edge for modification (Table 6.35). Another selection is observed in the face of retouch, where a distinct and well-defined preference for the dorsal face is evident (Table 6.35).

With regard to the form of the retouched edge, for several tool types this is dictated by their definition (i.e. concave notches, pointed borers). For the categories of side-scrapers and retouched flakes, the form of the active edge may reflect stylistic preferences. Such preferences are apparently recorded in the distribution of convex edges, which seem to be more frequent in both tool types, followed by straight edges (Table 6.35).

6.2 Typology of Flint Flake Tools and Core Tools

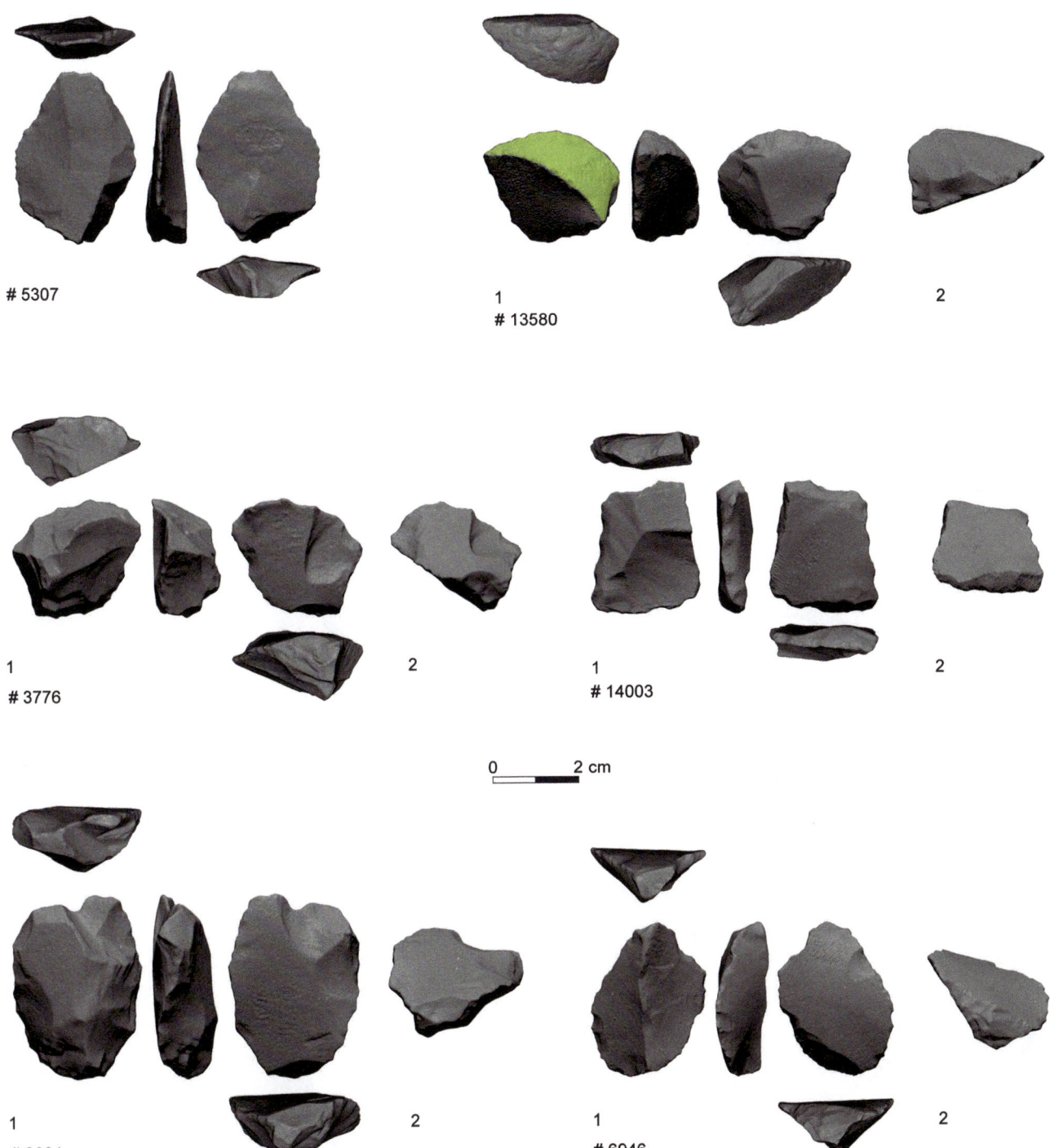

Fig. 6.39 Flint flakes with removed striking platforms **1** multiple views **2** detailed view of proximal end (#5307 Layer V-5 with signs of utilization, #3776 Layer II-6 Level 3 end-scraper, #2621 Layer II-6 Level 4 retouched flake with signs of utilization and ventral removals, #13580 Layer II-6 Level 7 naturally backed knife with signs of utilization, #14003 Layer II-6 Level 7 end-notch and denticulate, #6946 Layer II-6 Level 2 side-scraper on ventral face and retouched flake)

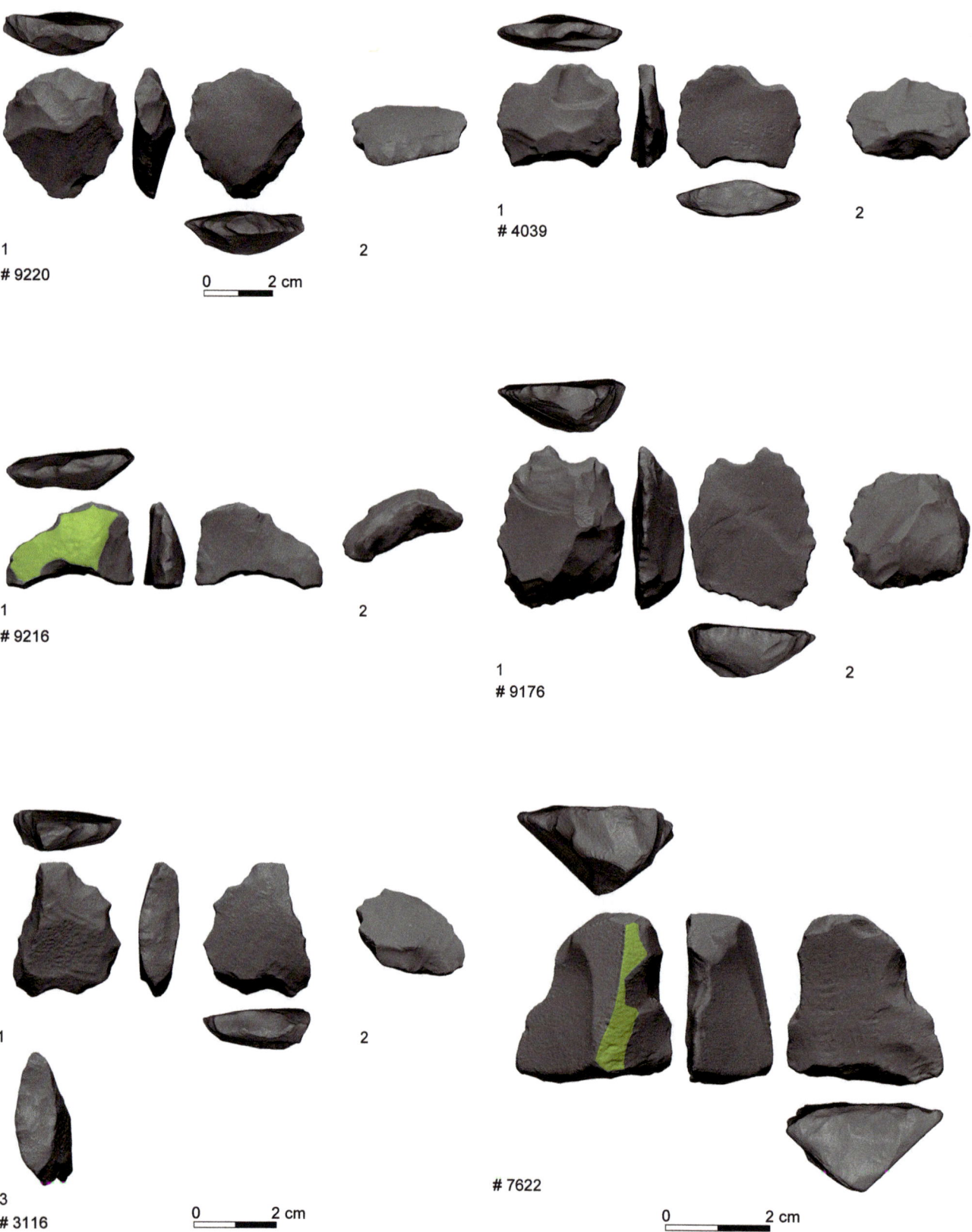

Fig. 6.40 Flint flakes with removed striking platforms **1** multiple views **2** detailed view of proximal end (#9220 Layer II-6 Level 6 side-scraper on ventral face with signs of utilization, #9216 Layer II-6 Level 6 abruptly retouched side-scraper with signs of utilization and ventral removals, #3116 Layer II-6 Level 4 denticulate)

Fig. 6.41 Flint flakes with removed striking platforms **1** multiple views **2** detailed view of proximal end (#4039 Layer II-6 Level 5 retouched flake, #9176 Layer II-6 Level 6 truncation and notch, #7622 Layer II-6 Level 7 notch with ventral removals)

6.2 Typology of Flint Flake Tools and Core Tools

Table 6.34 Descriptive statistics of different flint flake tool categories and of unretouched flint flakes (complete artifacts only; size in mm)

Type	Notch	Denticulate	Side-scraper	Borer	End-scraper	Knife	Truncation	Burin	Retouched flake	Unretouched flake
N	280	112	135	171	70	14	38	6	223	1,411
Maximum length										
Maximum	68.00	51.00	71.00	68.00	66.00	95.00	52.00	42.00	66.00	85.00
Median	28.00	30.00	31.00	27.00	28.50	47.00	29.00	35.00	27.00	26.00
Minimum	16.00	21.00	21.00	20.00	20.00	27.00	18.00	23.00	20.00	16.00
Mean	29.81	30.88	32.87	28.85	29.09	51.43	30.97	33.00	28.77	28.89
Std dev	7.78	7.99	8.98	7.10	6.80	18.09	7.93	7.24	6.92	8.52
Std err mean	0.46	0.76	0.77	0.54	0.81	4.83	1.29	2.96	0.46	0.23
Length										
Maximum	67.00	51.00	68.00	66.00	64.00	85.00	50.00	40.00	59.00	71.00
Median	25.00	26.00	27.00	24.00	26.00	44.50	27.00	29.50	25.00	23.00
Minimum	10.00	10.00	11.00	13.00	15.00	27.00	16.00	21.00	13.00	9.00
Mean	26.78	27.52	29.13	25.47	26.67	48.14	29.08	29.83	25.96	25.53
Std dev	8.68	8.25	9.51	7.56	7.58	16.57	8.14	7.63	7.41	8.85
Std err mean	0.52	0.78	0.82	0.58	0.91	4.43	1.32	3.11	0.50	0.24
Width										
Maximum	51.00	43.00	53.00	52.00	39.00	37.00	42.00	37.00	50.00	79.00
Median	23.00	24.00	26.00	22.00	22.00	26.50	23.00	24.00	22.00	22.00
Minimum	9.00	10.00	11.00	9.00	14.00	16.00	13.00	16.00	8.00	1.00
Mean	23.47	24.58	25.95	22.95	23.69	26.36	24.03	24.83	22.57	22.70
Std dev	6.87	7.66	8.04	6.43	5.78	6.23	6.33	7.25	6.69	7.33
Std err mean	0.41	0.72	0.69	0.49	0.69	1.67	1.03	2.96	0.45	0.20
Thickness										
Maximum	23.00	28.00	21.00	19.00	20.00	21.00	18.00	16.00	21.00	90.00
Median	9.00	9.00	10.00	9.00	9.50	11.00	9.00	13.00	9.00	9.00
Minimum	4.00	4.00	2.00	4.00	5.00	1.00	4.00	8.00	4.00	2.00
Mean	9.65	9.62	10.03	9.11	10.10	11.93	9.37	12.33	9.37	9.46
Std dev	3.29	3.60	3.16	2.97	3.08	4.83	3.60	3.20	3.12	4.42
Std err mean	0.20	0.34	0.27	0.23	0.37	1.29	0.58	1.31	0.21	0.12
L/W	(N=280)	(N=112)	(N=135)	(N=171)	(N=70)	(N=14)	(N=38)	(N=6)	(N=223)	(N=1,412)
Maximum	4.22	2.64	2.48	2.47	2.29	2.50	1.93	1.74	3.88	20.00
Median	1.16	1.19	1.16	1.16	1.08	1.75	1.22	1.30	1.17	1.15
Minimum	0.48	0.50	0.37	0.51	0.60	1.34	0.70	0.70	0.41	0.34
Mean	1.21	1.19	1.19	1.17	1.18	1.82	1.24	1.26	1.22	1.21
Std dev	0.45	0.41	0.39	0.40	0.39	0.35	0.31	0.35	0.43	0.66
Std err mean	0.03	0.04	0.03	0.03	0.05	0.09	0.05	0.14	0.03	0.02

Table 6.35 Counts and frequencies (%) of retouch characteristics of different flint flake tool categories

Type	Notch	Denticulate	Side-scraper	Borer	End-scraper	Knife	Truncation	Burin	Retouched flake
N	465	159	242	240	121	15	53	14	380
Location of retouch	(N=464)				(N=120)				
Distal	120	23	36	67	85	1	42	2	64
	25.86	14.47	14.88	27.92	70.83	6.67	79.25	14.29	16.84
Truncation	-	1	-	-	1	-	2	-	-
		0.63			0.83		3.77		
Proximal	37	4	13	16	6	-	3	1	14
	7.97	2.52	5.37	6.67	5.00		5.66	7.14	3.68
Left	120	42	78	37	7	9	3	5	93
	25.86	26.42	32.23	15.42	5.83	60.00	5.66	35.71	24.47
Right	155	24	88	26	9	4	2	-	120
	33.41	15.09	36.36	10.83	7.50	26.67	3.77		31.58

Table 6.35 (cont.)

Type	Notch	Denticulate	Side-scraper	Borer	End-scraper	Knife	Truncation	Burin	Retouched flake
N	465	159	242	240	121	15	53	14	380
Both edges	19	9	6	16	2	1	1	3	29
	4.09	5.66	2.48	6.67	1.67	6.67	1.89	21.43	7.63
Distal & both edges	1	4	-	2	-	-	-	1	6
	0.22	2.52		0.83				7.14	1.58
Distal & right	4	6	3	21	3	-	-	1	7
	0.86	3.77	1.24	8.75	2.50			7.14	1.84
Distal & left	1	6	4	16	3	-	-	-	7
	0.22	3.77	1.65	6.67	2.50				1.84
Proximal & both edges	1	1	-	-	1	-	-	-	1
	0.22	0.63			0.83				0.26
Proximal & left	-	-	1	4	-	-	-	-	1
			0.41	1.67					0.26
Proximal & right	-	-	-	-	-	-	-	-	1
									0.26
Convergent	1	9	12	33	1	-	-	1	4
	0.22	5.66	4.96	13.75	0.83			7.14	1.05
Circumference	4	28	1	2	1	-	-	-	28
	0.86	17.61	0.41	0.83	0.83				7.37
Indeterminate	1	2	-	-	1	-	-	-	5
	0.22	1.26			0.83				1.32
Type of retouch				(N=235)			(N=51)	(N=11)	(N=378)
Regular	6	2	3	18	2	1	7	2	70
	1.29	1.26	1.24	7.66	1.65	6.67	13.73	18.18	18.52
End-scraper	1	1	6	2	87	-	3	-	3
	0.22	0.63	2.48	0.85	71.90		5.88		0.79
Side-scraper	7	2	157	6	5	1	-	-	27
	1.51	1.26	64.88	2.55	4.13	6.67	-		7.14
Notch	417	140	14	158	15	2	3	1	21
	89.68	88.05	5.79	67.23	12.40	13.33	5.88	9.09	5.56
Invasive	-	-	9	3	1	-	2	-	10
			3.72	1.28	0.83		3.92		2.65
Semi-Quina	-	-	-	-	-	-	-	-	1
									0.26
Quina	-	1	4	1	-	-	-	-	1
		0.63	1.65	0.43					0.26
Raclette	-	-	-	-	-	-	1	-	4
							1.96		1.06
Semi-abrupt	2	-	6	1	-	-	4	-	9
	0.43		2.48	0.43			7.84		2.38
Abrupt	4	3	34	17	2	4	24	-	37
	0.86	1.89	14.05	7.23	1.65	26.67	47.06		9.79
Bipolar	-	-	-	-	-	1	1	-	-
						6.67	1.96		-
Bifacial	-	-	1	-	-	-	-	-	3
			0.41						0.79
Thinning	1	-	-	-	-	-	-	-	2
	0.22								0.53
Mixed	-	1	1	3	2	-	1	-	12
		0.63	0.41	1.28	1.65	-	1.96		3.17
Nahr Ibrahim	-	1	2	-	-	-	-	-	4
		0.63	0.83						1.06
Burin blow	-	-	-	-	-	-	-	3	-
								27.27	
Irregular	7	2	1	12	1	-	2	-	91
	1.51	1.26	0.41	5.11	0.83	-	3.92		24.07
Signs of use	18	2	2	7	5	5	1	2	46
	3.87	1.26	0.83	2.98	4.13	33.33	1.96	18.18	12.17
Indeterminate	2	4	2	7	1	1	2	3	37
	0.43	2.52	0.83	2.98	0.83	6.67	3.92	27.27	9.79

6.2 Typology of Flint Flake Tools and Core Tools

Table 6.35 (cont.)

Type	Notch	Denticulate	Side-scraper	Borer	End-scraper	Knife	Truncation	Burin	Retouched flake
N	465	159	242	240	121	15	53	14	380
Form of retouched edge	(N=427)	(N=157)	(N=234)	(N=235)	(N=116)	(N=12)	(N=51)	(N=4)	(N=337)
Straight	15	9	56	6	7	1	16	-	82
	3.51	5.73	23.93	2.55	6.03	8.33	31.37		24.33
Convex	53	9	98	8	83	7	14	2	104
	12.41	5.73	41.88	3.40	71.55	58.33	27.45	50.00	30.86
Concave	314	15	46	16	9	2	9	-	59
	73.54	9.55	19.66	6.81	7.76	16.67	17.65		17.51
Convergent	2	3	10	34	3	-	-	-	7
	0.47	1.91	4.27	14.47	2.59				2.08
Wavy	15	23	10	2	5	1	2	-	54
	3.51	14.65	4.27	0.85	4.31	8.33	3.92		16.02
Denticulate	18	91	2	3	6	1	2	1	8
	4.22	57.96	0.85	1.28	5.17	8.33	3.92	25.00	2.37
One tooth	8	5	6	164	2	-	1	1	4
	1.87	3.18	2.56	69.79	1.72		1.96	25.00	1.19
Oblique	-	-	4	1	-	-	6	-	3
			1.71	0.43			11.76		0.89
Indeterminate	2	2	2	1	1	-	1	-	16
	0.47	1.27	0.85	0.43	0.86		1.96		4.75
Face of retouch	(N=446)	(N=150)	(N=233)	(N=234)	(N=118)	(N=15)	(N=52)	(N=13)	(N=363)
Dorsal	286	84	161	146	96	4	33	4	220
	64.13	56.00	69.10	62.39	81.36	26.67	63.46	30.77	60.61
Ventral	102	22	37	27	10	3	8	4	53
	22.87	14.67	15.88	11.54	8.47	20.00	15.38	30.77	14.60
Dorsal & ventral	27	42	21	51	7	4	1	2	69
	6.05	28.00	9.01	21.79	5.93	26.67	1.92	15.38	19.01
Side	31	2	14	10	5	4	10	3	21
	6.95	1.33	6.01	4.27	4.24	26.67	19.23	23.08	5.79

Summary

The flint flake tools of GBY are less standardized than those of younger technocomplexes (e.g. Mousterian), as reflected in the length of the retouched edge, the characteristics of the retouch, and the location of the retouched edge. The GBY flint flake tool types also differ in their intensive retouch from those of many other Acheulian sites (e.g. Olorgesailie: Isaac 1977; Kalambo Falls: Clark 2001). Additional support for this view is provided by the moderate to high frequencies of double (multiple) and, more rarely, triple tools (e.g. Tables 6.9, 6.16, 6.21–6.22, 6.26). Again, this trait is limited to the flint assemblages and is entirely absent from the basalt and limestone artifacts. Even if our tool classification is somewhat arbitrary and subjective and could perhaps have been organized differently, or if some of the observed retouch is not intentional but derives from use, the GBY assemblages are characterized by extensive and diverse retouch.

The fact that only a minimal number of artifacts were classified as varia is an indication that the tools are not a simple ad hoc phenomenon but rather the result of repeated and consistent patterns of planning, modification, and use.

The flint flake tool assemblages differ radically and significantly from those of basalt and limestone in size, richness, and tool diversity, as well as in their consistently high frequencies, which range between a minimum of 9.00% (Layer V-6) and a maximum of 50.97% (Layer II-6 Level 5) throughout the stratigraphic sequence.

6.2.2 Typology of Core Tools

This section presents the flint tools that are not modified on flakes (artifacts lacking a ventral face). Bifaces, considered by many to be core tools (Isaac 1977) although they are often modified on flakes, form the largest component of the core tools and are described below (Chapter 7). Chopping tools, modified on pebbles, form the second largest group in this tool category. After this, but in lower frequencies, are tools whose retouched edges display characteristics that are usually observed on flake tools (e.g. notches).

The following discussion is limited to the typological characteristics of these tool types, while their technological traits are discussed below (Section 6.4)

Table 6.36 presents the typological inventory of core tools

Table 6.36 Counts and typological frequencies (%) of flint core tools

	Layer															
									II-6 Levels							
N	JB	V-5	V-6	I-4	I-5	II-2	II-3	II-5	II-5/6	L1	L2	L3	L4	L4b	L5	L6	L7
	2	4	1		2	2	1	5	1	36	18	23	23	2	1	10	23
Chopping tool	-	1 / 25.00	1 / 100.00	-	-	1 / 50.00	1 / 100.00	1 / 20.00	1 / 100.00	10 / 27.78	10 / 55.56	10 / 43.48	11 / 47.83	-	-	2 / 20.00	9 / 39.13
Notch	2 / 100.00	3 / 75.00	-	-	1 / 50.00	-	-	2 / 40.00	-	10 / 27.78	4 / 22.22	3 / 13.04	3 / 13.04	-	1 / 100.00	4 / 40.00	9 / 39.13
Denticulate	-	-	-	-	1 / 50.00	-	-	1 / 20.00	-	2 / 5.56	1 / 5.56	1 / 4.35	-	1 / 50.00	-	1 / 10.00	3 / 13.04
Side-scraper	-	-	-	-	-	-	-	1 / 20.00	-	1 / 2.78	2 / 11.11	8 / 34.78	6 / 26.09	1 / 50.00	-	2 / 20.00	1 / 4.35
Borer	-	-	-	-	-	1 / 50.00	-	-	-	5 / 13.89	1 / 5.56	1 / 4.35	1 / 4.35	-	-	1 / 10.00	1 / 4.35
End-scraper	-	-	-	-	-	-	-	-	-	6 / 16.67	-	-	-	-	-	-	-
Burin	-	-	-	-	-	-	-	-	-	2 / 5.56	-	-	2 / 8.70	-	-	-	-

throughout the cultural sequence. Generally, they occur in low frequencies throughout the sequence, mainly in Layer II-6.

Chopping Tools

Although chopping tools (Fig. 6.42) form a large part of the flint core tool category, they occur in relatively low frequencies in each of the layers and levels (Table 6.36). Combining these tools into a single group provides a sufficiently large sample for analyses and identification of their particular traits.

The taphonomic features of chopping tools are presented in Table 6.37. In this table we also included the amount of cortical coverage, which seems to vary extensively, far more than the coverage that typically characterizes chopping tools. There are specimens with minimal cortical coverage, while others are almost fully cortical.

The chopping tools are extremely small and are modified on relatively flat pebbles. The working edges form only a small portion of the entire circumference, as in typical chopping tools (Table 6.38). With only one exception, the chopping tools are characterized by a single working edge and two platforms (Table 6.38).

The morphology of the working edges is highly variable. Convex working edges are the most common (22.64%), followed by wavy (15.09%) and pointed (13.21%) forms. For 18.87% of the chopping tools, the form of the working edges could not be defined. The mean angle of the working edges is 80.70°. Signs of use were recorded on the working edges of 85.00% of the chopping tools. Three artifacts have battering marks on their cortical surfaces, suggesting their previous use as percussors.

The general aspect of the chopping tools is a non-standardized one. This is expressed in the high degree of variability of both categorical and continuous attributes (e.g. high standard deviations in the circumference and working edge length). The modification seen on these artifacts is a configuration of a few removals, mostly on two platforms/faces, which fits that of a chopping tool – a cortical pebble with a minimal number of working edges and often with minimal removals. However, the fact that the GBY chopping tools are extremely small (3.5 cm long on average) does not support their interpretation as objects aimed at the production of flakes or as objects designed for massive chopping activity, a function often attributed to classic Lower Paleolithic chopping tools.

Other Core Tools

The following section is concerned with tools whose retouched edges display characteristics that are usually observed on flake tools; these characteristics, however, occur on modified artifacts, angular fragments, and cores of different types (Fig. 6.43). This group includes denticulates, end-scrapers (only in Layer II-6 Level 1; Table 6.36), and burins (Fig. 6.43), but the most abundant types in descending order of frequency are notches, side-scrapers (e.g. transversal scraper Fig. 6.44), and borers (Fig. 6.43). This pattern of frequencies resembles that of the flint flake tools described above.

Table 6.39 presents the characteristics of the most common tool types of all layers and levels in order to provide a suitable sample for discussion. These tools are generally patinated, with relatively high frequencies of double

6.2 Typology of Flint Flake Tools and Core Tools

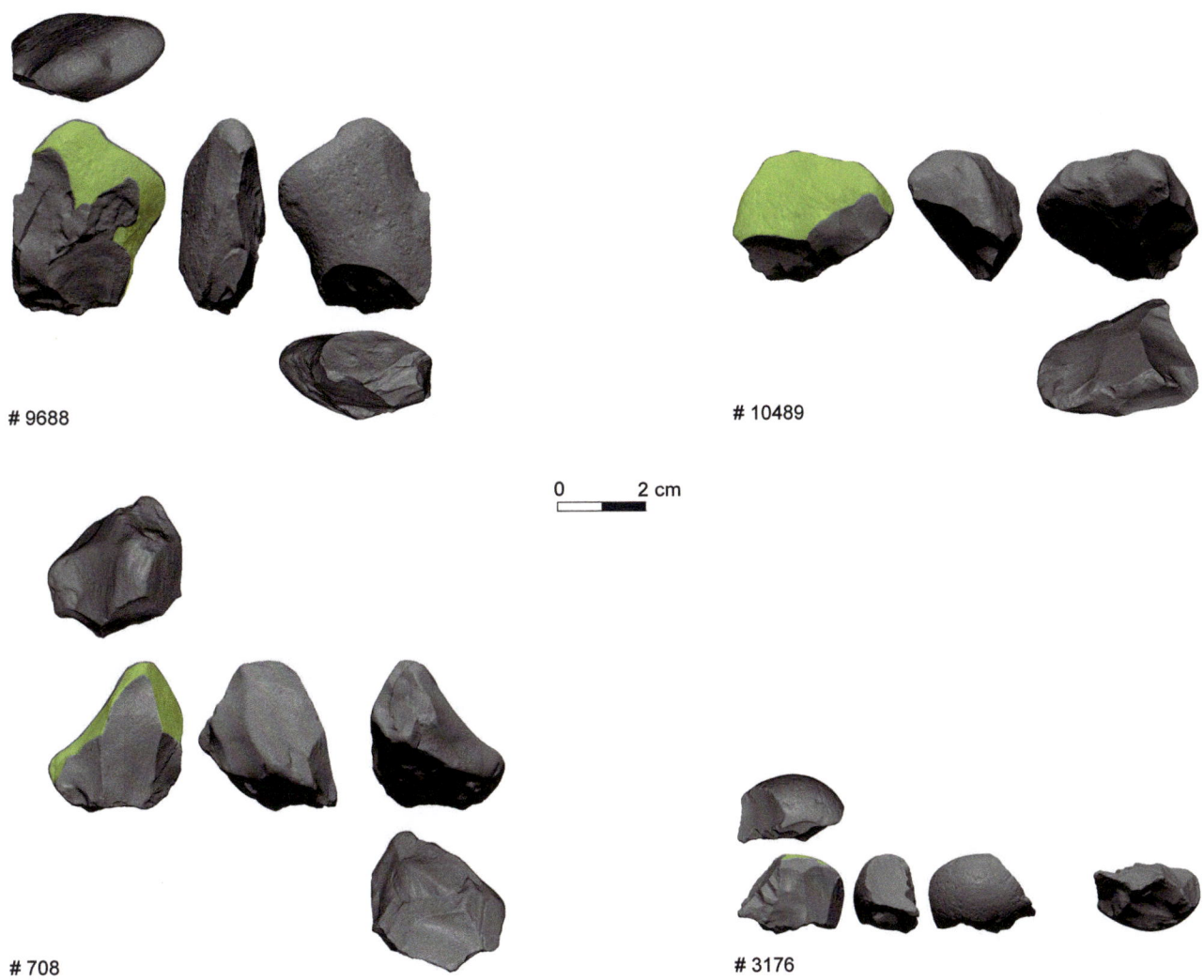

Fig. 6.42 Flint chopping tools (#9688 Layer V-6, #10489 Layer II-6 Level 1 with signs of utilization, #708 Layer II-6 Level 4 with signs of utilization, #3176 Layer II-6 Level 4)

Table 6.37 Counts and frequencies (%) of characteristics of flint chopping tools

	N=58	%
Preservation		
Fresh	15	25.86
Slightly abraded	34	58.62
Abraded	9	15.52
Patination		
Patinated	49	84.48
Double patinated	9	15.52
Breakage		
Complete	56	96.55
Lateral	2	3.45
Cortex		
1–25%	5	8.62
26–50%	20	34.48
51–75%	16	27.59
76–100%	17	29.31

patinated notches. They are mostly complete and display varied cortical coverage.

The small dimensions of these core tools (Table 6.40) are similar in all parameters to those of chopping tools and of cores in general (see below). There seems to be no real size preference for particular tool types. Chopping tools were found to be significantly wider (t ratio=2.27; df=158, $p(tt)$=0.0242) and thicker (t ratio=3.44; df=158, $p(tt)$=0.0007) than other core tool types, but not significantly longer. The differences, however, are of very low values (width: 2.95 mm; thickness: 4.05 mm).

To sum up, flint core tools, despite their typological diversity, exhibit variability in the blanks on which they are modified. There seems to be no selection of a particular blank type. The size of these blanks resembles the average size of flint CCT in general and of flint FFT as well, determined by the original small size of the flint nodules.

Table 6.38 Descriptive statistics of flint chopping tools (size in mm)

	Maximum length	Length	Width	Thickness	Circumference	Working edge length	N scars	N working edges	N platforms
N	58	58	58	58	55	44	57	54	55
Maximum	70.00	53.00	64.00	36.00	212.00	127.00	15.00	5.00	5.00
Median	32.00	29.00	26.50	17.50	96.00	32.50	5.00	1.00	2.00
Minimum	25.00	18.00	12.00	11.00	67.00	13.00	3.00	1.00	1.00
Mean	34.27	30.56	27.98	18.46	98.98	34.65	6.63	1.20	2.18
Std dev	8.33	7.55	8.52	5.06	25.80	19.83	3.40	0.62	0.64
Std err mean	1.09	0.99	1.11	0.66	3.47	2.98	0.45	0.08	0.08

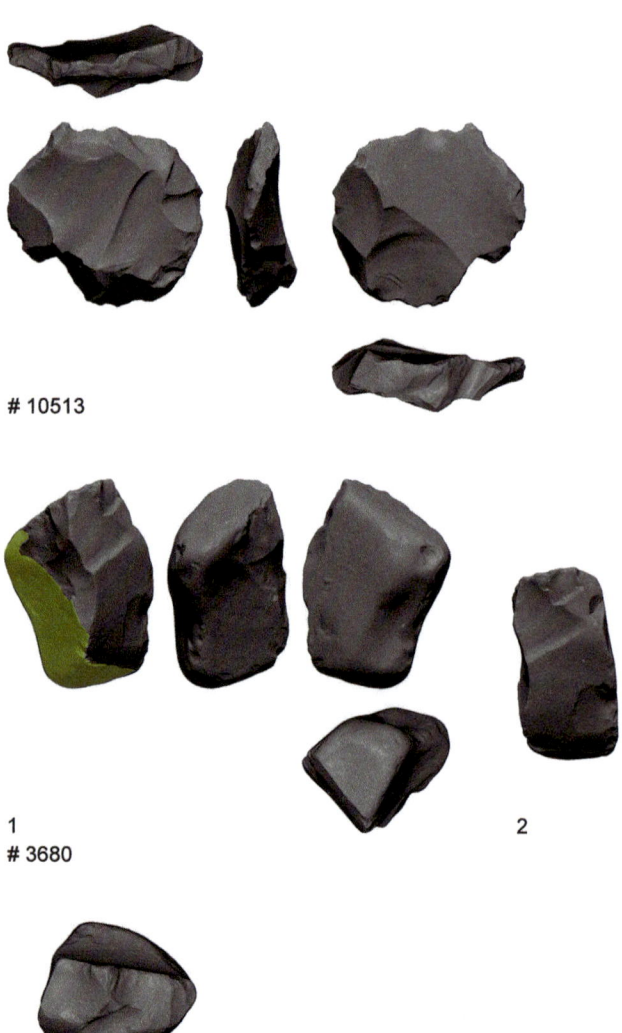

Fig. 6.43 Flint core tools (#10513 Layer II-6 Level 1 borer on core on flake, #3680 Layer II-6 Level 4 **1** burin on modified artifact **2** detailed view of burin, #17437 Layer II-6 Level 6 borer on modified artifact)

Fig. 6.44 Flint transversal scraper on nodule (#378 Layer II-6 Level 4)

Table 6.39 Counts and frequencies (%) of characteristics of flint core tools

	Notch	Side-scraper	Borer
N	42	22	11
Patination			
No patina	2	1	-
	4.76	4.55	
Patinated	25	17	9
	59.52	77.27	81.82
Double patinated	15	4	2
	35.71	18.18	18.18
Breakage			
Complete	39	18	10
	92.86	81.82	90.91
Broken	3	4	1
	7.14	18.18	9.09
Cortex	(N=41)	(N=21)	(N=10)
No cortex	10	5	3
	24.39	23.81	30.00
1–25%	8	2	2
	19.51	9.52	20.00
26–50%	6	4	1
	14.63	19.05	10.00
51–75%	5	7	1
	12.20	33.33	10.00
76–100%	12	3	3
	29.27	14.29	30.00

6.3 Technology of Flint Flakes

Table 6.40 Descriptive statistics of flint core tools (size in mm)

Type/Size	Maximum length	Length	Width	Thickness	Circumference
Notches (N=42)					(N=38)
Maximum	52.00	52.00	49.00	80.00	151.00
Median	30.50	28.50	24.00	13.00	91.00
Minimum	20.00	18.00	11.00	6.00	54.00
Mean	32.16	30.02	24.83	15.38	93.50
Std dev	8.54	8.28	8.07	10.96	25.64
Std err mean	1.31	1.27	1.24	1.69	4.15
Side-scrapers (N=22)					(N=21)
Maximum	46.00	44.00	34.00	24.00	121.00
Median	32.50	29.50	24.50	12.50	89.00
Minimum	21.00	16.00	15.00	8.00	60.00
Mean	31.81	28.95	24.22	13.54	90.19
Std dev	6.64	7.10	5.69	4.73	18.10
Std err mean	1.41	1.51	1.21	1.01	3.95
Borers (N=11)					(N=10)
Maximum	35.00	33.00	34.00	20.00	100.00
Median	31.00	29.00	22.00	11.00	85.50
Minimum	22.00	20.00	16.00	8.00	63.00
Mean	29.00	27.00	22.90	12.00	83.90
Std dev	5.01	4.83	5.46	3.82	13.48
Std err mean	1.51	1.45	1.65	1.15	4.26

In the previous section (6.2), several characteristics of the unretouched flint flakes were discussed. The data were presented for the entire assemblage of unretouched flakes, disregarding layer and level, for the purpose of comparison with the retouched component. Here, we present the detailed technological data of the individual unretouched flint flake assemblages by layer and level. This presentation encompasses assemblages with a minimum number of 30 unretouched flint flakes.

Within the definition of unretouched flakes we have included flakes, blades (Figs. 6.45–6.46), and *éclats de taille de biface* (*édtdb*) (Table 6.41), but excluded naturally backed knives (Section 6.2) and core management pieces (discussed below).

Table 6.41 demonstrates that the flakes are dominant in the assemblages, with a negligible representation of blades. Although flint handaxes are scarce at GBY, the products of their modification are present throughout the sequence and are represented by *édtdb*. Layers V-5 and V-6, in each of which a single flint handaxe was found, are the richest assemblages in terms of *édtdb* (Sharon and Goren-Inbar 1999; Goren-Inbar and Sharon 2006; see also discussion below).

6.3.1 Unretouched Flint Flakes

Table 6.42 presents the size variables of unretouched flint flakes. It clearly demonstrates an extensive degree of similarity in size among the different assemblages (Fig. 6.47). The standard deviations show the homogeneity within each layer and level. The only assemblage that is slightly larger is that of Layer V-6. The technological similarity is further expressed in the values of the measured angle of the striking platform.

Table 6.43 details further characteristics of unretouched flint flakes. Analyses of breakage patterns show that in general about half of the flakes are complete. The most frequent breakage pattern is that of the distal end, followed in descending order by proximal and lateral breaks. Some deviations from this pattern are observed in Layer II-6 Level 3, where proximal breaks are the most frequent, followed by lateral and then distal breaks. Another deviation from the general breakage pattern is observed on fragmented flakes, which occur in higher frequencies in Layer II-2, Layer II-6 Levels 5, 6, and 7, and Layer V-5.

When the direction of blow is examined (Table 6.43), the dominant mode in all layers and levels is that of end-struck flakes. Other directions of blow (side-struck and special side-struck) occur in lower frequencies. Two outliers are Layer II-5/6, where side-struck flakes amount to ca. 30%, and Layer V-6, where special side-struck flakes amount to ca. 15%.

Analyses of the amount of cortical coverage demonstrate that in all layers and levels the majority of flakes are non-cortical. Particularly high frequencies of non-cortical flakes are recorded in Layers V-5 and V-6, where ca. 80% of the unretouched flakes have no cortical coverage (Table 6.43). However, knapping of flint clearly took place at the site, as demonstrated by several layers and levels in which the frequency of cortical flakes with 76–100% cortex on their dorsal face is as high as that of flakes that retain only 1–25% cortex on their dorsal face (Table 6.43).

The extent of cortical coverage is also recorded in the analyses of the scar patterns. For many of the flint flakes (generally over 35% of each assemblage) the scar pattern could not be determined (Table 6.44). In cases where the scar pattern could be defined, the along axis and radial patterns are most frequent, while other patterns occur in lower percentages. As observed for other technological attributes, the assemblages of Layers V-5 and V-6 differ from other assemblages of the GBY sequence, with the occurrence of the along axis and radial patterns being much more frequent in these two layers. Similarly, in these two assemblages the mean number of scars on the dorsal face is the highest recorded among the flint assemblages. In general, however, the mean number of dorsal scars is similar throughout the sequence (Table 6.44).

Further evidence of the exceptional nature of the

Fig. 6.45 Flint blades with signs of utilization (#4408 Layer V-5, #4691 Layer V-5, #4723 Layer V-5 borer and *éclat de taille de biface*)

6.3 Technology of Flint Flakes

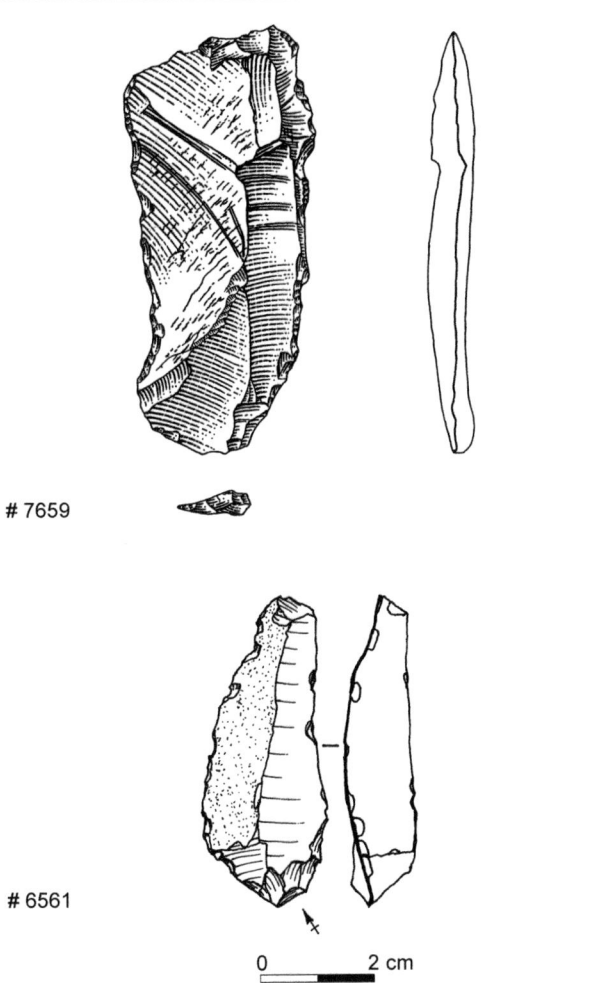

Fig. 6.46 Flint blades with signs of utilization (#7659 Layer V-6 *éclat de taille de biface*, #6561 Layer II-6 Level 1)

the final manufacturing stage of bifaces results in thin and hence easily breakable striking platforms (Herzlinger et al. 2015). Nevertheless, the dominant striking platform in all assemblages is the plain one, which in all layers and levels attains values of over 20%. In general there is a large degree of variability in the frequencies of striking platforms among the different layers and levels.

Flint knapping was performed at GBY with both hard and soft percussors (Section 6.4.2, Chapters 7, 8). However, during the analyses of the flint flakes there was no reliable way in which we could differentiate between their products. Assemblages such as those originating from Layers V-5 and V-6 have previously been described as resulting from thinning and finishing (the final modification phases of flint bifaces, after Newcomer 1971; Sharon and Goren-Inbar 1999). The current analysis furnishes further data illustrating this (Table 6.46). The identification of a hard percussor is more difficult and relies on several attributes such as scar on bulb of percussion, large (pronounced) cone, and double cone. Our observations regarding percussor type include all available flint flakes, retouched and unretouched, and are presented in Table 6.46.

The technological attributes presented in Table 6.46 illustrate another marked difference between the assemblages of Layers V-5 and V-6 and those of the rest of the site: the frequencies of lipped striking platforms are much higher. The other technological traits do not show a clear pattern (note that during the first years of the study some of the attributes, e.g. lipping or scar on the bulb of percussion, were not analyzed; Chapter 4).

To conclude, the flint assemblages of GBY were modified by the use of at least two types of techniques, those employing hard and soft percussors. Although several criteria are regarded as indicative of the use of a soft percussor (e.g. flake thinness, diffused bulbs of percussion, facetted and crushed or abraded striking platforms), we consider lipped striking platforms to be the most indicative. However, the analysis of *édtdb* (see below) revealed additional traits that may also be associated with the use of soft percussors.

assemblages of Layers V-5 and V-6 is provided by the relatively high frequencies of facetted striking platforms, as well as the very high frequencies of broken striking platforms (Table 6.45). These proximal breaks most probably derive from the fact that

Table 6.41 Counts and frequencies (%) of unretouched flint flakes by blank type

	Layer									II-6 Levels							
	V-5	V-6	JB	I-4	I-5	II-2	II-2/3	II-5	II-5/6	L1	L2	L3	L4	L4b	L5	L6	L7
N	197	258	122	26	28	52	66	103	65	513	188	222	239	69	77	135	396
Flake	118	78	39	26	28	50	65	97	64	488	184	216	227	68	75	127	366
	59.90	30.23	31.97	100.00	100.00	96.15	98.48	94.17	98.46	95.13	97.87	97.30	94.98	98.55	97.40	94.07	92.42
Blade	5	4	3	-	-	-	-	2	-	5	-	1	3	-	-	2	5
	2.54	1.55	2.46	-	-	-	-	1.94	-	0.97	-	0.45	1.26	-	-	1.48	1.26
Éclat de taille de biface	74	176	80	-	-	2	1	4	1	20	4	5	9	1	2	6	25
	37.56	68.22	65.57	-	-	3.85	1.52	3.88	1.54	3.90	2.13	2.25	3.77	1.45	2.60	4.44	6.31

Table 6.42 Descriptive statistics of unretouched flint flakes (size in mm; angle in degrees)

	Layer								
	JB	V-5	V-6	I-4	I-5	II-2	II-2/3	II-5	II-5/6
N	122	197	258	26	28	52	66	103	65
Maximum length									
Maximum	69.00	61.00	79.00	69.00	48.00	36.00	38.00	55.00	42.00
Median	31.00	26.00	30.00	23.00	23.00	24.00	23.00	25.00	24.00
Minimum	20.00	18.00	20.00	20.00	21.00	20.00	19.00	19.00	16.00
Mean	34.59	28.36	33.15	25.62	24.40	25.02	24.39	26.06	25.25
Std dev	12.43	8.53	10.71	9.41	5.16	3.61	4.38	5.40	4.72
Std err mean	1.12	0.61	0.67	1.85	0.98	0.50	0.54	0.53	0.59
Length									(N=64)
Maximum	69.00	56.00	71.00	62.00	40.00	34.00	38.00	55.00	40.00
Median	26.00	22.00	26.00	22.00	20.00	20.00	21.50	23.00	22.00
Minimum	12.00	10.00	8.00	14.00	13.00	9.00	11.00	12.00	14.00
Mean	30.11	24.93	28.75	23.19	20.29	20.19	21.95	23.94	22.39
Std dev	12.57	8.92	10.95	9.22	5.42	5.20	5.28	6.22	5.09
Std err mean	1.14	0.64	0.68	1.81	1.02	0.72	0.65	0.61	0.64
Width									(N=64)
Maximum	60.00	45.00	79.00	58.00	36.00	31.00	46.00	34.00	31.00
Median	25.50	20.00	24.00	17.50	19.00	21.00	20.00	20.00	19.00
Minimum	10.00	6.00	5.00	9.00	10.00	9.00	12.00	7.00	6.00
Mean	26.52	20.82	25.08	19.19	19.86	20.06	20.77	20.21	19.52
Std dev	9.26	7.79	9.14	9.02	6.12	4.94	6.18	5.05	5.32
Std err mean	0.84	0.55	0.57	1.77	1.15	0.69	0.76	0.50	0.67
Thickness									
Maximum	18.00	16.00	80.00	14.00	15.00	11.00	13.00	90.00	17.00
Median	7.00	6.00	5.50	7.00	7.50	7.50	9.00	9.00	9.00
Minimum	1.00	2.00	2.00	4.00	3.00	4.00	4.00	3.00	3.00
Mean	6.72	6.67	6.45	7.31	7.75	7.44	8.67	10.11	9.38
Std dev	3.09	2.64	5.57	2.81	3.22	1.63	2.19	8.87	3.02
Std err mean	0.28	0.19	0.35	0.55	0.61	0.23	0.27	0.87	0.37
L/W									(N=64)
Maximum	2.36	4.00	3.08	2.56	2.20	2.13	2.13	3.71	4.00
Median	1.15	1.27	1.19	1.22	1.11	1.00	1.12	1.12	1.17
Minimum	0.46	0.34	0.30	0.64	0.55	0.35	0.42	0.57	0.50
Mean	1.19	1.32	1.23	1.32	1.13	1.09	1.13	1.26	1.25
Std dev	0.42	0.54	0.49	0.50	0.48	0.44	0.38	0.47	0.53
Std err mean	0.04	0.04	0.03	0.10	0.09	0.06	0.05	0.05	0.07
Angle of striking platform	(N=64)	(N=57)	(N=136)	(N=3)	(N=3)	(N=18)	(N=24)	(N=54)	(N=13)
Maximum	127.00	133.00	142.00	142.00	120.00	132.00	138.00	145.00	140.00
Median	112.00	114.00	112.00	115.00	114.00	116.00	120.00	118.00	115.00
Minimum	91.00	95.00	92.00	107.00	111.00	98.00	91.00	93.00	94.00
Mean	112.09	113.07	111.76	121.30	115.00	115.28	117.79	116.93	116.38
Std dev	7.12	8.59	8.72	18.34	4.56	8.82	13.19	9.23	13.18
Std err mean	0.89	1.14	0.75	10.59	2.65	2.08	2.69	1.26	3.65

6.3 Technology of Flint Flakes

Table 6.42 (cont.)

	Layer II-6 Levels							
	L1	L2	L3	L4	L4b	L5	L6	L7
N	510	188	222	239	69	77	135	396
Maximum length								
Maximum	96.00	73.00	74.00	67.00	47.00	42.00	88.00	51.00
Median	25.00	25.00	27.00	26.00	26.00	24.00	25.00	24.00
Minimum	20.00	20.00	20.00	20.00	20.00	20.00	20.00	20.00
Mean	27.05	26.80	28.81	27.65	27.09	25.40	27.44	26.23
Std dev	7.21	6.24	7.56	6.22	5.44	4.73	7.85	5.50
Std err mean	0.32	0.46	0.51	0.40	0.65	0.54	0.68	0.28
Length	(N=508)	(N=187)						
Maximum	82.00	50.00	70.00	56.00	45.00	38.00	72.00	46.00
Median	23.00	23.00	24.00	24.00	23.00	22.00	22.00	22.00
Minimum	9.00	12.00	11.00	9.00	14.00	12.00	9.00	10.00
Mean	24.02	23.55	25.37	24.36	24.32	22.19	23.73	22.70
Std dev	7.48	6.22	8.35	6.77	6.12	4.75	8.07	5.86
Std err mean	0.33	0.45	0.56	0.44	0.74	0.54	0.69	0.29
Width	(N=508)	(N=187)						
Maximum	86.00	71.00	65.00	56.00	43.00	40.00	64.00	51.00
Median	21.00	20.00	23.00	21.00	21.00	21.00	21.00	21.00
Minimum	8.00	9.00	10.00	11.00	11.00	9.00	10.00	8.00
Mean	21.41	20.93	23.68	22.06	21.10	20.22	21.70	21.02
Std dev	7.23	6.68	7.24	6.23	5.90	5.74	6.79	6.39
Std err mean	0.32	0.49	0.49	0.40	0.71	0.65	0.58	0.32
Thickness								
Maximum	34.00	24.00	30.00	33.00	26.00	25.00	47.00	30.00
Median	9.00	9.00	9.00	9.00	9.00	9.00	9.00	8.00
Minimum	3.00	4.00	4.00	3.00	5.00	4.00	3.00	3.00
Mean	9.77	9.39	10.20	9.76	9.90	8.86	9.47	8.85
Std dev	3.90	3.67	4.08	4.19	3.80	3.45	4.37	3.21
Std err mean	0.17	0.27	0.27	0.27	0.46	0.39	0.38	0.16
L/W	(N=508)	(N=187)						
Maximum	3.00	2.50	3.10	2.63	2.42	2.56	2.59	2.92
Median	1.15	1.19	1.09	1.17	1.16	1.17	1.09	1.10
Minimum	0.41	0.46	0.42	0.35	0.58	0.57	0.42	0.31
Mean	1.20	1.21	1.14	1.17	1.24	1.19	1.17	1.17
Std dev	0.44	0.41	0.41	0.40	0.45	0.44	0.43	0.43
Std err mean	0.02	0.03	0.03	0.03	0.05	0.05	0.04	0.02
Angle of striking platform	(N=144)	(N=40)	(N=59)	(N=56)	(N=20)	(N=37)	(N=51)	(N=147)
Maximum	145.00	134.00	133.00	140.00	134.00	140.00	141.00	143.00
Median	118.00	115.50	116.00	117.00	117.00	115.00	116.00	114.00
Minimum	79.00	94.00	87.00	86.00	104.00	94.00	90.00	1.00
Mean	117.40	114.68	115.00	116.07	116.50	115.46	115.22	112.76
Std dev	9.22	8.91	9.00	9.68	7.19	10.47	10.22	14.14
Std err mean	0.77	1.41	1.17	1.29	1.61	1.72	1.43	1.17

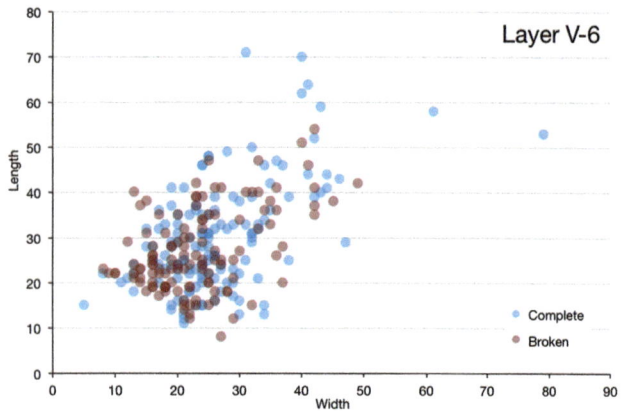

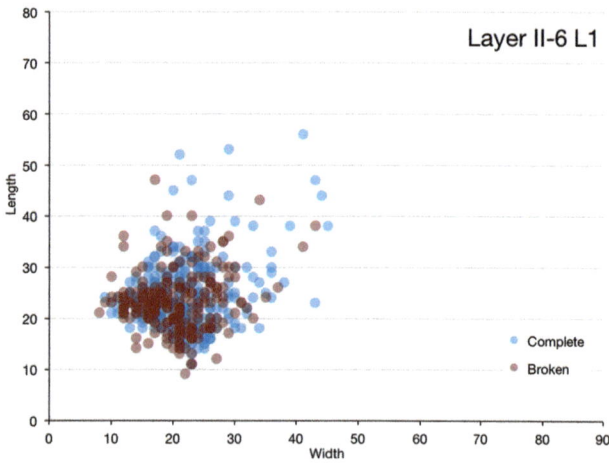

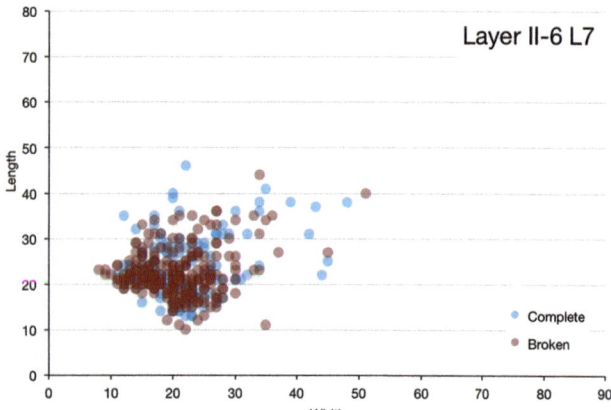

Fig. 6.47 Length and width of complete and broken unretouched flakes from Layer: V-6, II-6/L1, and II-6/L7

6.3.1.1 *Éclats de Taille de Biface*

The following analysis relates to those flint flakes classified as *édtdb* (in either Typology A or Typology B and C, Figs. 6.48–6.54) and those classified as unretouched flakes and blades. In order to arrive at sufficiently large samples, we have assembled artifacts of the above two categories from Layers V-5 and V-6 (which includes the JB assemblage) and from Layer II-6 Levels 1–7. In the following discussion these two samples are referred to as Layers V-5 and V-6 and Layer II-6.

Tables 6.47–6.51 provide the results of this comparative analysis, in which different technological attributes are examined.

Secure identification of *édtdb* is possible primarily when the artifacts are not broken. Each assemblage at GBY includes many thin flint flakes that cannot be assigned to *édtdb* because they could be fragments of the extremities of larger flakes. The same holds true for thin basalt flakes, which seem to shatter more frequently than those of flint. Hence, despite the fact that some of the basalt bifaces were undoubtedly modified in the final finishing stages, we lack the complete inventory of basalt *édtdb*. As with flint, there are many thin flake fragments that most probably originated in the finishing stages of basalt bifaces. Clearly, many products of the finishing stage of biface modification, on both flint and basalt, are items that are smaller than 2 cm and are hence included in the category of microartifacts.

The descriptive statistics of the two categories of flakes (Table 6.47) show that the only significant difference between them is in their thickness. *Édtdb* are significantly thinner than other unretouched flakes and blades (t ratio=-19.02; df=2989; $p(tt)$<0.0001; difference: 3.94 mm). In terms of length and width, while there is no clear difference between the two categories of flakes, it seems that they exhibit slightly larger sizes in the sample of Layers V-5 and V-6 in comparison with that of Layer II-6. This difference may result from a particular flake removal pattern associated with the modification of bifacial tools. In the sample of Layers V-5 and V-6, the dominant scar pattern is that of along axis removals, while in that of Layer II-6 the radial pattern is most common (Table 6.48). Other observations on the scar pattern suggest that for *édtdb* in the sample of Layers V-5 and V-6, the patterns of along axis and along axis and side are markedly more frequent (ca. 40%). In contrast, the dominant pattern on *édtdb* in the sample of Layer II-6 is the radial one (ca. 33%). In both samples the pattern of convergent flaking is significantly higher on *édtdb* than on other unretouched flakes (Table 6.48). Additional differences can be found in the mean number of scars on the dorsal face. In both categories of flakes, the mean number of scars is larger in the sample of Layers V-5–6 than in the sample of Layer II-6, and in both samples the mean number of dorsal scars for *édtdb* is higher (a single scar) than that of unretouched flakes (Table 6.47). When tested statistically,

6.3 Technology of Flint Flakes

Table 6.43 Counts and frequencies (%) of characteristics of unretouched flint flakes

	Layer																
										II-6 Levels							
	JB	V-5	V-6	I-4	I-5	II-2	II-2/3	II-5	II-5/6	L1	L2	L3	L4	L4b	L5	L6	L7
N	122	197	258	26	28	52	66	103	65	510	188	222	239	69	77	135	396
Breakage					(N=27)												
Complete	63	96	136	9	17	22	41	59	32	276	111	110	120	36	39	68	156
	51.64	48.73	52.71	34.62	62.96	42.31	62.12	57.28	49.23	54.12	59.04	49.55	50.21	52.17	50.65	50.37	39.39
Distal	20	29	46	3	5	8	7	15	8	52	18	21	32	9	11	24	56
	16.39	14.72	17.83	11.54	18.52	15.38	10.61	14.56	12.31	10.20	9.57	9.46	13.39	13.04	14.29	17.78	14.14
Lateral	9	9	17	2	1	5	1	4	3	33	17	28	10	6	6	9	34
	7.38	4.57	6.59	7.69	3.70	9.62	1.52	3.88	4.62	6.47	9.04	12.61	4.18	8.70	7.79	6.67	8.59
Proximal	12	16	23	6	3	-	7	6	8	66	10	32	27	5	3	11	47
	9.84	8.12	8.91	23.08	11.11		10.61	5.83	12.31	12.94	5.32	14.41	11.30	7.25	3.90	8.15	11.87
Distal & lateral	-	8	10	1	-	1	-	4	1	9	7	3	13	3	3	3	14
		4.06	3.88	3.85		1.92		3.88	1.54	1.76	3.72	1.35	5.44	4.35	3.90	2.22	3.54
Distal & proximal	3	9	6	2	-	1	3	4	3	19	5	7	9	4	2	3	15
	2.46	4.57	2.33	7.69		1.92	4.55	3.88	4.62	3.73	2.66	3.15	3.77	5.80	2.60	2.22	3.79
Fragment	12	26	15	2	1	10	4	5	5	44	16	17	22	5	11	15	59
	9.84	13.20	5.81	7.69	3.70	19.23	6.06	4.85	7.69	8.63	8.51	7.66	9.21	7.25	14.29	11.11	14.90
Proximal & lateral	3	4	5	1	-	5	2	5	5	10	4	4	6	1	2	2	15
	2.46	2.03	1.94	3.85		9.62	3.03	4.85	7.69	1.96	2.13	1.80	2.51	1.45	2.60	1.48	3.79
Indeterminate	-	-	-	-	-	-	1	1	-	1	-	-	-	-	-	-	-
							1.52	0.97		0.20							
Cortex		(N=195)					(N=65)			(N=509)							
No cortex	93	156	207	20	19	37	35	57	39	262	99	134	141	37	49	70	220
	76.23	80.00	80.23	76.92	67.86	71.15	53.85	55.34	60.00	51.47	52.66	60.36	59.00	53.62	63.64	51.85	55.56
1–25%	22	14	30	1	5	2	12	10	9	89	34	26	40	12	5	17	43
	18.03	7.18	11.63	3.85	17.85	3.85	18.46	9.71	13.85	17.49	18.09	11.71	16.74	17.39	6.49	12.59	10.86
26–50%	4	9	9	3	1	4	6	11	6	36	10	18	15	5	4	11	31
	3.28	4.62	3.49	11.54	3.57	7.69	9.23	10.68	9.23	7.07	5.32	8.11	6.28	7.25	5.19	8.15	7.83
51–75%	-	6	8	2	2	3	5	12	4	25	12	12	14	5	5	5	14
		3.08	3.10	7.69	7.14	5.77	7.69	11.65	6.15	4.91	6.38	5.41	5.86	7.25	6.49	3.70	3.54
76–100%	3	9	4	-	1	5	7	12	5	92	24	31	22	10	12	24	50
	2.46	4.62	1.55		3.57	9.62	10.77	11.65	7.69	18.07	12.77	13.96	9.21	14.49	15.58	17.78	12.63
Indeterminate	-	1	-	-	-	1	-	1	2	5	9	1	7	-	2	8	38
		0.51				1.92		0.97	3.08	0.98	4.79	0.45	2.93		2.60	5.93	9.60
Direction of blow	(N=106)	(N=189)	(N=257)		(N=27)	(N=51)	(N=56)	(N=101)	(N=61)	(N=491)	(N=180)	(N=220)	(N=238)	(N=67)	(N=76)	(N=127)	(N=385)
End-struck	49	81	103	13	17	11	27	46	28	214	76	99	102	33	25	43	108
	46.23	42.86	40.08	50.00	62.96	21.57	48.21	45.54	45.90	43.58	42.22	45.00	42.86	49.25	32.89	33.86	28.05
Side-struck	26	22	47	7	5	6	10	17	18	104	38	52	43	15	15	26	75
	24.53	11.64	18.29	26.92	18.52	11.76	17.86	16.83	29.51	21.18	21.11	23.64	18.07	22.39	19.74	20.47	19.48
Special side-struck	6	6	39	-	-	5	1	8	0	28	15	10	17	6	5	9	38
	5.66	3.17	15.18			9.80	1.79	7.92	0.00	5.70	8.33	4.55	7.14	8.96	6.58	7.09	9.87
Indeterminate	25	80	68	6	5	29	18	30	15	145	51	59	76	13	31	49	164
	23.58	42.33	26.46	23.08	18.52	56.86	32.14	29.70	24.59	29.53	28.33	26.82	31.93	19.40	40.79	38.58	42.60

Table 6.44 Counts and frequencies (%) of dorsal scar pattern and descriptive statistics of number of dorsal scars on unretouched flint flakes

	Layer																
										II-6 Levels							
	JB	V-5	V-6	I-4	I-5	II-2	II-2/3	II-5	II-5/6	L1	L2	L3	L4	L4b	L5	L6	L7
N	120	194	256	26	28	51	62	97	61	488	178	214	237	69	75	120	371
Scar pattern					(N=27)												
Cortical	4	14	12	-	2	8	13	19	8	89	27	34	20	12	9	26	44
	3.33	7.22	4.69		7.41	15.69	20.97	19.59	13.11	18.24	15.17	15.67	8.44	17.39	12.00	21.67	11.86
Plain	4	10	11	1	1	1	3	2	2	16	3	6	8	-	2	-	5
	3.33	5.15	4.30	3.85	3.70	1.96	4.84	2.06	3.28	3.28	1.69	2.76	3.38		2.67		1.35
Along axis	34	44	78	7	6	1	13	11	5	63	19	27	45	7	8	15	62
	28.33	22.68	30.47	26.92	22.22	1.96	20.97	11.34	8.20	12.91	10.67	12.44	18.99	10.14	10.67	12.50	16.71
Parallel	-	2	-	-	-	1	-	1	1	2	-	1	2	-	2	-	1
		1.03				1.61		1.64		0.41		0.46	0.84		2.67		0.27
Convergent	4	4	19	1	-	1	-	3	1	4	6	1	5	1	2	5	7
	3.33	2.06	7.42	3.85		1.96		3.09	1.64	0.82	3.37	0.46	2.11	1.45	2.67	4.17	1.89
Opposed	-	2	-	1	-	-	2	2	-	5	3	4	4	3	1	1	-
		1.03		3.85			3.23	2.06		1.02	1.69	1.84	1.69	4.35	1.33	0.83	
Radial	25	27	36	-	3	5	8	9	3	42	12	26	19	2	6	9	35
	20.83	13.92	14.06		11.11	9.80	12.90	9.28	4.92	8.61	6.74	11.98	8.02	2.90	8.00	7.50	9.43
Ridged	-	3	3	-	-	-	1	6	8	10	6	8	4	-	3	2	6
		1.55	1.17				1.61	6.19	13.11	2.05	3.37	3.69	1.69		4.00	1.67	1.62
Side	2	4	5	1	3	-	-	1	-	18	8	14	12	10	3	-	7
	1.67	2.06	1.95	3.85	11.11			1.03		3.69	4.49	6.45	5.06	14.49	4.00		1.89
Along axis & side	12	17	31	1	1	-	-	2	2	17	8	11	11	5	8	3	22
	10.00	8.76	12.11	3.85	3.70			2.06	3.28	3.48	4.49	5.07	4.64	7.25	10.67	2.50	5.93
Along axis & opposed	1	-	9	2	3	-	2	1	2	11	8	8	12	2	3	8	12
	0.83		3.52	7.69	11.11		3.23	1.03	3.28	2.25	4.49	3.69	5.06	2.90	4.00	6.67	3.23
Opposed & side	1	3	2	-	-	1	-	2	3	4	2	4	2	-	1	1	4
	0.83	1.55	0.78			1.96		2.06	4.92	0.82	1.12	1.84	0.84		1.33	0.83	1.08
Along axis & radial	2	-	-	-	-	-	-	-	-	-	-	-	1	-	-	-	-
	1.67												0.42				
Indeterminate	31	62	50	12	8	34	19	39	26	207	76	73	92	27	27	50	166
	25.83	31.96	19.53	46.15	29.63	66.67	30.65	40.21	42.62	42.42	42.70	33.64	38.82	39.13	36.00	41.67	44.74
N of scars	(N=117)	(N=189)	(N=254)	(N=25)	(N=23)	(N=49)	(N=58)	(N=83)	(N=48)	(N=386)	(N=141)	(N=185)	(N=206)	(N=60)	(N=59)	(N=99)	(N=297)
Maximum	21	16	16	8	11	8	8	11	9	12	12	14	12	9	9	15	15
Median	6	4	5	3	4	4	3	3	4	3	3	4	4	3	4	4	4
Minimum	1	1	1	1	1	1	1	1	1	1	1	1	1	1	1	1	1
Mean	6.67	4.72	5.32	3.44	4.17	3.92	3.40	4.17	4.25	3.41	3.73	4.03	3.95	3.67	4.02	4.41	4.13
Std dev	4.11	2.55	2.77	2.04	2.33	1.95	1.97	2.44	1.93	1.93	2.03	2.44	2.36	1.99	1.97	2.39	2.27
Std err mean	0.38	0.19	0.17	0.41	0.49	0.28	0.26	0.27	0.28	0.10	0.17	0.18	0.16	0.26	0.26	0.24	0.13

édtdb have significantly more scars than unretouched flakes and blades (t ratio=12.12; df=2989; $p(tt)<0.0001$; difference: 1.57 scar number).

Analysis of the identifiable striking platforms suggests that the same types dominate both *édtdb* and unretouched flakes. Thus, plain and facetted striking platforms are the most common in both categories and in both samples (Table 6.49). In the sample of Layer II-6, however, facetted striking platforms occur in significantly higher frequencies on *édtdb* and exhibit similar values to those of the facetted striking platforms in the sample of Layers V-5-6. Differences between the two categories of flakes are recorded mostly in the frequencies of punctiform and crushed striking platforms, which are significantly more frequent in the *édtdb* category in both samples.

Table 6.50 presents different technological observations relating to flake production. In contrast to other attributes presented and discussed above, the number of such

6.3 Technology of Flint Flakes

Table 6.45 Counts and frequencies (%) of striking platform types of unretouched flint flakes

	Layer									II-6 Levels							
	JB	V-5	V-6	I-4	I-5	II-2	II-2/3	II-5	II-5/6	L1	L2	L3	L4	L4b	L5	L6	L7
N	114	190	257	26	28	51	58	98	61	486	179	217	236	69	75	124	382
Striking platform					(N=27)												
Cortical	3	4	2	1	1	4	7	6	-	40	16	12	18	4	9	9	22
	2.63	2.11	0.78	3.85	3.70	7.84	12.07	6.12		8.23	8.94	5.61	7.63	5.80	12.00	7.26	5.76
Punctiform	2	11	17	2	2	1	2	6	1	20	8	9	3	5	7	9	9
	1.75	5.79	6.61	7.69	7.41	1.96	3.45	6.12	1.64	4.12	4.47	4.21	1.27	7.25	9.33	7.26	2.36
Plain	27	38	72	5	9	11	14	35	22	141	49	55	68	17	17	40	100
	23.68	20.00	28.02	19.23	33.33	21.57	24.14	35.71	36.07	29.01	27.37	25.70	28.81	24.64	22.67	32.26	26.18
Dihedral	13	5	25	-	-	1	2	3	1	9	2	5	3	2	2	1	6
	11.40	2.63	9.73			1.96	3.45	3.06	1.64	1.85	1.12	2.34	1.27	2.90	2.67	0.81	1.57
Facetted	17	38	58	2	-	5	3	10	1	32	14	20	19	6	10	8	25
	14.91	20.00	22.57	7.69		9.80	5.17	10.20	1.64	6.58	7.82	9.35	8.05	8.70	13.33	6.45	6.54
Crushed	8	13	16	-	-	3	-	2	-	5	4	2	7	6	1	6	11
	7.02	6.84	6.23			5.88	0.00	2.04	0.00	1.03	2.23	0.93	2.97	8.70	1.33	4.84	2.88
Removed	3	3	4	1	5	2	4	4	3	11	11	9	17	4	4	4	21
	2.63	1.58	1.56	3.85	18.52	3.92	6.90	4.08	4.92	2.26	6.15	4.21	7.20	5.80	5.33	3.23	5.50
Broken	29	66	50	11	7	16	15	17	20	156	40	71	73	15	15	28	124
	25.44	34.74	19.46	42.30	25.93	31.37	25.86	17.35	32.79	32.10	22.35	33.18	30.93	21.74	20.00	22.58	32.46
Indeterminate	12	12	13	4	3	8	11	15	13	72	35	31	28	10	10	19	64
	10.53	6.32	5.06	15.38	11.11	15.69	18.97	15.31	21.31	14.81	19.55	14.49	11.86	14.49	13.33	15.32	16.75

Table 6.46 Counts of technological features associated with percussor type on flint FFT

		Layer									II-6 Levels							
		V-5	V-6	JB	I-4	I-5	II-2	II-5	II-5/6	II-6/7	L1	L2	L3	L4	L4b	L5	L6	L7
Total FFT N		311	291	158	46	45	108	190	115	28	761	305	385	433	124	156	274	740
Hard	Scar on bulb of percussion	4	14	2	-	-	-	3	-	-	1	3	1	6	-	6	3	14
	Large cone	3	3	-	1	-	-	-	-	-	4	1	-	-	1	1	1	1
	Two cones	4	7	1	-	-	1	1	-	-	8	4	1	4	-	2	2	7
	Hinged flake	11	28	18	-	-	4	3	1	-	10	6	6	9	-	4	3	13
	Siret	1	5	-	-	-	-	2	-	1	1	-	-	1	-	1	-	2
Total		23	57	21	-	-	5	9	1	1	24	14	8	20	1	14	9	37
Soft	Lipped	40	63	27	-	-	4	2	-	2	16	5	3	3	7	5	7	12
	Not lipped	28	70	30	-	-	-	-	-	-	6	26	4	5	2	14	13	46
	Possibly lipped	21	39	15	-	-	-	-	-	-	2	5	3	1	-	2	2	8
	Total soft	89	172	72	-	-	4	2	-	2	24	36	10	9	9	21	22	66
Total		178	344	144	1	-	8	4	-	4	48	72	20	18	18	42	44	132

technological observations is minimal. As a result, the *édtdb* component of the sample of Layer II-6, which is small *a priori*, provides a small number of observations. Nevertheless, some technological attributes can apparently be associated with the production of *édtdb*. In the sample of Layers V-5–6, relatively high frequencies of hinged items occur on both categories of flakes, while steps (described in Sharon and Goren-Inbar 1999) are recorded in high frequencies for *édtdb* in this sample (Table 6.50; Figs. 6.48–6.49) as well as in the sample of Layer II-6.

The occurrence of *débordant* flakes (Section 6.4.1) in the sample of Layers V-5–6 in both categories of flakes may indicate a possible technological association between these products and the production of bifacial tools, particularly in the maintenance and correction of the working edge (Chapter 7).

Interestingly, although *édtdb* are often regarded as products of the soft percussor technique, our analysis shows that technological features associated with hard percussors (e.g. scar on bulb of percussion; Table 6.50) also occur among the *édtdb* samples.

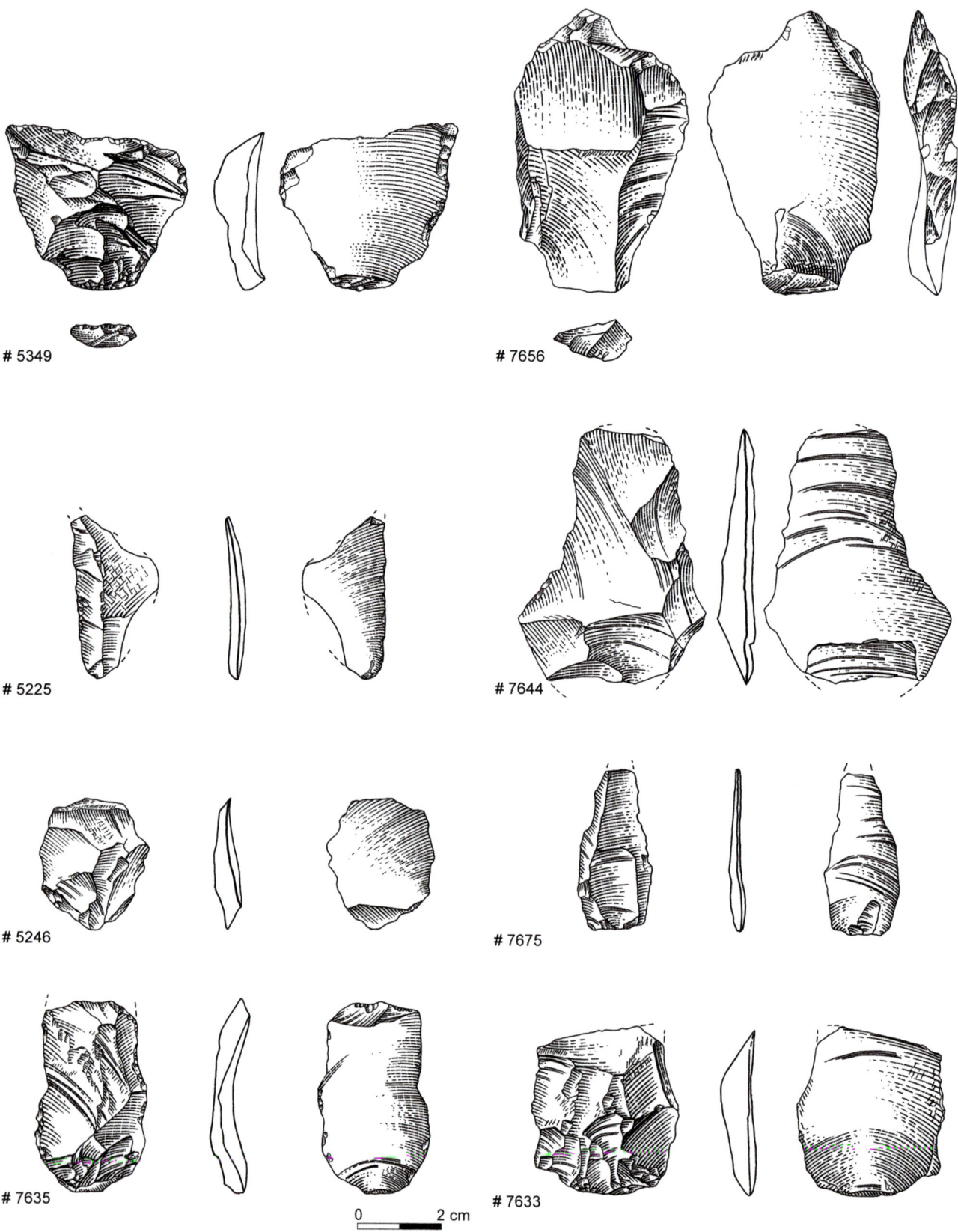

Fig. 6.48 Flint *éclats de taille de biface* (#5349 Layer V-5 borer with signs of utilization, note steps on proximal end, #5225 Layer V-5, #5246 Layer V-5, #7635 Layer V-5, note steps on proximal end, #7656 Layer V-5, #7644 Layer V-6, #7675 Layer V-6, #7633 Layer V-6, note steps on proximal end)

6.3 Technology of Flint Flakes

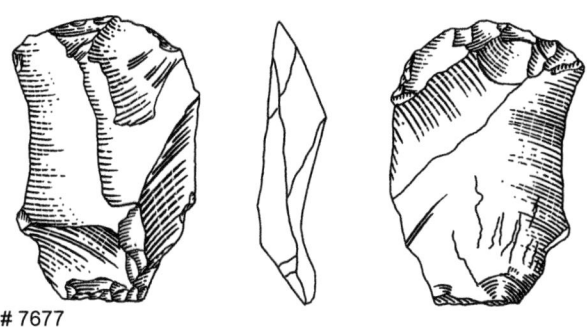

#7677

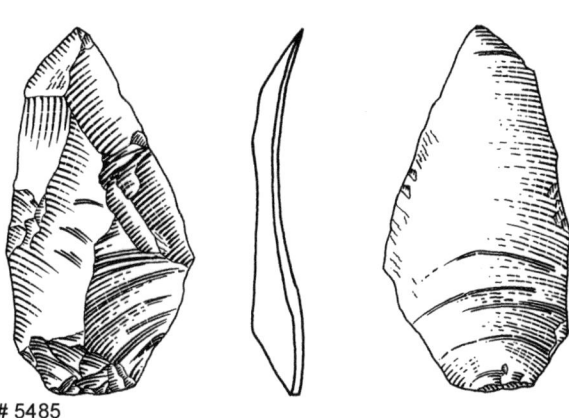

#5485

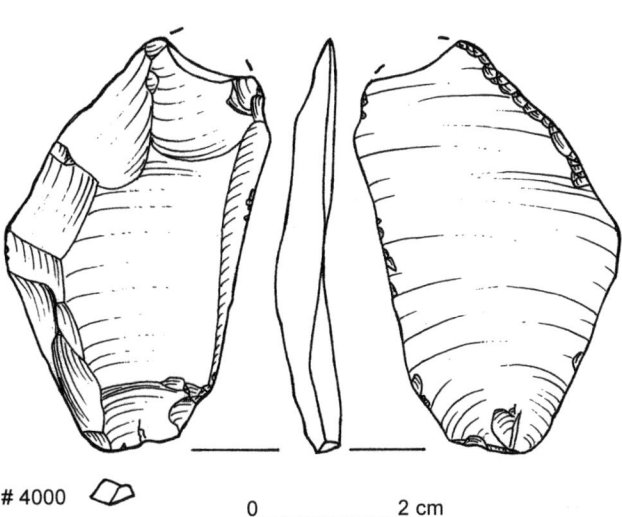

#4000

Fig. 6.49 Flint *éclats de taille de biface* (#7677 Layer V-6, #5485 Layer V-6, note steps on proximal end, #4000 JB retouched flake)

However, lipped striking platforms, a major characteristic of the soft percussor technique, occur in significantly higher frequencies on *édtdb* than on other unretouched flakes, in both samples (Table 6.51).

The above comparison between *édtdb* and unretouched flakes furnishes insights into different aspects of biface production, related both to technology and to intra-site variability. *Édtdb* exhibit distinct technological characteristics that include thinner flakes, steps, and lipped striking platforms, which are more often facetted, crushed, or punctiform. In the sample of Layers V-5–6, there is a high degree of similarity between the two flake categories. This resemblance probably indicates that both categories originate from the production of bifacial tools, those classified as *édtdb* representing the finishing stage of biface production (after Newcomer 1971).

Further insight into the technological production of flint bifaces derives from *édtdb*, which result from the very last stages of maintenance and regularization of the biface. These flakes exhibit diverse morphologies that are dictated by the angle between the striking axis and the biface. These include miniature side-struck flakes whose widest dimension is at the distal end while the proximal edge is pointed, struck perpendicular to the biface edge (Figs. 6.55, 6.57). Another example is that of non-perpendicular striking, which resulted in small elongated flakes bearing remnants of the bifacially worked edge on their lateral edge (Fig. 6.56). The maintenance of the biface surface and edge also produced an array of other forms, assigned on account of their small size to the final stages of biface modification (Fig. 6.56). Other such flakes are undoubtedly present within the smallest fraction of flint microartifacts, which is not discussed here (Sharon and Goring-Morris 2004; Malinsky-Buller et al. 2011).

Clear differences were recorded between the sample of Layers V-5–6 and that of Layer II-6. Each sample apparently displays different patterns of dorsal scars produced during the modification of flint bifaces. While in the sample of Layers V-5–6, the dominant scar pattern is that of along axis, in the sample of II-6, the radial pattern dominates.

6.3.1.2 Dorsally Plain (Kombewa) Flakes

A small number of dorsally plain flakes were identified (Fig. 6.57) and defined on the basis of an experimental study (Dag and Goren-Inbar 2001 and references therein). These are flakes that resemble Kombewa flakes but were not predetermined to be devoid of dorsal scars, as were those produced for the basalt bifacial tools (Chapter 7). Dorsally plain flakes (DPF) are viewed as by-products of modifications carried out on cores on flakes. These are the outcome of flaking that removed one or more flakes from the

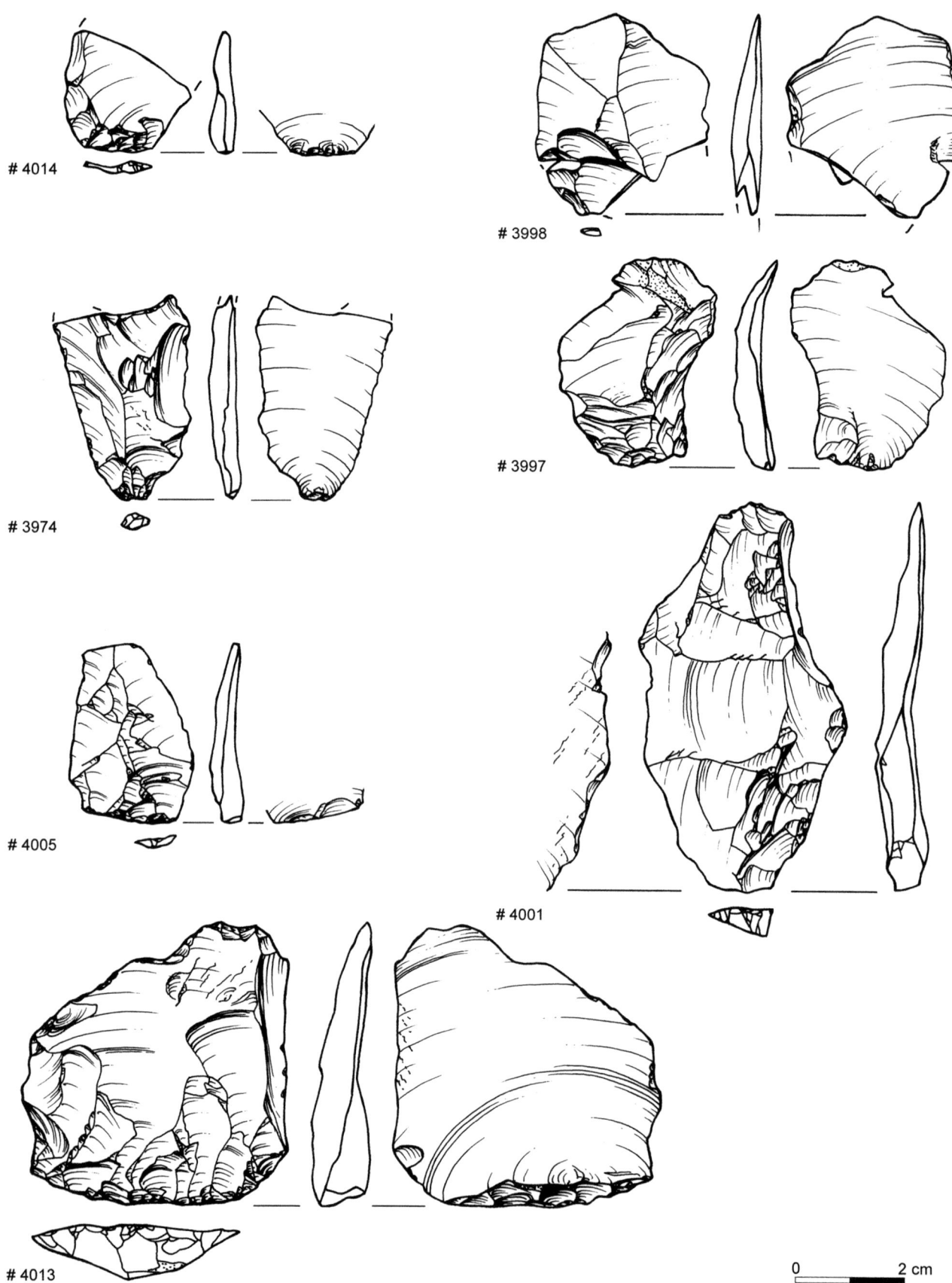

Fig. 6.50 Flint *éclats de taille de biface*, note steps on proximal end (#4014 JB, note crushing on proximal end, #3974 JB, #4005 JB, note crushing on proximal end, #3998 JB, #3997 JB, #4001 JB *débordant* flake, #4013 JB)

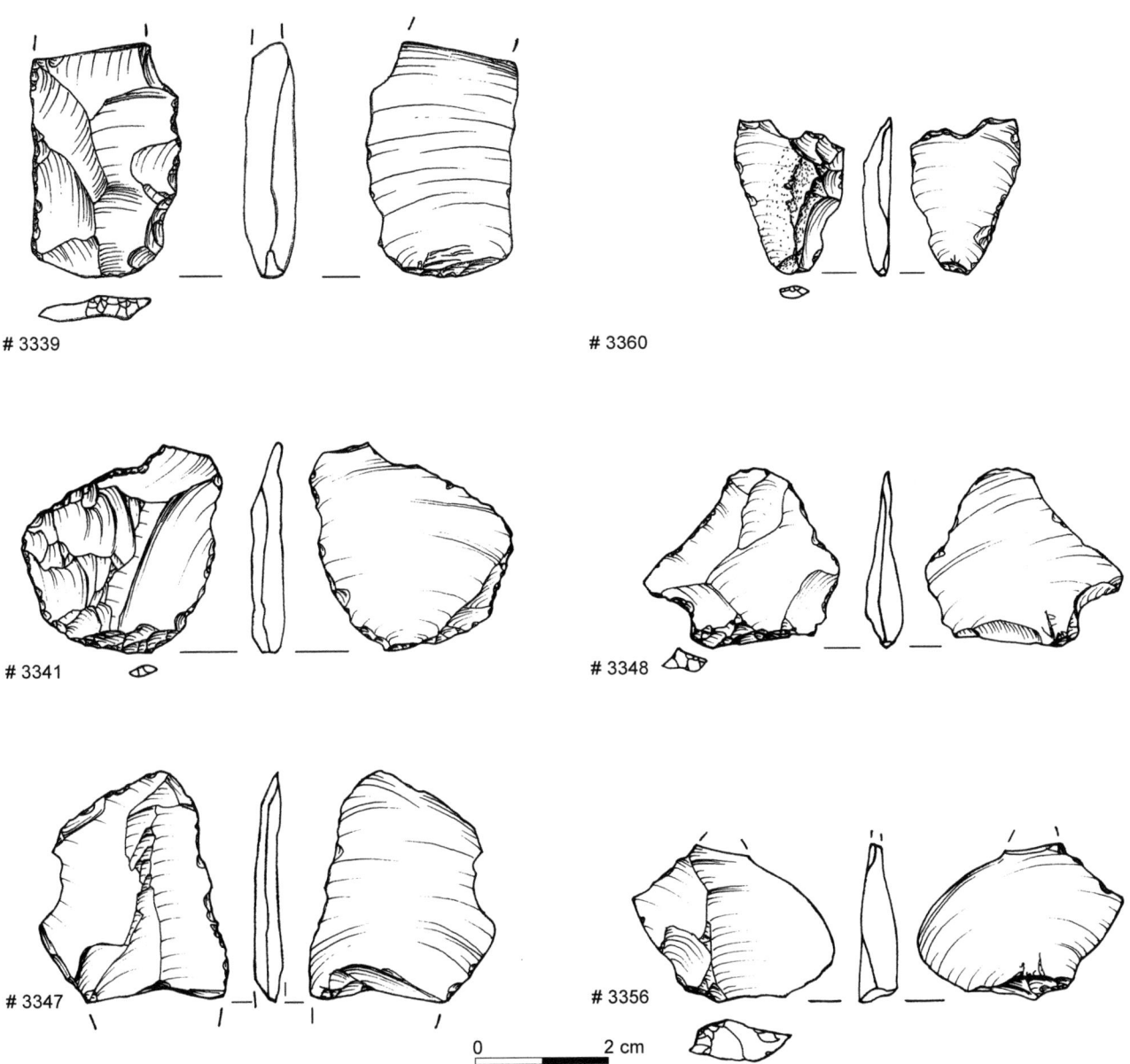

Fig. 6.51 Flint *éclats de taille de biface* with signs of utilization (#3339 JB, #3341 JB, note steps on proximal end, #3347 JB, #3360 JB, #3348 JB, note steps on proximal end, #3356 JB)

ventral face of these cores. Any such removal will produce a flake characterized by two ventral faces. The cross-section of the resulting flake will not necessarily be biconvex. Two alternative actions that may produce unintentional DPF are the flaking or modification of large flakes into bifaces and the spontaneous flaking of the bulbar scar. The first is low in probability, in view of the scarcity of flint bifaces at GBY. Furthermore, large flakes do not exist in the flint assemblage and the few handaxes that are present seem to be made on chunks rather than flakes (although the bifacially produced scars cover the technological features of the blanks). Clearly, this option is more feasible for the basalt assemblage, where flake sizes are much larger than those of the flint (Chapter 7). The second alternative is even less likely since such flakes, if they derived from the flint flake assemblage described above, would be present within the microartifacts category (>2 cm). Yet another possible source of flint DPF at GBY may be associated with the presence of "removals on the ventral face". This category, as described by Hovers (2009), comprises one or two removals that leave flake scars on the ventral faces of a flake. Thus, such removals could be also the source of flakes of this type. DPF have been identified in other Lower Paleolithic assemblages as well (e.g. Barkai et al. 2010; Malinsky-Buller et al. 2011). One way or another,

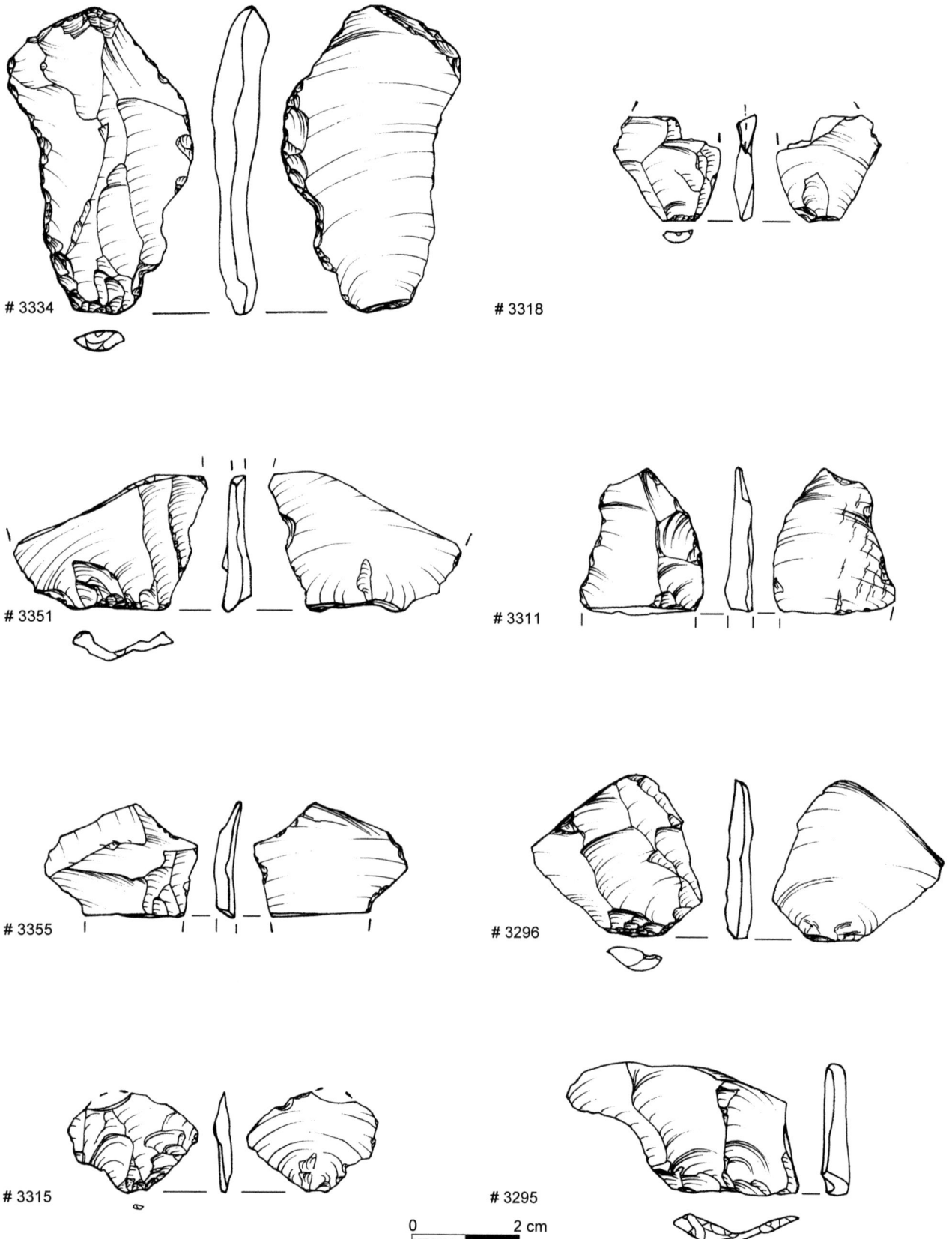

Fig. 6.52 Flint *éclats de taille de biface* (#3334 JB with signs of utilization, note steps on proximal end, #3351 JB, note steps on proximal end, #3355 JB with signs of utilization, #3315 JB, #3318 JB, #3311 JB with signs of utilization, #3296 JB hinged flake, note steps on proximal end, #3295 JB hinged flake)

6.3 Technology of Flint Flakes

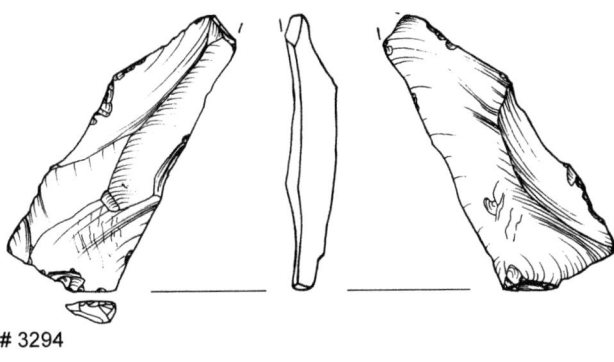

3294

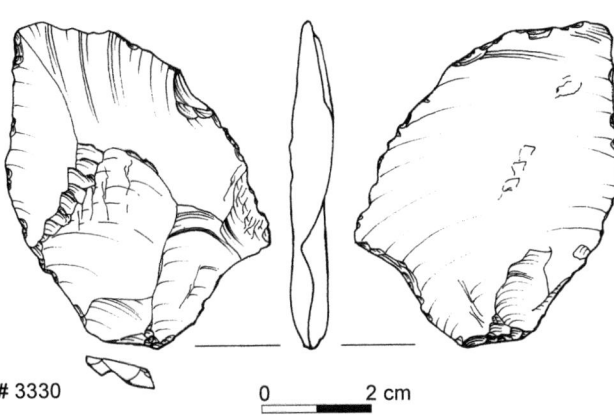

3330

Fig. 6.53 Flint *éclats de taille de biface* with signs of utilization (#3294 JB, #3330 JB)

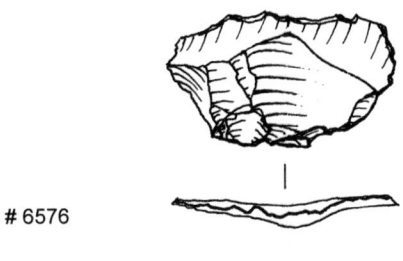

6576

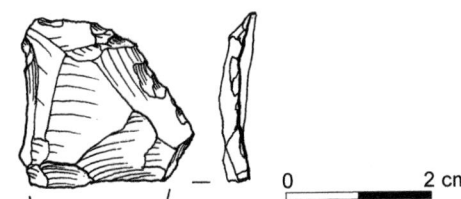

1240

Fig. 6.54 Flint *éclats de taille de biface* (#6576 Layer II-6 Level 1, #1240 Layer II-6 Level 4 with signs of utilization)

Table 6.47 Descriptive statistics of flint *éclats de taille de biface* and unretouched flint flakes of Layers V-5–6 and Layer II-6 (Levels 1–7) (size in mm; angle in degrees)

	Layer			
	Layers V-5 and V-6		Layer II-6	
	Flake	Édtdb	Flake	Édtdb
N	247	333	1,764	91
Maximum length				
Maximum	72.00	79.00	96.00	42.00
Median	27.00	29.00	25.00	26.00
Minimum	20.00	18.00	20.00	20.00
Mean	30.95	32.41	27.13	26.78
Std dev	10.79	10.61	6.70	4.65
Std err mean	0.69	0.58	0.16	0.49
Length			(N=1,761)	
Maximum	71.00	70.00	82.00	39.00
Median	24.00	25.00	23.00	23.00
Minimum	11.00	8.00	9.00	14.00
Mean	26.85	28.34	23.83	23.60
Std dev	10.71	10.93	7.06	5.50
Std err mean	0.68	0.60	0.17	0.58
Width			(N=1,761)	
Maximum	61.00	79.00	86.00	42.00
Median	21.00	23.00	21.00	21.00
Minimum	5.00	8.00	8.00	13.00
Mean	22.67	24.86	21.59	22.11
Std dev	9.05	8.85	6.82	5.49
Std err mean	0.58	0.48	0.16	0.58
Thickness				
Maximum	23.00	80.00	47.00	12.00
Median	8.00	5.00	9.00	5.00
Minimum	1.00	2.00	3.00	3.00
Mean	8.03	5.50	9.70	5.24
Std dev	3.18	4.63	3.79	1.60
Std err mean	0.20	0.25	0.09	0.17
Number of scars	(N=235)	(N=328)	(N=1,361)	
Maximum	21.00	20.00	15.00	12.00
Median	4.00	5.00	3.00	5.00
Minimum	1.00	1.00	1.00	1.00
Mean	5.02	5.69	3.79	5.00
Std dev	2.99	3.16	2.18	2.39
Std err mean	0.19	0.17	0.06	0.25
Angle of striking platform	(N=102)	(N=157)	(N=531)	(N=31)
Maximum	133.00	142.00	145.00	143.00
Median	115.00	112.00	116.00	115.00
Minimum	91.00	93.00	79.00	94.00
Mean	112.87	111.69	115.54	114.81
Std dev	8.24	8.42	9.73	10.97
Std err mean	0.82	0.67	0.42	1.97

Table 6.48 Counts and frequencies (%) of dorsal scar pattern of flint *éclats de taille de biface* and unretouched flint flakes of Layers V-5–6 and Layer II-6 (Levels 1–7)

	Layer			
	Layers V-5 and V-6		Layer II-6	
	Flake	Édtdb	Flake	Édtdb
N	244	329	1,683	91
Cortical	23	7	261	-
	9.43	2.13	15.51	
Plain	11	14	40	-
	4.51	4.26	2.38	
Along axis	56	100	237	11
	22.95	30.40	14.08	12.09
Parallel	2	1	7	2
	0.82	0.30	0.42	2.20
Convergent	7	21	27	7
	2.87	6.38	1.60	7.69
Opposed	1	1	20	1
	0.41	0.30	1.19	1.10
Radial	29	60	124	34
	11.89	18.24	7.37	37.36
Ridged	4	2	38	1
	1.64	0.61	2.26	1.10
Side	2	9	71	1
	0.82	2.74	4.22	1.10
Along axis & side	22	38	80	7
	9.02	11.55	4.75	7.69
Along axis & opposed	3	9	63	3
	1.23	2.74	3.74	3.30
Opposed & side	2	4	18	-
	0.82	1.22	1.07	
Along axis & radial	0	2	1	-
	0.00	0.61	0.06	
Indeterminate	82	61	696	24
	33.61	18.54	41.35	26.37

Table 6.49 Counts and frequencies (%) of striking platform types of flint *éclats de taille de biface* and unretouched flint flakes of Layers V-5–6 and Layer II-6 (Levels 1–7)

	Layer			
	Layers V-5 and V-6		Layer II-6	
	Flake	Édtdb	Flake	Édtdb
N	237	327	1,693	91
Cortical	8	1	130	-
	3.38	0.31	7.68	
Punctiform	4	26	64	8
	1.69	7.95	3.78	8.79
Plain	51	88	471	19
	21.52	26.91	27.82	20.88
Dihedral	18	24	29	2
	7.59	7.34	1.71	2.20
Facetted	45	67	118	22
	18.99	20.49	6.97	24.18
Crushed	8	29	32	14
	3.38	8.87	1.89	15.38
Removed	10	5	80	1
	4.22	1.53	4.73	1.10
Broken	73	70	510	24
	30.80	21.41	29.59	26.37
Indeterminate	20	17	268	1
	8.44	5.20	15.83	1.10

Table 6.50 Counts and frequencies (%) of technological observations on flint *éclats de taille de biface* and unretouched flint flakes of Layers V-5–6 and Layer II-6 (Levels 1–7)

	Layer			
	Layers V-5 and V-6		Layer II-6	
	Flake	Édtdb	Flake	Édtdb
N	241	328	1,746	91
Technological observations		(N=327)		(N=89)
Outrepassé	3	4	16	-
	1.24	1.22	0.92	
Débordant	15	16	58	5
	6.22	4.89	3.32	5.62
Débordant & hinge	-	-	1	-
			0.06	
Débordant & Siret	1	-	-	-
	0.41			
Débordant & steps	-	-	1	1
			0.06	1.12
Hinge	22	35	45	8
	9.13	10.70	2.58	8.99
Steps	4	30	6	20
	1.66	9.17	0.34	22.47
Siret	2	3	3	2
	0.83	0.92	0.17	2.25
Ventrally curved	3	6	21	1
	1.24	1.83	1.20	1.12
None	191	233	1,595	52
	79.25	71.25	91.35	58.43
Proximal end characteristics	(N=240)		(N=1,625)	
Large cone	3	2	11	-
	1.25	0.61	0.68	
Two cones	5	6	25	1
	2.08	1.83	1.54	1.10
Two cones & large cone	1	-	1	1
	0.42		0.06	1.10
Scar on bulb of percussion	9	11	16	21
	3.75	3.35	0.98	23.08
None	222	309	1,572	68
	92.50	94.21	96.74	74.73

the scarcity of flint DPF at GBY (Table 6.52) indicates that this type was not a predetermined target flake, as it was in the basalt flake assemblage, but a by-product that may have originated in both the cores on flakes and in ventral removals.

6.3.1.3 Ventral Removals

Presented in Table 6.53, ventral removals (isolated flake scars observed on the ventral faces of flakes, Fig. 6.58) occur in higher frequencies than those of the DPF discussed above. Ventral removals are more common in the complex of Layer II-6 than in any of the other layers. Further analyses of the FFT component suggested that ventral removals are associated

6.3 Technology of Flint Flakes

Table 6.51 Counts and frequencies (%) of lipped striking platform on flint *éclats de taille de biface* and unretouched flint flakes of Layers V-5–6 and Layer II-6 (Levels 1–7)

	Layer			
	Layers V-5 and V-6		Layer II-6	
	Flake	Édtdb	Flake	Édtdb
N	111	223	158	49
Lipped	35	95	46	17
	31.53	42.60	29.11	34.69
Possibly lipped	22	53	18	6
	19.82	23.77	11.39	12.24
Not lipped	54	75	94	26
	48.65	33.63	59.49	53.06

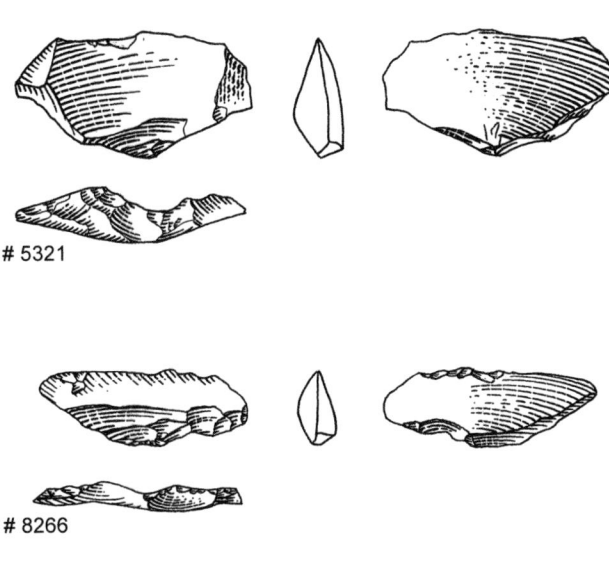

5321

8266

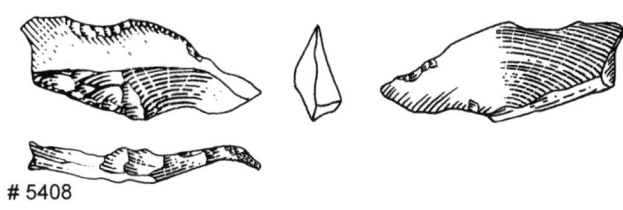

5408

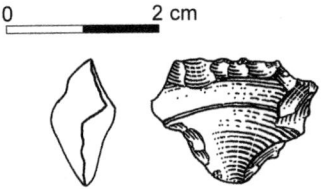

4711

Fig. 6.55 Flint *éclats de taille de biface*: finishing flakes struck perpendicular to biface edge (#5321 Layer V-5, #8266 Layer V-5, #5408 Layer V-6, #4711 Layer V-5)

with an innovative typological interpretation presented below (Section 6.5.1.4).

6.3.2 Core Management Pieces

In addition to the unretouched categories discussed above, the following classes were identified: core trimming elements (N=47; 0.93% of the entire flint assemblage; flakes that bear remnants of the cores' striking platform, Fig. 6.59); angular fragments (N=9; 0.18% of the entire flint assemblage; flakes with no clear dorsal arrangement but rather with a bulky, angular, dorsal face that likely originates in cores); and biface sharpening flakes (Figs. 6.8, 6.20; N=26; 0.52% of the entire flint assemblage; see Chapter 7). These three categories occur in very low frequencies (Table 6.54) but along the entire stratigraphic sequence. Their scarcity may indicate that their presence is occasional or even random, rather than being a result of systematic core/biface maintenance.

In addition to these categories, the presence of other types of core management pieces was recorded (see Chapter 4). These include *débordant* flakes (Figs. 6.10, 6.17, 6.27, 6.60) and *outrepassé* flakes (Fig. 6.20). The terminology of these two types is drawn from the definition of the Levallois method. However, although they display morphological traits similar to those of the Levallois method, they are not necessarily intentional products. Rather, in the context of GBY, they may result from biface production and maintenance.

For example, the abundance of flint flakes in Layers V-5 and V-6 enabled in-depth observations that concluded that these assemblages are the outcome of the final stages of biface modification and thinning (Sharon and Goren-Inbar 1999; Goren-Inbar and Sharon 2006). In these two layers, *débordant* flakes are the most common core waste product, while the most common type of unretouched flake is the *édtdb*. The high frequencies of these two waste products are clearly associated with biface production.

The other two categories of waste products (core trimming elements and angular fragments) occur in very low frequencies, suggesting that there was no systematic management of flint cores.

6.4 Classification and Technology of Cores and Core Tools

This section includes the detailed technological analyses of all cores and other non-flake waste products (e.g. angular

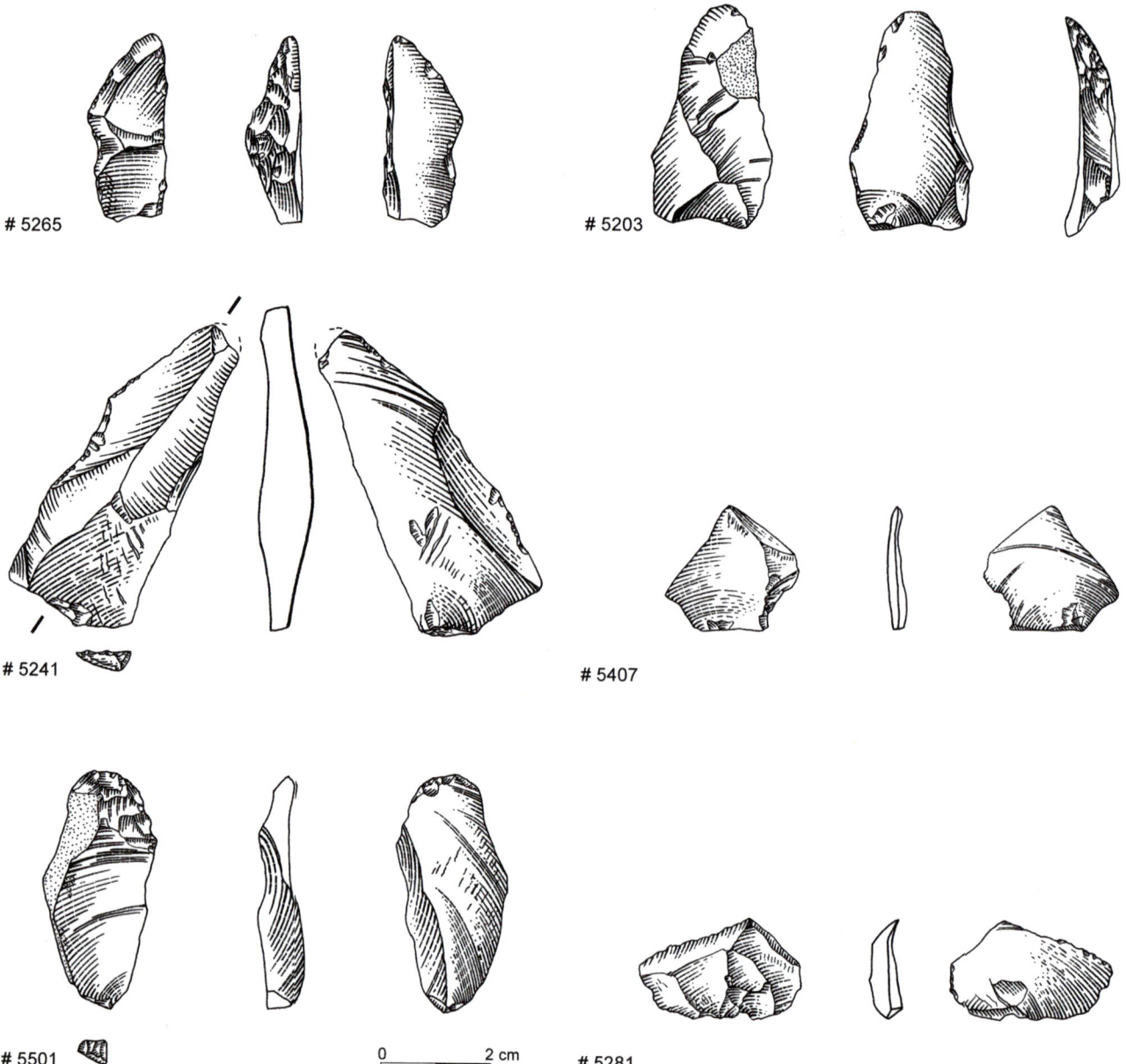

Fig. 6.56 Flint *éclats de taille de biface* (finishing flakes struck obliquely to biface edge: #5265 Layer V-5, #5241 Layer V-5, #5501 Layer V-6, #5203 Layer V-5; finishing flakes struck perpendicularly to biface edge: #5407 Layer V-6, #5281 Layer V-5)

fragments). Core tools (Section 6.2.2) are included within this discussion, since they are modified on non-flake waste products and/or cores; the present section, however, focuses on their technological characteristics. Chopping tools are excluded from the presentation, as they are a well-defined type that has been discussed in detail above in terms of technology (Section 6.2.2).

The technological classification of the entire CCT component (by layer and level) is presented in Table 6.55.

In the detailed discussion of the different categories and their characteristics, we have in general considered only layers and levels that yielded a minimum number of 50 artifacts. Exceptions to this are two levels of Layer II-6 that, although they yielded smaller samples (Table 6.55), are essential for the understanding of the extent of variability observed in the sequential cultural package of Layer II-6.

6.4 Classification and Technology of Cores and Core Tools

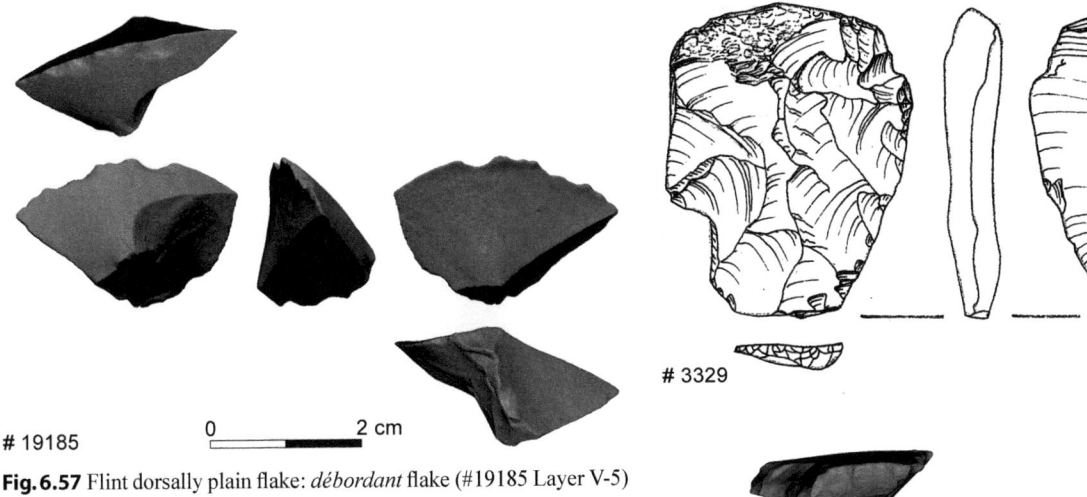

#19185

Fig. 6.57 Flint dorsally plain flake: *débordant* flake (#19185 Layer V-5)

Fig. 6.58 Flint flakes with ventral removals and signs of utilization (#3329 JB, #11425 Layer II-6 Level 1 borer with removed striking platform)

6.4.1 Cores

Cores are a major component within the flint assemblage and occur in relatively high frequencies throughout the cultural sequence at GBY (Table 6.55). Although the assemblage of cores is highly variable and composed of diverse morphologies and hence types, it represents the parent material from which most of the FFT originate. The high frequencies of flint cores contrast with the minimal occurrence of cores on other raw materials (i.e. basalt and limestone) at GBY. The following paragraphs

Table 6.52 Counts and frequencies (%) of dorsally plain flakes on flint FFT

	Layer												
				II-6 Levels									
	JB	V-5	V-6	L1	L2	L3	L4	L4b	L5	L6	L7	II-7
N	158	311	291	764	305	385	433	124	156	274	740	32
Dorsally plain flake	1	3	2	8	-	4	6	2	-	1	2	-
	0.63	0.96	0.69	1.05		1.04	1.39	1.61		0.36	0.27	
Possibly dorsally plain flake	1	2	4	2	-	4	-	1	1	3	9	-
	0.63	0.64	1.37	0.26		1.04		0.81	0.64	1.09	1.22	
None	156	306	285	754	305	377	427	121	155	270	729	32
	98.73	98.39	97.94	98.69	100.00	97.92	98.61	97.58	99.36	98.54	98.51	100.00

Table 6.53 Counts and frequencies (%) of ventral removals on flint FFT

	Layer															
								II-6 Levels								
	JB	V-5	V-6	II-2	II-2/3	II-5	II-5/6	L1	L2	L3	L4	L4b	L5	L6	L7	II-7
Ventral removals	3	11	13	6	-	14	11	47	23	40	41	17	23	27	84	7
	1.90	3.54	4.47	5.56		7.37	9.57	6.15	7.54	10.39	9.47	13.71	14.74	9.85	11.35	21.88
None	155	300	278	102	99	176	104	717	282	345	392	107	133	247	656	25
	98.10	96.46	95.53	94.44	100.00	92.63	90.43	93.85	92.46	89.61	90.53	86.29	85.26	90.15	88.65	78.13
Total	158	311	291	108	99	190	115	764	305	385	433	124	156	274	740	32

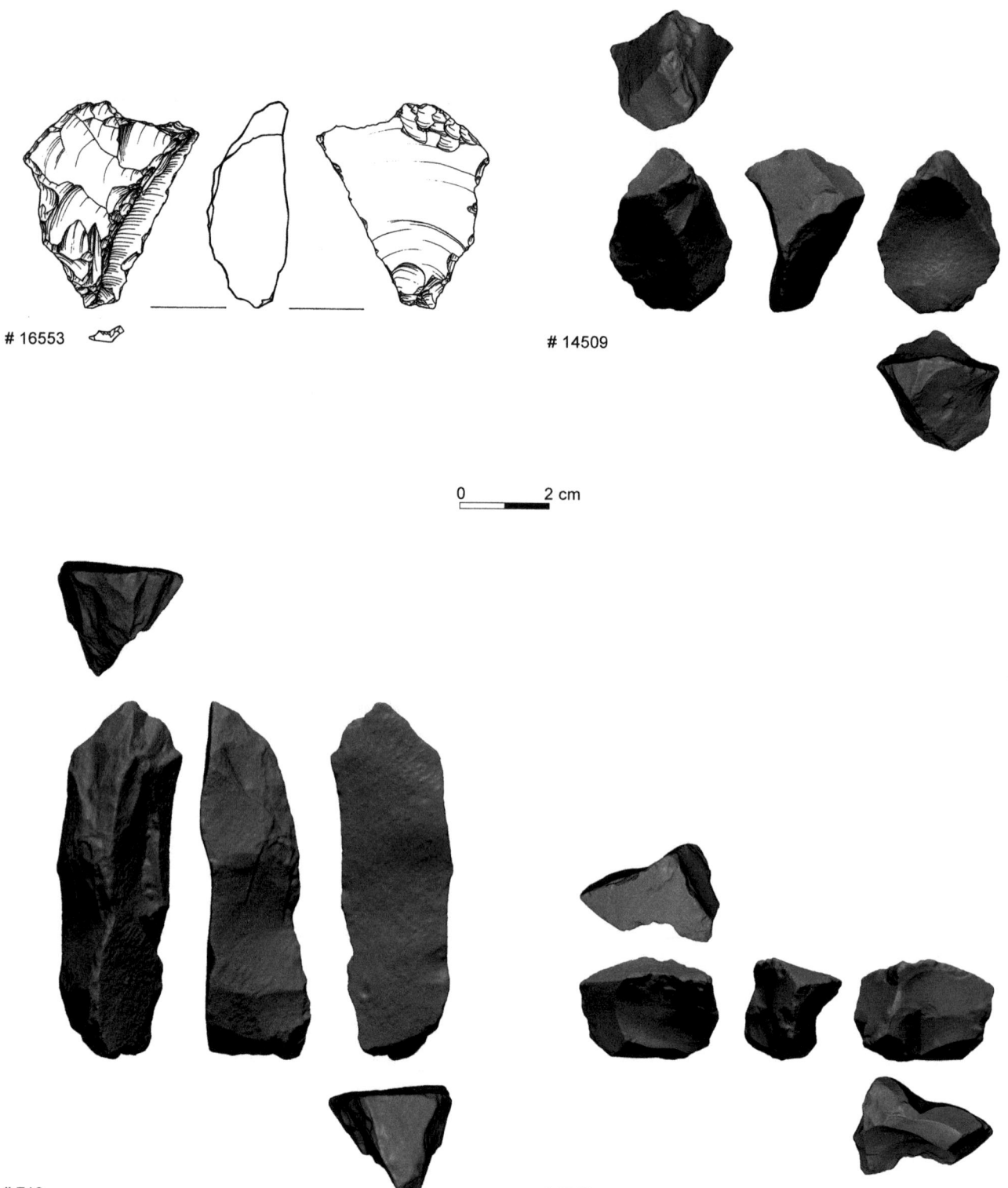

Fig. 6.59 Flint core trimming elements (#16553 Layer II-5 side-scraper on ventral face with removed striking platform, #710 Layer II-6 Level 1, #14509 Layer II-6 Level 2 with signs of utilization, #9845 Layer II-6 Level 7 with signs of utilization, note two cones)

6.4 Classification and Technology of Cores and Core Tools

Table 6.54 Counts of core management pieces on flint FFT*

	Layer																		
										II-6 Levels									
	JB	V-5	V-6	I-4	I-5	II-2	II-2/3	II-5	II-5/6	L1	L2	L3	L4	L4b	L5	L6	L7	II-6/7	II-7	Trench II green
Core trimming element	1	5	2		1	-	-	1	1	10	3	9	2	-	-	3	14	-	-	-
Biface sharpening flake	1	2	5		-	-	-	-	-	1	1	4	4	1	2	4	12	1	1	-
Angular fragment	-	2	-		-	2	1	-	-	2	1	-	-	-	-	-	-	-	-	-
Outrepassé flake	1	2	6		2	-	-	-	1	4	2	10	6	3	1	3	7	-	1	-
Débordant flake	12	21	20		2	1	-	7	1	7	4	16	14	2	13	12	43	-	-	3
Total	15	32	33		-	3	1	8	3	24	11	39	26	6	16	22	76	1	2	3

* including multiple tools (classified as core management pieces in Typology B or Typology C)

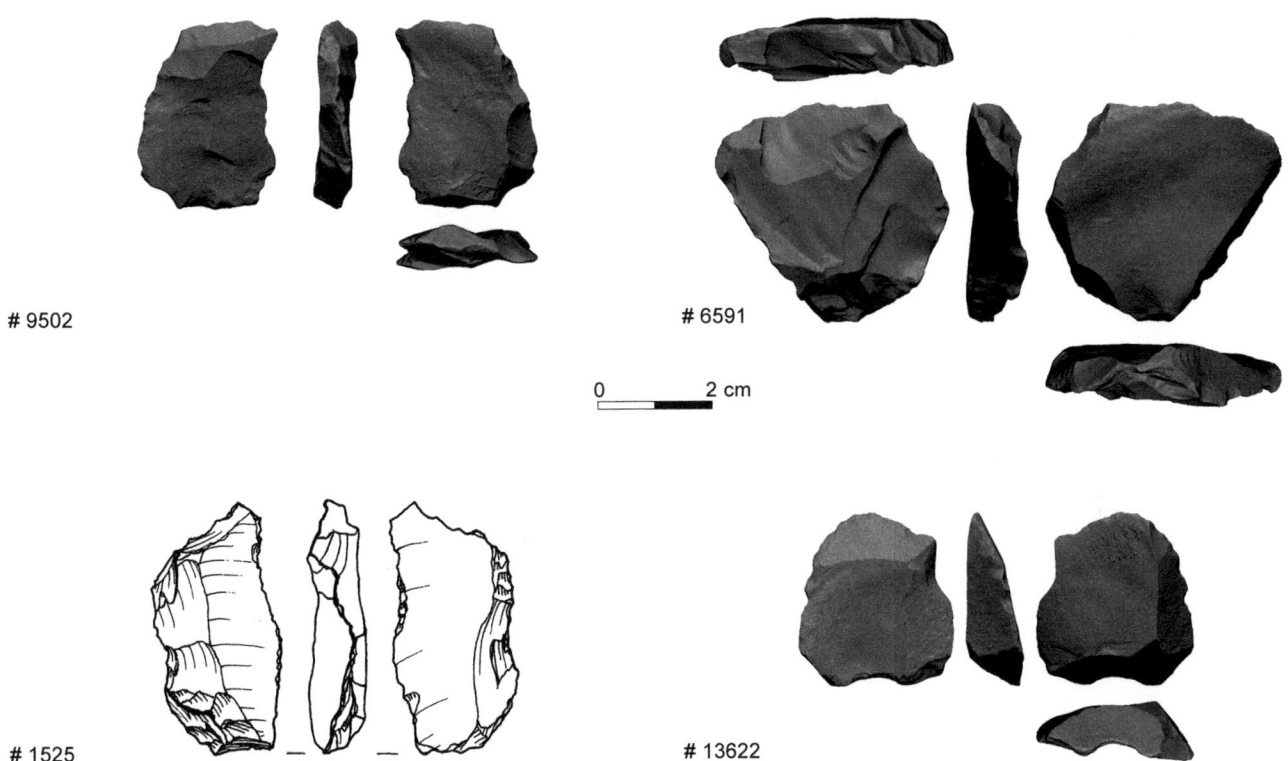

Fig. 6.60 Flint *débordant* flakes (#9502 Layer II-6 Level 3 notch, #1525 Layer II-6 Level 4 with signs of utilization, #6591 Layer II-6 Level 6 with signs of utilization, #13622 Layer II-6 Level 7 with signs of utilization)

provide the technological and typological descriptions of the different flint core categories.

Levallois Cores

Several cores were classified as Levallois cores (N=53), and they occur throughout the cultural sequence (Table 6.56, Figs. 6.61–6.64). The criteria for assigning them to this category were based on the following observations: 1) these cores predominantly have two hierarchical surfaces of which the upper debitage surface is flaked more intensively; 2) the debitage surface is generally convex; 3) the two surfaces are separated by a continuous working edge (striking platform) that separates the surfaces; 4) the acute angle of the working edge enables continuous flake removals along the circumference of the core. Since only complete or near-complete cores exhibiting the combination of all the above criteria could be

Table 6.55 Counts and typological frequencies (%) of flint CCT

N	Layer										
	V-4	V-4/5	V-5	V-6	JB	I-5	II-2	II-2/3	II-4	II-5	II-5/6
	6	6	23	8	10	11	6	18	4	49	38
Angular fragment	2	-	3	-	-	8	1	5	1	14	10
	33.33	-	13.04	-	-	72.73	16.67	27.78	25.00	28.57	26.32
Modified	4	2	7	1	2	1	1	4	1	9	8
	66.67	33.33	30.43	12.50	20.00	9.09	16.67	22.22	25.00	18.37	21.05
Percussor	-	-	1	-	-	-	-	-	-	1	-
	-	-	4.35	-	-	-	-	-	-	2.04	-
Core waste	-	1	-	1	1	-	-	1	-	1	2
	-	16.67	-	12.50	10.00	-	-	5.56	-	2.04	5.26
Amorphous core	-	2	1	-	-	-	-	-	-	-	1
	-	33.33	4.35	-	-	-	-	-	-	-	2.63
Core on flake	-	-	4	1	-	1	1	-	-	3	2
	-	-	17.39	12.50	-	9.09	16.67	-	-	6.12	5.26
Levallois core	-	-	1	3	1	1	1	2	-	2	2
	-	-	4.35	37.50	10.00	9.09	16.67	11.11	-	4.08	5.26
Discoidal core	-	-	1	-	-	-	-	-	-	-	-
	-	-	4.35	-	-	-	-	-	-	-	-
Globular core	-	-	-	-	-	-	-	-	-	1	-
	-	-	-	-	-	-	-	-	-	2.04	-
Pyramidal core	-	-	-	-	-	-	-	-	-	1	-
	-	-	-	-	-	-	-	-	-	2.04	-
Varia core	-	1	4	2	3	-	1	4	2	12	9
	-	16.67	17.39	25.00	30.00	-	16.67	22.22	50.00	24.49	23.68
Formless core	-	-	1	-	3	-	1	2	-	5	4
	-	-	4.35	-	30.00	-	16.67	11.11	-	10.20	10.53

classified as Levallois cores, most of these cores are complete (Table 6.57).

Although Levallois cores occur in varying frequencies throughout the GBY cultural sequence, their frequencies do not depend on the sample size of CCT. For example, while the largest core assemblage among the eight levels of Layer II-6 is that of Level 1, this level exhibits only four Levallois cores, while the largest number of Levallois cores (N=12) occurs in Layer II-6 Level 4, where the core assemblage is smaller.

The majority of Levallois cores are slightly abraded, although the frequencies of fresh artifacts are higher than in other core categories (Table 6.57). The bulk of the items are patinated and double patina occurs less frequently than in other core categories (Table 6.57). Most of these cores exhibit no cortical coverage and when recorded it is minimal (Table 6.57).

Levallois cores often exhibit a single working edge and two platforms (Table 6.58), although several cores deviate from this pattern (e.g. Fig. 6.65). The angle between the two platforms (working edge angle) is sharp (ca. 77° degrees on average; Table 6.58). Of the identifiable transversal sections of the Levallois cores, 55.81% exhibit a flat morphology and 44.18% a triangular one. The working edge length forms the bulk of the circumference of these cores, so that on average nearly 88% of the circumference is exploited (Table 6.58). For 56.86% of these cores, the entire circumference is worked, and in the remaining cores the dominant plan form of the working edge is wavy (50%) or convex (30%). These characteristics are very similar to those observed on cores on flakes (see below). Only three of the cores classified as Levallois cores are modified on flakes (Fig. 6.61).

6.4 Classification and Technology of Cores and Core Tools

Table 6.55 (cont.)

					Layer						
					II-6 Levels						
N	VI-14	L1	L2	L3	L4	L4b	L5	L6	L7	II-6/7	II-7
	2	303	136	185	238	25	42	63	128	2	2
Angular fragment	-	124 40.92	48 35.29	37 20.00	74 31.09	4 16.00	13 30.95	6 9.52	24 18.75	-	-
Modified	-	68 22.44	36 26.47	58 31.35	52 21.85	9 36.00	6 14.29	8 12.70	42 32.81	-	-
Percussor	-	-	-	-	-	-	-	3 4.76	-	-	-
Core waste	-	15 4.95	8 5.88	4 2.16	13 5.46	-	-	-	-	-	-
Amorphous core	-	3 0.99	6 4.41	5 2.70	6 2.52	-	-	-	-	-	-
Core on flake	1 50.00	10 3.30	6 4.41	13 7.03	16 6.72	1 4.00	2 4.76	15 23.81	13 10.16	2 100.00	-
Core on biface	-	-	-	-	1 0.42	-	-	-	-	-	-
Levallois core	1 50.00	4 1.32	4 2.94	5 2.70	12 5.04	1 4.00	2 4.76	3 4.76	8 6.25	-	-
Discoidal core	-	3 0.99	2 1.47	1 0.54	2 0.84	-	-	1 1.59	-	-	-
Globular core	-	-	1 0.74	2 1.08	1 0.42	-	1 2.38	3 4.76	-	-	1 50.00
Prismatic core	-	-	-	1 0.54	1 0.42	-	-	-	-	-	-
Pyramidal core	-	-	-	-	-	-	-	-	2 1.56	-	-
Varia core	-	56 18.48	16 11.76	35 18.92	37 15.55	9 36.00	13 30.95	17 26.98	28 21.88	-	1 50.00
Formless core	-	20 6.60	9 6.62	24 12.97	23 9.66	1 4.00	5 11.90	7 11.11	11 8.59	-	-

The measurements of Levallois cores suggest that their length and width dimensions are larger than those of other CCT, while their thickness is similar. We can assume that the original dimensions of these cores were larger, considering that the final debitage surface exhibits the removals of about 10 scars on average (Table 6.58). The size variability expressed in the standard deviation may indicate that these cores represent different degrees of exhaustion.

The size of the last measurable scar on the debitage surface suggests that the final removals from these Levallois cores were of small flakes, with an average length of 25 mm (Table 6.58); these scars, however, are larger than those observed on any other core type.

The scar pattern could be identified on all but three of the Levallois cores. Among the identifiable scar patterns (Table 6.59), the dominant one is the radial debitage surface (36.73%). Other common patterns that seem to characterize the debitage surfaces of the Levallois cores are the horseshoe and simple dominant patterns (Fig. 6.61). In both of these an entire or partial circumferential preparation of the debitage

Table 6.56 Counts of Levallois flint cores throughout the GBY sequence

Layer	N
V-5	1
V-6	3
JB	1
I-5	1
II-2	1
II-2/3	2
II-5	2
II-5/6	2
VI-14	1
II-6/L1	4
II-6/L2	4
II-6/L3	5
II-6/L4	12
II-6/L4b	1
II-6/L5	2
II-6/L6	3
II-6/L7	8
Total	53

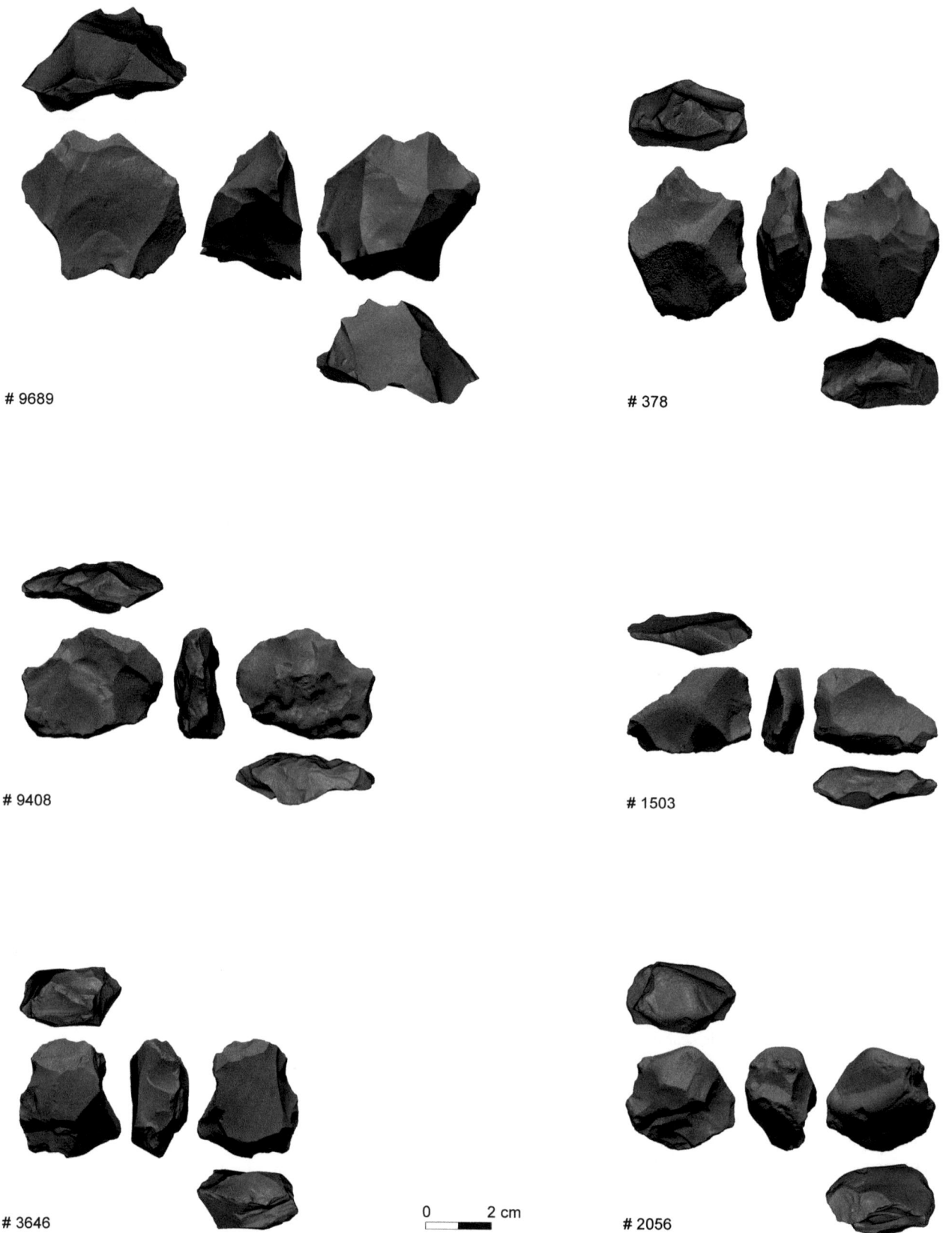

Fig. 6.61 Flint Levallois cores (#9689 Layer V-6, #378 Layer II-5, #9408 Layer II-6 Level, #1503 Layer II-6 Level 4, #3646 Layer II-6 Level 4, #2056 Layer II-6 Level 4)

6.4 Classification and Technology of Cores and Core Tools

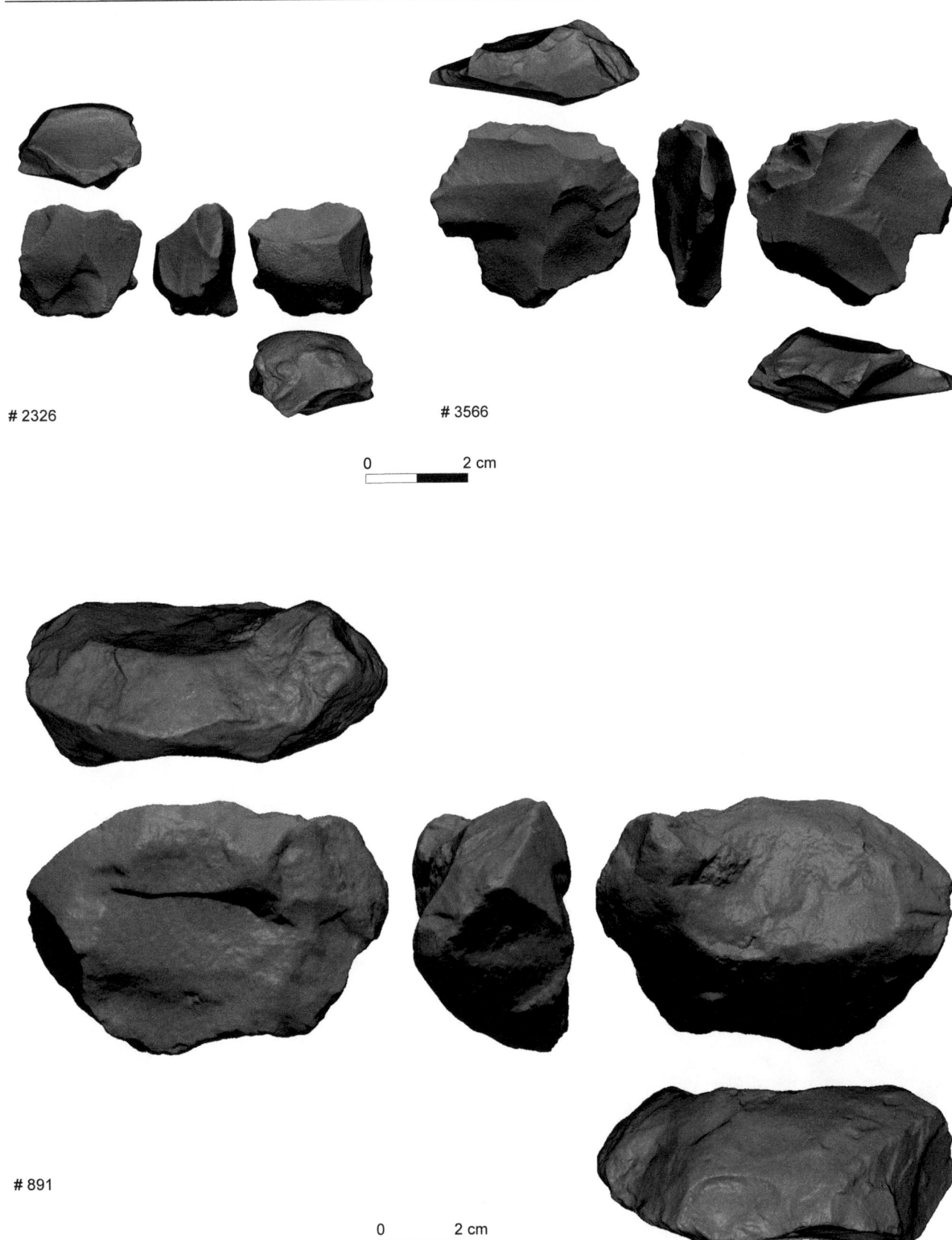

Fig. 6.62 Flint Levallois cores (#2326 Layer II-6 Level 4, #3566 Layer II-6 Level 4, #891 Layer II-6 Level 4)

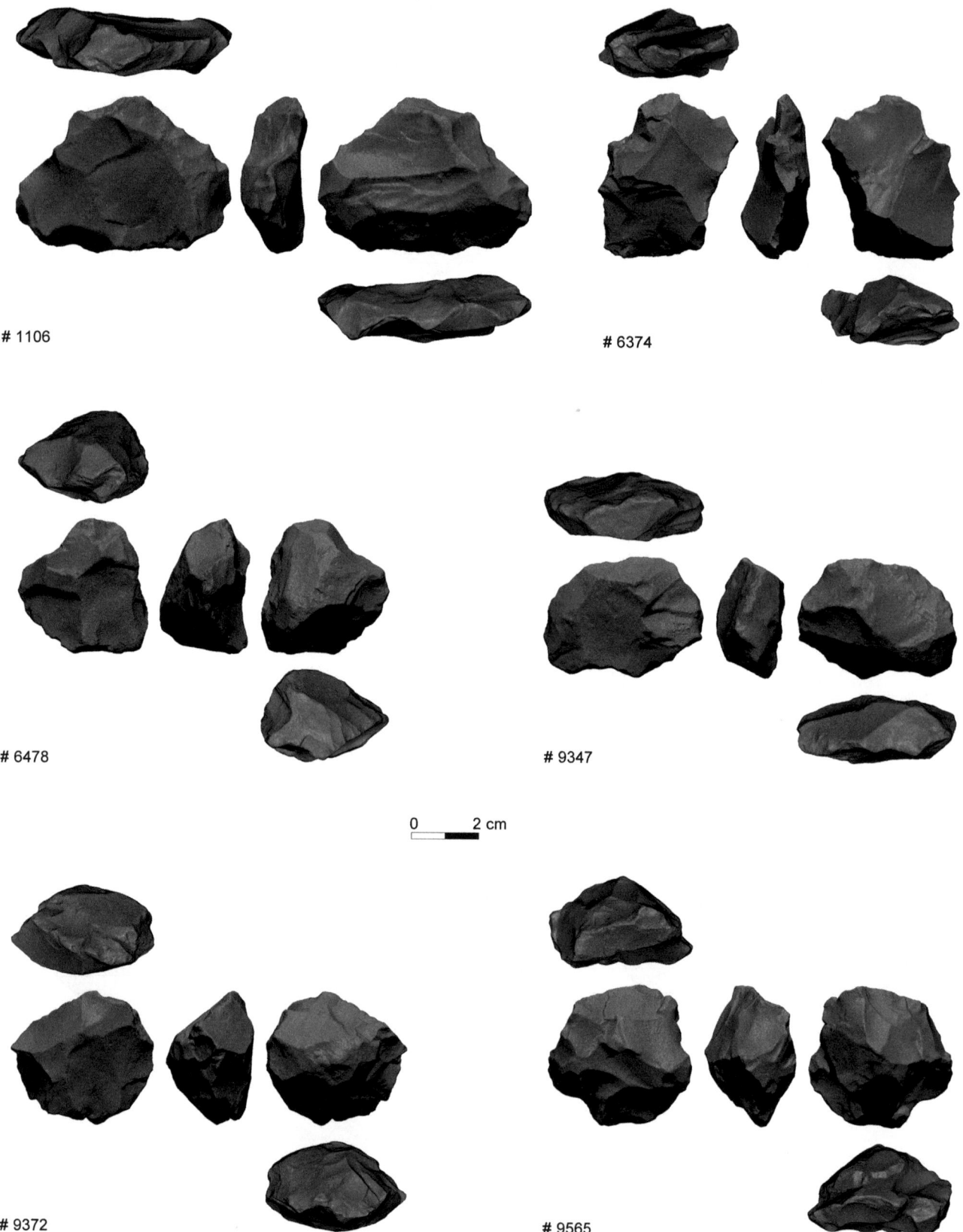

Fig. 6.63 Flint Levallois cores (#1106 Layer II-6 Level 4b, #6374 Layer II-6 Level 5, #6478 Layer II-6 Level 6, #9347 Layer II-6 Level 7, #9372 Layer II-6 Level 7, #9565 Layer II-6 Level 7)

6.4 Classification and Technology of Cores and Core Tools

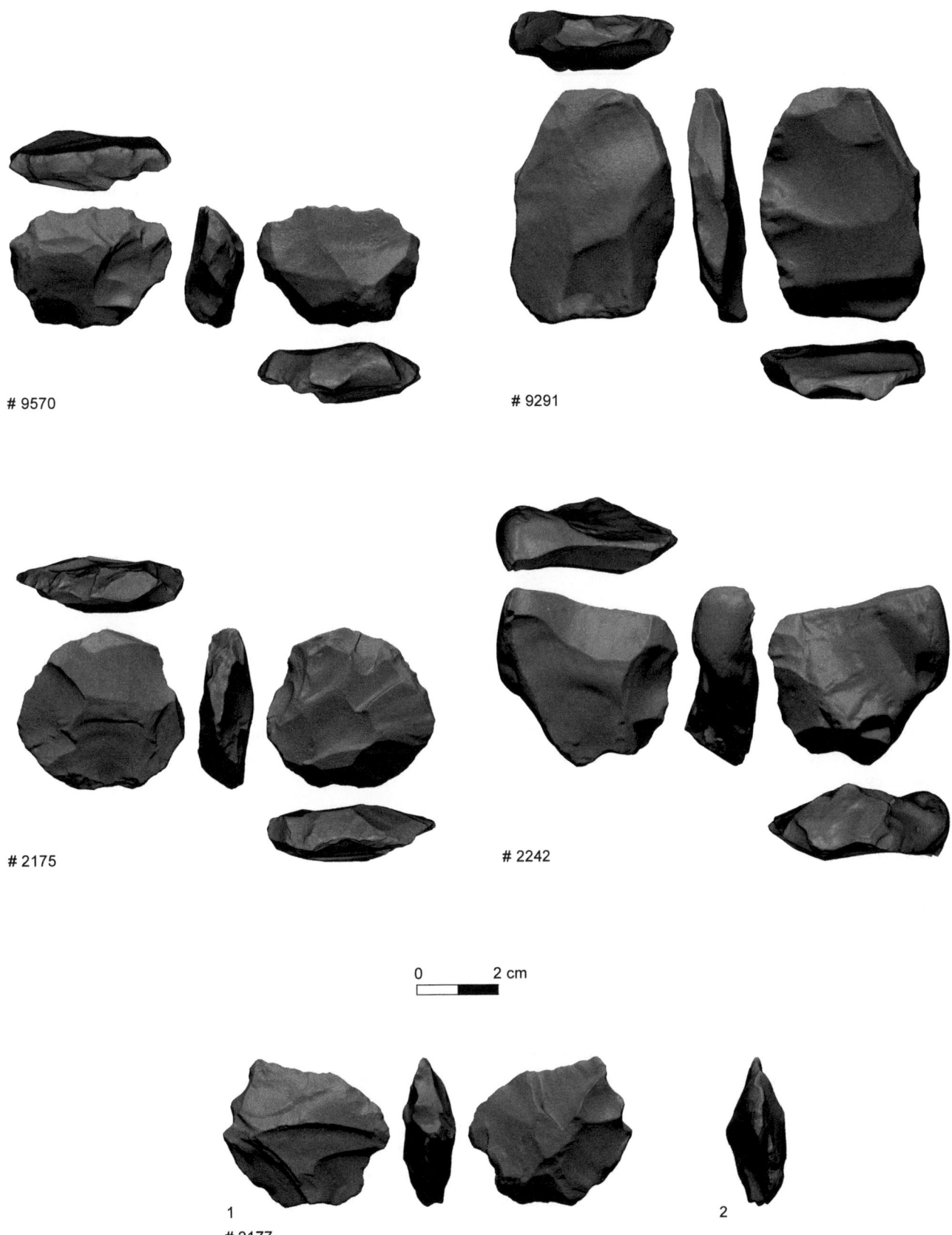

Fig. 6.64 Flint Levallois cores (#9570 Layer II-6 Level 7, #9291 Layer II-6 Level 7, #2175 Trench II, #2242 Trench II, #2177 Trench II **1** multiple views **2** detailed view of working edge)

Table 6.57 Counts and frequencies (%) of characteristics of Levallois flint cores

	N	%
Preservation (N=53)		
Fresh	19	35.85
Slightly abraded	26	49.06
Abraded	7	13.21
Heavily abraded	1	1.89
Patination (N=53)		
Patinated	47	88.68
Double patinated	6	11.32
Breakage (N=53)		
Complete	47	88.68
Distal	1	1.89
Lateral	5	9.43
Cortex (N=53)		
No cortex	35	66.04
1–25%	9	16.98
26–50%	5	9.43
51–75%	2	3.77
76–100%	2	3.77

surface allows the predetermination and removal of a single dominant flake. These removals (as determined by the last measurable scar) are on average 10 mm longer and 5 mm wider than the removals observed on other Levallois cores. Together, these patterns characterize some 40% of the Levallois cores (Table 6.59). Other patterns, such as the bipolar or convergent one (Fig. 6.65), also occur but in lower frequencies.

Cores on Flakes

Cores on flakes (N=76) (Figs. 6.65–6.68) are recorded throughout the cultural sequence of GBY (Table 6.55). Their varying frequencies depend on the sample size of CCT, but may also be the result of other functional/technological aspects.

Most of these cores are slightly abraded and patinated (Table 6.60). The occurrence of double patina reflects the use of patinated flakes on which later scar removals, bearing a different, younger patina, are seen. The majority of flakes on which these cores are modified exhibit no cortical coverage, and when recorded it is minimal (Table 6.60). Over 80% of these cores are complete (Table 6.60).

In the modification of flakes into cores, a single working edge and two platforms were often used (Table 6.61). The angle between the two platforms (working edge angle) is sharp on average and reflects the characteristics of the flakes on which these cores are modified. Of the identifiable transversal sections of these cores, 59.45% exhibit a flat morphology and 40.54% a triangular one. Measurement of the working edge length shows that it forms a substantial part of the circumference of these cores, so that on average nearly 70% of the circumference is exploited (Table 6.61). In some 40% of these cores the entire circumference is worked. For the remaining cores, the dominant plan form of the working edge is wavy (32.43%) or convex (24.32%).

Other measurements suggest that the dimensions of cores on flakes are slightly larger than those of flint FFT; for example, cores on flakes are on average 3 mm thicker than FFT. However, these cores are thinner than their original blanks due to scar removals (average of 6.32 scars; Table 6.61), so we can assume that thicker flakes were selected for cores on flakes. This assumption is supported by statistical testing of the dimensions of these cores; cores on flakes are significantly longer (t ratio=5.30; df=3073; $p(tt)$<0.0001; difference: 4.70 mm), wider (t ratio=4.50; df=3073; $p(tt)$<0.0001; difference: 3.67 mm), and thicker (t ratio=8.62; df=3073; $p(tt)$<0.0001; difference: 4.04 mm) than unretouched flakes.

The size of measurable scars on the debitage surface suggests that very small flakes were removed from these cores (Table 6.61). Their average length of 16.59 mm implies that only 57% of the length of the debitage surface was exploited.

For some 38% of the cores on flakes, the scar pattern could not be identified. Of the remaining identifiable scar patterns, the dominant one is the radial debitage pattern (36.11%), followed by the along axis pattern (13.88%). Worth noting is the fact that there is a high variability of patterns (see below), each represented by minimal occurrences.

Globular Cores and Discoidal Cores

Globular (N=10) and discoidal (N=10) (Fig. 6.69) cores are represented in small numbers within the GBY cultural sequence (Table 6.55). The characteristics of these two classes derive from their general volume and morphology. Table 6.62 presents the characteristics of these two classes, including all artifacts regardless of layer or level.

While globular cores are more abraded and lack a fresh component, both core categories exhibit equal frequencies of patinated and double patinated artifacts and in both all cores are complete. Globular cores exhibit more cortical coverage (Table 6.62).

Globular and discoidal cores are similar in size, except for the higher mean thickness of globular cores and their slightly longer circumference, both dictated by the spherical morphology of this core category. Globular cores are characterized by the exploitation of several platforms and thus demonstrate a relatively high number of flake scars (Table 6.63).

6.4 Classification and Technology of Cores and Core Tools

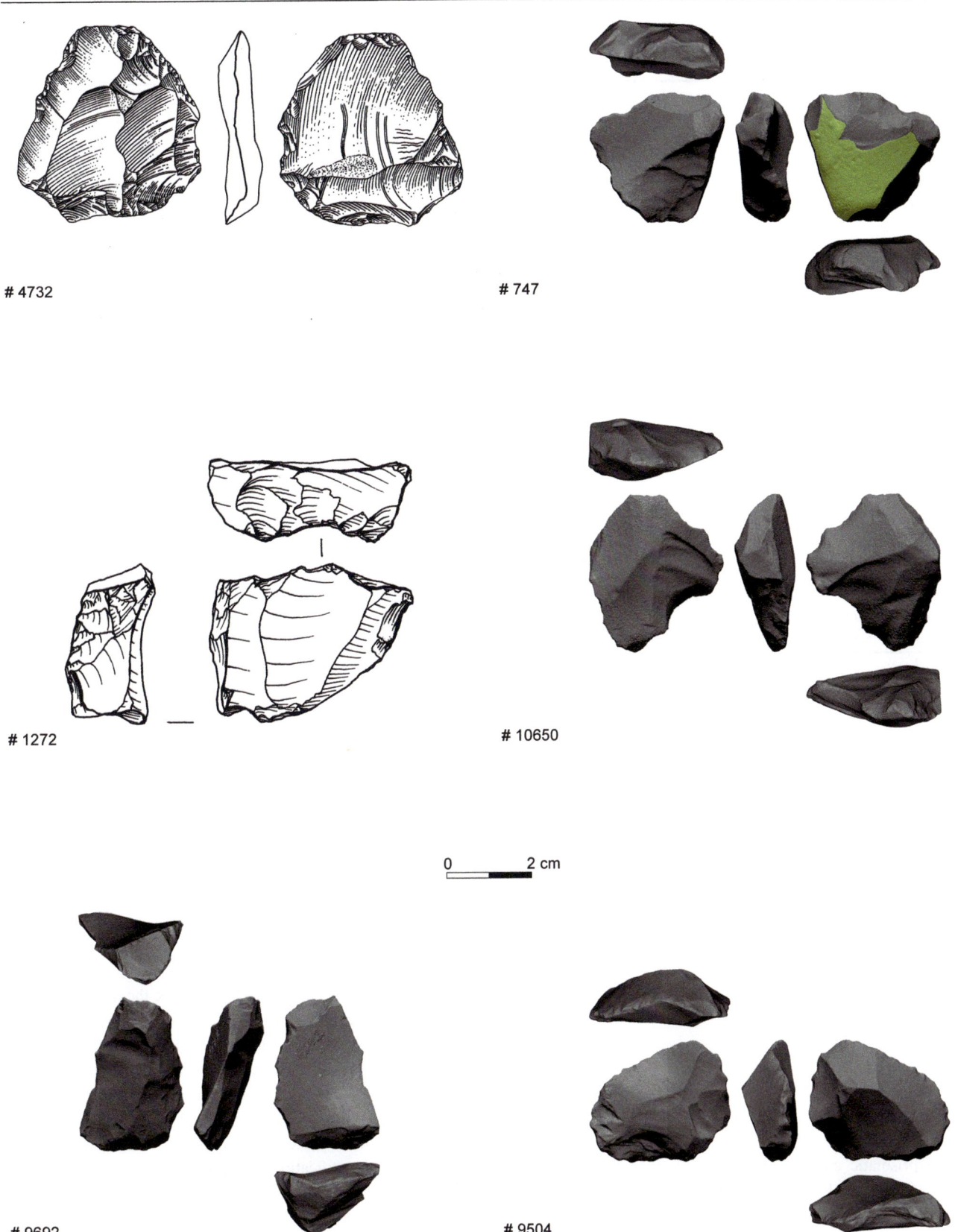

Fig. 6.65 Flint cores: Levallois cores (#4732 Layer V-5, #1272 Layer II-6 Level 4); cores on flakes (#9692 Layer V-6, #747 Layer II-5, #10650 Layer II-6 Level 1, #9504 Layer II-6 Level 3)

Table 6.58 Descriptive statistics of Levallois flint cores (size in mm; angle in degrees)

	Maximum length	Length	Width	Thickness	Circumference	Working edge length
N	53	53	53	53	51	44
Maximum	67.00	62.00	64.00	28.00	193.00	193.00
Median	38.00	35.00	32.00	15.00	108.00	108.00
Minimum	25.00	24.00	20.00	8.00	78.00	22.00
Mean	39.49	36.19	34.05	16.18	116.82	102.68
Std dev	10.07	9.66	9.81	5.43	28.13	39.60
Std err mean	1.38	1.33	1.34	0.74	3.94	5.97
	Working edges	Platforms	Scars	Working edge angle	Scar length	Scar width
N	53	53	53	30	35	35
Maximum	3.00	5.00	22.00	88.00	46.00	36.00
Median	1.00	2.00	9.00	79.00	25.00	22.00
Minimum	1.00	2.00	3.00	60.00	10.00	10.00
Mean	1.19	2.25	10.06	76.90	25.00	21.91
Std dev	0.48	0.68	4.81	7.45	8.80	6.56
Std err mean	0.07	0.09	0.66	1.36	1.49	1.11

Table 6.59 Counts and frequencies (%) of identifiable scar pattern of Levallois flint cores

Scar pattern (N=49)	N	%
Radial	18	36.73
Horseshoe	14	28.57
Simple dominant	6	12.24
Bipolar	5	10.20
Along axis & opposed	3	6.12
Along axis & side	1	2.04
Convergent	2	4.08

Discoidal cores exhibit much longer working edges, which on average stretch around 85.06% of the circumference. This is also expressed in the low values of number of working edges of these cores and the peripheral form of their working edges (Table 6.63). The dominant flaking pattern is the radial one (Table 6.64).

Other Cores

Two prismatic cores (Fig. 6.70) and three pyramidal cores (Fig. 6.70) are also included in the inventory of flint cores. Their minimal occurrence does not permit detailed analysis. In addition, a single core on a biface is recorded from Layer II-6 Level 4 (Fig. 6.70) (Table 6.55). The phenomenon of cores on bifaces is widely known from the later Acheulian (Revadim, Tabun, Ma'ayan Baruch, to mention but a few (Stekelis and Gilead 1967; deBono and Goren-Inbar 2001; Marder et al. 2006) but is rare at GBY. Three additional cores were registered as modified on broken bifaces (one classified as a Levallois core and two as varia cores, discussed and illustrated below under varia cores), the use of technological elements that derive from biface modification (e.g. biface sharpening flakes) is recorded for nine other cores (three cores on flakes, two Levallois cores, and four varia cores). This minimal representation of cores on bifaces may be due to the fact that most of the bifaces at GBY are modified on basalt, which is hardly used as raw material for systematic flake production but is limited to the production sequence aimed at manufacture of bifaces (Chapter 7).

Varia Cores and Formless Cores

These, particularly varia cores (N=211), are the most numerous of the flint core types (Table 6.55). Varia cores are cores with identifiable working edges and platforms on which the organization of scars is clear. However, they do not conform typologically to any of the other core classes and they are hence classified as an independent group (Figs. 6.71–6.74, 6.76). On some of these varia cores, additional observations resulted in the identification of a core with a single convex side-scraper (Fig. 6.75), a core on a biface (Fig. 6.75), and cores on biface sharpening flakes (Fig. 6.75). Similarly, formless cores (N=100) do not conform to any core typology but are even less standardized than varia cores, in that they exhibit no organization of scar removals and their various attributes are often classified as indeterminate (Fig. 6.76).

6.4 Classification and Technology of Cores and Core Tools

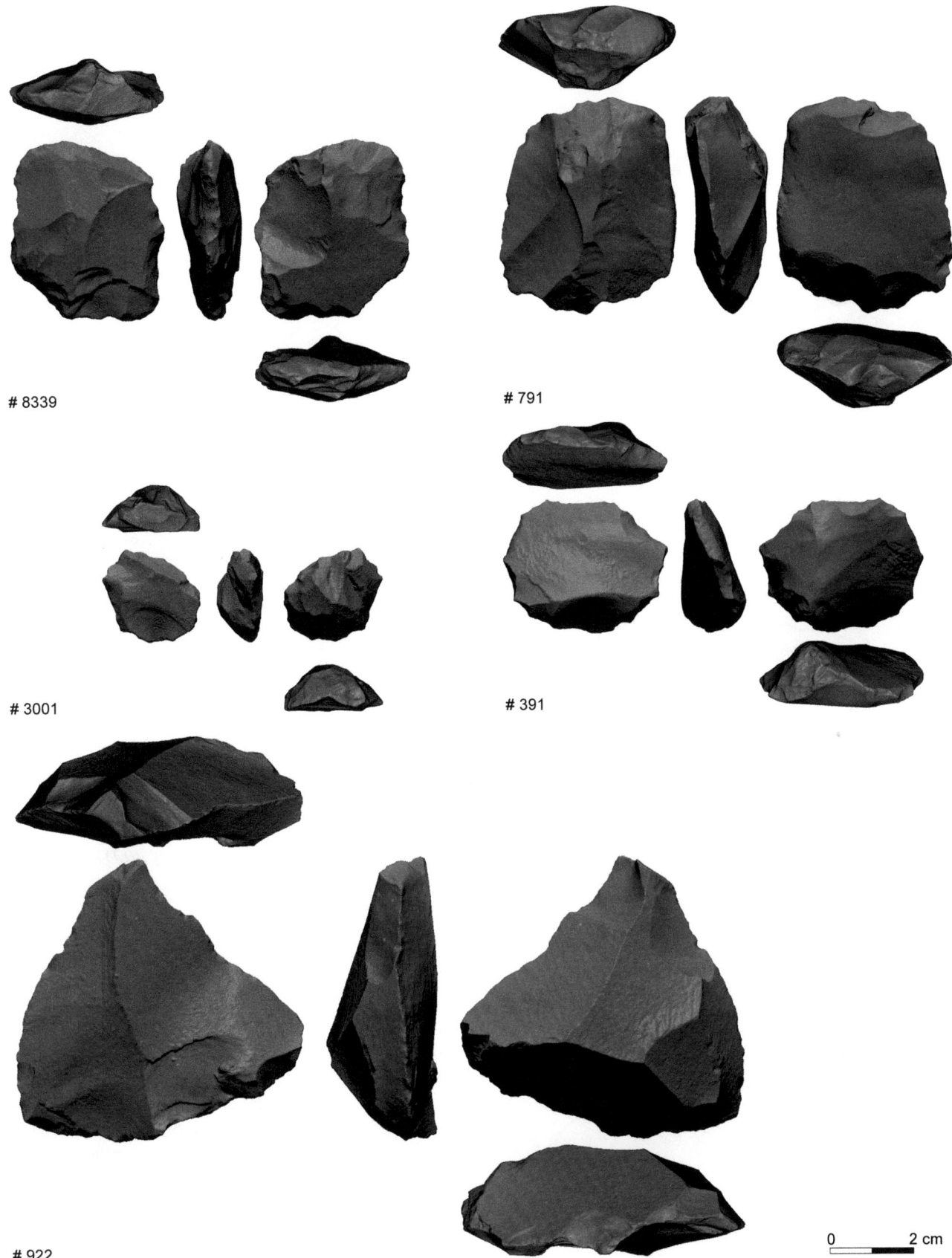

Fig. 6.66 Flint cores on flakes (#8339 Layer II-6 Level 3 on biface sharpening flake, #3001 Layer II-6 Level 4 on biface sharpening flake, #791 Layer II-6 Level 4 with Nahr Ibrahim technique, #391 Layer II-6 Level 4 with retouch, #922 Layer II-5/6 on biface sharpening flake)

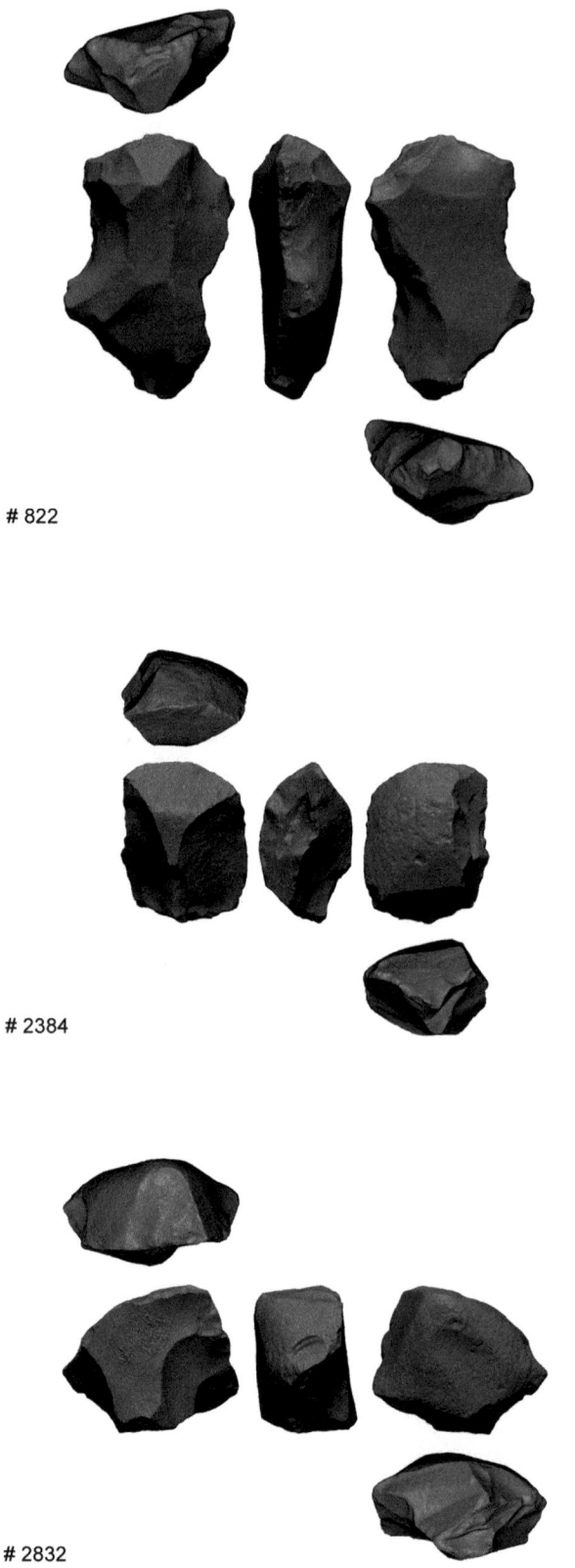

Fig. 6.67 Flint cores on flakes (#822 Layer II-6 Level 4 notch on both edges, #2384 Layer II-6 Level 4, #2832 Layer II-6 Level 4)

Fig. 6.68 Flint core on flake (#734 Layer II-6 Level 4)

Table 6.60 Counts and frequencies (%) of characteristics of flint cores on flakes of Layer II-6

	N	%
Preservation (N=76)		
Fresh	13	17.11
Slightly abraded	49	64.47
Abraded	13	17.11
Heavily abraded	1	1.32
Patination (N=76)		
Patinated	49	64.47
Double patinated	27	35.53
Breakage (N=76)		
Complete	62	81.58
Distal	2	2.63
Lateral	6	7.89
Proximal	4	5.26
Distal & lateral	1	1.32
Indeterminate	1	1.32
Cortex (N=76)		
No cortex	53	69.74
1–25%	12	15.79
26–50%	7	9.21
51–75%	1	1.32
76–100%	3	3.95

For the following analyses we have grouped together all varia cores and formless cores of Layer II-6. Table 6.65 presents the characteristics of these two core categories. There are general similarities between the two groups, although there are somewhat higher values of abraded artifacts among formless cores. Both core groups exhibit high values of patination and similar values of double patinated artifacts. Complete cores are the most common in both core categories and the breakage patterns are similar. In both categories the frequencies of non-cortical cores are similar and the majority of cores exhibit some cortical coverage. The degree of cortical

6.4 Classification and Technology of Cores and Core Tools

Table 6.61 Descriptive statistics of flint cores on flakes of Layer II-6 (size in mm; angle in degrees)

	Maximum length	Length	Width	Thickness	Circumference	Working edge length
N	75	76	76	76	71	44
Maximum	55.00	55.00	50.00	25.00	157.00	151.00
Median	32.00	30.00	23.50	12.00	92.00	68.00
Minimum	22.00	7.00	14.00	7.00	61.00	10.00
Mean	32.40	29.79	25.21	12.86	92.52	62.91
Std dev	7.46	7.43	6.58	3.66	20.12	38.11
Std err mean	0.86	0.85	0.75	0.42	2.39	5.74
	N working edges	N platforms	N scars	Working edge angle	Scar length	Scar width
N	69	69	75	35	29	29
Maximum	4.00	7.00	15.00	92.00	43.00	26.00
Median	1.00	2.00	6.00	78.00	15.00	15.00
Minimum	1.00	1.00	1.00	63.00	9.00	7.00
Mean	1.43	2.49	6.32	77.66	16.59	15.38
Std dev	0.83	1.20	3.07	6.89	7.07	4.63
Std err mean	0.10	0.14	0.35	1.17	1.31	0.86

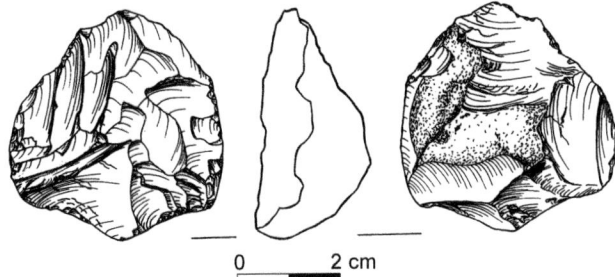

Fig. 6.69 Flint discoidal core (#6613 Layer II-6 Level 1)

Table 6.62 Counts and frequencies (%) of characteristics of globular and discoidal flint cores

	Globular		Discoidal	
N	10		10	
	N	%	N	%
Preservation				
Fresh	-	-	3	30.00
Slightly abraded	7	70.00	6	60.00
Abraded	3	30.00	1	10.00
Patination				
Patinated	7	70.00	7	70.00
Double patinated	3	30.00	3	30.00
Breakage				
Complete	10	100.00	10	100.00
Cortex				
No cortex	3	30.00	5	50.00
1–25%	3	30.00	4	40.00
26–50%	2	20.00	-	-
51–75%	2	20.00	-	-
76–100%	-	-	1	10.00

coverage, however, varies between the two core groups (Table 6.65).

Table 6.66 presents the size measurements of the two core groups, which are on average very similar but show a slightly higher diversity (expressed in the standard deviation) among the formless cores. The average length and width of the two groups are very similar and for both the length is slightly greater than the width. The only observed differences are in the longer circumference of the varia cores and in their slightly longer identifiable scars (Table 6.66). Despite the difference in circumference, in both core groups the ratio between the working edge length and the circumference is nearly identical (0.51/0.52).

In terms of design and exploitation, both core groups display highly similar values of number of working edges, number of platforms, the angle between them, and the number of scars (Table 6.66). The average number of scars is slightly higher for formless cores.

Table 6.67 presents some technological characteristics of the two core groups. It clearly shows that attributes of formless cores are more often indeterminate, emphasizing their unstandardized nature. This is particularly evident in the shape of the cores' section, as well as in the scar pattern of the analyzed artifacts. When identifiable, the transversal section of both core groups is more often flat than triangular. The dominant working edge form in the varia cores is peripheral, followed by convex and straight. In formless cores, the most frequent working edge form is peripheral as well, while wavy and convex occur in identical frequencies. Despite the general dominance of the peripheral form in both core groups, it is higher in varia cores (Table 6.67).

The scar pattern on both core groups is often unidentifiable (Table 6.67). In the case of formless cores, half of the recorded

Table 6.63 Descriptive statistics of globular and discoidal flint cores (size in mm; angle in degrees)

	Maximum length	Length	Width	Thickness	Circumference	Working edge length
Globular cores						
N	10	10	10	10	10	5
Maximum	45.00	42.00	39.00	26.00	134.00	99.00
Median	32.50	30.00	27.50	22.50	97.00	35.00
Minimum	30.00	26.00	24.00	19.00	87.00	17.00
Mean	34.20	31.10	28.30	22.60	100.70	47.00
Std dev	4.37	4.93	4.35	1.96	13.57	33.34
Std err mean	1.38	1.56	1.37	0.62	4.29	14.91
Discoidal cores						
N	10	10	10	10	10	9
Maximum	44.00	41.00	42.00	22.00	138.00	138.00
Median	32.50	26.50	26.50	16.50	92.00	78.00
Minimum	25.00	19.00	18.00	6.00	64.00	36.00
Mean	33.00	28.40	27.80	15.60	93.00	79.11
Std dev	6.45	6.40	7.39	4.74	19.89	28.14
Std err mean	2.04	2.02	2.34	1.50	6.29	9.38
	N working edges	N platforms	N scars	Working edge angle	Scar length	Scar width
Globular cores						
N	9	9	10	4	2	2
Maximum	4.00	7.00	20.00	85.00	18.00	19.00
Median	2.00	3.00	12.00	81.00	18.00	15.50
Minimum	1.00	2.00	3.00	73.00	18.00	12.00
Mean	2.33	3.56	11.20	80.00	18.00	15.50
Std dev	1.22	1.74	6.03	5.10	0.00	4.95
Std err mean	0.41	0.58	1.91	2.55	0.00	3.50
Discoidal cores						
N	10	10	10	6	1	1
Maximum	2.00	4.00	16.00	88.00	11.00	9.00
Median	1.00	2.00	9.00	83.00	11.00	9.00
Minimum	1.00	1.00	4.00	76.00	11.00	9.00
Mean	1.10	2.10	9.40	82.17	11	9
Std dev	0.32	0.74	4.27	4.49	-	-
Std err mean	0.10	0.23	1.35	1.83	-	-

Table 6.64 Counts and frequencies (%) of technological characteristics of globular and discoidal flint cores

	Globular		Discoidal	
	N	%	N	%
Form of section				
Triangular	2	25.00	4	50.00
Flat	-	-	2	25.00
Indeterminate	6	75.00	2	25.00
Total	8	100.00	8	100.00
Working edge form				
Straight	3	33.33	-	-
Convex	1	11.11	1	11.11
Wavy	2	22.22	-	-
Indeterminate	2	22.22	-	-
Peripheral	1	11.11	8	88.89
Total	9	100.00	9	100.00
Scar pattern				
Simple dominant	-	-	1	11.11
Radial	2	33.33	8	88.89
Along axis & opposed	1	16.67	-	-
Indeterminate	3	50.00	-	-
Total	6	100.00	9	100.00

attributes are indeterminate, a much higher value than that of the varia cores. As a result, we are left with a very small sample of formless cores when attempting to describe the scar patterns of this core group. Nevertheless, the dominant pattern within the formless cores is the parallel pattern, while for varia cores it is that of along axis (Table 6.67). In both groups the patterns of exploitation of the debitage surfaces show similar frequencies: unipolar patterns are dominant, followed by radial patterns and bipolar patterns. The radial pattern occurs in very similar frequencies. However, there are marked differences in the frequencies of unipolar vs. bipolar patterns between the two groups. The latter are observed on some 15% of the varia cores and only 7% of the formless cores.

6.4.2 Diverse Waste Artifacts

Angular Fragments

Within the large category of CCT, angular fragments are the most common waste artifacts and they occur throughout the cultural sequence in relatively high frequencies (Table 6.55). These artifacts, which are considered to be fragments of core waste, lack any consistent morphology and only rarely exhibit the remains of working edges or platforms. Their angularity is due to extensive flaking with no clearly defined orientation of the scars. These characteristics are the outcome of the long life-histories of these artifacts.

Most of the angular fragments are slightly abraded, while abraded artifacts occur in lower frequencies (Table 6.68). They are generally patinated, with presence of double patina on close to 30% of the artifacts (Table 6.68) showing that flaking took place on already patinated surfaces. Thus, cortex rarely occurs (artifacts with no cortical coverage comprise some 50% of the angular fragments) and when it is present the cortical coverage is minimal (Table 6.68). As far as the breakage pattern is concerned (Table 6.68), angular fragments are fragmented by definition and are characterized by high frequencies of items that are defined as "fragmented" (to the extent that all edges are broken). "Complete" angular fragments are artifacts with undamaged edges that display the original morphology of the waste artifact. Their breakage pattern is markedly different from that of the modified artifacts (discussed below).

Table 6.69 presents the descriptive statistics of angular fragments. It shows that, despite their extremely small dimensions, these waste artifacts exhibit a relatively high number of scar removals.

Modified Artifacts

The term "modified" (Fig. 6.43) refers to non-flake artifacts on which there is no evidence for preparation of a striking platform and there are no more than three flake scars. These artifacts are present throughout the cultural sequence and are considered waste. They most likely represent artifacts used as percussors or tested nodules.

Modified artifacts are more abraded than other lithic categories of FFT and of CCT (Table 6.70).

Like the entire flint component, modified artifacts are mostly patinated. The frequency of double patina is generally high, higher than that observed on unretouched flakes. This reflects the fact that the flake scars (mean=2.19; Table 6.71) were removed from patinated surfaces of the original nodule/chunk. These small nodules often exhibit extensive cortical coverage and are mostly complete (Table 6.70).

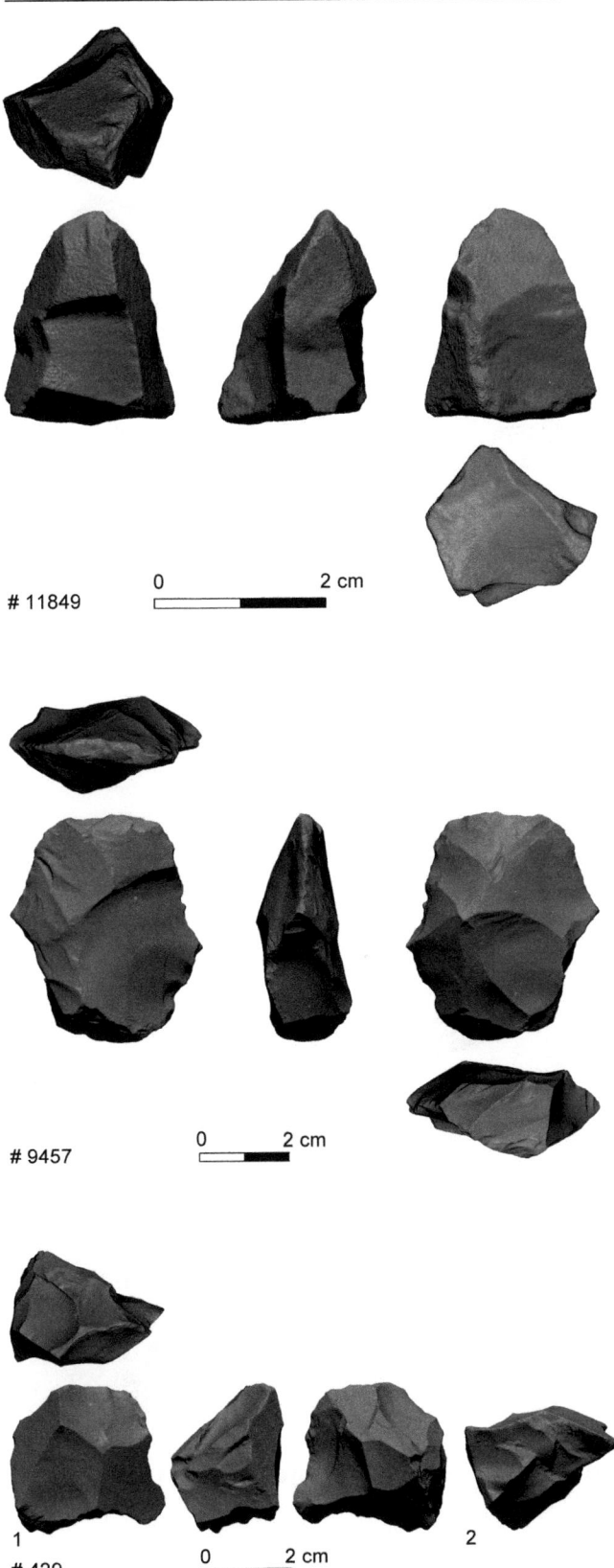

Fig. 6.70 Flint cores (#11849 Layer II-6 Level 3 prismatic core, #9457 Layer II-6 Level 7 pyramidal core, #429 Layer II-6 Level 4 core on broken biface)

Fig. 6.71 Flint varia cores (#5142 Layer V-5, #10570 Layer II-6 Level 1, #11183 Layer II-6 Level 1, #10752 Layer II-6 Level 1)

6.4 Classification and Technology of Cores and Core Tools

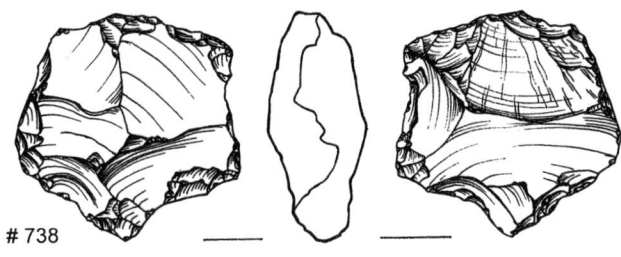

The "modified" artifacts are small, rounded (with similar length and width dimensions), and flat. The size of these artifacts is highly variable, demonstrated by the values of the standard deviations (Table 6.71).

There is extensive variability in the occurrence of modified artifacts in different levels of the Layer II-6 cultural sequence (Table 6.55). This variability is partially related to the size of the samples, but may well be a result of other (technological or functional) aspects as well, e.g. availability of flint or particular activities.

Core Waste and Amorphous Waste

The category of core waste includes artifacts that bear a core signature expressed by residues of working edges and a discernible scar pattern that identifies their origin in cores. In contrast, the group of amorphous waste assembles artifacts that exhibit no ventral faces, many scar removals, and no identifiable scar orientation. They do not conform to any of the core and waste categories (CCT types) and most likely represent waste products originating from the final stages of unknown reduction sequences. As both categories of waste artifacts occur in very low frequencies, we present their entire samples regardless of their stratigraphic assignment.

Both categories of waste are frequently complete and patinated. They are both more often slightly abraded and abraded, though fresh artifacts occur more frequently within core waste. Both categories of waste display varying degrees of cortical coverage, although core waste artifacts are generally less cortical (Table 6.72) and hence exhibit a higher mean number of scars (Table 6.73).

Core waste and amorphous waste products share similar values in their different size attributes, and are also very similar in size to angular fragments. They are all generally small in comparison to cores (Table 6.73).

Percussors

Flint percussors are minimally represented in comparison to those of basalt and limestone (Chapters 7 and 8). Only five flint percussors were identified in the entire GBY cultural sequence. These percussors are particularly small nodules (Table 6.74) and differ significantly from those of other raw materials in their dimensions. All five percussors exhibit signs of battering and pounding on their surfaces, and two bear scars (two and four scars).

The flint percussors are patinated, with one bearing double patina. They are all abraded and, apart from a single broken one, complete.

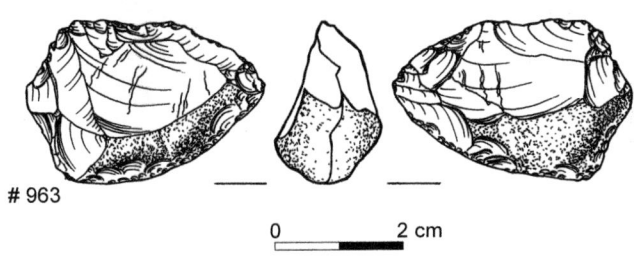

Fig. 6.72 Flint varia cores (#738 Layer II-5, #915 Layer II-6 Level 1, #557 Layer II-6 Level 1, #963 Layer II-6 Level 1)

Fig. 6.73 Flint varia cores (#2382 Layer II-6 Level 1, #8348 Layer II-6 Level 3, #15596 Layer II-6 Level 3, #860 Layer II-6 Level 4)

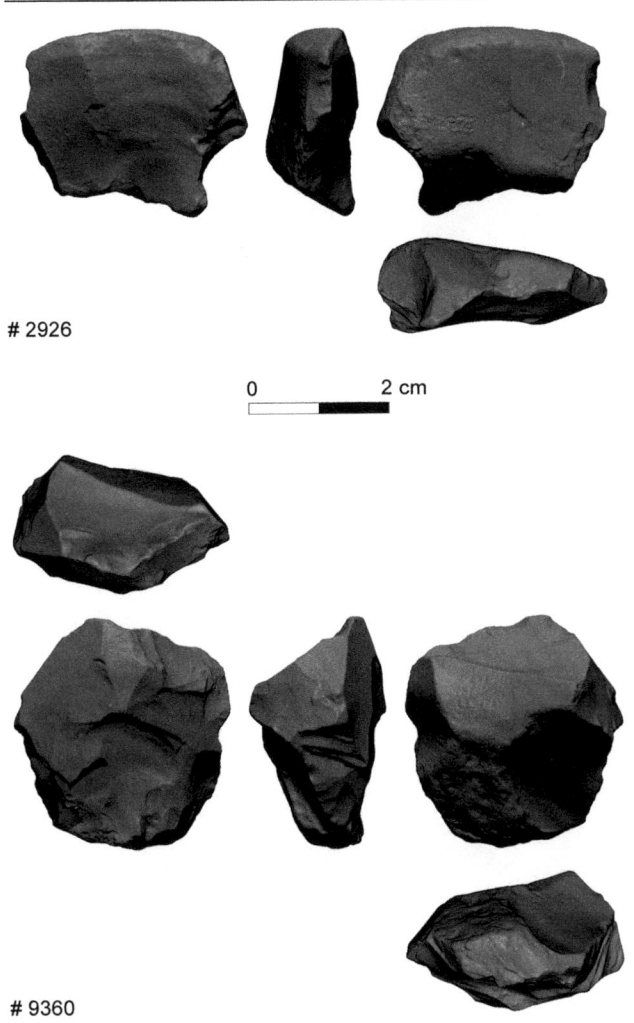

Fig. 6.74 Flint varia cores (#2926 Layer II-6 Level 4, #9360 Layer II-6 Level 7)

6.5 Discussion and Summary of the Flint Component

The analyses of the flint component provided detailed information on both the typology and the technology of flint flakes and flake tools as well as cores and core tools. In addition, they demonstrated that there are no significant differences between different archaeological layers. Thus, the homogeneity discernible throughout the cultural sequence allows us to analyze the GBY flint assemblage as a whole for the purposes of this discussion.

The following summary attempts to reconstruct varied aspects of the *chaînes opératoires* manifested in two major reduction sequences, those of flint bifaces and flint flake tools. Although flint bifaces are only a small component of the flint assemblages and are discussed elsewhere (Chapter 7), various by-products of this reduction sequence are discussed here as part of the overall aim of describing flint flakes and flake tools. Nevertheless, the reduction sequence of flint flakes and flake tools seems to have been the principal purpose of flint knapping at GBY.

6.5.1 The Flint Reduction Sequences

Fig. 6.77 presents the proposed scheme for the reduction sequence of flint at GBY. It is evident that two distinct size categories were involved in the selection of flint raw material. The dominant mode is the use of very small flint nodules in the production of cores and core tools, and particularly that of flakes and flake tools. Rarely, larger flint nodules were obtained and used solely for the production of handaxes. The fact that only handaxes were modified on larger flint nodules suggests that, although it was feasible to obtain such nodules, their acquisition required special investment. Most of the flint nodules used for the production of flake tools were very small. We do not suggest that these nodules were purposely selected for their small size, but rather that the small size of these nodules was not an important factor, since they were suitable for the production of the blanks desired for flake tools.

The selection of these nodules reflects the availability of this particular size range in the vicinity of the lake margin, either on the terraces of rivers flowing from the west and northwest into the Hula Valley (Chapter 4) or in exposures further away.

Signs of Use on CCT

An interesting feature emerging from the analyses of the cores and core waste is the fact that a substantial portion of them exhibits signs of use or retouch (Table 6.75). Although these artifacts were not classified as core tools but rather as cores and core waste, these signs were recorded. The highest frequencies are those of signs of use, followed by notches and denticulates, while other types of retouch occur in much lower frequencies (Table 6.75). Table 6.75 provides the data on the combined frequencies of signs of use and retouch for each of the core and core waste categories. These data show that the highest frequency occurs within the categories of cores on flakes and varia cores. This suggests that cores and core waste were manipulated beyond their technological role in the production of flakes. Furthermore, the high frequencies of signs of use and different types of retouch on these artifacts shows that, regardless of their technological determination or morphology, they were considered suitable blanks for use as tools.

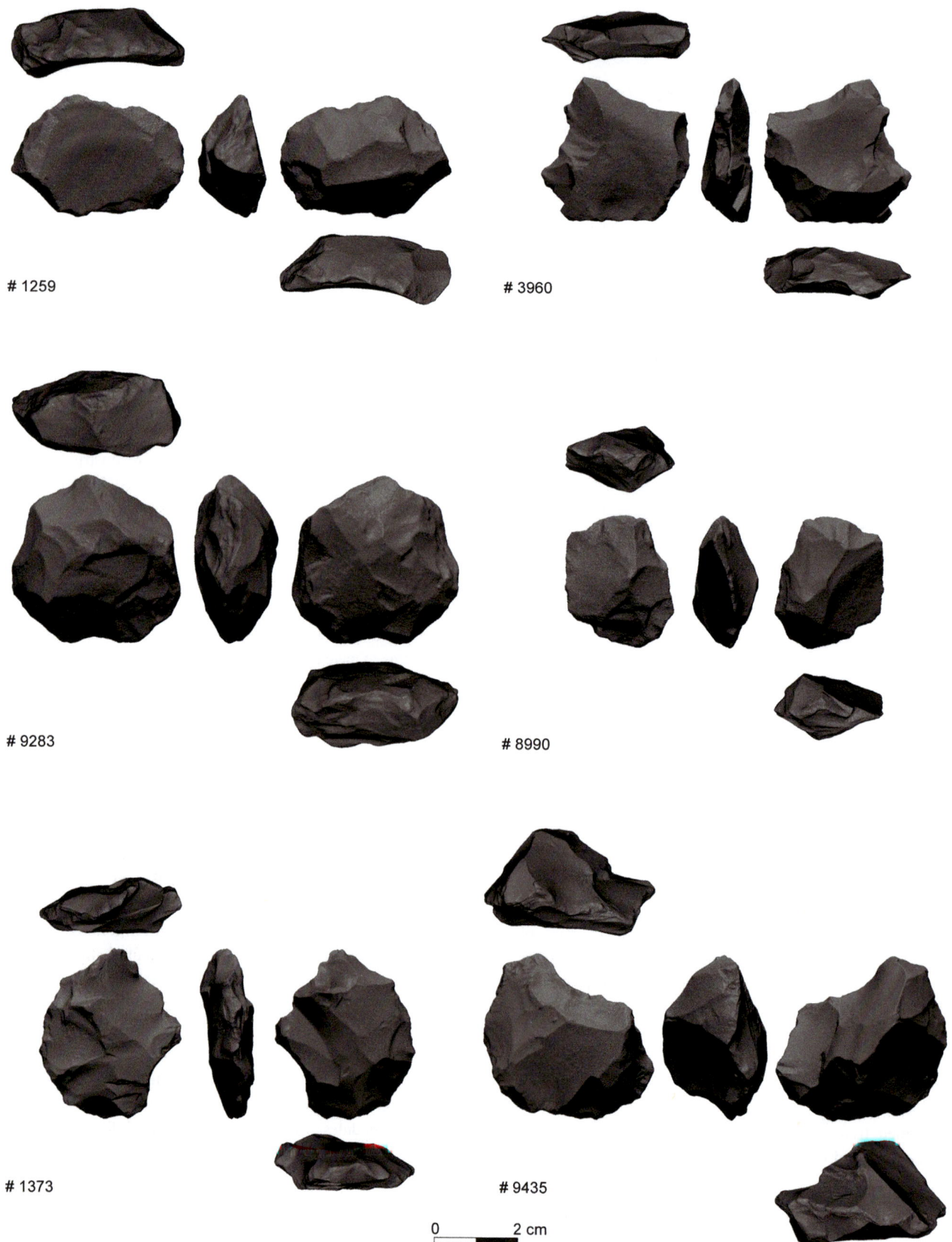

Fig. 6.75 Flint varia cores: with a single convex side-scraper (#1259 Layer II-6 Level 4), cores on broken bifaces (#9283 Layer II-6 Level 7, #1373 Layer II-6 Level 4), cores on biface sharpening flakes (#3960 JB, #8990 Layer II-6 Level 6, #9435 Layer II-6 Level 7)

6.5 Discussion and Summary of the Flint Component

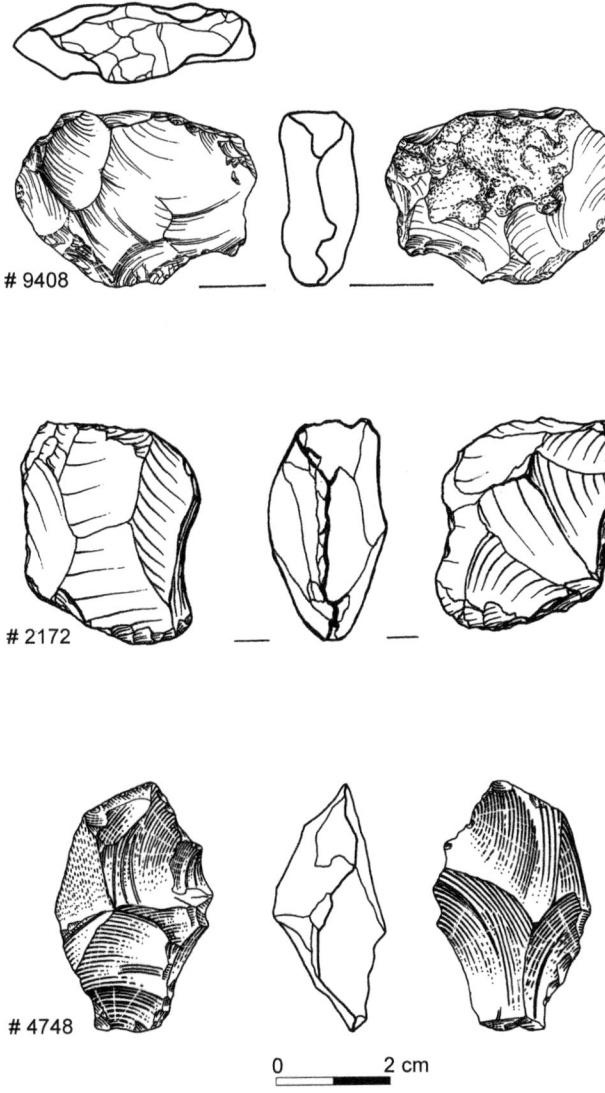

Fig. 6.76 Flint cores: varia cores (#9408 Layer II-6 Level 3, #2172 Layer II-6 Level 4), formless core (#4748 Layer V-5)

Table 6.65 Counts and frequencies (%) of characteristics of varia and formless flint cores of Layer II-6

	Varia		Formless	
	N	%	N	%
Preservation				
Fresh	28	13.27	13	13.00
Slightly abraded	139	65.88	58	58.00
Abraded	42	19.91	27	27.00
Heavily abraded	2	0.95	2	2.00
Total	211	100.00	100	100.00
Patination				
No patina	2	0.95	-	-
Patinated	150	71.43	70	70.00
Double patinated	58	27.62	30	30.00
Total	210	100.00	100	100.00
Breakage				
Complete	184	87.62	85	85.86
Distal	6	2.86	3	3.03
Lateral	14	6.67	4	4.04
Proximal	2	0.95	4	4.04
Fragment	2	0.95	3	3.03
Indeterminate	2	0.95	-	-
Total	210	100.00	99	100.00
Cortex				
No cortex	80	37.91	35	35.35
1–25%	41	19.43	24	24.24
26–50%	41	19.43	20	20.20
51–75%	30	14.22	9	9.09
76–100%	19	9.00	11	11.11
Total	211	100.00	99	100.00

The small size of the available nodules is probably the result of fluvial sorting processes. Any procurement mode would have necessitated transporting the flint from its depositional environment to the lake margin.

The small size of the flint nodules collected may also explain the paucity of bifaces made on flint. Flint bifaces (36 handaxes) occur in some of the layers and levels (Table 6.1) and are generally smaller than the basalt bifaces. The few flint handaxes found display a high degree of craftsmanship and sometimes retain partial cortical cover. This indicates that the approach to their reduction differed significantly from that used in exploitation of basalt for the production of bifaces. In this particular case, one can see the logic in the selection of the larger and more available type of raw material. We assume that if larger flint nodules were more readily available in the form of conglomerates with large clasts (resulting from changes in drainage patterns, shifts in fluvial velocity, etc.), the GBY hominins would most likely have used them.

The flint assemblages of GBY comprise all the components of the reduction sequence – small nodules, different types of cores and core management pieces, unretouched flakes, flake tools, and core tools, as well as a vast amount of microartifacts (Table 6.1). The presence of all these components is seen through the whole cultural sequence and furnishes evidence that flint knapping was carried out on the lake margin and encompassed the entire reduction sequence.

6.5.1.1 Percussor Techniques

Different characteristics of the flint assemblage suggest that FFT and CCT were in general produced with hard percussors. An exception to this is the reduction sequence of the bifacial tools, which involved the use of soft percussors.

Table 6.66 Descriptive statistics of varia and formless flint cores of Layer II-6 (size in mm; angle in degrees)

	Maximum length	Length	Width	Thickness	Circumference	Working edge length
Varia cores						
N	208	208	208	208	201	138
Maximum	65.00	57.00	48.00	32.00	187.00	131.00
Median	32.00	29.00	25.00	16.00	93.00	37.00
Minimum	20.00	8.00	14.00	6.00	60.00	13.00
Mean	32.77	29.30	26.33	16.31	95.00	49.71
Std dev	6.84	7.08	6.36	4.82	20.43	30.90
Std err mean	0.47	0.49	0.44	0.33	1.44	2.63
Formless cores						
N	99	99	99	99	86	45
Maximum	80.00	80.00	76.00	32.00	153.00	114.00
Median	30.00	27.00	22.00	15.00	85.00	40.00
Minimum	21.00	14.00	10.00	6.00	59.00	11.00
Mean	31.33	27.98	23.58	15.74	88.09	45.24
Std dev	8.91	9.04	7.83	5.36	20.84	26.02
Std err mean	0.90	0.91	0.79	0.54	2.25	3.88
	N working edges	N platforms	N scars	Working edge angle	Scar length	Scar width
Varia cores						
N	75	75	100	34	27	28
Maximum	4.00	6.00	24.00	99.00	97.00	43.00
Median	1.00	2.00	5.00	81.00	19.00	17.00
Minimum	1.00	1.00	1.00	61.00	8.00	6.00
Mean	1.47	2.35	5.80	80.11	20.68	17.54
Std dev	0.72	0.93	3.31	7.15	10.33	6.23
Std err mean	0.05	0.07	0.23	0.67	1.04	0.64
N	196	196	211	114	98	96
Formless cores						
Maximum	4.00	5.00	21.00	109.00	32.00	37.00
Median	1.00	2.00	6.00	82.00	17.00	16.00
Minimum	1.00	1.00	2.00	65.00	6.00	9.00
Mean	1.49	2.25	6.52	82.09	17.26	16.82
Std dev	0.72	0.86	3.64	8.05	6.02	7.30
Std err mean	0.08	0.10	0.36	1.38	1.16	1.38

An attempt has been made to compare the flint FFT of GBY with experimentally produced biface flakes to test the hypothesis that the former were produced with soft percussors (Sharon and Goren-Inbar 1999). In that study, the following attributes, considered by many to be indicative of soft percussor production, were selected: 1) diffused bulb of percussion, 2) lipped striking platform, 3) facetted and crushed or abraded striking platform, and 4) thinness of flake. The hypothesis was tested on 113 flint flakes from Layers V-5 and V-6 and on 99 flakes from Layer II-6 Level 1. The authors concluded that the former samples from Area C are indicative of biface production and contrasted this conclusion with Layer II-6 Level 1, where only minimal biface flint production took place. The study identified several trends, among which features of striking platform (particularly the lipped type) and proximal end (crushing, steps, angles) were considered to be good indicators for the use of soft percussors that could be useful in future studies. Despite this, the authors concluded that the material at hand did not provide firm proof of the presence of soft percussors in the process of bifacial production. With the completion of the GBY lithic analysis, larger samples are available and intra-site comparison has become possible. Hence, we are now able to present the full extent of the analysis and make another attempt to examine the issue of whether the biface production process employed soft percussors. As described above (Section 6.3.1.1), flakes classified as *édtdb* are significantly thinner than other flakes, suggesting the use of soft percussors for their production (Table 6.47). In addition, significantly higher frequencies of lipped striking platforms were recorded on *édtdb* (Table 6.51).

Apart from the extension of the flint samples, two new lines of evidence are introduced here. The first is the

6.5 Discussion and Summary of the Flint Component

Table 6.67 Counts and frequencies (%) of technological characteristics of varia and formless flint cores of Layer II-6

	Varia		Formless	
	N	%	N	%
Form of section				
Triangular	42	25.93	15	20.83
Flat	63	38.89	19	26.39
Indeterminate	57	35.19	38	52.78
Total	162	100.00	72	100.00
Working edge form				
Straight	30	16.48	7	10.77
Convex	35	19.23	8	12.31
Concave	9	4.95	6	9.23
Wavy	21	11.54	8	12.31
Denticulate	2	1.10	2	3.08
One tooth	4	2.20	3	4.62
Peripheral	44	24.18	10	15.38
Indeterminate	37	20.33	21	32.31
Total	182	100.00	65	100.00
Scar pattern				
Unipolar	-	-	-	-
Simple dominant	12	6.78	3	5.56
Horse shoe	5	2.82	1	1.85
Along axis	29	16.38	5	9.26
Parallel	16	9.04	8	14.81
Convergent	16	9.04	2	3.70
Opposed	1	0.56	-	-
Along axis & opposed	5	2.82	1	1.85
Bipolar	14	7.91	1	1.85
Radial	15	8.47	5	9.26
Along axis & side	13	7.34	1	1.85
Indeterminate	51	28.81	27	50.00
Total	177	100.00	54	100.00

Table 6.68 Counts and frequencies (%) of characteristics of flint angular fragments of Layer II-6

	N	%
Preservation (N=330)		
Fresh	9	2.73
Slightly abraded	215	65.15
Abraded	105	31.82
Heavily abraded	1	0.30
Patination (N=330)		
No patina	2	0.61
Patinated	234	70.91
Double patinated	94	28.48
Breakage (N=329)		
Complete	197	59.88
Distal	1	0.30
Lateral	14	4.26
Proximal	3	0.91
Distal & proximal	1	0.30
Fragment	106	32.22
Indeterminate	7	2.13
Cortex (N=323)		
No cortex	165	51.08
1–25%	72	22.29
26–50%	55	17.03
51–75%	19	5.88
76–100%	12	3.72

discovery of a soft percussor made of antler at the site (Fig. 6.78). This specimen clearly shows a wear pattern that is similar to that obtained by knapping experiments (Goren-Inbar 2011a) and provides direct evidence for the use of soft percussors at the site (see Stout et al. 2014). The second type of supporting evidence for the use of soft percussors originates in the limestone assemblage. We have recently identified a component of limestone percussors that occurs throughout the GBY cultural record (Alperson-Afil and Goren-Inbar 2015). Their presence contributes an additional dimension to the debate, since limestone percussors have been shown experimentally to produce flakes similar to those produced by antler percussors (Pelegrin 2000).

6.5.1.2 Cores

We are able to suggest that the original size of the flint nodules was small *a priori*, on the basis of the cores' small dimensions together with the remains of cortical coverage, which allows reconstruction of the initial size and morphology of the nodule (Fig. 6.79). Fig. 6.79 further illustrates that the life-history of these cores was short and did not include multiple stages of core reduction–core rejuvenation–core reduction. This is reflected

Table 6.69 Descriptive statistics of flint angular fragments of Layer II-6 (size in mm)

	Maximum length	Length	Width	Thickness	Circumference	N scars
N	330	330	330	330	126	327
Maximum	47.00	47.00	38.00	90.00	123.00	19.00
Median	25.00	23.00	18.00	13.00	72.00	7.00
Minimum	20.00	10.00	9.00	7.00	50.00	1.00
Mean	26.54	24.43	18.61	14.28	75.45	6.99
Std dev	4.93	5.13	4.73	5.86	15.40	3.28
Std err mean	0.27	0.28	0.26	0.32	1.37	0.18

Table 6.70 Counts and frequencies (%) of characteristics of flint modified artifacts of Layer II-6

	N	%
Preservation (N=279)		
Fresh	21	7.53
Slightly abraded	142	50.90
Abraded	112	40.14
Heavily abraded	4	1.43
Patination (N=278)		
No patina	3	1.08
Patinated	216	77.70
Double patinated	59	21.22
Breakage (N=278)		
Complete	255	91.73
Distal	3	1.08
Lateral	11	3.96
Distal & proximal	1	0.36
Fragment	7	2.52
Indeterminate	1	0.36
Cortex (N=274)		
No cortex	48	17.52
1–25%	20	7.30
26–50%	40	14.60
51–75%	53	19.34
76–100%	105	38.32
Indeterminate	8	2.92

Table 6.72 Counts and frequencies (%) of characteristics of flint core waste and amorphous waste

	Core waste		Amorphous waste	
	N	%	N	%
Preservation				
Fresh	12	25.53	3	12.50
Slightly abraded	28	59.57	11	45.83
Abraded	7	14.89	10	41.67
Total	47	100.00	24	100.00
Patination				
No patina	1	2.13	-	-
Patinated	37	78.72	19	82.61
Double patinated	9	19.15	4	17.39
Total	47	100.00	23	100.00
Breakage				
Complete	33	71.74	20	86.96
Broken	13	28.26	3	13.04
Total	46	100.00	23	100.00
Cortex				
No cortex	32	69.57	9	40.91
1–25%	11	23.91	4	18.18
26–50%	2	4.35	6	27.27
51–75%	1	2.17	2	9.09
76–100%	-	-	1	4.55
Total	46	100.00	22	100.00

not only in the consistently small size of the cores but also in the minimal representation of core management pieces. The latter are unstandardized and demonstrate the fact that there was no systematic management of the extremities of the flint cores (Section 6.3.2). In addition, most of the core management pieces, particularly *débordant* flakes and biface sharpening flakes, are in fact by-products of handaxe modification and are not the result of core maintenance (Section 6.3.2).

In addition to cores modified on nodules (particularly formless cores and varia cores), cores were modified on two other types of blanks: biface rejects and flakes that were modified into cores.

Cores on flakes form a substantial component within the flint cores of GBY (Section 6.4.1). This is a sub-set of artifacts that were selected from the flake assemblage, apparently since they had somewhat larger dimensions. As cores on flakes belong to a more advanced stage in the reduction sequence than other flint cores, one would expect their dimensions (Goren-Inbar 1988; Hovers 2007) to be smaller than those of other cores. Nevertheless, they exhibit similar dimensions. Among the products of these cores are the dorsally plain flakes (Section 6.3.1.2).

As illustrated in Fig. 6.77, the reduction sequence of bifaces inadvertently produced (as *accidents de travail*) broken

Table 6.71 Descriptive statistics of flint modified artifacts of Layer II-6 (size in mm)

	Maximum length	Length	Width	Thickness	Circumference	N scars
Maximum	63.00	61.00	52.00	30.00	167.00	3.00
Median	31.00	28.00	23.00	16.00	88.00	2.00
Minimum	20.00	11.00	12.00	5.00	60.00	1.00
Mean	32.63	29.70	23.69	16.10	92.47	2.19
Std dev	8.26	8.40	6.39	4.94	22.43	0.76
Std err mean	0.49	0.50	0.38	0.30	1.68	0.05
N	279	279	279	279	178	273

6.5 Discussion and Summary of the Flint Component

Table 6.73 Descriptive statistics of flint core waste and amorphous waste (size in mm)

	Maximum length	Length	Width	Thickness	Circumference	N scars
Core waste						
N	47.00	47.00	47.00	47.00	18.00	46.00
Maximum	46.00	44.00	33.00	27.00	120.00	18.00
Median	28.00	24.00	19.00	11.00	76.00	6.00
Minimum	21.00	16.00	13.00	6.00	61.00	2.00
Mean	28.45	26.06	19.57	12.70	77.28	6.52
Std dev	6.12	6.88	4.01	4.33	15.11	3.15
Std err mean	0.89	1.00	0.59	0.63	3.56	0.46
Amorphous waste						
N	24.00	24.00	24.00	24.00	10.00	21.00
Maximum	48.00	48.00	26.00	19.00	83.00	13.00
Median	26.50	24.50	19.00	12.00	70.50	6.00
Minimum	20.00	12.00	11.00	6.00	56.00	2.00
Mean	27.04	25.13	19.21	12.63	69.90	5.67
Std dev	5.96	6.66	3.68	3.98	10.74	2.71
Std err mean	1.22	1.36	0.75	0.81	3.40	0.59

Table 6.74 Descriptive statistics of flint percussors (size in mm)

	Maximum length	Length	Width	Thickness	Circumference
N	5	5	5	5	5
Maximum	59.00	59.00	48.00	42.00	153.00
Median	48.00	46.00	34.00	27.00	135.00
Minimum	36.00	33.00	26.00	17.00	116.00
Mean	48.40	47.40	34.60	27.80	137.40
Std dev	8.56	9.71	8.47	9.04	15.18
Std err mean	3.83	4.34	3.79	4.04	6.79

Table 6.75 Counts and frequencies (%) of retouch/signs of use on flint CCT

Type	Modified		Angular fragments		Formless cores		Varia cores		Cores on flakes		Levallois cores		Discoidal cores		Globular cores	
N	50		60		23		73		31		11		3		2	
Type of retouch	N	%	N	%	N	%	N	%	N	%	N	%	N	%	N	%
Signs of use	25	50.00	51	85.00	13	56.52	45	61.64	11	35.48	5	45.45	2	66.67	2	100.00
Regular	-	-	2	3.33	1	4.35	6	8.22	3	9.68	2	18.18	-	-	-	-
End-scraper	4	8.00	-	-	-	-	2	2.74	1	3.23	-	-	1	33.33	-	-
Side-scraper	4	8.00	-	-	1	4.35	4	5.48	5	16.13	-	-	-	-	-	-
Notch & denticulate	12	24.00	4	6.67	4	17.39	12	16.44	7	22.58	4	36.36	-	-	-	-
Indeterminate	1	2.00	-	-	2	8.70	2	2.74	2	6.45	-	-	-	-	-	-
Irregular	1	2.00	1	1.67	1	4.35	1	1.37	1	3.23	-	-	-	-	-	-
Abrupt	2	4.00	1	1.67	-	-	1	1.37	-	-	-	-	-	-	-	-
Semi-abrupt	1	2.00	1	1.67	-	-	-	-	1	3.23	-	-	-	-	-	-
Bifacial	-	-	-	-	1	4.35	-	-	-	-	-	-	-	-	-	-
% of category	17.92% (50 of 279)		18.18% (60 of 330)		23% (23 of 100)		34.59% (73 of 211)		40.78% (31 of 76)		28.20% (11 of 39)		30% (3 of 10)		20% (2 of 10)	

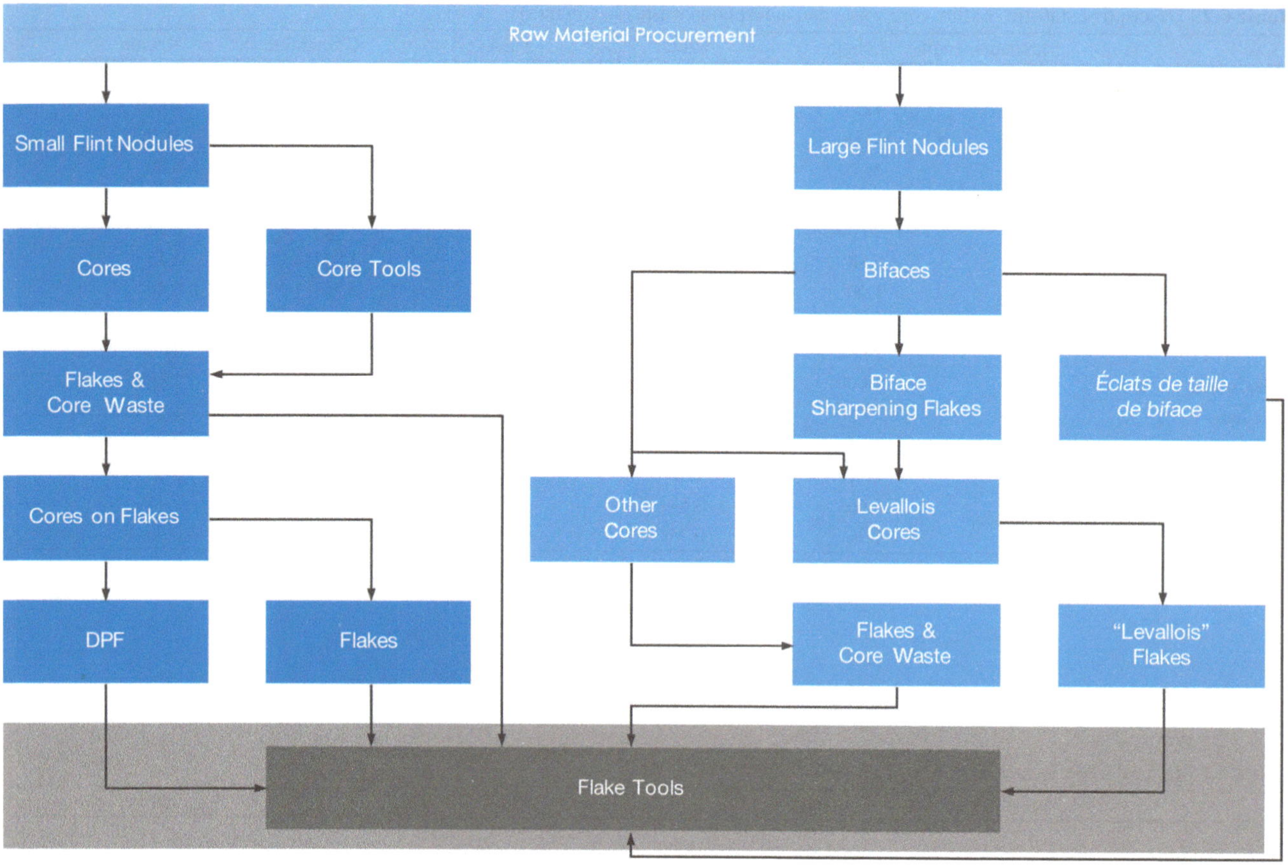

Fig. 6.77 Proposed scheme for the flint reduction sequence

Fig. 6.78 A damaged cervid (*Dama* sp.) antler from GBY (left) and an experimental antler percussor that was used to replicate basalt bifacial tools (right)

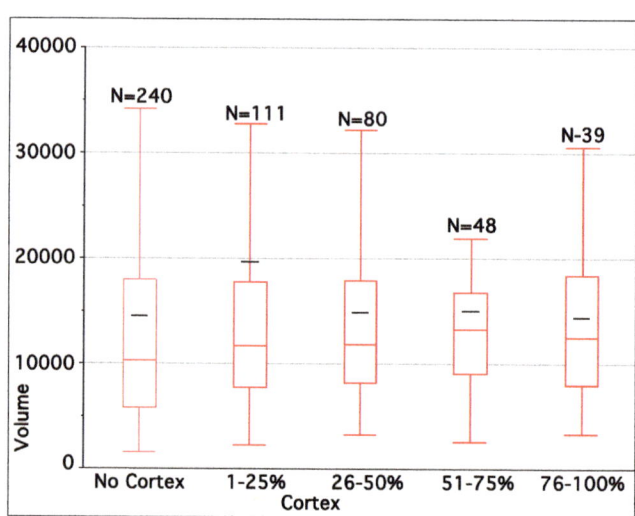

Fig. 6.79 Volume and cortex residue of the entire flint core assemblage (N=524) (volume in mm³=length x width x thickness)

handaxes and other products (e.g. biface sharpening flakes) whose morphologies enabled their use as cores. Within this group of cores we have identified a component assigned to the Levallois method (e.g. Fig. 6.80).

6.5.1.3 The Levallois Method

A small portion of the GBY flint assemblage is characterized by cores that have a single working edge and two platforms

6.5 Discussion and Summary of the Flint Component

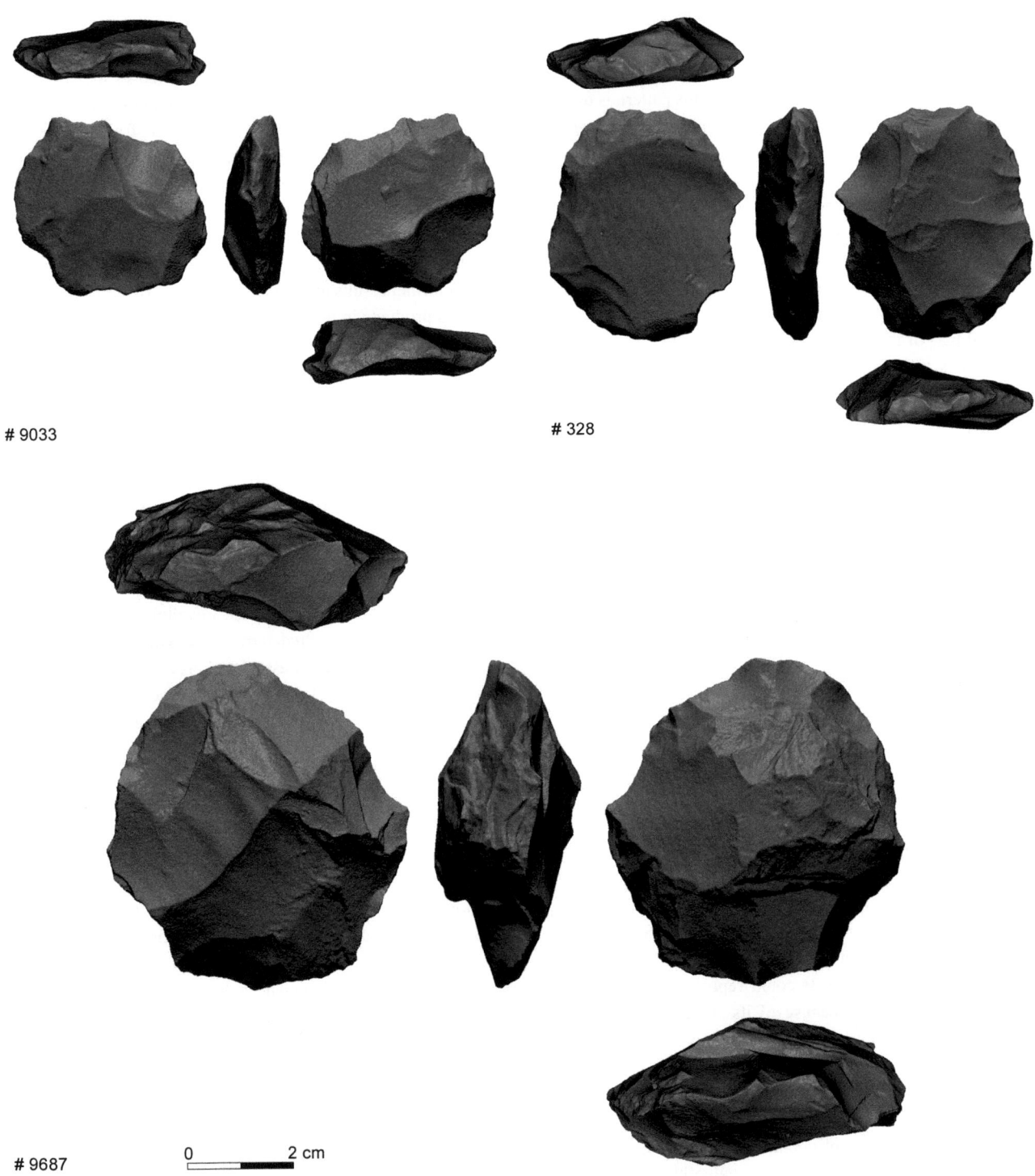

Fig. 6.80 Flint Levallois cores: on biface sharpening flakes (#9033 Layer II-6 Level 6, #328 Layer II-5/6) and on a biface (#9687 Layer V-6)

that are worked around the entire circumference or a substantial part of it. The platforms are usually hierarchical, resulting in one face of the core being more dominant or more extensively flaked than the other, considered here as a debitage surface. Thus, these cores display some similarities to discoidal cores or discoids but are generally flatter, possibly representing final stages of production and hence exhaustion.

Some of these cores are modified on handaxes, perhaps broken ones (Fig. 6.80). Other cores are modified on by-products of biface reduction, such as biface sharpening flakes (Fig. 6.80). This is evident from their edges, which are straighter (at least in part of the circumference) than those of the regular cores knapped with hard percussors. The biface phase of these cores' life-history is also seen in the typical bifacial scar pattern that is visible on the surface of the cores. This pattern is usually characterized by many small scars, frequently smaller than those common on cores, and are very flat, typical of the soft percussor technique. This clearly contrasts with other flake removals representing an earlier phase of the object's life-history, which were clearly achieved with a hard percussor. The sizes of these Levallois cores do not deviate from those of other core types (Tables 6.58, 6.61, 6.63, 6.66).

The typical unintentional breakage of handaxes occurs in the distal part of the artifacts, where it is thinner. Hence, some of the cores that are made on these blanks are already thin in their initial production phase. As a result, these cores are exceptionally thin and resemble discs of the Middle Paleolithic industries (Goren-Inbar 1990).

Flakes resulting from accidental breakage of bifaces bear on their edges a typical scar pattern (particularly of the last phases of thinning) characterized by intensive miniature scars, sometimes resulting from straightening the edge (rubbing) before inflicting a well-aimed blow. These miniature facets were exploited during the transformation of the broken biface into a core and were used as striking platforms for the Levallois core.

Exceptions to the above pattern are two cores that are made on larger chunks than those described above and seem to be associated with blanks made during earlier phases of biface modification. These are larger and have a smaller number of flake removals that were possibly achieved with a hard percussor. Fig. 6.61 illustrates one of these examples, a core of pyramidal shape originating from Layer V-6 in Area C, an archaeological horizon that yielded abundant *édtdb*, thousands of artifacts, and a single handaxe made of flint (Chapter 7; Sharon and Goren-Inbar 1999; Goren-Inbar and Sharon 2006). The second object of interest is another core (Fig. 6.62) that is larger than the average. It could represent an initial stage of handaxe production because of its straight edge, but was classified as a Levallois core due to its final phase of production.

As presented above, we classify some of the cores as Levallois cores (Tables 6.56–6.59). However, examination of the entire flake and flake tool inventory resulted in the conclusion that it does not contain typical Levallois products similar to those defined for Middle Paleolithic assemblages. Nevertheless, we do observe some features that could be assigned to the Levallois flaking method. These include radial scar patterning, facetted striking platforms, and characteristic waste products that are usually associated with the renewal of Levallois core convexity and include *débordant* and *outrepassé* flakes (Table 6.54). However, these products could result from biface production and modification processes, particularly the facetted striking platforms, radial scar patterns, and thin blanks, as well as the occurrence of dorsal face convexity. These traits are frequently observed on *édtdb* (Section 6.3.1.1) and on biface sharpening flakes.

While we have classified some of the cores as Levallois, we are aware of some particular differences that separate the Acheulian Levallois flint cores of GBY from the classic Levantine Middle Paleolithic Levallois. The main difference involves the extraction of the individual flakes. In most cases, the scars on the GBY Levallois cores do not extend over more than half of the length of the debitage surface. This results in a pattern of many flake scars in comparison to classic Levallois cores. It further results in production of flakes that are not typical Levallois products in that their dorsal faces are not convex and they are thicker and smaller in size than the typical Middle Paleolithic Levallois cores.

On the cores, the scar pattern is almost exclusively that of flakes and rarely that of points or blades. In addition, it is mostly radial; none of the planning that allows predetermination of the desired blank form (point/blade/flake) through preparation of the debitage surface is evident, and this holds true for both the recurrent and the lineal mode. No characteristic Levallois target flakes (predetermined flakes) were obtained from the cores at the site.

From the above it is evident that Levallois in the true sense does not exist at GBY. Predetermined Levallois flakes/blades/points are absent, as are the associated products (*débordant* and *outrepassé* flakes) that in Levallois assemblages are the means by which the convexities of the lateral and distal proximities of the cores are managed. Some flakes at GBY are indeed artifacts that fit the definitions, but they are a random result of knapping and not an intentional one (for a similar case of Acheulian assemblages with very low frequencies of these artifacts, see Malinsky-Buller et al. 2011). Our conclusion from the above is that all of the characteristics identified on the GBY cores that resemble Levallois cores (hierarchical surfaces, continuous working edge, exploitation of blanks with previous multiple bifacial scars, as well as the facetted edges of the latter that serve as striking platforms) illustrate an incipient understanding of the Levallois method. This adds to the evidence of the Levallois concept that has been identified on the giant basalt cores (Chapter 7; Goren-Inbar 2011; Goren-Inbar et al. 2011).

6.5.1.4 Tools

The GBY flint core tools exhibit comparable morphologies and sizes, suggesting that they similarly originate in the small nodules used for core knapping and flake production. Core tools include a variety of types that conform to conventional morpho-technological criteria. However, their meager representation and particularly their small dimensions indicate that they do not originate from an intentional and separate reduction sequence, but rather from ad hoc exploitation of an available resource.

6.5 Discussion and Summary of the Flint Component

This interpretation provides an explanation of their very low frequencies and for their typological composition, which resembles that of flake tools (Table 6.36).

Flake Tools

The desired products of the flint reduction sequence were flake tools. This is supported by the fact that flake tools occur in high frequencies in each of the layers and levels (Table 6.76). Generally, the same typological breakdown characterizes all the assemblages, with a dominance of retouched flakes, notches, borers, and side-scrapers. An additional typological feature is flakes with removed striking platforms, which are abundant throughout the GBY cultural sequence (Section 6.2.1.1, Table 6.76).

Analyses of the different flake tool categories and their comparison with the unretouched flakes did not reveal any consistent selection of particular blanks in most cases. However, retouched flint tools are often significantly longer than unretouched flint flakes (Appendix 7). Retouch reduces, to some extent, the dimensions of the tools, suggesting that their original size was even larger (Tables 6.32–6.34). The main difference is expressed by significantly higher frequencies of radial scar pattern on flake tools, while unretouched flakes mainly bear an along axis scar pattern (Section 6.2.1.1; Table 6.33). Examination of these two scar patterns in the entire sample of flint FFT (retouched and unretouched) shows higher frequencies of facetted striking platforms on radial flakes than on along axis flakes, and higher numbers of dorsal scars on radial flakes (Table 6.77). From these data, we conclude that if indeed there was intentional selection of blanks for tool production, it involved radial flakes with higher numbers of scars.

Removal of Striking Platform

We have demonstrated that the most frequent typological feature of flint flake tools is the removal of striking platforms. In addition, we have established a preference for debitage of the radial scar pattern for the modification of flake tools, as they exhibit higher numbers of dorsal scars. In an attempt to understand this triple association (removal of striking platform, high number of scars, and radial scar pattern), we compared FFT with striking platforms to those in which the striking platforms were removed (regardless of their typological classification). The results of these analyses demonstrate that:

Table 6.76 Counts and frequencies (%) of different typological groups and of artifacts with removed striking platforms (calculated out of the entire FFT component)

	Layer									
			II-6 Levels							
	V-5	V-6	L1	L2	L3	L4	L4b	L5	L6	L7
Typological group										
Notch	12	4	48	19	20	24	4	8	13	46
	8.51	2.68	12.57	12.10	10.87	11.16	6.78	9.20	9.35	14.42
Denticulate	11	1	15	5	9	9	4	3	10	14
	7.80	0.67	3.93	3.18	4.89	4.19	6.78	3.45	7.19	4.39
Side-scraper	7	5	13	5	10	15	1	6	13	29
	4.96	3.36	3.40	3.18	5.43	6.98	1.69	6.90	9.35	9.09
Borer	11	1	16	11	12	24	7	6	19	29
	7.80	0.67	4.19	7.01	6.52	11.16	11.86	6.90	13.67	9.09
End-scraper	1	2	11	5	9	7	4	3	3	8
	0.71	1.34	2.88	3.18	4.89	3.26	6.78	3.45	2.16	2.51
Knife	3	2	-	-	-	1	-	1	1	3
	2.13	1.34	-	-	-	0.47	-	1.15	0.72	0.94
Truncation	2	-	4	4	4	7	-	2	2	7
	1.42	-	1.05	2.55	2.17	3.26	-	2.30	1.44	2.19
Burin	1	-	-	-	-	1	-	-	1	1
	0.71	-	-	-	-	0.47	-	-	0.72	0.31
Retouched flake	11	1	31	8	26	19	3	19	13	34
	7.80	0.67	8.12	5.10	14.13	8.84	5.08	21.84	9.35	10.66
Unretouched flake	82	133	244	100	94	108	36	39	64	148
	58.16	89.26	63.87	63.69	51.09	50.23	61.02	44.83	46.04	46.39
Total	141	149	382	157	184	215	59	87	139	319
Striking platform removal										
N	135	148	373	149	181	210	59	86	136	314
Removed striking platform	26	4	29	19	24	38	9	19	21	51
	19.26	2.70	7.77	12.75	13.26	18.10	15.25	22.09	15.44	16.24

Table 6.77 Counts and frequencies (%) of scar pattern by striking platform and mean number of dorsal scars of flint FFT*

	Cortical	Punctiform	Plain	Dihedral	Facetted	Crushed	Total	Mean number of scars
N	157	105	558	79	262	71	1,232	5.07 (1,099)
Cortical	94	19	103	6	10	7	239	2.86 (101)
	39.33	7.95	43.10	2.51	4.18	2.93		
Plain	5	1	24	-	5	-	35	2.19 (26)
	14.29	2.86	68.57	-	14.29	-		
Radial	5	25	89	20	77	19	235	7.64 (230)
	2.12	10.63	37.87	8.51	32.76	8.08		
Along axis	16	28	139	30	65	23	301	3.90 (301)
	5.32	9.30	46.18	9.97	21.59	7.64		
Along axis & side	5	8	55	11	48	6	133	5.90 (134)
	3.76	6.02	41.35	8.27	36.09	4.51		
Along axis & opposed	4	4	29	3	12	5	57	5.28 (57)
	7.02	7.02	50.88	5.26	21.05	8.77		
Parallel	1	1	5	-	4	-	11	4.00 (11)
	9.09	9.09	45.45	-	36.36	-		
Convergent	-	4	23	6	16	4	53	5.63 (54)
	-	7.55	43.40	11.32	30.19	7.55		
Opposed	11	1	12	-	5	1	30	3.52 (29)
	36.67	3.33	40.00	-	16.67	3.33		
Ridged	3	6	30	2	10	2	53	7.04 (53)
	5.66	11.32	56.60	3.77	18.87	3.77		
Side	10	4	32	1	9	4	60	3.36 (61)
	16.67	6.67	53.33	1.67	15.00	6.67		
Opposed & side	3	4	17	-	1	-	25	3.80 (25)
	12.00	16.00	68.00	-	4.00	-		

* FFT (retouched and unretouched); complete only; excluding indeterminate; excluding removed striking platform

1) Both groups are similar in dimensions (length, width, thickness).
2) Non-cortical flakes occur somewhat more frequently within the group of removed striking platforms (65.60% vs. 61.03%).
3) Artifacts with removed striking platform are significantly more double patinated ($\chi^2=10.59$; $df=2$; $p=0.005$). The double patinated areas of these artifacts are often not the result of breakage but are an integral part of the tool, occurring either on the retouched edges or on the area of striking platform removal.
4) Clear differences in the typological composition of these two groups are observed (Table 6.78), whereby retouched tools occur significantly more frequently in the group of removed striking platforms ($\chi^2=34.07$; $df=1$; $p<0.0001$).
5) The radial scar pattern is significantly more frequent in the group of removed striking platforms (Table 6.79) ($\chi^2=31.43$; $df=13$; $p=0.0029$).
6) Removals on the ventral face (Section 6.3.1.3) occur in significantly higher frequencies among flakes with removed striking platform (Table 6.80) ($\chi^2=86.96$; $df=1$; $p<0.00001$).

Based on the above observations, we suggest that the removal of the striking platform is in fact a typological feature. Its occurrence in high frequencies and its consistent presence throughout the cultural sequence indicate that these artifacts were a product of special attention within the flint reduction process. The removal of the striking platform necessitated great precision, enforced by the small size of the flint flakes. The presence of these tools provides an indication that the Acheulians already possessed the physical skills and dexterity required for this precision.

It seems to us that the removal of the striking platform, and the flow of planning and implementation that was embedded within this process, are not restricted to the typological realm but are much more meaningful. We suggest that the precise and difficult task of removing a topographical high relief from the ventral face of a small flake was a specific planned goal that was aimed at acquiring a suitable surface, in this case even and flat. The precision and repetition required for this, as well as the morphological diversity of the proximal ends, calls for these modifications to be interpreted as preparation for hafting (Alperson-Afil and Goren-Inbar 2016).

Some additional artifacts displaying partial modification (but not complete removal) of their striking platform (e.g. Fig. 6.81) were not included in the category of removed striking platform. Moreover, some tools are characterized by modification that narrows a portion of their lateral edges (e.g. Fig. 6.81).

If hafting is a procedure that was practiced at GBY, one can assume that these hafted artifacts had longer life-histories

6.5 Discussion and Summary of the Flint Component

Table 6.78 Counts and typological frequencies (%) of flint FFT by striking platform (removed/not removed)

	Removed	Not removed
N	436	2,362
Retouched		
Side-scraper	54	125
	12.40	5.27
End-scraper	30	49
	6.88	2.08
Burin	3	6
	0.69	0.25
Borer	57	118
	13.07	4.99
Backed knife	4	9
	0.92	0.38
Truncation	15	24
	3.44	1.02
Notch	46	184
	10.55	7.79
Denticulate	37	77
	8.49	3.26
End-notched piece	11	51
	2.52	2.16
Retouched flake	52	188
	11.93	7.96
Varia	9	9
	2.07	0.39
Unretouched		
Flake	108	1,151
	24.77	48.73
Blade	-	26
		1.10
Éclat de taille de biface	7	295
	1.61	12.49
Biface sharpening flake	1	17
	0.23	0.72
Core waste	2	33
	0.46	1.39

Table 6.79 Counts and frequencies (%) of scar pattern of flint FFT by striking platform (removed/not removed)

	Removed	Not removed
N	434	2,338
Cortical	35	301
	8.06	12.87
Plain	11	55
	2.53	2.35
Along axis	64	464
	14.75	19.85
Parallel	-	15
	-	0.64
Convergent	9	78
	2.07	3.34
Opposed	4	35
	0.92	1.50
Radial	101	300
	23.27	12.83
Ridged	15	62
	3.46	2.65
Side	17	95
	3.92	4.06
Along axis & side	28	181
	6.45	7.74
Along axis & opposed	19	84
	4.38	3.59
Opposed & side	6	29
	1.38	1.24
Along axis & radial	1	8
	0.23	0.34
Indeterminate	124	631
	28.57	26.99

Table 6.80 Counts and frequencies (%) of ventral removals on flint FFT by striking platform (removed/not removed)

	Removed (N=436)	Not removed (N=2,362)
Yes	92	169
	21.10	7.15
No	344	2,193
	78.90	92.85

#8428   #735

0 2 cm

Fig. 6.81 Flint flakes with partial modification of the striking platforms (#8428 Layer II-6 Level 3 convex transversal scraper and a borer with a tang-like proximal end, #735 Layer II-6 Level 4 abruptly retouched side-scraper with narrowing of the lateral edges)

that probably involved maintenance, as indicated by the significantly higher frequencies of double patination on these items.

It is beyond the scope of this presentation to speculate on the function or tasks for which these tools were hafted. But if we accept the hypothesis that these tools were indeed hafted, we can perhaps test a possible breakage pattern of the artifacts to support this hypothesis. We examined the breakage patterns of the entire FFT sample and compared FFT with striking platforms to FFT with removed striking platforms. The pattern that emerged from this comparison is statistically significant ($\chi2=69.35$; $df=7$; $p<0.0001$) and shows that among the broken FFT with removed striking platforms 54.05% bear distal breaks, as opposed to 27.16% distal breaks on FFT with striking platforms (Table 6.81).

In summary, flake tools were the target products of the flint reduction sequence. With regard to these target products, we suggest that an important component among them consists of small flint flakes that were produced for the purpose of hafting.

Table 6.81 Counts and frequencies (%) of breakage pattern of flint FFT* by striking platform (removed/not removed)

	Removed	Not removed	
Distal	60	562	622
	54.05	27.16	
Lateral	28	299	327
	25.23	14.45	
Proximal	12	510	522
	10.81	24.65	
Distal & lateral	3	103	106
	2.70	4.98	
Distal & proximal	4	146	150
	3.60	7.06	
Fragment	4	342	346
	3.60	16.53	
Proximal & lateral	–	106	106
		5.12	
Indeterminate	–	1	1
		0.05	
Total	111	2069	2,180

* FFT (retouched and unretouched); broken only; excluding indeterminate striking platform

Chapter 7
The Basalt Component

Abstract Chapter 7 aims to provide a comprehensive description of the basalt assemblages, a major component of the archaeological horizons of Gesher Benot Ya'aqov. Analyses consist of taphonomic, morphological, technological, and typological observations, which enable characterization and reconstruction of the operational sequences of basalt. These involve percussive tools (anvils, percussors, and pitted stones) and the use of giant cores for the production of large flakes, which required minimal modification for their transformation into bifaces.

Keywords anvils · basalt · bifaces · handaxes · cleavers · giant cores · large flakes · percussors · pitted stones

Basalt is an important cultural aspect of the Gesher Benot Ya'aqov (GBY) Acheulian assemblages. It is the primary raw material for the production of bifaces, shaped on large flakes detached from giant basalt cores. The use of basalt characterizes the GBY assemblages technologically (e.g. the knapping of large flakes from giant cores), typologically (e.g. the production of cleavers), and stylistically (e.g. the design of bifacial tools). Basalt was infrequently used as raw material in Levantine Acheulian lithic assemblages (see, however, 'Ubeidiya: Bar-Yosef and Goren-Inbar 1993) and rarely appears in Levantine Acheulian assemblages postdating GBY (Stekelis 1966; Goren-Inbar 1985; Sharon 2007; Alperson-Afil 2015).

This chapter presents the complete inventory of the basalt artifacts within the GBY lithic assemblages (the few flint and limestone bifaces are discussed with the basalt biface assemblage presented here). The chapter presents the technology and typology of flakes and flake tools (FFT), followed by those of cores and core tools (CCT) and of bifaces.

Our observations of the GBY basalt component identified three distinct reduction sequences, two associated with percussive activities and the third, which produced the majority of basalt flakes, related to the production of bifacial tools.

Thus, although the FFT component is presented before the interpretation of the reduction sequence, we take the liberty of classifying different categories of flakes according to their role within the biface reduction sequence.

7.1 Inventory and Taphonomy of the Basalt Component

7.1.1 Inventory of the Basalt Component

The complete basalt inventory of GBY, presented in Table 7.1, indicates that basalt was used for the production of artifacts throughout the entire cultural sequence. In all tables the data are given in chronological order from the youngest (top) to the oldest (bottom) layer. In some of the assemblages, primarily those originating in the rich complex of Layer II-6, basalt is clearly the dominant raw material in terms of rock mass. Basalt FFT is always the most frequent artifact category followed by CCT, while bifaces are minimally represented in most of the layers (Table 7.1).

In some of the following presentations, we include only assemblages that are composed of a minimum number of 30 basalt artifacts and are hence suitable samples for analyses.

7.1.2 Taphonomy of the Basalt Component

The taphonomy of basalt at GBY is studied through observations on the state of preservation, patination, and breakage patterns. Basalt artifacts weather differently from other types of raw material at GBY and exhibit extreme differences in their degree of preservation. While some

Table 7.1 Counts of basalt artifacts and microartifacts throughout the GBY cultural sequence

Area	Layer	Artifacts					Microartifacts
		CCT	FFT	Handaxe	Cleaver	Total	
C	V-2	-	3	3	2	8	-
	V-3	-	1	1	-	2	2
	V-3/4	-	-	4	8	12	-
	V-4	1	5	-	-	6	54
	V-4/5	-	5	2	-	7	68
	V-5	8	76	-	2	86	5,702
	V-6	19	56	-	6	81	2,020
	JB	34	43	-	8	85	101
A	I-4	1	10	1	-	12	224*
	I-5	4	1	-	-	5	106
B	II-2	1	16	-	-	17	76
	II-2/3	15	34	7	-	56	44
	II-3	1	-	-	-	1	-
	II-4	1	4	-	-	5	-
	II-4/5	-	1	-	-	1	-
	II-5	7	100	4	2	113	79
	II-5/6	4	43	1	1	49	619
	Trench II	11	9	10	8	38	-
	Trench II green	7	61	24	8	100	-
	II-6/L1	117	906	54	17	1,094	2,897
	II-6/L2	122	835	19	10	986	3,930
	II-6/L3	66	711	12	12	801	2,751
	II-6/L4	104	785	165	60	1,114	7,895
	II-6/L4b	44	491	58	18	611	1,083
	II-6/L5	16	178	11	8	213	2,570
	II-6/L6	44	317	19	11	391	522
	II-6/L7	60	526	20	12	618	1,970
	II-6/7	-	24	-	-	24	53
	II-7	-	3	-	-	3	-
	VI-14	3	83	9	-	95	1
	Unconformity B	21	206	48	6	281	-
	Unconformity C	6	12	1	1	20	-
	Total	717	5,545	473	200	6,935	32,767

* including microartifacts that are not assigned to a particular layer within Area A

artifacts are very fresh, others have deteriorated completely, losing their shape and becoming small lumps of clay. The most extreme manifestation of weathering of basalt is the exfoliation of the artifact in an "onion-like" manner (Chapter 4.3). Two main agents are apparently responsible for this particular situation: the structure of the basalt flow exploited and the specific micro-environmental conditions of the waterlogged sediments at the site (Chapter 4.3). As a result, the analysis of basalt artifacts presents difficulties, as many of the technological and morphological features are difficult to "read", identify, and quantify.

The state of preservation, as reflected by the different weathering stages of the basalt artifacts, also varies substantially among layers along the cultural sequence of the site (Table 7.2), and hence is of great importance for the reconstruction of the environmental-accumulation conditions as well as postdepositional processes.

Table 7.2 presents the state of preservation recorded for the

7.1 Inventory and Taphonomy of the Basalt Component

Table 7.2 Frequencies (%) of state of preservation of basalt artifacts

Layer		Fresh %	Slightly abraded %	Abraded %	Heavily abraded %	Exfoliated %	Total N	Layer		Fresh %	Slightly abraded %	Abraded %	Heavily abraded %	Exfoliated %	Total N
V-5	Cleaver	-	100.00	-	-	-	2	II-6/L2	Cleaver	-	50.00	40.00	10.00	-	10
	Handaxe	-	-	-	-	-	-		Handaxe	-	36.84	42.11	21.05	-	19
	FFT	5.26	18.42	69.74	3.95	2.63	76		FFT	6.47	25.87	47.19	17.60	2.87	835
	CCT	-	87.50	12.50	-	-	8		CCT	0.82	37.70	57.38	2.46	1.64	122
V-6	Cleaver	-	83.33	16.67	-	-	6	II-6/L3	Cleaver	-	66.67	25.00	-	8.33	12
	Handaxe	-	-	-	-	-	-		Handaxe	-	54.55	18.18	18.18	9.09	11
	FFT	7.14	39.29	46.43	1.79	5.36	56		FFT	17.46	26.76	40.00	14.65	1.13	710
	CCT	-	63.16	36.84	-	-	19		CCT	1.52	37.88	56.06	4.55	-	66
JB	Cleaver	-	25.00	25.00	12.50	37.50	8	II-6/L4	Cleaver	-	56.67	20.00	8.33	15.00	60
	Handaxe	-	-	-	-	-	-		Handaxe	-	42.42	21.21	15.76	20.61	165
	FFT	2.33	30.23	51.16	16.28	-	43		FFT	17.58	23.44	41.15	13.25	4.59	785
	CCT	2.94	64.71	32.35	-	-	34		CCT	1.92	38.46	48.08	3.85	7.69	104
I-4	Cleaver	-	-	-	-	-	-	II-6/L4b	Cleaver	5.56	44.44	22.22	11.11	16.67	18
	Handaxe	-	100.00	-	-	-	1		Handaxe	-	43.10	22.41	8.62	25.86	58
	FFT	20.00	40.00	40.00	-	-	10		FFT	4.68	11.61	31.16	39.71	12.83	491
	CCT	-	-	100.00	-	-	1		CCT	-	2.27	84.09	6.82	6.82	44
I-5	Cleaver	-	-	-	-	-	-	II-6/L5	Cleaver	37.50	-	37.50	12.50	12.50	8
	Handaxe	-	-	-	-	-	-		Handaxe	-	-	18.18	18.18	63.64	11
	FFT	-	-	100.00	-	-	1		FFT	1.12	3.93	25.28	58.99	10.67	178
	CCT	-	25.00	75.00	-	-	4		CCT	-	18.75	68.75	6.25	6.25	16
II-2/3	Cleaver	-	-	-	-	-	-	II-6/L6	Cleaver	-	27.27	9.09	-	63.64	11
	Handaxe	-	28.57	28.57	-	42.86	7		Handaxe	-	15.79	36.84	5.26	42.11	19
	FFT	2.94	26.47	52.94	8.82	8.82	34		FFT	0.63	3.79	29.97	49.53	16.09	317
	CCT	-	20.00	80.00	-	-	15		CCT	-	18.18	59.09	6.82	15.91	44
II-5	Cleaver	-	-	-	100.00	-	2	II-6/L7	Cleaver	-	25.00	33.33	8.33	33.33	12
	Handaxe	25.00	75.00	-	-	-	4		Handaxe	-	5.00	20.00	30.00	45.00	20
	FFT	7.00	15.00	52.00	20.00	6.00	100		FFT	0.38	2.28	39.54	40.87	16.92	526
	CCT	14.29	14.29	71.43	-	-	7		CCT	-	20.00	46.67	10.00	23.33	60
II-5/6	Cleaver	-	-	-	100.00	-	1	VI-14	Cleaver	-	-	-	-	-	-
	Handaxe	-	-	100.00	-	-	1		Handaxe	-	11.11	11.11	22.22	55.56	9
	FFT	9.30	20.93	55.81	11.63	2.33	43		FFT	1.20	7.23	24.10	50.60	16.87	83
	CCT	-	-	100.00	-	-	4		CCT	-	-	33.33	-	66.67	3
II-6/L1	Cleaver	5.88	41.18	41.18	5.88	5.88	17	Trench II green	Cleaver	-	12.50	37.50	-	50.00	8
	Handaxe	-	40.74	35.19	16.67	7.41	54		Handaxe	4.17	8.33	20.83	25.00	41.67	24
	FFT	19.98	37.75	36.98	4.42	0.88	906		FFT	21.31	18.03	52.46	3.28	4.92	61
	CCT	0.85	35.04	57.26	5.98	0.85	117		CCT	-	-	85.71	14.29	-	7

artifacts of different archaeological horizons. Most of the GBY basalt artifacts are weathered to some degree and the patterns vary among the different artifact categories. There are similar low frequencies of fresh basalt artifacts and heavily exfoliated ones. The most common state is abraded, followed by slightly abraded (Table 7.2).

The basalt artifacts are more heavily weathered in some levels than in others, as in the case of Layer II-6, in which Levels 4b to 7 show higher frequencies of abraded artifacts and more exfoliated flakes and bifaces. Exfoliation generally occurs on bifaces more frequently than on any of the other artifact categories, and it seems not to be correlated to the size of the artifacts (see the discussion of giant cores and bifacial tools below, Section 7.6).

Table 7.3 presents the breakage and patination data recorded for the basalt artifacts. The differences in breakage pattern between the basalt FFT and CCT are difficult to discern, as the sample sizes of the CCT are always much smaller than those of the FFT. Inspection of assemblages with large sample sizes shows that over 60% of the FFT are broken, and when sample sizes of CCT are sufficiently large they show that the CCT are less broken than the FFT. The lowermost levels of Layer II-6 display a somewhat

Table 7.3 Frequencies (%) of breakage and patination of basalt artifacts

Layer		Breakage Complete %	Broken %	Patination No patina %	Patinated %	Double patinated %	N*	Layer		Breakage Complete %	Broken %	Patination No patina %	Patinated %	Double patinated %	N*
V-5	Cleaver	50.00	50.00	-	100.00	-	2	II-6/L2	Cleaver	80.00	20.00	-	100.00	-	10
	Handaxe	-	-	-	-	-	-		Handaxe	73.68	26.32	-	100.00	-	19
	FFT	34.67	65.33	-	96.00	4.00	75		FFT	31.26	68.74	1.92	93.89	4.19	835
	CCT	75.00	25.00	-	100.00	-	8		CCT	51.67	48.33	0.85	96.61	2.54	120; 118
V-6	Cleaver	66.67	33.33	-	100.00	-	6	II-6/L3	Cleaver	75.00	25.00	-	100.00	-	12
	Handaxe	-	-	-	-	-	-		Handaxe	50.00	50.00	-	100.00	-	12
	FFT	52.73	47.27	-	100.00	-	55; 56		FFT	34.37	65.63	1.97	94.79	3.24	710
	CCT	47.37	52.63	-	94.74	5.26	19		CCT	60.61	39.39	-	100.00	-	66
JB	Cleaver	62.50	37.50	-	85.71	14.29	8; 7	II-6/L4	Cleaver	73.33	26.67	1.72	98.28	-	60; 58
	Handaxe	-	-	-	-	-	-		Handaxe	52.12	47.88	1.84	96.93	1.23	165; 163
	FFT	55.81	44.19	-	97.67	2.33	43		FFT	39.36	60.64	1.66	97.96	0.38	785
	CCT	58.82	41.18	-	100.00	-	34		CCT	51.46	48.54	0.96	99.04	-	103; 104
I-4	Cleaver	-	-	-	-	-	-	II-6/L4b	Cleaver	55.56	44.44	-	100.00	-	18
	Handaxe	100.00	-	-	100.00	-	1		Handaxe	29.31	70.69	3.51	94.74	1.75	58; 57
	FFT	60.00	40.00	-	100.00	-	10		FFT	31.98	68.02	0.41	97.35	2.24	491
	CCT	100.00	-	-	100.00	-	1		CCT	27.91	72.09	-	100.00	-	43; 44
I-5	Cleaver	-	-	-	-	-	-	II-6/L5	Cleaver	50.00	50.00	-	100.00	-	8
	Handaxe	-	-	-	-	-	-		Handaxe	20.00	80.00	-	100.00	-	10
	FFT	-	100.00	-	100.00	-	1		FFT	26.97	73.03	-	98.88	1.12	178
	CCT	75.00	25.00	-	100.00	-	4		CCT	20.00	80.00	-	100.00	-	15; 16
II-2/3	Cleaver	-	-	-	-	-	-	II-6/L6	Cleaver	18.18	81.82	-	100.00	-	11; 10
	Handaxe	57.14	42.86	-	85.71	14.29	7		Handaxe	15.79	84.21	-	100.00	-	19
	FFT	47.06	52.94	11.76	88.24	-	34		FFT	31.23	68.77	1.58	96.85	1.58	317
	CCT	66.67	33.33	33.33	60.00	6.67	15		CCT	29.55	70.45	-	100.00	-	44
II-5	Cleaver	-	100.00	-	100.00	-	2	II-6/L7	Cleaver	41.67	58.33	-	100.00	-	12
	Handaxe	50.00	50.00	-	100.00	-	4		Handaxe	60.00	40.00	-	100.00	-	20
	FFT	44.00	56.00	10.00	88.00	2.00	100		FFT	25.10	74.90	0.19	96.20	3.61	526
	CCT	42.86	57.14	-	100.00	-	7		CCT	50.00	50.00	-	98.33	1.67	60
II-5/6	Cleaver	-	100.00	-	100.00	-	1	VI-14	Cleaver	-	-	-	-	-	-
	Handaxe	100.00	-	-	100.00	-	1		Handaxe	22.22	77.78	-	100.00	-	9
	FFT	34.88	65.12	6.98	90.70	2.33	43		FFT	31.71	68.29	-	97.59	2.41	82; 83
	CCT	33.33	66.67	-	100.00	-	4		CCT	66.67	33.33	-	66.67	33.33	3
II-6/L1	Cleaver	70.59	29.41	-	100.00	-	17	Trench II green	Cleaver	50.00	50.00	-	100.00	-	8; 7
	Handaxe	72.22	27.78	-	100.00	-	54		Handaxe	41.67	58.33	-	100.00	-	24
	FFT	41.72	58.28	7.17	91.72	1.10	906		FFT	54.10	45.90	16.39	81.97	1.64	61
	CCT	57.26	42.74	9.40	84.62	5.98	117		CCT	71.43	28.57	-	100.00	-	7

* in cases where N differs between the two attributes, the first value refers to counts of breakage and the second to those of patination

different pattern in which more artifacts, both FFT and CCT, are broken (similar to the variations in state of preservation mentioned above). Clearly, these observations are very general and will be detailed in the following sections, but nevertheless they reflect the taphonomic situation of the various assemblages.

Analysis of patination, another element in the taphonomic life-histories of the basalt artifacts, shows that most of the basalt artifacts are patinated. This kind of patination differs from that of the flint and is not necessarily biogenic (Chapter 4.3).

Observations on the pattern of breakage of basalt FFT show that in most of the assemblages complete FFT form the majority, followed by fragmented breaks (Table 7.4). Similar breakage patterns, of bimodality between complete and fragmented flakes, were also recorded for the flint FFT (Chapter 6). Somewhat different are a few levels of Layer II-6 (the deepest ones), which display higher fragmentation than the others; a similar trend was encountered for weathering, discussed above. The presence of relatively high frequencies of fragments should be analyzed with caution, as this may be a result of biface knapping rather than a postdepositional process. Further elaboration of this point will be presented in other sections of this chapter.

7.2 Typology of Basalt Flake Tools

Table 7.4 Frequencies (%) of breakage patterns of basalt FFT

	Complete	Distal	Proximal	Lateral	Distal & lateral	Distal & proximal	Proximal & lateral	Fragment	Indet.	N*
V-5	34.67	5.33	10.67	4.00	-	1.33	-	42.67	1.33	75
V-6	52.73	7.27	5.45	9.09	-	3.64	-	20.00	1.82	55
JB	55.81	2.33	2.33	4.65	2.33	-	2.33	30.23	-	43
II-5	44.00	13.00	5.00	5.00	9.00	1.00	3.00	19.00	1.00	100
II-5/6	34.88	16.28	6.98	13.95	2.33	4.65	2.33	18.60	-	43
II-6/L1	41.72	13.69	9.71	11.48	3.09	2.10	2.54	13.80	1.88	906
II-6/L2	31.26	12.34	7.07	12.94	7.19	4.55	4.31	18.08	2.28	835
II-6/L3	34.37	9.44	15.77	10.56	3.38	2.96	3.80	17.61	2.11	710
II-6/L4	39.36	12.74	9.04	9.81	5.99	3.69	4.97	13.12	1.27	785
II-6/L4b	31.98	11.61	9.78	10.18	6.31	3.05	3.46	18.33	5.30	491
II-6/L5	26.97	7.30	5.06	10.67	8.43	2.81	2.81	30.34	5.62	178
II-6/L6	31.23	10.41	6.62	11.05	6.31	4.10	3.47	23.97	2.84	317
II-6/L7	25.10	10.46	6.65	9.13	8.37	3.99	3.99	30.61	1.71	526
VI-14	31.71	14.63	6.10	6.10	8.54	-	3.66	29.27	-	82
Trench II green	54.10	6.56	13.11	9.84	1.64	3.28	1.64	3.28	6.56	61

* only layers with a minimum number of 30 FFT are included

Table 7.5 Typological frequencies (%) of basalt flake tools of Layer II-6*

Type/Layer	II-6/L1	II-6/L2	II-6/L3	II-6/L4	II-6/L4b	II-6/L5	II-6/L6	II-6/L7	VI-14
N	78	57	33	61	28	7	19	16	4
% of tools**	8.60	6.82	4.64	7.77	5.70	3.93	5.99	3.04	4.81
Single straight side-scraper	1.28	-	-	1.64	3.57	-	-	-	-
Single convex side-scraper	5.13	-	3.03	-	-	42.86	-	-	-
Single concave side-scraper	1.28	-	-	4.92	-	-	-	-	-
Double convex side-scraper	1.28	-	-	-	-	-	-	-	-
Convergent convex- side-scraper	-	-	-	1.64	-	-	-	-	-
Offset side-scraper	1.28	-	3.03	-	-	-	-	-	-
Straight transverse side-scraper	1.28	-	-	-	-	-	-	-	-
Convex transverse side-scraper	-	-	-	-	-	-	-	6.25	-
Side-scraper on ventral face	2.56	3.51	-	1.64	3.57	-	-	-	-
Abruptly retouched side-scraper	1.28	-	3.03	-	-	-	-	-	-
Thinned back side-scraper	2.56	-	-	-	-	-	-	-	-
Bifacially retouched side-scraper	-	-	-	1.64	3.57	-	-	-	-
Atypical end-scraper	2.56	-	-	-	-	-	-	-	-
Atypical burin	-	5.26	-	-	3.57	-	-	-	-
Atypical borer	1.28	-	9.09	-	-	-	5.26	6.25	-
Typical backed knife	-	-	-	-	-	14.29	-	-	-
Atypical backed knife	-	1.75	-	-	7.14	-	-	-	-
Notch	24.36	22.81	30.30	31.15	21.43	14.29	26.32	25.00	25.00
Denticulate	15.38	8.77	15.15	6.56	7.14	14.29	21.05	12.50	-
Varia	2.56	7.01	-	4.92	3.57	-	5.26	-	-
Retouched flake	15.38	31.58	27.27	27.87	21.43	-	15.79	43.75	25.00
Massive scraper	20.51	19.30	9.09	18.03	25.00	14.29	26.32	6.25	50.00

* only layers with a minimum number of 10 basalt flake tools are included; Layer VI-14 and the other levels of the Layer II-6 complex are included even if the sample is less than 10 tools
** % of flake tools out of FFT

The typological inventory of each of the GBY layers is presented in Chapter 5, where the basalt component is detailed. The number of basalt artifacts varies among the different layers at the site, and some layers do not provide large enough samples for detailed statistical analyses. Thus, in the following typological discussion we include only assemblages with a minimum number of 10 basalt flake tools (Table 7.5). Clearly, the number of basalt flake tools is very small, despite the fact that the number of basalt flakes is very high in all the GBY assemblages. Interestingly, there is no correlation between the overall number of basalt flakes and the number of basalt flake tools. This particular phenomenon will be further discussed in the summary of this chapter (Section 7.6). Nevertheless, the larger the basalt tool sample, the greater the typological diversification (similarly to flint flake tools, Chapter 6.2.1). In addition, high inter-layer variability is encountered (a high number of basalt tools in Layer II-6 Level 1), a phenomenon not related to sample size.

Table 7.5 shows that the highest frequencies of basalt tools are those of notches, denticulates, retouched flakes, and massive scrapers. These four tool categories are not only abundant in numbers but continuously represented throughout the cultural sequence. If notches and denticulates are combined, they form the largest group of tools.

As for basalt flake tools, descriptive and statistical constraints are imposed by their scarce occurrence. Consequently, the following presentation and discussion are limited to the combined data of all basalt flake tools from Layer II-6 (Levels 1–7) including those of Layer VI-14, which is the lateral continuation of Layer II-6 Level 1 (for convenience, these two layers will be referred to below simply as "Layer II-6").

The taphonomic characteristics of the basalt flake tools (Table 7.6) are similar to those of the rest of the basalt assemblage (Section 7.1.2). The majority of tools are fresh or slightly abraded with a negligible quantity of abraded items and most of them are patinated. Where breakage is concerned,

Table 7.6 Counts and frequencies (%) of taphonomic and technological characteristics of basalt flake tools of Layer II-6*

	Notch	Denticulate	Side-scraper	Massive scraper	Retouched flake	Varia		Notch	Denticulate	Side-scraper	Massive scraper	Retouched flake	Varia
N	69	36	33	57	71	10	N	69	36	33	57	71	10
Preservation							Distal & proximal	4	-	-	1	1	-
Fresh	6	9	10	15	16	3		5.80			1.75	1.41	
	8.70	25.00	30.30	26.32	22.54	30.00	Fragment	3	2	2	-	5	-
Slightly abraded	22	13	11	18	24	3		4.35	5.56	6.06		7.04	
	31.88	36.11	33.33	31.58	33.80	30.00	Proximal & lateral	7	1	1	2	3	-
Abraded	27	8	9	13	20	2		10.14	2.78	3.03	3.51	4.23	
	39.13	22.22	27.27	22.81	28.17	20.00	Indeterminate	2	1	-	1	-	-
Heavily abraded	13	6	2	4	8	2		2.90	2.78		1.75		
	18.84	16.67	6.06	7.02	11.27	20.00	**Direction of blow**						
Exfoliated	1	-	1	7	3	-	End-struck	19	16	16	27	25	6
	1.45		3.03	12.28	4.23			27.54	44.44	48.48	47.37	35.71	60.00
Patination							Side-struck	16	9	10	12	17	4
No patina	1	2	5	2	3	-		23.19	25.00	30.30	21.05	24.29	40.00
	1.45	5.56	15.15	3.51	4.23		Special side-struck	9	2	4	8	15	-
Patinated	66	28	28	53	67	10		13.04	5.56	12.12	14.04	21.43	
	95.65	77.78	84.85	92.98	94.37	100.00	Indeterminate	25	9	3	10	13	-
Double patinated	2	6	-	2	1	-		36.23	25.00	9.09	17.54	18.57	
	2.90	16.67		3.51	1.41		**Cortex**						
Breakage							No cortex	51	30	23	40	50	5
Complete	22	10	20	33	31	6		73.91	83.33	69.70	70.18	70.42	50.00
	31.88	27.78	60.61	57.89	43.66	60.00	1–25%	2	2	4	6	5	1
Distal	10	6	5	7	14	2		2.90	5.56	12.12	10.53	7.04	10.00
	14.49	16.67	15.15	12.28	19.72	20.00	26–50%	3	2	-	2	3	2
Lateral	12	7	2	7	9	1		4.35	5.56		3.51	4.23	20.00
	17.39	19.44	6.06	12.28	12.68	10.00	51–75%	6	-	1	4	5	-
Proximal	6	8	2	6	7	-		8.70		3.03	7.02	7.04	
	8.70	22.22	6.06	10.53	9.86		76–100%	4	2	3	4	4	2
Distal & lateral	3	1	1	-	1	1		5.80	5.56	9.09	7.02	5.63	20.00
	4.35	2.78	3.03		1.41	10.00	Indeterminate	3	-	2	1	4	-
								4.35		6.06	1.75	5.63	

* only layers with a minimum number of 10 basalt flake tools are included; Layer VI-14 and the other levels of the Layer II-6 complex are included even if the sample is less than 10 tools

---

the highest frequency is for complete tools, although different breakage variations are recorded. Technologically, in the direction of blow end-struck is the dominant mode, followed by side-struck. Nonetheless, the variability is high and includes some tools that are classified as special side-struck. Basalt tools are predominantly devoid of cortex, and when cortical coverage is present it is often minimal. An interesting point emerges from Table 7.6, which shows that, despite the definition of the tools as distinct morphotypes, there are similarities in the frequencies of their taphonomic and technological attribute states. This may be related to issues of blank availability and selection modes, a point that will be further touched upon in Section 7.3, where sample sizes allow significant statistical conclusions. In addition, various observations concerning the characteristics of retouch of basalt flake tools (location, face, and edge morphology) demonstrate an overall similarity among tool types (Table 7.7).

A detailed description of individual types of basalt flake tools follows, including those layers that are presented in Table 7.5.

Notches

Although notches (N=72; Layer II-6 N=69) are the most frequent basalt flake tool type, they form a very small group (Table 7.5). The frequencies are very low in all the archaeological horizons but are somewhat higher in the four uppermost levels of Layer II-6 (Figs. 7.1, 7.2; Table 7.5). Despite the meagerness of the samples, we present the main taphonomic and technological characteristics of this tool type in Table 7.6. Taphonomically, most of the notches are abraded or slightly abraded and the bulk are patinated. Notches are most commonly complete or laterally broken, with a near absence of fragments. Technologically,

7.2 Typology of Basalt Flake Tools

Table 7.7 Counts and frequencies (%) of retouch characteristics of basalt flake tools of Layer II-6*

	Notch	Denticulate	Side-scraper	Massive scraper	Retouched flake	Varia		Notch	Denticulate	Side-scraper	Massive scraper	Retouched flake	Varia
N	69	36	33	57	71	10	N	69	36	33	57	71	10
Location of retouch	(N=63)	(N=36)	(N=32)	(N=57)	(N=63)	(N=9)	Bifacial	-	-	3 / 9.68	-	3 / 4.84	2 / 22.22
Distal	10 / 15.87	11 / 30.56	6 / 18.75	14 / 24.56	11 / 17.46	1 / 11.11	Thinning	-	-	-	-	3 / 4.84	-
Proximal	3 / 4.76	-	-	2 / 3.51	2 / 3.17	-	Mixed	-	-	-	-	1 / 1.61	-
Left	19 / 30.16	9 / 25.00	11 / 34.38	9 / 15.79	18 / 28.57	2 / 22.22	Nahr Ibrahim	-	-	-	-	1 / 1.61	-
Right	28 / 44.44	8 / 22.22	11 / 34.38	20 / 35.09	22 / 34.92	1 / 11.11	Irregular	-	-	-	1 / 1.75	14 / 22.58	3 / 33.33
Both edges	1 / 1.59	3 / 8.33	1 / 3.13	-	3 / 4.76	2 / 22.22	Signs of use	1 / 1.61	-	-	1 / 1.75	2 / 3.23	-
Distal & both edges	-	1 / 2.78	-	-	-	-	Indeterminate	1 / 1.61	-	1 / 3.23	4 / 7.02	13 / 20.97	1 / 11.11
Distal & right	-	2 / 5.56	-	1 / 1.75	3 / 4.76	1 / 11.11	**Form of retouched edge**	(N=55)	(N=36)	(N=31)	(N=55)	(N=57)	(N=7)
Distal & left	1 / 1.59	-	1 / 3.13	2 / 3.51	2 / 3.17	2 / 22.22	Straight	1 / 1.82	5 / 13.89	5 / 16.13	2 / 3.64	11 / 19.30	2 / 28.57
Proximal & left	-	-	1 / 3.13	-	-	-	Convex	8 / 14.55	3 / 8.33	16 / 51.61	27 / 49.09	21 / 36.84	1 / 14.29
Proximal & right	-	-	-	1 / 1.75	-	-	Concave	41 / 74.55	2 / 5.56	4 / 12.90	4 / 7.27	4 / 7.02	1 / 14.29
Convergent	-	1 / 2.78	1 / 3.13	1 / 1.75	1 / 1.59	-	Convergent	-	1 / 2.78	1 / 3.23	1 / 1.82	1 / 1.75	-
Circumference	-	1 / 2.78	-	7 / 12.28	1 / 1.59	-	Wavy	3 / 5.45	2 / 5.56	3 / 9.68	11 / 20.00	16 / 28.07	2 / 28.57
Indeterminate	1 / 1.59	-	-	-	-	-	Denticulate	2 / 3.64	23 / 63.89	2 / 6.45	10 / 18.18	2 / 3.51	-
Type of retouch	(N=62)	(N=36)	(N=31)	(N=57)	(N=62)	(N=9)	One tooth	-	-	-	-	2 / 3.51	1 / 14.29
Regular	-	-	3 / 9.68	4 / 7.02	15 / 24.19	-	**Face of retouch**	(N=57)	(N=33)	(N=29)	(N=56)	(N=58)	(N=9)
Side-scraper	-	-	21 / 67.74	37 / 64.91	4 / 6.45	-	Dorsal	38 / 66.67	18 / 54.55	16 / 55.17	45 / 80.36	36 / 62.07	3 / 33.33
Notch	58 / 93.55	36 / 100.00	-	4 / 7.02	3 / 4.84	1 / 11.11	Ventral	11 / 19.30	6 / 18.18	9 / 31.03	6 / 10.71	10 / 17.24	-
Invasive	-	-	2 / 6.45	5 / 8.77	2 / 3.23	2 / 22.22	Dorsal & ventral	5 / 8.77	9 / 27.27	4 / 13.79	4 / 7.14	11 / 18.97	6 / 66.67
Semi-Quina	-	-	-	-	1 / 1.61	-	Side	3 / 5.26	-	-	1 / 1.79	1 / 1.72	-
Abrupt	2 / 3.23	-	1 / 3.23	1 / 1.75	-	-							

* only layers with a minimum number of 10 basalt flake tools are included; Layer VI-14 and the other levels of the Layer II-6 complex are included even if the sample is less than 10 tools

most of the notches are non-cortical and their axis of striking displays high variability and include all modes (end-struck, side-struck, and special side-struck), with a substantial number of indeterminate direction (Table 7.6). The location of retouch category includes items retouched on any edge, but the dominant locations are the left and the right edge, together forming more than 75% of the cases, followed by the distal edge. Obviously, notches are characterized by the notch/denticulate type of retouch (95% of the cases) and the most common form of the retouched edge is convex, occurring predominantly on the dorsal face (Table 7.7). The dominance of convex retouched edges reflects the fact that the notches, unlike the deep concave notches of the flint tools (Chapter 6.2.1), did not alter the original edge form of basalt flakes.

Table 7.8 gives the descriptive statistics of basalt notches. The data in this table are limited to notches originating in Layer

Table 7.8 Descriptive statistics of different basalt flake tool categories of Layer II-6*

	Notch	Denticulate	Side-scraper	Massive scraper	Retouched flake	Varia
N	69	36	33	57	71	10
Maximum length						
Maximum	160	178	153	192	134	170
Median	71	81	88	97	87	86.5
Minimum	25	27	30	50	20	59
Mean	74.01	81.33	87.70	98.93	82.08	97.20
Std dev	30.36	32.89	25.30	25.44	28.42	31.68
Std err mean	3.65	5.48	4.41	3.37	3.37	10.02
Length						
Maximum	132	178	153	162	133	170
Median	61	64	78	74	71	78
Minimum	17	24	23	44	20	44
Mean	62.20	71.92	74.73	85.74	71.62	87.50
Std dev	26.43	35.12	26.74	29.61	29.01	37.31
Std err mean	3.18	5.85	4.66	3.92	3.44	11.80
Width						
Maximum	135	130	121	122	124	140
Median	58	64	64	75	67	66
Minimum	21	20	30	51	17	39
Mean	61.09	63.44	68.64	77.07	65.80	74.50
Std dev	25.84	23.66	19.83	16.44	24.43	29.88
Std err mean	3.11	3.94	3.45	2.18	2.90	9.45
Thickness						
Maximum	45	44	45	60	68	61
Median	22	24	27	32	26	29
Minimum	7	8	14	18	7	18
Mean	23.43	25.00	27.67	31.56	27.00	34.30
Std dev	9.52	9.79	7.52	9.07	11.94	14.51
Std err mean	1.15	1.63	1.31	1.20	1.42	4.59
Angle of retouch	(N=42)	(N=26)	(N=31)	(N=51)	(N=37)	(N=8)
Maximum	110	105	93	103	105	112
Median	82.5	82	81	82	84	89
Minimum	72	42	72	68	73	78
Mean	83.52	79.38	82.19	82.12	84.68	90.88
Std dev	7.60	11.73	5.20	7.02	7.44	9.79
Std err mean	1.17	2.30	0.93	0.98	1.22	3.46

* only layers with a minimum number of 10 basalt flake tools are included; Layer VI-14 and the other levels of the Layer II-6 complex are included even if the sample is less than 10 tools

II-6, as other assemblages contain only a few notches (see Chapter 5). Two aspects characterize the notch assemblages. The first is the size of the items, which are much larger than notches made on flint. The second is the extensive variability within each of the notch assemblages, expressed in the large range of sizes in comparison to flint notches (Chapter 6.2.1). The mean size of notches originating from Layer II-6 is the smallest in all dimensions in comparison to all other basalt tool types and is characterized by high values of standard deviations (Table 7.8).

Denticulates

The basalt denticulates (N=38; Layer II-6 N=36) are a substantially smaller group than basalt notches. Most of

7.2 Typology of Basalt Flake Tools

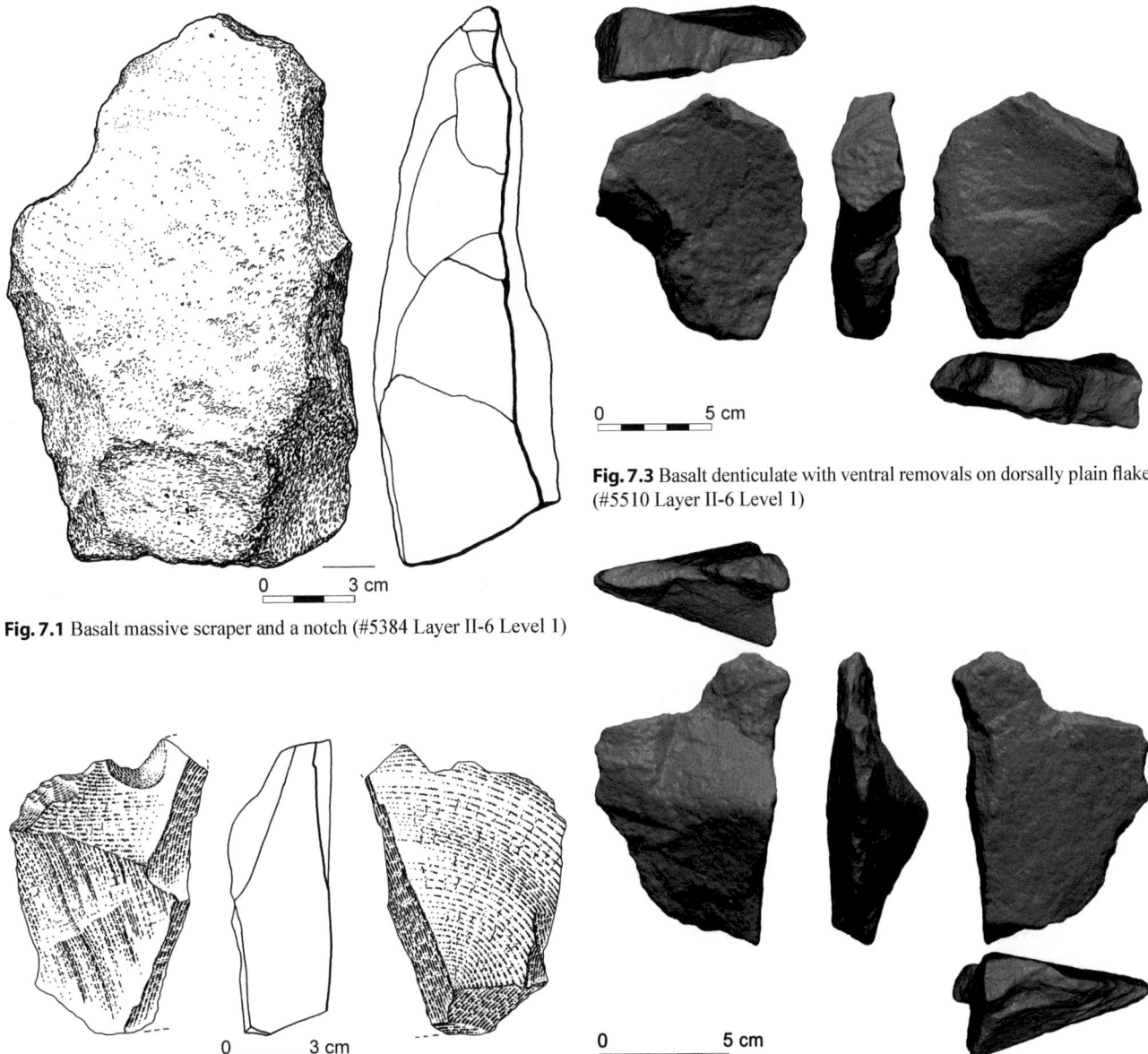

Fig. 7.1 Basalt massive scraper and a notch (#5384 Layer II-6 Level 1)

Fig. 7.2 Basalt end-notch on Kombewa flake, biface thinning flake (#155 Layer II-6 Level 4)

Fig. 7.3 Basalt denticulate with ventral removals on dorsally plain flake (#5510 Layer II-6 Level 1)

Fig. 7.4 Basalt denticulate (#5570 Layer II-6 Level 1)

the denticulates are slightly abraded, fresh, and abraded in descending order of frequency (Figs. 7.3, 7.4). All tools are patinated; some 27% are complete, while the rest are broken in different locations (Table 7.6). The denticulates' blanks are predominantly end-struck, but all the other directions of blow are observed as well. Most of these tools (more than 80%) lack cortex. The distal edge is the most frequent location of retouch, followed by both left and right edge, all formed by retouch of notch type and clearly characterized by a denticulated edge (Table 7.7). The retouch is most commonly located on the dorsal face, followed by both dorsal and ventral faces. Denticulates are larger than notches in all dimensions but exhibit smaller (sharper) values for the angle of retouch (Table 7.8).

Side-Scrapers

Side-scrapers (N=38; Layer II-6 N=33) (Table 7.6) are represented by 33 tools in Layer II-6. The most common types are single convex side-scrapers (Fig. 7.5), side-scrapers on the ventral face (Fig. 7.6), and single concave side-scrapers. Other types occur as well, but in very low frequencies (e.g. thinned backed side-scrapers: Figs. 7.7–7.9) (for the complete typological inventory, see Chapter 5). Most of the side-scrapers are slightly abraded or fresh and all are patinated. The frequency of complete side-scrapers is particularly high in comparison with other tool types, but when they are broken, distal breakages are the most common. The side-scrapers' blanks were struck in

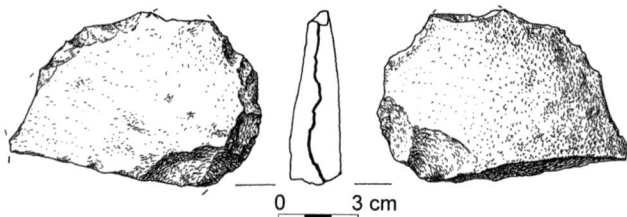

Fig. 7.5 Basalt single convex side-scraper (#5545 Layer II-6 Level 1)

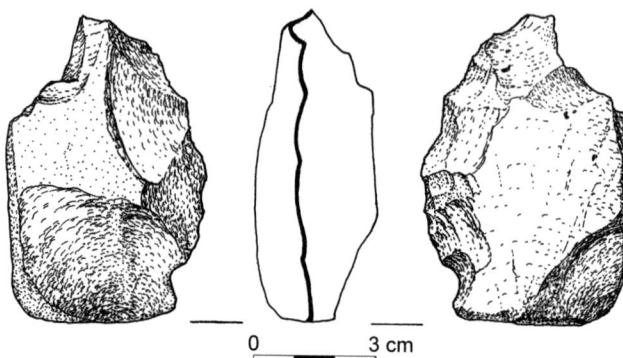

Fig. 7.7 Basalt thinned back side-scraper with ventral removals (#5532 Layer II-6 Level 1)

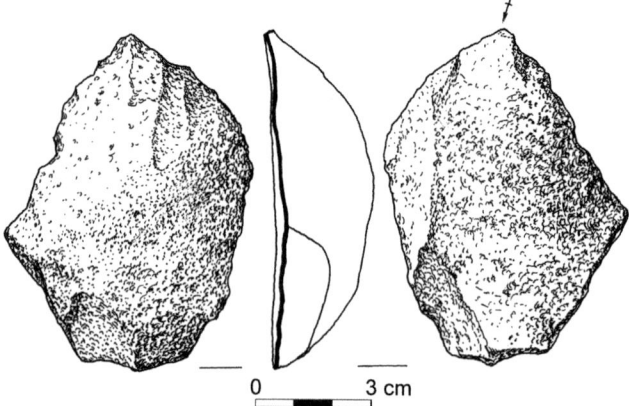

Fig. 7.6 Basalt side-scraper on ventral face (#5586 Layer II-6 Level 1)

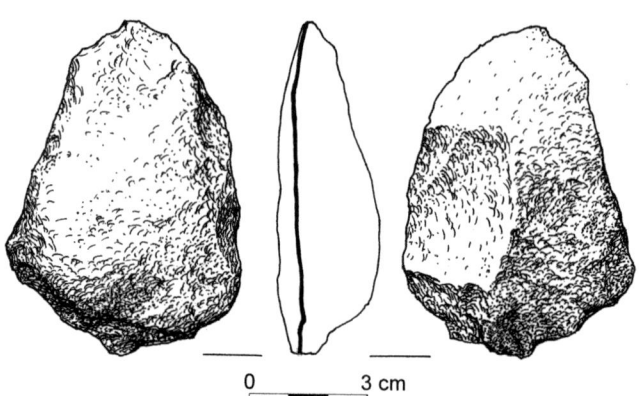

Fig. 7.8 Basalt thinned back side-scraper (#5572 Layer II-6 Level 1)

different ways, with a dominance of the end-struck mode, and nearly 70% of the tools are devoid of cortex. Equal frequencies of retouch were observed on the left and right edges of the tools, but other locations, including the distal edge, are also seen (Table 7.7). Over 67% of the artifacts display retouch of typical side-scraper type, but a variety of other types are observed as well. The dominant form of the retouched edge is convex (over 51% of the cases) and the retouch is predominantly located on the dorsal face, followed by the ventral face (Table 7.7). The average size of the side-scrapers is similar to that of denticulates and retouched flakes, and the mean angle of retouch is about 82° (Table 7.8).

Massive Scrapers

Of the 61 tools identified as massive scrapers, 57 were found in the complex of Layer II-6 (Table 7.6). Massive scrapers are relatively large (up to 192 mm) and thick and are characterized by a particular type of retouch that was applied continuously, intensively, and invasively, using a substantial part of the edge's thickness. The term "massive scraper" is adopted here to differentiate it from the common terms "heavy-duty scraper" and "core scraper" used for the East African Acheulian.

Massive scrapers are slightly abraded or abraded, most of

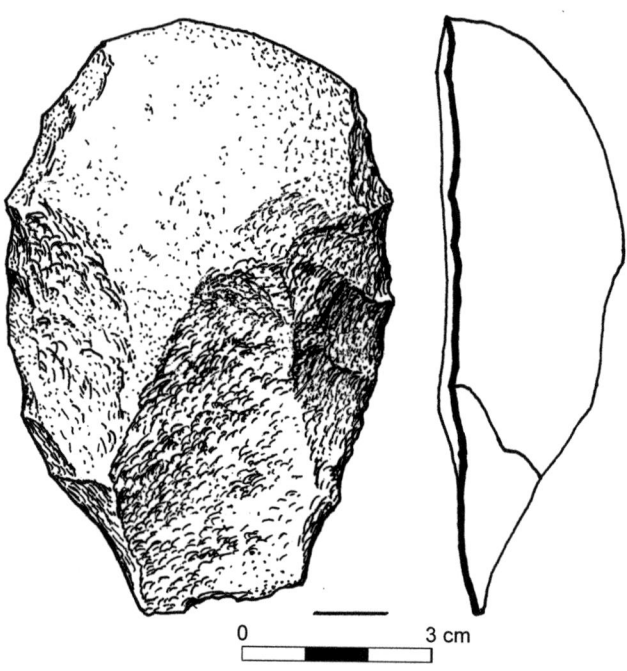

Fig. 7.9 Basalt thinned back side-scraper (#5523 Layer II-6 Level 1)

7.2 Typology of Basalt Flake Tools

them are patinated, and over half are complete (Table 7.6) (Figs. 7.10–7.15). Most massive scrapers are modified on end-struck blanks, followed by side-struck ones. Over 70% are devoid of cortex, followed by tools exhibiting 1–25% cortex. In some 35% of the massive scrapers the retouch is located on the right edge, followed by the distal edge, but combinations of edges occur, clearly demonstrating the variability of the retouched edge (Table 7.7). One should also note the cases (12.28%) in which the retouch is along the circumference, a trait only minimally present in other tool types. The most frequent retouch types of the massive scrapers are side-scraper and invasive. Massive scrapers are commonly characterized by a single dominant retouched edge (for exceptions see Fig. 7.16:4, 10, 14, 18) and a primarily convex morphology, followed by a wavy one, and in over 80% of the massive scrapers the retouch is on the dorsal face (Table 7.7). Some of the metrical characteristics of the massive scrapers of GBY are presented in Table 7.8; the mean maximal length of massive scrapers is greater than that of any of the other tool types originating in Layer II-6. The data reflect the extremely unstandardized nature of these tools, whether pertaining to size range, working edge morphology, or other technological modifications. Fig. 7.16 illustrates the morphology and location of the working edge, as well as the planform. The lack of standardization in the massive scrapers is most strikingly expressed by the tools' irregular planform (Fig. 7.16).

Several basalt flake tool types are minimally represented. These include borers, backed knives, burins, and end-scrapers. Since the same types occur in high frequencies in the flint tool assemblage, we view them as well-established types rather than as accidental phenomena. We provide below some observations about these rare cases. The description does not differentiate between different archaeological layers (the complete inventory of these types is provided in Chapter 5).

Borers

Borers (N=7) were found mostly in Layer II-6 and only a single one in Layer V-4. Borers are overwhelmingly made on non-cortical flakes (N=6) and only a single tool has cortical coverage (1–25%). The working edge is located on the distal end (N=3), followed by other edges or their combination. The mean dimensions of the borers are length: 38.28 mm, width: 33.71 mm, and thickness: 13 mm.

Backed Knives

Of the four tools classified as backed knives, three are identified as atypical and a single tool as typical. They are all devoid of cortex.

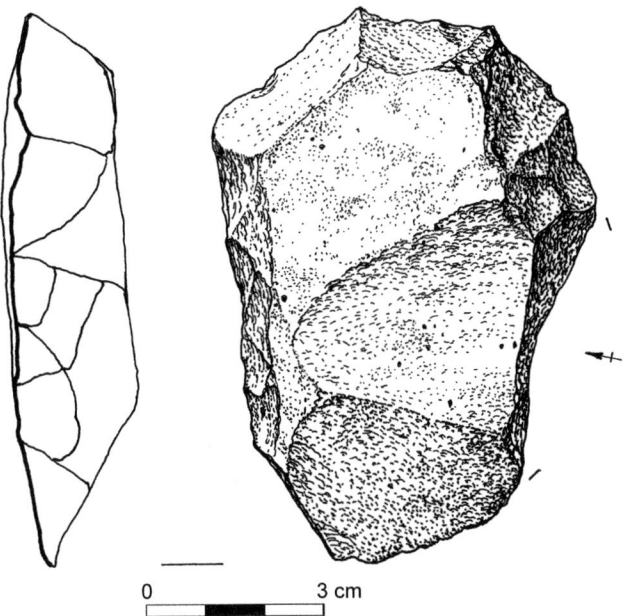

Fig. 7.10 Basalt massive scraper (#5682 Layer II-6 Level 1)

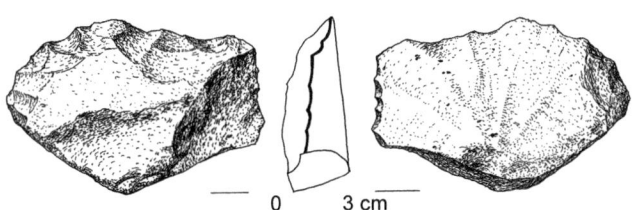

Fig. 7.11 Basalt massive scraper (#5338 Layer II-6 Level 1)

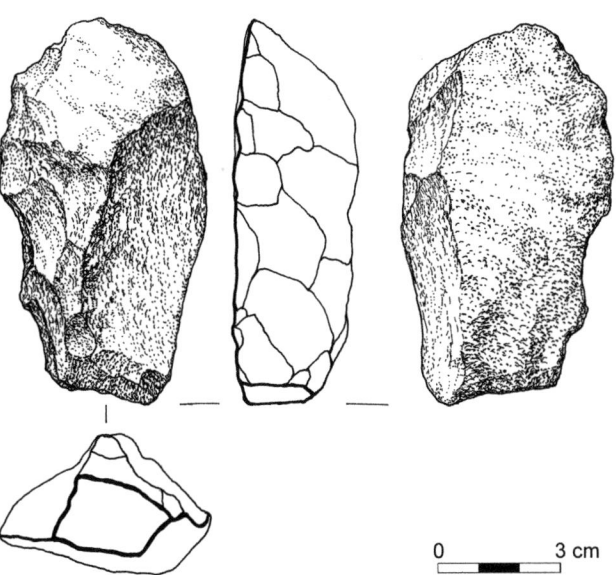

Fig. 7.12 Basalt massive scraper with ventral retouch (#5852 Layer II-6 Level 1)

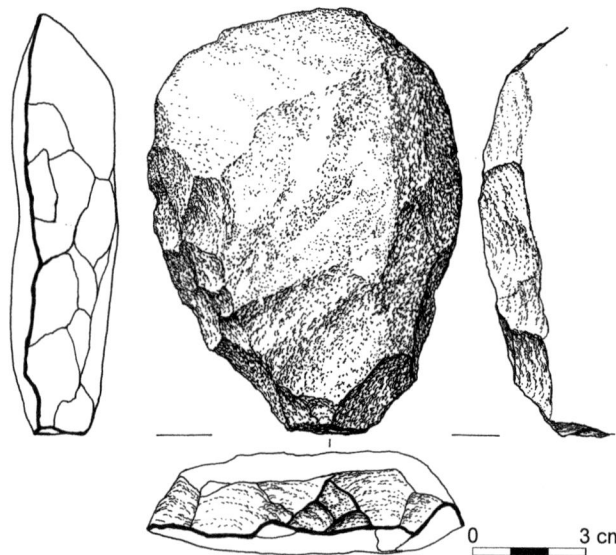

Fig. 7.13 Basalt massive scraper with ventral retouch (#5542 Layer II-6 Level 1)

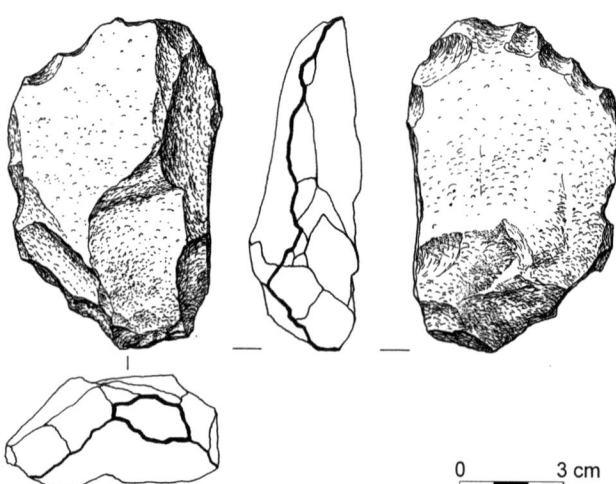

Fig. 7.14 Basalt massive scraper (#5587 Layer II-6 Level 1)

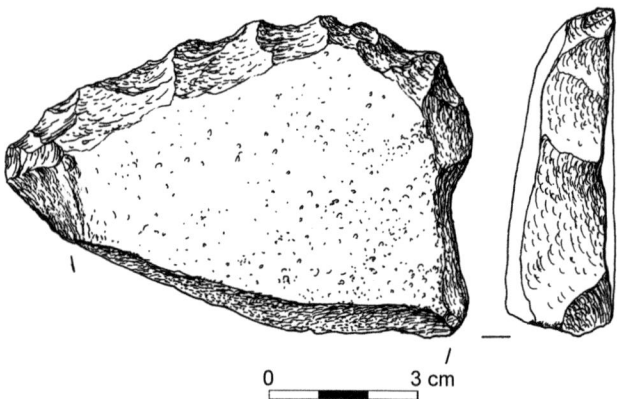

Fig. 7.15 Basalt denticulated massive scraper (#5453 Layer II-6 Level 1)

Burins

All four tools classified as atypical burins originate in Layer II-6. The working edge of these tools is located on different edges.

End-scrapers

Two end-scrapers are recorded, one classified as typical and the other as atypical; the retouch on the latter is located on the distal end and on the former on the proximal end.

Varia Tools

This tool category (N=11; Layer II-6 N=10) includes tools that are classified as varia as well as a single tool that was classified as an alternately retouched bec (beak). Within this tool category, four tools are worked by a scraper-like retouch forming a pointed edge. In addition, some of the varia tools are characterized by alternating or steep side-scraper retouch, while on others the length of the retouched edge is minimal.

Taphonomically, the tools exhibit three states of preservation, of which fresh and slightly abraded form the majority. The tools are patinated and over 55% are complete. Most of the tools are end-struck and some 44% are devoid of cortex. The retouch appears on all edges and the dominant form of the retouched edge is straight. The mean size of the varia tools is length: 91.66 mm, width: 76.22 mm, and thickness: 36.11 mm. Like other retouched basalt tools, the varia tools correspond in size with the large unretouched flakes, discussed below.

Retouched Flakes

Retouched flakes are the largest group of basalt flake tools (N=71). They originate mainly from the complex of Layer II-6, within which three levels, Level 1 (N=11), Level 2 (N=18), and Level 4 (N=17) are the richest (Figs. 7.17, 7.18). As in other tool types, most of the tools are slightly abraded, abraded, and fresh in descending order of frequency, and the bulk are patinated. Over 43% of the retouched flakes are complete, while distal, lateral, or proximal breaks occur on some 41% of the retouched flakes. As in other tool types, fragments of retouched flakes are minimally represented (N=5). Most of the tools are end-struck, but other directions of blow are represented as well. Some 70% of the retouched flakes are non-cortical, while the rest bear different degrees of cortex.

The predominant edge of retouch is the right edge (N=22), followed closely by the left edge (N=18). Only 24% of the

7.2 Typology of Basalt Flake Tools

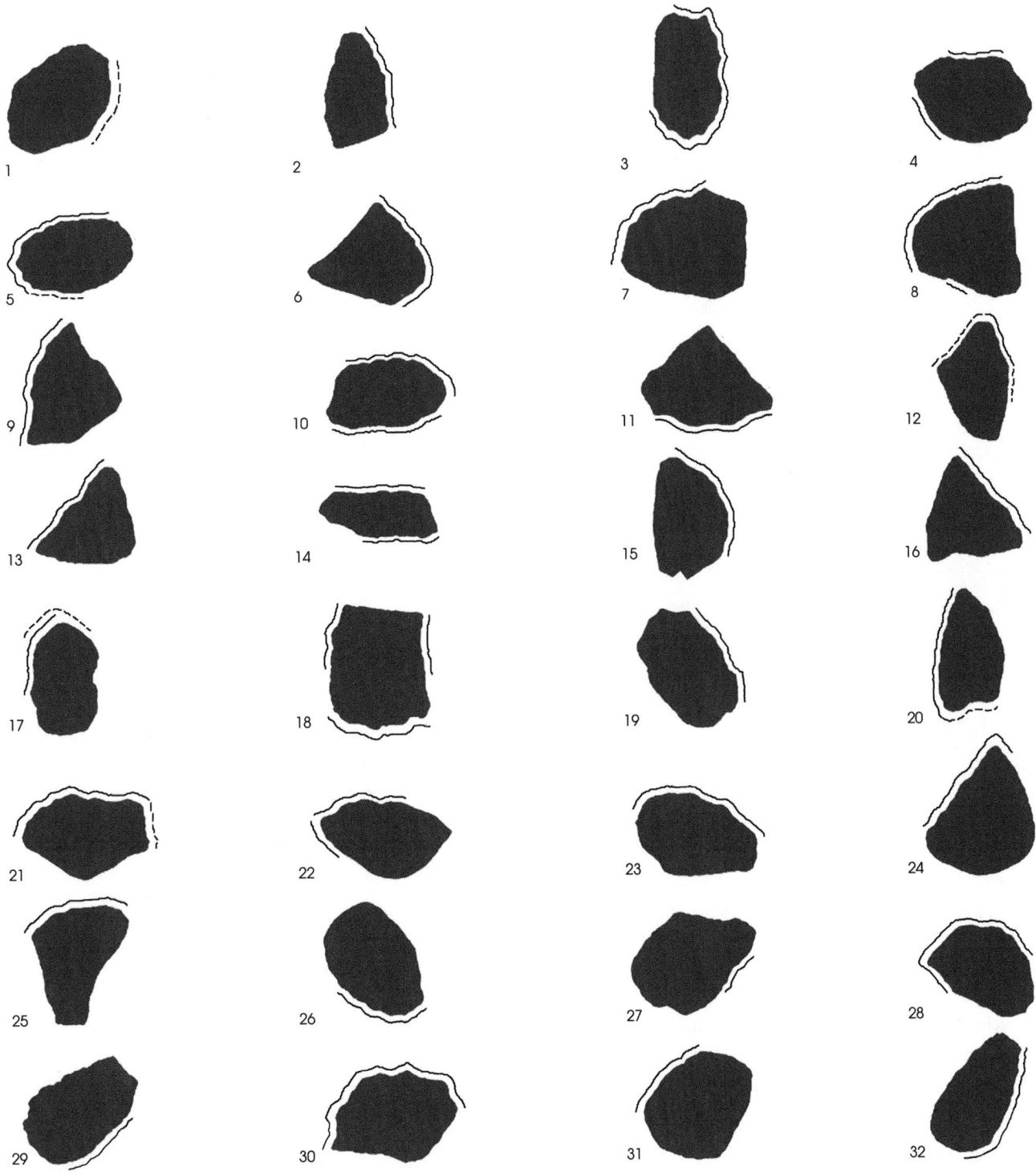

Fig. 7.16 Planform of selected of basalt massive scrapers from GBY. Lines mark the location of a retouched edge (a continuous line marks retouch on the dorsal face; a broken line marks retouch on the ventral face). (1) #13399, Layer II-6 Level 1; (2) #13411, Layer II-6 Level 4; (3) #13538, Layer II-6 Level 2; (4) #14190, Layer II-6 Level 2; (5) #14482, Layer II-6 Level 2; (6) #14739, Layer II-6 Level 2; (7) #16138, Trench II; (8) #2029, Layer II-6 Level 1; (9) #2032, Layer II-6 Level 1; (10) #2065, Layer II-6 Level 1; (11) #2077, Layer II-6 Level 1; (12) #29, Layer II-6 Level 4; (13) #4852, Layer II-5 Level 4; (14) # 4854, Layer II-6 Level 4; (15) #4887, Layer II-6 Level 4b; (16) #4913, Layer II-6 Level 4b; (17) #4920, Layer II-6 Level 4b; (18) #4943, II-6 Level 4; (19) #4959, Layer II-6 Level 4; (20) #5004, Layer II-6 Level 4; (21) #5338, Layer II-6 Level 1; (22) #5486, Layer V-6; (23) #5594, Layer II-6 Level 1; (24) #5608, Layer II-6 Level 4b; (25) #5671, Layer II-6 Level 6; (26) #5704, Layer II-6 Level 6; (27) #5726, Layer II-6 Level 4b; (28) #5739, Layer II-6 Level 4b; (29) #5838, Layer II-6 Level 2; (30) #8609, Layer II-6 Level 2; (31) #9247, Layer II-6 Level 6; (32) #4998, Layer II-6 Level 4

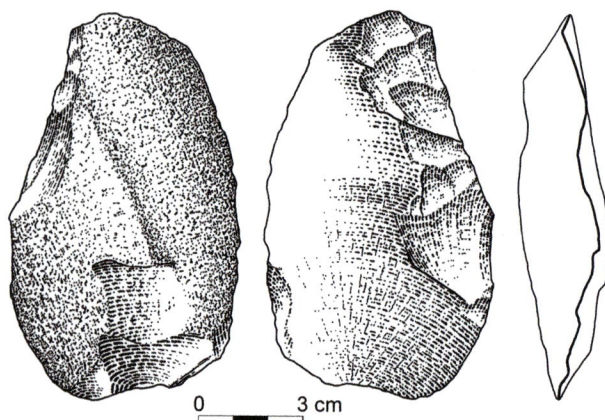

Fig. 7.17 Basalt retouched flake (#330 Layer II-6 Level 4)

Fig. 7.18 Basalt retouched wedge flake (#256 Layer II-6 Level 4)

retouched flakes bear regular retouch, followed by irregular retouch and then indeterminate retouch. The forms of the retouched edges are convex, wavy, or straight (36.84%, 28.07%, and 19.30% respectively) and the mean angle of retouch is 84.68°. In the majority of cases the retouched face is the dorsal face.

The characteristics of the retouched basalt flakes presented above illustrate their lack of standardization. This is further illustrated by their unstandardized size, which is presented in Table 7.8 and is particularly obvious from the extent of the standard deviations.

Multiple Tools

Although the number of basalt flake tools is very small, some of the characteristics that were observed for the flint flake tools are found here, such as the presence of multiple tools. Table 7.9 illustrates the presence of multiple tools, their specific tool associations, and their paucity. Worth noting is the co-occurrence of pitted stones (FFT) and retouched tools, combining two distinctly different tasks on the same tool.

Table 7.9 Counts of multiple tools on basalt flake tools

Typology A	Typology B						
	Naturally backed knife	Notch	Denticulate	Varia	Retouched flake	Modified	Pitted stone
Single concave side-scraper	-	1	-	-	-	-	1
Side-scraper on ventral face	-	-	1	-	-	1	-
Abruptly retouched side-scraper	-	2	-	-	-	-	-
Thinned back side-scraper	-	-	-	1	-	-	-
Massive scraper	-	2	-	-	-	-	-
Atypical burin	-	-	-	-	1	-	-
Atypical borer	-	-	-	-	3	-	-
Atypical backed knife	-	-	-	-	1	-	-
Notch	-	1	-	-	2	-	2
Denticulate	1	-	-	-	1	-	1
Varia	-	-	-	-	-	-	1
Retouched flake	-	-	-	-	1	1	3

7.3 Technology of Basalt Flakes

In contrast with the minimal amount of basalt flake tools, basalt flakes are numerous (N=5,221) and form the largest group of basalt artifacts at GBY. In the following sections we present the entire assemblage of unretouched basalt flakes. The presentation is divided into three sections, each describing a category of basalt flakes. The first category (Section 7.3.1) comprises all unretouched flakes according to two size groups. Within this category we also include artifacts with particular technological or morphological features (e.g. *éclats de taille de biface*); however, these distinct features are discussed in detail in the two following categories (Sections 7.3.2–7.3.3), where the distinction between different classes of flakes is based on technological features, size, and morphology (Fig. 7.19).

7.3 Technology of Basalt Flakes

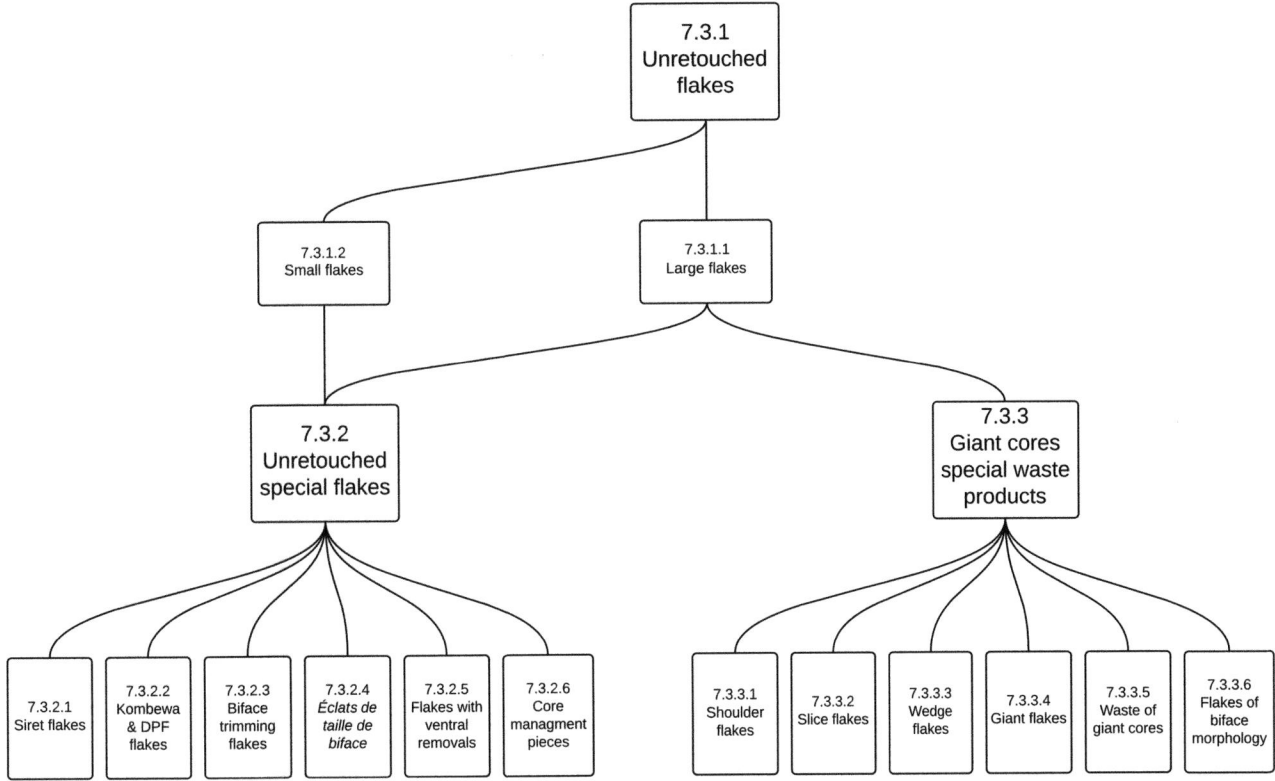

Fig. 7.19 Flow chart illustrating the order of presentation of different classes of basalt flakes section 7.3

Three additional groups of basalt flakes are present at GBY but are described separately under cores and core tools (Section 7.4). These include split percussors (N=88; discussed under basalt percussors), angular fragments (N=8), amorphous flakes (N=2), thin anvils (N=9), pitted stones (N=100), and giant cores (N=2).

7.3.1 Unretouched Flakes

Unretouched basalt flakes form the bulk of the basalt flake component. For the purpose of presentation we have divided this assemblage into two groups according to their size: large flakes that are 70–200 mm long and small flakes that are 20–69 mm long. Basalt flakes larger than 200 mm in length (N=5) are discussed below as a separate group of giant flakes (Section 7.3.3.4). The rationale for this separation lies in a number of previously published GBY studies that focus on the reduction sequence of basalt bifaces. The assumption is that flakes measuring 70 mm and above could have been suitable blanks for the production of bifaces (Section 7.5.1.1). This assumption is partially based on the size of the smallest bifaces of GBY, which serves here as the dividing line between the two groups of basalt flakes. In this we differ from Kleindienst (1961), who chose a dividing line of 100 mm on the basis of the East African Late Acheulian assemblages. In the following presentations we include only assemblages containing a minimum number of 30 unretouched flakes.

7.3.1.1 Large Flakes

The category of large basalt flakes comprises flakes measuring 70–200 mm and includes 1,638 artifacts. The size characteristics of this group are presented in Table 7.10, which shows a general trend of similarity in size among different assemblages, although somewhat larger flakes occur in the lowermost levels of Layer II-6. The angles of the striking platforms show minimal variation among the different levels.

Taphonomically, many of the large flakes are complete (Table 7.11). The percentage of complete flakes varies from level to level but can reach about half of the assemblage. As shown by Table 7.11, there is variability between assemblages, but in all layers complete is the dominant mode, followed by distally and laterally broken in similar frequencies. The category of fragments is represented in varying frequencies and, as for other attributes, particularly in the lowermost levels of Layer II-6 (Table 7.11). Like the flake tools, most of the large unretouched flakes are devoid of cortex. However, although all the different categories of cortical coverage are present, 75–100% is the second most frequent, a pattern that differs from that observed on flake tools

Table 7.10 Descriptive statistics of large basalt flakes (size in mm; angle in degrees) of Layer II-6*

	II-6/L1	II-6/L2	II-6/L3	II-6/L4	II-6/L4b	II-6/L5	II-6/L6	II-6/L7	VI-14
N	251	193	142	291	195	30	78	124	42
Maximum length	(N=250)								
Maximum	158	142	156	151	161	147	141	142	142
Median	87	82	85	92	87	83.5	85	88	92
Minimum	70	70	70	70	70	70	70	70	70
Mean	90.20	86.94	89.39	94.94	89.98	86.17	90.56	91.55	92.64
Std dev	18.01	15.18	16.57	18.42	16.08	15.68	17.24	16.95	16.46
Std err mean	1.14	1.09	1.39	1.08	1.15	2.86	1.95	1.52	2.54
Length	(N=250)								
Maximum	158	142	134	151	129	147	130	133	138
Median	76	74	74	78	76	69	76	74.5	74.5
Minimum	20	38	37	37	40	33	54	37	17
Mean	78.61	75.97	77.49	82.57	78.38	69.20	80.45	77.30	75.17
Std dev	21.03	17.57	19.17	20.17	18.46	20.45	17.43	18.06	20.84
Std err mean	1.33	1.26	1.61	1.18	1.32	3.73	1.97	1.62	3.22
Width	(N=249)								
Maximum	150	137	123	151	162	115	123	141	111
Median	68	65	70	72	68	71	66	75.5	78.5
Minimum	34	30	32	37	25	35	38	33	44
Mean	71.17	67.62	70.44	75.28	69.84	73.20	70.42	77.19	77.60
Std dev	21.07	18.37	19.74	21.75	19.77	19.08	21.02	20.22	18.09
Std err mean	1.33	1.32	1.66	1.27	1.42	3.48	2.38	1.82	2.79
Thickness	(N=250)								
Maximum	79	59	67	80	58	87	88	82	52
Median	31.5	31	31	32	31	30	35.5	37	32
Minimum	12	11	8	12	14	16	20	19	16
Mean	32.20	31.55	31.88	33.40	31.48	33.87	36.91	39.28	32.45
Std dev	9.81	8.47	9.22	10.42	9.20	14.81	12.57	10.94	8.20
Std err mean	0.62	0.61	0.77	0.61	0.66	2.70	1.42	0.98	1.26
L/W	(N=249)								
Maximum	2.72	2.53	2.31	2.24	2.92	1.71	1.89	2.24	2.14
Median	1.19	1.18	1.16	1.18	1.18	0.91	1.24	1.00	0.92
Minimum	0.39	0.50	0.49	0.39	0.47	0.46	0.56	0.53	0.24
Mean	1.19	1.21	1.19	1.18	1.21	0.99	1.22	1.07	1.03
Std dev	0.41	0.42	0.44	0.40	0.42	0.33	0.36	0.36	0.39
Std err mean	0.03	0.03	0.04	0.02	0.03	0.06	0.04	0.03	0.06
Angle of striking platform	(N=89)	(N=75)	(N=52)	(N=134)	(N=81)	(N=13)	(N=23)	(N=39)	(N=8)
Maximum	138	135	138	137	132	126	130	132	123
Median	118	115	120	117	115	109	114	112	111.5
Minimum	73	79	90	79	92	93	93	93	95
Mean	116.61	114.09	118.56	116.60	112.19	107.38	113.43	111.87	112.13
Std dev	11.68	10.27	8.94	10.04	9.16	10.07	8.61	10.57	9.60
Std err mean	1.24	1.19	1.24	0.87	1.02	2.79	1.80	1.69	3.39
Number of scars	(N=170)	(N=150)	(N=100)	(N=224)	(N=134)	(N=21)	(N=47)	(N=64)	(N=31)
Maximum	8	8	8	17	9	7	9	12	7
Median	2	2.5	3	3	3	3	2	3	3
Minimum	1	1	1	1	1	1	1	1	1
Mean	2.80	2.84	2.75	3.61	2.92	3.66	3.02	3.32	3.38
Std dev	1.61	1.69	1.38	2.40	1.67	1.46	1.81	1.81	1.92
Std err mean	0.12	0.13	0.13	0.16	0.14	0.31	0.26	0.22	0.34

* only layers with a minimum number of 30 large basalt flakes are included; excluding flakes of special morphology (core waste, angular fragment, wedge-like flake, shoulder/corner flake, and slice flake)

7.3 Technology of Basalt Flakes

Table 7.11 Counts and frequencies (%) of characteristics of large basalt flakes of Layer II-6*

	II-6/L1	II-6/L2	II-6/L3	II-6/L4	II-6/L4b	II-6/L5	II-6/L6	II-6/L7	VI-14
N	251	193	142	291	195	30	78	124	42
Breakage									(N=41)
Complete	127	89	71	148	71	9	34	38	12
	50.60	46.11	50.00	50.86	36.41	30.00	43.59	30.65	29.27
Distal	41	29	17	36	23	2	9	22	5
	16.33	15.03	11.97	12.37	11.79	6.67	11.54	17.74	12.20
Lateral	28	22	18	29	21	7	8	22	3
	11.16	11.40	12.67	9.97	10.77	23.33	10.26	17.74	7.32
Proximal	20	9	12	18	16	2	2	5	1
	7.97	4.66	8.45	6.19	8.21	6.67	2.56	4.03	2.44
Distal & lateral	7	14	2	16	19	4	8	11	5
	2.79	7.25	1.41	5.50	9.74	13.33	10.26	8.87	12.20
Distal & proximal	2	4	6	9	3	-	4	5	-
	0.80	2.07	4.23	3.09	1.54		5.13	4.03	
Fragment	14	16	9	23	25	5	9	15	12
	5.58	8.29	6.34	7.90	12.82	16.67	11.54	12.10	29.27
Proximal & lateral	5	5	3	8	7	-	3	3	3
	1.99	2.59	2.11	2.75	3.59		3.85	2.42	7.32
Indeterminate	7	5	4	4	10	1	1	3	-
	2.79	2.59	2.82	1.37	5.13	3.33	1.28	2.42	
Cortex	(N=247)	(N=191)	(N=141)	(N=290)	(N=194)				
No cortex	170	109	96	197	122	22	46	75	31
	68.83	57.07	67.61	67.93	62.89	73.33	58.97	60.48	73.81
1–25%	12	7	6	11	4	-	-	4	1
	4.86	3.66	4.23	3.79	2.06			3.23	2.38
26–50%	12	8	3	11	12	-	-	2	1
	4.86	4.19	2.11	3.79	6.19			1.61	2.38
51–75%	9	10	4	12	9	2	2	1	1
	3.64	5.24	2.82	4.14	4.64	6.67	2.56	0.81	2.38
76–100%	30	31	18	33	26	3	13	9	2
	12.15	16.23	12.68	11.38	13.40	10.00	16.67	7.26	4.76
Indeterminate	14	26	15	26	21	3	17	33	6
	5.67	13.61	10.56	8.97	10.82	10.00	21.79	26.61	14.29
Direction of blow	(N=243)	(N=191)	(N=141)	(N=276)	(N=190)		(N=73)	(N=113)	(N=41)
End-struck	126	92	67	129	86	8	31	22	2
	53.85	48.17	47.52	46.74	45.26	26.67	42.47	19.47	4.88
Side-struck	56	33	31	51	36	11	15	21	4
	23.93	17.28	21.99	18.48	18.95	36.67	20.55	18.58	9.76
Special side-struck	14	15	7	20	13	2	4	9	3
	5.98	7.85	4.96	7.25	6.84	6.67	5.48	7.96	7.32
Indeterminate	38	51	36	76	55	9	23	61	32
	16.24	26.70	25.53	27.54	28.95	30.00	31.51	53.98	78.05

* only layers with a minimum number of 30 large basalt flakes are included; excluding flakes of special morphology (core waste, angular fragment, wedge-like flake, shoulder/corner flake, and slice flake)

(Table 7.11). It is worth mentioning the substantial number of cases in which the extent of cortical coverage could not be identified, particularly in the lower levels of Layer II-6.

Most of the large basalt flakes are end-struck, followed by side-struck. Special side-struck flakes are represented by values lower than 8% in each of the excavated assemblages. One should note that on a substantial number of large flakes the direction of blow is indeterminate, including assemblages in which this is the case for over 50% of the items (Table 7.11).

On slightly less than half of the large flakes the scar pattern is indeterminate. Apart from these, the discernible patterns are primarily cortical, plain, and along axis. Each appears in different frequencies and additional patterns are present in low frequencies. An interesting feature is the radial scar pattern, which reaches 10% in some levels (Table 7.12). The mean number of scars on basalt flakes in

Table 7.12 Counts and frequencies (%) of technological characteristics of large basalt flakes of Layer II-6*

	II-6/L1	II-6/L2	II-6/L3	II-6/L4	II-6/L4b	II-6/L5	II-6/L6	II-6/L7	VI-14
N	251	193	142	291	195	30	78	124	42
Scar pattern	(N=230)	(N=190)	(N=138)	(N=286)	(N=192)		(N=73)	(N=101)	(N=40)
Cortical	31 13.48	30 15.79	16 11.59	27 9.44	26 13.54	5 16.67	13 17.81	10 9.90	4 10.00
Plain	20 8.70	14 7.37	13 9.42	32 11.19	12 6.25	6 20.00	5 6.85	13 12.87	9 22.50
Along axis	26 11.30	21 11.05	12 8.70	35 12.24	23 11.98	1 3.33	3 4.11	8 7.92	2 5.00
Parallel	1 0.43	-	-	-	-	-	-	-	-
Convergent	-	-	-	1 0.35	-	1 3.33	-	-	-
Opposed	4 1.74	5 2.63	2 1.45	6 2.10	3 1.56	-	2 2.74	-	-
Radial	13 5.65	14 7.37	5 3.62	28 9.79	6 3.13	1 3.33	5 6.85	5 4.95	4 10.00
Ridged	-	1 0.53	3 2.17	6 2.10	3 1.56	1 3.33	-	-	-
Side	10 4.35	6 3.16	3 2.17	10 3.50	10 5.21	-	1 1.37	2 1.98	-
Along axis & side	13 5.65	1 0.53	10 7.25	14 4.90	10 5.21	2 6.67	-	6 5.94	1 2.50
Along axis & opposed	6 2.61	4 2.11	3 2.17	10 3.50	1 0.52	1 3.33	3 4.11	2 1.98	-
Opposed & side	1 0.43	4 2.11	1 0.72	8 2.80	1 0.52	-	2 2.74	1 0.99	-
Along axis & radial	1 0.43	-	-	-	-	-	-	-	-
Indeterminate	104 45.22	90 47.37	70 50.72	109 38.11	97 50.52	12 40.00	39 53.42	54 53.47	20 50.00
Striking platform	(N=230)	(N=190)	(N=136)	(N=283)	(N=191)	(N=28)	(N=71)	(N=105)	(N=40)
Cortical	5 2.17	5 2.63	4 2.94	6 2.12	3 1.57	-	1 1.41	3 2.86	-
Punctiform	-	-	-	-	1 0.52	1 3.57	1 1.41	-	-
Plain	121 52.61	81 42.63	55 40.44	120 42.40	75 39.27	11 39.29	27 38.03	40 38.10	9 22.50
Dihedral	3 1.30	1 0.53	2 1.47	5 1.77	6 3.14	1 3.57	-	1 0.95	-
Facetted	11 4.78	8 4.21	3 2.21	13 4.59	7 3.66	3 10.71	3 4.23	2 1.90	-
Crushed	-	-	-	1 0.35	1 0.52	-	-	1 0.95	-
Removed	7 3.04	7 3.68	9 6.62	16 5.65	8 4.19	2 7.14	2 2.82	1 0.95	3 7.50
Broken	35 15.22	35 18.42	30 22.06	66 23.32	53 27.75	7 25.00	17 23.94	22 20.95	15 37.50
Indeterminate	48 20.87	53 27.89	33 24.26	56 19.79	37 19.37	3 10.71	20 28.17	35 33.33	13 32.50

* only layers with a minimum number of 30 large basalt flakes are included; excluding flakes of special morphology (core waste, angular fragment, wedge-like flake, shoulder/corner flake, and slice flake)

7.3 Technology of Basalt Flakes

all assemblages is larger than 2 and smaller than 4 scars per flake (Table 7.10).

Table 7.12 provides data on the type of striking platform in large unretouched basalt flakes. The most frequent type is plain, which in some of the layers characterizes half of the artifacts. It should be noted, however, that many of the striking platforms are either indeterminate or broken. Those that are broken are significantly less common ($\chi^2=59.29$; df=8; p<0.0001) than the breakage values observed for the small basalt flakes (Section 7.3.1.2). It is interesting that in a number of cases, albeit minimal, the striking platforms were removed, a phenomenon similar to that observed for the small flint flakes (Chapter 6). Only 5.61% of the large basalt flakes feature a lipped striking platform, a value that is very similar to that of flint flakes (Chapter 6).

Some particular technological features, such as *outrepassé* (Fig. 7.20), *débordant*, hinge (Figs. 7.21, 7.22), and steps (Fig. 7.23), are recorded on the basalt flakes (both large and small). Due to their scarcity in each of the two size groups, we present their types and frequencies together in Table 7.13. This scarcity indicates that they are not a repetitive component of the assemblage but rather a random result of knapping.

7.3.1.2 Small Flakes

This group of basalt flakes (20–69 mm in length) comprises 3,311 basalt flakes and is the largest group of artifacts in the complex of Layer II-6. With the exception of Layer V-5 (and the relatively small sample of Layer II-5/6), the mean sizes of flakes are generally very similar (Table 7.14). The data in Table 7.14 illustrate the high similarity of values of different size attributes among the different assemblages (with the same reservations as those expressed above). Layer VI-14 (the spatial continuation of Layer II-6 Level 1) displays somewhat larger sizes, expressed by slightly higher width values. The similarities among the

Fig. 7.21 Basalt flake with hinge (#437 Layer II-6 Level 4)

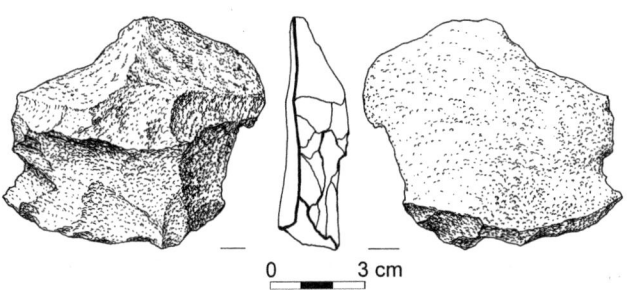

Fig. 7.22 Basalt flake with hinge (#5583 Layer II-6 Level 1)

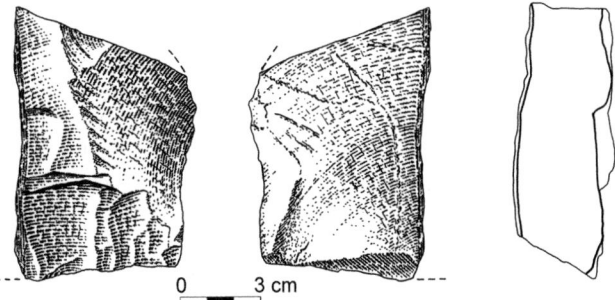

Fig. 7.23 Basalt Siret flake with steps (#5047 Layer II-6 Level 4)

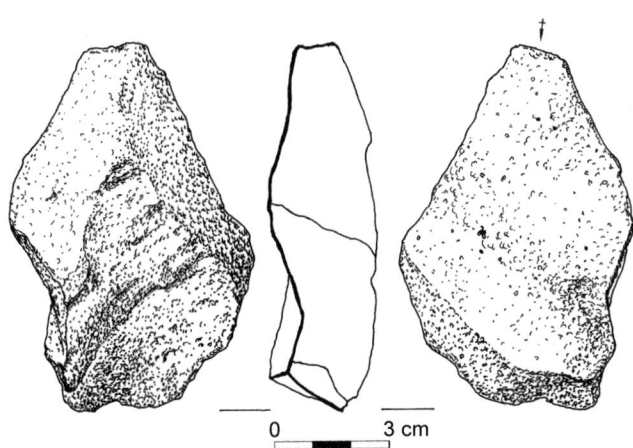

Fig. 7.20 Basalt *outrepassé* flake (#5740 Layer II-6 Level 1)

Table 7.13 Counts and frequencies (%) of special technological features of large and small basalt flakes

Technology	Large flakes		Small flakes	
	N	%	N	%
Outrepassé	5	4.24	2	2.02
Débordant	21	17.80	26	26.21
Hinge	11	9.32	5	5.05
Steps	6	5.08	6	6.06
Siret*	75	63.56	60	60.62
Total	118	100.00	99	100.00

* this technological feature is further described below (Section 7.3.2)

Table 7.14 Descriptive statistics of small basalt flakes (size in mm; angle in degrees)*

	V-5	II-5	II-5/6	II-6/L1	II-6/L2	II-6/L3	II-6/L4	II-6/L4b	II-6/L5	II-6/L6	II-6/L7	VI-14
N	67	67	30	522	543	522	374	246	127	209	360	36
Maximum length	(N=66)			(N=521)								
Maximum	63	69	68	69	69	69	69	69	69	69	69	69
Median	25	51	30	39	44	43	43	49	39	38	36	50
Minimum	20	20	22	20	21	20	21	20	21	21	20	22
Mean	28.55	47.79	36.67	41.07	44.48	43.16	43.95	47.17	40.48	40.94	39.80	46.78
Std dev	9.91	14.38	14.45	14.55	14.38	13.26	14.76	13.50	13.79	13.56	14.04	14.09
Std err mean	1.22	1.76	2.64	0.64	0.62	0.58	0.76	0.86	1.22	0.94	0.74	2.35
Length	(N=66)			(N=520)	(N=541)	(N=521)						
Maximum	55	66	68	69	69	69	69	69	68	69	69	64
Median	22.5	42	29.5	33	38	36	37	41	33	34	33	36
Minimum	14	15	15	11	12	13	12	13	16	12	15	15
Mean	25.06	43.69	33.27	35.38	39.32	37.56	38.60	41.88	36.75	36.14	35.77	36.14
Std dev	8.43	15.18	14.15	13.87	14.48	12.96	14.29	13.37	13.72	13.73	13.23	12.77
Std err mean	1.04	1.85	2.58	0.61	0.62	0.57	0.74	0.85	1.22	0.95	0.70	2.13
Width	(N=66)			(N=519)	(N=541)	(N=521)						
Maximum	51	67	60	69	68	76	68	69	86	69	72	69
Median	18.5	38	27	32	34	35	34	38	29	30	28	44
Minimum	10	13	13	11	11	11	13	13	8	14	11	21
Mean	20.53	38.06	29.53	34.02	36.31	35.95	35.82	37.95	32.30	32.79	31.38	42.75
Std dev	8.13	12.82	12.73	13.63	13.16	13.26	14.08	13.08	13.22	12.70	12.91	13.88
Std err mean	1.00	1.57	2.32	0.60	0.57	0.58	0.73	0.83	1.17	0.88	0.68	2.31
Thickness	(N=66)			(N=521)								
Maximum	36	43	30	52	56	51	39	40	38	37	42	34
Median	9	18	11	15	16	16	16	17	14	15	14	17.5
Minimum	4	4	6	4	5	4	4	5	4	5	4	8
Mean	10.18	19.12	13.97	15.85	17.32	16.55	16.69	17.70	16.44	16.75	16.26	18.72
Std dev	4.18	8.81	6.63	7.53	7.15	7.00	7.73	7.07	8.05	7.32	7.67	7.40
Std err mean	0.52	1.08	1.21	0.33	0.31	0.31	0.40	0.45	0.71	0.51	0.40	1.23
L/W	(N=66)			(N=519)	(N=541)	(N=521)						
Maximum	2.64	3.38	1.92	2.75	3.17	2.64	3.29	2.52	2.50	2.59	2.25	1.66
Median	1.32	1.24	1.17	1.05	1.11	1.05	1.12	1.15	1.19	1.14	1.16	0.80
Minimum	0.48	0.56	0.61	0.43	0.40	0.42	0.46	0.47	0.54	0.44	0.40	0.52
Mean	1.30	1.21	1.18	1.09	1.14	1.12	1.15	1.18	1.21	1.16	1.20	0.87
Std dev	0.37	0.44	0.38	0.34	0.37	0.38	0.38	0.39	0.37	0.38	0.35	0.27
Std err mean	0.05	0.05	0.07	0.01	0.02	0.02	0.02	0.02	0.03	0.03	0.02	0.05
Angle of striking platform	(N=9)	(N=18)	(N=4)	(N=146)	(N=153)	(N=122)	(N=101)	(N=69)	(N=32)	(N=34)	(N=73)	(N=8)
Maximum	118	135	108	147	132	133	140	134	131	128	130	133
Median	116	116.5	99	113	113	115	114	112	112.5	107.5	107	114.5
Minimum	102	93	92	87	79	84	90	90	94	88	1	93
Mean	112.78	112.89	99.50	113.19	111.78	114.72	113.25	111.26	111.75	107.32	104.71	111.75
Std dev	5.87	12.03	7.33	10.24	10.71	8.54	10.17	8.43	9.48	10.15	16.45	12.86
Std err mean	1.96	2.84	3.66	0.85	0.87	0.77	1.01	1.01	1.68	1.74	1.93	4.55
Number of scars	(N=52)	(N=31)	(N=23)	(N=369)	(N=348)	(N=360)	(N=259)	(N=140)	(N=49)	(N=82)	(N=147)	(N=23)
Maximum	6	6	8	7	10	11	8	8	10	8	6	
Median	2	3	2	2	2	2	2	2	2	2	2	2
Minimum	1	1	1	1	1	1	1	1	1	1	1	1
Mean	2.17	2.90	3.30	2.29	2.58	2.31	2.53	2.66	2.40	2.84	2.50	2.52
Std dev	1.21	1.42	2.34	1.10	1.43	1.43	1.38	1.52	1.47	1.72	1.30	1.34
Std err mean	0.16	0.25	0.48	0.05	0.07	0.07	0.08	0.12	0.21	0.19	0.10	0.28

* only layers with a minimum number of 30 small basalt flakes are included; excluding flakes of special morphology (core waste, angular fragment, wedge-like flake, shoulder/corner flake, and slice flake)

7.3 Technology of Basalt Flakes

Table 7.15 Counts and frequencies (%) of characteristics of small basalt flakes*

	V-5	II-5	II-5/6	II-6/L1	II-6/L2	II-6/L3	II-6/L4	II-6/L4b	II-6/L5	II-6/L6	II-6/L7	VI-14
N	67	67	30	522	543	522	374	246	127	209	360	36
Breakage	(N=66)											
Complete	18	27	9	188	127	159	98	61	33	56	80	22
	27.27	40.30	30.00	36.02	23.39	30.46	26.20	24.80	25.98	26.79	22.22	30.56
Distal	4	6	3	67	64	43	46	26	6	19	24	6
	6.06	8.96	10.00	12.84	11.79	8.24	12.30	10.57	4.72	9.09	6.67	16.67
Lateral	2	2	4	51	65	49	30	25	11	22	20	1
	3.03	2.99	13.33	9.77	11.97	9.39	8.02	10.16	8.66	10.53	5.56	2.78
Proximal	8	5	3	56	45	94	46	28	4	18	27	4
	12.12	7.46	10.00	10.73	8.29	18.01	12.30	11.38	3.15	8.61	7.50	11.11
Distal & lateral	-	4	1	17	41	22	28	10	10	9	30	2
		5.97	3.33	3.26	7.55	4.21	7.49	4.07	7.87	4.31	8.33	5.56
Distal & proximal	1	1	2	16	30	13	16	11	4	8	15	-
	1.52	1.49	6.67	3.07	5.52	2.49	4.28	4.47	3.15	3.83	4.17	
Fragment	32	19	7	104	132	112	76	64	47	64	144	12
	48.48	28.36	23.33	19.92	24.31	21.46	20.32	26.02	37.01	30.62	40.00	33.33
Proximal & lateral	-	2	1	13	28	21	30	7	5	5	15	-
		2.99	3.33	2.49	5.16	4.02	8.02	2.85	3.94	2.39	4.17	
Indeterminate	1	1	-	10	11	9	4	14	7	8	5	-
	1.52	1.49		1.92	2.03	1.72	1.07	5.69	5.51	3.83	1.39	
Cortex	(N=66)			(N=519)	(N=541)	(N=517)	(N=372)	(N=244)				
No cortex	45	58	26	405	321	411	279	182	73	129	170	32
	68.18	86.57	86.67	78.03	59.33	79.50	75.00	74.59	57.48	61.72	47.22	88.89
1–25%	-	1	-	18	17	6	8	3	-	3	2	-
		1.49		3.47	3.14	1.16	2.15	1.23		1.44	0.56	
26–50%	-	-	1	9	11	10	12	4	-	6	3	-
			3.33	1.73	2.03	1.93	3.23	1.64		2.87	0.83	
51–75%	-	-	-	5	18	14	15	5	3	4	5	-
				0.96	3.33	2.71	4.03	2.05	2.36	1.91	1.39	
76–100%	5	5	1	50	85	33	30	18	10	15	27	1
	7.58	7.46	3.33	9.63	15.71	6.38	8.06	7.38	7.87	7.18	7.50	2.78
Indeterminate	16	3	2	32	89	43	28	32	41	52	153	3
	24.24	4.48	6.67	6.17	16.45	8.32	7.53	13.11	32.28	24.88	42.50	8.33
Direction of blow	(N=63)	(N=59)		(N=477)	(N=528)	(N=509)	(N=372)	(N=241)	(N=125)	(N=172)	(N=327)	
End-struck	10	28	12	162	163	173	117	72	25	52	68	-
	15.87	47.46	40.00	33.96	30.87	33.99	31.45	29.88	20.00	30.23	20.80	
Side-struck	2	13	4	95	83	102	49	33	18	29	38	1
	3.17	22.03	13.33	19.92	15.72	20.04	13.17	13.69	14.40	16.86	11.62	2.78
Special side-struck	3	1	2	15	27	13	10	4	2	7	9	2
	4.76	1.69	6.67	3.14	5.11	2.55	2.69	1.66	1.60	4.07	2.75	5.56
Indeterminate	48	17	12	205	255	221	196	132	80	84	212	33
	76.19	28.81	40.00	42.98	48.30	43.42	52.69	54.77	64.00	48.84	64.83	91.67

* only layers with a minimum number of 30 small basalt flakes are included; excluding flakes of special morphology (core waste, angular fragment, wedge-like flake, shoulder/corner flake, and slice flake)

small basalt flakes of the different layers are also expressed in the angles of the striking platforms.

From a taphonomic perspective, the pattern of breakage (Table 7.15) shows that the highest frequencies are of fragments. The frequencies of complete flakes, however, range between 20% and 40% in all assemblages. The lowest values for complete flakes are in Layer V-5, in which the highest values of fragments occur. Within Layer II-6, the lowermost part, particularly Level 7, displays the highest frequencies of fragments (30–40%). Although this could have been interpreted as resulting from extensive taphonomic processes, but nevertheless the frequencies of complete flakes in these levels do not diverge from those that generally characterize other levels. It is possible that the fragmentation of small basalt flakes is not a taphonomic effect but rather results from spontaneous shattering during the production and manufacture of bifaces (Section 7.6). The majority of small basalt flakes are non-cortical, but when cortical coverage does occur it occupies 75–100% of the flakes (Table 7.15), resembling in this to the large unretouched basalt flakes and differing from basalt flake tools. The cortical coverage on basalt flakes is an enigmatic issue discussed in Chapter 4. The relatively high frequencies observed in the indeterminate category may be explained by our inability to differentiate between the cortical and non-cortical surfaces of some of the basalt flakes. The high values of indeterminate cortical coverage (Layer II-6 Levels 5–7) might well be interpreted (as shown in Chapter 6) as resulting from postdepositional processes. However, the variability in the amount of basalt flakes that are devoid of cortex or are fully cortical is clearly not associated with taphonomic agents but rather reflects particular modes of knapping that vary among the archaeological horizons.

The manner in which the flakes were detached from the cores is partially evident from the attribute describing the

Table 7.16 Counts and frequencies (%) of technological characteristics of small basalt flakes*

	V-5	II-5	II-5/6	II-6/L1	II-6/L2	II-6/L3	II-6/L4	II-6/L4b	II-6/L5	II-6/L6	II-6/L7	VI-14
N	67	67	30	522	543	522	374	246	127	209	360	36
Scar pattern	(N=62)	(N=57)	(N=27)	(N=467)	(N=528)	(N=502)	(N=362)	(N=239)	(N=123)	(N=153)	(N=313)	
Cortical	5	3	1	41	76	33	28	15	10	17	20	-
	8.06	5.26	3.70	8.78	14.39	6.57	7.73	6.28	8.13	11.11	6.39	
Plain	8	9	1	54	26	41	39	23	14	13	23	7
	12.90	15.79	3.70	11.56	4.92	8.17	10.77	9.62	11.38	8.50	7.35	19.44
Along axis	-	4	1	58	40	84	48	29	8	10	24	2
		7.02	3.70	12.42	7.58	16.73	13.26	12.13	6.50	6.54	7.67	5.56
Parallel	-	1	-	-	-	-	-	-	-	-	-	-
		1.75										
Convergent	-	1	-	1	-	1	2	1	1	2	1	-
		1.75		0.21		0.20	0.55	0.42	0.81	1.31	0.32	
Opposed	-	-	1	3	6	5	6	2	-	4	1	-
			3.70	0.64	1.14	1.00	1.66	0.84		2.61	0.32	
Radial	-	2	2	11	15	25	8	1	1	4	3	2
		3.51	7.41	2.36	2.84	4.98	2.21	0.42	0.81	2.61	0.96	5.56
Ridged	1	-	-	4	1	3	-	2	2	1	-	-
	1.61			0.86	0.19	0.60		0.84	1.63	0.65		
Side	1	2	1	20	13	19	4	4	-	3	1	-
	1.61	3.51	3.70	4.28	2.46	3.78	1.10	1.67		1.96	0.32	
Along axis & side	1	1	1	3	10	15	10	3	1	3	2	-
	1.61	1.75	3.70	0.64	1.89	2.99	2.76	1.26	0.81	1.96	0.64	
Along axis & opposed	-	1	3	9	8	9	6	3	1	2	3	-
		1.75	11.11	1.93	1.52	1.79	1.66	1.26	0.81	1.31	0.96	
Opposed & side	-	-	-	3	3	6	4	3	-	-	-	-
				0.64	0.57	1.20	1.10	1.26				
Indeterminate	46	33	16	260	330	261	207	153	85	94	235	25
	74.19	57.89	59.26	55.67	62.50	51.99	57.18	64.02	69.11	61.44	75.08	69.44
Striking platform	(N=58)	(N=48)	(N=25)	(N=463)	(N=522)	(N=489)	(N=355)	(N=238)	(N=121)	(N=161)	(N=318)	(N=34)
Cortical	-	-	-	5	13	2	5	1	2	2	3	-
				1.08	2.49	0.41	1.41	0.42	1.65	1.24	0.94	
Punctiform	-	1	-	2	2	1	2	1	-	3	2	-
		2.08		0.43	0.38	0.20	0.56	0.42		1.86	0.63	
Plain	4	19	10	154	148	153	105	67	27	40	64	6
	6.90	39.58	40.00	33.26	28.35	31.29	29.58	28.15	22.31	24.84	20.13	17.65
Dihedral	1	2	1	6	10	5	1	4	1	-	2	-
	1.72	4.17	4.00	1.30	1.92	1.02	0.28	1.68	0.83		0.63	
Facetted	1	1	2	13	15	3	18	1	1	3	6	1
	1.72	2.08	8.00	2.81	2.87	0.61	5.07	0.42	0.83	1.86	1.89	2.94
Crushed	1	-	-	-	1	-	1	-	-	-	1	-
	1.72				0.19		0.28				0.31	
Removed	-	1	1	6	6	8	4	5	-	1	1	-
		2.08	4.00	1.30	1.15	1.64	1.13	2.10		0.62	0.31	
Broken	35	11	8	160	208	190	150	97	49	63	164	18
	60.34	22.92	32.00	34.56	39.85	38.85	42.25	40.76	40.50	39.13	51.57	52.94
Indeterminate	16	13	3	117	119	127	69	62	41	49	75	9
	27.59	27.08	12.00	25.27	22.80	25.97	19.44	26.05	33.88	30.43	23.58	26.47

* only layers with a minimum number of 30 small basalt flakes are included; excluding flakes of special morphology (core waste, angular fragment, wedge-like flake, shoulder/corner flake, and slice flake)

7.3 Technology of Basalt Flakes

direction of blow. The small basalt flakes discussed here are characterized by very high frequencies of indeterminate direction of blow. While the values of this category vary greatly, in the largest samples they form about half of the cases, a situation that is probably partially associated with the frequent occurrence of fragmented flakes. Within the identifiable cases, the end-struck mode is the most common, followed by side-struck at about half the frequency. The special side-struck mode forms a negligible percentage, reaching 5% in only two layers (Table 7.15). This pattern is similar to that observed for the large basalt flakes.

With regard to the scar pattern, in over 50% of the artifacts this is indeterminate in all layers (Table 7.16). Again, much of this uncertainty is caused by the fragmented nature of many of the flakes. In the rest, the dominant pattern alternates between plain and along axis, with a single assemblage (Layer II-6 Level 2) in which over 14% are cortical flakes (Table 7.16). The small basalt flakes bear 2–3 scars on average on their dorsal surface (Table 7.14). Analysis of the striking platforms shows that when the high frequencies of broken and indeterminate striking platforms are excluded, the dominant type is plain; there is a minimal representation of all the other striking platform types, and removed platforms occur in low frequencies (Table 7.16). Additionally, 2.98% of the small basalt flakes have lipped striking platforms (N=93).

There is a general resemblance between the unretouched basalt flakes of the two size categories. A variety of technological attributes, such as cortical coverage, scar pattern, and striking platform, exhibit similar distributions in both categories. This may suggest that the flakes of the two size categories result from the same reduction sequence.

7.3.2 Unretouched Special Flakes

This section presents the analyses of several special technological features observed on basalt flakes. Table 7.17 presents the occurrence of these technological features

Table 7.17 Counts and frequencies of unretouched special basalt flakes

Layer	Siret	DPF	Kombewa	Biface sharpening flake	Éclat de taille de biface	Ventral removal	Core management piece
V-5	1 0.68	1 0.52	1 0.65	-	6 6.38	-	-
V-6	2 1.35	1 0.52	1 0.65	-	4 4.26	-	-
JB	-	1 0.52	1 0.65	-	3 3.19	-	-
II-2	-	-	-	-	1 1.06	-	-
II-5	-	4 2.09	-	-	3 3.19	6 1.75	-
II-5/6	1 0.68	1 0.52	2 1.31	-	-	5 1.46	-
II-6/L1	21 14.19	25 13.09	24 15.69	32 10.46	6 6.38	38 11.08	3 27.28
II-6/L2	24 16.22	26 13.61	14 9.15	67 21.90	7 7.45	66 19.24	2 18.17
II-6/L3	12 8.11	55 28.80	21 13.73	82 26.80	6 6.38	26 7.58	3 27.28
II-6/L4	42 28.38	39 20.42	52 33.99	77 25.16	21 22.34	72 20.99	1 9.09
II-6/L4b	22 14.86	24 12.57	13 8.50	30 9.80	5 5.32	50 14.58	-
II-6/L5	1 0.68	3 1.57	3 1.96	6 1.96	14 14.89	12 3.50	-
II-6/L6	6 4.05	6 3.14	7 4.58	3 0.98	7 7.45	25 7.29	1 9.09
II-6/L7	11 7.43	5 2.62	8 5.23	8 2.61	10 10.64	31 9.04	1 9.09
VI-14	1 0.68	-	2 1.31	1 0.33	1 1.06	6 1.75	-
Trench II green	4 2.70	-	4 2.61	-	-	6 1.75	-
Total	148 100.00	191 100.00	153 100.00	306 100.00	94 100.00	343 100.00	11 100.00

throughout the cultural sequence. It is worth noting that a single artifact can bear several technological features that are counted independently under each of the features. Each of these special flake types is analyzed and discussed below.

7.3.2.1 Siret Flakes (Brut)

Éclats Siret (Siret 1933; Inizan et al. 1999) is the name given to flakes that during knapping were spontaneously split longitudinally into two parts in accordance with the knapping axis. In classical cases the striking platform and the bulb of percussion are symmetrically split. Although there are many deviations from the above, in the present work we have attempted, despite the difficulties caused by the preservation state of the basalt, to include only classical cases. Siret flakes are considered to be the result of a very forceful blow and are recorded in various frequencies as early as the Lower Paleolithic, as well as later (e.g.

Fig. 7.24 Basalt Siret flake (#2029 Layer II-6 Level 1)

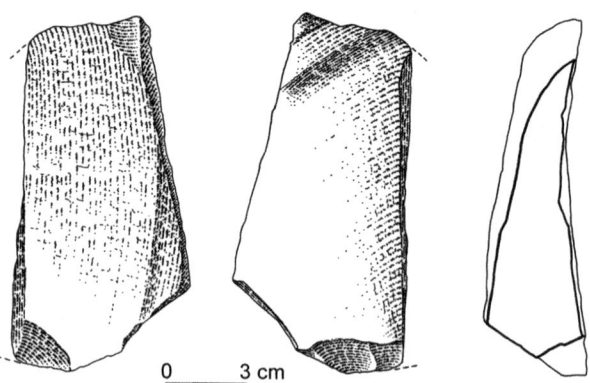

Fig. 7.27 Basalt Siret flake (#4960 Layer II-6 Level 4)

Fig. 7.25 Basalt Siret flake (#13252 Layer II-6 Level 1)

Fig. 7.28 Basalt Siret flake (#5575 Layer II-6 Level 4b)

Table 7.18 Counts and frequencies (%) of basalt Siret flakes on tools and blanks

Typology	N	%
Atypical end-scraper	1	0.68
Atypical burin	1	0.68
Notch	3	2.03
End-notched piece	1	0.68
Flake	129	87.16
Blade	1	0.68
Éclat de taille de biface	3	2.03
Retouched flake	1	0.68
Massive scraper	8	5.41
Total	148	100.00

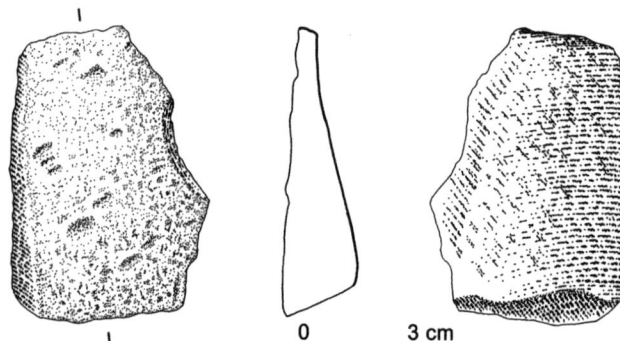

Fig. 7.26 Basalt Siret flake (#854 Layer II-6 Level 4)

7.3 Technology of Basalt Flakes

Malinsky Buller et al. 2011; for a relatively high frequency, see Zaidner 2010).

Basalt Siret flakes are present along the entire GBY cultural sequence but are particularly common in the upper levels of Layer II-6 (Table 7.17). Like other technological phenomena described in this section, the bulk of the basalt Siret flakes are not retouched and tools on Siret flakes occur very rarely (Figs. 7.23–7.28), except perhaps in the case of massive scrapers (made on thicker flakes) (Table 7.18). In sum, it seems that these flakes, which were not predetermined, were not selected for further modification. The size of the basalt Siret flakes varies substantially, as seen in Table 7.19, especially when the

Table 7.19 Descriptive statistics of basalt Siret flakes of Layer II-6 (size in mm; angle in degrees)

	II-6/L1	II-6/L2	II-6/L3	II-6/L4	II-6/L4b	II-6/L5	II-6/L6	II-6/L7
N	21	24	12	42	22	1	6	11
Maximum length								
Maximum	137	96	133	145	107	82	100	135
Median	71	62	80	78	74.5	82	71.5	63
Minimum	22	32	50	25	40	82	41	49
Mean	70.62	65.75	84.08	77.74	77.23	82.00	72.00	72.91
Std dev	27.19	20.58	26.06	28.09	17.54	-	18.83	25.89
Std err mean	5.93	4.20	7.52	4.33	3.74	-	7.69	7.80
Length								
Maximum	126	88	127	145	104	82	95	133
Median	63	54	77.5	73.5	72	82	71	63
Minimum	21	28	47	25	40	82	41	46
Mean	64.90	56.00	78.92	74.43	73.05	82.00	69.33	70.09
Std dev	27.16	19.44	25.02	28.81	16.63	-	17.18	26.04
Std err mean	5.93	3.97	7.22	4.45	3.55	-	7.01	7.85
Width								
Maximum	88	96	86	82	76	48	66	83
Median	51	45.5	55.5	51	53	48	48.5	46
Minimum	13	27	36	14	16	48	22	24
Mean	49.29	49.33	57.92	50.98	49.91	48.00	47.83	48.00
Std dev	18.31	18.80	15.74	16.51	14.60	-	15.87	16.54
Std err mean	4.00	3.84	4.54	2.55	3.11	-	6.48	4.99
Thickness								
Maximum	45	43	43	47	43	28	30	50
Median	23	20	24.5	23	22	28	25.5	23
Minimum	6	9	15	6	11	28	14	14
Mean	24.24	21.33	27.75	23.90	25.55	28.00	24.00	27.00
Std dev	10.44	8.61	9.93	9.29	9.52	-	5.76	11.12
Std err mean	2.28	1.76	2.87	1.43	2.03	-	2.35	3.35
L/W								
Maximum	2.00	2.00	1.94	2.11	2.57	1.71	1.86	2.25
Median	1.47	1.33	1.38	1.48	1.54	1.71	1.53	1.67
Minimum	0.63	0.55	0.85	0.67	0.89	1.71	1.16	0.82
Mean	1.37	1.21	1.39	1.48	1.56	1.71	1.51	1.53
Std dev	0.39	0.43	0.36	0.32	0.45	-	0.27	0.45
Std err mean	0.09	0.09	0.10	0.05	0.10	-	0.11	0.14
Angle of striking platform	(N=13)	(N=14)	(N=7)	(N=28)	(N=17)		(N=4)	(N=7)
Maximum	128	124	138	125	132	109	130	130
Median	120	115	125	117.5	115	109	122	117
Minimum	73	93	100	85	94	109	106	101
Mean	115.15	113.36	122.29	114.32	113.29	109.00	120.00	114.57
Std dev	15.23	9.32	13.43	10.60	8.80	-	10.20	10.37
Std err mean	4.22	2.49	5.07	2.00	2.13	-	5.10	3.92
Number of scars	(N=19)	(N=20)	(N=11)	(N=35)	(N=19)			(N=7)
Maximum	8	7	4	10	7	7	5	5
Median	2	3	3	3	2	7	2	2
Minimum	1	1	1	1	1	7	1	1
Mean	2.74	3.00	2.82	3.31	2.47	7.00	2.50	2.43
Std dev	1.59	3.00	0.87	1.98	1.39	-	1.64	1.27
Std err mean	0.37	0.36	0.26	0.33	0.32	-	0.67	0.48

standard deviations are considered. In general, the width and thickness of these flakes are less variable than their length. Taphonomically, in the larger samples of Siret flakes about half are complete, although a substantial proportion are laterally broken (Table 7.20). Most of these flakes are devoid of cortex and their dominant direction of blow is end-struck, a fact to be expected in view of the origin of these flakes (Table 7.20). The dominant scar pattern on the Siret flakes is along axis followed by indeterminate, but other patterns are present as well in very small frequencies (Table 7.21). The mean number of scars is between 2 and 3 (Table 7.19). Despite the breakage (split) of the striking platform of these flakes, the majority could be identified as plain, followed by indeterminate, although other patterns are discernible as well (Table 7.21).

Table 7.20 Counts and frequencies (%) of taphonomic and technological characteristics of basalt Siret flakes of Layer II-6

	II-6/L1	II-6/L2	II-6/L3	II-6/L4	II-6/L4b	II-6/L5	II-6/L6	II-6/L7
N	21	24	12	42	22	1	6	11
Breakage								
Complete	8 / 38.10	12 / 50.00	4 / 33.33	18 / 42.86	12 / 54.55	1 / 100.00	4 / 66.67	5 / 45.45
Distal	2 / 9.52	3 / 12.50	2 / 16.67	6 / 14.29	2 / 9.09	-	1 / 16.67	1 / 9.09
Lateral	8 / 38.10	5 / 20.83	5 / 41.67	11 / 26.19	6 / 27.27	-	1 / 16.67	4 / 36.36
Proximal	1 / 4.76	-	-	2 / 4.76	-	-	-	-
Distal & lateral	1 / 4.76	2 / 8.33	1 / 8.33	4 / 9.52	-	-	-	1 / 9.09
Distal & proximal	-	-	-	1 / 2.38	-	-	-	-
Proximal & lateral	1 / 4.76	2 / 8.33	-	-	1 / 4.55	-	-	-
Indeterminate	-	-	-	-	1 / 4.55	-	-	-
Cortex								
No cortex	17 / 80.95	18 / 75.00	10 / 83.33	40 / 95.24	13 / 59.09	1 / 100.00	6 / 100.00	7 / 63.64
1–25%	2 / 9.52	-	-	-	1 / 4.55	-	-	-
26–50%	-	1 / 4.17	1 / 8.33	-	1 / 4.55	-	-	-
51–75%	2 / 9.52	1 / 4.17	-	-	1 / 4.55	-	-	-
76–100%	-	2 / 8.33	-	2 / 4.76	2 / 9.09	-	-	1 / 9.09
Indeterminate	-	2 / 8.33	1 / 8.33	-	4 / 18.18	-	-	3 / 27.27
Direction of blow		(N=23)		(N=41)				
End-struck	11 / 52.38	11 / 47.83	8 / 66.67	25 / 60.98	18 / 81.82	1 / 100.00	5 / 83.33	3 / 27.27
Side-struck	1 / 4.76	3 / 13.04	1 / 8.33	1 / 2.44	-	-	-	2 / 18.18
Special side-struck	1 / 4.76	-	2 / 16.67	-	2 / 9.09	-	1 / 16.67	-
Indeterminate	8 / 38.10	9 / 39.13	1 / 8.33	15 / 36.59	2 / 9.09	-	-	6 / 54.55

7.3 Technology of Basalt Flakes

Table 7.21 Counts and frequencies (%) of technological characteristics of basalt Siret flakes of Layer II-6

	II-6/L1	II-6/L2	II-6/L3	II-6/L4	II-6/L4b	II-6/L5	II-6/L6	II-6/L7
N	21	24	12	42	22	1	6	11
Scar pattern		(N=23)						(N=10)
Cortical	2 9.52	-	-	3 7.14	3 13.64	-	-	1 10.00
Plain	3 14.29	2 8.70	-	8 19.05	2 9.09	-	-	2 20.00
Along axis	4 19.05	9 39.13	4 33.33	7 16.67	8 36.36	-	2 33.33	3 30.00
Convergent	-	-	-	1 2.38	-	-	-	-
Opposed	1 4.76	-	-	-	-	-	-	-
Radial	3 14.29	1 4.35	1 8.33	3 7.14	-	-	1 16.67	-
Side	1 4.76	-	1 8.33	-	-	-	-	-
Along axis & side	1 4.76	2 8.70	2 16.67	5 11.90	-	1 100.00	-	-
Along axis & opposed	-	-	-	2 4.76	-	-	1 16.67	-
Opposed & side	-	-	1 8.33	1 2.38	-	-	-	-
Indeterminate	6 28.57	9 39.13	3 25.00	12 28.57	9 40.91	-	2 33.33	4 40.00
Striking platform		(N=23)						(N=10)
Cortical	-	-	-	1 2.38	-	-	-	-
Plain	16 76.19	9 39.13	7 58.33	23 54.76	19 86.36	1 100.00	3 50.00	4 40.00
Dihedral	-	-	-	1 2.38	-	-	-	-
Facetted	2 9.52	5 21.74	-	6 14.29	-	-	1 16.67	2 20.00
Removed	-	3 13.04	-	2 4.76	1 4.55	-	-	1 10.00
Broken	2 9.52	2 8.70	1 8.33	6 14.29	1 4.55	-	-	-
Indeterminate	1 4.76	4 17.39	4 33.33	3 7.14	1 4.55	-	2 33.33	3 30.00

7.3.2.2 Kombewa and Dorsally Plain Flakes (DPF)

In this study we use two different terms to define flakes with two ventral faces. Small flakes are termed dorsally plain flakes (DPF), a terminology that was defined by Dag and Goren-Inbar (2001), and large flakes are termed Kombewa flakes (Owen 1938). Dorsally plain flakes were not produced intentionally but are products of large flake modification, cores on flakes, and spontaneous detachment of the bulbar face (*esquilles bulbaires*) (Dag and Goren-Inbar 2001 and references therein). Kombewa flakes were predetermined and produced intentionally. The GBY basalt flake inventory includes both variants, which are described separately below. The DPF group belongs to the realm of small basalt flakes, while the Kombewa flakes are a component of the large flakes and comprise a group of predetermined blanks suitable for the production of bifaces.

In the following paragraphs we describe first the DPF group and then Kombewa flakes. Section 7.5.3.2 provides additional data on the Kombewa component in relation to biface technology.

DPF

The analysis of basalt flakes resulted in the identification of DPF. The categorization was at times very difficult, and hence artifacts that were not considered typical for different reasons were classified as possibly DPF. Table 7.22 presents the detailed breakdown of the DPF, which includes both the uncertain ones and some that are additionally classified as Siret flakes. As sample sizes are very small, the analyses were carried out on all three categories lumped together.

Table 7.17 shows that the frequencies of DPF vary substantially in the complex of Layer II-6. Some levels that are rich in basalt artifacts (e.g. Level 7) have very few DPF. One can also see the opposite trend, in which levels that are not particularly rich in basalt artifacts (e.g. Level 4b) have relatively high values. The frequencies of DPF may be correlated with biface production, a topic that is dealt with in Section 7.5. As would be expected by their definition, DPF are small in size (Table 7.23) but are characterized by very high standard deviations, demonstrating a lack of systematic or intentional production. The size of DPF does not vary extensively among the levels of Layer II-6. Some variation is encountered in the angles of striking platforms, where the standard deviations again indicate high variation within the levels (Table 7.23).

The breakage patterns of DPF, presented in Table 7.24, show extreme variability. This variability and the relative high occurrence of fragments in comparison with other flake categories is probably the result of biface modification that produced DPF with spontaneous breakages, including fragments. When the direction of blow is discernible on DPF, the highest frequencies are for end-struck and side-struck, with noticeable frequencies of indeterminate direction. With regard to striking platforms, the most common type is plain, as is so common at GBY, but the categories of broken and indeterminate show higher frequencies than in other flake categories.

There seem to be higher frequencies of DPF on unretouched basalt flakes than on retouched ones (Figs. 7.29, 7.30). Table 7.25 presents the counts and frequencies of basalt tools on DPF (N=27 in the entire basalt assemblage). It is evident that there is no systematic selection of DPF for the production of tools, as demonstrated by the very low frequencies of these items in the basalt assemblage (Fig. 7.3). It is also evident that there is no particular selection of tool type in the use of DPF, since the highest frequencies of DPF are distributed among the different tool types.

Kombewa

Kombewa flakes occur throughout the GBY cultural sequence (Table 7.17) but, like DPF, are much more frequent in the levels of Layer II-6; again, their presence is dependent on the sample size. The large majority of Kombewa flakes are unretouched and a minimal number were modified into tools (Table 7.26). Some of the Kombewa flakes can be classified as particular morphological types that are often associated with biface modification (Table 7.27). In addition, different technological features were recorded on these flakes (Table 7.28). A substantial number of artifacts are classified as possibly Kombewa, partially due to the difficulties posed by weathering of the artifacts and partially due to our strict classification procedure that tended to minimize rather than maximize the occurrence. Worth noting is the combination of Kombewa and Siret, which amounts to over 16% in Level 4 of Layer II-6 (Fig. 7.31). From a morphological perspective, the Kombewa flakes are diverse and include a variety of forms dominated by flakes (Figs. 7.32–7.37). Kombewa flakes of biface morphology form just over 9% of the entire GBY Kombewa assemblage (Figs. 7.38–7.40).

In view of the small samples of Kombewa flakes in some of the layers, we focus in the following analysis on the different levels of Layer II-6. The highest frequency of Kombewa flakes occurs in Level 4, the level that is richest in bifaces (Section 7.5). Despite this, there seems to be no correlation between the frequency of bifaces and that of Kombewa flakes, as Level 3 is richer in Kombewa flakes than Level 1, whose biface assemblage is larger.

The Kombewa flakes, while all placed within the category of large basalt flakes, vary substantially in size (Table 7.29). There is, however, great similarity in the mean width and thickness among the different levels. The lack of samples of suitable size rules out any generalizations about the angle of the striking platform. There is also extensive variability in the taphonomic condition of the Kombewa flakes (Table 7.30), where complete is generally the dominant mode, followed by laterally broken. As for direction of blow, Kombewa flakes were predominantly end-struck, followed by side-struck and indeterminate. Although other types are present, Kombewa flakes are predominantly characterized by plain striking platforms, especially when broken and indeterminate striking platforms are excluded.

Table 7.22 Counts and frequencies (%) of basalt dorsally plain flake categories of Layer II-6

	II-6/L1	II-6/L2	II-6/L3	II-6/L4	II-6/L4b	II-6/L5	II-6/L6	II-6/L7
N	25	26	55	39	24	3	6	5
DPF	16	12	22	18	4	-	-	1
	64.00	46.15	40.00	46.15	16.67			20.00
Possibly DPF	9	13	33	19	20	3	6	3
	36.00	50.00	60.00	48.72	83.33	100.00	100.00	60.00
DPF & Siret	-	1	-	2	-	-	-	1
		3.85		5.13				20.00

7.3 Technology of Basalt Flakes

Table 7.23 Descriptive statistics of basalt dorsally plain flakes of Layer II-6 (size in mm; angle in degrees)

	II-6/L1	II-6/L2	II-6/L3	II-6/L4	II-6/L4b	II-6/L5	II-6/L6	II-6/L7
N	25	26	55	39	24	3	6	5
Maximum length								
Maximum	68	68	66	69	69	45	66	66
Median	42	50.5	48	48	42.5	42	39	63
Minimum	23	27	27	25	20	27	23	44
Mean	44.52	47.92	46.67	47.15	42.58	38.00	44.50	59.80
Std dev	15.71	14.54	11.16	14.81	16.47	9.64	17.48	9.04
Std err mean	3.14	2.85	1.50	2.37	3.36	5.57	7.14	4.04
Length								
Maximum	65	67	65	67	69	45	59	66
Median	37	41.5	38	43	37.5	29	35.5	59
Minimum	16	16	16	18	18	27	22	39
Mean	39.08	42.38	37.89	41.51	38.63	33.67	36.17	55.00
Std dev	15.60	14.44	12.31	14.58	15.78	9.87	13.64	11.25
Std err mean	3.12	2.83	1.66	2.34	3.22	5.70	5.57	5.03
Width								
Maximum	68	68	70	68	59	39	64	65
Median	35	35.5	43	39	36	26	30.5	43
Minimum	11	15	16	18	15	24	15	30
Mean	38.60	40.27	42.33	39.87	33.38	29.67	37.33	47.20
Std dev	15.61	15.64	12.36	14.02	12.76	8.14	21.30	14.13
Std err mean	3.12	3.07	1.67	2.25	2.60	4.70	8.70	6.32
Thickness								
Maximum	43	29	38	31	28	16	26	30
Median	12	15.5	15	15	13	16	14.5	22
Minimum	7	7	5	7	5	9	8	11
Mean	15.20	16.88	15.91	15.44	13.75	13.67	15.33	21.20
Std dev	8.91	5.88	6.42	6.31	6.15	4.04	6.68	7.19
Std err mean	1.78	1.15	0.87	1.01	1.26	2.33	2.73	3.22
L/W								
Maximum	2.00	1.89	2.00	1.82	1.79	1.73	2.00	2.20
Median	1.00	1.07	0.86	1.04	1.22	1.13	1.13	1.02
Minimum	0.63	0.70	0.43	0.55	0.53	0.74	0.57	0.74
Mean	1.05	1.11	0.94	1.09	1.18	1.20	1.17	1.29
Std dev	0.30	0.32	0.34	0.35	0.27	0.50	0.56	0.60
Std err mean	0.06	0.06	0.05	0.06	0.05	0.29	0.23	0.27
Angle of striking platform	(N=10)	(N=8)	(N=20)	(N=14)	(N=8)	(N=2)	(N=1)	(N=2)
Maximum	121	126	126	140	119	114	115	122
Median	108	112.5	115.5	116.5	106.5	104	115	113
Minimum	100	100	91	96	90	94	115	104
Mean	108.40	113.13	113.95	114.57	106.38	104.00	115.00	113.00
Std dev	6.90	8.08	8.52	10.78	10.25	14.14	-	12.73
Std err mean	2.18	2.86	1.90	2.88	3.63	10.00	-	9.00
Number of scars	(N=12)	(N=14)	(N=28)	(N=15)	(N=11)	-	(N=1)	(N=2)
Maximum	2	4	8	6	5	-	1	1
Median	1	1	2	2	2	-	1	1
Minimum	1	1	1	1	1	-	1	1
Mean	1.25	1.71	2.07	1.93	2.27	-	1.00	1.00
Std dev	0.45	1.14	1.46	1.33	1.27	-	-	-
Std err mean	0.13	0.30	0.28	0.34	0.38	-	-	-

Table 7.24 Counts and frequencies (%) of taphonomic and technological characteristics of basalt dorsally plain flakes of Layer II-6

	II-6/L1	II-6/L2	II-6/L3	II-6/L4	II-6/L4b	II-6/L5	II-6/L6	II-6/L7
N	25	26	55	39	24	3	6	5
Breakage								
Complete	11	7	12	13	8	-	1	2
	44.00	26.92	21.82	33.33	33.33		16.67	40.00
Distal	1	2	5	3	-	1	-	1
	4.00	7.69	9.09	7.69		33.33		20.00
Lateral	2	5	11	5	3	2	2	-
	8.00	19.23	20.00	12.82	12.50	66.67	33.33	
Proximal	4	4	5	5	2	-	-	-
	16.00	15.38	9.09	12.82	8.33			
Distal & lateral	3	3	5	4	2	-	1	1
	12.00	11.54	9.09	10.26	8.33		16.67	20.00
Distal & proximal	-	-	-	-	2	-	-	-
					8.33			
Fragment	2	4	12	8	4	-	1	-
	8.00	15.38	21.82	20.51	16.67		16.67	
Proximal & lateral	1	-	4	1	1	-	-	1
	4.00		7.27	2.56	4.17			20.00
Indeterminate	1	1	1	-	2	-	1	-
	4.00	3.85	1.82		8.33		16.67	
Cortex								
No cortex	24	23	55	38	22	3	6	5
	96.00	88.46	100.00	97.44	91.67	100.00	100.00	100.00
1–25%	1	1	-	-	-	-	-	-
	4.00	3.85						
51–75%	-	-	-	-	1	-	-	-
					4.17			
76–100%	-	1	-	1	1	-	-	-
		3.85		2.56	4.17			
Indeterminate	-	1	-	-	-	-	-	-
		3.85						
Direction of blow								(N=4)
End-struck	9	8	11	9	8	1	2	1
	36.00	30.77	20.00	23.08	33.33	33.33	33.33	25.00
Side-struck	6	8	20	10	5	1	1	2
	24.00	30.77	36.36	25.64	20.83	33.33	16.67	50.00
Special side-struck	1	3	1	3	-	-	-	-
	4.00	11.54	1.82	7.69				
Indeterminate	9	7	23	17	11	1	3	1
	36.00	26.92	41.82	43.59	45.83	33.33	50.00	25.00
Striking platform		(N=25)		(N=37)				(N=4)
Cortical	-	1	-	-	1	-	-	-
		4.00			4.17			
Plain	13	11	29	18	8	2	2	3
	52.00	44.00	52.73	48.65	33.33	66.67	33.33	75.00
Dihedral	-	1	2	-	1	-	-	-
		4.00	3.64		4.17			
Facetted	1	-	-	4	-	-	1	-
	4.00			10.81			16.67	
Crushed	-	-	-	1	-	-	-	-
				2.70				
Removed	1	-	-	-	-	-	-	-
	4.00							
Broken	7	6	13	12	10	-	1	-
	28.00	24.00	23.64	32.43	41.67		16.67	
Indeterminate	3	6	11	2	4	1	2	1
	12.00	24.00	20.00	5.41	16.67	33.33	33.33	25.00

7.3 Technology of Basalt Flakes

Table 7.24 (cont.)

	II-6/L1	II-6/L2	II-6/L3	II-6/L4	II-6/L4b	II-6/L5	II-6/L6	II-6/L7
N	25	26	55	39	24	3	6	5
Scar pattern	(N=24)	(N=25)	(N=54)	(N=38)				(N=4)
Cortical	-	1 / 4.00	-	1 / 2.63	1 / 4.17	-	-	1 / 25.00
Plain	18 / 75.00	13 / 52.00	24 / 44.44	23 / 60.53	15 / 62.50	3 / 100.00	5 / 83.33	3 / 75.00
Along axis	1 / 4.17	2 / 8.00	18 / 33.33	7 / 18.42	4 / 16.67	-	-	-
Side	1 / 4.17	-	1 / 1.85	1 / 2.63	-	-	-	-
Along axis & side	-	1 / 4.00	1 / 1.85	2 / 5.26	1 / 4.17	-	-	-
Indeterminate	4 / 16.67	8 / 32.00	10 / 18.52	4 / 10.53	3 / 12.50	-	1 / 16.67	-

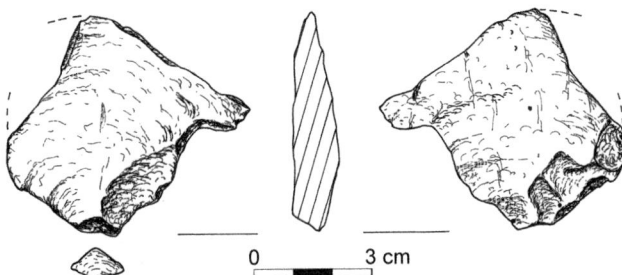

Fig. 7.29 Basalt dorsally plain flake (#5474 Layer II-6 Level 1)

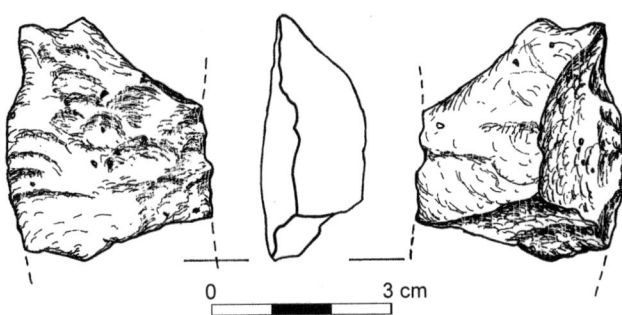

Fig. 7.30 Basalt dorsally plain flake (#5516 Layer II-6 Level 1)

Table 7.25 Typological counts and frequencies (%) of basalt dorsally plain flakes on tools and blanks of Layer II-6

	N	%
Abruptly retouched side-scraper	1	0.52
Notch	3	1.57
Flake	177	92.67
Éclat de taille de biface	9	4.71
Retouched flake	1	0.52
Total	191	100.00

Table 7.26 Typological counts and frequencies (%) of Kombewa flakes

	N	%
Single straight side-scraper	1	0.65
Single convex side-scraper	1	0.65
Single concave side-scraper	1	0.65
Side-scraper on ventral face	1	0.65
Atypical burin	1	0.65
Denticulate	2	1.31
End-notched piece	1	0.65
Varia	1	0.65
Flake	130	84.97
Blade	1	0.65
Éclat de taille de biface	1	0.65
Retouched flake	8	5.23
Massive scraper	3	1.96
Biface preform	1	0.65
Total	153	100.00

Table 7.27 Counts and frequencies (%) of special large basalt flake categories on Kombewa flakes

Flake type	N	%
Handaxe-shaped flake	12	26.09
Biface thinning flake	17	36.96
Roughing-out flake	2	4.35
Wedge-like flake	1	2.17
Cleaver-shaped flake	4	8.70
Shoulder/corner flake	2	4.35
Éclat de taille de biface	1	2.17
Waste of giant core	4	8.70
Large flake	3	6.52
Total	46	100.00

Table 7.28 Counts and frequencies (%) of different categories of basalt Kombewa flakes of Layer II-6

	II-6/L1	II-6/L2	II-6/L3	II-6/L4	II-6/L4b	II-6/L5	II-6/L6	II-6/L7	VI-14	Trench II green
N	24	14	21	52	13	3	7	8	2	4
Kombewa	18 75.00	7 50.00	14 66.67	24 46.15	7 53.85	2 66.67	1 14.29	2 25.00	-	2 50.00
Possibly Kombewa	3 12.50	7 50.00	5 23.81	18 34.62	6 46.15	1 33.33	6 85.71	6 75.00	2 100.00	2 50.00
Kombewa & hinge	1 4.17	-	-	1 1.92	-	-	-	-	-	-
Débordant & Kombewa	-	-	1 4.76	-	-	-	-	-	-	-
Kombewa & Siret	2 8.33	-	1 4.76	9 17.31	-	-	-	-	-	-

Fig. 7.31 Basalt massive scraper on Kombewa biface thinning flake (#25 Layer II-6 Level 4)

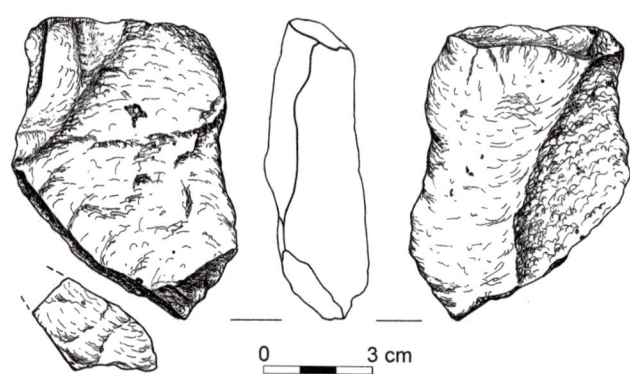

Fig. 7.34 Basalt Kombewa flake (#5314 Layer II-6 Level 1)

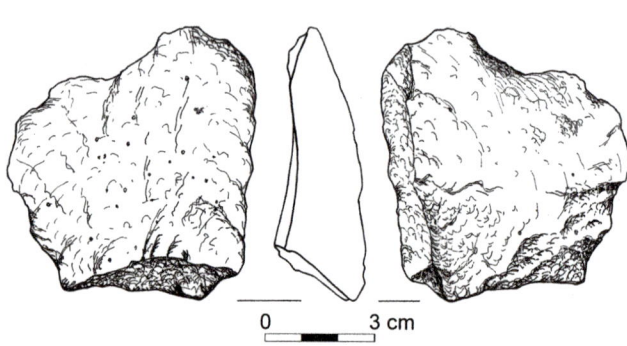

Fig. 7.32 Basalt Kombewa biface thinning flake (#5573 Layer II-6 Level 1)

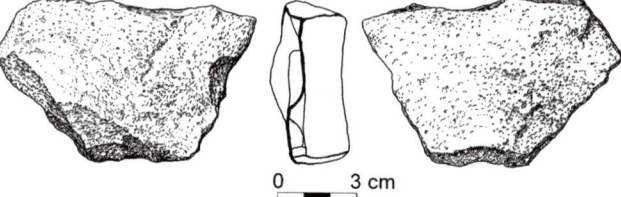

Fig. 7.35 Basalt Kombewa flake (#5544 Layer II-6 Level 1)

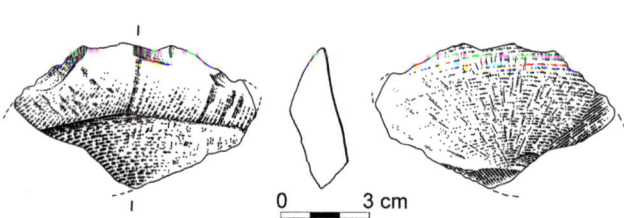

Fig. 7.33 Basalt Kombewa biface thinning flake (#721 Layer II-6 Level 4)

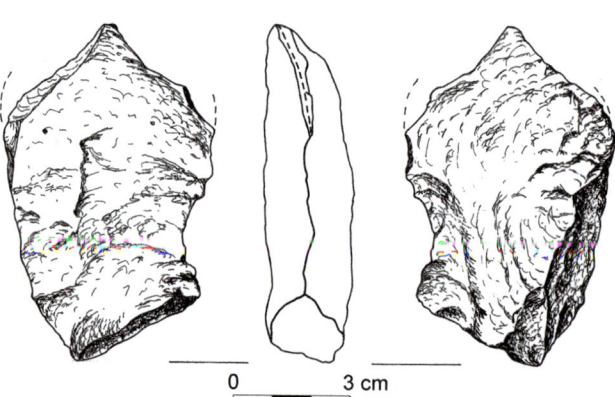

Fig. 7.36 Basalt Kombewa flake (#5525 Layer II-6 Level 1)

7.3 Technology of Basalt Flakes

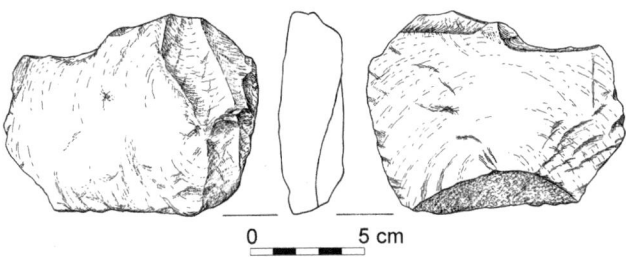

Fig. 7.37 Basalt Kombewa flake (#3332 JB)

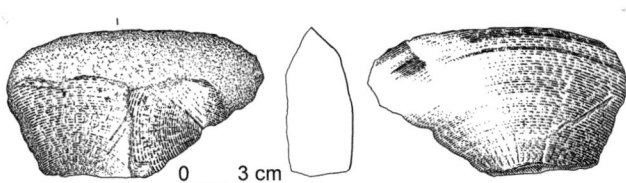

Fig. 7.39 Basalt handaxe-shaped Kombewa flake (#107 Layer II-6 Level 4)

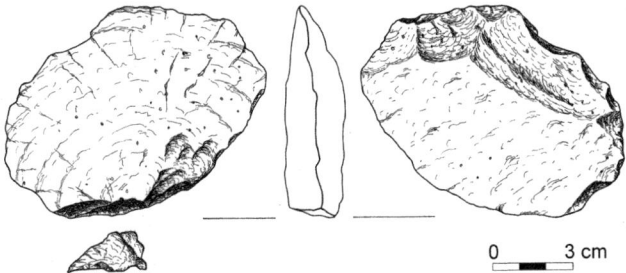

Fig. 7.38 Basalt handaxe-shaped Kombewa flake (#5362 Layer II-6 Level 1)

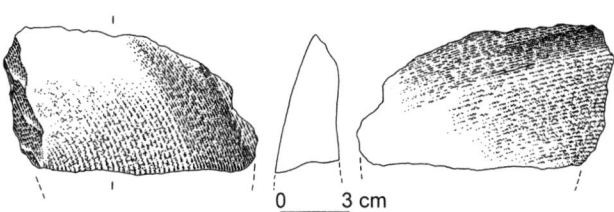

Fig. 7.40 Basalt cleaver-shaped Kombewa flake (#4850 Layer II-6 Level 4)

As mentioned above, our data on Kombewa technology and flakes originate mostly from Kombewa blanks used for the production of bifaces, rather than from flakes or cores. Based on experimental knapping data (Madsen and Goren-Inbar 2004), it is evident that the blank intended to be modified into a biface could not be detached from the same direction as that of the large parent flake (the Kombewa core), since the angle between the striking platform and the ventral face of the Kombewa core is too large. This is why the striking axes of the two visible platforms of each Kombewa flake are aligned at ca. 90° to each other. These features are recorded primarily on bifaces and are discussed in detail below (Section 7.5)

7.3.2.3 Biface Sharpening Flakes

Biface sharpening flakes belong to the shaping process (*façonnage*) of biface production. They are characteristically larger than the category of *éclats de taille de biface* and are quite diagnostic at GBY. Table 7.17 presents the occurrence of these flakes in the different layers of GBY, showing that they are much more common than *éclats de taille de biface*. Biface sharpening flakes were used at times, but very infrequently, for the modification of tools (Table 7.31), but clearly most of them remained unretouched (Figs. 7.41–7.50). The size of these flakes (Table 7.32) varies extensively, as reflected by their standard deviations and as expected from their place within the biface reduction sequence. The biface sharpening flakes originating in Layer II-6 Level 4 are the largest.

The assemblages that contain sufficiently large samples of biface sharpening flakes have a sizeable number of complete flakes followed by distal breaks, with the exception of Level 3, which has a relatively high frequency of proximal and lateral breaks (Table 7.33). As expected, cortical coverage is minimally observed and the direction of blow, when discernible, is predominantly end-struck, followed closely by side-struck (Table 7.33). When the scar pattern on the dorsal face is identifiable, the most common pattern in most levels is along axis and plain (Table 7.34), with a mean scar count of between 3 and 4 scars per artifact (Table 7.32). Most biface sharpening flakes have plain striking platforms, but there are high frequencies of broken platforms as well (Table 7.34).

7.3.2.4 *Éclats de Taille de Biface*

Éclats de taille de biface (*édtdb*) appear in the assemblages of all the excavated layers at GBY (Figs. 7.51–7.55). From Table 7.17, which presents the frequencies of these basalt flakes, it is evident that, although basalt bifaces are abundant at the site, the frequency of *édtdb* is extremely low. Furthermore, even in assemblages containing many basalt bifaces (e.g. Layer II-6 Level 1) the frequency of basalt *édtdb* is minimal. This feature is even more striking when the occurrence of basalt *édtdb* is compared with that of flint (Section 6.3.1.1), as the latter are

Table 7.29 Descriptive statistics of basalt Kombewa flakes of Layer II-6 (size in mm; angle in degrees)

	II-6/L1	II-6/L2	II-6/L3	II-6/L4	II-6/L4b	II-6/L5	II-6/L6	II-6/L7	VI-14	Trench II green
N	24	14	21	52	13	3	7	8	2	4
Maximum length										
Maximum	136	137	123	153	137	112	109	141	108	134
Median	94.5	89	91	95.5	102	100	80	104	101	88
Minimum	70	72	71	70	75	75	73	83	94	75
Mean	95.63	91.29	91.67	96.48	98.69	95.67	83.71	107.75	101.00	96.25
Std dev	18.07	16.75	16.65	19.87	17.03	18.88	12.00	21.17	9.90	26.71
Std err mean	3.69	4.48	3.63	2.76	4.72	10.90	4.53	7.49	7.00	13.36
Length										
Maximum	136	97	103	153	126	112	76	130	94	74
Median	77.5	73	76	78	71	89	67	76.5	93.5	66.5
Minimum	57	53	47	42	56	75	58	61	93	64
Mean	84.29	73.93	74.81	83.37	76.85	92.00	68.00	82.00	93.50	67.75
Std dev	22.29	12.52	13.31	21.62	22.81	18.68	6.51	23.78	0.71	4.50
Std err mean	4.55	3.35	2.90	3.00	6.33	10.79	2.46	8.41	0.50	2.25
Width										
Maximum	149	137	123	145	114	100	105	141	94	134
Median	68.5	78	82	70	89	66	78	102	69	79
Minimum	34	30	41	39	55	45	47	73	44	62
Mean	77.29	76.21	79.95	76.12	85.92	70.33	71.57	102.25	69.00	88.50
Std dev	25.01	25.35	25.68	23.84	19.33	27.75	21.35	21.08	35.36	33.68
Std err mean	5.11	6.77	5.60	3.31	5.36	16.02	8.07	7.45	25.00	16.84
Thickness										
Maximum	79	47	41	80	40	29	49	63	45	38
Median	27	30	27	30	26	28	33	37	34.5	26.5
Minimum	17	22	16	13	19	19	20	31	24	19
Mean	31.88	31.50	27.48	30.40	26.92	25.33	30.57	41.13	34.50	27.50
Std dev	13.80	7.72	7.19	10.10	6.20	5.51	10.26	11.26	14.85	9.47
Std err mean	2.82	2.06	1.57	1.40	1.72	3.18	3.88	3.98	10.50	4.73
L/W										
Maximum	2.00	2.53	1.88	2.00	1.53	1.70	1.62	1.13	2.14	1.10
Median	1.11	0.99	0.90	1.24	0.89	1.67	0.86	0.76	1.56	0.85
Minimum	0.52	0.50	0.55	0.39	0.53	0.89	0.70	0.56	0.99	0.55
Mean	1.18	1.11	1.04	1.20	0.93	1.42	1.04	0.81	1.56	0.84
Std dev	0.43	0.53	0.41	0.42	0.30	0.46	0.37	0.20	0.81	0.26
Std err mean	0.09	0.14	0.09	0.06	0.08	0.26	0.14	0.07	0.57	0.13
Angle of striking platform	(N=9)	(N=4)	(N=12)	(N=34)	(N=8)	(N=1)	(N=4)	(N=3)	(N=1)	(N=1)
Maximum	138	135	135	135	128	112	125	123	106	126
Median	125	117.5	118	120	121.5	112	114	120	106	126
Minimum	107	112	100	98	102	112	106	102	106	126
Mean	122.33	120.50	117.25	118.26	118.50	112.00	114.75	115.00	106.00	126.00
Std dev	8.66	10.21	9.08	8.17	9.93	-	7.97	11.36	-	-
Std err mean	2.89	5.11	2.62	1.40	3.51	-	3.99	6.56	-	-
Number of scars	(N=13)	(N=10)	(N=10)	(N=36)	(N=5)	(N=1)	(N=5)	(N=5)		(N=1)
Maximum	4	3	8	6	5	3	3	4	6	1
Median	2	1.5	2	2	1	3	2	2	3.5	1
Minimum	1	1	1	1	1	3	1	1	1	1
Mean	1.77	1.80	2.50	2.42	2.00	3.00	1.80	2.20	3.50	1.00
Std dev	0.93	0.92	2.17	1.27	1.73	-	0.84	1.30	3.54	-
Std err mean	0.26	0.29	0.69	0.21	0.77	-	0.37	0.58	2.50	-

7.3 Technology of Basalt Flakes

Table 7.30 Counts and frequencies (%) of taphonomic and technological characteristics of basalt Kombewa flakes of Layer II-6

	II-6/L1	II-6/L2	II-6/L3	II-6/L4	II-6/L4b	II-6/L5	II-6/L6	II-6/L7	VI-14	Trench II green
N	24	14	21	52	13	3	7	8	2	4
Breakage										
Complete	12 / 50.00	2 / 14.29	10 / 47.62	27 / 51.92	6 / 46.15	-	1 / 14.29	3 / 37.50	1 / 50.00	3 / 75.00
Distal	6 / 25.00	3 / 21.43	1 / 4.76	5 / 9.62	2 / 15.38	-	2 / 28.57	3 / 37.50	1 / 50.00	-
Lateral	2 / 8.33	5 / 35.71	6 / 28.57	10 / 19.23	2 / 15.38	-	2 / 28.57	2 / 25.00	-	-
Proximal	2 / 8.33	-	1 / 4.76	2 / 3.85	-	1 / 33.33	1 / 14.29	-	-	-
Distal & lateral	-	2 / 14.29	-	2 / 3.85	3 / 23.08	-	1 / 14.29	-	-	-
Distal & proximal	-	-	2 / 9.52	-	-	-	-	-	-	-
Fragment	1 / 4.17	1 / 7.14	-	4 / 7.69	-	2 / 66.67	-	-	-	1 / 25.00
Proximal & lateral	1 / 4.17	1 / 7.14	1 / 4.76	2 / 3.85	-	-	-	-	-	-
Cortex										
No cortex	21 / 87.50	12 / 85.71	18 / 85.71	50 / 96.15	13 / 100.00	3 / 100.00	6 / 85.71	6 / 75.00	2 / 100.00	3 / 75.00
1–25%	1 / 4.17	-	2 / 9.52	-	-	-	-	-	-	1 / 25.00
26–50%	-	1 / 7.14	-	1 / 1.92	-	-	-	-	-	-
51–75%	1 / 4.17	-	-	-	-	-	1 / 14.29	-	-	-
76–100%	-	1 / 7.14	-	-	-	-	-	-	-	-
Indeterminate	1 / 4.17	-	1 / 4.76	1 / 1.92	-	-	-	2 / 25.00	-	-
Direction of blow	(N=23)			(N=50)			(N=5)			
End-struck	12 / 52.17	5 / 35.71	6 / 28.57	25 / 50.00	4 / 30.77	2 / 66.67	2 / 40.00	1 / 12.50	-	-
Side-struck	5 / 21.74	3 / 21.43	5 / 23.81	11 / 22.00	7 / 53.85	-	2 / 40.00	3 / 37.50	-	-
Special side-struck	2 / 8.70	1 / 7.14	4 / 19.05	3 / 6.00	1 / 7.69	-	1 / 20.00	-	1 / 50.00	3 / 75.00
Indeterminate	4 / 17.39	5 / 35.71	6 / 28.57	11 / 22.00	1 / 7.69	1 / 33.33	-	4 / 50.00	1 / 50.00	1 / 25.00
Striking platform	(N=23)			(N=51)		(N=2)	(N=5)	(N=7)		
Punctiform	-	-	-	-	-	-	1 / 20.00	-	-	-
Plain	14 / 60.87	6 / 42.86	12 / 57.14	27 / 52.94	8 / 61.54	1 / 50.00	4 / 80.00	2 / 28.57	1 / 50.00	3 / 75.00
Dihedral	-	-	1 / 4.76	2 / 3.92	-	-	-	-	-	-
Facetted	1 / 4.35	2 / 14.29	-	2 / 3.92	2 / 15.38	-	-	-	-	-
Removed	-	-	2 / 9.52	5 / 9.80	-	-	-	-	-	-
Broken	4 / 17.39	3 / 21.43	4 / 19.05	7 / 13.73	1 / 7.69	1 / 50.00	-	1 / 14.29	1 / 50.00	1 / 25.00
Indeterminate	4 / 17.39	3 / 21.43	2 / 9.52	8 / 15.69	2 / 15.38	-	-	4 / 57.14	-	-

Table 7.30 (cont.)

	II-6/L1	II-6/L2	II-6/L3	II-6/L4	II-6/L4b	II-6/L5	II-6/L6	II-6/L7	VI-14	Trench II green
N	24	14	21	52	13	3	7	8	2	4
Scar pattern								(N=7)		
Cortical	1 4.17	-	-	1 1.92	-	-	-	1 14.29	-	-
Plain	17 70.83	6 42.86	12 57.14	22 42.31	10 76.92	2 66.67	2 28.57	3 42.86	1 50.00	4 100.00
Along axis	3 12.50	4 28.57	1 4.76	12 23.08	2 15.38	1 33.33	1 14.29	-	-	-
Opposed	-	-	1 4.76	1 1.92	-	-	-	-	-	-
Radial	-	1 7.14	-	-	-	-	-	-	-	-
Side	-	-	-	4 7.69	1 7.69	-	-	-	-	-
Along axis & side	1 4.17	-	-	3 5.77	-	-	-	1 14.29	-	-
Along axis & opposed	-	-	-	1 1.92	-	-	-	-	-	-
Opposed & side	-	-	1 4.76	2 3.85	-	-	-	-	-	-
Indeterminate	2 8.33	3 21.43	6 28.57	6 11.54	-	-	4 57.14	2 28.57	1 50.00	-

Table 7.31 Typological counts and frequencies (%) of biface sharpening flakes

Typology	N	%
Single convex side-scraper	1	0.33
Offset side-scraper	1	0.33
Side-scraper on ventral face	1	0.33
Notch	5	1.63
Denticulate	4	1.31
End-notched piece	2	0.65
Flake	272	88.89
Blade	4	1.31
Biface sharpening flake	5	1.63
Retouched flake	8	2.61
Massive scraper	3	0.98
Total	306	100.00

much more frequent. This is a remarkable phenomenon, since the basalt bifaces are much more numerous overall at the site than the flint and limestone bifaces. The paucity of basalt *édtdb* is clearly enigmatic, but an explanation can perhaps be found within the realm of microartifacts. Methodologically (Chapter 4), all basalt flakes smaller than 2 cm were considered microartifacts. Thus, if for some reason the production of basalt bifaces resulted in particularly small *édtdb*, they are invisible for the purposes of this presentation. Further discussion of this phenomenon is provided at the end of this chapter. It is evident that the assemblage that is richest in basalt bifaces (Layer II-6 Level 4) is also the richest in basalt *édtdb* (Table 7.17) and basalt microartifacts (Table 7.1). Nevertheless, in the combined assemblage of Layer V-5, Layer V-6, and the JB (under Area C in Table 7.35) the sample of *édtdb* is larger than that of most levels of Layer II-6, despite the fact that only 16 basalt bifaces were found in these layers (Section 7.5). The size of *édtdb* is small, and the observed variability within the Layer II-6 complex is probably a result of the small samples (Table 7.35).

Taphonomically, breaks occur on different edges and complete *édtdb* do not dominate in most cases (Table 7.36). Technologically, the flakes are non-cortical and end-struck, but there is a noticeable frequency of indeterminate artifacts (Table 7.36). As mentioned above, the samples of *édtdb* are too small for in-depth analysis; however, we present the scar pattern and direction of blow (Table 7.36).

7.3.2.5 Flakes with Ventral Removals

The phenomenon of removals on the ventral face is described in detail in Chapter 6 for the flint assemblage, where it occurs in large numbers. Here we describe the same phenomenon, which occurs throughout the cultural sequence, for basalt flakes and flake tools (Table 7.17). Ventral removals are observed on both flakes and tools but are most frequent on unretouched flakes (Figs. 7.2, 7.56, 7.57, 7.7). Tools with ventral removals include massive scrapers, notches, and split percussors, but these are not frequent (less than 11% of the basalt flake tools). In contrast, unretouched flakes with ventral removals are very common and form over 80% of this assemblage (Table 7.37). Ventral removals occur on both small and large basalt flakes, although 61% of the flakes with this feature are large. This

7.3 Technology of Basalt Flakes

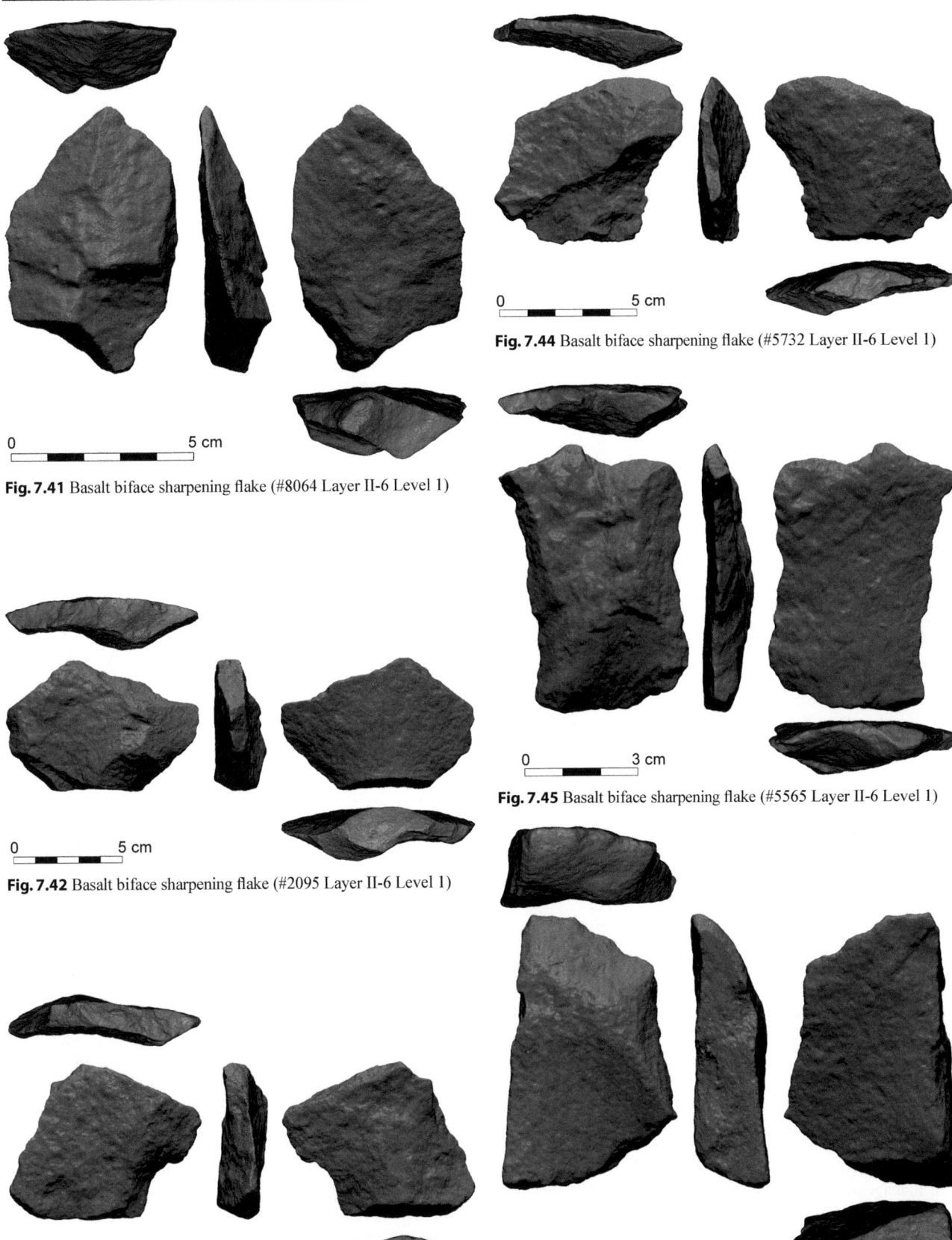

Fig. 7.41 Basalt biface sharpening flake (#8064 Layer II-6 Level 1)

Fig. 7.42 Basalt biface sharpening flake (#2095 Layer II-6 Level 1)

Fig. 7.43 Basalt biface sharpening flake (#2102 Layer II-6 Level 1)

Fig. 7.44 Basalt biface sharpening flake (#5732 Layer II-6 Level 1)

Fig. 7.45 Basalt biface sharpening flake (#5565 Layer II-6 Level 1)

Fig. 7.46 Basalt biface sharpening flake (#5589 Layer II-6 Level 1)

Fig. 7.47 Basalt biface sharpening flake (#7492 Layer II-6 Level 1)

Fig. 7.49 Basalt biface sharpening flake (#147 Layer II-6 Level 4)

Fig. 7.48 Basalt biface sharpening flake (#5555 Layer II-6 Level 1)

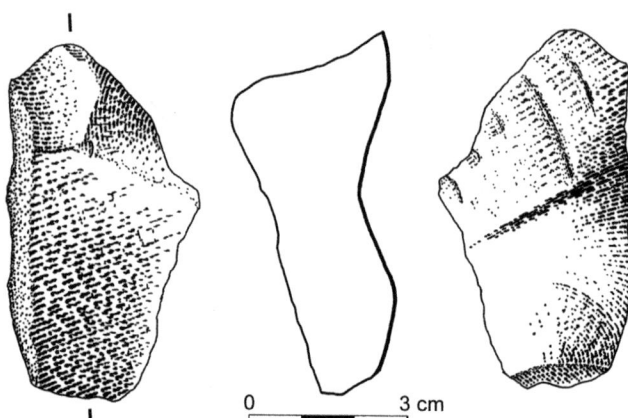

Fig. 7.50 Basalt biface sharpening flake (#783 Layer II-6 Level 4)

may be the result of thinning large biface blanks in the thickest area of the striking platform, as part of blank modification. Technologically, the dominant striking platform mode is plain, followed in significantly lower frequencies by removed (Table 7.38).

7.3.2.6 Core Management Pieces

Core management pieces include a variety of flakes that are diverse in form but all result from the reduction process of cores. There are very few basalt core management pieces at the site. Table 7.17 presents the excavated inventory, which is restricted to Area B. The very low frequency of waste products is clearly associated with the very low occurrence of basalt cores at the site (see below). The size attributes of the core waste are presented in Table 7.39, where, despite the fact that all levels of Layer II-6 are lumped together, they form an extremely small group of artifacts. Naturally, due to the small sample size, there is very high variability within the group. More than half of these artifacts are complete and have plain striking platforms, and their frequent scar patterns are ridge and indeterminate, with a mean of 4 scars per artifact.

7.3.3 Giant Core Special Waste Products

The prolonged study of the basalt component of GBY revealed that slabs were exploited for the production of large blanks for biface modification. This reduction sequence produced characteristic large waste products. Each of these has been identified and previously published (Madsen and Goren-Inbar 2004; Goren-Inbar 2011; Goren-Inbar et al. 2011). The following section describes the flake products of this reduction

7.3 Technology of Basalt Flakes

Table 7.32 Descriptive statistics of basalt biface sharpening flakes of Layer II-6 (size in mm; angle in degrees)

	II-6/L1	II-6/L2	II-6/L3	II-6/L4	II-6/L4b	II-6/L5	II-6/L6	II-6/L7
N	32	67	82	77	30	6	3	8
Maximum length								
Maximum	103	94	121	124	88	100	80	93
Median	60.5	52	52	70	63	83	72	62
Minimum	25	22	27	35	32	55	45	48
Mean	60.22	56.19	53.95	72.10	63.93	82.00	65.67	63.25
Std dev	19.47	19.52	19.17	21.80	14.47	16.24	18.34	14.85
Std err mean	3.44	2.38	2.12	2.48	2.64	6.63	10.59	5.25
Length								
Maximum	101	89	94	123	88	89	67	93
Median	50.5	48	42	62	52.5	62.5	58	52.5
Minimum	23	15	20	25	25	44	45	38
Mean	53.72	47.37	45.18	64.53	51.97	63.83	56.67	57.00
Std dev	18.92	18.24	16.92	22.85	16.11	14.47	11.06	17.66
Std err mean	3.34	2.23	1.87	2.60	2.94	5.91	6.39	6.24
Width								
Maximum	88	93	121	109	87	100	80	75
Median	48	45	45	56	53.5	71	53	51
Minimum	15	18	23	19	21	55	30	27
Mean	50.16	47.75	47.05	57.01	53.90	73.17	54.33	50.13
Std dev	16.32	18.97	17.61	18.66	15.41	16.14	25.03	15.55
Std err mean	2.89	2.32	1.94	2.13	2.81	6.59	14.45	5.50
Thickness								
Maximum	30	27	34	33	23	38	25	27
Median	14.5	15	15	19	15.5	21.5	20	18.5
Minimum	7	6	7	7	8	16	10	14
Mean	15.50	15.81	15.50	19.55	15.77	24.17	18.33	18.75
Std dev	5.59	5.32	5.89	6.24	4.32	7.81	7.64	4.59
Std err mean	0.99	0.65	0.65	0.71	0.79	3.19	4.41	1.62
L/W								
Maximum	1.80	2.06	2.16	2.15	2.52	1.00	1.50	1.63
Median	1.05	0.98	0.93	1.20	1.00	0.85	1.26	1.22
Minimum	0.59	0.52	0.58	0.55	0.50	0.79	0.73	0.67
Mean	1.11	1.05	1.00	1.18	1.03	0.88	1.16	1.19
Std dev	0.27	0.36	0.31	0.37	0.42	0.09	0.40	0.32
Std err mean	0.05	0.04	0.03	0.04	0.08	0.04	0.23	0.11
Angle of striking platform	(N=20)	(N=37)	(N=40)	(N=46)	(N=19)			(N=6)
Maximum	130	135	133	135	122	122	125	122
Median	118	112	117.5	116	115	109.5	120	114
Minimum	97	95	98	94	97	96	97	90
Mean	116.80	113.81	115.68	116.17	111.37	109.33	114.00	110.33
Std dev	8.84	8.76	8.62	7.56	7.90	8.43	14.93	12.14
Std err mean	1.98	1.44	1.36	1.11	1.81	3.44	8.62	4.96
Number of scars		(N=60)	(N=69)	(N=67)	(N=27)	(N=5)		(N=7)
Maximum	10	10	9	9	6	6	2	6
Median	3	3	3	3	3	3	2	4
Minimum	1	1	1	1	1	3	1	1
Mean	3.72	3.15	2.94	3.66	2.74	4.00	1.67	3.57
Std dev	2.08	1.70	1.64	1.98	1.40	1.41	0.58	1.72
Std err mean	0.37	0.22	0.20	0.24	0.27	0.63	0.33	0.65

Table 7.33 Counts and frequencies (%) of taphonomic and technological characteristics of basalt biface sharpening flakes of Layer II-6

	II-6/L1	II-6/L2	II-6/L3	II-6/L4	II-6/L4b	II-6/L5	II-6/L6	II-6/L7
N	32	67	82	77	30	6	3	8
Breakage								
Complete	15 / 46.88	30 / 44.78	27 / 32.93	32 / 41.56	10 / 33.33	3 / 50.00	1 / 33.33	2 / 25.00
Distal	7 / 21.88	11 / 16.42	6 / 7.32	12 / 15.58	4 / 13.33	-	1 / 33.33	3 / 37.50
Lateral	4 / 12.50	5 / 7.46	14 / 17.07	10 / 12.99	2 / 6.67	-	1 / 33.33	1 / 12.50
Proximal	1 / 3.13	4 / 5.97	18 / 21.95	8 / 10.39	4 / 13.33	-	-	-
Distal & lateral	2 / 6.25	8 / 11.94	3 / 3.66	5 / 6.49	4 / 13.33	2 / 33.33	-	2 / 25.00
Distal & proximal	-	3 / 4.48	3 / 3.66	4 / 5.19	2 / 6.67	-	-	-
Fragment	2 / 6.25	1 / 1.49	8 / 9.76	3 / 3.90	3 / 10.00	1 / 16.67	-	-
Proximal & lateral	1 / 3.13	5 / 7.46	3 / 3.66	3 / 3.90	1 / 3.33	-	-	-
Cortex								
No cortex	28 / 87.50	58 / 86.57	81 / 98.78	73 / 94.81	26 / 86.67	6 / 100.00	3 / 100.00	8 / 100.00
1–25%	1 / 3.13	4 / 5.97	1 / 1.22	3 / 3.90	1 / 3.33	-	-	-
26–50%	1 / 3.13	1 / 1.49	-	-	1 / 3.33	-	-	-
51–75%	2 / 6.25	2 / 2.99	-	-	-	-	-	-
76–100%	-	1 / 1.49	-	1 / 1.30	1 / 3.33	-	-	-
Indeterminate	-	1 / 1.49	-	-	1 / 3.33	-	-	-
Direction of blow		(N=66)	(N=81)					(N=7)
End-struck	15 / 46.88	22 / 33.33	22 / 27.16	28 / 36.36	9 / 30.00	1 / 16.67	1 / 33.33	1 / 14.29
Side-struck	8 / 25.00	16 / 24.24	24 / 29.63	11 / 14.29	7 / 23.33	4 / 66.67	1 / 33.33	-
Special side-struck	3 / 9.38	7 / 10.61	6 / 7.41	13 / 16.88	5 / 16.67	-	1 / 33.33	1 / 14.29
Indeterminate	6 / 18.75	21 / 31.82	29 / 35.80	25 / 32.47	9 / 30.00	1 / 16.67	-	5 / 71.43

sequence, while the cores (giant cores) are described in detail elsewhere (Section 7.4.7.2).

7.3.3.1 Shoulder/Corner Flakes

These large flakes (N=78) are of particular morphology and dimensions as they all originate from the corners of the large basalt slabs. The geometry of the natural thick basalt slabs (Section 7.4.7.2) includes two surfaces (a flat base and a flat/sloping top). The thickness of the slab is not homogeneous along the entire circumference, and the shoulder/corner flakes were removed from the thickest part of the slab, where it often forms a right angle (Figs. 7.57–7.61). Their location on the slab explains the cortical coverage, morphology, and robustness of these flakes. The removal of these flakes was clearly aimed at producing an acute angle, an initial stage in the formation of a

7.3 Technology of Basalt Flakes

Table 7.34 Counts and frequencies (%) of technological characteristics of basalt biface sharpening flakes of Layer II-6

	II-6/L1	II-6/L2	II-6/L3	II-6/L4	II-6/L4b	II-6/L5	II-6/L6	II-6/L7
N	32	67	82	77	30	6	3	8
Scar pattern		(N=65)	(N=81)					(N=7)
Plain	1 3.13	6 9.23	15 18.52	13 16.88	5 16.67	4 66.67	-	1 14.29
Along axis	11 34.38	15 23.08	8 9.88	13 16.88	8 26.67	-	1 33.33	3 42.86
Parallel	-	1 1.54	-	-	-	-	-	-
Convergent	-	-	1 1.23	1 1.30	-	-	-	-
Opposed	-	2 3.08	-	-	-	-	-	-
Radial	4 12.50	7 10.77	12 14.81	9 11.69	-	-	-	-
Ridged	-	1 1.54	1 1.23	-	-	-	-	-
Side	4 12.50	4 6.15	8 9.88	3 3.90	4 13.33	-	-	-
Along axis & side	3 9.38	3 4.62	4 4.94	6 7.79	2 6.67	1 16.67	-	2 28.57
Along axis & opposed	1 3.13	3 4.62	6 7.41	6 7.79	1 3.33	1 16.67	-	-
Opposed & side	-	-	2 2.47	1 1.30	1 3.33	-	-	-
Indeterminate	8 25.00	23 35.38	24 29.63	25 32.47	9 30.00	-	2 66.67	1 14.29
Striking platform		(N=66)						(N=7)
Cortical	1 3.13	1 1.52	-	-	-	-	-	-
Punctiform	1 3.13	-	-	-	-	-	1 33.33	-
Plain	19 59.38	40 60.61	43 52.44	36 46.75	15 50.00	5 83.33	2 66.67	5 71.43
Dihedral	1 3.13	2 3.03	2 2.44	2 2.60	4 13.33	1 16.67	-	-
Facetted	2 6.25	2 3.03	2 2.44	10 12.99	2 6.67	-	-	1 14.29
Removed	1 3.13	2 3.03	3 3.66	4 5.19	-	-	-	-
Broken	4 12.50	13 19.70	27 32.93	19 24.68	7 23.33	-	-	-
Indeterminate	3 9.38	6 9.09	5 6.10	6 7.79	2 6.67	-	-	1 14.29

striking platform, which requires an extremely powerful blow, an "opening" of the core. There are shoulder/corner flakes in all levels of Layer II-6, with the highest frequencies occurring in Levels 1, 3, and 4 (Table 7.40). Table 7.40 presents the data for their size in the different archaeological levels of Layer II-6. While some variability is observed in the size measurements, the variability is minimal when the largest samples are compared. One would expect some size variability, as the slabs themselves are not homogeneous in their thickness (and most probably in other size attributes).

Shoulder/corner flakes display different taphonomic states. Complete artifacts are frequent and the most common breaks

Fig. 7.51 Basalt *éclat de taille de biface* (#687 Layer II-6 Level 1)

Fig. 7.54 Basalt *éclat de taille de biface* (#12956 Layer II-6 Level 1)

Fig. 7.52 Basalt *éclat de taille de biface* (#5876 Layer II-6 Level 1)

Fig. 7.55 Basalt *éclat de taille de biface* (#13253 Layer II-6 Level 1)

Fig. 7.53 Basalt *éclat de taille de biface* (#6446 Layer II-6 Level 1)

are of the distal and lateral edges. Many flakes are classified as indeterminate; this is most probably the result of weathering of the dorsal faces (Table 7.41). End-struck is the most common direction of blow, but other directions occur in lower frequencies (Table 7.41). A noticeable feature is the high variability of cortical coverage, showing that, regardless of origin, some of the flakes are more extensively flaked than others (Table 7.41), yet most of the dorsal surfaces are cortical. The most common scar pattern is cortical, followed by indeterminate and along axis (Table 7.42), averaging between 2 and 4 scars per flake (Table 7.40). The most frequent type of striking platform is plain, followed by indeterminate and broken (Table 7.42). The largest sample of shoulder/corner flakes (Layer II-6 Level 4) shows that on a minimal number of flakes there are indications

7.3 Technology of Basalt Flakes

Table 7.35 Descriptive statistics of basalt *éclats de taille de biface* of Area C and Layer II-6 (size in mm; angle in degrees)

	Area C	II-6/L1	II-6/L2	II-6/L3	II-6/L4	II-6/L4b	II-6/L5	II-6/L6	II-6/L7
N	13	6	7	6	21	5	14	7	10
Maximum length		(N=5)							
Maximum	80	33	49	36	88	65	46	41	49
Median	45	27	32	25	34	41	30	29	34.5
Minimum	20	26	23	21	21	22	22	21	22
Mean	43.69	27.80	33.43	27.33	38.71	42.00	33.00	30.43	33.40
Std dev	19.20	2.95	8.62	6.19	16.57	15.41	9.44	7.55	8.67
Std err mean	5.33	1.32	3.26	2.53	3.62	6.89	2.52	2.85	2.74
Length									
Maximum	64	27	44	35	86	65	45	39	39
Median	35	25	28	22	26	38	29	22	25.5
Minimum	17	20	15	18	15	19	18	16	15
Mean	38.08	24.17	27.86	25.33	32.76	38.60	30.64	24.86	26.70
Std dev	15.60	2.71	10.19	7.28	17.37	17.64	9.45	8.97	8.01
Std err mean	4.33	1.11	3.85	2.97	3.79	7.89	2.53	3.39	2.53
Width		(N=5)							
Maximum	77	28	40	36	70	60	45	41	43
Median	25	23	28	24	32	27	27	25	27.5
Minimum	13	20	20	18	16	18	15	21	19
Mean	31.92	23.60	28.86	24.33	32.38	32.20	26.86	26.86	27.50
Std dev	18.20	3.36	7.17	6.62	13.54	17.06	8.75	7.15	7.53
Std err mean	5.05	1.50	2.71	2.70	2.95	7.63	2.34	2.70	2.38
Thickness		(N=5)							
Maximum	26	7	9	12	30	19	12	9	11
Median	9	6	7	5	9	9	7	5	6
Minimum	6	4	5	4	4	5	4	5	5
Mean	10.69	5.80	6.86	6.83	10.00	10.00	7.79	6.00	7.10
Std dev	5.59	1.30	1.68	3.31	6.02	5.39	2.58	1.53	2.23
Std err mean	1.55	0.58	0.63	1.35	1.31	2.41	0.69	0.58	0.71
L/W		(N=5)							
Maximum	2.27	1.24	1.46	1.42	2.13	2.20	1.54	1.86	1.80
Median	1.25	1.04	0.88	1.07	0.84	1.08	1.14	0.80	1.01
Minimum	0.74	0.96	0.54	0.69	0.62	0.75	0.67	0.57	0.54
Mean	1.33	1.07	0.98	1.07	1.03	1.30	1.18	0.97	1.02
Std dev	0.48	0.11	0.34	0.27	0.40	0.55	0.29	0.43	0.36
Std err mean	0.13	0.05	0.13	0.11	0.09	0.25	0.08	0.16	0.11
Angle of striking platform	(N=3)	(N=1)	(N=4)	(N=1)	(N=9)	(N=2)	(N=7)	(N=3)	(N=3)
Maximum	118	114	124	126	123	105	126	128	127
Median	110	114	114	126	105	103.5	116	111	108
Minimum	102	114	93	126	96	102	100	106	107
Mean	110.00	114.00	111.25	126.00	107.33	103.50	114.00	115.00	114.00
Std dev	8.00	-	13.45	-	8.94	2.12	10.08	11.53	11.27
Std err mean	4.62	-	6.73	-	2.98	1.50	3.81	6.66	6.51

Table 7.36 Counts and frequencies (%) of taphonomic and technological characteristics of basalt *éclats de taille de biface* of Area C and Layer II-6

	Area C	II-6/L1	II-6/L2	II-6/L3	II-6/L4	II-6/L4b	II-6/L5	II-6/L6	II-6/L7
N	13	6	7	6	21	5	14	7	10
Breakage									
Complete	7	2	3	1	4	1	4	1	4
	53.85	33.33	42.86	16.67	19.05	20.00	28.57	14.29	40.00
Distal	1	1	1	1	5	1	-	1	1
	7.69	16.67	14.29	16.67	23.81	20.00		14.29	10.00
Lateral	1	1	1	1	1	-	4	-	2
	7.69	16.67	14.29	16.67	4.76		28.57		20.00
Proximal	-	-	-	2	2	1	-	1	1
				33.33	9.52	20.00		14.29	10.00
Distal & lateral	-	1	1	-	3	-	1	1	-
		16.67	14.29		14.29		7.14	14.29	
Distal & proximal	-	-	-	-	-	-	-	1	1
								14.29	10.00
Fragment	4	1	-	1	4	1	5	2	1
	30.77	16.67		16.67	19.05	20.00	35.71	28.57	10.00
Proximal & lateral	-	-	1	-	2	-	-	-	-
			14.29		9.52				
Indeterminate	-	-	-	-	-	1	-	-	-
						20.00			
Cortex		(N=5)							
No cortex	9	5	6	6	20	4	11	7	9
	69.23	100.00	85.71	100.00	95.24	80.00	78.57	100.00	90.00
51–75%	1	-	-	-	-	-	-	-	-
	7.69								
76–100%	2	-	-	-	-	1	-	-	-
	15.38					20.00			
Indeterminate	1	-	1	-	1	-	3	-	1
	7.69		14.29		4.76		21.43		10.00
Direction of blow	(N=11)	(N=5)							
End-struck	5	2	3	1	4	2	4	-	2
	45.45	40.00	42.86	16.67	19.05	40.00	28.57		20.00
Side-struck	2	-	1	1	5	-	2	2	2
	18.18		14.29	16.67	23.81		14.29	28.57	20.00
Special side-struck	-	-	1	1	-	-	1	-	-
			14.29	16.67			7.14		
Indeterminate	4	3	2	3	12	3	7	5	6
	36.36	60.00	28.57	50.00	57.14	60.00	50.00	71.43	60.00
Striking platform	(N=11)	(N=5)		(N=5)	(N=18)			(N=6)	
Punctiform	-	-	-	-	-	-	-	-	1
									10.00
Plain	1	1	3	1	8	2	4	2	3
	9.09	20.00	42.86	20.00	44.44	40.00	28.57	33.33	30.00
Dihedral	1	-	1	-	-	-	1	-	-
	9.09		14.29				7.14		
Facetted	1	2	2	-	1	-	-	1	1
	9.09	40.00	28.57		5.56			16.67	10.00
Crushed	2	-	-	-	-	-	-	-	1
	18.18								10.00
Removed	-	-	-	-	1	-	-	-	-
					5.56				
Broken	3	-	1	2	7	3	4	3	4
	27.27		14.29	40.00	38.89	60.00	28.57	50.00	40.00
Indeterminate	3	2	-	2	1	-	5	-	-
	27.27	40.00		40.00	5.56		35.71		
Scar pattern	(N=12)	(N=5)						(N=6)	
Cortical	2	-	-	-	-	1	-	-	-
	16.67					20.00			

7.3 Technology of Basalt Flakes

Table 7.36 (cont.)

	Area C	II-6/L1	II-6/L2	II-6/L3	II-6/L4	II-6/L4b	II-6/L5	II-6/L6	II-6/L7
N	13	6	7	6	21	5	14	7	10
Plain	3 25.00	2 40.00	-	-	6 28.57	-	2 14.29	-	3 30.00
Along axis	1 8.33	-	3 42.86	1 16.67	4 19.05	-	4 28.57	1 16.67	-
Convergent	-	-	-	1 16.67	-	-	-	1 16.67	-
Radial	1 8.33	-	2 28.57	1 16.67	1 4.76	-	-	-	1 10.00
Side	1 8.33	-	-	-	-	-	-	-	-
Along axis & side	-	-	-	-	2 9.52	-	-	-	-
Indeterminate	4 33.33	3 60.00	2 28.57	3 50.00	8 38.10	4 80.00	8 57.14	4 66.67	6 60.00

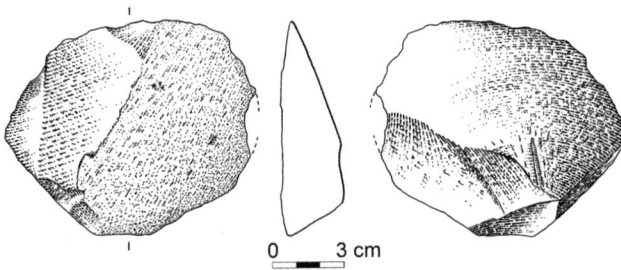

Fig. 7.56 Basalt biface thinning flake with ventral removals (#18 Layer II-6 Level 4)

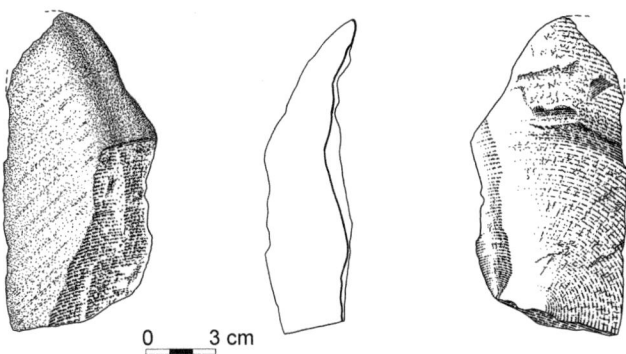

Fig. 7.57 Basalt shoulder flake with ventral removals (#170 Layer II-6 Level 4)

that, in addition to plain striking platforms, some were removed (Table 7.42).

7.3.3.2 Slice Flakes

The method of core slicing (Sharon 2007), defined at the Indian Acheulian site of Hunsgi V (Paddayya 1982), is represented at GBY by four slice flakes, all originating from Layer II-6 (Levels 2, 4b, 5, and 6) (Figs. 7.62–7.63). Apart from these four slice flakes, the use of this method at GBY is recorded on the giant cores and is discussed and illustrated below (Section 7.4.7.2). The flakes are complete, large, and thick (Figs. 7.62–7.63), and represent different amounts of cortex. Technologically, they are

Table 7.37 Typological counts and frequencies (%) of basalt flakes and tools with ventral removals

Typology	N	%
Single straight side-scraper	1	0.28
Single convex side-scraper	1	0.28
Single concave side-scraper	1	0.28
Thinned back side-scraper	1	0.28
Atypical backed knife	2	0.56
Notch	10	2.82
Denticulate	2	0.56
Retouch on ventral face	1	0.28
End-notched piece	2	0.56
Varia	1	0.28
Flake	300	84.75
Éclat de taille de biface	1	0.28
Retouched flake	8	2.26
Massive scraper	12	3.39
Split percussor	10	2.82
Biface preform	1	0.28
Total	354	100.00

Table 7.38 Counts and frequencies (%) of types of striking platforms on basalt flakes with ventral removals

Striking platform	N	%
Cortical	14	4.01
Punctiform	1	0.29
Plain	163	46.70
Dihedral	5	1.43
Facetted	14	4.01
Removed	25	7.16
Broken	50	14.33
Indeterminate	77	22.06
Total	349	100.00

Table 7.39 Descriptive statistics of basalt core management pieces of Layer II-6 (size in mm; angle in degrees)

N	Maximum length	Length	Width	Thickness	L/W	Number of scars	Angle of striking platform
	11	11	11	11	11	11	6
Maximum	104	92	104	50	1.86	7	113
Median	61	53	38	22	1.37	4	109
Minimum	27	15	16	12	0.54	2	100
Mean	58.73	52.00	45.00	26.09	1.31	4.09	108.17
Std dev	25.72	22.43	26.73	12.13	0.48	1.76	4.40
Std err mean	7.76	6.76	8.06	3.66	0.15	0.53	1.80

Fig. 7.58 Basalt shoulder flake (#5379 Layer II-6 Level 1)

Fig. 7.60 Basalt shoulder flake (#5075 Layer II-6 Level 4)

Fig. 7.59 Basalt shoulder flake (#5841 Layer II-6 Level 1)

Fig. 7.61 Basalt shoulder flake (#9343 Layer II-6 Level 7)

7.3 Technology of Basalt Flakes

Table 7.40 Descriptive statistics of basalt shoulder/corner flakes of Layer II-6 (size in mm; angle in degrees)

	II-6/L1	II-6/L2	II-6/L3	II-6/L4	II-6/L4b	II-6/L5	II-6/L6	II-6/L7
N	17	13	4	31	7	1	1	4
Maximum length								
Maximum	139	157	178	159	140	82	181	150
Median	125	121	135	113	118	82	181	129.5
Minimum	98	83	101	69	91	82	181	119
Mean	123.29	121.77	137.25	116.13	116.71	82.00	181.00	132.00
Std dev	13.49	23.27	31.54	23.33	16.10	-	-	14.72
Std err mean	3.27	6.46	15.77	4.19	6.09	-	-	7.36
Length								
Maximum	138	150	178	159	140	59	179	143
Median	120	94	101	102	75	59	179	117.5
Minimum	85	52	81	55	54	59	179	69
Mean	116.00	99.31	115.25	101.48	82.86	59.00	179.00	111.75
Std dev	17.50	31.78	42.88	27.15	27.82	-	-	31.26
Std err mean	4.24	8.81	21.44	4.88	10.52	-	-	15.63
Width								
Maximum	135	140	135	159	130	82	140	146
Median	88	93	132.5	91	112	82	140	125.5
Minimum	59	13	87	47	64	82	140	93
Mean	90.06	92.69	121.75	90.84	106.71	82.00	140.00	122.50
Std dev	19.20	36.23	23.29	27.20	24.14	-	-	22.55
Std err mean	4.66	10.05	11.64	4.88	9.13	-	-	11.27
Thickness								
Maximum	80	56	52	81	50	28	93	69
Median	47	38	38	45	39	28	93	55.5
Minimum	33	26	30	16	31	28	93	41
Mean	51.47	40.77	39.50	47.26	38.86	28.00	93.00	55.25
Std dev	14.73	10.27	10.38	15.23	6.62	-	-	15.33
Std err mean	3.57	2.85	5.19	2.74	2.50	-	-	7.66
L/W								
Maximum	1.82	5.69	1.37	2.40	1.17	0.72	1.28	1.20
Median	1.26	0.98	0.95	1.14	0.74	0.72	1.28	0.96
Minimum	0.89	0.65	0.60	0.51	0.45	0.72	1.28	0.58
Mean	1.32	1.50	0.97	1.21	0.81	0.72	1.28	0.92
Std dev	0.24	1.44	0.36	0.48	0.29	-	-	0.26
Std err mean	0.06	0.40	0.18	0.09	0.11	-	-	0.13
Angle of striking platform	(N=6)	(N=6)	(N=4)	(N=14)	(N=2)	(N=1)		(N=2)
Maximum	123	131	130	130	125	113	-	126
Median	119.5	116.5	123.5	119.5	116.5	113	-	115
Minimum	112	100	119	97	108	113	-	104
Mean	118.83	115.17	124.00	117.00	116.50	113.00	-	115.00
Std dev	4.07	10.76	4.69	10.64	12.02	-	-	15.56
Std err mean	1.66	4.39	2.35	2.84	8.50	-	-	11.00
Number of scars	(N=16)	(N=12)		(N=21)				
Maximum	5	6	5	5	9	2	-	3
Median	3	3	5	2	3	2	-	3
Minimum	1	1	2	1	1	2	-	2
Mean	2.56	2.67	4.25	2.29	3.86	2.00	-	2.75
Std dev	1.15	1.44	1.50	1.42	2.61	-	-	0.50
Std err mean	0.29	0.41	0.75	0.31	0.99	-	-	025

Table 7.41 Counts and frequencies (%) of taphonomic and technological characteristics of shoulder/corner flakes of Layer II-6

	II-6/L1	II-6/L2	II-6/L3	II-6/L4	II-6/L4b	II-6/L5	II-6/L6	II-6/L7
N	17	13	4	31	7	1	1	4
Breakage								
Complete	6	9	1	16	6	1	-	3
	35.29	69.23	25.00	51.61	85.71	100.00		75.00
Distal	4	2	1	6	-	-	1	-
	23.53	15.38	25.00	19.35			100.00	
Lateral	5	1	-	3	1	-	-	-
	29.41	7.69		9.68	14.29			
Proximal	1	-	1	2	-	-	-	1
	5.88		25.00	6.45				25.00
Distal & lateral	-	-	-	2	-	-	-	-
				6.45				
Distal & proximal	-	1	-	2	-	-	-	-
		7.69		6.45				
Fragment	-	-	-	-	-	-	-	-
Proximal & lateral	1	-	-	-	-	-	-	-
	5.88							
Indeterminate	-	-	1	-	-	-	-	-
			25.00					
Cortex								
No cortex	4	5	3	3	3	-	1	-
	23.53	38.46	75.00	9.68	42.86		100.00	
1–25%	-	1	-	2	-	-	-	-
		7.69		6.45				
26–50%	2	2	-	1	1	-	-	-
	11.76	15.38		3.23	14.29			
51–75%	8	2	1	3	3	1	-	-
	47.06	15.38	25.00	9.68	42.86	100.00		
76–100%	3	3	-	19	-	-	-	3
	17.65	23.08		61.29				75.00
Indeterminate	-	-	-	3	-	-	-	1
				9.68				25.00
Direction of blow	(N=16)			(N=30)				
End-struck	11	6	2	16	2	-	-	-
	68.75	46.15	50.00	53.33	28.57			
Side-struck	2	5	1	4	2	1	-	3
	12.50	38.46	25.00	13.33	28.57	100.00		75.00
Special side-struck	-	-	-	4	3	-	-	-
				13.33	42.86			
Indeterminate	3	2	1	6	-	-	1	1
	18.75	15.38	25.00	20.00			100.00	25.00

end-struck (N=3) and side-struck (N=1) and display extensive variability in their dimensions and different types of striking platform, while the mean number of scars is 1.5.

7.3.3.3 Wedge-Like Flakes

Wedge-like flakes (N=36) are large basalt flakes of a particular and hence identifiable morphology. These are elongated, triangular flakes ("frontal flakes" in Madsen and Goren-Inbar 2004:14, fig. 5) that were struck from cortical flat surfaces of the slabs. The initial removals from these slabs included those of at least two deep flakes, forming a ridge between them; the next blow, located above the ridge, thus resulted in a triangular flake of wedge morphology (Figs. 7.18, 7.64–7.67). Another alternative for the formation of such flakes is an entirely cortical wedge in which the ridge was part of the original topography of the basalt slab. It should be noted that some of the large flakes of handaxe shape (Section 7.3.3.6) are also of wedge morphology (Figs. 7.68–7.69).

Basalt wedge-like flakes were found, in very small frequencies, only in Layer II-6. Level 1 is the richest assemblage,

7.3 Technology of Basalt Flakes

Table 7.42 Counts and frequencies (%) of technological characteristics of basalt shoulder/corner flakes of Layer II-6

	II-6/L1	II-6/L2	II-6/L3	II-6/L4	II-6/L4b	II-6/L5	II-6/L6	II-6/L7
N	17	13	4	31	7	1	1	4
Scar pattern	(N=16)		(N=3)					
Cortical	6 37.50	4 30.77	-	15 48.39	1 14.29	-	-	1 25.00
Plain	1 6.25	-	-	1 3.23	-	-	1 100.00	-
Along axis	1 6.25	2 15.38	-	6 19.35	-	1 100.00	-	2 50.00
Opposed	1 6.25	-	-	3 9.68	1 14.29	-	-	-
Radial	1 6.25	2 15.38	1 33.33	1 3.23	1 14.29	-	-	-
Ridged	-	-	-	-	-	-	-	-
Side	1 6.25	1 7.69	-	-	1 14.29	-	-	-
Along axis & side	-	-	2 66.67	1 3.23	-	-	-	-
Along axis & opposed	-	1 7.69	-	-	-	-	-	-
Indeterminate	5 31.25	3 23.08	-	4 12.90	3 42.86	-	-	1 25.00
Striking platform	(N=16)		(N=3)					
Cortical	-	-	-	6 19.35	1 14.29	-	-	-
Plain	8 50.00	6 46.15	1 33.33	10 32.26	5 71.43	-	1 100.00	2 50.00
Dihedral	-	-	1 33.33	-	-	1 100.00	-	-
Facetted	1 6.25	1 7.69	-	-	-	-	-	-
Removed	1 6.25	1 7.69	-	2 6.45	-	-	-	-
Broken	3 18.75	1 7.69	1 33.33	4 12.90	-	-	-	2 50.00
Indeterminate	3 18.75	4 30.77	-	9 29.03	1 14.29	-	-	-

Fig. 7.62 Basalt slice flake (#9230 Layer II-6 Level 5)

Fig. 7.63 Basalt slice flake (#9154 Layer II-6 Level 6)

and even this sample is small (Table 7.43). The artifacts display great intra-level and inter-level variability in dimensions, with the exception of homogeneity in the angles of striking platforms. Taphonomically, most of these flakes are complete, followed by distal breaks and minimal representation of other types of breaks (Table 7.44). Most of the wedge-like flakes are devoid of cortex, but there are individual cases, as described above, in which cortex covers most of the surface (Table 7.44). Clearly, these flakes have no scar pattern, but the majority of the artifacts show an along axis pattern (Table 7.45). There is extensive variability with respect to the number of scars, the mean ranging between 2 and 5 (Table 7.43). Their dominant type of striking platform is plain and none are cortical (Table 7.45; see Section 7.6 for discussion).

7.3.3.4 Giant Flakes

Giant flakes are classified by their particularly large dimensions (201 mm and above) and are rare at GBY (N=5). The longest flake reaches 242 mm in maximal length and is 177 mm wide, and the widest flake is 120 mm long and 225 mm wide (Fig. 7.70).

Of the five giant flakes, three originate from Layer II-6 and two from Layer II-1. Typologically, they are all classified as unretouched flakes, and two were identified as pitted stones. Of the two identified striking platforms, one was plain and the other facetted.

7.3.3.5 Waste of Giant Cores

Flakes classified as waste of giant cores (N=72) are thick flakes of large dimensions with a general similarity between the length and width axis (Table 7.46). The dorsal face of these flakes

Fig. 7.65 Basalt wedge flake (#5588 Layer II-6 Level 1)

Fig. 7.64 Basalt wedge flake (#5366 Layer II-6 Level 1)

Fig. 7.66 Basalt wedge flake (#4818 Layer II-6 Level 4)

7.3 Technology of Basalt Flakes

Fig. 7.67 Basalt wedge flake (#4948 Layer II-6 Level 4)

reduction sequence, these flakes all originate from Layer II-6 and are rarely associated with retouched tools (Table 7.47).

The waste flakes of giant cores are often complete, followed by distal and lateral breakages (Table 7.48), and are mostly non-cortical; when cortex occurs it is minimal (Table 7.48). Technologically, these waste flakes frequently exhibit plain striking platforms and are mostly end-struck, followed by side-struck and only rarely special side-struck (Table 7.49). Unlike other basalt flakes, the dominant scar pattern of these flakes is not along axis; rather, varied patterns occur, including radial and along axis and side (Table 7.49).

7.3.3.6 Flakes of Biface Morphology

As in other Large Flake Acheulian assemblages, GBY is characterized by the presence of bifaces that were modified on flakes. This particular tradition is also characterized by minimal retouch in the final stages of biface modification (thinning, shaping, and finishing following Newcomer 1971), as seen in many sites (e.g. Kilombe: Gowlett 1991; Olorgesailie: Isaac 1977; Sharon 2007). Thus, a blank with morphology similar to that of a biface can be considered an artifact with potential to be modified into a biface. As expected, the large basalt flake assemblage of GBY indeed includes items whose morphology is that of a blank that with a minimal investment could be

has a protruding and uneven morphology that associates them with the reduction sequence and maintenance of giant cores (Figs. 7.4, 7.71–7.74). The morphology of these flakes is varied (Figs. 7.75–7.76) and includes some elongated artifacts (Figs. 7.77–7.79). Like some other waste products of the giant core

Fig. 7.68 Basalt wedge flake of handaxe morphology (#4998 Layer II-6 Level 4)

Fig. 7.69 Basalt wedge flake of handaxe morphology (#194 Layer II-6 Level 4)

Table 7.43 Descriptive statistics of basalt wedge-like flakes of Layer II-6 (size in mm; angle in degrees)

	II-6/L1	II-6/L2	II-6/L3	II-6/L4	II-6/L4b	II-6/L5	II-6/L6	II-6/L7
N	16	2	1	6	4	3	1	3
Maximum length								
Maximum	141	97	142	158	110	134	67	119
Median	99.5	96.5	142	115	80.5	88	67	112
Minimum	80	96	142	92	69	75	67	103
Mean	103.81	96.50	142.00	115.50	85.00	99.00	67.00	111.33
Std dev	19.52	0.71	-	24.02	19.88	31.00	-	8.02
Std err mean	4.88	0.50	-	9.81	9.94	17.90	-	4.63
Length								
Maximum	141	88	142	158	110	134	63	118
Median	99.5	70.5	142	109.5	79.5	75	63	99
Minimum	70	53	142	90	66	45	63	63
Mean	102.25	70.50	142.00	112.33	83.75	84.67	63.00	93.33
Std dev	20.19	24.75	-	24.60	20.50	45.28	-	27.93
Std err mean	5.05	17.50	-	10.04	10.25	26.14	-	16.13
Width								
Maximum	99	96	59	81	70	85	46	113
Median	57	89.5	59	65	49.5	53	46	67
Minimum	47	83	59	52	35	41	46	63
Mean	60.50	89.50	59.00	65.17	51.00	59.67	46.00	81.00
Std dev	13.79	9.19	-	9.95	14.40	22.74	-	27.78
Std err mean	3.45	6.50	-	4.06	7.20	13.13	-	16.04
Thickness								
Maximum	64	33	56	49	30	41	22	38
Median	35	29	56	30	27.5	38	22	37
Minimum	23	25	56	26	27	31	22	35
Mean	36.44	29.00	56.00	35.33	28.00	36.67	22.00	36.67
Std dev	11.97	5.66	-	10.33	1.41	5.13	-	1.53
Std err mean	2.99	4.00	-	4.22	0.71	2.96	-	0.88
L/W								
Maximum	2.11	1.06	2.41	1.95	2.57	2.53	1.37	1.76
Median	1.70	0.81	2.41	1.70	1.48	1.83	1.37	1.57
Minimum	1.16	0.55	2.41	1.45	1.35	0.53	1.37	0.56
Mean	1.72	0.81	2.41	1.72	1.72	1.63	1.37	1.30
Std dev	0.29	0.36	-	0.21	0.58	1.01	-	0.65
Std err mean	0.07	0.25	-	0.08	0.29	0.59	-	0.37
Angle of striking platform	(N=10)			(N=5)	(N=3)			
Maximum	127	-	120	131	124	-	126	128
Median	112.5	-	120	120	123	-	126	122
Minimum	97	-	120	113	119	-	126	114
Mean	112.80	-	120.00	120.40	122.00	-	126.00	121.33
Std dev	9.46	-	-	7.13	2.65	-	-	7.02
Std err mean	2.99	-	-	3.19	1.53	-	-	4.06
Number of scars	(N=13)	(N=1)						
Maximum	6	5	2	7	3	3	5	5
Median	3	5	2	4.5	2.5	3	5	2
Minimum	1	5	2	1	2	2	5	2
Mean	2.77	5.00	2.00	4.50	2.50	2.67	5.00	3.00
Std dev	1.36	-	-	2.51	0.58	0.58	-	1.73
Std err mean	0.38	-	-	1.02	0.29	0.33	-	1.00

7.3 Technology of Basalt Flakes

Table 7.44 Counts and frequencies (%) of taphonomic and technological characteristics of wedge-like flakes of Layer II-6

	II-6/L1	II-6/L2	II-6/L3	II-6/L4	II-6/L4b	II-6/L5	II-6/L6	II-6/L7
N	16	2	1	6	4	3	1	3
Breakage								
Complete	11 68.75	1 50.00	-	5 83.33	2 50.00	1 33.33	-	1 33.33
Distal	3 18.75	-	-	1 16.67	2 50.00	-	1 100.00	2 66.67
Lateral	-	1 50.00	-	-	-	-	-	-
Proximal	1 6.25	-	-	-	-	1 33.33	-	-
Distal & lateral	1 6.25	-	-	-	-	-	-	-
Distal & proximal	-	-	-	-	-	1 33.33	-	-
Indeterminate	-	-	1 100.00	-	-	-	-	-
Cortex								
No cortex	11 68.75	2 100.00	1 100.00	5 83.33	3 75.00	3 100.00	1 100.00	3 100.00
1–25%	-	-	-	-	1 25.00	-	-	-
26–50%	1 6.25	-	-	-	-	-	-	-
51–75%	1 6.25	-	-	1 16.67	-	-	-	-
76–100%	2 12.50	-	-	-	-	-	-	-
Indeterminate	1 6.25	-	-	-	-	-	-	-
Direction of blow								
End-struck	16 100.00	-	1 100.00	6 100.00	4 100.00	1 33.33	1 100.00	2 66.67
Side-struck	-	1 50.00	-	-	-	1 33.33	-	-
Special side-struck	-	-	-	-	-	-	-	1 33.33
Indeterminate	-	1 50.00	-	-	-	1 33.33	-	-

transformed into a biface. In order to verify the assumption that this particular morphological group is indeed a separate entity, we carried out a series of tests aimed at examining the relationship between the two (large flakes of biface morphology and large flakes). For the following statistical examination, we included all large basalt flakes from Layer II-6. Flakes with the morphology of bifaces (both handaxes and cleavers) are significantly longer (maximum length: difference: -30.36 mm; t-ratio: -15.059; df: 1,326; $p(tt) < .0001$) and significantly wider (width: difference: -25.10 mm; t-ratio: -10.27; df: 1,325; $p(tt) < .0001$) than large (7–20 mm) unretouched flakes. However, there is no significant difference in their thickness (thickness: difference: -1.95 mm; t-ratio: -1.55; df: 1,326; $p(tt) = 0.1196$).

In addition, flakes of biface morphology are significantly more complete than large unretouched flakes (N=1,328; $\chi 2=9.30$; $df=1$; $p=0.0023$), and occur on special side-struck flakes significantly more often than large unretouched flakes (N=922; $\chi 2=12.97$; $df=2$; $p=0.0015$). As for scar patterns, these are significantly different for flakes of biface morphology, which are less cortical and more along axis and side in comparison to large unretouched flakes (N=685; $\chi 2=42.85$; $df=12$; $p<0.0001$). In terms of cortical coverage, flakes of biface morphology do not differ significantly from large unretouched flakes. The number of scars on the dorsal face, however, is significantly different, with significantly more scars on flakes of biface morphology than on other large

Table 7.45 Counts and frequencies (%) of taphonomic and technological characteristics of wedge-like flakes of Layer II-6

	II-6/L1	II-6/L2	II-6/L3	II-6/L4	II-6/L4b	II-6/L5	II-6/L6	II-6/L7
N	16	2	1	6	4	3	1	3
Scar pattern								
Cortical	3 / 18.75	-	-	-	-	-	-	-
Plain	-	1 / 50.00	-	-	1 / 25.00	-	-	-
Along axis	4 / 25.00	-	-	4 / 66.67	1 / 25.00	-	-	-
Convergent	-	-	-	1 / 16.67	-	-	-	-
Radial	1 / 6.25	-	-	-	-	-	1 / 100.00	1 / 33.33
Ridged	-	-	-	-	-	-	-	2 / 66.67
Along axis & opposed	1 / 6.25	-	-	-	-	-	-	-
Indeterminate	7 / 43.75	1 / 50.00	1 / 100.00	1 / 16.67	2 / 50.00	3 / 100.00	-	-
Striking platform	(N=15)							
Plain	9 / 60.00	-	1 / 100.00	3 / 50.00	4 / 100.00	-	1 / 100.00	2 / 66.67
Dihedral	-	-	-	2 / 33.33	-	-	-	-
Removed	1 / 6.67	1 / 50.00	-	1 / 16.67	-	-	-	-
Broken	1 / 6.67	-	-	-	-	1 / 33.33	-	-
Indeterminate	4 / 26.67	1 / 50.00	-	-	-	2 / 66.67	-	1 / 33.33

Fig. 7.70 Basalt giant flake (#13397 Layer II-6 Level 2)

7.3 Technology of Basalt Flakes

Table 7.46 Descriptive statistics of basalt waste of giant cores of Layer II-6 (size in mm; angle in degrees)

	II-6/L1	II-6/L2	II-6/L3	II-6/L4	II-6/L4b	II-6/L5	II-6/L6	II-6/L7
N	16	6	9	23	2	2	3	11
Maximum length								
Maximum	192	141	156	151	113	105	130	142
Median	104	112	112	100	109	94	128	113
Minimum	84	104	63	71	105	83	97	96
Mean	111.94	118.50	108.67	103.13	109.00	94.00	118.33	114.00
Std dev	28.44	15.58	28.65	20.88	5.66	15.56	18.50	13.05
Std err mean	7.11	6.36	9.55	4.35	4.00	11.00	10.68	3.93
Length								
Maximum	162	135	134	133	111	83	130	114
Median	99.5	107.5	96	93	90.5	75.5	103	91
Minimum	30	56	49	51	70	68	82	74
Mean	96.81	104.17	95.00	92.87	90.50	75.50	105.00	89.64
Std dev	32.80	26.16	28.25	20.10	28.99	10.61	24.06	12.99
Std err mean	8.20	10.68	9.42	4.19	20.50	7.50	13.89	3.92
Width								
Maximum	150	115	112	151	105	105	123	139
Median	79.5	95	70	76	86.5	79.5	94	102
Minimum	51	82	36	43	68	54	76	77
Mean	83.63	98.17	74.67	81.87	86.50	79.50	97.67	101.64
Std dev	26.58	13.92	25.10	25.01	26.16	36.06	23.71	19.27
Std err mean	6.64	5.68	8.37	5.21	18.50	25.50	13.69	5.81
Thickness								
Maximum	62	60	60	67	53	41	88	82
Median	48	47	43	42	47.5	38.5	50	55
Minimum	35	35	30	28	42	36	48	43
Mean	47.25	48.00	43.67	44.30	47.50	38.50	62.00	58.82
Std dev	7.43	10.49	10.42	11.33	7.78	3.54	22.54	13.53
Std err mean	1.86	4.28	3.47	2.36	5.50	2.50	13.01	4.08
L/W								
Maximum	1.92	1.28	2.12	1.77	1.63	1.54	1.71	1.48
Median	1.39	1.18	1.34	1.23	1.15	1.09	0.87	0.89
Minimum	0.39	0.49	0.78	0.51	0.67	0.65	0.84	0.53
Mean	1.23	1.08	1.36	1.22	1.15	1.09	1.14	0.93
Std dev	0.47	0.29	0.49	0.37	0.68	0.63	0.49	0.29
Std err mean	0.12	0.12	0.16	0.08	0.48	0.44	0.29	0.09
Angle of striking platform	(N=7)	(N=5)	(N=3)	(N=14)			(N=2)	(N=2)
Maximum	133	128	130	130	132	103	115	112
Median	120	122	110	118	118.5	99.5	107.5	110
Minimum	117	104	107	88	105	96	100	108
Mean	122.29	119.40	115.67	115.79	118.50	99.50	107.50	110.00
Std dev	5.79	9.10	12.50	11.35	19.09	4.95	10.61	2.83
Std err mean	2.19	4.07	7.22	3.03	13.50	3.50	7.50	2.00
Number of scars	(N=15)			(N=22)			(N=1)	(N=8)
Maximum	16	11	7	10	6	6	8	4
Median	4	5.5	4	5	3.5	5.5	8	3
Minimum	1	4	2	2	1	5	8	1
Mean	5.07	6.17	3.89	4.91	3.50	5.50	8.00	2.88
Std dev	4.28	2.64	1.45	2.45	3.54	0.71	-	1.25
Std err mean	1.11	1.08	0.48	0.52	2.50	0.50	-	0.44

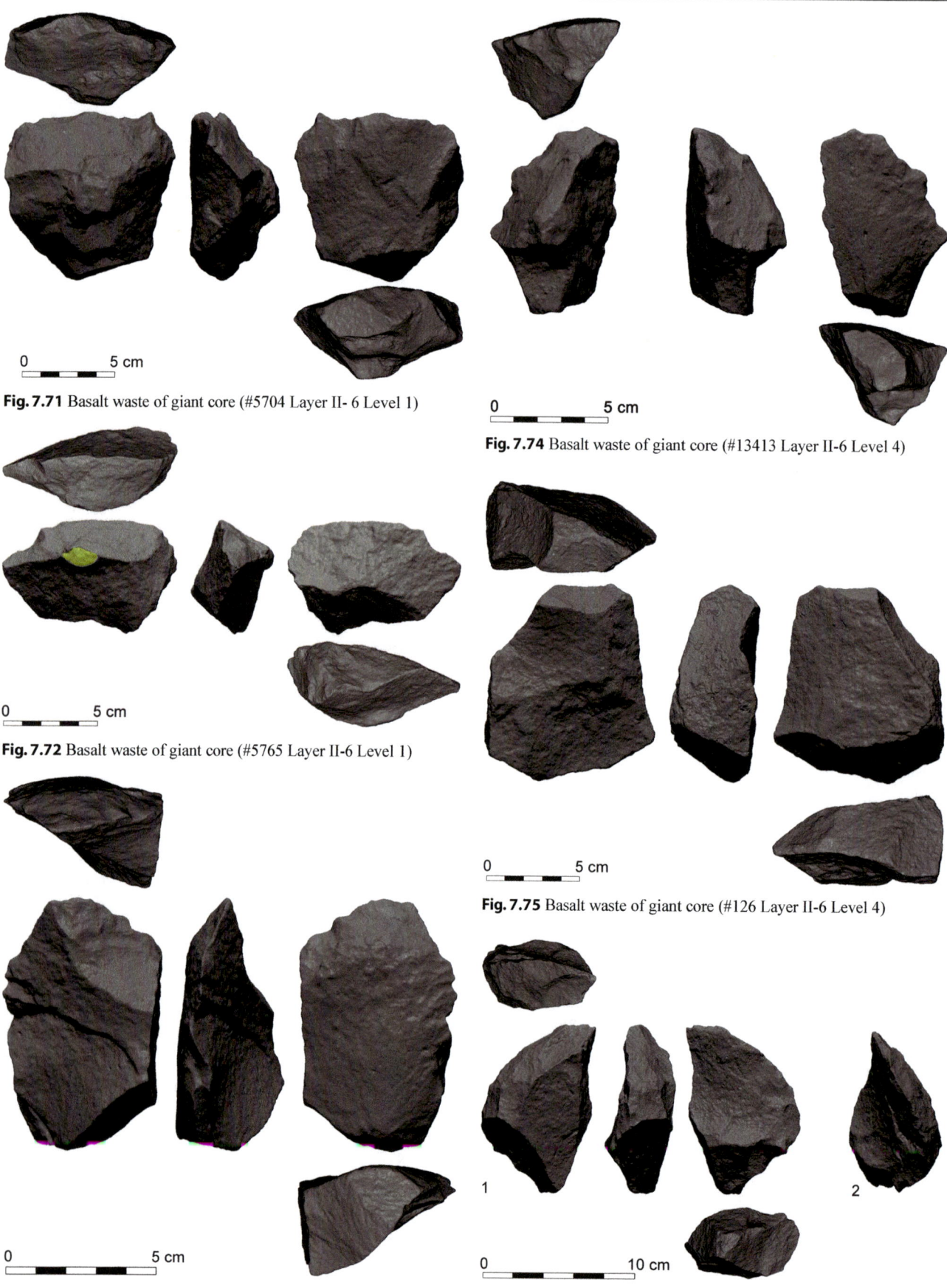

Fig. 7.71 Basalt waste of giant core (#5704 Layer II-6 Level 1)

Fig. 7.72 Basalt waste of giant core (#5765 Layer II-6 Level 1)

Fig. 7.73 Basalt waste of giant core (#5639 Layer II-6 Level 4)

Fig. 7.74 Basalt waste of giant core (#13413 Layer II-6 Level 4)

Fig. 7.75 Basalt waste of giant core (#126 Layer II-6 Level 4)

Fig. 7.76 Basalt waste of giant core (#268 Layer II-6 Level 4)

7.3 Technology of Basalt Flakes

Fig. 7.77 Basalt waste of giant core (#2014 Layer II-6 Level 1)

unretouched flakes (difference:-2.09 scars; t-ratio: -8.35; *df*: 931; *p(tt)*<.0001). The striking platforms of these flakes also differ significantly, with more plain and more removed striking platforms on biface-shaped flakes than on large unretouched flakes (N=971; χ2=20.84; *df*=7; *p*=0.004).

These results clearly indicate that biface-shaped flakes are technologically and morphologically different from large unretouched flakes, hence supporting our decision to present this morphotype in detail. Two categories of flakes of biface morphology were identified: those of handaxe morphology and those of cleaver morphology. Below we present the main characteristics of each of these groups.

Flakes of Handaxe Morphology

Flakes of handaxe morphology are large basalt flakes with a pointed morphology. They are thickest at the proximal end and thin towards the pointed distal end (Figs. 7.68–7.69, 7.80–

Table 7.47 Counts and frequencies (%) of the typology of basalt waste of giant cores

Typology	N	%
Abruptly retouched side-scraper	1	1.39
Bifacially retouched side-scraper	1	1.39
Typical backed knife	1	1.39
Notch	2	2.78
Denticulate	1	1.39
Core trimming element	7	9.72
Flake	51	70.83
Retouched flake	6	8.33
Massive scraper	1	1.39
Pitted stone	1	1.39
Total	72	100.00

Fig. 7.78 Basalt waste of giant core (#5301 Layer II-6 Level 4)

Fig. 7.79 Basalt waste of giant core (#56 Layer II-6 Level 4)

7.85). These large basalt flakes occur (in low frequencies) mainly in Layer II-6; their highest occurrence is in Layer II-6 Level 4, which is also the level that is richest in handaxes. The metrical characteristics of these flakes are presented in Table 7.50. In the largest sample, Layer II-6 Level 4, more than half of these flakes are complete and most are devoid of cortex

Table 7.48 Counts and frequencies of taphonomic characteristics of basalt waste of giant cores of Layer II-6

	II-6/L1	II-6/L2	II-6/L3	II-6/L4	II-6/L4b	II-6/L5	II-6/L6	II-6/L7
N	16	6	9	23	2	2	3	11
Breakage								
Complete	12 75.00	5 83.33	5 55.56	17 73.91	-	1 50.00	3 100.00	5 45.45
Distal	1 6.25	-	2 22.22	3 13.04	1 50.00	-	-	2 18.18
Lateral	2 12.50	1 16.67	-	2 8.70	-	1 50.00	-	1 9.09
Proximal	-	-	2 22.22	-	-	-	-	-
Distal & lateral	-	-	-	-	1 50.00	-	-	-
Distal & proximal	-	-	-	1 4.35	-	-	-	1 9.09
Fragment	1 6.25	-	-	-	-	-	-	1 9.09
Proximal & lateral	-	-	-	-	-	-	-	1 9.09
Cortex								
No cortex	11 68.75	6 100.00	8 88.89	19 82.61	2 100.00	2 100.00	1 33.33	8 72.73
1–25%	3 18.75	-	1 11.11	1 4.35	-	-	-	-
26–50%	2 12.50	-	-	-	-	-	-	1 9.09
76–100%	-	-	-	2 8.70	-	-	1 33.33	1 9.09
Indeterminate	-	-	-	1 4.35	-	-	1 33.33	1 9.09

Table 7.49 Counts and frequencies of technological characteristics of basalt waste of giant cores of Layer II-6

	II-6/L1	II-6/L2	II-6/L3	II-6/L4	II-6/L4b	II-6/L5	II-6/L6	II-6/L7
N	16	6	9	23	2	2	3	11
Direction of blow	(N=15)			(N=22)	(N=1)			
End-struck	10 66.67	1 16.67	6 66.67	11 50.00	-	1 50.00	-	3 27.27
Side-struck	4 26.67	2 33.33	1 11.11	6 27.27	1 100.00	-	2 66.67	2 18.18
Special side-struck	-	3 50.00	1 11.11	1 4.55	-	1 50.00	-	-
Indeterminate	1 6.67	-	1 11.11	4 18.18	-	-	1 33.33	6 54.55
Scar pattern	(N=15)							(N=10)
Cortical	-	-	-	1 4.35	-	-	1 33.33	-
Plain	1 6.67	-	-	-	1 50.00	-	-	1 10.00
Along axis	2 13.33	-	2 22.22	3 13.04	-	-	-	-
Radial	3 20.00	3 50.00	-	6 26.09	-	-	1 33.33	1 10.00
Ridged	2 13.33	1 16.67	-	4 17.39	-	1 50.00	-	-
Side	1 6.67	-	-	1 4.35	-	-	-	-

7.3 Technology of Basalt Flakes

Table 7.49 (cont.)

	II-6/L1	II-6/L2	II-6/L3	II-6/L4	II-6/L4b	II-6/L5	II-6/L6	II-6/L7
N	16	6	9	23	2	2	3	11
Along axis & side	1 6.67	-	4 44.44	1 4.35	-	1 50.00	-	1 10.00
Along axis & opposed	-	1 16.67	2 22.22	1 4.35	-	-	-	-
Opposed & side	1 6.67	-	-	-	1 50.00	-	-	1 10.00
Indeterminate	4 26.67	1 16.67	1 11.11	6 26.09	-	-	1 33.33	6 60.00
Striking platform				(N=22)				(N=10)
Cortical	-	-	-	2 9.09	-	-	-	1 10.00
Punctiform	-	-	1 11.11	-	-	-	-	-
Plain	12 75.00	5 83.33	3 33.33	14 63.64	2 100.00	1 50.00	1 33.33	2 20.00
Dihedral	-	1 16.67	-	1 4.55	-	-	-	-
Facetted	-	-	1 11.11	-	-	1 50.00	1 33.33	-
Crushed	-	-	-	-	-	-	-	1 10.00
Removed	3 18.75	-	1 11.11	1 4.55	-	-	-	1 10.00
Broken	1 6.25	-	1 11.11	1 4.55	-	-	1 33.33	-
Indeterminate	-	-	2 22.22	3 13.64	-	-	-	5 50.00

Fig. 7.80 Basalt flake of handaxe morphology (#2120 Layer II-6 Level 1)

Fig. 7.81 Basalt flake of handaxe morphology (#5450 Layer II-6 Level 1)

Fig. 7.82 Basalt flake of handaxe morphology (#5540 Layer II-6 Level 1)

Fig. 7.83 Basalt flake of handaxe morphology (#5004 Layer II-6 Level 4)

Fig. 7.84 Basalt flake of handaxe morphology (#5125 Layer II-6 Level 4)

Fig. 7.85 Basalt flake of handaxe morphology (#5726 Layer II-6 Level 4)

(Table 7.51). The dominant direction of blow is end-struck (Table 7.51) and the striking platforms are predominantly plain (Table 7.52). The minimal size of the sample rules out in-depth understanding of the scar pattern, but it is clear that there is extensive variability with the most common pattern being radial, followed by along axis and side and then by side (Table 7.52).

Flakes of Cleaver Morphology

There is only a meager occurrence of basalt large flakes of cleaver morphology, once again in Layer II-6. Nevertheless, as we consider this component an important feature of the basalt assemblage, the relevant data are presented below. There are 13 flakes of cleaver morphology (Level 4: N=11;

7.3 Technology of Basalt Flakes

Table 7.50 Descriptive statistics of basalt large flakes of biface morphology (size in mm; angle in degrees)

	Handaxe morphology				Cleaver morphology
	II-6/L1	II-6/L2	II-6/L4	II-6/L4b	II-6/L4 & L7
N	6	3	18	5	13
Maximum length					
Maximum	158	112	146	124	133
Median	117	90	122	117	120
Minimum	94	86	99	112	105
Mean	121.33	96.00	123.78	117.20	117.77
Std dev	23.14	14.00	14.75	4.44	9.08
Std err mean	9.45	8.08	3.48	1.98	2.52
Length					
Maximum	158	84	137	124	130
Median	103.5	82	97.5	87	101
Minimum	75	76	65	64	60
Mean	108.83	80.67	97.11	94.60	95.00
Std dev	35.02	4.16	19.26	25.59	26.51
Std err mean	14.30	2.40	4.54	11.44	7.35
Width					
Maximum	149	112	146	117	127
Median	104.5	87	106.5	107	93
Minimum	58	71	66	70	60
Mean	102.67	90.00	106.17	99.80	93.31
Std dev	29.12	20.66	26.91	19.77	23.02
Std err mean	11.89	11.93	6.34	8.84	6.39
Thickness					
Maximum	79	45	44	46	43
Median	33.5	43	33	37	33
Minimum	23	32	24	31	27
Mean	38.83	40.00	33.11	38.80	34.77
Std dev	21.15	7.00	5.69	6.14	5.10
Std err mean	8.63	4.04	1.34	2.75	1.41
L/W					
Maximum	2.72	1.18	1.70	1.69	1.85
Median	0.90	0.94	0.88	0.81	1.10
Minimum	0.73	0.68	0.54	0.55	0.55
Mean	1.19	0.93	1.00	1.02	1.14
Std dev	0.77	0.25	0.40	0.49	0.54
Std err mean	0.31	0.15	0.09	0.22	0.15
Angle of striking platform	(N=3)		(N=15)	(N=4)	(N=9)
Maximum	119	-	132	122	122
Median	114	-	124	117.5	106
Minimum	112	-	115	107	102
Mean	115.00	-	122.80	116.00	108.67
Std dev	3.61	-	5.73	6.68	7.33
Std err mean	2.08	-	1.48	3.34	2.44
Number of scars	(N=5)		(N=14)		(N=12)
Maximum	5	4	12	7	9
Median	3	3	6	4	4
Minimum	1	1	1	3	1
Mean	2.80	2.67	5.71	4.60	5.00
Std dev	1.48	1.53	3.41	1.52	2.37
Std err mean	0.66	0.88	0.91	0.68	0.69

Table 7.51 Counts and frequencies (%) of taphonomic and technological characteristics of basalt large flakes of biface morphology

	Handaxe morphology				Cleaver morphology
	II-6/L1	II-6/L2	II-6/L4	II-6/L4b	II-6/L4 & L7
N	6	3	18	5	13
Breakage					
Complete	6 / 100.00	1 / 33.33	12 / 66.67	2 / 40.00	10 / 76.92
Distal	-	-	2 / 11.11	-	3 / 23.08
Lateral	-	-	1 / 5.56	1 / 20.00	-
Proximal	-	1 / 33.33	2 / 11.11	-	-
Distal & lateral	-	-	-	2 / 40.00	-
Distal & proximal	-	-	1 / 5.56	-	-
Indeterminate	-	1 / 33.33	-	-	-
Cortex					
No cortex	5 / 83.33	2 / 66.67	15 / 83.33	5 / 100.00	11 / 84.62
1–25%	-	1 / 33.33	-	-	1 / 7.69
26–50%	-	-	1 / 5.56	-	1 / 7.69
76–100%	1 / 16.67	-	2 / 11.11	-	-
Direction of blow				(N=5)	
End-struck	1 / 20.00	1 / 33.33	9 / 50.00	3 / 60.00	7 / 58.33
Side-struck	-	1 / 33.33	7 / 38.89	-	5 / 41.67
Special side-struck	4 / 80.00	-	2 / 11.11	1 / 20.00	-
Indeterminate	-	1 / 33.33	-	1 / 20.00	-

Level 7: N=2) (Figs. 7.40, 7.86). Table 7.50 presents their metrical characteristics. They resemble the flakes of handaxe morphology but differ (not significantly) in being slightly thicker, as specified above. The artifacts are mostly complete and non-cortical (Table 7.51) and technologically their direction of blow is predominantly end-struck, although other directions are present as well. The dominant scar patterns are along axis and along axis and side (Table 7.52), and the mean number of scars is 5.5 (Table 7.50). The most frequent type of striking platform is plain, followed by removed (Table 7.52). Clearly, the small sample size rules out thorough study of these traits.

The presence of large basalt flakes of biface morphology, though in low frequencies, sheds light on an important stage in the reduction sequence of bifaces at GBY. Their similarity to bifaces is expressed mainly in their size, which

Table 7.52 Counts and frequencies (%) of technological characteristics of basalt large flakes of biface morphology

	Handaxe morphology			Cleaver morphology	
	II-6/L1	II-6/L2	II-6/L4	II-6/L4b	II-6/L4 & L7
	6	3	18	5	13
Striking platform	(N=5)				
Plain	3	1	14	3	8
	60.00	33.33	77.78	60.00	61.54
Dihedral	1	-	-	1	-
	20.00			20.00	
Facetted	1	-	1	1	-
	20.00		5.56	20.00	
Broken	-	1	2	-	-
		33.33	11.11		
Removed					3
					23.08
Indeterminate	-	1	1	-	2
		33.33	5.56		15.38
Scar pattern					
Cortical	1	-	2	-	-
	16.67		11.11		
Plain	1	-	2	-	-
	16.67		11.11		
Along axis	-	1	2	1	-
		33.33	11.11	20.00	
Opposed	1	-	-	-	-
	16.67				
Radial	1	1	6	-	-
	16.67	33.33	33.33		
Along axis & side	1	-	4	3	-
	16.67		22.22	60.00	
Indeterminate	1	1	2	1	-
	16.67	33.33	11.11	20.00	

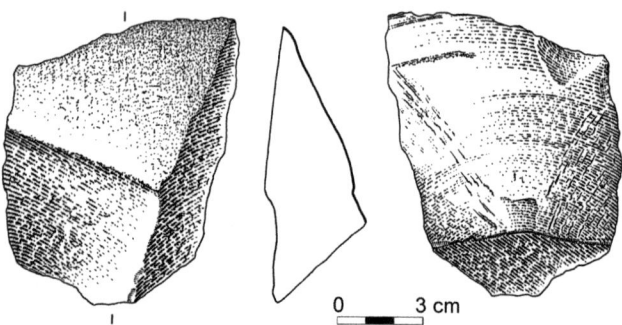

Fig. 7.86 Basalt flake of cleaver morphology (#282 Layer II-6 Level 4)

will be discussed in detail below (Section 7.5.3.1). It is interesting that, despite the clear difference in morphology between handaxe-shaped flakes and cleaver-shaped flakes, there is no significant difference in any of their dimensions (max. length: difference: -3.51 mm; t-ratio: -.811; df: 68; $p(tt)$=.4198; length: difference: -7.29 mm; t-ratio: -1.11; df: 68; $p(tt)$=.2673; width: difference: -4.96 mm; t-ratio: -.78; df: 68; $p(tt)$=.4375; thickness: difference: -.98 mm; t-ratio: -.46; df: 68; $p(tt)$=.6452). This may indicate that both flake categories originate in the same reduction sequence targeted at the production of large flakes of similar dimensions. While the products of this reduction sequence were all suitable for biface modification, some could be selected later for modification as handaxes and others as cleavers, according to their predetermined morphology.

7.4 Typology and Technology of Basalt Cores and Core Tools

Basalt cores and core tools (CCT) form an impressive mass within the lithic assemblages of GBY, the bulk of this mass comprising giant cores and bifacial tools. Other CCT categories, surprisingly, are minimally represented, particularly those of a formal typology (e.g. cores, side-scrapers, notches, and denticulates). Instead, higher frequencies are recorded for unflaked basalt objects that were transported into the site and were manipulated differently into diverse types of percussive tools (e.g. percussors, anvils, pitted stones).

Table 7.53 presents the inventory of basalt cores and core tools within the cultural sequence of GBY and the different tool categories are discussed in the following sections.

7.4.1 Percussive Tools: Percussors, Pitted Stones, and Thin Anvils

Percussive tools form a significant component of the basalt assemblage of GBY. These artifacts, which until recently have often been ignored in descriptions of prehistoric assemblages, are actually elementary working tools used by the knapper while shaping, retouching, and maintaining stone artifacts, as well as in other percussive activities. Percussors, pitted stones, and anvils were utilized for many additional tasks, including processing of food (both vegetal and faunal) and many other activities (see discussion in Section 7.6). While flint and limestone percussive tools form only a very small component of the GBY assemblages (Chapters 6, 8), those made on basalt are many and varied.

7.4.1.1 Percussors

The identification of percussors in Lower Paleolithic assemblages is a difficult and at times somewhat subjective task,

7.4 Typology and Technology of Basalt Cores and Core Tools

Table 7.53 Typological counts of basalt CCT by layer

Typology	V-4	V-5	V-6	JB	I-4	I-5	II-2	II-2/3	II-3	II-4	II-5	II-5/6	II-6/L1	II-6/L2	II-6/L3	II-6/L4	II-6/L4b	II-6/L5	II-6/L6	II-6/L7	VI-14	Trench II green	Total
Percussive tools																							
Percussor	-	2	12	19	1	3	1	5	-	1	5	1	25	28	14	28	23	5	20	28	3	2	226
Pitted stone	-	-	-	-	-	-	-	-	-	-	-	-	-	1	-	-	-	-	-	-	-	-	1
Thin anvil	-	-	-	2	-	-	-	-	-	-	-	-	4	7	5	6	2	1	6	2	-	-	35
Modified	-	4	2	3	1	1	-	2	-	-	-	2	26	40	29	29	2	3	7	14	-	3	168
Core tools*																							
Chopping tool	-	-	-	2	-	-	-	-	-	-	-	-	-	2	1	2	-	-	-	-	-	-	7
Massive scraper	-	-	-	-	-	-	-	-	-	-	-	-	1	1	-	-	-	-	-	-	-	-	2
Atypical burin	-	-	-	-	-	-	-	-	-	-	-	-	1	-	-	-	-	-	-	-	-	-	1
Denticulate	-	-	-	-	-	-	-	-	-	-	-	-	1	-	-	-	-	-	-	-	-	-	1
Denticulated point	-	-	-	-	-	-	-	-	-	-	-	-	1	-	-	-	-	-	-	-	-	-	1
Subspheroid	-	-	-	-	-	-	-	-	-	-	-	-	-	-	-	-	-	-	1	-	-	-	1
Cores																							
Levallois core for flakes	-	-	-	-	-	-	-	-	-	-	-	-	1	-	-	-	-	-	-	-	-	-	1
Globular core	-	-	-	-	-	-	-	-	-	-	-	-	-	-	-	-	-	2	-	-	-	-	2
Varia core	-	-	1	-	-	-	-	-	-	-	-	-	3	-	2	1	-	-	1	1	-	-	9
Amorphous core	-	-	1	-	-	-	-	-	-	-	-	-	1	4	-	2	-	-	1	-	-	1	10
Core on flake	-	-	-	-	-	-	-	-	-	1	-	1	1	1	1	-	-	-	-	-	-	5
Giant core	-	-	1	-	-	-	-	1	-	-	-	-	4	7	3	5	-	2	3	7	-	-	33
Angular fragment	1	2	2	8	-	-	-	6	1	-	1	1	45	29	10	28	17	3	5	8	-	-	167
Core waste	-	-	-	-	-	-	-	-	-	-	-	-	1	-	-	1	-	-	-	-	-	-	2
Manuport	-	-	-	-	-	-	-	-	-	-	-	-	-	-	1	-	-	-	-	-	-	1	2
Bifaces*																							
Handaxe	-	-	-	-	1	-	-	7	-	-	4	1	54	19	12	165	58	11	19	20	9	24	404
Cleaver	-	2	6	8	-	-	-	-	-	-	2	1	17	10	12	60	18	8	11	12	-	8	175
Partial biface	-	-	-	-	-	-	-	-	-	-	-	-	1	-	-	1	-	-	-	-	-	-	2
Biface preform	-	-	-	-	-	-	-	-	-	-	-	-	1	2	-	-	-	-	-	-	-	-	3
Total	1	10	25	34	1	4	1	14	1	1	7	4	117	122	66	104	44	16	44	60	3	7	1258

* the presentation of bifaces is not included within that of core tools, but is discussed separately in Section 7.5

depending on the researchers' experience and the preservation of this category of artifacts. The identification of percussors in the Acheulian is confronted by many difficulties. Percussors of various rock types were used during the different stages of the reduction process, and in addition organic materials such as antler were applied in some of the diverse stages of tool modification and maintenance. Each of the rock types used had different properties of hardness, durability, and fracture mechanics, thus showing different impact marks resulting from its use. These damage traces can be observed on the knapped object and, more relevantly to the current discussion, on the percussor itself. Hence, we cannot predict the morphology of basalt percussors or the battering marks left on them by studying, for example, flint percussors. At GBY, rocks of three types (flint, basalt, and limestone) were used as percussors and each of them displays different typical battering and damage marks (Chapters 6 and 8).

Clearly, the preservation condition has a significant effect on the visibility and identification of battering and other damage marks. The basalt percussors, like all of the other basalt artifacts at GBY, frequently show postdepositional weathering. As a result, the identification of typical battering marks is more difficult for basalt than for limestone and flint.

The identification of percussors at GBY, based on varied morphological attributes, reflects observations and data obtained during experimental knapping (Sharon and Goren-Inbar 1999; Madsen and Goren-Inbar 2004). During the analyses of the GBY percussors, the morphology of the artifact was used as a primary criterion for definition. Rounded nodules, somewhat elongated and usually of cobble size, are the most obvious candidates. In addition, a suitable size was considered a criterion. Nevertheless, at GBY shape and size are not sufficient in themselves to assign a basalt artifact to the percussor class, and tools are identified as percussors only if battering damage

Table 7.54 Counts and frequencies (%) of basalt percussors and split percussors

Layer	Percussors N	Percussors %	Split percussors N	Split percussors %
V-5	2	0.89	1	1.14
V-6	12	5.33	-	-
JB	19	8.44	1	1.14
I-4	-	-	-	-
I-5	3	1.33	-	-
II-2/3	5	2.22	-	-
II-4	1	0.44	-	-
II-5	5	2.22	-	-
II-5/6	1	0.44	-	-
II-6/L1	25	11.11	12	13.64
II-6/L2	28	12.44	17	19.32
II-6/L3	14	6.22	3	3.41
II-6/L4	28	12.44	18	20.45
II-6/L4b	23	10.22	7	7.95
II-6/L5	5	2.22	9	10.23
II-6/L6	20	8.89	7	7.95
II-6/L7	28	12.44	13	14.77
VI-14	3	1.33	-	-
Trench II green	3	1.33	-	-
Total	225	99.95	88	100.00

Fig. 7.87 Basalt percussor (#16326 Layer II-2/3); note battering on surface and damage on both edges

Fig. 7.88 Basalt percussor (#5717 Layer II-6 Level 1); note spontaneous flaking on surface and damage on both edges

marks are visible on their surfaces. Battering marks are difficult to identify on the basalt artifacts, particularly cobbles; thus, despite their high frequencies, basalt percussors may actually be under-represented in the GBY assemblages

The morphological criteria for the identification of percussors at GBY included 1) the presence of battering or small pitting damage on any part of the tool's surface; 2) the presence around the edges (the thickness) of the cobble of a typical flat surface, visible as a narrow band, resulting from repeated use; and 3) flake scars (in most cases small) not reflecting organized or planned, controlled knapping. The latter are most probably a result of accidents during the knapping process (see Chapter 8.1.1 for limestone analogies).

Table 7.54 presents the counts and frequencies of basalt percussors in the GBY lithic assemblages. An additional group of percussor-related artifacts consists of basalt flakes identified as split percussors (see below).

High frequencies of percussors were found in the complex of Layer II-6 (Figs. 7.87–7.94), as well as in Layer V-6 and the JB (Table 7.54). Interestingly, Table 7.54 also shows that split percussors generally occur in high frequencies in the same layers and levels. With the exception of Layer II-6 Levels 3 and 5, the frequencies of basalt percussors are quite similar among the different levels, with a mean of around 25 per archaeological horizon. In addition, the frequencies of potential percussors (see below) are somewhat similar. The archaeological horizons with abundant percussors are interpreted as representing some intensive activity (such as the butchering of an elephant in Level 1 and the exceptionally rich assemblage of bifacial tools in Level 4). We are not able to identify all of the activities that involved percussors, but the typical edges were clearly caused by knapping, as was frequently demonstrated by the experiments (Figs. 7.95–7.97). In addition, the large quantities of fractured bones of medium-

7.4 Typology and Technology of Basalt Cores and Core Tools

Fig. 7.89 Basalt percussor (#8073 Layer II-6 Level 1); note spontaneous flaking and battering damage on the same surface

Fig. 7.91 Basalt percussor (#7030 Layer II-6 Level 2); note battering and pitting damage on the flat surface

Fig. 7.90 Basalt percussor (#831 Layer II-6 Level 1)

Fig. 7.92 Basalt percussor (#13706 Layer II-6 Level 2)

sized to large mammals are an indication that percussors were used for tasks other than knapping.

The breakage patterns observed for the GBY percussors are presented in Table 7.55, where the three most common categories are complete, laterally broken, and substantially broken. The bulk of the percussors are broken, clearly a typical result of the knapping procedure. In addition, some percussor have been damaged by postdepositional effects. By definition, most of these artifacts are cortical, with only a few artifacts displaying smaller amounts of cortex.

There are large differences in the size of percussors in each of the archaeological horizons (Table 7.56), which could be explained by the diverse tasks for which these artifacts were intended. Percussors were used for an extremely wide range of knapping activities, from the extraction of large basalt flakes struck off giant cores to the well-controlled removal of small, thin flakes during the thinning of bifaces (handaxes

Fig. 7.93 Basalt percussor (#14887 Layer II-6 Level 3)

Fig. 7.95 Basalt experimental percussor (#P01); note battering and breakage damage on proximal edge

Fig. 7.94 Basalt percussor (#5064 Layer II-6 Level 4); note battering and pitting damage on the flat surface and lateral edge

Fig. 7.96 Basalt experimental percussor (#P02)

and cleavers). We have no information on the use of basalt percussors in the knapping of flint and limestone, but both could have taken place in the archaeological horizons. Another factor in the observed variability of percussors could be the individual preferences of the knapper; this factor is well known from many actualistic knapping sessions, in which a knapper carries around his "favorite" percussors. These individual preferences are also associated with the extent of the knapper's dexterity, level of expertise, etc. A skilled knapper can carry out very intense knapping with a relatively small percussor. Experimental work has shown that during the process of gaining experience in knapping, the size of the percussor used for the heavy tasks of large flake production

7.4 Typology and Technology of Basalt Cores and Core Tools

Fig. 7.97 Basalt experimental percussor (#P03); **1** note breakage on surface **2** detailed view of percussive damage

is dramatically reduced (Jones 1994; Madsen and Goren-Inbar 2004). During the experimental project of replicating the GBY bifacial tools, the very large percussors (up to 30 kg each) that were applied to the detaching of large flakes from giant cores were eventually replaced by much smaller ones (ca. 1.5–3 kg).

The size of percussors in the different archaeological horizons (Tables 7.56) shows a negligible difference among the different assemblages. There is an overall size similarity across the entire cultural sequence at the site. The size range of the percussors is quite homogeneous, between ca. 60 mm and 120 mm. Most of the percussors are not large, a fact that to the modern eye makes them unsuitable for knapping giant cores or for smashing the very large bones, such as those of elephants, hippos, or large bovids, that were found at the site (Rabinovich et al. 2012). Exceptions were found in Layer II-6 Level 1, where a very large boulder was found under an elephant skull adjacent to a giant core (Goren-Inbar et al. 1994), along with exceptionally large percussors (Figs. 7.98–7.99). Based on previous analyses of the GBY lithic assemblages, it seems likely that some of the primary knapping activities that are associated with basalt, e.g. the extraction of giant slabs, their fragmentation, and parts of the extraction of giant and large flakes, were not carried out at the site. It is therefore only natural that the largest percussors, those used in the initial phases of core preparation, were not found at the site, since they were most probably left in the vicinity of the basalt flows from which the material was quarried.

Table 7.56 also provides the weight of the percussors for the various levels of Layer II-6. It is evident (as for the dimensions) that the weight of the percussors varies greatly within each layer and level, as reflected by the values of the standard deviations. The heaviest percussor was close to 4 kg and the lowest recorded weight is 66 g. Hence, the range is very large and reflects the use of percussors in the multiple tasks suggested above.

Split Percussors and Percussor Fragments

Many of the percussors, as presented in Table 7.54, are split. These are represented in the assemblage as split cobbles or as large, thick cortical flakes of distinctive morphology. This breakage pattern is a typical result of the use of percussors in the massive knapping of very large and durable materials (e.g. giant cores and anvils) or the smashing of very large bones. Typically, these artifacts show a surface morphology that is

Table 7.55 Counts and frequencies (%) of breakage patterns and cortical coverage of basalt percussors

	V-6	JB	II-2/3	II-5	II-6/L1	II-6/L2	II-6/L3	II-6/L4	II-6/L4b	II-6/L5	II-6/L6	II-6/L7
N	12	19	5	5	25	28	14	28	23	5	20	28
Breakage						(N=27)						
Complete	4	10	3	1	11	10	6	10	9	-	3	16
	33.33	52.63	60.00	20.00	44.00	37.04	42.86	35.71	39.13		15.00	57.14
Distal	-	-	-	-	1	1	1	-	-	-	-	-
					4.00	3.70	7.14					
Lateral	5	2	-	-	6	10	4	6	7	-	9	7
	41.67	10.53			24.00	37.04	28.57	21.43	30.43		45.00	25.00
Proximal	1	2	-	2	-	1	1	2	1	-	1	-
	8.33	10.53		40.00		3.70	7.14	7.14	4.35		5.00	
Distal & lateral	1	-	-	-	-	2	-	-	-	-	-	-
	8.33					7.41						
Distal & proximal	-	-	-	-	1	-	-	-	2	-	-	-
					4.00				8.70			
Fragment	-	-	-	-	-	-	-	3	-	3	3	1
								10.71		60.00	15.00	3.57
Proximal & lateral	-	1	-	-	-	-	-	3	-	-	-	-
		5.26						10.71				
Substantially broken	1	4	2	2	6	2	2	3	4	2	4	4
	8.33	21.05	40.00	40.00	24.00	7.41	14.29	10.71	17.39	40.00	20.00	14.29
Indeterminate	-	-	-	-	-	1	-	1	-	-	-	-
						3.70		3.57				
Cortex	(N=9)					(N=26)						(N=27)
No cortex	-	-	3	-	-	1	-	3	-	-	1	2
			60.00			3.85		10.71			5.00	7.41
1–25%	-	1	-	-	-	1	-	-	-	-	1	-
		5.26				3.85					5.00	
26–50%	1	2	-	1	9	4	4	7	1	1	1	2
	11.11	10.53		20.00	36.00	15.38	28.57	25.00	4.35	20.00	5.00	7.41
51–75%	-	2	-	1	7	4	3	1	2	-	-	1
		10.53		20.00	28.00	15.38	21.43	3.57	8.70			3.70
76–100%	5	1	2	2	8	10	5	12	18	2	10	16
	55.56	5.26	40.00	40.00	32.00	38.46	35.71	42.86	78.26	40.00	50.00	59.26
Indeterminate	3	13	-	1	1	6	2	5	2	2	7	6
	33.33	68.42		20.00	4.00	23.08	14.29	17.86	8.70	40.00	35.00	22.22

rough and uneven and lacks features of the ventral face, such as identifiable point of percussion, bulb of percussion, or conchoidal waves (Fig. 7.100). The ventral surfaces differ from the relatively smooth ones of flakes removed during intentional knapping. The rough surfaces probably result from the repeated high-impact percussion that is typical of percussor battering. Such surfaces are one of the criteria for the identification of a split cobble or thick flake within the categories of broken percussors or percussor fragments.

Table 7.57 presents the characteristics of the split percussors. Evidently, these data are very similar to those for the percussors (Table 7.55). The direction of the blow that split the percussors resulted more frequently in end-struck flakes (Table 7.57), providing additional information on the manner in which the percussors were applied, using the length axis more often.

Table 7.58 provides data on different size attributes of split percussors. These data demonstrate a high variability that is well documented in the high values of the standard deviations. There are several cases in which the average length and width of the split percussors is larger than that of the percussors (Tables 7.56 and 7.58). This may result from the fact that the larger the percussor, the more likely it is to split during massive blows.

Natural Cobbles as Potential Percussors

As shown above, the percussors of GBY all share similar morphology and all show battering and surface damage marks that suggests their use as percussors. However, we must take into consideration the possibility that minimal use of percussors may have left no damage marks. Thus, it is important to discuss the group of natural basalt cobbles that could have been used as percussors. While the identification of a percussor is based on

7.4 Typology and Technology of Basalt Cores and Core Tools

Table 7.56 Descriptive statistics of basalt percussors (size in mm; weight in g)

	V-6	JB	II-2/3	II-5	II-6/L1	II-6/L2	II-6/L3	II-6/L4	II-6/L4b	II-6/L5	II-6/L6	II-6/L7
N	12	19	5	5	25	28	14	28	23	5	20	28
Maximum length						(N=27)						
Maximum	114	136	126	123	182	163	112	141	135	107	123	126
Median	104.5	98	111	75	91	79	72.5	88	90	83	87	94
Minimum	75	47	76	70	74	56	45	55	59	73	49	61
Mean	100.67	89.11	101.00	83.20	96.80	85.37	75.21	88.96	90.35	86.80	85.30	92.82
Std dev	14.34	23.36	21.86	22.51	22.62	21.65	18.51	19.90	19.89	14.46	16.78	17.61
Std err	4.14	5.36	9.78	10.07	4.52	4.17	4.95	3.76	4.15	6.47	3.75	3.33
Length						(N=27)						
Max	112	112	123	123	135	108	110	140	125	102	123	120
Median	102	90	101	74	90	75	70	78.5	86	75	81	93
Minimum	73	47	76	61	56	44	45	55	47	66	44	58
Mean	97.50	84.11	97.80	81.00	92.28	78.04	73.36	83.64	83.00	81.20	79.05	89.79
Std dev	14.49	21.73	20.95	24.24	18.24	19.15	19.11	19.98	21.09	16.33	18.78	19.31
Std err	4.18	4.98	9.37	10.84	3.65	3.69	5.11	3.78	4.40	7.30	4.20	3.65
Width						(N=27)						
Maximum	110	136	103	109	144	116	88	89	117	76	95	100
Median	73.5	73	99	54	68	66	58.5	66.5	67	68	68	71
Minimum	49	39	62	37	50	37	38	35	45	53	33	38
Mean	79.33	72.32	86.20	60.80	70.80	64.04	59.29	64.64	68.48	65.80	64.90	69.32
Std dev	18.54	22.40	20.36	27.94	19.74	15.69	15.14	15.45	16.72	8.98	15.89	14.09
Std err	5.35	5.14	9.11	12.50	3.95	3.02	4.05	2.92	3.49	4.02	3.55	2.66
Thickness					(N=24)	(N=27)						
Maximum	72	66	69	81	102	66	77	71	71	57	65	84
Median	53.5	43	58	52	46	42	37.5	49	45	43	46	51
Minimum	41	28	31	27	33	26	30	29	36	33	26	35
Mean	54.25	44.84	55.60	53.20	50.67	42.30	41.36	47.32	47.87	46.00	46.75	52.54
Std dev	8.57	9.68	15.01	25.09	14.57	10.93	13.14	11.85	9.75	9.95	10.75	11.34
Std err	2.47	2.22	6.71	11.22	2.97	2.10	3.51	2.24	2.03	4.45	2.40	2.14
Circumference		(N=18)		(N=1)	(N=4)	(N=16)	(N=4)	(N=15)		(N=19)	(N=27)	
Maximum	353	376	360	220	480	385	319	345	418	314	333	362
Median	290.5	269.5	335	220	286	229	228.5	275	246	242	247	270
Minimum	205	136	225	220	216	169	138	163	174	125	129	189
Mean	292.08	254.56	301.20	220.00	317.00	239.00	228.50	256.93	259.65	234.80	238.79	263.30
Std dev	49.81	63.17	68.90	-	114.32	51.17	79.68	55.12	60.86	69.81	47.14	42.12
Std err	14.38	14.89	30.81	-	57.16	12.79	39.84	14.23	12.69	31.22	10.81	8.11
Number of scars	(N=7)	(N=15)			(N=23)	(N=16)	(N=11)	(N=23)	(N=16)	(N=3)	(N=17)	(N=24)
Maximum	4	6	5	3	4	4	6	7	6	5	6	7
Median	2	2	3	2	3	2	2	3	2	5	3	2
Minimum	1	1	1	1	1	1	1	1	1	2	1	1
Mean	2.29	2.67	2.80	2.20	2.52	2.06	2.55	3.13	2.75	4.00	2.65	2.58
Std dev	1.11	1.76	1.48	0.84	1.08	0.85	1.86	1.46	1.65	1.73	1.41	1.32
Std err	0.42	0.45	0.66	0.37	0.23	0.21	0.56	0.30	0.41	1.00	0.34	0.27
Weight*					(N=11)	(N=6)	(N=5)	(N=8)	(N=5)	(N=1)	(N=3)	(N=16)
Maximum	-	-	-	-	2,812	1,278	1,028	571	1,354	3,846	451	772
Median	-	-	-	-	467	332	335	344	540	3,846	178	383
Minimum	-	-	-	-	209	151	66	105	300	3,846	137	141
Mean	-	-	-	-	630.45	448.50	403.00	333.88	641.60	3,846.00	255.33	430.56
Std dev	-	-	-	-	734.40	414.30	373.99	192.94	429.46	-	170.69	179.18
Std err	-	-	-	-	221.43	169.14	167.25	68.21	192.06	-	98.55	44.80

* the weight variable includes only complete artifacts

Fig. 7.98 Basalt very large percussor; 2.81 kg (#5278 Layer II-6 Level 1)

Fig. 7.99 Basalt very large percussor; 2.18 kg (#2312 Layer II-6 Level 1)

an observed damage pattern that has resulted from repeated use, objects that lack such a pattern may have been used in the same manner but not to the same extent.

The natural objects (pebbles, cobbles, and boulders) that were unearthed during the excavations and which show no evidence of human modification were analyzed and their size, weight, and roundness and sphericity (following Krumbein 1941; Fig. 7.101) were recorded. From these data we extracted potential percussors that fall within the size and morphology (roundness) range of the identified basalt percussors. These are artifacts that were suitable for use as percussors due to their similarity in size and morphology to well-defined percussors (Fig. 7.102). On the other hand, they show no unequivocal evidence of the typical damage that would allow them to be defined as percussors. This lack of damage may be due to minimal use of the artifact, possibly for tasks that were less forceful or less frequently repeated, or to our failure to identify such marks on the "hard to read" GBY basalt. Table 7.59 presents the frequency of unmodified artifacts from GBY (all raw materials) larger than 65 mm (see discussion of percussor size) and for which a roundness value of >7 was recorded.

Of the 242 potential percussors, the great majority (N=218) are basalt, reflecting the relative frequency of percussors in each of the raw materials (Chapters 6–8). These potential percussors are observed in almost all archaeological horizons, with the highest frequencies occurring in Layer II-6 Levels 4 and 4b.

Fig. 7.100 Basalt split percussor (#6877 Layer II-6 Level 2)

7.4.2 Pitted Stones

Pitted stones are often unflaked nodules or slabs that bear small depressions on their surfaces resulting from repeated percussive activity (Figs. 7.103–7.107). The character of this artificially produced damage depends on the type of raw material, the object used to inflict the damage, and the intensity of the use. Common terminology includes names such as "pitted anvils," "pitted

7.4 Typology and Technology of Basalt Cores and Core Tools

Table 7.57 Counts and frequencies (%) of taphonomic and technological characteristics of basalt split percussors

	II-6/L1	II-6/L2	II-6/L3	II-6/L4	II-6/L4b	II-6/L5	II-6/L6	II-6/L7
N	12	17	3	18	7	9	7	13
Breakage								
Complete	5	13	2	13	4	-	2	7
	41.67	76.48	66.67	72.22	57.13		28.57	53.84
Distal	1	1	-	-	-	4	-	-
	8.33	5.88				44.44		
Lateral	1	1	-	2	-	1	1	2
	8.33	5.88		11.11		11.11	14.29	15.38
Distal & lateral	1	-	-	-	1	1	2	2
	8.33				14.29	11.11	28.57	15.38
Distal & proximal	-	1	-	1	-	-	-	-
		5.88		5.56				
Fragment	2	-	1	1	-	1	1	1
	16.67		33.33	5.56		11.11	14.29	7.70
Proximal & lateral	2	-	-	-	1	-	1	-
	16.67				14.29		14.29	
Indeterminate	-	1	-	1	1	2	-	1
		5.88		5.56	14.29	22.22		7.70
Cortex								
No cortex	3	2	1	5	-	5	2	2
	25.00	11.76	33.33	27.78		55.56	28.57	15.38
1–25%	-	-	-	1	-	-	-	-
				5.56				
26–50%	-	-	-	1	-	-	-	1
				5.56				7.69
51–75%	2	-	-	-	1	-	1	-
	16.67				14.29		14.29	
76–100%	6	12	2	8	6	-	3	3
	50.00	70.59	66.67	44.44	85.71		42.86	23.08
Indeterminate	1	3	-	3	-	4	1	7
	8.33	17.65		16.67		44.44	14.29	53.85
Direction of blow		(N=16)		(N=16)			(N=5)	(N=11)
End-struck	8	7	2	7	2	3	1	4
	66.67	43.75	66.67	43.75	28.57	33.33	20.00	36.36
Side-struck	3	5	1	2	1	1	1	2
	25.00	31.25	33.33	12.50	14.29	11.11	20.00	18.18
Special side-struck	1	2	-	-	-	-	1	1
	8.33	12.50					20.00	9.09
Indeterminate	-	2	-	7	4	5	2	4
		12.50		43.75	57.14	55.56	40.00	36.36

stones," and "nutting stones" (for different nomenclatures, chimpanzees' use of these objects, and ethnographic data see Goren-Inbar et al. 2002a). Pitted stones occur at GBY on all three types of raw material. They are minimally present in flint (Chapter 6), more frequent in limestone (Chapter 8), and most abundant in basalt.

Characterizing the damage patterns of the basalt pitted stones of GBY is a very difficult task, due primarily to the fact that basalt underwent *in situ* weathering at the site (Section 7.1.2). In addition, the damage caused to their surfaces following their use in battering and pounding activities may have induced weathering.

Although we have made a previous attempt to characterize

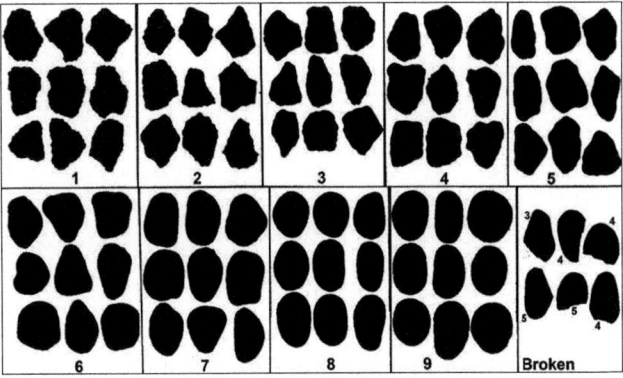

Fig. 7.101 Roundness index for GBY natural objects (after Krumbein 1941)

Table 7.58 Descriptive statistics of basalt split percussors of Layer II-6 (size in mm)

	II-6/L1	II-6/L2	II-6/L3	II-6/L4	II-6/L4b	II-6/L5	II-6/L6	II-6/L7
N	12	17	3	18	7	9	7	13
Maximum length								
Maximum	118	128	130	167	111	123	131	122
Median	91	95	128	102.5	77	97	94	80
Minimum	61	64	70	68	64	62	69	60
Mean	89.67	94.41	109.33	103.61	84.00	94.78	95.00	88.00
Std dev	20.10	16.23	34.08	26.48	17.88	18.08	19.18	21.32
Std err	5.80	3.94	19.68	6.24	6.76	6.03	7.25	5.91
Length								
Maximum	116	128	128	167	111	98	130	122
Median	79	83	103	97.5	76	84	85	75
Minimum	61	46	63	47	60	52	65	55
Mean	80.92	84.41	98.00	97.17	80.57	79.00	87.86	79.23
Std dev	15.37	18.68	32.79	31.37	20.48	14.70	22.35	18.53
Std err	4.44	4.53	18.93	7.39	7.74	4.90	8.45	5.14
Width								
Maximum	113	106	74	104	72	121	95	115
Median	71.5	66	69	74.5	62	78	67	79
Minimum	42	56	58	61	44	58	56	36
Mean	75.00	72.35	67.00	78.83	58.00	85.67	73.14	75.46
Std dev	24.43	15.66	8.19	12.72	11.20	23.13	15.68	23.19
Std err	7.05	3.80	4.73	3.00	4.23	7.71	5.93	6.43
Thickness								
Maximum	49	50	47	60	37	68	58	57
Median	41	36	40	44.5	35	43	54	44
Minimum	20	13	33	24	26	30	32	26
Mean	38.92	33.71	40.00	44.22	32.86	45.56	50.00	42.31
Std dev	9.04	9.11	7.00	10.81	4.22	11.54	9.50	9.60
Std err	2.61	2.21	4.04	2.55	1.60	3.85	3.59	2.66
L/W								
Maximum	1.55	1.82	1.78	1.84	2.52	1.45	1.65	1.58
Median	1.21	1.26	1.73	1.20	1.15	0.90	1.18	1.13
Minimum	0.70	0.65	0.91	0.66	0.83	0.64	0.71	0.70
Mean	1.15	1.21	1.47	1.25	1.49	0.97	1.24	1.11
Std dev	0.28	0.35	0.49	0.38	0.67	0.27	0.34	0.28
Std err	0.08	0.09	0.28	0.09	0.25	0.09	0.13	0.08

Table 7.59 Counts of unmodified cobbles (potential percussors) with maximum size greater than 65 mm and roundness value greater than or equal to 7

		Raw material		
Layer	Flint	Limestone	Basalt	Total
V-6	-	1	2	3
JB	-	-	17	17
II-2/3	3	-	-	3
II-5	-	1	4	5
II-5/6	-	-	1	1
II-6/L1	4	2	37	43
II-6/L2	3	-	5	8
II-6/L3	6	2	7	15
II-6/L4	-	-	84	84
II-6/L4b	-	1	17	18
II-6/L5	1	-	6	7
II-6/L6	-	-	5	5
II-6/L7	-	-	33	33
Total	17	7	218	242

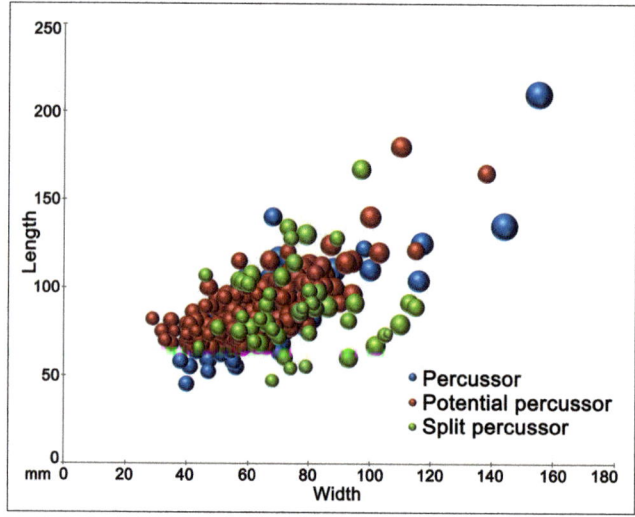

Fig. 7.102 Length, width, and thickness (bubble size) of Layer II-6 basalt complete percussors, split percussors, and potential percussors

7.4 Typology and Technology of Basalt Cores and Core Tools

Fig. 7.103 Basalt pitted stone (#13525 Layer II-6 Level 2)

Fig. 7.105 Basalt pitted stone (#5520 Layer V-5)

Fig. 7.106 Basalt pitted stone (#5508 Layer II-6 Level 6)

Fig. 7.104 Basalt pitted stone (#13387 Layer II-6 Level 3)

Fig. 7.107 Basalt pitted stone (#8907 II-6 Level 4b)

these artifacts (Goren-Inbar et al. 2002a), we believe that such a characterization should be carried out in the context of an exhaustive and comprehensive experimental study. As the latter has not yet been carried out, we refrain here from providing a detailed description of the pits and limit the scope to an attempt to characterize the artifacts in a more general manner. Here we describe the blank types, size, and morphology of pitted stones. This presentation includes the complete assemblage of basalt pitted stones, a much larger sample than previously published. In addition, we also present a sample of pitted stones for which the pits' morphology, location, and clustering were analyzed in depth, data that have been published previously (Goren-Inbar et al. 2002a).

Leakey suggested that the pitted anvil (pitted stone) is in fact only one component of a pair (Leakey and Roe 1994:7). The pair consists of an immobile element (an anvil) and a mobile one (a percussor) (Chavaillon 1979). We use the term "pitted stones" to describe any stone that bears pitting. The basalt pitted stones of GBY are made on two different types of blank: FFT (lighter and smaller) and CCT (heavier and larger). Table 7.60 presents the entire inventory of the basalt pitted stones found in the archaeological horizons of GBY and distinguishes between those made on FFT and those made on CCT. As with other tool types, the largest samples of pitted stones occur within the complex of Layer II-6. Some of the pitted stones have additional tool types (Table 7.61). The most common additional tool types within the CCT are other percussive tools, while within the FFT the dominant blank are unretouched flakes, though percussors and retouched flakes occur as well, with a minimal representation of other retouched tools (Table 7.61). Only one pitted stone has two additional typological classifications: it is also classified as a thin anvil (see below, Section 7.4.3) and as a percussor.

The dominant breakage patterns of pitted stones are complete and laterally broken, while other breakage types occur in lower frequencies (Tables 7.62–7.63). Clearly, broken cobbles or broken thick flakes were suitable for use as pitted stones in different tasks (e.g. food preparation) if they were of the requisite size and weight. Pitted stones on CCT blanks are usually covered by 76–100% cortex and rarely lesser amounts; as is usual with basalt, there are some indeterminate cases (Table 7.62). FFT pitted stones are not as cortical as those on CCT and are often classified as indeterminate with regard to their cortical coverage (Table 7.63).

Tables 7.64 and 7.65 as well as Fig. 7.108 present the size of pitted stones. Those on CCT are significantly longer (t ratio=-2.39; df=187; $p(tt)$=0.0175) and thicker (t ratio=-4.56; df=187; $p(tt)$<0.0001), but not significantly wider than those on FFT. There is some intra-layer and inter-layer variability, as can be seen in the descriptive statistics of Tables 7.64–7.65. Most pitted stones on CCT are larger than 90 mm in length and up to 140 mm in maximal length. Most of the pitted stones seem to have been of a hand-held size and thus easy to handle. The considerable variability in the size of the pitted stones suggests that size was only to an extent a factor in selection, whereas the preponderance of flat surfaces on the artifacts seems to reflect trait-specific selection.

As with other artifacts described and discussed in this chapter, the pitted stones on CCT of the upper part of Layer II-6 seem to be slightly larger than those of the lower part of this layer. Thus, the pitted stones of Level 1 are somewhat larger than those of the other layers, albeit by only a few centimeters (Tables 7.64). Interestingly, the Level 1 pitted stones are larger than those of other layers in both CCT and FFT assemblages, a

Table 7.60 Counts and frequencies (%) of basalt pitted stones on CCT and FFT

Layer	CCT N	CCT %	FFT N	FFT %
I-5	1	1.05	-	-
V-4	1	1.05	-	-
V-5	2	2.11	1	1.05
V-6	2	2.11	1	1.05
JB	6	6.32	-	-
II-2/3	1	1.05	1	1.05
II-5	1	1.05	1	1.05
II-5/6	-	-	1	1.05
II-6/L1	8	8.42	10	10.53
II-6/L2	14	14.74	8	8.42
II-6/L3	6	6.32	6	6.32
II-6/L4	14	14.74	20	21.05
II-6/L4b	17	17.89	6	6.32
II-6/L5	2	2.11	3	3.16
II-6/L6	13	13.68	16	16.84
II-6/L7	7	7.37	18	18.95
VI-14	-	-	2	2.11
Trench II green	-	-	1	1.05
Total	95	100.00	95	100.00

Table 7.61 Counts and frequencies of co-occurrence of basalt pitted stones and other typological categories

Typology	CCT N	CCT %	FFT N	FFT %
Single concave side-scraper	-	-	1	1.05
Notch	-	-	2	2.11
Denticulate	-	-	1	1.05
Varia	1	1.05	1	1.05
Flake	-	-	75	78.95
Retouched flake	-	-	3	3.16
Percussor	47	49.47	8	8.42
Anvil	26	27.37	4	4.21
Angular fragment	11	11.59	-	-
Modified	9	9.47	-	-
Manuport	1	1.05	-	-
Total	95	100.00	95	100.00

7.4 Typology and Technology of Basalt Cores and Core Tools

Table 7.62 Counts and frequencies (%) of breakage patterns and cortical coverage of basalt CCT pitted stones of Layer II-6

	II-6/L1	II-6/L2	II-6/L3	II-6/L4	II-6/L4b	II-6/L5	II-6/L6	II-6/L7
N	8	14	6	14	17	2	13	7
Breakage								
Complete	3 37.50	6 42.86	1 16.67	6 42.86	4 23.53	-	3 23.08	4 57.14
Distal	1 12.50	-	-	-	1 5.88	-	-	-
Lateral	1 12.50	5 35.71	2 33.33	2 14.29	7 41.18	-	6 46.15	2 28.57
Proximal	-	1 7.14	1 16.67	-	1 5.88	-	1 7.69	-
Distal & lateral	-	-	-	1 7.14	-	-	-	-
Distal & proximal	1 12.50	-	-	-	-	-	-	-
Fragment	2 25.00	1 7.14	1 16.67	2 14.29	1 5.88	-	2 15.38	-
Proximal & lateral	-	-	1 16.67	1 7.14	-	-	-	-
Split	-	1 7.14	-	1 7.14	3 17.65	1 50.00	1 7.69	1 14.29
Indeterminate	-	-	-	1 7.14	-	1 50.00	-	-
Cortex		(N=13)						
No cortex	1 12.50	4 30.77	-	3 21.43	-	1 50.00	1 7.69	-
1–25%	-	-	1 16.67	-	-	-	1 7.69	-
26–50%	2 25.00	1 7.69	-	2 14.29	1 5.88	-	-	1 14.29
51–75%	-	2 15.38	1 16.67	1 7.14	2 11.76	-	-	-
76–100%	4 50.00	6 46.15	3 50.00	4 28.57	11 64.71	1 50.00	6 46.15	3 42.86
Indeterminate	1 12.50	-	1 16.67	4 28.57	3 17.65	-	5 38.46	3 42.86

fact that may be attributed to a particular type of activity carried out in this level.

Some of the large basalt slabs recovered at GBY, weighing up to 22.5 kg, show signs of battering, and a single example also exhibit pits of different types (Fig. 7.109). This suggests that the GBY giant cores served not only as a source of raw material for tool production but also as dormant large anvils, perhaps being used as working tables in the processing of different materials.

Detailed analyses of the pits were carried out on a majority of the basalt pitted stones. The sample comprises 154 basalt artifacts and the analyses provide data on the sizes of the pits (large/small), their depth (deep/shallow), and their counts and clustering. In addition, an indication of the spatial location of the pits on the surface of the artifacts is provided.

Most of the pits are quite shallow, and those that are large and deep and thus easily defined are less frequent. Over 17% of the pits are organized in clusters of small pits located on one of the artifact's faces. On some 15% of the pitted stones the pits are less distinct and seem to resemble battering marks in restricted areas, although these marks differ from the typical percussor marks observed at GBY (Table 7.66).

The location of the pits in regard to the overall shape of the pitted stone is described in Table 7.67. The majority of pits are located in the central part of the artifact and none of the other options seems to be quantitatively significant. This observation is not surprising, as the center of the artifact is the most likely location considering its use as an anvil or a percursor. Pitted stones on FFT seem to show more pitting on their edges, and

Table 7.63 Counts and frequencies (%) of breakage patterns and corticas coverage of basalt FFT pitted stones of Layer II-6

	II-6/L1	II-6/L2	II-6/L3	II-6/L4	II-6/L4b	II-6/L5	II-6/L6	II-6/L7
N	10	8	6	20	6	3	16	18
Breakage								
Complete	4 40.00	1 12.50	1 16.67	5 25.00	2 33.33	2 66.67	8 50.00	7 38.89
Distal	1 10.00	1 12.50	1 16.67	3 15.00	-	-	1 6.25	4 22.22
Lateral	2 20.00	-	2 33.33	3 15.00	1 16.67	-	1 6.25	5 27.78
Proximal	1 10.00	1 12.50	-	2 10.00	-	-	-	-
Distal & lateral	-	1 12.50	-	4 20.00	-	-	3 18.75	-
Distal & proximal	-	1 12.50	-	1 5.00	-	-	1 6.25	1 5.56
Fragment	1 10.00	1 12.50	1 16.67	2 10.00	2 33.33	1 33.33	2 12.50	-
Proximal & lateral	-	1 12.50	-	-	1 16.67	-	-	1 5.56
Split	1 10.00	1 12.50	-	-	-	-	-	-
Indeterminate	-	-	1 16.67	-	-	-	-	-
Cortex								
No cortex	5 50.00	2 25.00	4 66.67	5 25.00	3 50.00	3 100.00	7 43.75	12 66.67
1–25%	1 10.00	-	-	-	-	-	-	1 5.56
26–50%	-	-	-	2 10.00	-	-	-	1 5.56
51–75%	1 10.00	1 12.50	-	3 15.00	-	-	-	-
76–100%	1 10.00	2 25.00	1 16.67	4 20.00	1 16.67	-	6 37.50	1 5.56
Indeterminate	2 20.00	3 37.50	1 16.67	6 30.00	2 33.33	-	3 18.75	3 16.67
Striking platform		(N=7)		(N=19)		(N=2)	(N=15)	(N=17)
Cortical	1 10.00	-	-	1 5.26	-	-	-	-
Plain	3 30.00	2 28.57	2 33.33	6 31.58	-	1 50.00	5 33.33	10 58.82
Dihedral	-	-	-	-	-	-	-	1 5.88
Facetted	1 10.00	-	1 16.67	-	-	-	-	-
Crushed	-	-	-	-	1 16.67	-	-	-
Removed	-	1 14.29	-	1 5.26	1 16.67	-	1 6.67	-
Broken	2 20.00	3 42.86	1 16.67	4 21.05	3 50.00	1 50.00	4 26.67	-
Indeterminate	3 30.00	1 14.29	2 33.33	7 36.84	1 16.67	-	5 33.33	6 35.29

7.4 Typology and Technology of Basalt Cores and Core Tools

Table 7.64 Descriptive statistics of basalt CCT pitted stones (size in mm) of Layer II-6

	II-6/L1	II-6/L2	II-6/L3	II-6/L4	II-6/L4b	II-6/L5	II-6/L6	II-6/L7
N	8	14	6	14	17	2	13	7
Maximum length								
Maximum	198	163	160	150	157	107	149	122
Median	137	113	129.5	105	101	81	96	99
Minimum	96	73	88	64	62	55	49	69
Mean	140.38	121.54	125.83	102.93	99.82	81.00	101.00	96.29
Std dev	40.33	30.07	31.86	24.16	29.46	36.77	30.85	18.81
Std err	14.26	8.34	13.01	6.46	7.14	26.00	8.56	7.11
Length		(N=13)						
Maximum	196	155	148	150	157	102	137	122
Median	129.5	104	112.5	96.5	89	74.5	86	98
Minimum	84	72	88	62	47	47	44	64
Mean	132.50	112.92	113.17	97.71	91.06	74.50	91.69	93.29
Std dev	41.62	26.47	21.39	28.56	28.92	38.89	31.71	20.35
Std err	14.72	7.34	8.73	7.63	7.01	27.50	8.80	7.69
Width		(N=13)						
Maximum	133	156	126	127	131	76	102	105
Median	91	76	91	74.5	73	61.5	85	82
Minimum	64	59	45	52	45	47	37	49
Mean	95.13	84.92	88.67	78.71	77.24	61.50	76.46	77.00
Std dev	23.04	26.21	32.36	18.05	24.23	20.51	20.59	17.66
Std err	8.15	7.27	13.21	4.82	5.88	14.50	5.71	6.68
Thickness		(N=13)						
Maximum	83	64	75	71	71	55	68	66
Median	48	46	45.5	50	45	43.5	51	48
Minimum	32	34	33	37	29	32	28	36
Mean	49.63	45.69	48.33	51.14	48.76	43.50	49.54	49.86
Std dev	16.22	8.29	15.02	10.43	11.91	16.26	11.69	11.57
Std err	5.74	2.30	6.13	2.79	2.89	11.50	3.24	4.37
Circumference	(N=5)	(N=9)	(N=5)	(N=10)	(N=16)			
Maximum	550	447	430	411	445	314	413	357
Median	302	352	410	319	298	241.5	291	300
Minimum	278	222	245	231	191	169	143	192
Mean	366.20	336.56	369.40	309.00	295.13	241.50	284.62	282.14
Std dev	117.28	70.94	77.63	57.55	81.59	102.53	78.74	57.44
Std err	52.45	23.65	34.72	18.20	20.40	72.50	21.84	21.71
Number of scars	(N=7)	(N=10)	(N=3)	(N=10)	(N=11)		(N=12)	
Maximum	7	9	5	6	6	4	6	4
Median	3	2.5	2	3	2	3	2.5	3
Minimum	1	1	2	1	1	2	1	2
Mean	3.43	3.00	3.00	3.20	2.82	3.00	2.83	2.86
Std dev	1.99	2.31	1.73	1.48	1.72	1.41	1.64	0.90
Std err	0.75	0.73	1.00	0.47	0.52	1.00	0.47	0.34

the occurrence of multiple pitting is higher than that observed for CCT (Table 7.67).

The shape characteristics of the pits observed on the GBY pitted stones are summarized in Table 7.68. Shape was difficult to determine due to weathering of the surfaces and particularly to the blurred nature of the pits' margins, resulting in high frequencies of indeterminate cases. The identifiable pits are rounded or elliptical in shape. It should be noted that no other shapes were observed and, in particular, that no narrow elliptical pits were recorded.

7.4.3 Thin Anvils

The thin anvils of GBY (N=42) are small, thin, flat slabs of non-vesicular basalt, at times slightly weathered (Fig. 7.110). The identification of the thin anvils is based not on pitting or other damage markings but on other, primarily morphological, characteristics. The thin anvils are found diachronically in the cultural sequence at the site (Table 7.69).

The thin anvils are characterized by two flat, parallel,

Table 7.65 Descriptive statistics of basalt FFT pitted stones (size in mm; angle in degrees) of Layer II-6

	II-6/L1	II-6/L2	II-6/L3	II-6/L4	II-6/L4b	II-6/L5	II-6/L6	II-6/L7
N	10	8	6	20	6	3	16	18
Maximum length								
Maximum	200	230	106	152	112	147	132	157
Median	129.5	84.5	70	103	73	65	84	95.5
Minimum	51	58	41	70	64	62	61	76
Mean	123.50	101.25	73.17	106.90	83.33	91.33	95.31	101.83
Std dev	48.54	56.24	22.11	19.61	20.61	48.23	24.52	23.30
Std err	15.35	19.89	9.03	4.39	8.41	27.85	6.13	5.49
Length								
Maximum	185	230	106	152	106	147	130	138
Median	89	66	59	98	73	62	76.5	74
Minimum	51	52	37	64	58	54	58	57
Mean	108.70	92.50	63.33	95.80	79.50	87.67	86.81	83.56
Std dev	49.56	58.87	22.91	22.27	20.22	51.54	25.45	24.75
Std err	15.67	20.81	9.35	4.98	8.25	29.76	6.36	5.83
Width								
Maximum	142	179	94	119	96	101	123	123
Median	95	69	68	78	48.5	50	76.5	84
Minimum	44	45	37	42	40	50	46	57
Mean	94.30	80.50	68.33	80.35	54.33	67.00	75.13	88.56
Std dev	38.79	43.85	20.56	17.93	20.85	29.44	23.06	19.52
Std err	12.27	15.50	8.39	4.01	8.51	17.00	5.76	4.60
Thickness	10	8	6	20	6	3	16	18
Maximum	68	69	33	66	40	58	88	82
Median	39.5	27.5	25.5	39	29.5	38	37.5	43
Minimum	20	20	19	25	23	26	21	24
Mean	44.60	32.88	25.50	43.10	30.50	40.67	39.88	46.06
Std dev	16.69	16.36	5.21	12.79	6.53	16.17	16.17	14.28
Std err	5.28	5.78	2.13	2.86	2.67	9.33	4.04	3.37
L/W								
Maximum	1.50	1.43	1.13	2.14	2.52	1.46	1.74	1.48
Median	1.21	1.21	0.98	1.16	1.48	1.24	1.20	0.87
Minimum	0.65	0.65	0.70	0.66	1.07	1.08	0.70	0.69
Mean	1.18	1.16	0.93	1.24	1.55	1.26	1.21	0.96
Std dev	0.28	0.26	0.15	0.34	0.51	0.19	0.35	0.26
Std err	0.09	0.09	0.06	0.08	0.21	0.11	0.09	0.06
Angle of striking platform	(N=5)	(N=1)	(N=3)	(N=7)		(N=1)	(N=6)	(N=9)
Maximum	130	125	130	132	-	123	115	132
Median	124	125	113	117	-	123	108	112
Minimum	114	125	110	100	-	123	102	93
Mean	122.60	125.00	117.67	118.00	-	123.00	107.50	110.00
Std dev	6.54	-	10.79	9.85	-	-	5.13	13.11
Std err	2.93	-	6.23	3.72	-	-	2.09	4.37
Number of scars	(N=7)	(N=6)	(N=5)	(N=13)	(N=1)		(N=5)	(N=13)
Maximum	4	7	5	5	3	4	6	7
Median	3	1.5	5	3	3	3	3	3
Minimum	1	1	2	1	3	2	1	2
Mean	3.14	2.5	3.8	2.84	3.00	3.00	2.80	3.84
Std dev	1.06	2.34	1.64	1.21	-	1.00	2.04	1.77
Std err	0.40	0.95	0.73	0.33	-	0.57	0.91	0.49

7.4 Typology and Technology of Basalt Cores and Core Tools

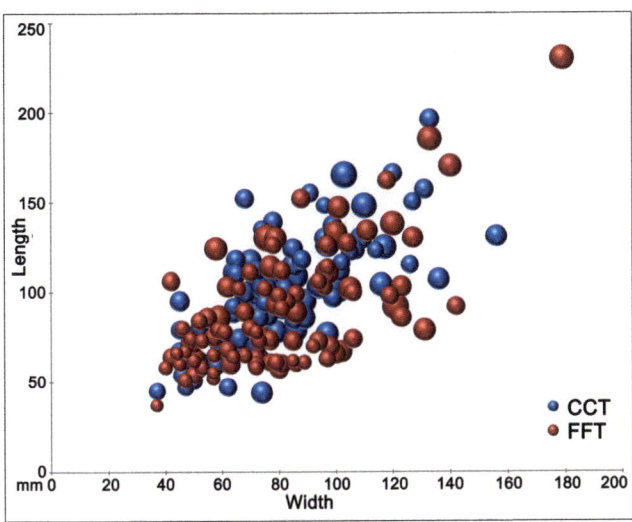

Fig. 7.108 Length, width, and thickness (bubble size) of Layer II-6 basalt pitted stones on FFT and on CCT

Fig. 7.109 Basalt giant core with a cluster of pits (#14177 Layer II-6 Level 2)

Table 7.66 Counts and frequencies (%) of pit characteristics of basalt pitted stones

	CCT		FFT	
	N	%	N	%
Flat	26	35.62	26	32.10
Easily defined and deep	9	12.33	17	20.99
Battering	13	17.81	11	13.58
Large and deep	10	13.70	13	16.05
Cluster of pits	15	20.55	14	17.28
Total	73	100.00	81	100.00

Table 7.67 Counts and frequencies (%) of spatial location of pits of basalt pitted stones

	CCT		FFT	
	N	%	N	%
Central	40	54.79	36	44.44
Marginal	7	9.59	19	23.46
Side	9	12.33	6	7.41
More than one pit (no primary one)	6	8.22	3	3.70
Multiple pits	10	13.70	16	19.75
Indeterminate	1	1.37	1	1.23
Total	73	100.00	81	100.00

Table 7.68 Counts and frequencies (%) of shape of pits of basalt pitted stones

	CCT		FFT	
	N	%	N	%
Rounded	22	30.14	38	46.91
Elliptical	4	5.48	8	9.88
Cluster of pits	16	21.92	14	17.28
Indeterminate	31	42.47	21	25.93
Total	73	100.00	81	100.00

unflaked surfaces. All these thin slabs are fragments of larger objects (Figs. 7.111–7.112) and some bear signs of heavy percussive impact that caused their fragmentation. The thin basalt slabs have a particular geometry that is similar to that of the massive basalt slabs from which giant cores were flaked (Section 7.4.7.2). As with the massive slabs, the original cross-section morphology of the thin slabs sometimes has a generally trapezoid form, with part of the upper surface occasionally sloping to form an acute angle with the lower surface.

The identification of anvils at GBY is a demanding task, since the acquired shape of the artifact is frequently a result of the intensity with which it was used. Thus, limited use of anvils may not have left a clear signature and they can easily be classified as natural or minimally used objects; other examples of such cases are percussors (Section 7.4.1.1) and pitted stones (Section 7.4.2).

In each of the layers the frequency of the anvils is too low for significant analysis and does not permit in-depth morphological study of their original planform (Table 7.69). In view of the conservative nature of the GBY cultural sequence, we have lumped together the items from all levels to form a single assemblage that is large enough for analysis.

Since the thin anvils are natural objects, a consistent system of positioning was necessary in order to achieve a significant analysis. Three planes (surfaces) were defined (Fig. 7.113) for each artifact following the methodology of Mora and de la Torre (2005). First, the two large opposite planes were termed A and B, Plane B being the flatter one; thus, Plane B is considered the one that was in contact with the ground. The third plane, the slab's profile, was termed Plane C. In the few cases in which the artifact was identified as a flake, its ventral face was considered Plane B. For the purposes of systematic recording and analysis, the artifact was rotated so that its maximal length

Fig. 7.110 Basalt thin anvil (#7695 Layer II-6 Level 6)

Table 7.69 Counts and frequencies of basalt thin anvils

Layer	N	%
JB	2	4.76
II-6/L1	8	19.05
II-6/L2	7	16.67
II-6/L3	5	11.90
II-6/L4	6	14.29
II-6/L4b	2	4.76
II-6/L6	6	14.29
II-6/L7	5	11.90
Trench II	1	2.38
Total	42	100.00

Fig. 7.111 Broken basalt thin anvil (#5889 Layer II-6 Level 2)

Fig. 7.112 Broken basalt thin anvil (#5883 Layer II-6 Level 2)

Fig. 7.113 Positioning method of basalt thin anvils and their A, B, and C planes (#5571 Layer II-6 Level 1)

was parallel to a fixed Y axis and its widest axis was placed proximally (Fig. 7.113). This systematic positioning is not related to the rock damage observed on the anvils but simply aims to generate an objective common denominator for their geometric documentation and analysis.

The systematic positioning of the anvils enabled efficient recording of their breakage patterns. A breakage is defined here as a fracture that extends through the entire thickness of the artifact. The breakage damage type is the result either of intentional fragmentation or of a massive accidental blow that caused the thin slab to break (longitudinally or transversally). The morphology of the anvils was recorded in accordance with their systematic positioning, using four measurements (length, width, thickness, and weight).

The analysis focused on the signs of use and other rock damage types identified on the three different planes of the anvils. We defined a number of damage categories that include pitting, flake scars, and damage caused by the use of the artifact as a percussor.

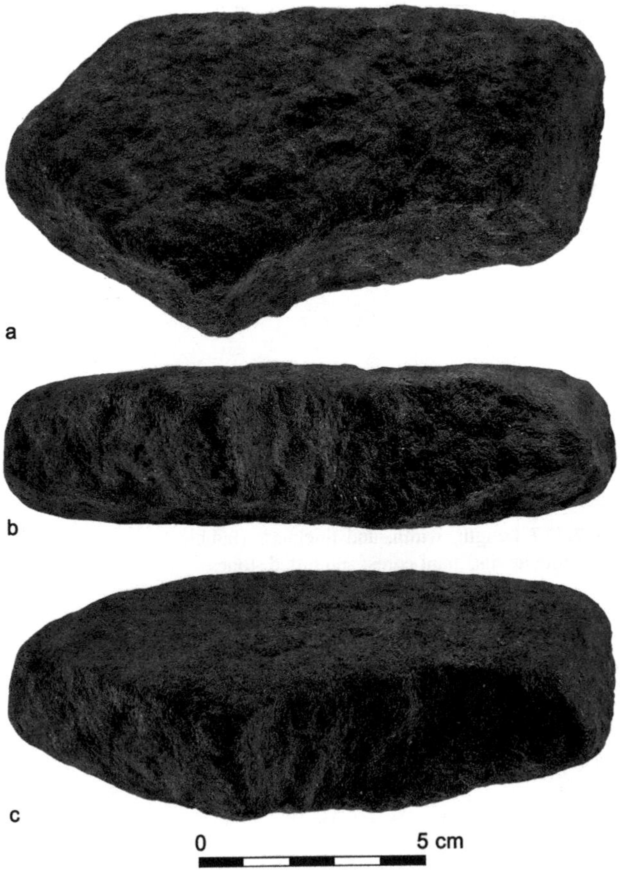

7.4 Typology and Technology of Basalt Cores and Core Tools

Fig. 7.114 Basalt thin anvil (#16123 Layer II-6 Level 2): a) Plane A with view of the pitted surface, b) Plane C – a break, c) another view of Planes A and C

Fig. 7.115 Basalt thin anvil (#7498 Layer II-6 Level 1); note damage in the form of flake scars

a) Pitting damage: defined as depressions on any of the surfaces of the artifact, caused by energy transferred to the artifact during its use as a passive (dormant) percussion object. A pit was determined when a small area, with distinct limits, was observed to be deeper than its surrounding natural surface. Pits appear on different locations of the artifacts and differ in depth, diameter, and concentration, all of which result from distinct types of functions and intensity (Fig. 7.114a). Intensity of damage was recorded for pitting on the horizontal Planes A and B, where it is most common. Given the scarcity of this damage type on Plane C, we registered only its presence or absence. These observations of signs of damage, and specifically those of pitting, were subjective and did not involve microscopic tools. The intensity of pitting was determined according to the number of pits, their depth, and their concentration on a specific plane.

b) Flake scars: caused by flaking and determined by scars on the surfaces of the artifacts (Fig. 7.115). Six types of flaking damage were defined and documented with regard to the direction of the scar removal and the plane from which it originates (Fig. 7.116):

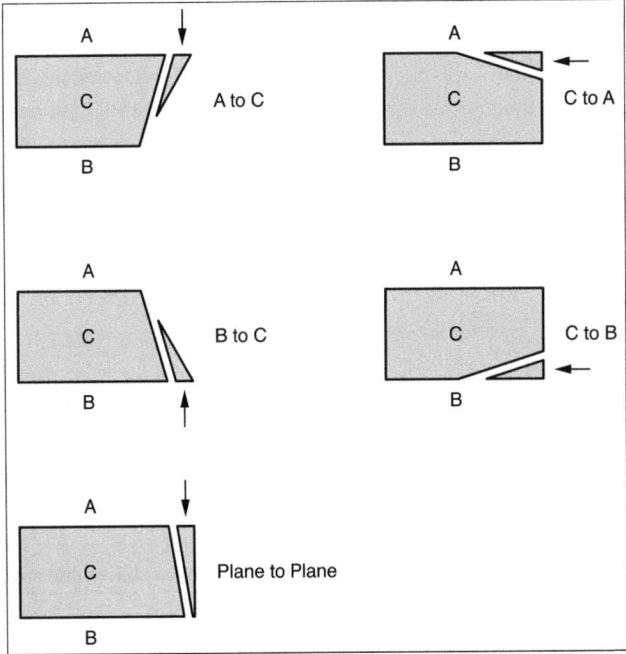

Fig. 7.116 Location and direction of types of flaking damage on basalt thin anvils

1. A to C flaking direction, documented by a scar on Plane C originating from Plane A.
2. B to C flaking direction, documented by a scar on Plane C originating from Plane B.
3–4. Directions C to A and C to B: in both cases Plane C serves as the striking platform and the scar is located on either Plane A or Plane B.
5. Plane to plane flaking. This is in practice a subtype of the first two categories, but differs from them in that it resulted from a powerful blow that was inflicted on either of the horizontal planes and removed a flake from the entire thickness of the slab, either from Plane A or from Plane B, somewhat similar to an overshot flake.
6. The last type of flaking damage is a longitudinal scar. This type appears only on Plane C and reflects a flake removal from one edge of Plane C to another.

c) Damage caused by the use of the artifact as a percussor: recorded in cases in which a restricted area on one or more of the surfaces exhibited the battering damage that is distinctive of use as a percussor. This type of damage was recorded as percussor damage.

The size of the thin anvils presents minimal variability, expressed in all three dimensions of length, width, and thickness (Table 7.70, Fig. 7.117). The fact that the large majority of anvils (95.2%) are broken on at least one edge and 11.9% are fragments broken on all sides makes their uniformity of size even more notable. The thickness presents an even greater homogeneity, as it is not influenced by breakage. As we have no knowledge of the original length and width of these slabs, we consider their observed thickness to represent the original feature selected by the GBY hominins.

Pitting damage is very common and occurs on 81% of Plane A, while on Plane B the frequency falls to 57.1% (Table 7.71). In both cases the most common pattern is low intensity, although frequencies vary between planes (50% for Plane A and 35.7% for Plane B) (Table 7.72). This variability supports the

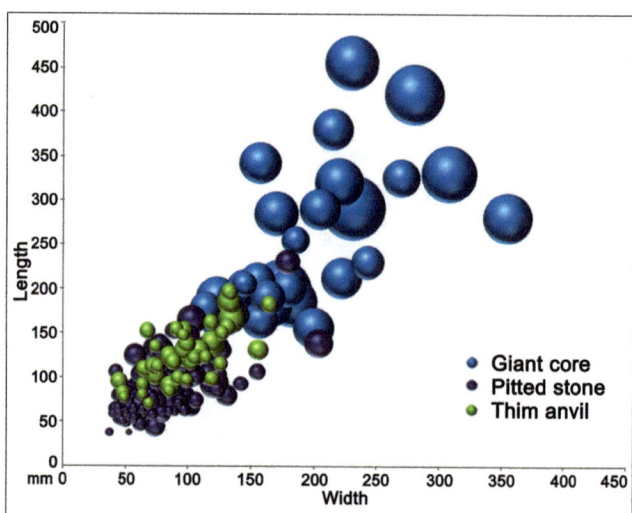

Fig. 7.117 Length, width, and thickness (bubble size) of Layer II-6 basalt thin anvils, giant cores, and pitted stones

Table 7.71 Damage types and location on basalt thin anvils

Damage types	N	%
Plane A pitting	34	80.95
Plane B pitting	24	57.14
Plane C pitting	8	19.05
Flaking	34	80.95
Plane to plane	16	38.10
Percussor	11	26.19

identification of Plane A as the upper surface, on which damage occurred more frequently and more intensively. Pitting damage on Plane C is rare, occurring on 19% of the artifacts.

Like pitting, damage resulting from flaking is very common and 81% of the artifacts bear flake scars on Plane C (Table 7.71). This type of damage appears primarily on one of the lateral edges of Plane C (66.7%) and is less abundant on its proximal and distal edges (23.8% and 21.4% respectively). On Plane C the flake scars usually appear on a single edge (52.4%), although items with two (26.2%) and three (2.4%) flaked edges are recorded as well. In over 64% of the anvils with flake scars the scars on Plane C originate in plane B, and in 38.1% the direction is the opposite one (Table 7.73). Longitudinal scars, occurring only on Plane C and dictated by the thinness of the slab, which guides the percussive energy, occur on 11.9% of the anvils. On 40.5% of the anvils we identified flake scars with multiple directions. Flaking damage is less common on Planes A and B, as only 11.9% of the items display it on Plane A and 2.4% (a single item) on Plane B.

Table 7.70 Descriptive statistics of basalt thin anvils (N=42) (size in mm; weight in g)

	Maximum Length	Length	Width	Thickness	Weight
Maximum	209	196	165	81	2,350
Median	132.5	123	96.5	43.5	649.5
Minimum	78	70	45	28	163
Mean	135.14	124.40	96.93	44.74	827.12
Std dev	31.33	30.47	28.41	11.98	553.18
Std err	4.83	4.70	4.38	1.85	85.36

Table 7.72 Location and intensity of pitting damage on basalt thin anvils

	No pitting		Low		Medium		High		Total	
	N	%	N	%	N	%	N	%	N	%
Plane A	8	19.05	21	50.00	4	9.52	9	21.43	42	100.00
Plane B	18	42.86	15	35.71	8	19.05	1	2.38	42	100.00

7.4 Typology and Technology of Basalt Cores and Core Tools

Table 7.73 Type and direction of flaking damage on basalt thin anvils

Type and direction of flaking	N	%
A to C	16	38.10
B to C	27	64.29
C to A	5	11.90
C to B	1	2.38
Longitudinal	5	11.90
Plane to plane	16	38.10

Plane to plane flaking damage, which extends through the entire thickness of the artifact and most likely results from a blow, occurs on 38.1% of the anvils.

It appears that some of the items were also used as active percussors, as 26.2% of them bear the typical buttering damage that is indicative of percussive action (Table 7.71).

7.4.4 Modified Artifacts

The typological classification of modified artifacts follows the definition of Leakey (1971) and refers to a group of artifacts that were minimally knapped or tested. We have included in this category artifacts bearing up to three flake scars. This definition does not rule out the possibility that the three removals were purposely made, despite the lack of obvious planning of knapping in the form of a well-delineated striking platform or a working edge.

Table 7.74 provides data on the occurrence and quantity of basalt modified artifacts in the different archaeological horizons. As in many other cases, the highest counts are in the lithic assemblages of Layer II-6, but this artifact appears in almost all of the site's archaeological horizons in different frequencies. Taphonomically, most of the modified artifacts are complete, followed by fragments (Table 7.75); this is relevant only to those levels that have samples of sufficient size. The high frequency of modified artifacts in Level 2 is interesting and can be attributed to the large variety of activities that were carried out in this horizon (Alperson-Afil et al. 2009).

A varied distribution of cortical coverage is observed for this class of artifacts, which displays different extents of cortex and a notable percentage of indeterminate coverage, the latter typical of the basalt artifacts (Table 7.75). The size of the modified artifacts is extremely varied in each of the lithic assemblages (Table 7.76). The largest sizes (and means) occur in Layer II-6 Levels 7, 6, and 1 in descending order. This extensive variability, indicating use in multiple tasks, is perhaps an outcome of our rigid definition. One should remember that each of the items was brought to the site and was not a component of the sedimentological setting. Yet despite the great variability, there is an inter-assemblage similarity that could be explained by use of a familiar source for a very long time by the occupants of the site.

Table 7.74 Counts and frequencies (%) of basalt modified artifacts

Layer	N	%
V-5	4	2.37
V-6	2	1.18
I-5	2	1.18
II-2/3	2	1.18
II-5/6	2	1.18
II-6/L1	26	15.38
II-6/L2	40	23.67
II-6/L3	30	17.75
II-6/L4	29	17.16
II-6/L4b	2	1.18
II-6/L5	3	1.78
II-6/L6	7	4.14
II-6/L7	14	8.28
JB	3	1.78
Trench II green	3	1.78
Total	169	100.00

Table 7.75 Counts and frequencies (%) of breakage patterns and cortical coverage characteristics of basalt modified artifacts of Layer II-6

	II-6/L1	II-6/L2	II-6/L3	II-6/L4	II-6/L6	II-6/L7
N	26	40	30	29	7	14
Breakage				(N=28)		
Complete	19	27	26	18	4	6
	73.08	67.50	86.67	64.29	57.14	42.86
Distal	-	2	-	-	-	1
		5.00				7.14
Lateral	3	1	3	1	1	4
	11.54	2.50	10.00	3.57	14.29	28.57
Proximal	1	1	-	2	-	-
	3.85	2.50		7.14		
Distal & lateral	-	-	1	1	-	-
			3.33	3.57		
Fragment	1	6	-	5	1	2
	3.85	15.00		17.86	14.29	14.29
Indeterminate	2	3	-	1	1	1
	7.69	7.50		3.57	14.29	7.14
Cortex		(N=34)	(N=28)	(N=28)	(N=5)	(N=13)
No cortex	12	10	7	9	3	5
	46.15	29.41	25.00	32.14	60.00	38.46
1–25%	2	2	1	4	-	-
	7.69	5.88	3.57	14.29		
26–50%	6	6	8	5	-	1
	23.08	17.65	28.57	17.86		7.69
51–75%	4	6	4	1	-	2
	15.38	17.65	14.29	3.57		15.38
76–100%	1	2	2	1	2	3
	3.85	5.88	7.14	3.57	40.00	23.08
Indeterminate	1	8	6	8	-	2
	3.85	23.53	21.43	28.57		15.38

Table 7.76 Descriptive statistics of basalt modified artifacts of Layer II-6 (size in mm)

	II-6/L1	II-6/L2	II-6/L3	II-6/L4	II-6/L6	II-6/L7
N	26	40	30	29	7	14
Maximum length				(N=28)		
Maximum	171	183	90	158	127	230
Median	80	66	56.5	75.5	68	83
Minimum	42	42	34	30	48	44
Mean	81.81	71.25	59.40	77.79	82.57	96.29
Std dev	27.31	26.21	13.02	30.44	33.31	45.28
Std err	5.36	4.14	2.38	5.75	12.59	12.10
Length						
Maximum	165	164	90	153	123	220
Median	79.5	58.5	56.5	65	62	79
Minimum	41	38	31	28	40	41
Mean	75.15	64.70	55.73	71.38	75.57	86.50
Std dev	26.51	26.50	14.47	30.92	32.92	41.22
Std err	5.20	4.19	2.64	5.74	12.44	11.02
Width						
Maximum	103	121	66	131	96	147
Median	49.5	47	44	54	61	65.5
Minimum	29	32	27	23	35	32
Mean	57.65	50.45	44.60	59.07	60.29	71.93
Std dev	18.32	16.41	10.02	24.93	23.19	33.63
Std err	3.59	2.59	1.83	4.63	8.77	8.99
Thickness						
Maximum	83	70	54	81	80	130
Median	39	32.5	27.5	39	41	43
Minimum	20	23	15	16	34	22
Mean	40.85	35.88	30.30	38.69	49.86	52.50
Std dev	12.55	10.42	8.75	14.67	16.01	26.71
Std err	2.46	1.65	1.60	2.72	6.05	7.14
Circumference	(N=4)	(N=19)	(N=5)	(N=13)		
Maximum	276	220	222	445	364	625
Median	208	172	165	222	203	232
Minimum	173	118	116	66	132	135
Mean	216.25	171.79	165.60	238.31	231.14	269.00
Std dev	48.80	26.14	40.32	94.74	89.12	120.28
Std err	24.40	6.00	18.03	26.27	33.69	32.15
Number of scars				(N=29)		
Maximum	3	3	3	3	3	3
Median	3	2	2	3	2	2.5
Minimum	1	1	1	1	1	1
Mean	2.58	2.35	1.93	2.41	2.14	2.36
Std dev	0.64	0.62	0.70	0.78	0.90	0.74
Std err	0.13	0.10	0.13	0.14	0.34	0.20

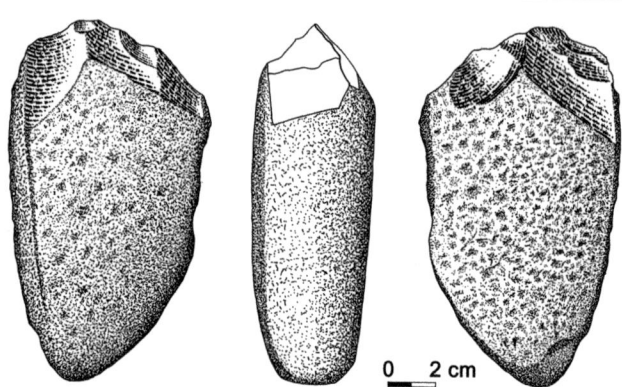

Fig. 7.118 Basalt chopping tool (#677 Layer II-6 Level 4)

7.4.5.1 Chopping Tools

Chopping tools on all types of raw material appear in very small numbers. Basalt chopping tools are extremely rare and number only seven (Table 7.53; Fig. 7.118). These tools were found in Layer II-6 Levels 2, 3, and 4 (five artifacts altogether) and in the JB (two artifacts). All but two are complete and they are covered by different amounts of cortex, including a single artifact with 1–25% cortical coverage. The latter is very atypical of chopping tools and in a sense represents the deviation of the GBY artifacts from the normative appearance of this tool type as recorded in sites such as 'Ubeidiya, Revadim, and others (Bar-Yosef and Goren-Inbar 1993; Malinski-Buller et al. 2011). While the number of working edges (1–3) and platforms (1–3) is normative for this class, the working edge forms include two cases that are concave and hence foreign to the general appearance of this type. Another aspect of the atypical nature of these artifacts is their dimensions (average length: 95.29 mm, width: 67 mm, and thickness: 45.71), all relatively small, particularly when compared to percussors and modified artifacts.

Basalt cobbles of suitable size, abundant in the GBY lithic assemblage and in the site's vicinity, were brought to the site; hence raw material and suitable blanks for the production of chopping tools were available to the GBY knappers. The almost complete lack of chopping tools in the lithic assemblages of the site shows that this tool type was not a regular component of the GBY lithic tradition.

7.4.5.2 Other Basalt Core Tools

The presence of other basalt core tool types at GBY, as presented in Table 7.53, is negligible. It can be argued that practically no tool types beyond the few types presented in the table (massive scraper, atypical burin, denticulate) were produced on non-flake basalt artifacts.

7.4.5 Core Tools

Like basalt flake tools (Section 7.2), basalt core tools are minimally represented. The core tools include chopping tools and a variety of tool types that mirror typologies encountered on flake tools.

7.4 Typology and Technology of Basalt Cores and Core Tools

Only a single subspheroid was identified. This contrasts with the situation revealed in the nearby Acheulian location of NBA, where a lithic assemblage similar to that of the study area included some ten spheroids shaped on both basalt and limestone (Sharon et al. 2002, 2010). This may suggest that spheroids were part of the GBY Acheulian toolkit but nevertheless were not produced or used within the range of activities represented in the GBY archaeological horizons.

7.4.6 Angular Fragments

Angular fragments are artifacts that were produced by knapping but lack any clear morphological/technological features that enable their classification in any of the typological groups. The name is somewhat misleading, as they are not necessarily broken, but they were given this name years ago at the beginning of the lithic analyses. These are artifacts resulting from different stages of knapping that lack any visible signs of flake or other type of blank organization (e.g. scar pattern, striking platforms) or any other morphological features in common.

Angular fragments are quite a common component at GBY and there are 167 basalt examples, which appear along the cultural sequence of the site (Table 7.77). The most frequent occurrence of angular fragments is in the levels of Layer II-6 and they are most abundant in its upper levels, particularly in Level 1. Although many of the angular fragments are broken, the second most common category is complete (Table 7.78). Most of the artifacts are devoid of cortex or bear minimal cortical coverage, a self-explanatory phenomenon, since the angularity is the result of intensive

Table 7.77 Counts and frequencies (%) of basalt angular fragments

Layer	N	%
V-4	1	0.60
V-5	2	1.20
V-6	2	1.20
JB	8	4.79
II-2/3	6	3.59
II-3	1	0.60
II-5	1	0.60
II-5/6	1	0.60
II-6/L1	45	26.95
II-6/L2	29	17.37
II-6/L3	10	5.99
II-6/L4	28	16.77
II-6/L4b	17	10.18
II-6/L5	3	1.80
II-6/L6	5	2.99
II-6/L7	8	4.79
Total	167	100.00

Table 7.78 Counts and frequencies (%) of breakage patterns and cortical coverage of basalt angular fragments of Layer II-6

	II-6/L1	II-6/L2	II-6/L3	II-6/L4	II-6/L4b	II-6/L5	II-6/L6	II-6/L7
N	45	29	10	28	17	3	5	8
Breakage		(N=28)			(N=16)			
Complete	20 / 44.44	9 / 32.14	2 / 20.00	14 / 50.00	1 / 6.25	-	1 / 20.00	1 / 12.50
Distal	-	-	-	1 / 3.57	-	-	-	-
Lateral	3 / 6.67	-	2 / 20.00	2 / 7.14	1 / 6.25	-	-	2 / 25.00
Proximal	1 / 2.22	-	-	-	-	-	-	-
Fragment	16 / 35.56	19 / 67.86	6 / 60.00	10 / 35.71	14 / 87.50	2 / 66.67	4 / 80.00	5 / 62.50
Indeterminate	5 / 11.11	-	-	1 / 3.57	-	1 / 33.33	-	-
Cortex	(N=43)	(N=25)						
No cortex	27 / 62.79	15 / 60.00	7 / 70.00	18 / 64.29	7 / 41.18	3 / 100.00	1 / 20.00	4 / 50.00
1–25%	5 / 11.63	2 / 8.00	-	4 / 14.29	1 / 5.88	-	-	-
26–50%	5 / 11.63	4 / 16.00	1 / 10.00	2 / 7.14	-	-	-	1 / 12.50
51–75%	3 / 6.98	-	-	-	-	-	-	-
76–100%	-	1 / 4.00	1 / 10.00	-	2 / 11.76	-	2 / 40.00	-
Indeterminate	3 / 6.98	3 / 12.00	1 / 10.00	4 / 14.29	7 / 41.18	-	2 / 40.00	3 / 37.50

Table 7.79 Descriptive statistics of basalt CCT angular fragments of Layer II-6 (size in mm)

	II-6/L1	II-6/L2	II-6/L3	II-6/L4	II-6/L4b	II-6/L5	II-6/L6	II-6/L7
N	45	29	10	28	17	3	5	8
Maximum length								
Maxmum	134	104	104	103	101	74	67	94
Median	62	42	49	55.5	63	55	64	65.5
Minimum	23	26	23	25	21	43	49	56
Mean	60.88	51.96	59.40	58.64	62.11	57.33	60.60	68.00
Std dev	30.39	23.77	25.67	19.39	19.38	15.63	7.30	12.03
Std err	4.53	4.41	8.12	3.66	4.70	9.02	3.26	4.25
Length								
Maxmum	134	96	102	103	95	66	62	84
Median	59	38	43.5	49	57	47	61	59.5
Minimum	21	20	23	25	21	42	45	54
Mean	58.35	47.41	55.20	54.14	56.52	51.66	57.00	63.37
Std dev	30.47	22.39	26.75	19.63	19.72	12.66	7.10	9.37
Std err	4.54	4.15	8.46	3.71	4.78	7.31	3.17	3.31
Width								
Maxmum	82	69	74	80	89	47	56	76
Median	40	29	38	41.5	40	47	47	52.5
Minimum	12	13	15	15	15	28	37	38
Mean	40.15	35.86	42.20	43.00	43.94	40.66	47.00	53.25
Std dev	20.04	16.57	18.44	16.86	16.31	10.96	7.44	11.86
Std err	2.98	3.07	5.830	3.180	3.95	6.33	3.33	4.19
Thickness								
Maxmum	71	56	53	54	66	39	38	54
Median	29	20	24.5	30	31	32	32	38.5
Minimum	11	10	12	15	9	23	28	32
Mean	31.53	27.55	29.00	31.50	32.41	31.33	32.80	38.50
Std dev	16.06	14.00	13.00	10.76	11.72	8.02	3.63	7.32
Std err	2.390	2.60	4.11	2.03	2.84	4.63	1.62	2.59
Circumference	(N=9)	(N=19)	(N=5)	(N=8)				
Maxmum	278	284	263	225	303	188	189	275
Median	112	109	129	169	175	169	183	186
Minimum	72	63	66	70	60	124	143	153
Mean	142.44	139.10	160.80	154.62	172.88	160.33	176.60	195.25
Std dev	74.07	70.51	83.22	54.74	54.71	32.86	19.04	37.72
Std err	24.69	16.17	37.21	19.35	13.27	18.97	8.51	13.33
Number of scars	(N=44)	(N=28)			(N=16)	(N=2)	(N=4)	
Maxmum	8	9	10	14	8	6	5	10
Median	5	5.5	4	5	4.5	5	4	5.5
Minimum	2	2	2	2	2	4	3	3
Mean	5.05	5.50	4.70	5.50	4.44	5.00	4.00	6.25
Std dev	1.48	1.90	2.31	2.62	1.75	1.41	0.82	2.25
Std err	0.22	0.36	0.73	0.49	0.44	1.00	0.41	0.80

7.4 Typology and Technology of Basalt Cores and Core Tools

knapping (Table 7.78). The size of the angular fragments is extremely varied in each of the archaeological horizons, with a large range of variation expressed by the high values of the standard deviation (Table 7.79). It is interesting that the average values of the size of angular fragments in each of the archaeological horizons exhibit intra-layer and inter-layer similarities (Table 7.79).

Angular fragments are common in archaeological horizons in which a large basalt assemblage is recorded and seem to indicate that their presence, at least to some extent, is a result of extensive knapping. Technologically, they seem to result from different stages of the reduction sequence, but due to their variability in size and extensive flaking (expressed by the number of scars) it is difficult to draw particular conclusions about the process from which they originate.

Fig. 7.119 Basalt core (#7531 Layer V-6)

7.4.7 Basalt Cores

The GBY basalt cores are divided into two groups: giant cores designed for the production of large flakes for biface modification and other cores intended for the production of small flakes. While the first group is a major characteristic of the Acheulian of GBY, the smaller basalt cores appear in low frequencies and are not a significant component of the lithic assemblages.

7.4.7.1 Cores

The occurrence of basalt cores is minimal and in most cases does not exceed five cores per archaeological horizon (Table 7.80). The total number of cores recorded for all of the GBY archaeological horizons is 31, and they are lumped together here in order to provide a sufficient sample for analyses and description. They are often devoid of cortex, and when cortical coverage occurs it is minimal and never exceeds 50% of the core's surface (Fig. 7.119). These cores are mostly informal in shape and seem to represent ad hoc knapping. This clearly contrasts with the systematic knapping of giant cores (discussed below).

Only three cores, a single Levallois core for flakes and two globular cores, can be classified within formal organized core types (Table 7.80). Their meager occurrence illustrates the fact that at GBY the production of flakes from basalt cores (excluding giant cores) does not represent a distinct, well-formulated reduction sequence. Additional support for this observation comes from the near-absence of core management pieces (Section 7.3.2.6) and the high variability observed in the different size measures of these cores (Table 7.81).

An interesting feature is the presence of basalt cores on flakes (Figs. 7.120–7.121). Despite its meager representation in the basalt assemblage, this type is abundant in the flint assemblage (Section 6.4)

7.4.7.2. Giant Cores

The giant cores of GBY are basalt artifacts of extremely large dimensions. They exhibit a clear sequence of scar removals that defines them as cores, intentionally shaped to produce large

Table 7.80 Typological counts and frequencies of basalt cores

Typology	V-6	II-5	II-6/L1	II-6/L2	II-6/L3	II-6/L4	II-6/L5	II-6/L6	II-6/L7	Trench II green
Core on flake	-	1 100.00	2 25.00	1 16.67	1 33.33	1 20.00	-	-	-	-
Levallois core for flakes	-	-	1 12.50	-	-	-	-	-	-	-
Globular core	-	-	-	-	-	-	2 100.00	-	-	-
Varia core	1 50.00	-	4 50.00	-	2 66.67	1 20.00	-	1 50.00	1 100.00	-
Formless core	1 50.00	-	1 12.50	5 83.33	-	3 60.00	-	1 50.00	-	1 100.00

Table 7.81 Descriptive statistics of basalt cores (N=31)

	Maximum length	Length	Width	Thickness	Circumference	Number of platforms
					(N=28)	(N=23)
Maximum	160	290	128	107	420	4
Median	79	78	54	36	233	2
Minimum	27	23	23	16	83	1
Mean	83.94	83.29	64.74	41.29	241.46	2.17
Std dev	32.48	48.89	29.49	20.92	90.89	0.89
Std err	5.83	8.78	5.30	3.76	17.18	0.18
	Number of working edges	Working edge length	Number of scars	Scar length	Scar width	
	(N=24)	(N=12)		(N=8)	(N=8)	
Maximum	3	420	14	94	101	
Median	1	213.5	4	41	49.5	
Minimum	1	61	1	18	17	
Mean	1.38	227.42	5.26	51.63	57.00	
Std dev	0.71	140.15	2.94	30.91	31.86	
Std err	0.15	40.46	0.53	10.93	11.26	

Fig. 7.120 Basalt core on flake (#2119 Layer II-5)

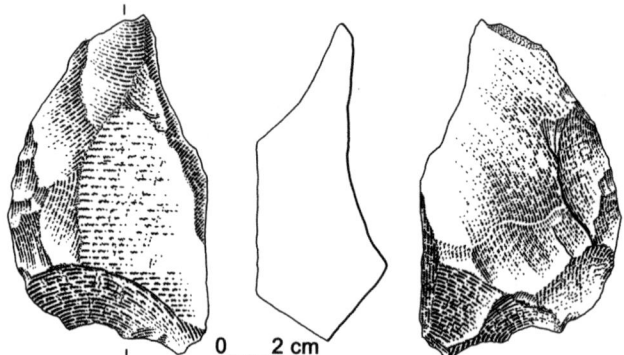

Fig. 7.121 Basalt core on flake (#268 Layer II-6 Level 4)

flakes that were suitable for the production of bifaces (e.g. handaxes and cleavers). At GBY, the category of giant cores includes all cores of sufficient dimensions for the production of large flakes (>10 cm following the criterion suggested by Kleindienst 1961).

The giant cores are all modified on basalt slabs, quarried from unknown quarry sites. These quarries were likely situated on basalt flows in the vicinity of the site, as the size and weight of these cores suggest that they were not transported over very long distances.

All slabs were selected from unweathered, compact, non-vesicular, and fine-grained basalts of the highest quality, a fact suggestive of profound knowledge of the characteristics of the basaltic bedrock and its fracture mechanics. Such slabs, a well-known volcanic phenomenon, occur in the middle part of the flow, where the basalt is most dense and is characterized by horizontal fissures (Green and Short 1971). The GBY knappers were also familiar with the bedding plane of these basalt slabs and took it into account while exploiting the giant cores (see below).

The quarrying of these slabs clearly involved different methods that we cannot reconstruct, but one method has apparently left its traces on the slabs in the form of flake scars or "notches" (Fig. 7.122). These scars are relatively small and isolated, without additional scars near them, and show a morphology that is not typical of intentional scar removals; they are ellipsoid, flat, and apparently detached from an unsuitable striking surface and angle. These scars may have resulted from a quarrying procedure that involved the use of levers or wedges for the detachment of the slabs from the outcrop.

Morphology of Giant Cores

The morphology of the slabs used for the knapping of giant cores at GBY is quite uniform (Fig. 7.123). Their cross-sections commonly show a flat base and frequently a double-sloping top. The sloping face (upper surface or surfaces) of the slab joins the flat base, forming an acute angle (Fig. 7. 124). The presence of

7.4 Typology and Technology of Basalt Cores and Core Tools

Fig. 7.122 Basalt giant cores showing notches, possibly the result of quarrying; (#5447 Layer II-6 Level 1; #5896 Layer II-6 Level 2; two faces of #7696 Layer II-6 Level 7)

this angle is of the utmost importance for the flaking process of the slabs. This homogeneity, expressed in shape and morphological characteristics, is shared by almost all of the GBY giant cores and can be summarized as follows: 1) all giant cores share a natural flat base; 2) in most of the giant cores the upper surface is also a natural surface, a remnant of the upper surface of the slab; even in heavily knapped cores, some of this original surface is present, enabling definition of the core blank as a slab; 3) in some cases the upper surface is naturally flat and is parallel to the base; 4)

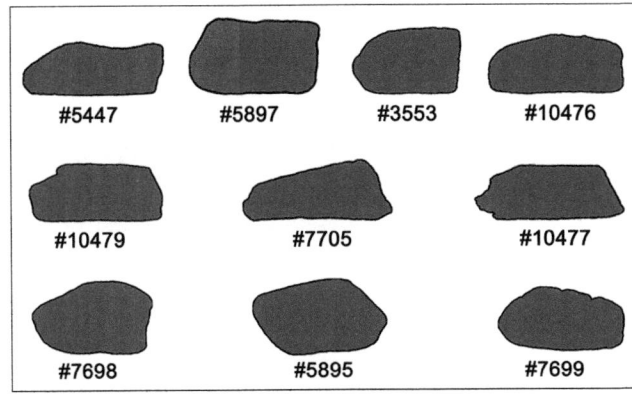

Fig. 7.123 Cross-sections of basalt giant cores from Layer II-6 (obtained by 3-D scanning) indicative of their slab origin

in other cases the upper natural surface slants, creating an acute angle with the base; in others, a ridge is observed from which the upper surface slants toward both lateral edges of the slab.

The combination of the slanting surface(s) and the acute angle that is formed with the base provided a suitable striking surface and angle and allowed easy knapping without previous preparation. However, this strategy is only one manner of exploiting the slabs, and the assemblages of both cores and flakes provide evidence that other, much more difficult, strategies were applied (Section 7.6).

Despite the large size variation of the GBY giant cores, they are all far larger than the cores made on flint and limestone that occur in the same archaeological horizons. Consequently, they are all included in the general category of giant cores.

At GBY, dimension-based classification had resulted in a previous subdivision of the giant core assemblage into different size categories (giant, very large, and large cores; Madsen and Goren-Inbar 2004). However, such a classification does not take into consideration the life-history of the core and the extent of mass reduction, or the amount of flaking that is "readable" from the scars and their visible orientation. Thus, within the category of giant cores we now define three sub-groups that represent different stages of reduction (Table 7.82). Each of these groups is analyzed (Table 7.83) and discussed independently below.

a) Fragmented slabs: with the exception of a few large slabs, most of the artifacts in the giant cores group are in fact fragmented slabs whose original size is unknown. Despite this lack of information, the geometry and morphology of the available slab fragments is apparent and the form of the original slabs could be partially reconstructed (Fig. 7.124). These fragmented slabs are the result of a procedure of fragmentation, the "opening" (initiation) and shaping of large slabs. They are thinner than the giant cores and the exhausted cores (Table 7.83), but they originate from the same exploitation mode. Although these fragmented slabs are not cores, they are included here as they provide additional information on the selection strategies for giant cores.

b) Giant cores: the giant cores of GBY vary in morphology,

Fig. 7.124 Cross-sections of basalt giant cores from Layer II-6 and Layer II-2/3 and their probable location within the basalt slab morphology (#5446 Level 1, #7701 Level 6, #3561 Layer II-2/3, #14213 Level 2, #10479 Level 1, #10478 Level 1)

Table 7.82 Counts of groups of basalt giant cores

Layer	Slab fragment	Giant core	Exhausted giant core	Total
V-6	1	-	-	1
II-2/3	-	1	-	1
II-6/L1	1	3	-	4
II-6/L2	4	2	2	8
II-6/L3	3	1	-	4
II-6/L4	2	1	1	4
II-6/L5	1	1	-	2
II-6/L6	2	1	-	3
II-6/L7	2	4	1	7
Total	16	14	4	34

size (Table 7.83), and reduction method, but nonetheless are pre-planned and prepared cores. Generally, their reduction methods were based on striking a flat or facetted platform, or a predominantly plain platform area placed next to, or among, facets. By using the scars of previous flakes, the individual flake as well as a series of well-spaced large and medium-sized flakes could be controlled. The use of serially overlapping flakes from cores increased the number of regular flakes with lateral or distal ready-for-use edges. These flakes are the target products of giant core reduction and could be used for the production of a variety of large basalt tools such as massive scrapers (Section 7.2), handaxes, and cleavers (Section 7.5).

c) Exhausted cores: these are classified as exhausted on the basis of the relatively large number of scars observed on their surfaces (Table 7.83), and particularly the fact that they are no longer suitable in form and size for the extraction of additional large flakes (Figs. 7.125–7.127). The basalt exhausted cores from GBY are mainly in the large and very large size classes, although a few were continuously flaked and not discarded until they reached medium size. These cores can be interpreted as having been reduced for as long as an individual core could yield flakes large enough for making bifaces. Thus, they were abandoned at the stage when production of further blanks for bifaces or other flake tools was no longer possible. Obviously, the smaller exhausted cores exhibit only residual features of the original slabs, usually identified by their very flat base. Other cores have a typical triangular cross-section, with one edge forming a right angle with the straight base and a slightly larger angle with the upper surface of the core/slab.

Methods of Giant Core Reduction

The reconstruction of giant core reduction methods is somewhat hypothetical, particularly since these cores are the outcome of a long reduction process and are the remnants of an exhaustive flaking sequence of large, uniform flakes. Thus, in order to reconstruct the different methods used in the reduction of giant cores at GBY, we draw on three primary lines of evidence: a)

7.4 Typology and Technology of Basalt Cores and Core Tools

Table 7.83 Descriptive statistics of groups of basalt giant cores (size in mm)

	Slab fragment	Giant core	Exhausted giant core
N	16	14	4
Length	(N=11)		(N=3)
Maximum	328	370	190
Median	202	268	176
Minimum	116	149	125
Mean	201.64	249.64	163.67
Std dev	62.35	63.17	34.21
Std err	18.80	16.88	19.75
Width	(N=11)		(N=3)
Maximum	286	454	168
Median	171	230.5	142
Minimum	135	172	132
Mean	183.91	278.50	147.33
Std dev	49.01	100.64	18.58
Std err	14.78	26.90	10.73
Thickness	(N=11)		(N=3)
Maximum	147	326	148
Median	108	152	111
Minimum	73	98	108
Mean	108.36	163.14	122.33
Std dev	17.90	57.73	22.28
Std err	5.40	15.43	12.86
Weight	(N=13)	(N=12)	(N=1)
Maximum	8,753	22,500	3,621
Median	4,023	13,139.5	3,621
Minimum	2,311	3,669	3,621
Mean	4,494.69	1,2997.83	3,621.00
Std dev	2,069.04	6,696.74	-
Std err	573.85	1,933.18	-
N platforms	(N=5)	(N=6)	(N=1)
Maximum	4	5	3
Median	2	1	3
Minimum	1	1	3
Mean	2.40	2.00	3.00
Std dev	1.14	1.67	-
Std err	0.51	0.68	-
N working edges	(N=5)	(N=6)	(N=2)
Maximum	3	5	2
Median	2	1.5	1.5
Minimum	1	1	1
Mean	2.00	2.00	1.50
Std dev	0.71	1.55	0.71
Std err	0.32	0.63	0.50
Number of scars	(N=13)	(N=13)	
Maximum	5	6	6
Median	2	1	5.5
Minimum	1	1	2
Mean	2.54	2.15	4.75
Std dev	1.61	1.57	1.89
Std err	0.45	0.44	0.95

Fig. 7.125 Basalt exhausted giant core (#3562 Layer II-6 Level 4)

Fig. 7.127 Basalt exhausted giant core (#14213 Layer II-6 Level 2)

Fig. 7.126 Basalt exhausted giant core (#9580 Layer II-6 Level 7)

Fig. 7.128 Basalt giant core of the slab slicing method (#10479 Layer II-6 Level 1)

study of the giant cores; b) study of the giant cores' products and flakes (Section 7.3.1.1) and the handaxes and cleavers (Section 7.5) produced on these large flakes; and c) experimental replication of the GBY artifacts (Madsen and Goren-Inbar 2004). Here we describe the core methods observed and illustrate these with actual cores, often only a single one. Within the general framework of the giant core reduction scheme, five different core methods are identified at the site.

Slab Slicing

This core method was defined at the Acheulian site of Hunsgi V (Paddayya 1982), where it is the dominant core method (Sharon 2007). At GBY, there are only a few examples of such slice flakes (Section 7.3.3.2). However, one of the cores (Fig. 7.128) was clearly sliced by a similar method. Although this method was not frequently used at GBY, the

7.4 Typology and Technology of Basalt Cores and Core Tools

volumetric principle of using the flat surface of a slab as a striking platform for flakes that "slice" the entire thickness of the slab was known and practiced by the GBY knappers (Fig. 7.129).

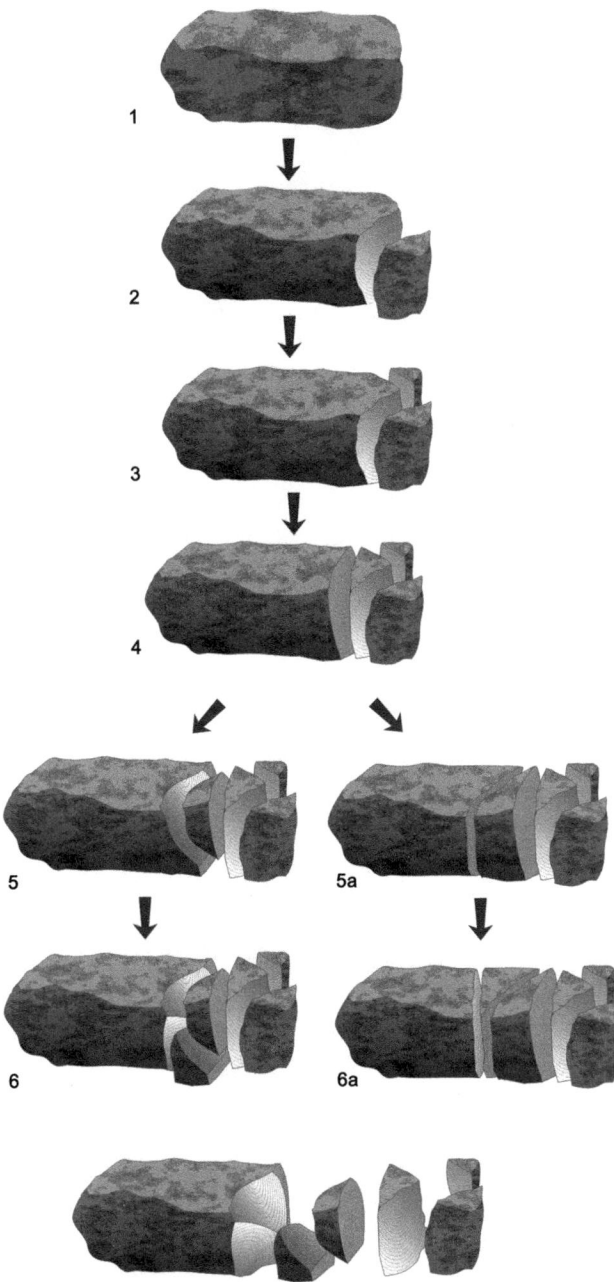

Fig. 7.130 Basalt giant core of the bifacial method (#5447 Layer II-6 Level 1)

Bifacial Cores

This method is once again represented at GBY by a single core (Fig. 7.130). The flakes were removed from both faces of the same striking platform by alternate removals in which each of the scars was used as a striking platform for the removal of the next flake. This core furnishes an indication that the knappers of GBY were familiar with the bifacial core method and applied it to the production of large flakes (Fig. 7.131).

Kombewa

A single giant core represents this method at GBY (Fig. 7.132). It is a very large amorphous flake from the ventral face of which a large flake was detached. This core is heavily weathered and its morphology is very difficult to "read." However, the use of the Kombewa core method for the production of bifacial tools at GBY is evident from the presence of Kombewa flakes among the bifacial tools and large flakes at the site (Fig. 7.133) (Sections 7.3.2.2 and 7.5).

Levallois

This method is represented by a single core (Fig. 7.134), found under the butchered elephant skull in Layer II-6 Level 1 (Goren-Inbar et al. 1994; Madsen and Goren-Inbar 2004). This is the most advanced core design at GBY and falls well within the definition of the recurrent Levallois core method (Boëda 1995). The flakes detached from this core were large and at least one of the scars points to the removal of a large side-struck flake that matches handaxe blanks from the site. All of the other GBY giant cores have lower scar counts.

Fig. 7.129 Hypothetical reconstruction of the slab slicing method from top to bottom according to the reduction sequence: stage 1) basalt slab; stages 2 and 3) removal of two opening shoulder flakes; stage 4) removal of a wedge flake creating a suitable debitage surface. Following this stage, two strategies are possible. The first (stages 5a and 6a) is to detach flakes that remove the entire thickness of the slab, resulting in flakes that have an inherent cleaver shape. The second strategy (stages 5 and 6) is to detach flakes that remove only half or more of the slab's thickness, creating a backed knife flake (after Goren-Inbar et al. 2011)

Ad hoc or Indeterminate Methods

A few of the GBY giant cores (Figs. 7.122, 7.135) are large slabs from which one or two flakes were removed without

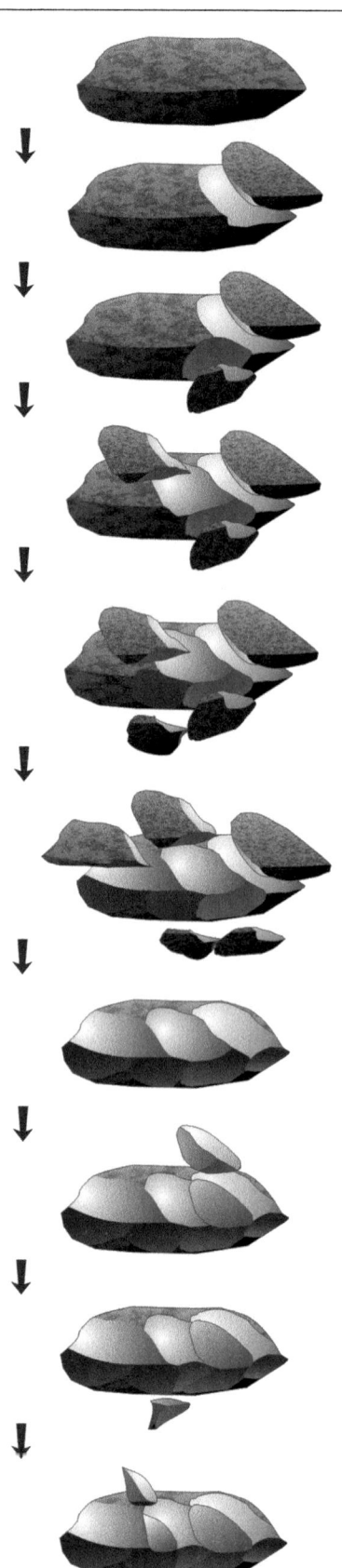

Fig. 7.132 Basalt giant core of the Kombewa method (#7704 Layer II-6 Level 7)

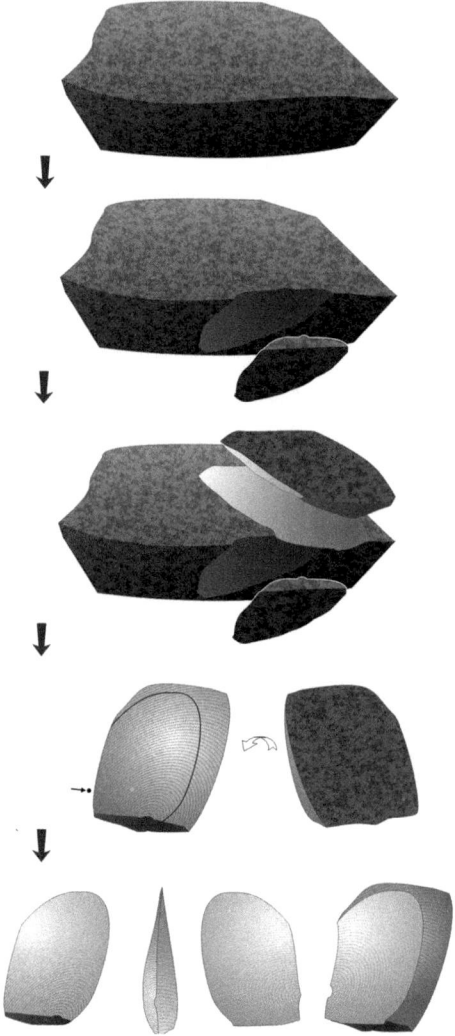

Fig. 7.131 Hypothetical reconstruction of basalt giant core reduction sequence of the bifacial method

Fig. 7.133 Hypothetical reconstruction of basalt giant core reduction sequence of the Kombewa method (after Goren-Inbar et al. 2011)

Fig. 7.134 Basalt giant core of the Levallois method (#10478 Layer II-6 Level 1)

Fig. 7.135 Basalt slab fragment (5897# Layer II-6 Level 2)

the basalt slabs is the key to efficient knapping. The GBY knappers employed the presence of a naturally sharp angle, which allowed immediate knapping without preparation of the core. The preparation of the giant core for the production of predetermined (target) flakes sometimes necessitated further steps in addition to the exploitation of the natural geometric properties of the slab. These are visible archaeologically through shoulder/corner flakes, exceptionally thick flakes that were extracted from the edges of the fragmented slab (Section 7.3.3.1).

7.5 Bifacial Tools

At GBY, as at many other Acheulian sites, bifacial tools were recognized as a component of the lithic assemblage from the very beginning of archaeological investigation. The uniqueness of the GBY bifaces, as expressed by their raw material, typology, and technology, was first recognized in the pioneering account of Stekelis (1960). It was Stekelis who first identified the presence of basalt cleavers and the use of large flakes as a dominant blank type in the production of bifaces, the first evidence of such an industry in the Levant. He also noted that the European Acheulian scheme could not serve as a basis for comparison in the study of this assemblage, an assessment that was later discussed in detail by Gilead (1970 and in Chapter 2). In his comprehensive account of the Lower Paleolithic industries of the Near East, Gilead placed the GBY bifaces as a stand-alone member of the Levantine "Middle Acheulian" (Gilead 1968, 1970).

Our excavations at GBY uncovered a large assemblage of bifaces, characterized by the following features that form the basis for their presentation and discussion below. These features place GBY as a unique assemblage in the Levant and highlight its significance to the study of the Old World Acheulian.

a) Bifaces occur in high frequencies in comparison with the limited extent of the excavation. Layer II-6 yielded 535 bifaces from a relatively small excavation volume of ca. 21 m^3 (Chapter 5). In Level 4 of Layer II-6, the density of bifacial tools is as high as 13.70 bifacial tools per m^2. Acheulian sites with very large bifacial assemblages are known mainly from Africa (Howell and Clark 1961; Isaac 1969, 1977; Gowlett 1991; Beaumont and Morris 1990; Leakey and Roe 1994; Clark 2001; Piperno 2001; Sharon 2007). In the Levant, rich Acheulian biface assemblages are known from younger sites such as Ma'ayan Barukh, where thousands of bifaces were collected from the surface (Stekelis and Gilead 1967), and Tabun Cave, where Layer F also yielded thousands of handaxes

any preparation. In other cases the method used cannot be reconstructed due to the fragmentary nature of the core and the weathering of the basalt. These cores are defined here as indeterminate.

To sum up, the characteristics of the basalt giant cores and their waste, present in many of GBY's archaeological horizons, suggest that the slabs were extracted from the dense middle section of basalt flows by unknown extraction techniques, one of which has left its "notch" markings on some of the giant cores. Following the extraction of the basalt slab from the basalt flow, it was deliberately fragmented into several smaller, workable fragments. The fragments were then worked into giant cores by a variety of core reduction methods. The control and exploitation of the particular geometry of

(Rollefson 1978). Large collections of many hundreds of handaxes are known from other sites such as el-Jafar in Jordan (Rollefson et al. 2005).

b) Cleavers are a common tool among the GBY bifaces. At no other Levantine site do the technology and frequencies of cleavers resemble those of GBY.

c) The primary raw material used for the production of the GBY bifacial tools was basalt. The use of basalt as the dominant raw material for biface production is not recorded at other Levantine Acheulian sites apart from 'Ubeidiya, where basalt was used for the production of handaxes, trihedrals, quadrihedrals, and picks (Bar-Yosef and Goren-Inbar 1993).

d) The primary technology practiced at GBY was the use of large flakes as blanks for biface production. The use of this technology is expressed in many technological features that will be discussed in detail below.

7.5.1 Methodology of Biface Analyses

While most researchers seem to agree on the definition of handaxes, cleavers are the subject of much debate. In this study we follow Kleindienst (1962), Bordes (1961), and Roe (1994) for the definition and analyses of bifaces. The definition of cleavers is somewhat more challenging, as some researchers see ovate handaxes as cleavers (for examples and discussion see Rollefson et al. 2005; White 2006). For the identification and analysis of cleavers we applied the definitions of Roe (1994) and followed Tixier (1957) for two additional criteria. The first is that cleavers are exclusively flake tools; hence bifaces with a transverse cutting edge that were made on non-flake blanks (such as cobbles or flat slabs) are not cleavers. The second is that a cleaver's cutting edge was never shaped by secondary retouch; the edge of a GBY cleaver is a portion of a scar that existed on the original large core from which the blank of the cleaver was extracted. Alternatively it was modified after the production of the flake (Section 7.5.6.4).

The methodology of the analyses applied to the study of the GBY bifacial tools was developed in order to integrate metrical measurements with technological and morphological observations. These were incorporated into a techno-typological attribute analysis. Generally, the metrical attributes are based on the systems developed by Bordes (1961) and Roe (1964, 1968; Leakey and Roe 1994; Roe in Clark 2001). Technological attributes were adopted from various sources, among which the works of M. Leakey (1971; Leakey and Roe 1994) and Isaac (1968, 1977) are notable. They were then further modified in order to suit them to the GBY assemblage (Appendices 5–6). In addition to the GBY attribute analysis,

other approaches and methods have been applied to the study of the bifacial tools from the site. A cleaver assemblage was studied by Goren-Inbar et al. (1991a) and Zohar (1993), and Saragusti studied the handaxes from the site as part of her comparative study of the morphological development of Levantine handaxes (Saragusti et al. 1998; Saragusti and Goren-Inbar 2001; Saragusti 2002; Saragusti et al. 2005).

Many of our technological observations and insights are derived from the experimental knapping project. The controlled knapping experiments, carried out by B. Madsen, focused on the production of large flakes from basalt giant cores and the technology of shaping bifacial tools from these large flakes, as well as other types of blanks. These experiments made a remarkable contribution to our ability to reconstruct the reduction sequnce of the GBY bifaces (Madsen and Goren-Inbar 2004).

7.5.2 Raw Materials and Taphonomic Characteristics

7.5.2.1 Raw Materials

The rock types exploited for the production of bifacial tools at GBY were basalt, flint, and limestone. Chapters 6 and 8 discuss the flint and limestone assemblages of GBY with the exception of the few flint and limestone bifaces, which are integrated in the current section. Here we focus on the frequencies of different raw materials used and the impact of raw material type on the preservation, size, technology, and morphology of the bifacial tools. Table 7.84 presents the counts and frequencies of the raw materials used in the production of handaxes and cleavers in the different archaeological assemblages.

As seen in Table 7.84, cleavers were made exclusively on basalt, the only exception being a single limestone cleaver (Fig. 7.136). Basalt is the dominant raw material used for the production of handaxes, although both flint and, to a lesser extent, limestone, were used as well (Figs. 7.137–7.140).

The frequencies of flint and limestone handaxes are very low, with a maximal occurrence of four flint handaxes each in Levels 1, 2, and 5 of Layer II-6. Hence, the raw material exploitation pattern for bifaces at GBY is primarily one of basalt, while flint and limestone occur minimally. This cannot be explained by raw material availability alone. While we did not find cleavers made on flint in the study area, we did find three of these in the BYF on the eastern slopes of the Crusader fortress (Chapters 1 and Section 3.3.2). Nevertheless, the lack of flint cleavers is most probably due to lack of appropriate size and/or availability of the raw material.

7.5 Bifacial Tools

Table 7.84 Counts and frequencies (%) of handaxes and cleavers by raw material

	Handaxes				Cleavers		
	Flint	Basalt	Limestone	Total	Basalt	Limestone	Total
N	28	414	9	451	185	1	186
V-2	-	3 100.00	-	3 100.00	2 100.00	-	2 100.00
V-3	-	1 100.00	-	1 100.00	-	-	-
V-3/4	-	4 100.00	-	4 100.00	8 100.00	-	8 100.00
V-4/5	-	2 100.00	-	2 100.00	-	-	-
V-5	-	-	-	-	2 100.00	-	2 100.00
V-6	1 100.00	-	-	1 100.00	6 100.00	-	6 100.00
JB	2 100.00	-	-	2 100.00	8 100.00	-	8 100.00
I-4	-	1 100.00	-	1 100.00	-	-	-
II-2	1 100.00	-	-	1 100.00	-	-	-
II-2/3	-	7 100.00	-	7 100.00	-	-	-
II-5	-	4 100.00	-	4 100.00	2 100.00	-	2 100.00
II-5/6	-	1 100.00	-	1 100.00	1 100.00	-	1 100.00
II-6/L1	4 6.35	54 85.72	5 7.93	63 100.00	17 100.00	-	17 100.00
II-6/L2	4 17.39	19 82.61	-	23 100.00	10 100.00	-	10 100.00
II-6/L3	1 7.70	12 92.30	-	13 100.00	12 100.00	-	12 100.00
II-6/L4	3 1.79	165 98.21	-	168 100.00	60 100.00	-	60 100.00
II-6/L4b	1 1.64	58 95.08	2 3.28	61 100.00	18 100.00	-	18 100.00
II-6/L5	4 26.67	11 73.33	-	15 100.00	8 100.00	-	8 100.00
II-6/L6	-	19 100.00	-	19 100.00	11 91.67	1 8.33	12 100.00
II-6/L7	3 12.50	20 83.33	1 4.17	24 100.00	12 100.00	-	12 100.00
VI-14	1 10.00	9 90.00	-	10 100.00	-	-	-
Trench II green	3 10.72	24 85.71	1 3.57	28 100.00	8 100.00	-	8 100.00
Total	28 6.20	414 91.80	9 2.00	451 100.00	185 99.46	1 0.54	186 100.00

As for the flint handaxes of GBY, some of them are of the highest craftsmanship. The GBY knappers clearly possessed the knowhow of procuring and using flint and had mastered the art of producing flint handaxes from generally very small nodules (Fig. 7.141). Lack of technological knowledge, therefore, is not the cause of the low frequency of flint handaxes in the GBY assemblage.

7.5.2.2 Preservation and Patination

As discussed above (Chapter 3), some of the basalt artifacts at GBY were subject to *in situ* chemical weathering that resulted in their exfoliation and finally deterioration into clay. The following presentation will address issues of the varying degrees of preservation among the different raw materials

Fig. 7.136 Limestone cleaver (#7717 Layer II-6 Level 6)

Fig. 7.139 Limestone handaxe (#1124 Layer II-6 Level 1)

Fig. 7.137 Limestone handaxe (#469 Layer II-6 Level 1)

Fig. 7.140 Flint handaxe (#8169 Layer II-6 Level 2)

Fig. 7.138 Limestone handaxe (#6609 Layer II-6 Level 1)

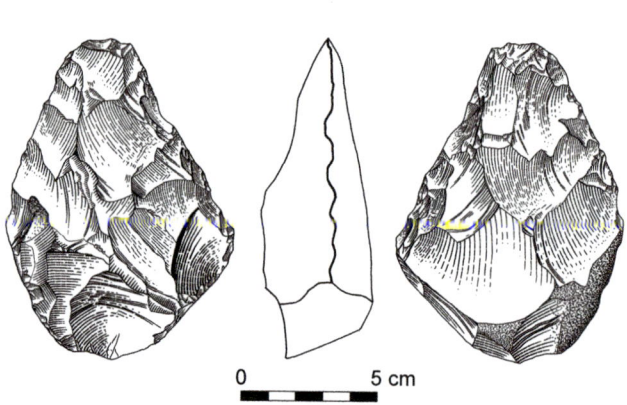

Fig. 7.141 Flint handaxe (#5785 Layer II-6 Level 4b)

7.5 Bifacial Tools

used for bifacial tool production, as well as specific aspects of the preservation of basalt bifaces and their significance.

While some of the GBY basalt handaxes and cleavers were found in pristine and unweathered condition and remain in that state today, others (alas, most) have undergone an exfoliation process that began when they were first exposed to atmospheric conditions during excavation. This process continued, with conservation treatment only slowing the tempo of decay of the basalt. The result is that many features observed when the tools were unearthed, and even later when they were analyzed in the laboratory, cannot now be identified (Chapter 3.5). Obviously, the observations presented in this volume refer to the state of preservation at the time of analysis. Table 7.85 presents the state of preservation of basalt handaxes and cleavers, whereas only assemblages with a minimum of 10 bifaces are included. The data indicate that the preservation is similar in most of the levels of Layer II-6. Well-preserved, fresh tools are quite rare, possibly due to some mild postdepositional weathering in the waterlogged environment. The common preservation mode is slightly abraded in both typological classes, but in the older part of Layer II-6 handaxes are somewhat more abraded and the frequencies of exfoliated handaxes can reach over 40%.

Among flint handaxes the most common preservation mode is slightly abraded as well (62.50%; N=15), but the fresh tools display higher values than those observed for basalt (20.83%; N=5). Among limestone handaxes there are no fresh tools and the common preservation mode is slightly abraded (66.67%; N=6). Clearly, different rock types show different degrees of weathering despite their prolonged burial in the same sediments, and basalt in particular is more prone to postdepositional weathering than other raw materials.

The issue of patination and the difficulties presented by this attribute, particularly with regard to basalt artifacts, have been discussed earlier in this volume (Chapter 3). We will mention here only that the bulk of bifaces are patinated and only a few were registered as unpatinated (Table 7.3).

Among flint handaxes the majority are patinated (75.00%; N=18), followed by double patinated and then unpatinated. The small sample size of flint handaxes does not enable further conclusions. In contrast, all the limestone handaxes (N=9) are patinated.

7.5.2.3 Breakage

Tables 7.86 and 7.87 present the breakage patterns of basalt handaxes and cleavers (respectively).

Most of the basalt bifaces are complete in most of the levels, and they always form at least half of the cases. The only deviations from this pattern are Levels 4b, 5, and 6, where exfoliated handaxes are more frequent. Worth noting also are the

Table 7.85 Counts and frequencies (%) of preservation of basalt handaxes and cleavers of Layer II-6

	Handaxes						Cleavers					
	Fresh	Slightly abraded	Abraded	Heavily abraded	Exfoliated	Total	Fresh	Slightly abraded	Abraded	Heavily abraded	Exfoliated	Total
N	1	137	96	63	93	390	5	69	41	11	30	156
II-6 L1	-	22	19	9	4	54	1	7	7	1	1	17
		40.74	35.19	16.67	7.41	100.00	5.88	41.18	41.18	5.88	5.88	100.00
II-6 L2	-	7	8	4	-	19	-	5	4	1	-	10
		36.84	42.11	21.05		100.00		50.00	40.00	10.00		100.00
II-6 L3	-	6	2	2	1	11	-	8	3	-	1	12
		54.55	18.18	18.18	9.09	100.00		66.67	25.00		8.33	100.00
II-6 L4	-	70	35	26	34	165	-	34	12	5	9	60
		42.42	21.21	15.76	20.61	100.00		56.67	20.00	8.33	15.00	100.00
II-6 L4b	-	25	13	5	15	58	1	8	4	2	3	18
		43.10	22.41	8.62	25.86	100.00	5.56	44.44	22.22	11.11	16.67	100.00
II-6 L5	-	-	2	2	7	11	3	-	3	1	1	8
			18.18	18.18	63.64	100.00	37.50		37.50	12.50	12.50	100.00
II-6 L6	-	3	7	1	8	19	-	3	1	-	7	11
		15.79	36.84	5.26	42.11	100.00		27.27	9.09		63.64	100.00
II-6 L7	-	1	4	6	9	20	-	3	4	1	4	12
		5.00	20.00	30.00	45.00	100.00		25.00	33.33	8.33	33.33	100.00
VI-14	-	1	1	2	5	9	-	-	-	-	-	-
		11.11	11.11	22.22	55.56	100.00						
Trench II green	1	2	5	6	10	24	-	1	3	-	4	8
	4.17	8.33	20.83	25.00	41.67	100.00		12.50	37.50		50.00	100.00
Total	1	137	96	63	93	390	5	69	41	11	30	156
	0.26	35.13	24.62	16.15	23.85	100.00	3.21	44.23	26.28	7.05	19.23	100.00

Table 7.86 Counts and frequencies (%) of breakage patterns of basalt handaxes of Layer II-6

	Complete	Slightly broken tip	Distal	Lateral	Proximal	Lateral & distal	Proximal & distal	Fragment	Exfoliated	Total
II-6/L1	39	4	4	1	4	1	-	1	-	54
	72.22	7.41	7.41	1.85	7.41	1.85		1.85		100.00
II-6/L2	14	4	1	-	-	-	-	-	-	19
	73.68	21.05	5.26							100.00
II-6/L3	6	2	-	1	2	-	-	-	1	12
	50.00	16.67		8.33	16.67				8.33	100.00
II-6/L4	86	19	34	3	6	2	2	2	11	165
	52.12	11.52	20.61	1.82	3.64	1.21	1.21	1.21	6.67	100.00
II-6/L4b	17	15	6	2	5	1	1	-	11	58
	29.31	25.86	10.34	3.45	8.62	1.72	1.72		18.97	100.00
II-6/L5	2	1	-	-	-	-	1	-	6	10
	20.00	10.00					10.00		60.00	100.00
II-6/L6	3	4	2	-	2	-	-	-	8	19
	15.79	21.05	10.53		10.53				42.11	100.00
II-6/L7	12	1	2	-	1	-	-	-	4	20
	60.00	5.00	10.00		5.00				20.00	100.00
VI-14	2	-	2	2	1	-	2	-	-	9
	22.22		22.22	22.22	11.11		22.22			100.00
Trench II green	10	1	7	1	-	-	2	-	3	24
	41.67	4.17	29.17	4.17			8.33		12.50	100.00
Total	191	51	58	10	21	4	8	3	44	390
	48.97	13.08	14.87	2.56	5.38	1.03	2.05	0.77	11.28	100.00

Table 7.87 Counts and frequencies (%) of breakage patterns of basalt cleavers of Layer II-6

	Complete	Slightly broken tip	Distal	Lateral	Proximal	Lateral & distal	Proximal & distal	Exfoliated	Total
II-6/L1	12	3	1	-	-	-	-	1	17
	70.59	17.65	5.88					5.88	100.00
II-6/L2	8	1	-	-	-	1	-	-	10
	80.00	10.00				10.00			100.00
II-6/L3	9	2	1	-	-	-	-	-	12
	75.00	16.67	8.33						100.00
II-6/L4	44	4	3	1	2	-	1	5	60
	73.33	6.67	5.00	1.67	3.33		1.67	8.33	100.00
II-6/L4b	10	4	-	1	1	-	-	2	18
	55.56	22.22		5.56	5.56			11.11	100.00
II-6/L5	4	4	-	-	-	-	-	-	8
	50.00	50.00							100.00
II-6/L6	2	1	2	-	-	1	-	5	11
	18.18	9.09	18.18			9.09		45.45	100.00
II-6/L7	5	2	-	-	2	-	-	3	12
	41.67	16.67			16.67			25.00	100.00
Trench II green	4	-	2	-	1	1	-	-	8
	50.00		25.00		12.50	12.50			100.00
Total	98	21	9	2	6	3	1	16	156
	62.82	13.46	5.77	1.28	3.85	1.92	0.64	10.26	100.00

cases of Level 4, Layer VI-14, and Trench II green, where some 20% of the tools are broken distally. Distal breaks result either from working accidents or from the deterioration of the basalt. Basalt is the only type of raw material that exfoliates and, since exfoliation attacks the extremities, it causes greater damage to the thinnest area; this explains the slight damage to the bifaces' tips.

While the picture emerging from the breakage analysis of cleavers is generally similar to that of the handaxes, there are some differences (Table 7.87). There are more complete tools than among handaxes and slightly broken tips are the second most frequent category. The breakage pattern is somewhat different in the lower levels of Layer II-6, where exfoliation, and hence breakage, of the cleavers is more common.

There are marked differences in the breakage pattern of the

7.5 Bifacial Tools

Table 7.88 Counts and frequencies (%) of breakage patterns of flint and limestone handaxes of Layer II-6

	Flint		Limestone	
	N	%	N	%
Complete	14	58.33	7	77.78
Distal	-	-	2	22.22
Proximal	7	29.17	-	-
Fragment	3	12.50	-	-
Total	24	100.0	9	100.00

three raw materials. While multiple types of breaks occur, as shown above, in the basalt, the flint and limestone handaxes provide a much simpler pattern (Table 7.88). The breakage patterns of flint and limestone bifaces may result from the different properties of their rock type or from their very small sample sizes.

From the breakage distributions, it is evident that slightly broken tips (distal ends) occur frequently. Taking this into consideration, we present here a comparison between the length of complete bifaces and that of bifaces with a slightly broken tip. The analysis included only basalt bifaces and was carried out separately for handaxes and for cleavers. We found that the difference among basalt handaxes from Layer II-6 (and Trench II green) amounts to only 3 mm and that there is no significant difference between the length of complete basalt handaxes and those with slightly damaged tips. Similarly, the analysis of cleavers showed that there is no significant difference between the two groups (they differ by 8 mm). However, when the maximal length was examined it showed a significant difference: cleavers with slightly damaged tips are longer (difference=11.2 mm; t-ratio=2.09; df=117; $p(tt)$=0.0384).

As no significant difference was found in the length of complete and slightly broken bifaces, the latter are included in the following analyses.

7.5.3 Technological Characteristics

The analysis of bifaces involved a variety of technological attributes. These included the different size measurements and particularly the type and characteristics (e.g. direction of blow, striking platform) of the blank selected for the production of the biface. Although the technological traits of the bifaces were observed primarily on Face 1, when data are available for Face 2 these are supplied as well (Chapter 4).

7.5.3.1 Size

The size of basalt handaxes and cleavers is presented in Tables 7.89–7.90, which include tools with complete and those with slightly broken tips but exclude all other types of broken bifaces. The variations in size are reflected in the length of basalt handaxes, among which the smallest is 69 mm long and the longest 193 mm long (Table 7.89). Similar ranges are recorded for basalt cleavers, with the smallest cleaver measuring 74 mm long and the longest 188 mm long (Table 7.90). Despite this, the variability of biface size both within and among the levels is limited, since the handaxes and cleavers are clustered in a relatively restricted size range. Moreover, the length of over 90% of the bifaces falls between 69 and 158 mm. Thus, the length of the great majority of complete bifaces (and those with a slightly broken tip) of Layer II-6 is found within a range of 9 cm. The width and thickness demonstrate an even more pronounced homogeneity, with the great majority of the tools clustering in a range of 55 and 23 mm respectively. Furthermore, the interquartile range of the length is 31 mm for handaxes and 29 mm for cleavers. This means that the length of 50% of the bifaces is within a range of 3 cm. A similar pattern is evident for other size attributes.

A similar trend of homogeneity emerges for the inter-level variability of the different size attributes. Although there are significant differences among some levels in some attributes, in the vast majority of cases the differences are statistically insignificant. Most of the significant differences concern aspects of length and they usually highlight higher values for Level 4 than for some of the other levels, such as Levels 1, 2, 4b, and 7 and Trench II green. The width and thickness aspects present a much more homogeneous pattern, and there are only a few isolated significant differences among the levels. In general, these observations support the notion of substantial similarity in the size of bifaces, both handaxes and cleavers, in the different levels.

Another observation that further strengthens the general pattern of similarity in size emerges from the comparison between handaxes and cleavers in the different levels. In almost all levels of Layer II-6 there are no statistically significant differences in any of the size attributes between these two biface types. The only exceptions are Level 1, where cleavers are significantly longer (difference: 23.4; t-ratio=3.91; df=56; $p(tt)$=0.0002), wider (difference: 19.05; t-ratio=4.58; df=56; $p(tt)$<0.0001), and thicker (difference: 6.3; t-ratio=2.98; df=56; $p(tt)$=0.0042) than the handaxes, and Level 2, where the cleavers are significantly wider (difference: 12; t-ratio=2.29; df=25; $p(tt)$=0.0307). This pattern indicates that the sizes of the handaxes and cleavers are on the whole very similar.

The general conclusion that can be drawn from these patterns is that the sizes of bifacial tools are homogeneous *within* each of the bifacial tool types, similarly to their morphology and general production technology and style. However, there is also a marked homogeneity in size *between* the two bifacial tool types, although they differ in their morphology and present some differences in their production technology and style. This

Table 7.89 Descriptive statistics of basalt handaxes of Layer II-6 (size in mm)

	II-6/L1	II-6/L2	II-6/L3	II-6/L4	II-6/L4b	II-6/L5	II-6/L6	II-6/L7	VI-14	Trench II green
N	43	18	8	105	32	4	7	13	2	11
Length										
Maximum	174	151	149	193	173	135	166	143	128	174
Median	108	115.5	111	134	131.5	119	142	114	114	124
Minimum	77	74	78	79	89	96	69	72	100	95
Mean	111.53	112.06	115.63	136.57	128.63	117.25	132.43	116.46	114.00	124.91
Std dev	21.97	19.73	24.76	22.30	21.79	20.19	32.71	19.34	19.80	20.95
Std err	3.35	4.65	8.75	2.18	3.85	10.09	12.36	5.36	14.00	6.32
Width										
Maximum	109	100	93	111	98	82	110	93	88	103
Median	71	76.5	75	83	79	75.5	87	75	79.5	89
Minimum	48	53	34	54	54	52	59	56	71	63
Mean	75.35	75.72	71.00	82.67	76.84	71.25	87.14	76.31	79.50	86.36
Std dev	14.85	12.71	19.05	12.13	12.44	13.50	16.77	11.87	12.02	12.31
Std err	2.26	3.00	6.73	1.18	2.20	6.75	6.34	3.29	8.50	3.71
Thickness										
Maximum	50	44	55	50	50	52	48	52	50	47
Median	35	38	35.5	39	35.5	45.5	37	43	45	40
Minimum	21	23	26	20	24	28	23	20	40	30
Mean	35.63	36.33	36.75	38.27	35.72	42.75	38.71	38.92	45.00	40.27
Std dev	7.40	6.16	8.45	5.90	6.94	10.50	8.60	9.44	7.07	5.42
Std err	1.13	1.45	2.99	0.58	1.23	5.25	3.25	2.62	5.00	1.64
Roe's refinement (T/W)										
Maximum	0.79	0.62	0.97	0.75	0.70	1.00	0.60	0.75	0.57	0.54
Median	0.47	0.48	0.49	0.46	0.47	0.59	0.44	0.50	0.56	0.46
Minimum	0.26	0.34	0.33	0.31	0.31	0.35	0.33	0.35	0.56	0.43
Mean	0.48	0.49	0.55	0.47	0.47	0.63	0.45	0.51	0.57	0.47
Std dev	0.09	0.09	0.19	0.08	0.08	0.27	0.09	0.12	0.00	0.04
Std err	0.01	0.02	0.07	0.01	0.01	0.13	0.03	0.03	0.00	0.01

pattern further highlights the general cultural conservatism and homogeneity of the GBY cultural sequence.

Despite the small numbers of flint and limestone handaxes, it is clear that they are generally smaller overall than the basalt handaxes (Table 7.91). Both flint and limestone handaxes exhibit a high variability of size (the smallest measure 53 mm in flint and 60 mm in limestone; the longest measure 133 mm in flint and 120 mm in limestone). Due to the constraints imposed by the small samples of flint and limestone handaxes, it is only in the handaxe assemblage of Layer II-6 that this size variation among the different raw materials is significant. The difference is clearly rooted in the originally small size of the available flint and limestone nodules (Chapters 6, 8). While some of the flint nodules were of sufficient size to produce handaxes that fall within the range of the basalt handaxes, in some cases (admittedly few) the small size of the flint nodules dictated the production of small handaxes.

We consider that basalt was the prevailing preference of the GBY hominins for the production of bifaces. Thus, flint and limestone were used only rarely for biface production (8.20% of the entire handaxe assemblage).

7.5.3.2 Blank Type

The counts and frequencies of different blank types used for the production of bifaces of all raw materials in Layer II-6 are presented in Table 7.92. It is evident that basalt bifaces are made primarily on flakes, as there is only one case of a handaxe made on a chunk. Where cleavers are concerned, it is evident that the knappers made exclusive use of large basalt flakes, a rigidity that is not observed in the handaxes where tools made on raw materials other than basalt were often made on non-flake blanks.

Some basalt bifaces were modified on Kombewa flakes. These blanks are more frequent in cleavers (Figs. 7.142–7.146) than in handaxes (there are no handaxes certainly made on Kombewa flakes, but there are a few possible cases). This meager representation is probably due primarily to the extensive scar coverage of handaxes, which masks the blank type.

There are substantial differences in the counts of Kombewa bifaces between the publication of Goren-Inbar and Saragusti (1996) and our present observations. While 105

7.5 Bifacial Tools

Table 7.90 Descriptive statistics of basalt cleavers of Layer II-6 (size in mm)

	II-6/L1	II-6/L2	II-6/L3	II-6/L4	II-6/L4b	II-6/L5	II-6/L6	II-6/L7	Trench II green
N	15	9	11	48	14	8	3	7	4
Maximum length									
Maximum	159	155	170	188	158	164	163	162	155
Median	138	133	138	136	130	135.5	155	133	121
Minimum	120	82	97	80	85	94	128	74	114
Mean	137.53	127.00	135.55	139.17	128.07	130.63	148.67	126.00	127.75
Std dev	13.18	21.53	23.45	25.06	20.00	24.86	18.34	28.17	18.48
Std err	3.40	7.18	7.07	3.62	5.34	8.79	10.59	10.65	9.24
Length									
Maximum	156	148	167	185	155	158	163	155	146
Median	137	115	138	133	127	132	153	132	114.5
Minimum	119	80	91	80	83	85	120	110	106
Mean	134.93	118.56	133.73	136.29	125.57	126.25	145.33	128.71	120.25
Std dev	11.98	23.38	23.49	24.62	20.09	26.64	22.50	16.66	17.67
Std err	3.09	7.79	7.08	3.55	5.37	9.42	12.99	6.30	8.84
Width									
Maximum	112	105	93	139	85	96	94	102	93
Median	92	88	82	82.5	71	87	93	80	79
Minimum	81	67	35	53	49	62	73	46	68
Mean	94.40	87.78	72.82	83.29	71.43	82.13	86.67	79.14	79.75
Std dev	10.43	13.25	18.77	15.62	9.10	13.40	11.85	17.48	10.28
Std err	2.69	4.42	5.66	2.26	2.43	4.74	6.84	6.61	5.14
Thickness									
Maximum	53	48	50	52	43	49	44	111	51
Median	41	40	45	38	33	40.5	40	38	41
Minimum	32	30	31	24	25	24	28	32	36
Mean	41.93	39.33	42.00	37.77	33.07	37.13	37.33	49.00	42.25
Std dev	5.86	6.04	7.38	7.07	5.53	8.48	8.33	28.21	6.50
Std err	1.51	2.01	2.22	1.02	1.48	3.00	4.81	10.66	3.25
Roe's refinement (T/W)									
Maximum	0.65	0.57	1.43	0.64	0.61	0.66	0.47	2.41	0.64
Median	0.46	0.45	0.54	0.43	0.46	0.44	0.43	0.45	0.52
Minimum	0.30	0.33	0.36	0.33	0.31	0.31	0.38	0.39	0.46
Mean	0.45	0.45	0.64	0.46	0.47	0.46	0.43	0.74	0.53
Std dev	0.09	0.07	0.31	0.09	0.08	0.11	0.04	0.74	0.09
Std err	0.02	0.02	0.09	0.01	0.02	0.04	0.02	0.28	0.04

basalt handaxes and 41 cleavers were analyzed previously, the present counts are 389 and 157 respectively. However, the absolute numbers as well as the frequencies of bifaces made on Kombewa blanks are much smaller in the present study. The main reason for this discrepancy is the fact that our current definition of Kombewa is much more rigid than in the former publication. Only definite cases where two ventral surfaces could be observed with certainty were included in this category. Furthermore, the continuous deterioration of the basalt bifaces (despite their conservation) presented additional difficulties for assigning items to the Kombewa category.

The relative high frequencies of indeterminate blanks is clearly related to the fact that the scars observed on both faces of many of the bifaces masks the identity of the blank, a feature that is common to all three raw materials. It is also evident that limestone and flint handaxes were more frequently shaped on cobbles (Figs. 7.147–7.148). From the small size of both flint and limestone bifaces it seems that the selection was dependent on the size of the raw material, and that when large enough nodules were available flakes were produced for the modification of handaxes. In accordance with the above, we note that the great majority of both handaxes and cleavers are devoid of cortex; frequencies of non-cortical bifaces are, for example, 78.72% (handaxes) and 83.33% (cleavers) in Layer II-6 Level 1, and 93.10% (handaxes) and 84.31% (cleavers) in Layer II-6 Level 4.

Table 7.91 Descriptive statistics of flint and limestone handaxes of Layer II-6 (size in mm)

	Flint N=24	Limestone N=7
Length		
Maximum	133	120
Median	98	100
Maximum	53	60
Mean	91.38	89.57
Std dev	21.66	23.75
Std err	4.42	8.98
Width		
Maximum	93	90
Median	60.5	57
Maximum	35	41
Mean	62.38	60.14
Std dev	15.92	15.75
Std err	3.25	5.95
Thickness		
Maximum	60	50
Median	35	36
Maximum	14	23
Mean	34.58	35.14
Std dev	10.22	9.15
Std err	2.09	3.46
Roe's refinement (T/W)		
Maximum	0.86	0.78
Median	0.55	0.56
Maximum	0.33	0.47
Mean	0.56	0.59
Std dev	0.13	0.11
Std err	0.03	0.04

Fig. 7.142 Basalt cleaver, possibly Kombewa (#5904 Layer V-5)

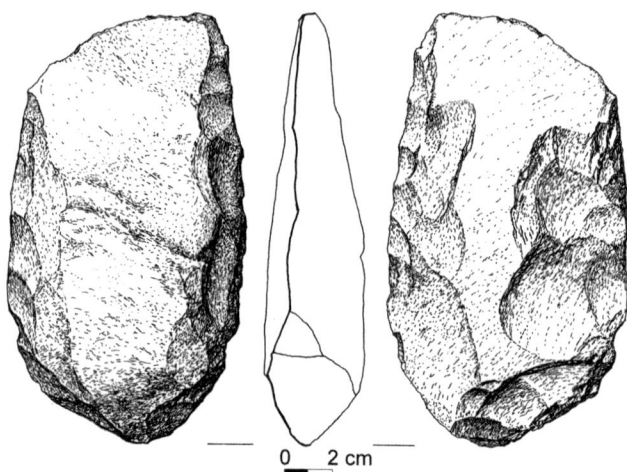

Fig. 7.143 Basalt cleaver, Kombewa (#223 Layer II-6 Level 4)

Table 7.92 Counts and frequencies (%) of blank type of bifaces of Layer II-6 by raw material

Handaxes						
	Basalt		Flint		Limestone	
	N	%	N	%	N	%
Flake	264	67.86	6	25.00%	2	22.22
Chunk	1	0.25	4	16.67%	2	22.22
Indeterminate	99	25.45	14	58.33%	4	44.44
Possibly Kombewa	11	2.82	-	-	1	11.11
Possibly flake	14	3.59	-	-	-	-
Total	389	100.00	24	100.00%	9	

Cleavers		
Basalt		
	N	%
Flake	119	75.80
Indeterminate	7	4.46
Kombewa	13	8.28
Possibly Kombewa	18	11.46
Total	157	100.00

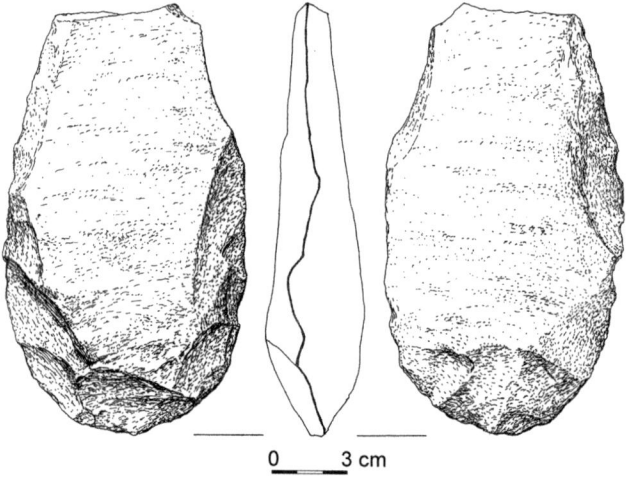

Fig. 7.144 Basalt cleaver, Kombewa (#331 Layer II-6 Level 4)

7.5 Bifacial Tools

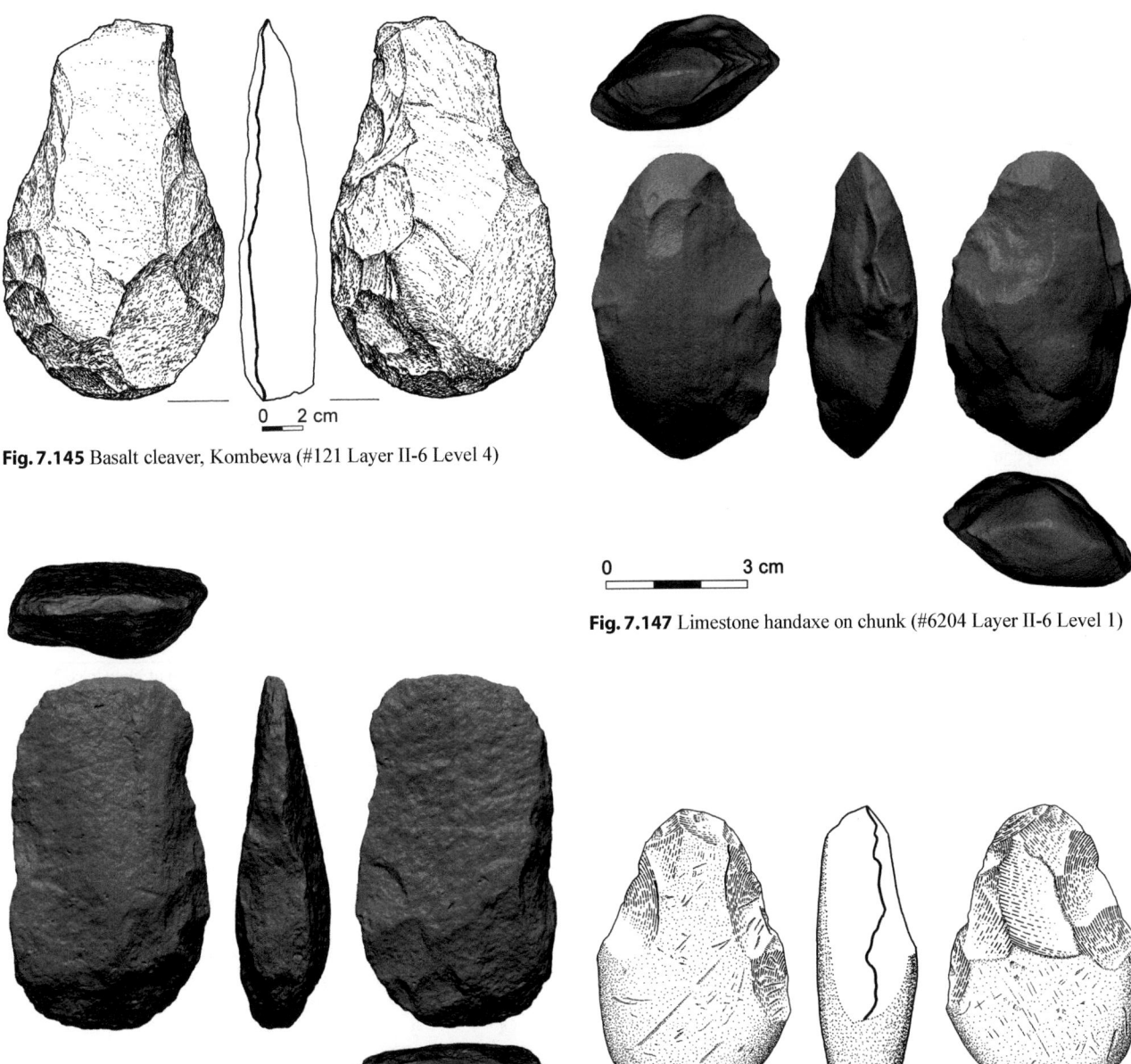

Fig. 7.145 Basalt cleaver, Kombewa (#121 Layer II-6 Level 4)

Fig. 7.146 Basalt cleaver, Kombewa (#11961 Layer II-6 Level 5)

Fig. 7.147 Limestone handaxe on chunk (#6204 Layer II-6 Level 1)

Fig. 7.148 Limestone handaxe on chunk (#5750 Layer II-6 Level 4b)

7.5.3.3 Direction of Blow

The direction of blow was recorded for the flakes on which bifaces were modified. The different blow directions were recorded in relation to the biface's axis of symmetry (Section 4.2.2.3).

Tables 7.93–7.94 present the direction of blow for bifaces modified on flakes from the different levels of Layer II-6. The primary difference observed between handaxes and cleavers is the higher frequency of indeterminate direction for the handaxes (Figs. 7.149–7.152). This clearly results from the intensive retouch on both surfaces of the handaxes (Figs. 7.153–7.156), especially on their ventral faces (see below). If we exclude indeterminate cases, the following observations can be suggested.

It is evident that the removal of large flakes for biface modification was carried out in several directions, particularly those positioned in the lateral (Figs. 7.157–7.159) and proximal zones (i.e. blow directions 3–7). Clearly, these zones reflect the striking platform area of the flake, which is thicker and hence

Table 7.93 Counts and frequencies (%) of direction of blow of basalt handaxes of Layer II-6*

	Direction of blow Face 1										Direction of blow Face 2#
	II-6 L1	II-6 L2	II-6 L3	II-6 L4	II-6 L4b	II-6 L5	II-6 L6	II-6 L7	VI-14	Trench II green	II-6
N	43	17	11	140	43	7	15	19	9	23	10
Direction of blow*	(N=17)	(N=4)	(N=11)	(N=90)	(N=19)	(N=3)	(N=7)	(N=3)	(N=1)	(N=3)	
1	-	-	-	-	-	-	-	-	-	-	
2	-	-	-	-	1 / 5.26	-	-	-	-	-	
3	5 / 29.41	1 / 25.00	2 / 33.33	17 / 18.89	1 / 5.26	-	2 / 28.57	-	-	-	3 / 30.00
4	1 / 5.88	-	1 / 16.67	11 / 12.22	5 / 26.32	-	2 / 28.57	-	-	2 / 66.67	
5	3 / 17.65	1 / 25.00	1 / 16.67	22 / 24.44	4 / 21.05	1 / 33.33	1 / 14.29	-	-	-	2 / 20.00
6	6 / 35.29	-	2 / 33.33	25 / 27.78	3 / 15.79	1 / 33.33	-	1 / 33.33	-	1 / 33.33	2 / 20.00
7	2 / 11.76	2 / 50.00	-	15 / 16.67	5 / 26.32	1 / 33.33	2 / 28.57	2 / 66.67	1 / 100.00	-	3 / 30.00

* excluding indeterminate category; differences in counts reflect lack of observations, resulting from scars masking the indicative traits
# cases where a second ventral face is recorded for which the direction of blow is visible

Table 7.94 Counts and frequencies (%) of direction of blow of basalt cleavers of Layer II-6*

	Direction of blow Face 1									Direction of blow Face 2#
	II-6/L1	II-6/L2	II-6/L3	II-6/L4	II-6/L4b	II-6/L5	II-6/L6	II-6/L7	Trench II green	II-6
N	13	10	9	39	12	8	8	7	6	14
Direction of blow*	(N=12)			(N=35)	(N=11)	(N=7)	(N=6)		(N=5)	
3	1 / 8.33	1 / 10.00	4 / 44.44	4 / 11.43	1 / 9.09	-	-	-	1 / 20.00	-
4	5 / 41.67	2 / 20.00	-	3 / 8.57	2 / 18.18	2 / 28.57	2 / 33.33	-	2 / 40.00	3 / 21.43
5	2 / 16.67	2 / 20.00	-	12 / 34.29	5 / 45.45	2 / 28.57	-	1 / 14.29	-	3 / 21.43
6	2 / 16.67	2 / 20.00	-	12 / 34.29	1 / 9.09	2 / 28.57	-	3 / 42.86	1 / 20.00	2 / 14.29
7	2 / 16.67	3 / 30.00	5 / 55.56	4 / 11.43	2 / 18.18	1 / 14.29	4 / 66.67	3 / 42.86	1 / 20.00	6 / 42.86

* excluding indeterminate category; differences in counts reflect lack of observations, resulting from scars masking the indicative traits
# cases where a second ventral face is recorded for which the direction of blow is visible

more suitable as a butt, unlike the thinner tip. Directions 1, 2, and 8 are represented by a single case each in the handaxes and are absent from the cleavers (Tables 7.93–7.94). Even when two ventral faces are observed (Fig. 7.160), directions 1, 2, and 8 are missing.

While the proximal and lateral (3, 4, 6, and 7) (side-struck: Figs. 7.161–7.162) directions of blow are the most common, the longitudinal direction (5) is less frequent (Figs. 7.163–7.164). The flake type most commonly used at GBY for the production of bifacial tools was apparently the special side-struck flake (Figs. 7.165–7.167) (Goren-Inbar and Saragusti 1996). These flakes are the result of the knapping method applied to some of the cores discussed above, in which the scar of a previous flake was used as the striking platform for the detachment of the next flake. This resulted in a special side-struck flake. As demonstrated experimentally, such a sequence of removals is typical of bifacial, as well as Kombewa, cores (Champault 1966; Madsen and Goren-Inbar 2004).

7.5.3.4 Striking Platforms

As many of the bifaces were modified on flakes, remnants of the original striking platform could be observed and classified. Tables 7.95–7.96 present the types and frequencies

7.5 Bifacial Tools

Fig. 7.149 Basalt handaxe (#8620 Layer II-6 Level 2)

Fig. 7.151 Basalt handaxe (#7747 Layer II-6 Level 5)

Fig. 7.150 Basalt handaxe (#5672 Layer II-6 Level 2)

Fig. 7.152 Basalt handaxe (#7724 Layer II-6 Level 7)

of striking platforms observed on handaxes and cleavers of Layer II-6.

The most common type recorded for striking platforms is removed (Figs. 7.168–7.170). The removal was carried out by intentional flaking, a particular type of retouch that is characterized by the presence of a few flat, shallow scars aimed at reducing the topographic high (thickness) formed by the bulb of percussion, the thickest part of the tool. In some cases, thinning of the bulb was all that was needed in order to transform the flake (blank) into the desired biface, as the lenticular shape was already present. When the bulb-thinning strategy is used, the ventral face is often minimally retouched either by a few bulb-thinning scars or by retouch of a small area along both lateral edges (Fig. 7.160).

The next most frequent type of striking platform is plain (Figs. 7.171–175). Experimental work has shown that plain striking platforms are the expected result of most of the

Fig. 7.153 Basalt handaxe (#2006 Layer II-6 Level 1)

Fig. 7.155 Basalt handaxe (#5655 Layer II-6 Level 2)

Fig. 7.154 Basalt handaxe (#2126 Layer II-6 Level 1)

Fig. 7.156 Basalt handaxe (#8614 Layer II-6 Level 2)

giant core methods (Jones 1994; Madsen and Goren-Inbar 2004; Sharon 2007). Plain striking platforms do not imply unsophisticated core technology or lack of core preparation/maintenance. The knapping properties needed for removal of large flakes (discussed above) dictate the use of thick, large and plain striking platforms. Small, thin platforms cannot withstand the powerful blow needed for large flake detachment and will result in small and broken flakes. All other types of striking platforms appear in very low frequencies.

The absence of cortical striking platforms is of special interest, as it indicates that the selected large flakes rarely derive from the initial stages of core reduction but represent a more advanced (and pre-planned) stage of knapping within the biface reduction sequence. The frequencies of striking platform types recorded for bifaces clearly show that similar knapping

7.5 Bifacial Tools

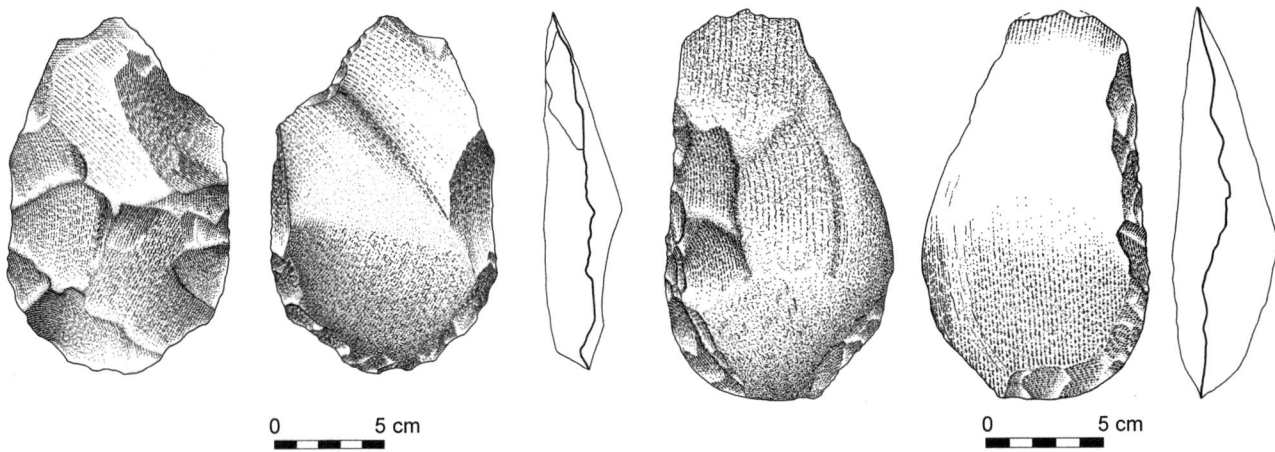

Fig. 7.157 Basalt handaxe on side-struck flake (#13764 Layer II-6 Level 1)

Fig. 7.160 Basalt handaxe on side-struck flake (#7720 Layer II-6 Level 6)

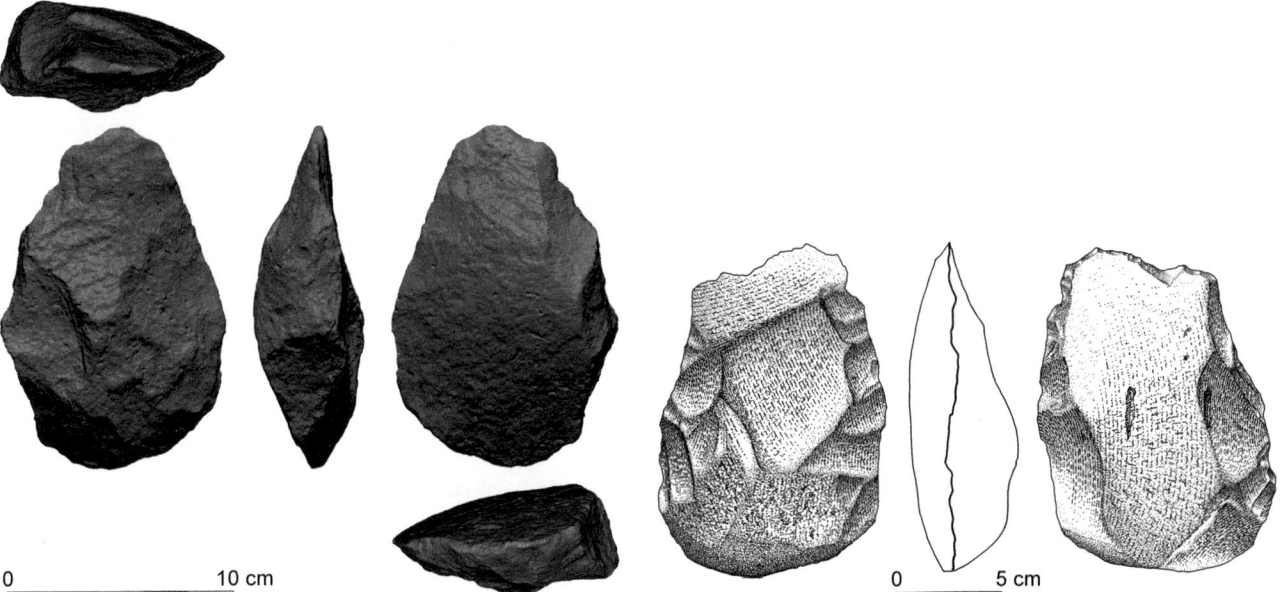

Fig. 7.158 Basalt handaxe on side-struck flake (#8623 Layer II-6 Level 2)

Fig. 7.161 Basalt cleaver on side-struck flake (#8666 Layer II-6 Level 3)

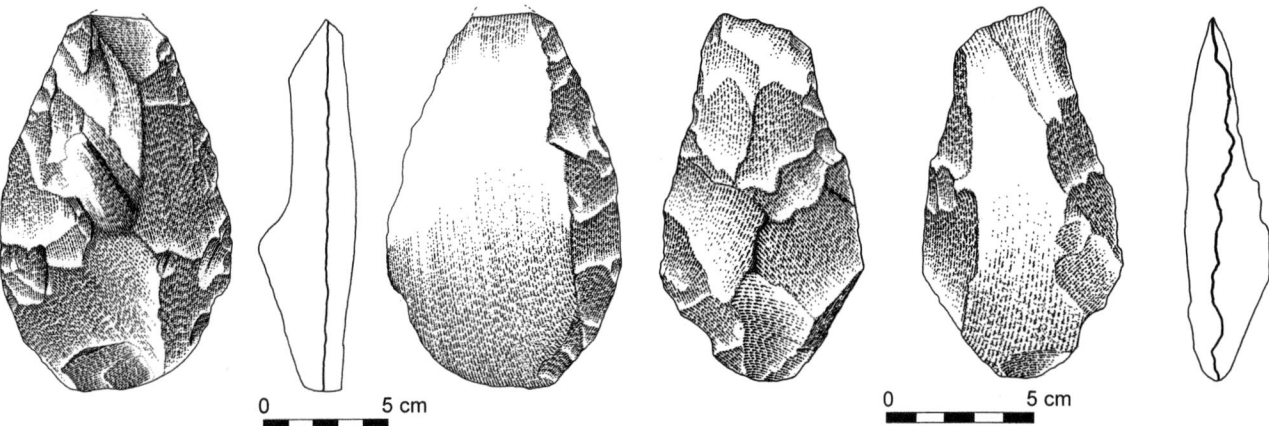

Fig. 7.159 Basalt handaxe on side-struck flake (#7719 Layer II-6 Level 7)

Fig. 7.162 Basalt cleaver on side-struck flake (#5778 Layer II-6 Level 4b)

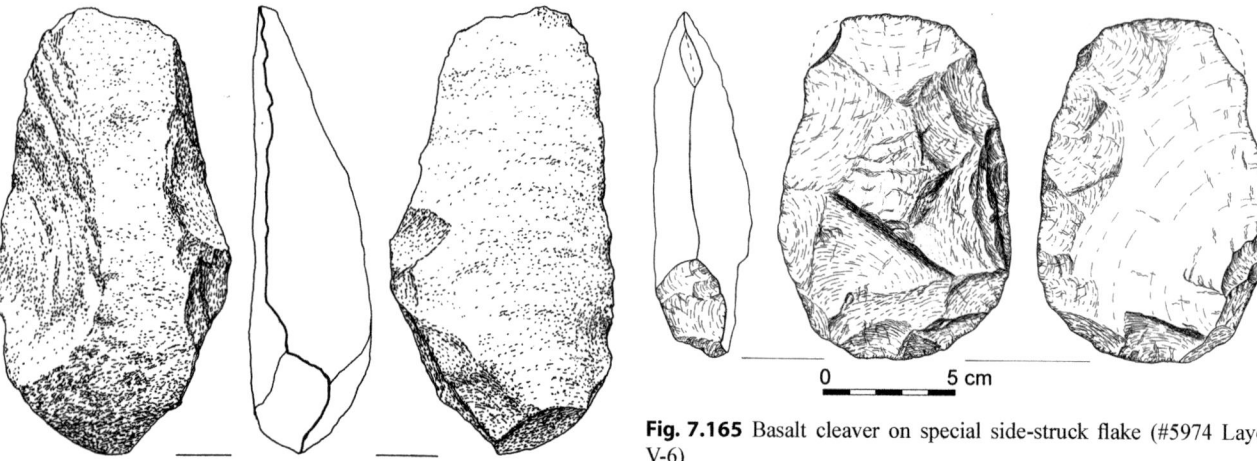

Fig. 7.163 Basalt cleaver on end-struck flake (#180 Layer II-6 Level 4)

Fig. 7.165 Basalt cleaver on special side-struck flake (#5974 Layer V-6)

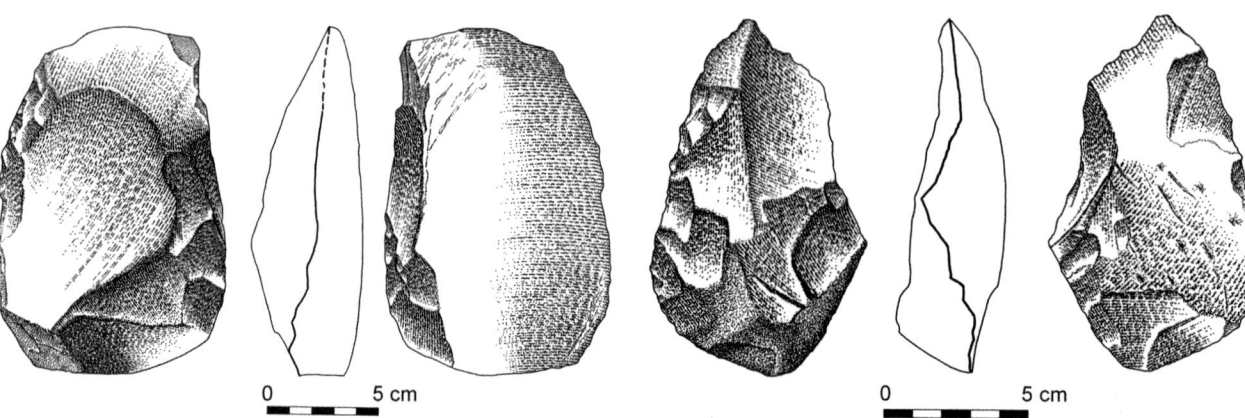

Fig. 7.164 Basalt cleaver on end-struck flake (#7737 Layer II-6 Level 7)

Fig. 7.166 Basalt cleaver on special side-struck flake (#5780 Layer II-6 Level 4b)

methods were applied to both handaxes and cleavers in all the archaeological horizons and during the entire time trajectory that they represent.

The pattern of abundant removals of the striking platform is also recorded in cases where two striking platforms were observed. While removal of the striking platform is not strictly speaking a technological feature, it is viewed here as a component of the much wider realm of the configuration and design of bifaces, discussed in the following section.

7.5.4 Modification and Design

The modification of the GBY bifaces involved a variety of approaches that integrated different aspects of retouch. These

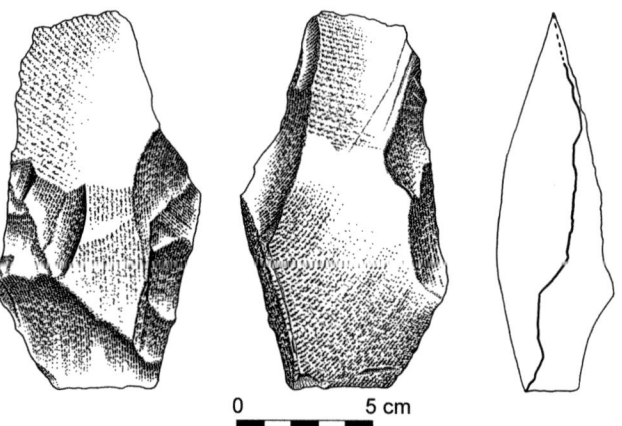

Fig. 7.167 Basalt cleaver on special side-struck flake (#5792 Layer II-6 Level)

7.5 Bifacial Tools

Table 7.95 Counts and frequencies (%) of striking platform type on basalt handaxes of Layer II-6

	Striking platform Face 1										Striking platform Face 2*
	II-6/L1	II-6/L2	II-6/L3	II-6/L4	II-6/L4b	II-6/L5	II-6/L6	II-6/L7	VI-14	Trench II green	
N	54	19	12	165	58	11	19	20	9	24	
	(N=16)	(N=10)	(N=5)	(N=98)	(N=34)	(N=7)	(N=11)	(N=6)	(N=1)	(N=3)	(N=12)
Punctiform	-	-	1 20.00	-	-	-	-	-	-	-	-
Plain	6 37.50	1 10.00	2 40.00	46 46.94	7 20.59	2 28.57	4 36.36	-	-	2 66.67	1 8.33
Dihedral	-	1 10.00	-	2 2.04	1 2.94	-	1 9.09	-	-	1 33.33	1 8.33
Removed	9 56.25	7 70.00	2 40.00	47 47.96	23 67.65	4 57.14	6 54.55	5 83.33	1 100.00	-	10 83.33
Missing	1 6.25	-	-	3 3.06	2 5.88	1 14.29	-	-	-	-	-
Indeterminate	-	1 10.00	-	-	1 2.94	-	-	1 16.67	-	-	-

* in Kombewa flakes where two striking platforms are recorded

Table 7.96 Counts and frequencies (%) of striking platform type on basalt cleavers of Layer II-6

	Striking platform Face 1									Striking platform Face 2*
	II-6/L1	II-6/L2	II-6/L3	II-6/L4	II-6/L4b	II-6/L5	II-6/L6	II-6/L7	Trench II green	
N	17	10	12	60	18	8	11	12	8	25
Striking platform 1	(N=15)		(N=11)	(N=52)	(N=16)	(N=7)	(N=8)		(N=7)	
Plain	5 33.33	-	1 9.09	16 30.77	3 18.75	6 85.71	1 12.50	-	2 28.57	6 24.00
Facetted	1 6.67	1 10.00	-	1 1.92	-	-	-	-	-	-
Removed	8 53.33	9 90.00	10 90.91	30 57.69	11 68.75	1 14.29	6 75.00	10 83.33	4 57.14	17 68.00
Missing	1 6.67	-	-	4 7.69	2 12.50	-	1 12.50	2 16.67	1 14.29	1 4.00

* in Kombewa flakes where two striking platforms are recorded

Fig. 7.168 Basalt handaxe with removed striking platform (#5776 Layer II-6 Level 4b)

Fig. 7.169 Basalt cleaver with removed striking platform (#5908 Layer II-6 Level 3)

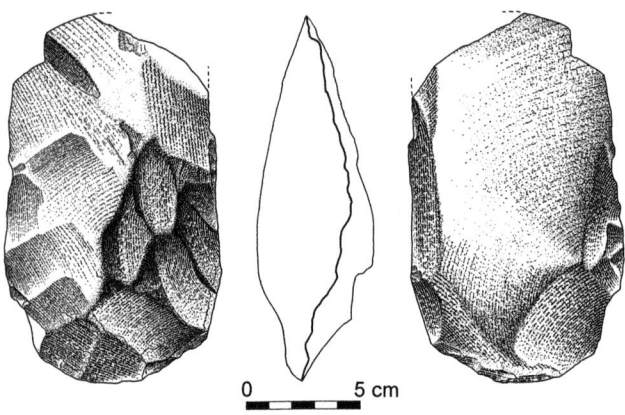

Fig. 7.170 Basalt cleaver with removed striking platform (#13762 Layer II-6 Level 3)

Fig. 7.173 Basalt handaxe with plain striking platform (#179 Layer II-6 Level 4)

Fig. 7.171 Basalt handaxe with plain striking platform (#152 Layer II-6 Level 4)

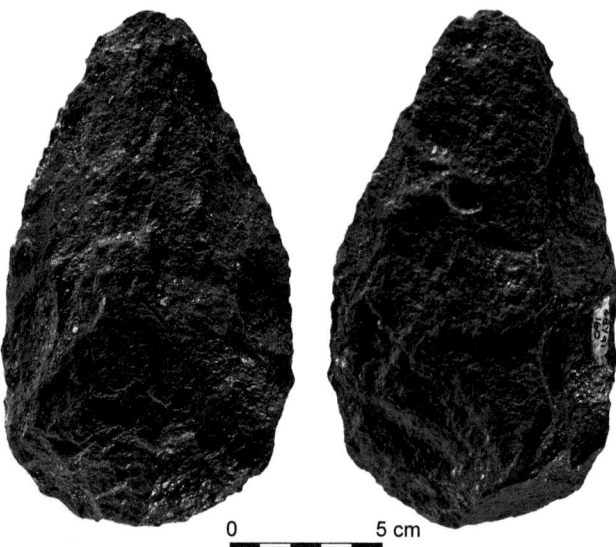

Fig. 7.171 Basalt handaxe with plain striking platform (#152 Layer II-6 Level 4)

resulted in bifaces of a particular morphology and design, which characterize the GBY lithic assemblages. The removal of striking platform, discussed above, is one of the modes used by knappers to design the bifaces in conformity with a suitable morphology. The following sections discuss other design methods, particularly those related to the type, quantity, and location of retouch.

The recording of retouch on bifaces relates to both faces of the tool, which were analyzed individually. The upper, dorsal face was termed Face 1 and the flat, ventral face was designated Face 2. These methodological definitions, as well as others related to the analysis of biface retouch, are provided in Section 4.2.2.3.

7.5.4.1 Type of Retouch

Table 7.97 presents the frequencies of different retouch types on basalt handaxes, primarily bifacial retouch (Fig. 7.176) and thinning (Figs. 7.177–7.178). In most cases both faces of the handaxes were retouched and the same types of retouch occur on both, though in different frequencies. The dominant retouch type used for the modification of basalt handaxes is bifacial. On Face 1 it forms the bulk of the cases and is even more frequent since it also occurs in association with thinning (Figs. 7.179–7.181) (Table 7.97). On Face 2, bifacial retouch is again most common but in lower frequencies than on Face 1; this is because Face 2 is more significantly modified by thinning and flat limited retouch (Table 7.97).

These patterns are highlighted by examination of the relationship between the two faces. Table 7.98 presents the combinations of retouch types that occur on Face 1 and Face 2 of the handaxes. It shows that the most common mode

7.5 Bifacial Tools

Fig. 7.174 Basalt cleaver with plain striking platform (#13765 Layer II-6 Level 1)

Fig. 7.175 Basalt cleaver with plain striking platform (#5906 Layer II-6 Level 3)

Table 7.97 Counts and frequencies (%) of type of retouch on basalt handaxes of Layer II-6

	II-6/L1	II-6/L2	II-6/L3	II-6/L4	II-6/L4b	II-6/L5	II-6/L6	II-6/L7	VI-14	Trench II green
N	52	19	11	164	57	11	19	20	9	24
Face 1	(N=48)		(N=9)	(N=144)	(N=48)	(N=7)	(N=14)	(N=17)	(N=5)	(N=18)
Bifacial	38	15	7	96	31	6	9	10	5	14
	79.17	78.95	77.78	66.67	64.58	85.71	64.29	58.82	100.00	77.78
Thinning	1	3	-	7	6	1	1	2	-	-
	2.08	15.79		4.86	12.50	14.29	7.14	11.76		
Bifacial & thinning	8	1	1	21	4	-	2	2	-	2
	16.67	5.26	11.11	14.58	8.33		14.29	11.76		11.11
Flat & limited	1	-	-	3	4	-	2	-	-	-
	2.08			2.08	8.33		14.29			
Indeterminate	-	-	1	17	3	-	-	3	-	2
			11.11	11.81	6.25			17.65		11.11
Face 2	(N=47)		(N=9)	(N=141)	(N=49)	(N=7)	(N=14)	(N=17)	(N=6)	(N=18)
Bifacial	24	6	2	69	15	1	1	6	3	14
	51.06	31.58	22.22	48.94	30.61	14.29	7.14	35.29	50.00	77.78
Thinning	10	7	3	41	17	3	7	4	2	-
	21.28	36.84	33.33	29.08	34.69	42.86	50.00	23.53	33.33	
Bifacial & thinning	6	3	3	19	5	-	1	5	-	4
	12.77	15.79	33.33	13.48	10.20		7.14	29.41		22.22
Flat & limited	7	3	-	10	12	2	5	2	-	-
	14.89	15.79		7.09	24.49	28.57	35.71	11.76		
Indeterminate	-	-	1	2	-	1	-	-	1	-
			11.11	1.42		14.29			16.67	

of handaxe modification is bifacial retouch on both faces (34.58%), followed by handaxes on which bifacial retouch was employed for Face 1 while Face 2 was modified by either thinning (17.45%) or flat limited retouch (7.79%) (Table 7.98). These patterns characterize the entire cultural sequence of Layer II-6.

In a few cases, in addition to the dominant retouch described above, a secondary retouch was also applied to the

Fig. 7.176 Basalt handaxe with bifacial retouch on both faces (#7202 Layer II-6 Level 1)

Fig. 7.179 Basalt handaxe with mixed retouch types (bifacial and thinning) (#8057 Layer II-6 Level 1)

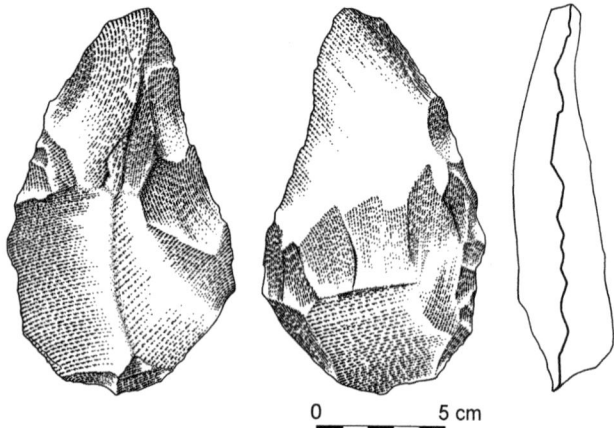

Fig. 7.177 Basalt handaxe with thinning retouch on ventral face (#5787 Layer II-6 Level 6)

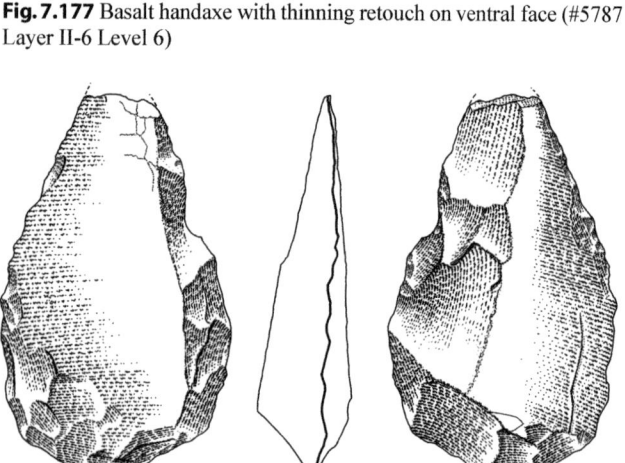

Fig. 7.178 Basalt handaxe with thinning retouch on ventral face (#1099 Layer II-6 Level 4b)

Fig. 7.180 Basalt handaxe with mixed retouch types (bifacial and thinning) (#5673 Layer II-6 Level 2)

same face of the handaxe. This was observed on both faces of the handaxes but is more common on Face 2 (N=48) than on Face 1 (N=29). In rare cases, a secondary retouch was observed on both faces, suggestive of a more intensive retouch (N=8) (Figs. 7.158–7.159).

In general, the secondary retouch is of the same types as the primary retouch. The analysis allows us to associate the primary retouch with various secondary retouch types. Thus, it

7.5 Bifacial Tools

when it occurs in association with thinning, forms the majority there (Table 7.99). For Face 2 of the cleavers, the dominant retouch type (excluding Layer II-6 Level 5) is thinning; when combined with the thinning that occurs in association with bifacial retouch, this amounts to the majority (Table 7.99). Scraper-like retouch (Fig. 7.182) was occasionally observed on cleavers, mostly on Face 1 (Table 7.99); this type of retouch was recorded on only one basalt handaxe, and there only as secondary retouch.

7.5.4.2 Quantity of Retouch

The quantity of retouch on bifaces is determined by estimation of the percentage of the overall surface that is covered by scars. Four categories, ranging from 0% to 100%, describe the recorded coverage of the surface by scars. The quantity of retouch was recorded for each of the faces of handaxes (Table 7.100) and cleavers (Table 7.101). In both handaxes and cleavers, Face 1 is much more intensively flaked than Face 2 (Fig. 7.183). The category of 76–100% retouch is much more frequent on Face 1, where it forms the majority of cases (Tables 7.100–7.101). Face 2 of handaxes exhibits similar frequencies of all four categories of retouch quantity. On Face 2 of cleavers, on the other hand, there are fewer cases of the highest amount of scar coverage, and 0–25% and 26–50% are the most frequent modes of quantity of retouch (Table 7.101).

Some of the GBY bifaces are only minimally retouched, as their blanks already provided the predetermined morphology and could be used as finished tools with no further secondary retouch (Figs. 7.142, 7.182, 7.184–7.185).

Number of Scars

The quantity of retouch is additionally expressed by the number of scars on each of the faces of the biface. The attribute of number of scars is not identical to that of quantity of retouch (discussed above), as it is the size of the scars and not necessarily their number that dictates the amount of surface that they cover. Only scars larger than 5 mm in maximal dimension were counted.

Fig. 7.181 Basalt handaxe with mixed retouch types (bifacial and thinning) (#5867 Layer II-6 Level 4b)

is interesting to note that in cases where a secondary retouch was observed on Face 2, over 90% were modified primarily by thinning (a significantly higher value than that observed for the entire handaxe assemblage: Tables 7.97–7.98) and the secondary retouch was frequently in the form of flat and limited retouch.

As for flint handaxes, bifacial retouch is similarly the dominant mode of modification, but in contrast with basalt handaxes it occurs in similarly high frequencies on both faces (88.89% of Face 1; 80.00% of Face 2). Thinning is rarely used for the modification of flint handaxes and was observed on Face 1 and Face 2 in similar frequencies (5.56% and 5.00% respectively). The few limestone handaxes exhibit similar tendencies, with bifacial retouch comprising the dominant mode and rare examples of the use of thinning.

In the case of basalt cleavers, the majority were modified on both faces and very few on a single face. The dominant retouch type is bifacial, which on Face 1 can amount to over 50% and,

Table 7.98 Counts and frequencies (%) of type of retouch on both faces of basalt handaxes of Layer II-6

	Face 1										Total Face 2	
	Bifacial		Thinning		Thinning & bifacial		Flat & limited		Indeterminate			
Face 2	N	%	N	%	N	%	N	%	N	%	N	%
Bifacial	111	34.58	3	0.93	11	3.43	2	0.62	11	3.43	138	42.99
Thinning	56	17.45	13	4.05	13	4.05	3	0.93	8	2.49	93	28.97
Bifacial & thinning	30	9.35	0	0.00	11	3.43	0	0.00	3	0.93	44	13.71
Flat & limited	25	7.79	4	1.25	6	1.87	5	1.56	1	0.31	41	12.77
Indeterminate	3	0.93	0	0.00	0	0.00	0	0.00	2	0.62	5	1.56
Total Face 1	225	70.09	20	6.23	41	12.77	10	3.12	25	7.79	321	100.00

Table 7.99 Counts and frequencies (%) of type of retouch on basalt cleavers of Layer II-6

	II-6/L1	II-6/L2	II-6/L3	II-6/L4	II-6/L4b	II-6/L5	II-6/L6	II-6/L7	Trench II green
N	17	10	12	60	18	8	11	12	8
Face 1	(N=14)		(N=11)	(N=53)			(N=10)		(N=6)
Bifacial	7	5	6	34	13	5	8	8	4
	50.00	50.00	54.55	64.15	72.22	62.50	80.00	66.67	66.67
Thinning	1	1	1	6	1	1	1	1	1
	7.14	10.00	9.09	11.32	5.56	12.50	10.00	8.33	16.67
Bifacial & thinning	4	3	3	11	2	2	1	3	1
	28.57	30.00	27.27	20.75	11.11	25.00	10.00	25.00	16.67
Flat & limited	-	-	-	-	1	-	-	-	-
					5.56				
Bifacial & scraper	1	1	-	2	1	-	-	-	-
	7.14	10.00		3.77	5.56				
Indeterminate	1	-	1	-	-	-	-	-	-
	7.14		9.09						
Face 2	(N=16)		(N=10)	(N=49)			(N=9)	(N=11)	(N=6)
Bifacial	4	-	-	7	7	3	2	1	1
	25.00			14.29	38.89	37.50	22.22	9.09	16.67
Thinning	7	6	5	20	6	-	3	8	1
	43.75	60.00	50.00	40.82	33.33		33.33	72.73	16.67
Bifacial & thinning	3	3	3	10	3	3	3	1	3
	18.75	30.00	30.00	20.41	16.67	37.50	33.33	9.09	50.00
Flat & limited	1	1	2	9	1	1	1	1	-
	6.25	10.00	20.00	18.37	5.56	12.50	11.11	9.09	
Bifacial & scraper	-	-	-	1	-	-	-	-	-
				2.04					
Indeterminate	1	-	-	2	1	1	-	-	1
	6.25			4.08	5.56	12.50			16.67

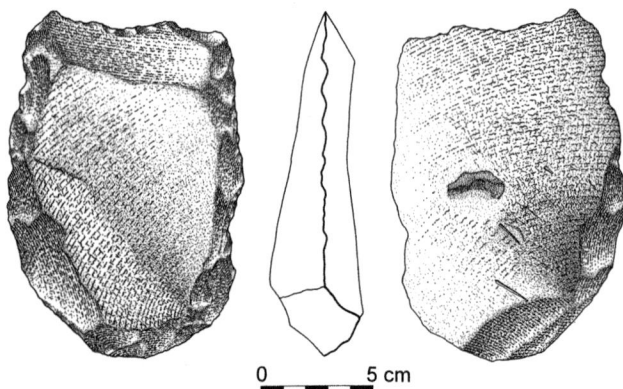

Fig. 7.182 Basalt cleaver with scraper-like retouch (#13761 Layer II-6 Level 2)

Table 7.102 presents the distribution of scar counts for the handaxes and cleavers of Layer II-6.

As for all other retouch characteristics described above, the number of scars recorded for Face 1 of the bifaces is higher than that for Face 2. The data also indicate that, while there is no significant difference between handaxes and cleavers in the number of scars on Face 1, a significant difference is observed for Face 2 (difference: 1.88; t-ratio=3.41; df=444; $p(tt)$=0.0007): in most levels of Layer II-6 the scar count on Face 2 of handaxes exceeds the count recorded for cleavers. However, in Level 4 and to a lesser extent in Level 4b cleavers and handaxes show a similar number of scars on their ventral face. The reason for this difference may lie in a different shaping technology that was applied to the cleavers in these layers.

7.5.4.3 Location of Retouch

This attribute describes the location of retouch on each of the two faces of the biface. Recording this attribute is important since its distribution reflects one of the primary aspects of biface modification at GBY, namely the investment during the final stages of modification. Table 7.103 shows the combinations of the location of retouch between the two faces as recorded for the entire assemblage of basalt handaxes from Layer II-6. It is evident that the most common category is handaxes on which both faces were entirely retouched, forming 20.74% of all cases. Nevertheless, there is a distinctly different approach to each of the faces. While the frequency of handaxes on which only Face 1 is entirely retouched is 41.79% (N=202) (Figs. 7.157–7.159), those on which only Face 2 is entirely retouched compose merely 4.64% (N=15) (Table 7.103). This pattern is in accordance with the quantity of retouch (discussed above). The

7.5 Bifacial Tools

Table 7.100 Counts and frequencies (%) of quantity of retouch on basalt handaxes of Layer II-6

	II-6/L1	II-6/L2	II-6/L3	II-6/L4	II-6/L4b	II-6/L5	II-6/L6	II-6/L7	VI-14	Trench II green
N	52	19	11	164	57	11	19	20	9	24
Face 1	(N=46)		(N=9)	(N=143)	(N=48)	(N=7)	(N=14)	(N=17)	(N=5)	(N=19)
0–25%	3	-	-	5	-	-	1	1	-	1
	6.52			3.50			7.14	5.88		5.26
26–50%	4	3	1	17	9	1	1	3	-	1
	8.70	15.79	11.11	11.89	18.75	14.29	7.14	17.65		5.26
51–75%	7	3	3	33	10	-	4	-	1	-
	15.22	15.79	33.33	23.08	20.83		28.57		20.00	
76–100%	32	13	5	88	29	6	8	13	4	17
	69.57	68.42	55.56	61.54	60.42	85.71	57.14	76.47	80.00	89.47
Face 2	(N=47)	(N=18)	(N=8)	(N=137)	(N=47)	(N=7)	(N=14)	(N=15)	(N=6)	(N=19)
0–25%	11	2	3	24	14	3	6	5	-	1
	23.40	11.11	37.50	17.52	29.79	42.86	42.86	33.33		5.26
26–50%	11	5	1	49	18	2	4	3	5	6
	23.40	27.78	12.50	35.77	38.30	28.57	28.57	20.00	83.33	31.58
51–75%	11	5	-	33	7	1	3	1	1	2
	23.40	27.78		24.09	14.89	14.29	21.43	6.67	16.67	10.53
76–100%	14	6	4	31	8	1	1	6	-	10
	29.79	33.33	50.00	22.63	17.02	14.29	7.14	40.00		52.63

Table 7.101 Counts and frequencies (%) of quantity of retouch on basalt cleavers of Layer II-6

	II-6/L1	II-6/L2	II-6/L3	II-6/L4	II-6/L4b	II-6/L5	II-6/L6	II-6/L7	Trench II green
N	17	10	12	60	18	8	11	12	8
Face 1	(N=13)		(N=11)	(N=51)	(N=16)		(N=9)	(N=11)	(N=6)
0–25%	1	1	-	2	-	1	-	1	-
	7.69	10.00		3.92		12.50		9.09	
26–50%	2	-	1	10	3	3	-	-	1
	15.38		9.09	19.61	18.75	37.50			16.67
51–75%	1	-	1	10	4	-	-	1	1
	7.69		9.09	19.61	25.00			9.09	16.67
76–100%	9	9	9	29	9	4	9	9	4
	69.23	90.00	81.82	56.86	56.25	50.00	100.00	81.82	66.67
Face 2	(N=14)		(N=11)	(N=53)	(N=16)		(N=8)	(N=10)	(N=6)
0–25%	4	8	5	9	2	3	2	6	3
	28.57	80.00	45.45	16.98	12.50	37.50	25.00	60.00	50.00
26–50%	6	1	4	24	10	3	2	4	1
	42.86	10.00	36.36	45.28	62.50	37.50	25.00	40.00	16.67
51–75%	4	1	2	16	2	-	4	-	1
	28.57	10.00	18.18	30.19	12.50		50.00		16.67
76–100%	-	-	-	4	2	2	-	-	1
				7.55	12.50	25.00			16.67

data in Table 7.103 illustrate that, though each combination of retouch location is represented in low frequencies, the diversity of such combinations is extensive.

The frequency of cleavers on which Face 1 is entirely retouched (Table 7.104) is similar to that of handaxes (62.96% and 62.54% respectively). However, only 4.44% of the cleavers exhibit two entirely retouched faces (Table 7.104), a significantly different pattern from that observed for handaxes ($\chi^2=23.70$; $df=2$; $p<0.0001$). In cleavers, the modification of Face 2 is restricted to the proximal area and the two edges, with emphasis on the right edge (the left edge of Face 1) (Table 7.104). In addition, a variety of combinations of retouch locations occur, though in low frequencies, illustrating flexibility in the modification of the cleavers (Table 7.104). However, combinations in the location of retouch are less diverse in cleavers than in handaxes.

7.5.5 Morphology of Bifaces

The morphology of the GBY bifaces has been studied previously (Brande and Saragusti 1996; Goren-Inbar and Saragusti 1996; Saragusti et al. 1998). Most of these studies, however, related

Fig. 7.183 Basalt handaxe with intensively flaked Face 1 (#7727 Layer II-6 Level 7)

Fig. 7.184 Basalt handaxe with minimal modification (#14195 Layer II-6 Level 1)

only to a portion of the entire GBY bifacial assemblage. Some studies (e.g. Sharon 2007) referred to morphology, but only briefly and very generally as part of a broader discussion of other topics. The morphology of bifaces has been related to various aspects of hominin behavior, such as patterns of raw material acquisition (Sharon 2008), functionality (Machin et al. 2007), and cognitive development (McNaab and Cole 2015), to mention but a few. Analysis of the morphology of the bifacial tools of GBY may provide insights on such aspects.

The term "morphology" refers to the general shape of the tool and should not be confused with size, which is represented by metrical measurements. As bifaces are stone tools made by hominins, they have complex and irregular three-dimensional shapes. Objective quantitative description and analysis of their shapes therefore poses a challenge. In order to describe the shapes of the GBY handaxes we have employed the method developed by Bordes (1961) for typological classification of handaxes. This method, which is based on calculation of ratios of various distance measurements, has several discrete stages. The first consists of the acquisition of several distance measurements in accordance with the standard positioning protocol for handaxes. The distance measurements include the maximal length, width, and thickness of the handaxe, as well as the length to the point of maximal width, the width at mid-length, and the width at three quarters of the length (note, however, that in the analysis of GBY we have used the width at four fifths of the length, following Roe 1968) (Appendix 4). The next stage is the application of these measurements to the calculation of a number of ratios (Table 7.105). These include flatness ratio, elongation index, location of maximal width, roundness of

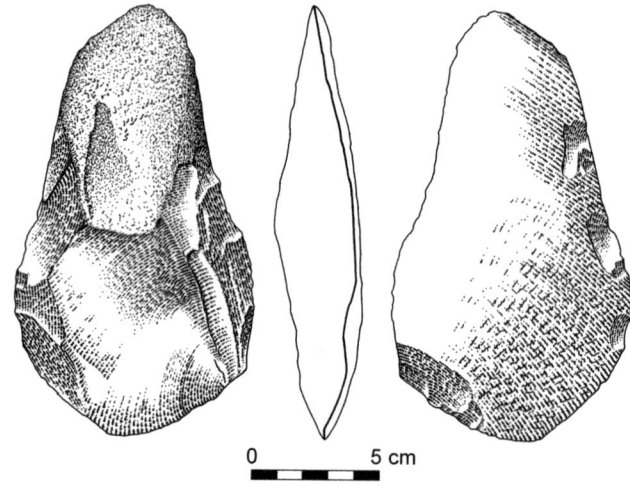

Fig. 7.185 Basalt handaxe with minimal modification (#5782 Layer II-6 Level 4b)

edges, and pointedness ratio. Each of these ratios quantitatively describes a morphological aspect of the handaxes.

The flatness ratio is calculated by dividing the maximal width by the maximal thickness for each handaxe. The result indicates the degree of flatness of the artifact. The higher the result, the flatter the tool; for example, a handaxe with a flatness ratio of 2.50 is flatter than a handaxe with a ratio of 2.00. The elongation index, calculated by dividing the maximal length by the maximal width, describes the degree of elongation of each handaxe. Again, the higher the result, the more elongated the handaxe. The location of the maximal width is calculated by dividing the maximal length by the length up to the point

7.5 Bifacial Tools

Table 7.102 Descriptive statistics of number of scars on basalt bifaces of Layer II-6

	II-6/L1	II-6/L2	II-6/L3	II-6/L4	II-6/L4b	II-6/L5	II-6/L6	II-6/L7	VI-14	Trench II green
Handaxes										
N	52	19	11	164	57	11	19	20	9	24
Face 1	(N=46)		(N=8)	(N=139)	(N=44)	(N=7)	(N=14)	(N=15)	(N=7)	(N=18)
Maximum	30	31	26	35	28	32	26	22	15	26
Median	16	15	12	12	11.5	17	14	12	9	15
Minimum	1	7	5	2	2	6	3	6	4	1
Mean	15.24	16.11	14.00	12.17	13.20	17.14	14.21	12.80	8.71	15.28
Std dev	6.49	6.57	8.35	5.55	5.83	8.55	6.53	5.33	3.55	5.93
Std err	0.96	1.51	2.95	0.47	0.88	3.23	1.75	1.38	1.34	1.40
Face 2	(N=47)		(N=7)	(N=140)	(N=43)	(N=7)	(N=13)	(N=15)	(N=7)	(N=18)
Maximum	31	36	18	28	22	13	14	30	14	30
Median	10	10	7	8	10	7	10	10	6	13
Minimum	3	5	3	1	3	2	4	3	3	6
Mean	12.00	13.00	8.29	9.19	9.93	7.71	9.69	10.33	6.86	14.17
Std dev	6.09	7.85	4.89	4.94	4.58	4.03	3.33	7.45	3.93	6.39
Std err	0.89	1.80	1.85	0.42	0.70	1.52	0.92	1.92	1.49	1.51
Cleavers										
N	17	10	12	60	18	8	11	12		8
Face 1	(N=16)		(N=11)	(N=49)	(N=17)		(N=10)	(N=11)		(N=7)
Maximum	24	32	20	25	20	17	27	20		21
Median	12.5	18.5	15	12	13	12.5	14.5	10		16
Minimum	5	6	6	4	8	5	8	6		5
Mean	12.75	17.50	14.00	12.18	12.71	11.38	15.50	11.91		14.57
Std dev	5.42	7.58	4.29	5.33	3.18	4.53	5.76	5.03		5.77
Std err	1.36	2.40	1.29	0.76	0.77	1.60	1.82	1.52		2.18
Face 2	(N=15)		(N=10)	(N=47)	(N=15)		(N=8)	(N=11)		(N=5)
Maximum	14	17	12	22	13	20	16	10		25
Median	6	5	8	8	8	8	8.5	7		9
Minimum	2	2	1	2	1	1	4	3		5
Mean	6.87	6.30	7.40	8.89	7.93	10.50	9.38	7.00		11.20
Std dev	3.34	4.11	3.50	4.47	3.33	7.23	4.72	2.57		7.95
Std err	0.86	1.30	1.11	0.65	0.86	2.56	1.67	0.77		3.56

of maximal width. This ratio has a minimal value of 1.00, which hypothetically describes a handaxe whose maximal width is located at the distal end. The higher this ratio, the closer the maximal width of the handaxe to the proximal end. The roundness of edges ratio is calculated by dividing the width at mid-length by the maximal width. This ratio has a maximal value of 1.00, which indicates that the width at mid-length is equal to the maximal width. Such a value reflects a planform with straight to rounded edges; as the value decreases, it signifies edges that are more convergent toward one of the ends (usually the distal end). The last ratio is the pointedness ratio, which is calculated by dividing the width at three quarters of the length (at GBY four fifths of the length) by the maximal width. This ratio has a maximal value of 1.00, which hypothetically indicates an item whose maximal width is at its upper fifth and hence has a very rounded distal end. As the value of this ratio decreases, the distal end of the handaxe will be more pointed.

An additional attribute describing the morphology of the handaxe is the shape of its cross-section. This qualitative and discrete attribute is obtained by direct observation and provides information on the general shape of the item as it appears in cross-section, an observation that is not expressed by the quantitative ratios described above (Appendix 4).

The description of the morphology of the cleavers is slightly different from that of the handaxes. Bordes (1961) did not provide quantitative ratios for the description of their shapes. Our description of their morphology, therefore, is based exclusively on qualitative attributes that include the shape of the base, lateral edges, cross-section, and working edge, as well as the design of the working edge. These observations provide a comprehensive description and permit analysis of the morphology of the GBY cleavers (Appendix 5).

7.5.5.1 Morphology of Basalt Handaxes

Table 7.106 presents the descriptive statistics of the quantitative morphological attributes of basalt handaxes from the various

Table 7.103 Counts and frequencies (%) of location* of retouch on both faces of basalt handaxes of Layer II-6

	Face 1																																
	Distal		Left		Right		Both edges		Convergent		Distal & both edges		Distal & left		Circumference		Proximal & both edges		Proximal & left		Proximal & right		Entirely covered		Distal & proximal & left		Distal & proximal & right		Total Face 2	
Face 2	N	%	N	%	N	%	N	%	N	%	N	%	N	%	N	%	N	%	N	%	N	%	N	%	N	%	N	%	N	%
Distal	1	0.31	-	-	-	-	-	-	-	-	-	-	-	-	-	-	1	0.31	-	-	-	-	1	0.31	-	-	-	-	3	0.93
Proximal	-	-	-	-	-	-	-	-	-	-	-	-	-	-	1	0.31	1	0.31	-	-	2	0.62	8	2.48	-	-	-	-	12	3.72
Left	1	0.31	1	0.31	4	1.24	1	0.31	-	-	-	-	-	-	5	1.55	3	0.93	2	0.62	-	-	18	5.57	1	0.31	-	-	36	11.15
Right	-	-	5	1.55	4	1.24	2	0.62	1	0.31	1	0.31	-	-	5	1.55	2	0.62	3	0.93	1	0.31	24	7.43	-	-	-	-	48	14.86
Both edges	-	-	1	0.31	2	0.62	3	0.93	2	0.62	-	-	1	0.31	-	-	-	-	-	-	1	0.31	19	5.88	-	-	-	-	29	8.98
Convergent	-	-	-	-	-	-	1	0.31	-	-	5	1.55	-	-	-	-	2	0.62	1	0.31	-	-	8	2.48	-	-	-	-	17	5.26
Distal & left	1	0.31	-	-	-	-	-	-	-	-	-	-	1	0.31	-	-	-	-	-	-	-	-	1	0.31	-	-	-	-	3	0.93
Distal & right	-	-	1	0.31	-	-	-	-	-	-	-	-	-	-	1	0.31	-	-	-	-	-	-	2	0.62	-	-	-	-	4	1.24
Circumference	-	-	1	0.31	1	0.31	-	-	1	0.31	-	-	-	-	5	1.55	2	0.62	-	-	-	-	15	4.64	-	-	1	0.31	26	8.05
Proximal & both edges	-	-	1	0.31	-	-	-	-	-	-	-	-	1	0.31	-	-	1	0.31	1	0.31	-	-	12	3.72	-	-	-	-	16	4.95
Proximal & left	-	-	1	0.31	1	0.31	-	-	2	0.62	-	-	-	-	2	0.62	-	-	-	-	1	0.31	9	2.79	-	-	-	-	16	4.95
Proximal & right	-	-	-	-	-	-	-	-	-	-	-	-	-	-	6	1.86	-	-	3	0.93	1	0.31	14	4.33	-	-	-	-	24	7.43
Entirely covered	-	-	2	0.62	2	0.62	2	0.62	2	0.62	1	0.31	-	-	3	0.93	-	-	3	0.93	-	-	67	20.74	-	-	-	-	82	25.39
Distal & proximal & left	-	-	-	-	-	-	-	-	-	-	-	-	-	-	-	-	-	-	-	-	-	-	2	0.62	1	0.31	-	-	3	0.93
Distal & proximal & right	-	-	-	-	-	-	-	-	-	-	-	-	-	-	-	-	-	-	-	-	-	-	2	0.62	-	-	-	-	2	0.62
Convergent & proximal	-	-	-	-	-	-	-	-	-	-	1	0.31	-	-	-	-	-	-	-	-	1	0.31	-	-	-	-	-	-	2	0.62
Total Face 1	3	0.93	13	4.02	15	4.64	8	2.48	14	4.33	3	0.93	2	0.62	30	9.29	11	3.41	12	3.72	7	2.17	202	62.54	2	0.62	1	0.31	323	10

* note that when the observations of the location of retouch were carried out, the analyst was facing each of the analyzed faces separately; thus the right edge of Face 1 is the left edge of Face 2

Table 7.104 Counts and frequencies (%) of location* of retouch on both faces of basalt cleavers of Layer II-6

	Face 1																												
	Left		Right		Both edges		Convergent		Distal & both edges		Circumference		Proximal & both edges		Proximal & left		Proximal & right		Entirely covered		Distal & proximal & right		Total Face 2		
Face 2	N	%	N	%	N	%	N	%	N	%	N	%	N	%	N	%	N	%	N	%	N	%	N	%
Distal	-	-	-	-	-	-	-	-	-	-	-	-	-	-	-	-	-	-	1	0.74	-	-	1	0.74
Proximal	-	-	-	-	-	-	-	-	-	-	-	-	2	1.48	-	-	-	-	4	2.96	-	-	6	4.44
Left	1	0.74	-	-	1	0.74	-	-	-	-	-	-	1	0.74	-	-	-	-	7	5.19	-	-	10	7.41
Right	-	-	1	0.74	1	0.74	-	-	-	-	2	1.48	3	2.22	1	0.74	1	0.74	8	5.93	-	-	17	12.59
Both edges	2	1.48	1	0.74	3	2.22	1	0.74	2	1.48	-	-	3	2.22	2	1.48	2	1.48	20	14.81	1	0.74	37	27.41
Distal & both edges	-	-	-	-	-	-	-	-	1	0.74	-	-	-	-	-	-	-	-	1	0.74	-	-	2	1.48
Distal & right	-	-	-	-	-	-	-	-	-	-	-	-	-	-	-	-	-	-	1	0.74	-	-	1	0.74
Circumference	-	-	-	-	-	-	-	-	-	-	-	-	-	-	-	-	-	-	5	3.70	-	-	5	3.70
Proximal & both edges	-	-	-	-	3	2.22	1	0.74	-	-	1	0.74	3	2.22	-	-	1	0.74	15	11.11	-	-	24	17.78
Proximal & left	-	-	-	-	1	0.74	-	-	-	-	1	0.74	-	-	-	-	1	0.74	5	3.70	-	-	8	5.93
Proximal & right	-	-	-	-	2	1.48	-	-	-	-	-	-	2	1.48	-	-	1	0.74	11	8.15	-	-	16	11.85
Entirely covered	-	-	-	-	1	0.74	-	-	-	-	-	-	-	-	-	-	-	-	6	4.44	-	-	7	5.19
Indeterminate	-	-	-	-	-	-	-	-	-	-	-	-	-	-	-	-	-	-	1	0.74	-	-	1	0.74
Total Face 1	3	2.22	2	1.48	12	8.89	2	1.48	3	2.22	4	2.96	14	10.37	3	2.22	6	4.44	85	62.96	1	0.74	135	100.00

* note that when the observations of the location of retouch were carried out, the analyst was facing each of the analyzed faces separately; thus the right edge of Face 1 is the left edge of Face 2

7.5 Bifacial Tools

Table 7.105 Bordes' morphological ratios and their calculations

Ratio	Calculation
Flatness ratio	maximal width / maximal thickness
Elongation index	maximal length / maximal width
Location of maximal width	maximal length / length to maximal width
Roundness of edges	width at half length / maximal width
Pointedness ratio	width at 3/4 of length / maximal width

levels of Layer II-6. In order to avoid morphological biases stemming from broken items, the ratios were calculated only for complete handaxes and those that have a slightly broken tip.

The flatness ratio presents a relatively uniform trend across the various levels. The mean values of the different assemblages range from a maximal value of 2.20 in Level 4b to a minimal value of 1.77 in Layer VI-14. This trend of homogeneity is even more pronounced in view of the fact that the two lowest mean values appear in the two smallest assemblages, which consist of only two and three tools each. If these two assemblages are excluded on the basis of sample size, the minimal value is 1.97 in Level 3, resulting in an even more homogeneous distribution of the mean flatness ratio across levels. Nevertheless, when each of the assemblages is examined individually a more variable trend can be seen, with a maximal value of 3.88 in Level 1 and a minimal value of 1.00 in Level 5. Furthermore, the standard deviations are relatively high, ranging from 0.17 to 0.57 (excluding the two small assemblages). The three largest assemblages (Levels 1, 4, and 4b) present quite similar

Table 7.106 Descriptive statistics of the morphological ratios of basalt handaxes of Layer II-6

	Layer									
	II-6/L1	II-6/L2	II-6/L3	II-6/L4	II-6/L4b	II-6/L5	II-6/L6	II-6/L7	VI-14	Trench II green
N	43	18	8	105	32	3	7	13	2	11
Flatness ratio										
Maximum	3.88	2.93	3.00	3.20	3.23	2.82	3.06	2.85	1.78	2.34
Median	2.09	2.06	2.03	2.14	2.10	1.67	2.27	1.98	1.77	2.15
Minimum	1.27	1.61	1.03	1.34	1.43	1.00	1.67	1.33	1.76	1.86
Mean	2.16	2.12	1.97	2.19	2.20	1.83	2.30	2.05	1.77	2.15
Std dev	0.45	0.41	0.57	0.36	0.42	0.92	0.44	0.48	0.01	0.17
Std err	0.07	0.10	0.20	0.04	0.07	0.53	0.17	0.13	0.01	0.05
Elongation index										
Maximum	1.80	1.79	2.68	2.22	2.13	1.86	1.71	2.02	1.45	1.81
Median	1.48	1.46	1.57	1.67	1.65	1.85	1.48	1.51	1.43	1.36
Minimum	1.16	1.39	1.37	1.14	1.39	1.71	1.17	1.29	1.41	1.32
Mean	1.49	1.48	1.70	1.66	1.68	1.81	1.51	1.53	1.43	1.45
Std dev	0.17	0.10	0.42	0.18	0.16	0.08	0.18	0.17	0.03	0.16
Std err	0.03	0.02	0.15	0.02	0.03	0.05	0.07	0.05	0.02	0.05
Location of max width	(N=42)			(N=104)				(N=9)		
Maximum	6.12	3.77	3.54	9.53	6.60	3.35	3.86	2.53	2.10	5.09
Median	2.32	2.24	2.55	2.60	2.78	3.00	2.44	2.33	1.99	2.30
Minimum	1.56	1.79	2.10	1.70	1.57	2.23	2.02	1.91	1.89	1.70
Mean	2.50	2.40	2.62	2.80	2.99	2.86	2.61	2.27	1.99	2.88
Std dev	0.79	0.54	0.44	0.93	0.91	0.57	0.70	0.21	0.15	1.16
Std err	0.12	0.13	0.15	0.09	0.16	0.33	0.26	0.07	0.11	0.35
Roundness of edges	(N=42)			(N=100)				(N=9)		
Maximum	1.00	1.00	1.00	1.00	1.00	0.96	1.00	0.97	1.00	1.00
Median	0.95	0.98	0.96	0.93	0.93	0.89	0.92	0.93	0.98	0.95
Minimum	0.82	0.83	0.74	0.72	0.80	0.89	0.86	0.89	0.97	0.90
Mean	0.94	0.96	0.94	0.92	0.93	0.91	0.93	0.94	0.98	0.95
Std dev	0.05	0.05	0.09	0.05	0.05	0.04	0.05	0.03	0.02	0.04
Std err	0.01	0.01	0.03	0.01	0.01	0.02	0.02	0.01	0.02	0.01
Pointedness ratio	(N=42)			(N=100)				(N=8)		
Maximum	0.80	0.83	0.68	0.75	0.85	0.69	0.71	0.70	0.85	0.70
Median	0.62	0.66	0.56	0.57	0.60	0.61	0.65	0.57	0.73	0.57
Minimum	0.21	0.49	0.48	0.30	0.45	0.49	0.52	0.47	0.61	0.47
Mean	0.59	0.66	0.58	0.57	0.61	0.60	0.62	0.58	0.73	0.58
Std dev	0.12	0.10	0.08	0.08	0.09	0.10	0.08	0.09	0.17	0.06
Std err	0.02	0.02	0.02	0.01	0.02	0.06	0.03	0.03	0.12	0.02

variability in the flatness ratio, with Level 4 being the least variable and Level 1 the most variable.

The elongation index presents a similar trend but even greater homogeneity. The mean values of the various assemblages range from 1.48 in Level 2 to 1.81 in Level 5. However, if we omit the latter because of its small sample size, the highest mean value is 1.70 in Level 3. These values of the elongation index are fairly high, signifying that the mean maximal length of the handaxes is 1.50 times greater than their mean maximal width. The intra-assemblage variability for this index seems to be smaller as well. The values range from a minimum of 1.14 in Level 4 to a maximum of 2.68 in Level 3. The standard deviations also reflect this trend, as all but one (0.42 in Level 3) are lower than 0.2.

The location of maximal width also shows a trend toward uniformity. In all assemblages (excluding Layer VI-14) the mean values are higher than 2.00. These values indicate that in all assemblages the mean location of maximal width is always in the lower half of the handaxe, that is, closer to the proximal end. The inter-assemblage variability of this ratio is also fairly low, as no assemblage reaches a mean value of 3.00 (indicating that the maximal width is located in the lower third of the item). However, similarly to the flatness ratio, higher levels of variability are observed within the individual assemblages. The lowest value in all assemblages is 1.57 in Level 4b, a value that places the maximal width almost at the distal quarter of the artifact. The highest value in all assemblages is 9.53 in Level 4, which locates the maximal width at the very proximal end of the handaxe. The standard deviations also point toward this trend, as they range from a minimum of 0.15 (0.21 if the small sample of Layer VI-14 is excluded) to a maximum of 1.16.

The roundness of edges ratio follows the general trend of inter-assemblage homogeneity but also presents the highest intra-assemblage homogeneity. The mean values of the different assemblages range from 0.91 to 0.98. These values, which are very close to the highest possible value for this index (1.00), indicate that for the majority of handaxes the width at mid-length is almost identical to the maximal width. The minimal value for this ratio is 0.72 (Level 4) and the maximal value is exactly 1.00 in all assemblages except Levels 5 and 7. The exceptionally high intra-assemblage homogeneity can also be seen in the standard deviations, which, with the exception of Level 3, do not surpass 0.05.

The last quantitative ratio is that of pointedness. This ratio too displays relatively low intra- and inter-assemblage variability. The mean values in the different assemblages range from a minimum of 0.57 in Level 4 to a maximum of 0.73 in Layer VI-14; if the latter is omitted due to small sample size, the next highest ratio is that of Level 2, with a value of 0.66. The lowest value in all assemblages is 0.21 in Level 1, while the highest is 0.85 in Level 4b. The standard deviations also point to the fact that the assemblages are very homogeneous, with a maximal value of 0.12.

It should be noted that in Layer II-6 there are a few handaxes (N=7) with a cleaver-like distal end (chisel-end) (Figs. 7.160, 7.186–7.189). These artifacts could not be classified as cleavers according to Roe's (1994) definition, as the length of their working edge is less than half of their maximal width. The classification of these bifaces as "cleaver-like handaxes" is based on qualitative observations, since this phenomenon cannot be efficiently detected by the pointedness ratio, which is based on the measurement of width at the upper fifth of the item and not at its distal end.

Table 7.107 shows the distribution of cross-section shapes across the different handaxe assemblages. As this attribute is not quantitative, the table also includes observations made on

Fig. 7.186 Basalt handaxe with cleaver-like distal end (#244 Layer II-6 Level 4)

Fig. 7.187 Basalt handaxe with cleaver-like distal end (#264 Layer II-6 Level 4)

7.5 Bifacial Tools

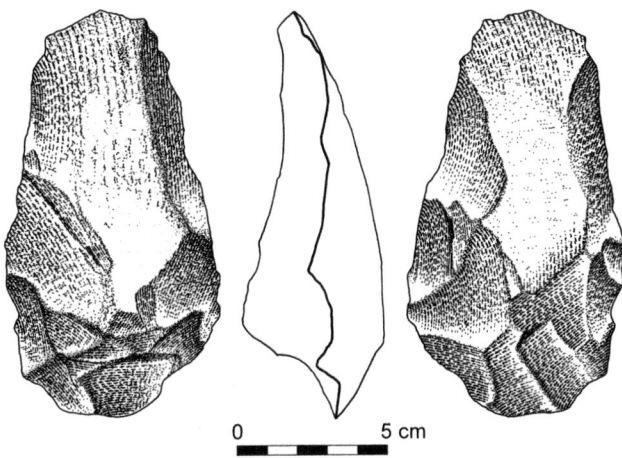

Fig. 7.188 Basalt handaxe with cleaver-like distal end (#5772 Layer II-6 Level 4b)

broken handaxes whose cross-section shape was not affected by the breakage.

The most common cross-section shape in all levels but Level 2 is clearly lenticular, followed in decreasing frequency by flat lenticular, plano-convex, and thick. Other cross-section shapes, such as angular, backed, triangular (Figs. 7.184, 7.188–7.190), and flat triangular, also appear but in substantially smaller frequencies.

The various morphological attributes point to relatively homogeneous shapes of handaxes in the different assemblages. Examination of the morphology of handaxes according to the metrical description and analysis method of Roe (1964, 1968) yields similar observations to the above. This is to be expected, given the fact that Roe's morphological ratios differ only slightly from those of Bordes (1961). Applying Roe's method (Fig.

Fig. 7.189 Basalt handaxe with cleaver-like distal end (#193 Layer II-6 Level 4)

7.191) shows that the vast majority of the handaxes of Layer II-6 are either pointed or oval, meaning that their maximal width is closer to their proximal end. The oval type is the more common, but pointed handaxes form a significant portion. Furthermore, Roe's diagram clearly illustrates that the handaxes of all types are tightly clustered in respect to pointedness and elongation,

Table 7.107 Distribution of cross-section shapes of basalt handaxes of Layer II-6

	II-6/L1	II-6/L2	II-6/L3	II-6/L4	II-6/L4b	II-6/L5	II-6/L6	II-6/L7	VI-14	Trench II green
N	(N=48)	(N=11)	(N=10)	(N=156)	(N=53)	(N=9)	(N=17)	(N=18)	(N=8)	(N=21)
Lenticular	18	1	3	59	24	-	10	8	3	8
	37.50	9.09	30.00	37.82	45.28		58.82	44.44	37.50	38.10
Plano-convex	7	2	2	21	5	3	1	1	-	3
	14.58	18.18	20.00	13.46	9.43	33.33	5.88	5.56		14.29
Thick	5	4	1	14	2	1	-	1	2	-
	10.42	36.36	10.00	8.97	3.77	11.11		5.56	25.00	
Angular	1	1	2	14	4	-	-	-	-	1
	2.08	9.09	20.00	8.97	7.55					4.76
Flat lenticular	10	2	2	22	10	2	4	4	2	6
	20.83	18.18	20.00	14.10	18.87	22.22	23.53	22.22	25.00	28.57
Backed	2	1	-	20	1	1	1	1	-	2
	4.17	9.09		12.82	1.89	11.11	5.88	5.56		9.52
Triangular	5	-	-	-	2	2	1	2	1	-
	10.42				3.77	22.22	5.88	11.11	12.50	
Flat triangular	-	-	-	-	2	-	-	-	-	-
					3.77					
Indeterminate	-	-	-	6	3	-	-	1	-	1
				3.85	5.66			5.56		4.76

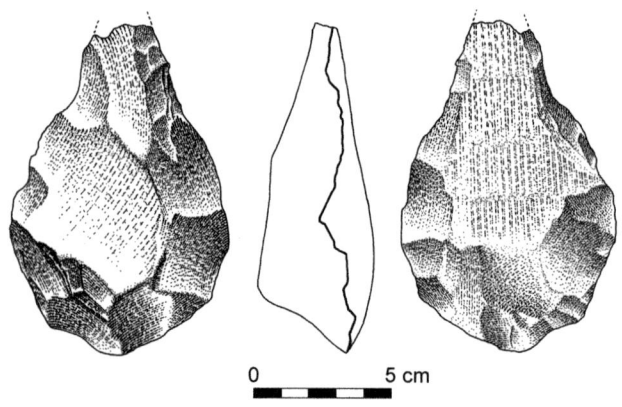

Fig. 7.190 Basalt handaxe with triangular cross-section (#7743 Layer II-6 Level 6)

supporting the morphological homogeneity discussed above.

To sum up, the most common handaxe shape at GBY is relatively thick, with its maximum width in the proximal half, lateral edges that range from straight and parallel to convex or distally converging, and a pointed distal end (Figs. 7.192–7.195). It will usually be lenticular in cross-section with a variable thickness ranging from flat to biconvex.

7.5.5.2 Morphology of Flint Handaxes

Flint handaxes make up only a minute fraction of the entire GBY handaxe assemblage. As with the basalt handaxes, the

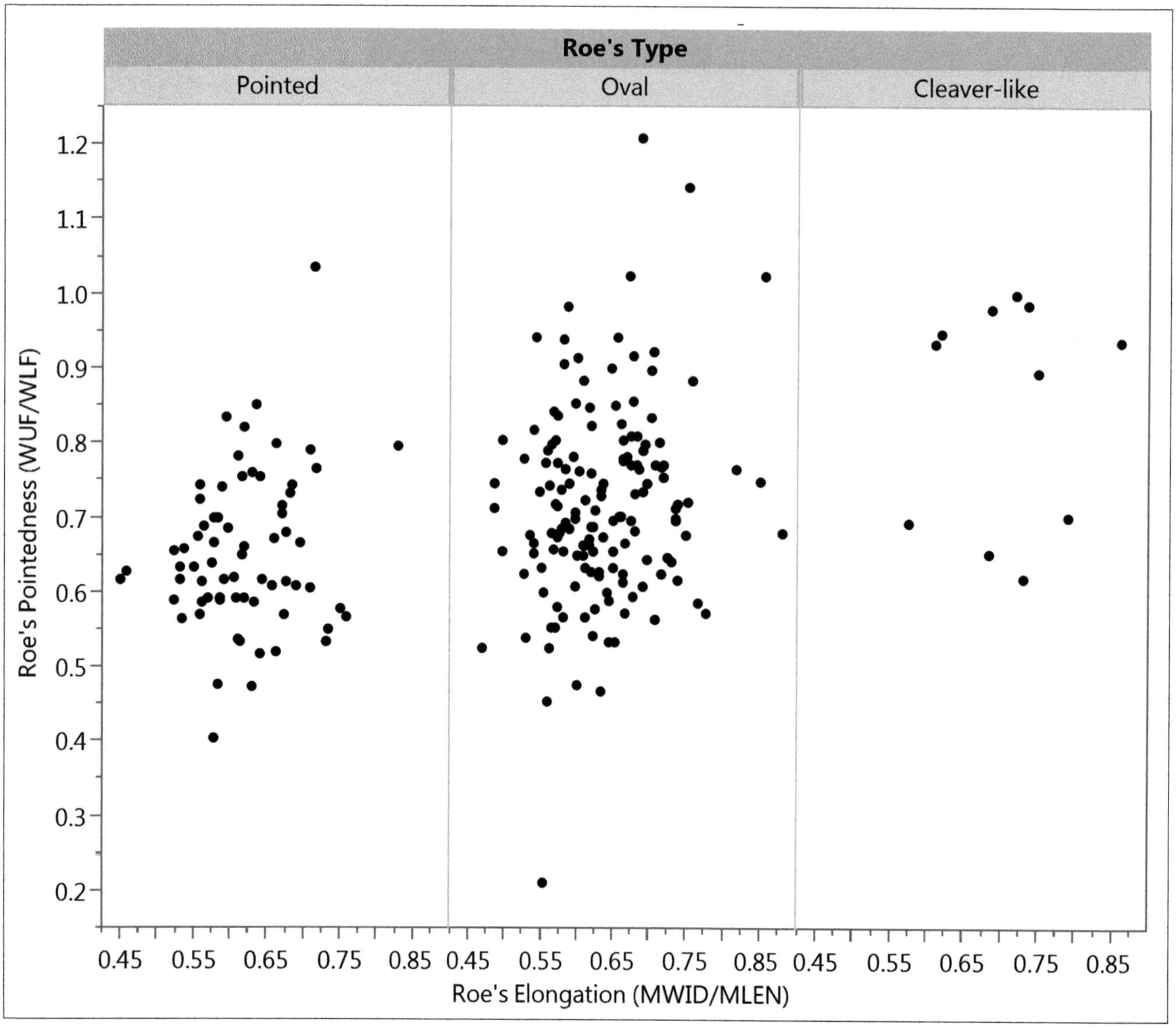

Fig. 7.191 A scatter plot presenting handaxe shape variability at GBY following Roe's method (1964)

7.5 Bifacial Tools

Fig. 7.192 Basalt handaxe with pointed distal end (#1023 Layer II-6 Level 3)

However, a difference can be seen between the assemblages of the two raw materials with respect to the cross-section. In the flint assemblage handaxes with a thick cross-section are the most common, in contrast to the basalt assemblages, where the thick cross-section is far less common. Flint handaxes with lenticular, flat lenticular, and triangular cross-sections are in joint second place. However, these data should be viewed with caution in light of the very small sample size.

Fig. 7.193 Basalt handaxe with pointed distal end (#5907 Layer II-6 Level 3)

Fig. 7.194 Basalt handaxe with pointed distal end (#81 Layer II-6 Level 4)

analysis and description of their morphologies are restricted here to complete artifacts, reducing even further the already small sample size. The distribution of complete flint handaxes across the different levels of Layer II-6 is presented in Table 7.108 (see Table 7.84 for the entire inventory of flint handaxes). Due to the extremely small sample sizes in each of the different levels, the flint handaxes are lumped together for the purposes of the following analysis.

Table 7.109 presents the descriptive statistics of the quantitative morphological attributes of flint handaxes and the distribution of the qualitative attribute of cross-section shape. The different values of the morphological attributes of flint handaxes all conform to the mean value ranges of basalt handaxes. This similarity is also reflected in the standard deviations and minimal and maximal values.

Fig. 7.195 Basalt handaxe with pointed distal end (#144 Layer II-6 Level 4)

Table 7.108 Distribution of complete flint handaxes by layer

Layer	N	%
II-6/L1	3	21.43
II-6/L2	3	21.43
II-6/L4	2	14.29
II-6/L4b	1	7.14
II-6/L5	2	14.29
Trench II green	3	21.43
Total	14	100.00

7.5.5.3 Morphology of Basalt Cleavers

Table 7.110 presents the morphological characteristics of basalt cleavers from the different levels of Layer II-6. Unlike the analysis of handaxes, which incorporated only complete tools, the analysis of cleavers was based on qualitative morphological observations that are not affected by breakage, and hence the analysis includes all basalt cleavers.

Three attributes were recorded to describe the planform shape of cleavers: the shapes of the lateral edges, base, and working edge. The most common shapes for lateral edges in all levels are straight and parallel, convex, or straight and convex. Combined, these shapes comprise more than half of the cleavers and are recorded in each of the levels. Next most common are the distally converging and irregular shapes, which appear in similar frequencies but are recorded in only about half of the levels. Proximally converging, hourglass (Figs. 7.196–7.197), concave and convex, and concave and straight lateral edges are very rare and appear on only a few isolated artifacts. While most of the levels are highly uniform, Level 4 is again more variable.

The proximal end (base) of the cleavers shows a more varied pattern. All three shapes (straight, pointed, and convex) appear in all assemblages but Level 6, where pointed bases are absent. In Levels 1 and 3 more than half of the cleavers have a convex base (Fig. 7.145), while in Level 2 the majority of cleavers are proximally pointed (Fig. 7.198). In Level 4 the three base shapes are almost evenly distributed. In all the levels below Level 4, at least 50% of proximal ends are straight.

The most common cross-section shapes of cleavers are trapeze, parallelogram, and lenticular, which appear in almost all levels in varying frequencies. In the largest assemblages, however, there seems to be a preference for the lenticular cross-section. The convex, rhombus, triangular, and flat triangular cross-sections are relatively uncommon and appear only sporadically.

A similar tendency can be seen in the shape of the working edge. In Levels 1–4 the most common cleavers are those with a straight working edge (Fig. 7.199) (in Level 3 this type comprises more than half of the cleavers). In Level 4b the straight and diagonal shapes are represented in equal frequencies (Figs. 7.200–7.202). In Levels 5 and 6 the most common shape is diagonal. In Level 7 and Layer VI-14 the most common shape is convex (Fig. 7.203, 7.146). These three types are present in most assemblages, along with the concave type, which also appears in almost all assemblages but in smaller frequencies. The pointed working edge is very rare and appears on only a few isolated cleavers.

The design of the working edge is very homogeneous across all assemblages. The pre-flake (designed prior to the detachment of the cleaver's flake blank) is clearly the most common in all levels of Layer II-6, since the pre-flake design conforms to the definition of the cleaver (Figs. 7.204–7.208). The shaping of the working edge by retouch, while rare, is sporadically present in most assemblages.

The cleaver assemblages of Layer II-6 present a morphological pattern that is generally similar to that of the handaxe assemblages. The differences between the levels are expressed in the frequencies of the different morphologies rather than their presence or absence. Consequently, cleavers of similar morphology appear throughout the sequence of Layer II-6. However, the difference in frequencies appears to be more pronounced in two specific morphological attributes: the shape of the base and that of the working edge. In these two attributes there seems to be a meaningful trend of differences in morphological preference over time, but these differences are in fact not statistically significant. For the three other

Table 7.109 Morphological characteristics of complete flint handaxes

	Descriptive statistics of quantitative morphological attributes (N=14)					Shape of cross-section (N=13)		
	Flatness ratio	Elongation index	Location of max. width	Roundness of edges	Pointedness ratio		N	%
Maximum	3.07	1.80	3.70	1.00	0.83	Lenticular	2	15.38
Median	1.88	1.50	2.37	0.94	0.59	Plano-convex	1	7.69
Minimum	1.17	1.26	1.66	0.89	0.43	Thick	4	30.77
Mean	1.94	1.51	2.40	0.94	0.60	Angular	1	7.69
Std dev	0.58	0.20	0.55	0.04	0.12	Flat lenticular	2	15.38
Std err	0.15	0.05	0.15	0.01	0.03	Triangular	2	15.38
						Indeterminate	1	7.69

7.5 Bifacial Tools

Table 7.110 Morphological characteristics of basalt cleavers of Layer II-6

	II-6/L1	II-6/L2	II-6/L3	II-6/L4	II-6/L4b	II-6/L5	II-6/L6	II-6/L7	Trench II green
N	17	10	12	60	18	8	11	12	8
Lateral edges	(N=14)			(N=59)		(N=7)	(N=10)		
Straight parallel	1 7.14	4 40.00	4 33.33	8 13.56	2 11.11	2 28.57	2 20.00	2 16.67	3 37.50
Distally converging	1 7.14	2 20.00	2 16.67	5 8.47	2 11.11	-	3 30.00	1 8.33	-
Proximally converging	-	1 10.00	1 8.33	2 3.39	-	-	-	-	-
Convex & straight	2 14.29	-	2 16.67	15 25.42	5 27.78	3 42.86	1 10.00	5 41.67	4 50.00
Hourglass	-	-	-	5 8.47	1 5.56	-	-	1 8.33	-
Convex	8 57.14	2 20.00	1 8.33	13 22.03	4 22.22	2 28.57	3 30.00	3 25.00	1 12.50
Irregular	2 14.29	1 10.00	1 8.33	3 5.08	1 5.56	-	-	-	-
Convex & concave	-	-	1 8.33	5 8.47	1 5.56	-	1 10.00	-	-
Concave & straight	-	-	-	-	2 11.11	-	-	-	-
Indeterminate	-	-	-	3 5.08	-	-	-	-	-
Base	(N=14)			(N=59)		(N=7)	(N=10)		
Straight	1 7.14	2 20.00	2 16.67	15 25.42	9 50.00	4 57.14	7 70.00	6 50.00	4 50.00
Pointed	2 14.29	6 60.00	4 33.33	20 33.90	6 33.33	1 14.29	-	4 33.33	2 25.00
Convex	10 71.43	2 20.00	6 50.00	21 35.59	3 16.67	2 28.57	3 30.00	2 16.67	1 12.50
Indeterminate	1 7.14	-	-	3 5.08	-	-	-	-	1 12.50
Cross-section	(N=16)	(N=9)	(N=9)	(N=45)	(N=16)	(N=6)	(N=8)	(N=10)	(N=6)
Trapeze	3 18.75	3 33.33	4 44.44	4 8.89	4 25.00	1 16.67	3 37.50	2 20.00	-
Parallelogram	1 6.25	2 22.22	2 22.22	7 15.56	1 6.25	4 66.67	-	4 40.00	1 16.67
Lenticular	5 31.25	1 11.11	-	22 48.89	6 37.50	1 16.67	3 37.50	1 10.00	2 33.33
Convex	2 12.50	-	-	2 4.44	-	-	-	-	-
Triangular	1 6.25	1 11.11	-	3 6.67	1 6.25	-	-	2 20.00	-
Rhombus	-	1 11.11	1 11.11	5 11.11	4 25.00	-	-	1 10.00	2 33.33
Flat triangular	-	-	-	-	-	-	-	-	1 16.67
Indeterminate	4 25.00	1 11.11	2 22.22	2 4.44	-	-	2 25.00	-	-
Working edge shape		(N=9)	(N=11)	(N=55)	(N=16)	(N=7)	(N=8)		(N=4)
Straight	6 35.29	4 44.44	6 54.55	24 43.64	6 37.50	1 14.29	1 12.50	2 16.67	1 25.00
Convex	2 11.76	2 22.22	-	19 34.55	1 6.25	2 28.57	1 12.50	5 41.67	3 75.00
Concave	4 23.53	2 22.22	2 18.18	4 7.27	1 6.25	-	1 12.50	2 16.67	-
Pointed	2 11.76	-	1 9.09	2 3.64	-	-	-	-	-
Diagonal	-	1 11.11	1 9.09	-	6 37.50	3 42.86	4 50.00	3 25.00	-
Indeterminate	3 17.65	-	1 9.09	6 10.91	2 12.50	1 14.29	1 12.50	-	-

Table 7.110 (cont.)

	II-6/L1	II-6/L2	II-6/L3	II-6/L4	II-6/L4b	II-6/L5	II-6/L6	II-6/L7	Trench II green
N	17	10	12	60	18	8	11	12	8
Working edge design		(N=8)	(N=11)	(N=55)	(N=15)	(N=7)	(N=8)		(N=4)
Pre-flake	13	5	10	40	10	4	3	6	4
	76.47	62.50	90.91	72.73	66.67	57.14	37.50	50.00	100.00
Retouch	2	3	-	1	1	1	-	3	-
	11.76	37.50		1.82	6.67	14.29		25.00	
Indeterminate	2	-	1	14	4	2	5	3	-
	11.76		9.09	25.45	26.67	28.57	62.50	25.00	

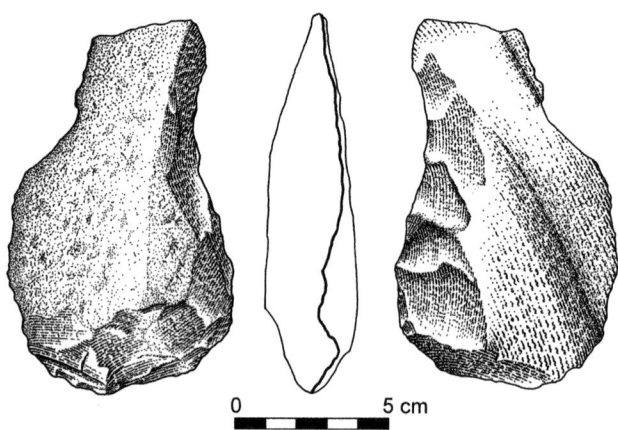

Fig. 7.196 Basalt cleaver of hourglass form (#1102 Layer II-6 Level 4b)

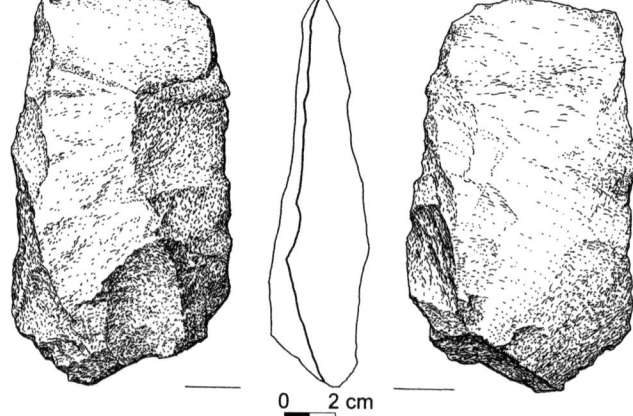

Fig. 7.199 Basalt cleaver with straight working edge (#220 Layer II-6 Level 4)

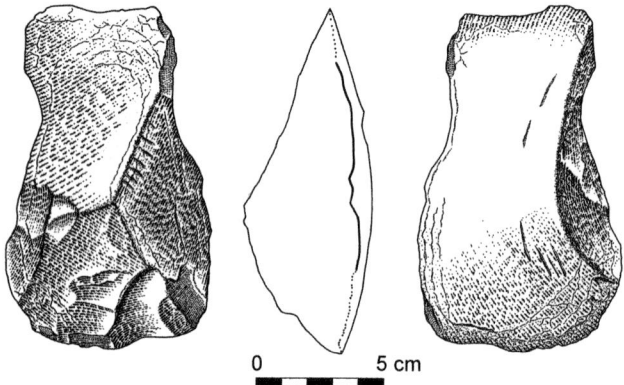

Fig. 7.197 Basalt cleaver of hourglass form (#7783 Layer II-6 Level 7)

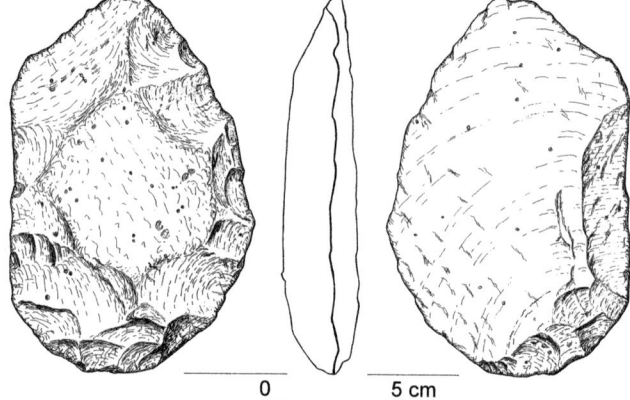

Fig. 7.200 Basalt cleaver with diagonal working edge (#5977 Layer V-6)

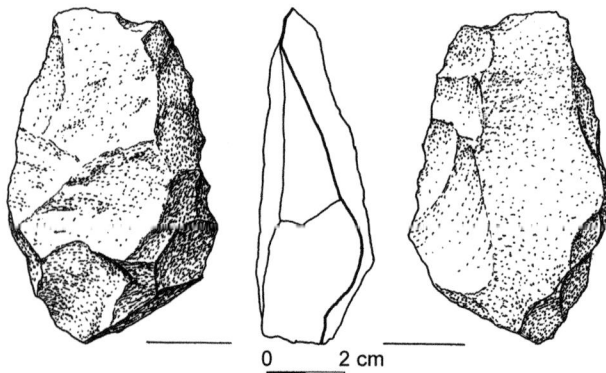

Fig. 7.198 Basalt cleaver with proximally pointed base (#269 Layer II-6 Level 4)

morphological attributes of lateral edges, cross-section, and working edge design, the differences in frequencies do not reflect a clear diachronic trend.

In general, the GBY cleavers have straight, convex, or pointed proximal ends, straight to convex lateral edges, and straight, convex, concave, or diagonal working edges. In cross-section their shapes are usually lenticular, trapezoid, or parallelogram. Thus, despite the intra-level variability, the different morphologies of cleavers occur in all levels but in changing frequencies (Fig. 7.209). It can be concluded that the cleaver assemblages of GBY also express the conservatism that is apparent in the morphology of handaxes.

7.5 Bifacial Tools

Fig. 7.201 Basalt cleaver with diagonal working edge (#5670 Layer II-6 Level 1)

Fig. 7.203 Basalt cleaver with convex working edge (#176 JB)

Fig. 7.202 Basalt cleaver with diagonal working edge (#5835 Layer II-6 Level 5)

7.5.6 Typology of Bifaces

The handaxes of GBY were typologically classified following Bordes (1961). This typological classification is based mainly on the morphology of the artifact and in several instances also on observations that are not strictly morphological (e.g. cortex). The values of the quantitative morphological ratios serve as the criteria for the classification of the handaxes into typological groups according to a number of threshold values (Table 7.111).

The flatness ratio serves as the primary criterion for the following typological classification, as it divides the handaxes into two main classes of flat and thick tools. The threshold determined by Bordes for a flat handaxe is a minimal value of 2.35. Each of the two flatness classes has corresponding, but different, typological groups. The other ratios used to determine the typological classification are the location of maximum width and the roundness of edges. These two ratios serve as x and y coordinates for each handaxe, which can then be plotted on a scatter plot. The scatter plot is divided into four shape zones separated by three straight-line functions, so that each shape zone in each flatness class represents a typological group (Table 7.112). There are two exceptions to this principle. The first is that in the thick handaxe class the two uppermost shape zones are not separated and items falling within both of them are classified as lanceolate. The second exception is that in the thin handaxe class the lower shape zone includes three different typological groups based on the elongation ratio. It should be noted that, while additional typological groups exist in this method, they are not determined by objective quantitative criteria but rather by subjective observations such as the refinement of edges. These were not taken into account in the analyses of the GBY handaxes, which followed only quantitative criteria.

Fig. 7.204 Basalt cleaver (#5976 Layer V-5)

Fig. 7.205 Basalt cleaver (#9896 JB)

Fig. 7.206 Basalt cleaver (#8646 Layer II-6 Level 3)

Fig. 7.207 Basalt cleaver (#13763 Layer II-6 Level 3)

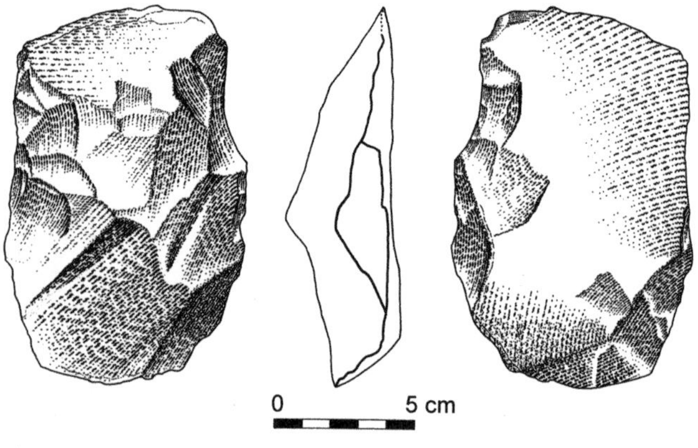

Fig. 7.208 Basalt cleaver (#7784 Layer II-6 Level 7)

7.5.6.1 Typology of Basalt Handaxes

As the typological classification is dependent on the morphology of the handaxe and broken artifacts could therefore bias the results, only complete handaxes and those with slightly broken tips were included in the analysis. In addition, samples of small size were excluded and only the handaxes from the different levels of Layer II-6 were included.

The studied sample displays several typological patterns. In terms of elongation, in most of the levels a majority of the handaxes are non-elongated (Table 7.113). However, in two of the richest levels (4 and 4b) there is a clear majority of elongated

7.5 Bifacial Tools

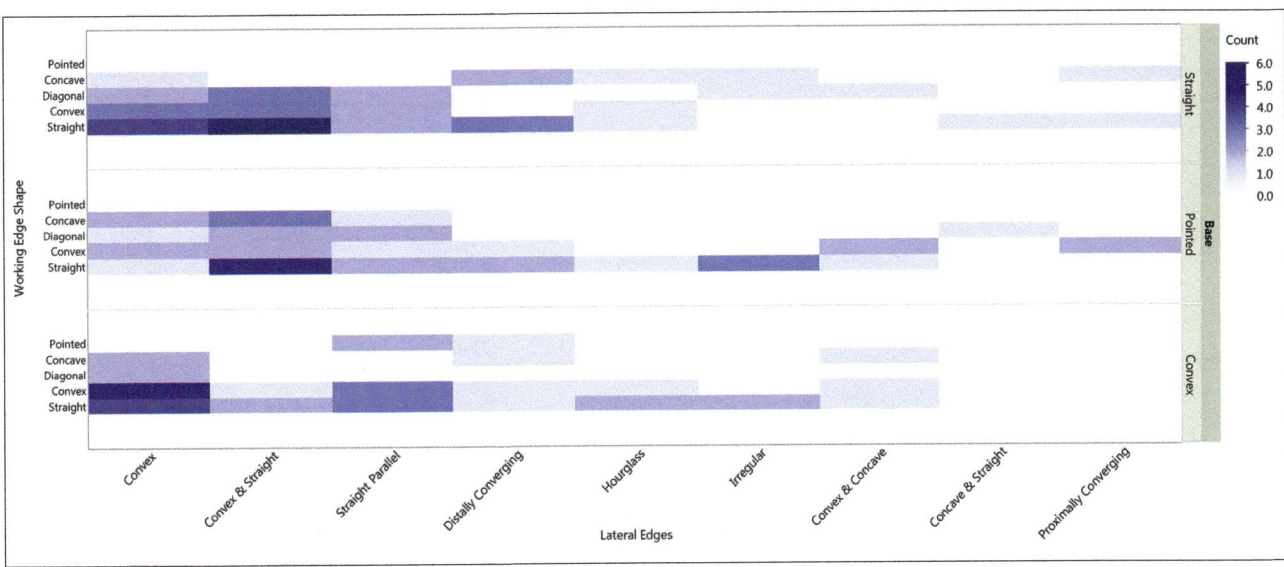

Fig. 7.209 A heat-map presenting the GBY morphological attributes of cleavers' planform

Table 7.111 Morphological ratios and threshold values

Ratio	Threshold values
Flatness ratio	*Thick* ≤ 2.35 < *Flat*
Elongation index	1.5 < *Elongated*
Location of maximum width	$L = 4.75R + 0.975$
	$L = 4.75R - 0.325$
Roundness of edges	$L = 4,75R - 1.625$

Table 7.112 Bordes' shape zones and corresponding typology

Zone	Upper limit	Lower limit	Flat category	Thick category
I	-	Y=4.75X+0.975	Triangular	Lanceolate
II	Y=4.75X+0.975	Y=4.75X-0.325	Subtriangular	
III	Y=4.75X-0.325	Y=4.75X-1.625	Cordiform	Amygdaloid
IV	Y=4.75X-1.625	-	Diverse (discoid, ovate, limande)	Core-like biface (fusiform)

handaxes. The flatness classes show a much more significant pattern, as in all levels the vast majority of handaxes fall within the thick class (Table 7.113). With respect to the distribution of shape zones in the different levels, it can be seen that in almost all levels the vast majority of items fall within shape zone IV (Table 7.114). Shape zone III is also represented in relatively high frequencies, while items that fall within shape zones I and II are very rare. Naturally, these patterns affect the typological distribution of the sample, so that in all levels the thick types are the most common (Table 7.115).

If we follow the typology of Bordes, the core-like handaxe is the most frequent in all levels but two, Levels 5 and 6, which are also two of the smallest assemblages. The core-like biface, sometimes referred to as fusiform when both of its ends are pointed, was very poorly defined by Bordes, due to the fact that this type appears very rarely in the European Acheulian assemblages studied by him. In general this type can be described as a thick handaxe of a relatively rounded planform. The next most common type is the amygdaloid, occurring in frequencies of more than 10% in most levels, while the lanceolate type is the

Table 7.113 Flatness class and elongation distribution of basalt handaxes of Layer II-6

	II-6/L1	II-6/L2	II-6/L3	II-6/L4	II-6/L4b	II-6/L5	II-6/L6	II-6/L7	II-6/LVI-14	Trench II green
N	43	18	8	105	32	3	7	13	2	11
Flatness class										
Flat	12	4	1	36	9	1	3	4	-	-
	27.91	22.22	12.50	34.29	28.13	33.33	42.86	30.77		
Thick	31	14	7	69	23	2	4	9	2	11
	72.09	77.78	87.50	65.71	71.88	66.67	57.14	69.23	100.00	100.00
Elongated										
Yes	12	2	3	71	22	3	2	2	-	2
	27.91	11.11	37.50	67.62	68.75	100.00	28.57	15.38		18.18
No	31	16	5	34	10	-	5	11	2	9
	72.09	88.89	62.50	32.38	31.25		71.43	84.62	100.00	81.82

Table 7.114 Shape zone distribution of basalt handaxes of Layer II-6

	II-6/L1	II-6/L2	II-6/L3	II-6/L4	II-6/L4b	II-6/L5	II-6/L6	II-6/L7	VI-14	Trench II green
N	43	18	8	105	32	3	7	13	2	11
	(N=42)			(N=100)			(N=9)			
Zone I	1 2.38	-	-	2 2.00	-	-	-	-	-	-
Zone II	1 2.38	-	1 12.50	3 3.00	2 6.25	-	-	-	-	2 18.18
Zone III	8 19.05	3 16.67	1 12.50	40 40.00	15 46.88	2 66.67	2 28.57	-	-	2 18.18
Zone IV	32 76.19	15 83.33	6 75.00	55 55.00	15 46.88	1 33.33	5 71.43	9 100.00	2 100.00	7 63.64

Table 7.115 Typological distribution of basalt handaxes of Layer II-6

	II-6/L1	II-6/L2	II-6/L3	II-6/L4	II-6/L4b	II-6/L5	II-6/L6	II-6/L7	VI-14	Trench II green
N	43	18	8	105	32	3	7	13	2	11
Type	(N=42)			(N=100)			(N=9)			
Lanceolate	1 2.38	-	1 12.50	4 4.00	1 3.13	-	-	-	-	2 18.18
Amygdaloid	5 11.90	3 16.67	1 12.50	28 28.00	9 28.13	1 33.33	2 28.57	-	-	2 18.18
Core-like	24 57.14	11 61.11	5 62.50	34 34.00	13 40.63	1 33.33	2 28.57	7 77.78	2 100.00	7 63.64
Triangular	-	-	-	1 1.00	-	-	-	-	-	-
Subtriangular	1 2.38	-	-	-	1 3.13	-	-	-	-	-
Cordiform	3 7.14	-	-	12 12.00	6 18.75	1 33.33	-	-	-	-
Discoid	2 4.76	-	-	1 1.00	-	-	1 14.29	1 11.11	-	-
Ovate	6 14.29	4 22.22	1 12.50	8 8.00	1 3.13	-	2 28.57	1 11.11	-	-
Limande	-	-	-	12 12.00	1 3.13	-	-	-	-	-

rarest among the thick forms. This distribution results from the fact that the vast majority of handaxes are thick and fall within the two lower shape zones. This pattern is valid for the thin forms as well: the most common types are cordiform, ovate, discoid, and limande, which correspond to the two lower shape zones. The subtriangular and triangular types, representing the two upper zones of the flat class, are represented by only three handaxes.

The typological distribution reflects a high level of homogeneity that conforms with the technological, morphological, and stylistic conservatism seen in the handaxe production sequence at GBY. However, the validity of this insight is limited by several problems stemming from the classification method. All the thresholds used as classification criteria were arbitrarily determined by Bordes (1961) based on his familiarity with the European Acheulian record. It can be seen from Fig. 7.210 that, while the spaces between the three boundaries could theoretically express the morphological variability that exists at GBY, their location prevents a better description of this variability. As shape zone IV has no lower boundary it could, by definition, comprise a more varied range of shapes than shape zones II and III. Furthermore, thick items that fall within shape zone IV do not have an additional classification criteria, in contrast to thin items from the same shape zone that are further classified by their elongation. An additional problem is the relatively crude manner in which this method builds upon the morphological attributes. One such example, which is especially relevant to the GBY assemblages, is the fact that the typological classification fails to consider the pointedness ratio. Consequently, when only Bordes' typology is considered, the impression formed is that the GBY assemblage consists mainly of handaxes with a very rounded planform, whereas in reality the artifacts are commonly pointed. Moreover, the relatively few measurements and ratios fail to take into consideration finer aspects of shape, such as the form and nature of edges and the distribution of thickness at different locations on the handaxes. Therefore, the data used in the typological classification of the handaxes exclude shape

7.5 Bifacial Tools

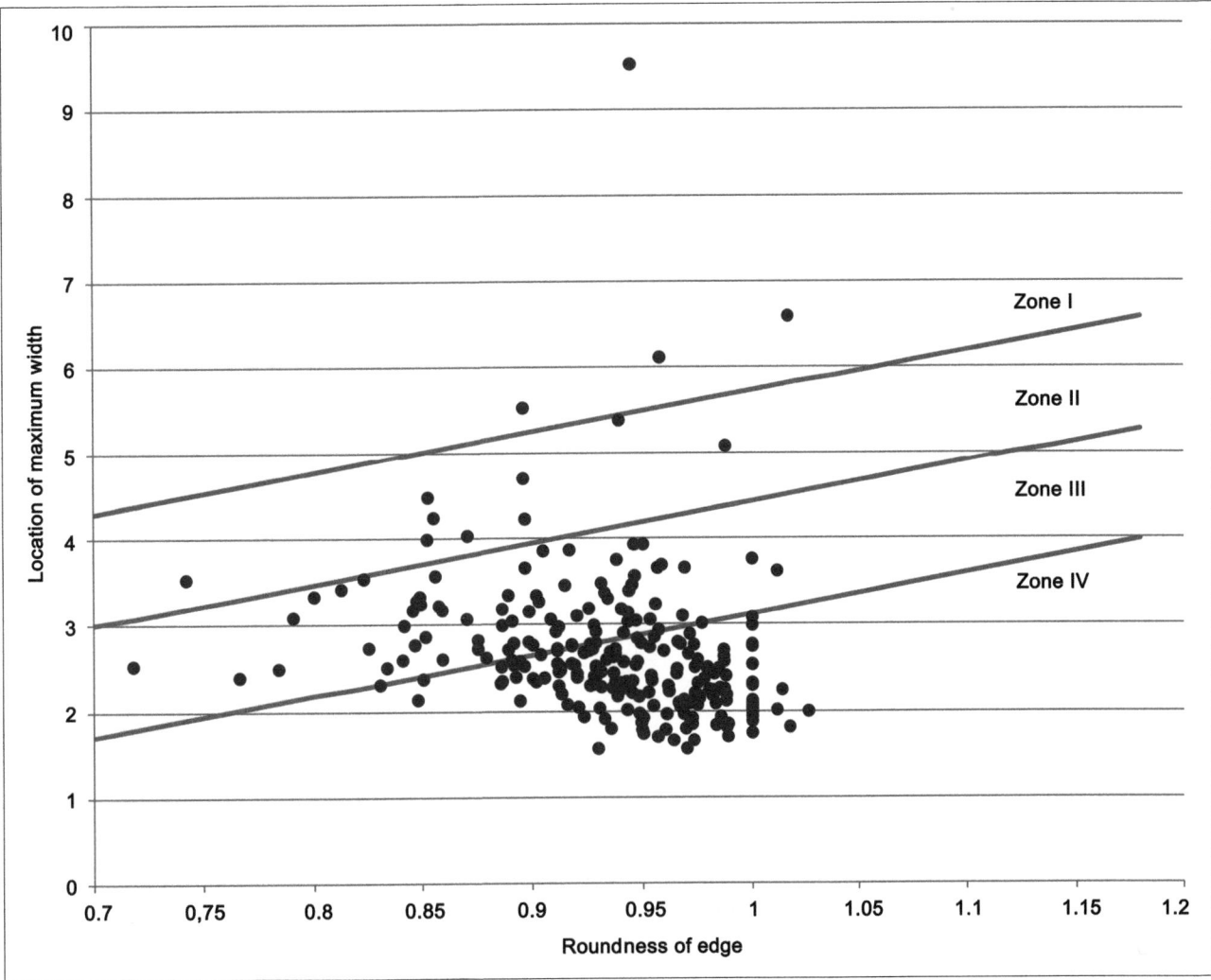

Fig. 7.210 A scatterplot presenting the distribution of GBY handaxes according to Bordes' (1961) shape zones

aspects that could be of significance for the description of the basalt handaxes at GBY.

7.5.6.2 Typology of Flint Handaxes

The small sample of flint handaxes was subjected to the typological analysis of Bordes (1961).

Table 7.116 presents the results of this typological classification, which suggest a morphological similarity between the flint and basalt handaxes. The majority of flint handaxes fall within the thick class and are not elongated. In terms of shape zone the pattern is almost identical to that of the basalt assemblage, as most of the items fall within shape zone IV and only two within shape zone III. The two upper shape zones are completely devoid of handaxes. Thus, most of the flint handaxes are of the core-like type, while the other types are represented by only a single artifact. These typological conclusions do not

Table 7.116 Typological classification data of flint handaxes

	N	%
Flatness class		Column
Flat	3	21.43
Thick	11	78.57
Total	14	100.00
Elongation		
Yes	6	42.86
No	8	57.14
Total	14	100.00
Shape zone		
Zone III	2	14.29
Zone IV	12	85.71
Total	14	100.00
Type		
Amygdaloid	1	7.14
Core-like	10	71.43
Cordiform	1	7.14
Discoid	1	7.14
Limande	1	7.14
Total	14	100.00

represent the true morphology of the pointed flint handaxes of GBY.

7.5.6.3 Typology of Cleavers

The typological classification of the GBY cleavers followed that of Tixier (1957), which is not based on the morphologies of the cleavers *per se* but rather on technological criteria such as blank type and working edge design. At GBY almost all cleavers fall within Type II, which is a cleaver made on a regular, non-cortical blank with the working edge shaped by a prior removal while the blank was still on the core. The other type, which occasionally appears in the assemblage, is Type VI, a cleaver made on a Kombewa flake (Figs. 7.143–7.144). Although the morphology of the GBY cleavers cannot be fully understood through the use of Tixier's typology, we present these data for the sake of comparison.

7.5.6.4 Predetermination of Cleavers

The concept of predetermination of cleavers is expressed in two different aspects: 1) obtaining predetermined flakes suitable for modification into cleavers and 2) forming a predetermined working edge (or cutting edge, as translated from the French *trenchant*) by using a particular scar configuration, a remnant of the core's debitage surface. These predeterminations provided a flake with a distinct distal morphology, expressed by 1) a wedge-like distal edge with an acute angle between the debitage surface and the ventral face of the blank, and 2) a straight working edge.

The flatness of the ventral face often did not require modification, and when modifications occur they are focused primarily on removing the striking platform and its relief by thinning the bulb of percussion. Thus, the conventional view of predetermination of cleavers focuses on the configuration of the scar forming the dorsal distal end (Fig. 7.210a). This scar, which exhibits many variations (and is shown in schematic illustrations) is documented in different lists of typological and technological attributes (Mourre 2003 and references therein). This view sees the sophistication of the predetermination as lying in the ability to design a particular scar pattern on the core's debitage surface, which will later form the distal end of the cleaver. This distal configuration is viewed as the main objective of cleaver production and as a requirement for the function of these tools (Mourre 2003 and references therein).

Through the study of the GBY cleavers we have gained a somewhat different perspective on hominin decision-making as

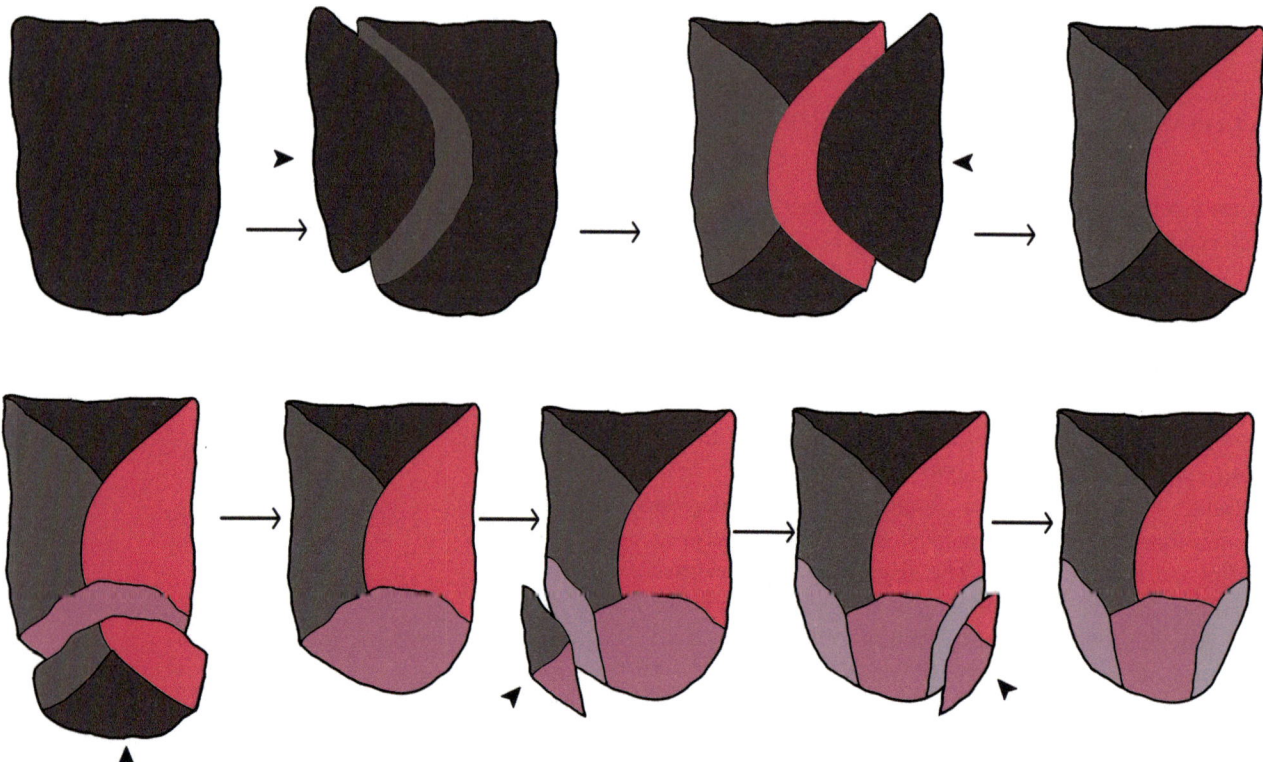

Fig. 7.210a Schematic illustration of cleaver modifications; working edge modification after detachment of the predetermined cleaver flake

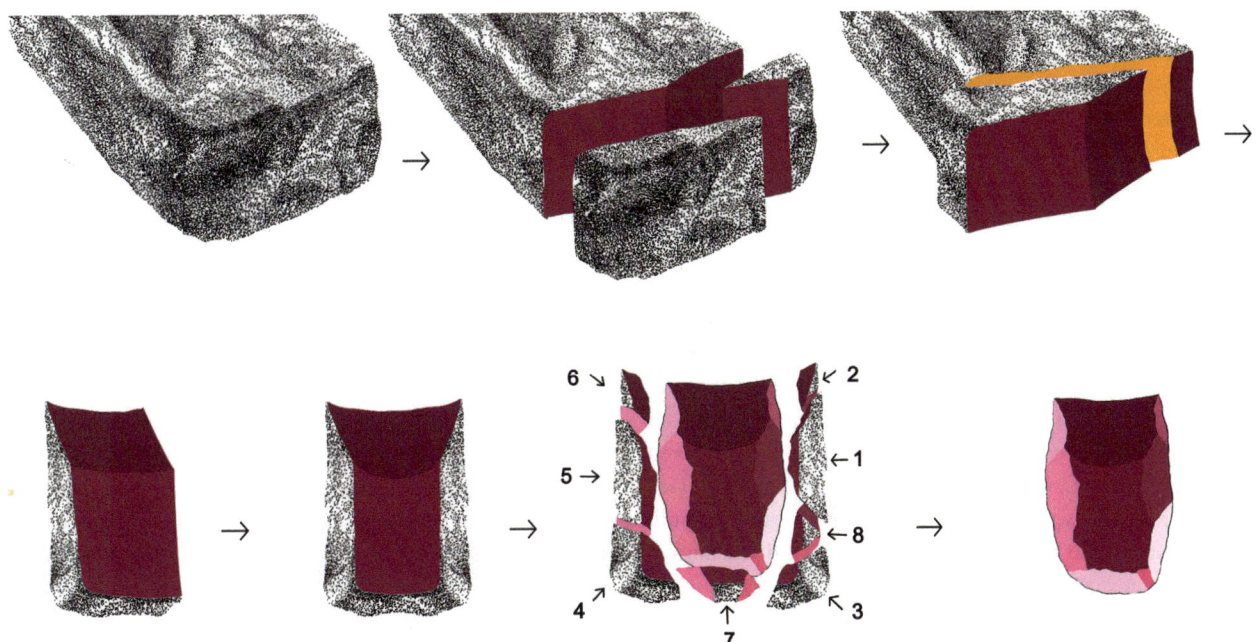

Fig. 7.210b Schematic illustration of cleaver modifications; hypothetical case of cleaver modification with a predetermined working edge (a scar of the giant core) and a series of flakes struck after the detachment of the predetermined cleaver flake

regards the production of cleavers. It is our view that for the efficient use of a cleaver and the achievement of controlled (functional) penetration, its distal end had to be formed in a way that provided an optimal shearing configuration.

As an addition to the traditional view, we suggest that the distal configuration of cleavers resulted not only from the core's debitage surface but also from the intentional shaping of the blank, which adhered to a desired cleaver design. According to this view, hominins were engaged mainly in volume/mass reduction by delineating the borders of the desired shearing surface. This delineated surface remains unflaked and its geometry has, in most cases, a somewhat triangular morphology; the triangle's base is the distal working edge and its opposing vertex is located in the middle of the cleaver, frequently at its thickest point (Figs. 7.210b, 7.175, 7.204).

In both methods, whether utilizing the original debitage surface or intentionally modifying it into a triangular shearing surface, the cleaver exhibits two intersecting surfaces, forming a wedge-like working edge.

In most of the GBY cleavers, the first stage, as at other LFA sites, was the extraction of a large flake with distal straight-edged planform morphology. The next stage was concerned with the reduction of the volume to produce a wedge-like section.

When the predetermined flake already had a cleaver-like configuration, the hominins made use of this. When this was not the case, they flaked around an "imaginary" triangular surface, reducing the volume to achieve the same configuration. Interestingly, at GBY the latter mode is more frequent (e.g. Figs. 7.174, 7.208).

7.6 Discussion and Summary of the Basalt Component

In different chapters of this volume we have demonstrated that the number of flint artifacts exceeds that of basalt artifacts and that limestone was selected by the GBY hominins mainly for use as percussive tools. The properties and sources of both flint and basalt were well known to the GBY knappers and they were used in sophisticated and skillful ways. Nevertheless, basalt artifacts form the bulk of the mass of the GBY lithic assemblages, which by far exceeds that of any other raw material.

In its use of basalt the GBY assemblage is unique among Levantine Acheulian sites, which (apart from the much older site of 'Ubeidiya) are characterized by an almost exclusive use of flint.

This chapter has presented detailed analyses of the basalt assemblages of GBY, analyses that encompass the taphonomic, typological, and technological characteristics of tools, flakes, cores, and bifaces. The results of these analyses show that there are no significant differences between the archaeological horizons and that when differences exist, they are concerned not with typology or technology but with varying frequencies of particular types of artifacts. This homogeneity, discernible throughout the cultural sequence, allows us to consider the GBY basalt assemblage as a whole for the purposes of the following discussion.

This discussion attempts to reconstruct varied aspects of the reduction sequences of basalt, which involve, among others, the relatively well-known production of bifaces. We have previously published papers on this subject, focusing mainly on bifaces (Goren-Inbar and Saragusti 1996), giant cores (Madsen and Goren-Inbar 2004; Goren-Inbar 2011; Goren-Inbar et al. 2011), massive scrapers (Goren-Inbar et al. 2008), and thin anvils (Goren-Inbar et al. 2015). Nevertheless, this chapter has provided various additional and original observations of these reduction sequences and illuminated others, enabling a comprehensive description of the reduction sequences of basalt.

Our analyses revealed the presence of more than one reduction sequence. Fig. 7.211 schematically illustrates the three reduction sequences that we have identified, which differ significantly from each other in the morphologies of their natural blanks. All of these distinct reduction sequences are present in each of the archaeological horizons, and each sequence involved a particular process of decision-making, dictated by strategies aimed at producing specific target artifacts. Although the initial morphologies of the blank and the target artifact differ among reduction sequences, some of the basalt artifacts identified as waste products (e.g. angular fragments, small flakes) could have resulted from any one of these sequences (Fig. 7.211).

Identification of the natural morphologies of the procured basalt is essential for understanding the production and modification of the basalt assemblage. The basalt morphologies selected by the hominins of GBY include: 1) rounded nodules of different sizes that were introduced to the site to be manipulated as percussors, modified artifacts, and pitted stones; 2) thin slabs that were used primarily as anvils; and 3) large, thick basalt slabs, selected for their dimensions, morphology, and geometry, as well as for their particular flaking properties; these were knapped into giant cores and were exploited in a long and complex reduction process aimed at the production of bifaces.

The procurement of basalt of different morphologies involved multiple constructs and environmental constraints, necessitating extensive knowhow on the part of the hominins. Each procurement mode was based on a particular plan (a function or a task) and on in-depth environmental knowledge of the location and availability (i.e. the potential for finding the appropriate raw material) of basalt. For example, the acquisition of basalt slabs (whether thick or thin) required knowledge of the location and particular characteristics of basalt flows, leading to the selection of dense, high-quality slabs of appropriate size and morphology. The selection of percussors was concerned with an entirely different blank morphology, which is primarily more spherical and rounded. The search for this morphology probably took place in a different type of terrain within the same volcanic landscape.

Each of the different reduction sequences that we identified involved a distinct set of actions and flaking intensities. These range from artifacts showing no intentional modification after acquisition, such as anvils and percussors, to the extremely long

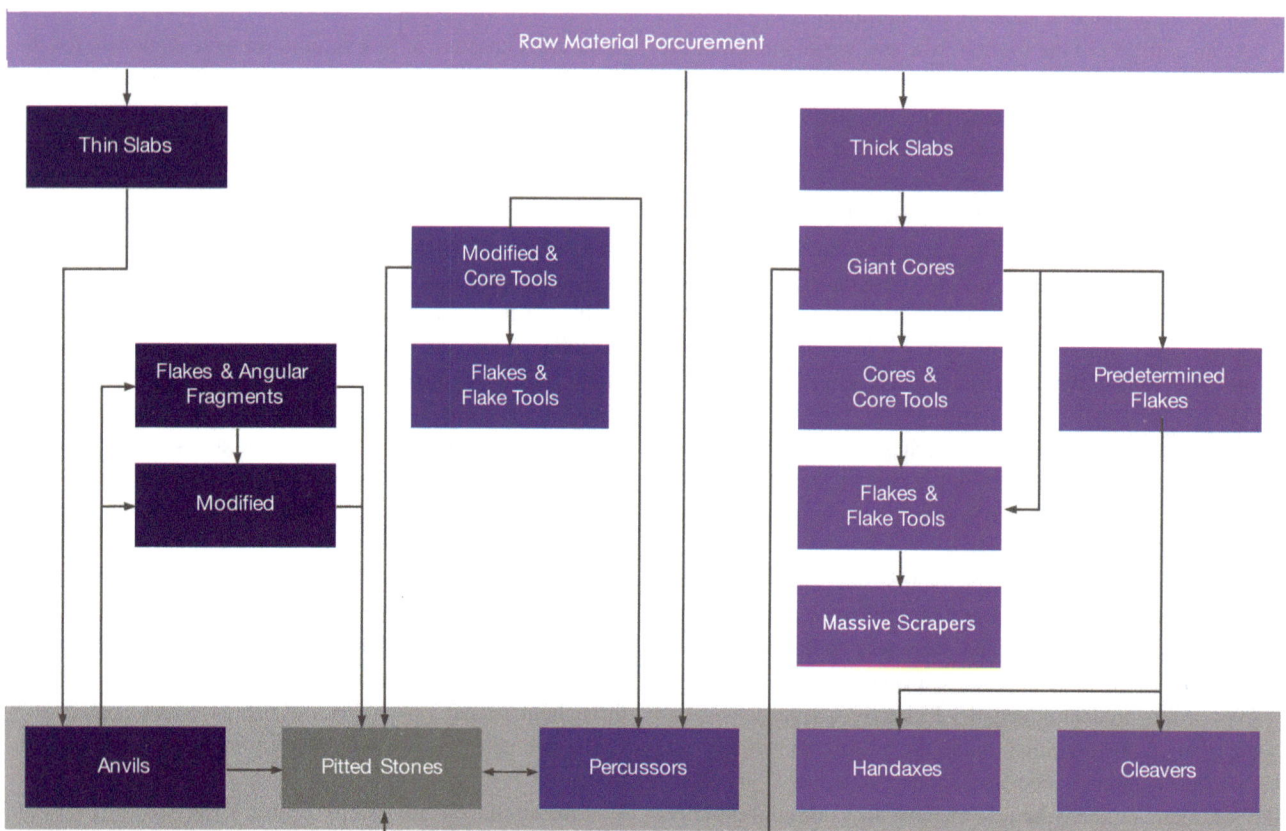

Fig. 7.211 Proposed scheme for the basalt reduction sequences

7.6 Discussion and Summary of the Basalt Component

reduction sequence aimed at the production of handaxes and cleavers.

As the reduction process of basalt bifaces was the main objective of basalt exploitation, it was also the producer of the bulk of the basalt flakes excavated at the site. In contrast with the uniformity of size that typifies the predetermined basalt flakes (large flakes and those modified into bifaces), an extensive variability is observed in the size of the other basalt FFT.

The presence of basalt tools such as massive scrapers, denticulates, notches, etc. is minimal and there is no evidence for selection of a particular type of blank for their production, except perhaps in the case of the massive scrapers. These tools are perhaps the best illustration of the exploitation of giant core waste products, selected purposefully for a particular trait – their thickness. However, the flake tools in other categories were also produced on blanks resulting from the reduction sequence that was intended to produce bifaces. It is interesting that basalt flake tools and core tools appear in very low frequencies in comparison to bifaces. Most of the basalt tool types can probably be considered as resulting from an ad-hoc need, and hence they represent only a small segment of the rich repertoire that characterizes the flint flake tools. This being said, it is also important to note that, despite the minimal size of the sample and the apparent lack of conscious planning, the analysis of the basalt flake tools presents indications of intentionality. Similarly to the assemblage of flint flake tools, a highly significant pattern emerges when types of striking platform are compared between retouched flake tools and (unmodified) flakes: the basalt flake tools have significantly more removed striking platforms than the unretouched flake assemblage ($\chi^2=36.17$; $df=6$; $p<0.0001$). This pattern highlights the cognitive flexibility and cultural conservatism reflected by the assemblage. While it is clear that there is a strong preference for the use of flint blanks to produce flake tools, in some cases similar tool types were made on basalt flakes. When this occurred, those tools were modified in a very similar manner that imposed identical production concepts on a different raw material. Thus, when the GBY hominins chose to produce a specific type of tool they applied similar methods to the blank, regardless of the reduction sequence from which it derived. Another aspect of the basalt flake tools concerns their breakage. The fact that most of the items in the basalt FFT category are broken does not necessarily indicate postdepositional activity. Rather, it is very possible that the frequent breakages were caused during the final stages of biface modification (Herzlinger et al. 2015).

7.6.1 Reduction Sequences of Basalt Percussive Tools

The basalt percussive tools of GBY belong to three categories: percussors, anvils, and pitted stones. They are all grouped under the same heading because of two primary aspects that characterize them. The first is that they all seem to have been used in percussive activities, applying force in order to process materials. These materials could have been bones, plant material, or stones (knapping). The other aspect that they have in common is that they required minimal modification in order to be functional. Rather, it seems that great care was invested in selecting raw material of suitable size and shape for different percussive tools.

The minimal modification observed on the GBY percussive tools does not mean that they are less "evolved" than other tool types. On the contrary, finding the right tool for a particular task required extensive knowledge of the location and properties of basalt resources in the vicinity of the site. Planning the acquisition had to take into consideration the morphology, size, weight, and density of the basalt needed for each of the planned tasks. Thus, the selection of percussors necessitated cobbles of particular size, weight, and roundness. Anvils, in contrast, necessitated flat surfaces of particular size and density, available as thin basalt slabs. Pitted stones occur in diverse morphologies and sizes, including among others percussors and thin anvils (Fig. 7.212).

Much of what we know about knapping percussors originates in the experimental study, in which many of the stages of the GBY basalt reduction sequence were replicated, using local basalt both as raw material and as percussors. The wear patterns and breakage modes of the excavated basalt percussors indicate that their primary function was knapping. The knapping was carried out by the direct

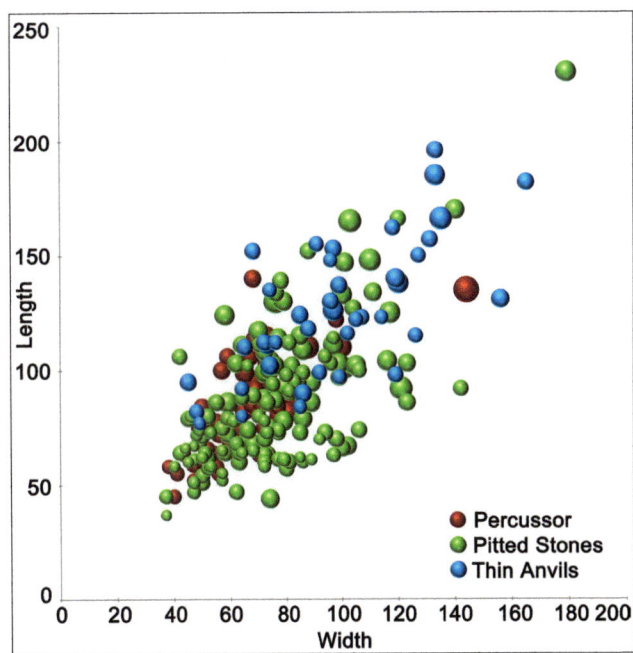

Fig. 7.212 Length, width, and thickness (bubble size) of basalt percussors, pitted stones, and thin anvils from Layer II-6

percussion technique (no evidence for the bipolar technique was identified at GBY), as suggested by the damage to the rock and by the shattered and broken percussors. It should be mentioned, however, that basalt was only one of many raw materials used for knapping percussors at GBY; others included limestone, flint, and soft organic materials (Section 6.5.1.1). A variety of other functions were probably performed, as suggested for the thin anvils based on their different damage marks. However, thin anvils, unlike percussors, served primarily as passive tools. One should bear in mind that stone tool production was not the sole activity in which percussors were used. At GBY, percussive tools could have been used for nut cracking (Goren-Inbar et al. 2002a), nut popping (Goren-Inbar et al. 2014), bone fragmentation for the extraction of marrow, which is very common at the site (Rabinovich and Biton 2011; Rabinovich et al. 2012), cracking open the shells of crabs, fruit smashing, and possibly the preparation of other plant foodstuffs such as underground storage organs (Melamed et al. 2016).

A wealth of percussive activities undoubtedly took place during the occupations of early sites, covering an extensive array of materials such as stone, bone, meat, vegetal material and various other foodstuffs. At GBY, different tool types in addition to the percussors and the thin anvils were employed in these activities and bear pitting signs. These occur primarily on basalt CCT and FFT that are classified as pitted stones. Their definition relies on their pitting damage rather than their morphology or size. For example, a single pitted surface occurs on a massive fragmented slab of basalt (Goren-Inbar 2011; Goren-Inbar et al. 2011) that, due to its great weight and morphology (two flat surfaces), was clearly used as a passive anvil.

Although the percussive tools of GBY were not products of a complex multi-stage reduction sequence, they do manifest long life-histories that are expressed in their multiple damage marks. The percussive tools were involved in a variety of tasks, making them an important component of the GBY Acheulian tool kit.

To sum up, percussive tools are mostly unmodified basalt manuports, including cobbles, thin slabs, and thick slab fragments, that were selected and transported into the site by hominins. Items of the FFT assemblage as well have been selected for use as percussive tools due to their morphology and size, which met the requirements of the percussive task.

The percussive tool assemblage provides an additional indication of the GBY hominins' knowledge and understanding of the structure, traits, and variability of basalt flows. The fact that these items were identified, quarried, and transported to the site shows that the hominins' powers of observation and thought processing went beyond a simple understanding of the properties of rocks. The knowledge involved in identifying their location and acquiring them, as well as their association with an array of different tasks, point to the unique importance of these items within the Acheulian of the upper Jordan Valley and the Levantine Corridor. The occurrence of different types of percussive tools is characteristic of the entire cultural sequence of GBY. The repetitive procurement of similar objects through time and for similar functions is indicative of a continuous tradition.

7.6.2 The Biface Reduction Sequence

The reduction sequence beginning in the extraction of a large basalt slab and ending in a predetermined biface is evidently the principal reduction sequence carried out for basalt by the GBY knappers. Most of the non-bifacial basalt flake tools are actually by-products of this reduction sequence, produced on flakes that were unsuitable for biface production. Most of the basalt waste at the site can also be attributed to this reduction sequence.

7.6.2.1 Reduction Strategies of Giant Cores

The initial step in the reduction sequence of giant cores was the acquisition of large, thick, unweathered, non-vesicular basalt slabs of particular compact quality that existed within basalt flows in the vicinity of the site. The actual processes and techniques of slab extraction are unknown, but perhaps involved a lever, fire, or the combination of both (Goren-Inbar 2011). Hard organic levers were not found at GBY, apart from a wooden log that was located under an elephant skull (Goren-Inbar et al. 1994; Goren-Inbar et al. 2002b; Section 5.4.9). Clearly, an unknown extraction technology has left its mark on the slabs in the form of indentations or notches that do not have the typical features of flake scars (Section 7.4.7.2; Fig. 7.122).

Following the extraction of the basalt slabs from the flow, they were transported to the site, where they were deliberately fragmented into several smaller, workable fragments. Alternatively, it is possible that this fragmentation took place at the quarry and that the slabs were transported to the site as fragmented slabs. The fact that cortical large flakes appear only in small numbers at the site may be an indication that the initial stage of decortication of the giant cores generally took place at the quarry. On the other hand, the occurrence of some cortical large flakes suggests that some giant cores were knapped at the site, starting at the very first stages of reduction. The great majority of giant cores found at the site belong to an early stage of the reduction sequence, with a few flakes having been removed from them.

The next step was concerned with the transformation of the slab into a core. This procedure involved their exploitation and/or preparation to produce suitable knapping angles. The control and exploitation of the particular geometry of the basalt slabs is

7.6 Discussion and Summary of the Basalt Component

the key to efficient knapping. The GBY knappers employed the natural presence of an acute angle, which allowed the immediate removal of flakes without preliminary preparation of the core. The exploitation of natural acute angles (Figs. 7.128, 7.135) on the one hand, and the formation of acute angles by knapping on the other hand, produces diagnostic waste products. These include shoulder flakes (Figs. 7.60–7.61), cortical large flakes, and wedges (Figs. 7.64–7.69). Both the fragmentation of the slabs and the formation of a suitable striking platform by the removal of shoulder flakes necessitated the application of very large percussors (Section 7.4.1.1; Fig. 7.98).

These procedures of slab fragmentation and preparation facilitated the modification of giant cores by diverse methods as described above (Section 7.4.7.2). We have identified the presence of the Kombewa, bifacial, slab-slicing, and Levallois methods. Regrettably, these core methods are minimally represented in the giant core assemblage, which includes cores exhausted by unidentified methods. Furthermore, the analysis of the basalt flake assemblages suggests that core reduction methods that are not represented in the giant core assemblage were used by the GBY hominins. One such example is the handaxe-shaped flakes, which are of a pointed morphology that could not have resulted from any of the above-mentioned methods.

Giant cores were reduced in order to produce large flakes suitable for biface modification, a reduction that involved a variety of methods, all practiced by the GBY knappers. The selection of the core method was dependent on the particular morphology and size of the slab and on the characteristics of the basalt's bedding as expressed by differences in mineralogy and structure (Fig. 7.213).

It is clear that a particular core method was selected prior to the modification of the slab. This is evident from the specific waste products, such as shoulder flakes, which required great force to reduce thickness and create a suitable angle, even though these same slabs could have been manipulated more easily if thickness were reduced from different edge morphologies. If the geometry of the slab allowed the immediate extraction of large flakes in the form of appropriate sharp angles, what could be the reason for detaching shoulder flakes? A possible answer may lie in the selection of a particular reduction sequence from several alternatives. If a knapper intends to adopt a strategy that exploits the entire circumference of the fragmented slab (as seen in Fig. 7.134b), then a core morphology that includes the thickest part of the slab may be an obstacle. Thinning the slab by removal of the thickest part will result in a longer working edge and will allow continuous exploitation of the core and its transformation into a predesigned core, whether bifacial, Levallois, or other.

Despite the variety of core methods, the large flakes that they produced were all of suitable size and morphology for modification into bifaces with minimal investment. The technological abilities of the GBY giant core knappers were at a remarkably high level and the efficiency of their work is demonstrated by the homogeneity of their products (see below).

Fig. 7.213 Visible bedding structure (marked by arrows) in basalt artifacts (#3562 and #9580 Layer II-6 Level 2; #7699 Layer II-6 Level 7)

The presence of diverse giant core methods points to a higher variability in core reduction strategies than in most of the studied Acheulian sites (Sharon 2009). This variability reflects the hominins' behavioral flexibility, technological dexterity, and excellent knowledge of raw material properties.

7.6.2.2 Biface Design Strategies

Blank Selection

The analysis of the GBY giant cores products provides insight into the criteria of blank selection. Our results show that the size and morphology of large flakes were pre-planned. The GBY knappers shaped and maintained the giant cores in a way that enabled them to dictate the morphology of the predetermined flake. When carried out successfully, the core methods used by the knappers resulted in the production of flakes that needed minimal additional investment in order to modify them into bifaces (the "minimal investment in shaping" strategy, following the terminology of Isaac 1981). Thus, the fact that most of the planning, energy, dexterity, and work were invested during blank production facilitated the process of blank selection.

Another decision-making procedure was involved in the selection of blanks: the differentiation between cleaver blanks and handaxe blanks. This differentiation was dictated mostly by the morphology of the blanks. Cleaver blanks were more carefully selected than those of handaxes, according to a more rigid set of morphological constraints. This was probably because cleavers received less investment during the shaping stage (below).

To evaluate the criteria that characterize blanks that are suitable for biface modification, we compared several characteristics of the bifaces with those of biface-shaped flakes (Section 7.3.3.6). The comparison focused on the levels of Layer II-6 and included only complete bifaces or those with a slightly broken tip. With respect to size, there appears to be no significant difference in length and thickness between handaxe-shaped flakes and handaxes. However, handaxe-shaped flakes are significantly wider (difference: 20.17 mm; t-ratio: 7.3; df: 244; $p(tt) < .0001$). When cleavers are compared to cleaver-shaped flakes, it is evident that cleavers are significantly longer and thicker than cleaver-shaped flakes (length, difference: 17.60 mm; t-ratio: -1.999; df: 133; $p(tt)=0.001$; thickness, difference: 4.49 mm; t-ratio: -3.37; df: 133; $p(tt)=0.0475$), while cleaver-shaped flakes are significantly wider (difference: 10 mm; t-ratio: 2.527; df:133; $p(tt)=0.0127$). For both handaxes and cleavers, it appears that a few potential blanks coincide in size with the main clusters of handaxes and cleavers. However, most of the biface-shaped flakes are shorter than the majority of bifaces, a fact that would have made it impossible to modify them into finished tools of appropriate size and proportions (Figs. 7.214–7.215). This observation may explain the fact that these particular biface-shaped flakes were not selected for modification into bifaces.

When the biface-shaped blanks were appropriate in term of size, they were selected for modification into bifaces regardless of technological variations such as their direction of blow or type

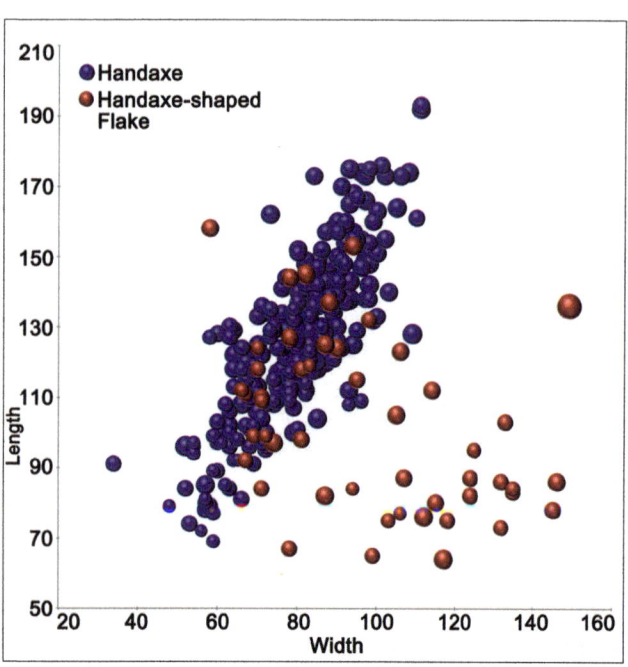

Fig. 7.214 Length, width, and thickness (bubble size) of Layer II-6 basalt handaxe and handaxe-shaped flakes

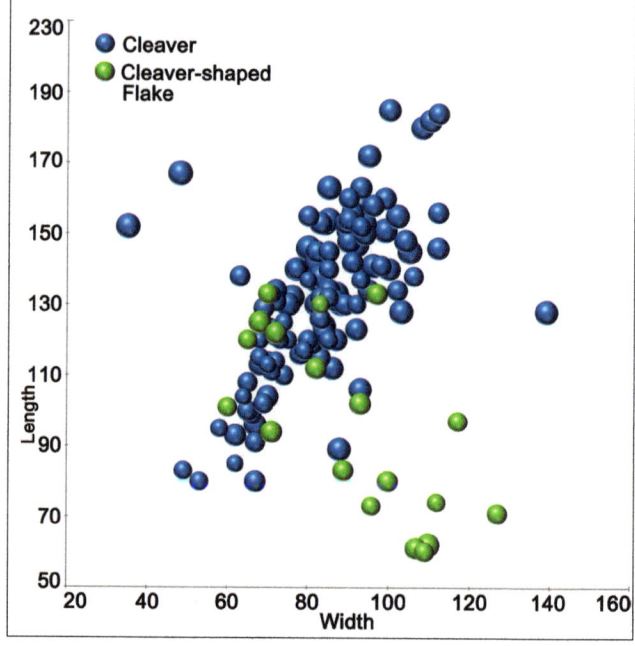

Fig. 7.215 Length, width, and thickness (bubble size) of Layer II-6 basalt cleavers and cleaver-shaped flakes

of striking platform. The stylistic modification was applied in a highly skilled manner to achieve the desired biface morphology. This modification was usually restricted to the thinning of the thickest area of the blank to create a symmetrical morphology and appropriate distribution of mass. Since the thickest parts of a flake are generally the striking platform and bulb of percussion, these areas were most commonly modified by retouch and the removal of the striking platform. Thus, our results indicate that the direction of blow and type of striking platform were not important criteria for the selection of blanks to be modified into bifaces. Rather, the size and general morphology of the blank were the criteria that dictated whether a specific blank would be selected for modification.

7.6.2.3 Biface Modification and Morphology

Different methodologies were applied to the analysis of the morphology of handaxes and cleavers. Consequently, it is difficult to compare the extent of variability of these types. Handaxe morphology is described by quantitative continuous ratios and that of cleavers by qualitative discrete ones, a discrepancy that prevents effective comparison between handaxes and cleavers in the degree of their variability and homogeneity. However, some conclusions are still valid, particularly those related to the technology, style, and size of these tools.

The most obvious pattern is the substantial intra- and inter-assemblage homogeneity of both handaxe and cleaver morphology. The low intra-assemblage morphological variability of handaxes is reflected in the low standard deviations of the various morphological ratios. This trend is somewhat less apparent in cleavers, as their qualitative description does not allow similar observations. Furthermore, the morphological attributes of the cleavers seem to be rather diverse in each of the levels of Layer II-6. However, it should be noted that the different morphological characteristics of the cleavers do not necessarily indicate a substantial difference. For example, cleavers with convex, straight, or convex and straight lateral edges may be in fact very similar in general planform. The fact that the combination of these three lateral edge morphologies makes up more than half of the cleavers in all levels highlights the intra-assemblage morphological homogeneity of this tool.

Similar conclusions may be drawn with respect to inter-assemblage homogeneity. For the handaxes, the homogeneity is evident in the similarities of the mean values and ranges of the different morphological ratios in all levels of Layer II-6. For the cleavers, however, such a morphological homogeneity is again more difficult to discern. As described in the morphological section (Section 7.5.5), there appears to be some preference for cleaver morphology that changed temporally across the different levels of Layer II-6. Nevertheless, it is also clear that all the different edge and base forms appear in the majority of levels and differences are expressed only in their frequencies, differences that are statistically insignificant. Thus, we may conclude that the morphology of the bifacial tools points to cultural conservatism. Handaxes at GBY tend to be relatively thick, with their maximal width in their proximal half, lateral edges that range from straight and parallel to convex and distally converging, a pointed distal end, and a lenticular cross-section. Cleavers generally possess a convex, straight, or pointed proximal end, straight to convex lateral edges, a straight, convex, or diagonal working edge, and a lenticular, trapezoid, or parallelogram cross-section. This is not to say, however, that all handaxes and cleavers are identical or even highly similar in shape, as in each of the levels there is a range of morphologies. This range is composed of the common shapes together with shapes that are less frequent and diverge from the mode; the latter include, for example, very thin triangular handaxes and thick, stubby and rounded handaxes and cleavers with irregular lateral edges and concave working edge. It is clear that the morphological uniformity of bifacial tools was important to the GBY hominins, and the common shapes of bifaces should be viewed as fitting the ideal "mental template" of the knappers. Nonetheless, this uniformity existed within a range of acceptable morphological variations.

In this regard, it should be noted that the current methods of morphological description and analysis of bifaces involve some difficulties. The methods of both Bordes (1961) and Roe (1964, 1968) use ratios that focus on very specific morphological aspects but ignore others, such as the shape and refinement of the edges and the distribution of thickness across the item's surface. The latter may very well be significant for the understanding of various behavioral issues. Furthermore, Bordes' typological classification system is biased towards the European record and is based on a very low-resolution morphological description of the handaxes that fails to consider the shape aspects mentioned above. Improved description and analysis of bifacial tool morphologies, which provide high-resolution data and consider shape in a more holistic manner, require the application of more sophisticated methods of analyzing shape. Such methods for the description and analysis of shape variability in handaxes have recently been published (Herzlinger 2014). These methods, which are based on high-resolution description of shape and multivariate statistical analyses, will surely provide better observations and enable further conclusions about the nature and significance of the morphological variability in biface assemblages.

The morphological homogeneity seen in the biface assemblages of GBY is even more notable when considering the technological and stylistic flexibility that characterizes their production. Examination of the relationships between the morphological attributes and the technological or stylistic attributes provides no significant patterns, either for handaxes or for cleavers. This means that even when differences in technological attributes such as blank type, striking platform, and direction of blow are observed, the various morphological attributes of the bifaces do not differ significantly. The pattern is similar for stylistic attributes such as location, type, and intensity of retouch. Thus, the same morphologies were repeatedly produced, both concurrently and over time, regardless of variations in stylistic modification and production technology.

However, some repeated and significant patterns do emerge from the relationships between several of the technological and stylistic attributes. There are significant differences in the retouch locations on Face 2 of handaxes in relation to the direction of blow of their blank removal. Handaxes produced on end-struck flakes have significantly more proximal retouch. Those produced on side-struck flakes are significantly more retouched on their left and right side and less on both edges and the proximal end. Handaxes whose blanks are classified as special side-struck are significantly more retouched on their proximal end and right side and less on their left side ($\chi^2=41.74$; $df=26$; $p=0.014$). These differences indicate that the GBY hominins systematically thinned the area of the striking platform and bulb of percussion, which is usually the thickest area of the flake, in accordance with its position on the blank. The other edges of the blank, however, were subjected to significantly less secondary modification. This pattern shows how technological and stylistic flexibility was used to achieve the desired handaxe morphology.

Although this particular pattern is not seen in the cleaver assemblage, a different but similarly important pattern can be discerned when the type of striking platform of the cleaver's blank is examined in relation to the direction of blow of blank removal. Cleavers that are end-struck have significantly more plain striking platforms, while those that are side-struck have significantly fewer plain and more removed striking platforms ($\chi^2=25.99$; $df=6$; $p=0.0002$). This pattern indicates that in general striking platforms were removed only when their location on the blank interfered with the desired morphology. This highlights the fact that, as with the handaxes, the GBY hominins applied flexible stylistic solutions to technological constraints in order to achieve and maintain the homogeneous morphology of the cleavers.

All of these observations point towards a substantial similarity in the way that the GBY hominins perceived handaxes and cleavers. It is clear that morphological homogeneity and conservatism, both within levels and along the time trajectory, played an important role. Furthermore, for both tool types the goals were achieved by generally similar means. These can be summarized as selective thinning of areas of the blank that interfered with the desired morphology and minimal investment in secondary shaping. Despite the fact that the significant patterns of modification of handaxes and cleavers occur in different attributes (direction of blow vs. location of ventral retouch in handaxes and direction of blow vs. striking platform for cleavers), they point toward the same concept. Thus, the GBY hominins took advantage of a technology that enabled the production of large flakes in order to obtain blanks that needed minimal secondary modification, as they already possessed morphological attributes similar to those of the desired end-product. Modification of the blanks' morphology was applied parsimoniously, for example in artifacts whose blanks were side-struck, as a result of which the striking platforms were positioned on one of the lateral edges. The behavioral implications of this pattern point to cognitive flexibility and selective application of technical solutions in order to maintain a common cultural preference, which was sustained over the entire period of the site's occupation.

To sum up, the reduction sequence aimed at the production of bifaces is a single path in which the goal (the predetermined target tool), once achieved, was either a handaxe or a cleaver and was most probably used for a single type of task. This is well documented in the pristine state or lack of use damage that is typical of these artifacts. Pitted stones, thin anvils, modified artifacts, and an assortment of other types show multiple typological or functional traits, which frequently co-occur on the same artifact. Clearly, while we can only guess at the different tasks or functions performed by these tools, it is evident that hominins at GBY used them either alternately or simultaneously. This flexibility, which has cognitive implications, is documented in an entirely different manner in the biface reduction sequence, where it is expressed by the exploitation of rejected blanks and their transformation into yet another tool type – the massive scraper. These patterns of exploitation and manipulation of the same raw material in diverse ways illustrate both overall cognitive planning abilities and the consistent expression of the complex Acheulian tradition, a tradition that displays flexibility side by side with conservatism throughout the cultural sequence at GBY.

Chapter 8
The Limestone Component

Abstract In-depth study of Acheulian limestone artifacts from Gesher Benot Ya'aqov (0.79 Ma) has revealed that limestone nodules procured from fluvial deposits were transported to the lake margin and exploited throughout the occupational sequence (ca. 50 ka). Analyses of the limestone assemblages show that individual artifacts went through several use-stages or complex life-histories within a single reduction sequence. This reduction sequence began with the targeting of nodules suitable for use as percussors. Use of the percussors sometimes resulted in breakage that produced flakes typical of working accidents. Broken percussors were shaped into a second morphotype, chopping tools, while cores comprise a third morphotype. These morphotypes are viewed as consecutive, interrelated options. Once a morphotype was inadequate for use it was transformed into another, resulting in gradual reduction of dimensions from one type to the next. The ability to renovate/recycle implies flexibility and contingency.

Keywords Gesher Benot Ya'aqov · limestone · *chaîne opératoire* · percussor · *percuteur de concassage*

8.1 Inventory and Taphonomy of the Limestone Component

8.1.1 Inventory of the Limestone Component

The limestone component is the smallest among the three raw materials used at Gesher Benot Ya'aqov (GBY), and limestone artifacts occur throughout the cultural sequence in particularly low frequencies (Table 8.1). In general, the richer assemblages are those originating in the archaeological horizons of Layer II-6, although even they rarely provide sample sizes sufficient for in-depth statistical analysis. Therefore, the following analysis is limited to layers that yielded a minimum number of 10 limestone artifacts. There are varying frequencies of different lithic categories, but FFT are the most frequent in all levels and substantial amounts of microartifacts appear throughout the sequence (Table 8.1).

Table 8.1 Counts of limestone artifacts and microartifacts throughout the GBY sequence

Area	Layer	Artifacts					Microartifacts
		FFT	CCT	HX	CL	Total	
C	V-3	-	-	-	-	-	4
	V-4	-	-	-	-	-	8
	V-4/5	-	-	-	-	-	9
	V-5	2	2	-	-	4	333
	V-6	2	-	-	-	2	96
	JB	8	2	-	-	10	4
A	I-4	-	-	-	-	-	103
	I-5	-	1	-	-	1	135
	II-2	10	2	-	-	12	170
	II-2/3	3	-	-	-	3	85
	II-4	1	-	-	-	1	-
	II-5	7	7	-	-	14	212
	II-5/6	2	2	-	-	4	1,055
	Trench II	-	4	1	-	5	-
	Trench II green	1	5	-	-	6	-
B	II-6/L1	22	14	5	-	41	1,506
	II-6/L2	17	6	-	-	23	2,154
	II-6/L3	24	9	-	-	33	1,570
	II-6/L4	21	21	-	-	42	4,727
	II-6/L4b	20	5	2	-	27	145
	II-6/L5	26	3	-	-	29	1,996
	II-6/L6	25	3	-	1	29	331
	II-6/L7	77	13	1	-	91	1,602
	II-6/7	5	-	-	-	5	37
	II-14	2	-	-	-	2	-
	Unconformity B	11	10	-	-	21	5
	Unconformity C	-	5	-	-	5	-
Total		286	114	9	1	410	16,287

Table 8.2 Counts and frequencies (%) of state of preservation of limestone CCT and FFT

Layer		Fresh	Slightly abraded	Preservation Abraded	Heavily abraded	Exfoliated	Total
JB	CCT	-	50.00	50.00	-	-	2
	FFT	37.50	37.50	25.00	-	-	8
II-2	CCT	-	-	100.00	-	-	2
	FFT	-	20.00	40.00	40.00	-	10
II-5	CCT	14.29	28.57	57.14	-	-	7
	FFT	14.29	14.29	57.14	14.29	-	7
II-6/L1	CCT	7.14	21.43	64.29	7.14	-	14
	FFT	13.64	13.64	59.09	13.64	-	22
II-6/L2	CCT	-	33.33	66.67	-	-	6
	FFT	5.88	23.53	52.94	17.65	-	17
II-6/L3	CCT	-	44.44	55.56	-	-	9
	FFT	8.33	29.17	45.83	16.67	-	24
II-6/L4	CCT	-	47.62	52.38	-	-	21
	FFT	19.05	28.57	38.10	14.29	-	21
II-6/L4b	CCT	-	40.00	60.00	-	-	5
	FFT	-	15.00	35.00	50.00	-	20
II-6/L5	CCT	-	-	33.33	33.33	33.33	3
	FFT	-	7.69	23.08	69.23	-	26
II-6/L6	CCT	-	-	66.67	-	33.33	3
	FFT	-	-	44.00	52.00	4.00	25
II-6/L7	CCT	-	30.77	61.54	7.69	-	13
	FFT	1.30	7.79	33.77	54.55	2.60	77

8.1.2 Taphonomy of the Limestone Component

Preservation state, patination, and breakage pattern are used to study the taphonomy of the limestone component. Similarly to other raw materials at the site, the states of preservation vary throughout the sequence, indicating different depositional environments (Table 8.2). The most common mode of preservation is that of abraded artifacts, followed by slightly abraded ones. Heavily abraded artifacts appear in higher frequencies in the lower levels of Layer II-6, in which at times they form more than half of the sample. Different preservation modes emerge when the CCT and FFT components are compared. In general, CCT tend to be less abraded than FFT, which are characterized by higher frequencies of heavily abraded artifacts. In the upper levels, FFT seem to show a more variable pattern in which all preservation modes are represented. In contrast, the CCT in these levels are generally either slightly abraded or abraded. In the lower levels, both classes are more abraded, with FFT showing higher frequencies of heavily abraded artifacts than CCT. These observations accord with the trends seen for other types of raw material (Chapters 6–7).

Patination of the limestone component probably resulted both from exposure to atmospheric conditions and from the waterlogged depositional environment. Artifacts that are not patinated appear only sporadically, and the most frequent patination mode is that of patinated artifacts, which form the majority of artifacts in all levels (Table 8.3). Double patinated artifacts, which reflect a more complex life-history, also appear in very low frequencies, although somewhat higher values occur in the lower levels of Layer II-6. No major differences can be discerned between the patination patterns of the CCT and FFT categories.

The breakage pattern of the limestone component (Table 8.3) displays a similar trend to those seen for preservation and patination. In the upper levels of Layer II-6 the majority of the artifacts are complete, in contrast to the lower levels, which show higher frequencies of broken artifacts. This trend, however, is significantly more pronounced in the FFT component. While in the upper levels FFT show generally equal frequencies of complete and broken artifacts, in the lower levels broken artifacts become substantially more common. CCT, on the other hand, display a marked predominance of complete artifacts in all levels of Layer II-6. This trend is consistent with that seen for the weathering (preservation attribute), as well as with the trends observed for the flint component (Section 6.1.2).

The taphonomic state of the limestone component as seen in the preservation, patination, and breakage attributes suggests a postdepositional history that is quite similar to that of the flint component. The distribution patterns of these attributes indicate that in the upper levels of Layer II-6 the artifacts were only very slightly affected by postdepositional processes. Although in the lower levels these processes were somewhat more pronounced, their effect on the artifacts is negligible. The higher degree of weathering seen for the FFT component could be the result of different factors affecting the *in situ* weathering of the artifacts, similarly to the other raw materials.

8.2 Typology and Technology of the Limestone Flakes and Flake Tools

Table 8.3 Counts and frequencies (%) of breakage and patination of limestone CCT and FFT

Layer		Breakage			Patination		
		Complete	Broken	No patina	Patinated	Double patinated	Total
JB	CCT	100.00	-	-	50.00	50.00	2
	FFT	37.50	62.50	-	100.00	-	8
II-2	CCT	50.00	50.00	-	50.00	50.00	2
	FFT	60.00	40.00	-	80.00	20.00	10
II-5	CCT	85.71	14.29	-	100.00	-	7
	FFT	71.43	28.57	28.57	57.14	14.29	7
	FFT	-	100.00	-	-	100.00	1
II-6/L1	CCT	85.71	14.29	-	92.86	7.14	14
	FFT	59.09	40.91	13.64	77.27	9.09	22
II-6/L2	CCT	83.33	16.67	-	66.67	33.33	6
	FFT	52.94	47.06	5.88	88.24	5.88	17
II-6/L3	CCT	100.00	-	-	100.00	-	9
	FFT	50.00	50.00	-	95.83	4.17	24
II-6/L4	CCT	95.24	4.76	-	100.00	-	21
	FFT	52.38	47.62	4.76	90.48	4.76	21
II-6/L4b	CCT	80.00	20.00	-	100.00	-	5
	FFT	55.00	45.00	5.00	85.00	10.00	20
II-6/L5	CCT	66.67	33.33	-	66.67	33.33	3
	FFT	42.31	57.69	-	80.77	19.23	26
II-6/L6	CCT	100.00	-	-	66.67	33.33	3
	FFT	24.00	76.00	12.00	80.00	8.00	25
II-6/L7	CCT	92.31	7.69	-	100.00	-	13
	FFT	35.06	64.94	5.19	92.21	2.60	77

Table 8.4 Counts and typological frequencies (%) of limestone FFT*

	II-2	II-6/L1	II-6/L2	II-6/L3	II-6/L4	II-6/L4b	II-6/L5	II-6/L6	II-6/L7
N	10	22	13	24	18	18	25	24	75
Typology									
Convex transverse side-scraper	-	4.55	-	-	5.56	-	-	-	-
Side-scraper on ventral face	-	-	-	-	-	5.56	-	-	-
Typical end-scraper	-	-	-	4.17	-	-	-	-	-
Atypical borer	-	-	-	-	-	-	-	-	1.33
Naturally backed knife	-	-	-	-	-	-	-	4.17	-
Notch	-	4.55	-	-	5.56	5.56	4.00	-	-
Denticulate	-	-	-	-	-	5.56	-	-	-
Varia	-	-	-	4.17	-	-	4.00	-	-
Retouched flake	-	9.09	-	-	-	5.56	-	8.33	1.33
Flake	100.00	81.82	100.00	91.67	88.89	77.78	92.00	87.50	96.00
Éclat de taille de biface	-	-	-	-	-	-	-	-	1.33

* only layers with a minimum number of 10 limestone FFT are included

8.2 Typology and Technology of the Limestone Flakes and Flake Tools

The typological inventory of each of the GBY layers has been presented in Chapter 5, where the limestone component is presented in detail. In this analysis we describe assemblages that comprise a minimum number of 10 artifacts. Furthermore, due to the extremely small number of limestone flake tools, we have lumped together all tool types from all layers and discuss them as a single assemblage.

Table 8.4 presents the typological composition of the limestone FFT component. Unretouched flakes comprise the vast majority of the assemblages, reaching over 77% in each of them. It should be noted that a single item was classified as *éclat de taille de biface*, indicating that limestone bifaces were in fact produced at the site, although evidently to a minimal extent (Chapter 7).

Retouched tools appear in significantly lower frequencies and are very sporadic in all layers, amounting to only 19 tools. The most frequent tool type is the retouched flake (N=6), followed by notches (N=4). Other types, such as side- and end-scrapers, denticulates, knives, and varia, are represented only by one or two artifacts. The very limited assemblage of flake tools suggests that this component was not a systematic and regular target of the reduction sequence. Rather, it reflects an ad-hoc approach to the modification of limestone flakes to be used as tools. This is further supported by the fact that retouched flakes and varia tools amount to almost 50% of all limestone flake tools. These are two of the more variable typological categories, as they include artifacts that cannot be assigned to any of the other, more formal, tool types. Thus, the typological frequencies further highlight the irregularity and inconsistency of this component.

The breakage pattern of limestone FFT reflects the general taphonomic state of the entire limestone component. As presented above, the majority of unretouched flakes are complete in the upper levels of Layer II-6, while in the lower levels there are higher frequencies of broken artifacts (Table 8.5). Among the broken

Table 8.5 Counts and frequencies (%) of characteristics of limestone FFT*

	Tools	Unretouched flakes								
		II-2	II-6/L1	II-6/L2	II-6/L3	II-6/L4	II-6/L4b	II-6/L5	II-6/L6	II-6/L7
N	19	10	18	13	22	16	14	23	21	73
Breakage										
Complete	12 / 63.16	6 / 60.00	9 / 50.00	9 / 69.23	11 / 50.00	10 / 62.50	8 / 57.14	10 / 43.48	5 / 23.81	26 / 35.62
Distal	2 / 10.53	-	6 / 33.33	-	1 / 4.55	1 / 6.25	1 / 7.14	3 / 13.04	2 / 9.52	10 / 13.70
Lateral	1 / 5.26	1 / 10.00	1 / 5.56	-	1 / 4.55	-	1 / 7.14	-	3 / 14.29	6 / 8.22
Proximal	2 / 10.53	-	-	1 / 7.69	4 / 18.18	1 / 6.25	2 / 14.29	1 / 4.35	3 / 14.29	7 / 9.59
Distal & lateral	-	2 / 20.00	-	-	1 / 4.55	-	-	-	-	8 / 10.96
Distal & proximal	1 / 5.26	1 / 10.00	1 / 5.56	-	-	-	-	-	1 / 4.76	1 / 1.37
Fragment	1 / 5.26	-	1 / 5.56	2 / 15.38	3 / 13.64	2 / 12.50	2 / 14.29	8 / 34.78	7 / 33.33	13 / 17.81
Proximal & lateral	-	-	-	1 / 7.69	1 / 4.55	1 / 6.25	-	-	-	2 / 2.74
Indeterminate	-	-	-	-	-	1 / 6.25	-	1 / 4.35	-	-
Cortex										
No cortex	10 / 52.63	5 / 50.00	8 / 44.44	6 / 46.15	9 / 40.91	4 / 25.00	5 / 35.71	9 / 39.13	6 / 28.57	22 / 30.14
1–25%	1 / 5.26	-	1 / 5.56	3 / 23.08	3 / 13.64	2 / 12.50	-	-	1 / 4.76	1 / 1.37
26–50%	1 / 5.26	-	-	-	3 / 13.64	2 / 12.50	3 / 21.43	-	1 / 4.76	4 / 5.48
51–75%	1 / 5.26	-	1 / 5.56	-	1 / 4.55	3 / 18.75	-	2 / 8.70	-	-
76–100%	3 / 15.79	4 / 40.00	8 / 44.44	4 / 30.77	4 / 18.18	4 / 25.00	4 / 28.57	7 / 30.43	-	17 / 23.29
Indeterminate	3 / 15.79	1 / 10.00	-	-	2 / 9.09	1 / 6.25	2 / 14.29	5 / 21.74	13 / 61.90	29 / 39.73
Direction of blow		(N=8)							(N=13)	(N=67)
End-struck	9 / 47.37	-	5 / 27.78	8 / 61.54	15 / 68.18	9 / 56.25	6 / 42.86	11 / 47.83	3 / 23.08	21 / 31.34
Side-struck	5 / 26.32	2 / 25.00	7 / 38.89	-	4 / 18.18	3 / 18.75	4 / 28.57	1 / 4.35	4 / 30.77	12 / 17.91
Special side-struck	1 / 5.26	1 / 12.50	-	2 / 15.38	-	-	-	1 / 4.35	-	2 / 2.99
Indeterminate	4 / 21.05	5 / 62.50	6 / 33.33	3 / 23.08	3 / 13.64	4 / 25.00	4 / 28.57	10 / 43.48	6 / 46.15	32 / 47.76

* only layers with a minimum number of 10 limestone FFT are included; flake tools from all layers are lumped together

8.2 Typology and Technology of the Limestone Flakes and Flake Tools

flakes, the most common mode is that of fragmented artifacts, again with higher frequencies in Layer II-6 Levels 5, 6, and 7. Proximal and distal breakages appear in most levels, although in lower frequencies. Other types of breakage appear more sporadically. Among the retouched tools the pattern is mostly similar.

Technologically, less than half of the unretouched components in all levels are completely devoid of cortex (Table 8.5). Most artifacts present at least some cortical coverage, with artifacts that have 75–100% coverage being the most frequent (Figs. 8.1–8.3). This pattern is somewhat unusual in comparison with the other raw materials, suggesting that a large portion of the limestone FFT originate in the initial stages of the reduction sequence. It should be noted, though, that significant numbers of unretouched flakes of unidentified cortical coverage were found in the lower levels of Layer II-6. This phenomenon is probably due to the unfavorable preservation conditions characterizing these levels.

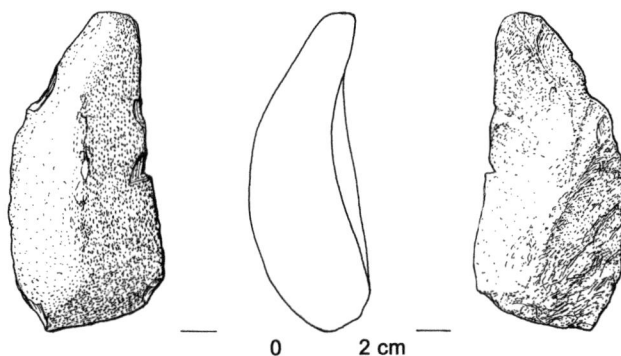

Fig. 8.1 Limestone cortical flake (#2100 Layer II-6 Level 4)

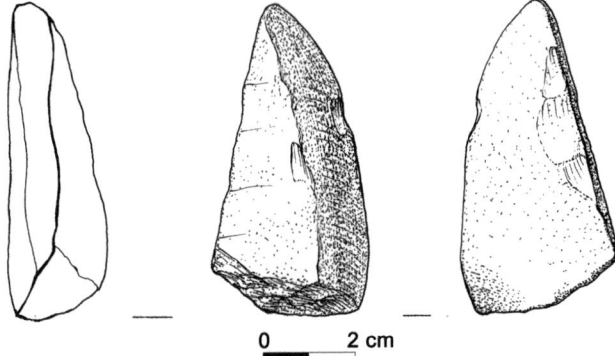

Fig. 8.2 Limestone cortical flake (#9171 Layer II-6 Level 3)

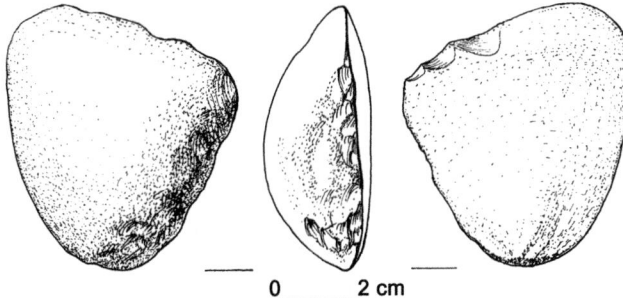

Fig. 8.3 Limestone cortical flake (#3440 Layer II-6 Level 4b)

The variability in cortical coverage demonstrates a marked difference between the retouched and unretouched components. Notwithstanding their small number, fully cortical limestone flake tools appear in significantly lower frequencies than unretouched flakes. Furthermore, there is a higher occurrence of non-cortical tools than unretouched flakes in all levels.

The direction of blow of limestone FFT is usually end-struck, while side- and special side-struck artifacts are less common (Table 8.5). Here again there are high frequencies of indeterminate artifact, especially in the lower levels of Layer II-6. The pattern of the retouched tools is similar to that of the unretouched ones.

The items of the limestone FFT component are generally small and thin (Table 8.6). Mean length and width values of the unretouched flakes are fairly similar and range between 2 and 3 cm (Fig. 8.4). In some cases, such as in Layer II-6 Level 4, the values are higher due to a small number of larger unretouched flakes, as can be seen from the maximal and median values. Somewhat higher size variability appears in Level 7, which yielded the largest sample of limestone FFT. In general, this component is homogeneous in terms of size, as indicated by the small standard deviations of most of the layers. This observation also holds true for the similarity of the angle of striking platforms along the archaeological sequence. The retouched tools demonstrate similar overall mean values to those of the unretouched flakes (Table 8.6). The maximal and minimal values are all within the ranges seen for the unretouched component. However, standard deviation values are somewhat higher than those seen in most levels for the unretouched component, indicating that the retouched tools are more variable in size. This suggests that there was no systematic size selection of blanks for the production of limestone flake tools. This observation further strengthens the notion that limestone flake tools did not constitute a systematic and regular part of the tool kit of the GBY hominins.

Other technological observations of the limestone FFT are presented in Table 8.7. The scar pattern of most artifacts, whether unretouched flakes or tools, is indeterminate, while the second most common mode is cortical. Other modes, such as plain, along axis, and others, rarely appear throughout the sequence, and when they do it is in very low frequencies. The tools component shows a relatively high frequency of radial scar pattern, which is very rare in the unretouched flakes. Nevertheless, due to their small number this observation is not necessarily a significant one. The number of scars for the flakes is generally low, never exceeding a mean value of 3.25 scars (Table 8.7). This attribute demonstrates homogeneity both between the levels and within them, as can be seen by the low values of standard deviation. Similarly to the attribute of cortical coverage presented above, this attribute displays a marked difference between the tools and the flakes: the scar count on tools shows a significantly higher mean value of 4 scars per item. This difference is supported by the standard deviation value, which is not higher than those of the flakes.

Table 8.6 Descriptive statistics of limestone FFT (size in mm; angle in degrees)[*]

	Tools	Unretouched flakes								
		II-2	II-6/L1	II-6/L2	II-6/L3	II-6/L4	II-6/L4b	II-6/L5	II-6/L6	II-6/L7
N	19	10	18	13	22	16	14	23	21	73
Maximum length										
Max	76	41	53	52	64	144	59	45	50	70
Median	30	27.5	32	28	32.5	32	25.5	26	28	28
Min	23	21	23	22	22	22	20	20	21	21
Mean	36.32	28.80	33.56	31.85	35.32	40.56	32.64	26.39	29.29	31.42
Std dev	15.78	7.28	9.48	8.75	12.30	29.57	12.86	4.91	7.48	10.29
Std err	3.62	2.30	2.23	2.43	2.62	7.39	3.44	1.02	1.63	1.20
Length										(N=71)
Max	65	38	47	50	64	144	51	45	43	59
Median	27	24.5	24.5	27	29	29.5	23	24	21	25
Min	15	15	18	21	20	15	17	17	15	12
Mean	32.26	26	27	29	33	38	29	25	24	27
Std dev	14.68	8.19	7.72	8.86	11.62	30.57	11.68	5.71	6.66	10.02
Std err	3.37	2.59	1.82	2.46	2.48	7.64	3.12	1.19	1.45	1.19
Width										(N=71)
Max	72	31	53	35	51	102	53	30	49	60
Median	26	23	27.5	25	25	24	22	18	23	23
Min	17	19	19	14	14	11	14	13	15	10
Mean	29.47	25	30	24	27	30	25	19	24	24
Std dev	13.61	4.65	9.78	7.49	9.60	20.66	10.41	4.67	8.32	9.41
Std err	3.12	1.47	2.31	2.08	2.05	5.16	2.78	0.97	1.82	1.12
Thickness										
Max	25	14	22	19	27	27	30	13	22	45
Median	12	9	12.5	11	11	10.5	10	9	9	10
Min	7	8	7	7	5	8	7	4	6	4
Mean	13.32	9.80	12.83	12.15	12.82	13.25	13.36	9.00	10.19	12.05
Std dev	5.86	2.10	4.16	4.65	6.37	6.01	7.38	2.41	4.30	6.74
Std err	1.34	0.66	0.98	1.29	1.36	1.50	1.97	0.50	0.94	0.79
Angle of striking platform	(N=7)	(N=3)	(N=4)	(N=1)	(N=2)		(N=5)	(N=4)	(N=1)	(N=19)
Max	136	122	120	85	115	-	127	130	116	123
Median	126	110	112.5	85	109	-	105	110	116	109
Min	102	103	108	85	103	-	94	90	116	89
Mean	122.57	111.67	113.25	85.00	109.00	-	109.60	110.00	116.00	107.21
Std dev	11.49	9.61	5.38	-	8.49	-	12.93	16.41	-	11.79
Std err	4.34	5.55	2.69	-	6.00	-	5.78	8.21	-	2.71

[*] only layers with a minimum number of 10 limestone FFT are included; flake tools from all layers are lumped together

The distribution of striking platform types (Table 8.7) suggests that the most common artifacts in most levels, especially in the lower levels of Layer II-6, are those whose type of striking platform could not be determined. Alongside these, the most frequent modes of striking platforms are broken, plain, and cortical, while other types appear sporadically. It is worth noting that, in contrast to the flint flake tools, there appears to be no systematic removal of the striking platform.

The patterns presented above suggest a very irregular and unsystematic production of limestone FFT. The small number of artifacts, together with the minimal number of modified tools, the high frequencies of indeterminate striking platforms and scar patterns, the low number of scars, and high degree of cortical coverage, tend to rule out a systematic and intentional reduction sequence whose main target was the production of limestone flakes and their subsequent modification into tools. Rather, limestone FFT seem to be by-products of the production and use of core tools. It is worth noting, though, that in the few instances in which flake tools were modified on limestone, there appears to be a preference towards a selection of less cortical blanks with higher scar counts, blanks that probably originated in more advanced stages of reduction.

8.3 Typology and Technology of the Limestone Cores and Core Tools

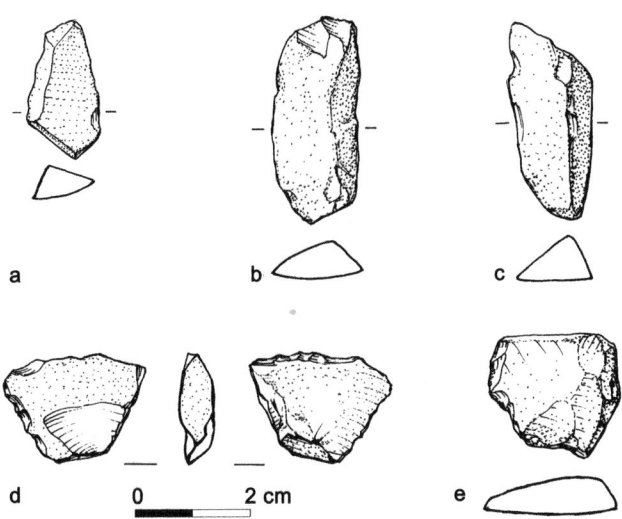

Fig. 8.4 Limestone flakes: a) small non-cortical limestone flake (#9633 Layer II-6 Level 5), b) limestone elongated flake (#13311 Layer II-6 Level 3), c) limestone elongated flake (#1693 Layer II-6 Level 4b), d) limestone cortical flake (#14322 Layer II-6 Level 7), e) small non-cortical limestone flake (#4329 Layer II-6 Level 6)

The typological classification of the limestone CCT component is presented in Table 8.8. The most common typological groups are percussors and modified artifacts. It should be noted that the category of split percussors, which by definition belongs to the FFT component as it consists of flakes, is described in this section, as it clearly belongs to the realm of percussive tools (Section 7.4.1.1). Chopping tools appear in lower frequencies and cores are even scarcer. Other types of core tools, such as massive scrapers, appear only sporadically. The very low sample size of limestone CCT precludes the possibility of discussing each level individually. Therefore, for the following detailed typological and technological presentation, all layers and levels are lumped together for each typological category.

Some 38 artifacts (33.93% of the entire limestone CCT assemblage) have a secondary typology. The secondary typologies consist of only three types – percussors, pitted stones, and a single case of a notch (Table 8.9). The percussors were defined on the basis of damage marks on artifacts not primarily classified as percussors. The presence of these artifacts in the limestone assemblage illustrates their exploitation as percussors in earlier stages of their life-histories. The definition of pitted stones is based on the identification of indicative pits, mainly on artifacts defined as percussors (Section 7.4.2). The distribution of the secondary typologies, indicating that 62% of the entire limestone CCT assemblage was used in some form of percussive activity, highlights the key role of these tasks within the limestone component. Two artifacts have a triple typology; one item is a chopping tool, a pitted stone, and a percussor, while another is modified, a percussor, and a notch.

8.3.1 Percussors

The typological classification of percussors is based on their general morphology and additional percussive signatures. Percussors are defined as unflaked cobbles ranging from flat to globular in form. Furthermore, the definition is also based on the presence of impact signs on their surfaces and/or edges. Two distinct groups were identified within the assemblage of limestone percussors: knapping percussors (*percuteurs de taille*) and *percuteurs de concassage* (de Beaune 2000).

The vast majority of percussors are complete (Table 8.10) and the few breaks that occur are located on their proximal and distal ends. As expected, they are mostly covered by cortex: more than 80% of the artifacts bear 75–100% cortex. Less than 7% of the percussors bear less than 50% cortex, and none are completely devoid of cortex. The scar count on these artifacts is relatively high, with a mean value of more than 4 scars. It is worth noting, however, that this value is somewhat biased due to the large difference in this attribute between the two groups of percussors, as can be seen by the wide ranges and high standard deviations (see below).

Percussors are relatively large in size, reaching a mean maximal length of almost 10 cm (Table 8.11). The width and thickness values are smaller, having mean values of 7.5 and 5.2 cm respectively. Their average circumference is 27 cm and their weight ranges between less than 200 g and almost 1 kg, with an average of 522 g. Thus, their size and weight ranges are similar to those of flint and basalt percussors (Sections 6.4 and 7.4.1.1).

Knapping percussors

This group (N=32) was defined on the basis of their classic rounded morphology and appropriate size, as well as damage marks indicative of knapping or retouching (Figs. 8.5–8.8). Artifacts of this group are often flat and rounded (6–9 in the roundness scale of 1–9: Krumbine 1941). Their flat morphology ranges from flat discs (44% of the cases) to plano-convex (sub-discoid) cobbles (52%). The impact signs on these knapping percussors vary in intensity from minimal battering traces to extensive weathering of the surface and/or edges. The damage marks are located mainly on the narrow edges (62%) and to a lesser extent on both the edges and the large semi-flat surfaces (38%) of the artifacts. The surface damage seems to be more

Table 8.7 Counts and frequencies (%) of technological characteristics and descriptive statistics of number of dorsal scars of limestone FFT*

	Tools	Unretouched flakes								
		II-2	II-6/L1	II-6/L2	II-6/L3	II-6/L4	II-6/L4b	II-6/L5	II-6/L6	II-6/L7
N	19	10	18	13	22	16	14	23	21	73
Scar pattern	(N=17)	(N=7)							(N=10)	(N=64)
Cortical	4	2	9	3	5	5	3	6	2	15
	23.53	28.57	50.00	23.08	22.73	31.25	21.43	26.09	20.00	23.44
Plain	1	1	-	1	1	-	3	-	1	1
	5.88	14.29		7.69	4.55		21.43		10.00	1.56
Along axis	1	-	1	2	-	3	1	-	-	4
	5.88		5.56	15.38		18.75	7.14			6.25
Opposed	-	-	-	-	2	1	-	1	-	1
					9.09	6.25		4.35		1.56
Radial	4	-	1	-	-	-	-	-	-	-
	23.53		5.56							
Ridged	-	-	-	-	1	-	-	-	-	-
					4.55					
Side	-	-	-	-	-	1	-	-	1	-
						6.25			10.00	
Along axis & side	1									1
	5.88									1.56
Along axis & opposed	-	-	-	-	-	1	2	-	-	-
						6.25	14.29			
Indeterminate	6	4	7	7	13	5	5	16	6	42
	35.29	57.14	38.89	53.85	59.09	31.25	35.71	69.57	60.00	65.63
Number of scars	(N=13)	(N=6)	(N=8)	(N=10)	(N=14)	(N=11)	(N=9)	(N=4)	(N=4)	(N=27)
Max	8	3	8	5	9	4	4	4	6	8
Median	4	1.5	3	2.5	2	2	3	3.5	2.5	3
Min	1	1	1	1	1	1	1	1	2	1
Mean	4.08	1.83	3.75	2.50	2.71	2.45	2.44	3.00	3.25	2.74
Std dev	1.98	0.98	2.71	1.35	2.40	1.21	1.24	1.41	1.89	1.61
Std err	0.55	0.40	0.96	0.43	0.64	0.37	0.41	0.71	0.95	0.31
Striking platform	(N=18)	(N=8)	(N=17)	(N=13)		(N=13)		(N=9)	(N=64)	
Cortical	1	1	1	1	1	5	-	-	1	2
	5.56	12.50	5.88	8.33	4.55	31.25			11.11	3.13
Punctiform	1	-	1	-	-	1	-	1	-	-
	5.56		5.88			6.25		4.35		
Plain	4	4	6	3	4	1	6	2	3	17
	22.22	50.00	35.29	25.00	18.18	6.25	46.15	8.70	33.33	26.56
Dihedral	-	-	-	-	-	1	-	-	-	1
						6.25				1.56
Facetted	2	-	-	1	-	-	1	-	-	2
	11.11			8.33			7.69			3.13
Crushed	-	-	-	-	-	-	-	-	-	1
										1.56
Removed	2	-	-	-	2	1	1	-	-	-
	11.11				9.09	6.25	7.69			
Broken	4	1	4	6	5	5	3	6	2	19
	22.22	12.50	23.53	50.00	22.73	31.25	23.08	26.09	22.22	29.69
Indeterminate	4	2	5	1	10	2	2	14	3	22
	22.22	25.00	29.41	8.33	45.45	12.50	15.38	60.87	33.33	34.38

* only layers with a minimum number of 10 limestone FFT are included; flake tools from all layers are lumped together

8.3 Typology and Technology of the Limestone Cores and Core Tools

Table 8.8 Counts and typological frequencies (%) of limestone CCT*

	JB	II-2	II-5	II-6/L1	II-6/L2	II-6/L3	II-6/L4	II-6/L4b	II-6/L5	II-6/L6	II-6/L7
N	2	2	7	14	10	9	24	7	4	4	15
Typology											
End-notched piece	-	-	28.57	-	-	-	-	-	-	-	-
Massive scraper	-	-	-	7.14	-	-	-	-	-	-	-
Core waste	-	-	-	-	-	-	4.17	-	-	-	-
Core varia#	-	-	-	7.14	-	-	4.17	-	-	-	6.67
Amorphous core#	-	-	-	-	-	11.11	-	-	-	-	-
Angular fragment	-	-	-	-	-	11.11	8.33	14.29	25.00	-	6.67
Chopper^	-	-	-	7.14	-	-	-	-	-	-	-
Chopping tool^	50.00	-	-	7.14	20.00	11.11	-	28.57	-	-	-
Modified	-	-	57.14	42.86	20.00	66.67	37.50	-	25.00	-	26.67
Percussor	50.00	100.00	14.29	28.57	20.00	-	33.33	28.57	25.00	75.00	46.67
Split percussor**	-	-	-	-	40.00	-	12.50	28.57	25.00	25.00	13.33

* only layers with a minimum number of 10 limestone artifacts are included
# grouped below under cores;
^ grouped below under chopping tools;
** counted here although they are FFT

Table 8.9 Counts of multiple typologies of limestone CCT

	Typology B		
Typology A	Notch	Percussor	Pitted stone
Chopper		1	
Chopping tool		4	1
Amorphous core			1
Percussor			14
Modified		13	1
Massive scraper			1
Split percussor	1		1

Table 8.10 Counts and frequencies (%) of characteristics and descriptive statistics of number of dorsal scars of limestone CCT from the entire cultural sequence of GBY by typological category

	Typological category				
	Percussor	Split percussor	Modified	Chopping tool	Core
N	35	13	35	11	7
Breakage					
Complete	32 / 91.43	13 / 100.00	32 / 91.43	11 / 100.00	7 / 100.00
Distal	1 / 2.86	-	-	-	-
Lateral	1 / 2.86	-	3 / 8.57	-	-
Distal & proximal	1 / 2.86	-	-	-	-
Cortex	(N=33)		(N=33)		
No cortex	-	-	2 / 6.06	-	2 / 28.57
1–25%	1 / 3.03	-	3 / 9.09	-	2 / 28.57
26–50%	1 / 3.03	1 / 7.69	4 / 12.12	2 / 18.18	1 / 14.29
51–75%	4 / 12.12	1 / 7.69	5 / 15.15	5 / 45.45	1 / 14.29
76–100%	27 / 81.82	11 / 84.62	19 / 57.58	4 / 36.36	1 / 14.29
Number of scars	(N=21)	(N=9)	(N=34)		
Max	25	6	4	12	20
Median	3	2	2	7	4
Min	1	1	1	2	3
Mean	4.81	2.67	2.18	6.91	6.71
Std dev	5.58	1.41	0.80	3.42	5.99
Std err	1.22	0.47	0.14	1.03	2.26

frequent on percussors of flat disc morphology (45%) than on plano-convex percussors (30%), for which edge damage is more frequent (69%). Taphonomically and technologically, the patterns of breakage and cortical coverage of this group of percussors are similar to those of the second group; however, they have significantly less dorsal scars (Table 8.12) and are slightly smaller and lighter (Table 8.13).

Percuteurs de concassage

The second group of percussors is very small and differs distinctly from the first. These artifacts are defined as *percuteurs de concassage* (de Beaune 2000; see also *percuteurs de fracturation* in Soressi 2011) rather than *percuteurs de taille*, based on the substantial damage signs that have altered their original morphology, resulting in rugged and uneven surfaces (Figs. 8.9–8.10). The group comprises five artifacts, two originating in the excavated archaeological horizons and the rest in the Unconformity of Area B (N=2) and in the excavation of Trench II (N=1). These percussors are characterized by signs of heavy wear such as battering,

Table 8.11 Descriptive statistics of limestone CCT from the entire cultural sequence of GBY by typological category

	Typological category				
	Percussor	Split percussor	Modified	Chopping tool	Core
N	35	13	35	11	7
Maximum length					
Max	123	123	125	108	109
Median	96	97	65	96	67
Min	64	72	33	55	31
Mean	96.23	98.62	70.63	89.64	69.86
Std dev	12.03	14.85	26.66	19.56	32.01
Std err	2.03	4.12	4.51	5.90	12.10
Length					
Max	120	114	206	106	103
Median	90	89	70	84	64
Min	55	60	30	25	25
Mean	91.03	83.69	71.66	80.73	62.43
Std dev	15.18	16.67	35.13	25.76	31.38
Std err	2.57	4.62	5.94	7.77	11.86
Width					
Max	95	117	96	100	90
Median	76	76	46	76	60
Min	53	57	23	52	27
Mean	75.40	82.92	53.09	74.09	56.29
Std dev	11.78	18.58	22.34	14.49	23.97
Std err	1.99	5.15	3.78	4.37	9.06
Thickness					
Max	79	61	68	65	49
Median	52	36	29	44	25
Min	37	22	15	37	18
Mean	52.23	37.62	36.00	48.09	32.00
Std dev	10.81	10.58	16.01	10.78	12.33
Std err	1.83	2.93	2.71	3.25	4.66
Circumference	(N=25)		(N=6)	(N=8)	
Max	335	339	349	325	303
Median	276	267	237	299.5	205
Min	98	210	113	154	91
Mean	271.72	272.62	233.83	265.25	198.00
Std dev	49.46	34.72	82.06	64.85	85.73
Std err	9.89	9.63	33.50	22.93	32.40
Weight	(N=29)	-	(N=27)	(N=8)	(N=5)
Max	978	-	1001	615	585
Median	522	-	106	373	29
Min	185	-	9	125	16
Mean	552.55	-	228.89	391.25	217.80
Std dev	218.07	-	258.65	184.19	272.38
Std err	40.49	-	49.78	65.12	121.81

crushing, and pounding, which clearly resulted from heavy and repeated blows inflicted on a hard surface. In addition, the damaged area is associated with intensive scars, typical of spontaneous flake removals. Such scars generally exhibit very shallow surfaces that on the one hand lack the distinctive depressions left by bulbs of percussion, but on the other hand are very rugged and unevenly broken. It is evident that both features (battering and shattering) derived from actions that used extensive, violent force, most probably blows repeatedly inflicted on the same hard surface.

Analysis of the few limestone *percuteurs de concassage* from GBY demonstrates that, like the knapping percussors, they exhibit a relatively rounded-flat morphology (either flat discs or plano-convex). The signs of damage, however, occur in all cases on both the edges and the surfaces of the percussors and result in significantly more scars than those observed on

8.3 Typology and Technology of the Limestone Cores and Core Tools

Fig. 8.5 Limestone knapping percussor (#705 Layer II-6 Level 4)

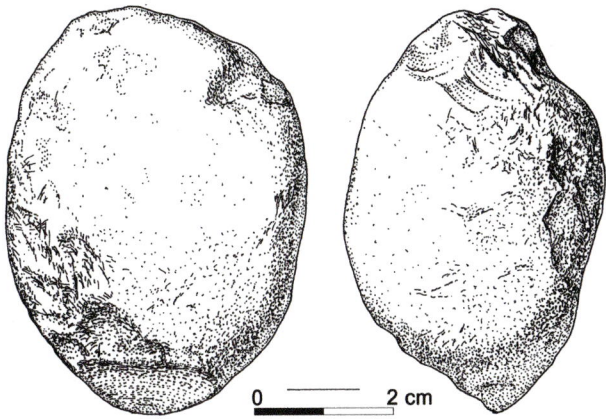

Fig. 8.6 Limestone knapping percussor (#413 Layer II-6 Level 4)

the knapping percussors (Table 8.12). Despite the general morphological similarity of the *percuteurs de concassage* to the knapping percussors, they are slightly larger and heavier on average, although still in their range (Table 8.13).

Clearly, the differences in morphology, size, and damage patterns seen between the two groups indicate that they were used for different functions. It can safely be assumed that the limestone percussors of the first group were used mainly for knapping and for other types of percussive activities identified at GBY, such as nut cracking (Section 7.4.1). *The percuteurs de concassage*, on the other hand, were probably used for a different and specific percussive activity. Building on the variety of finds excavated in the different archaeological horizons at GBY, we can attempt to identify their possible function. Stone knapping can safely be excluded, given the rugged and uneven surfaces caused by the

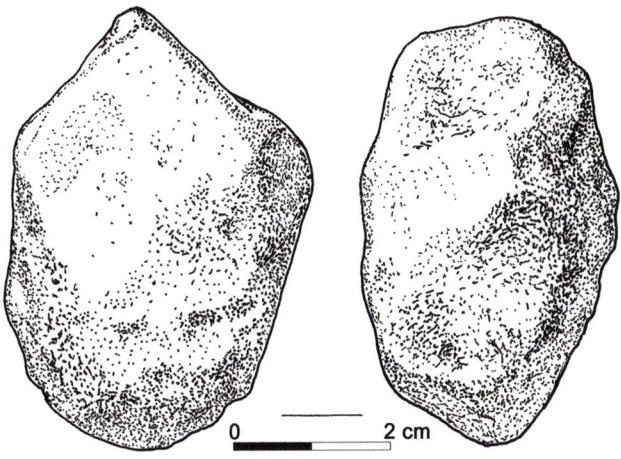

Fig. 8.7 Limestone knapping percussor (#844 Layer II-6 Level 4)

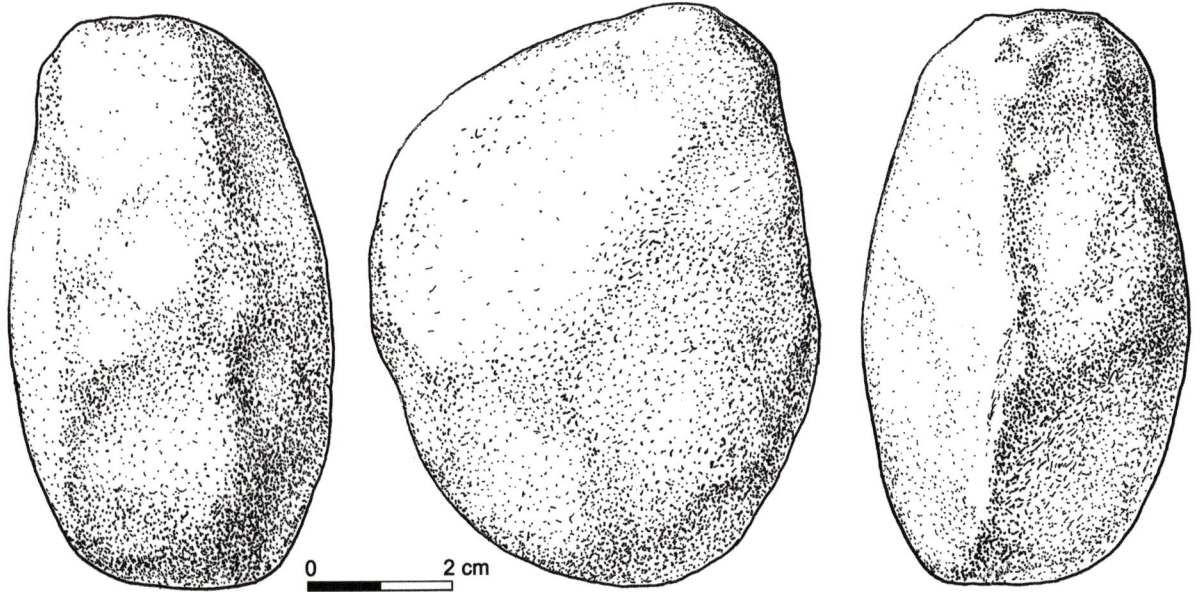

Fig. 8.8 Limestone knapping percussor (#2971 Layer I-5)

Table 8.12 Counts and frequencies (%) of characteristics and descriptive statistics of number of dorsal scars of limestone percussors from the entire stratigraphic sequence of GBY* by type

	Percussor type	
	Percuteur de concassage	Knapping percussor
N	5	32
Breakage		
Complete	5	29
	100.00	90.63
Distal	-	1
		3.13
Lateral	-	1
		3.13
Distal & proximal	-	1
		3.13
Cortex		(N=30)
1–25%	-	1
		3.33
26–50%	-	1
		3.33
51–75%	1	3
	20.00	10.00
76–100%	4	25
	80.00	83.33
Number of scars		(N=18)
Max	25	8
Median	10	2.5
Min	1	1
Mean	11.6	3.0
Std dev	8.62	2.14
Std err	3.85	0.50

* including 2 *percuteurs de concassage* from Unconformity B

Table 8.13 Descriptive statistics of limestone percussors from the entire stratigraphic sequence of GBY* by type

	Percussor type	
	Percuteur de concassage	Knapping percussor
N	5	32
Maximum length		
Max	112	123
Median	99	95.5
Min	97	64
Mean	101.80	95.63
Std dev	5.97	12.25
Std err	2.67	2.17
Length		
Max	109	120
Median	98	90
Min	86	55
Mean	97.80	90.41
Std dev	8.17	15.47
Std err	3.65	2.74
Width		
Max	95	94
Median	91	75
Min	75	53
Mean	87.20	74.03
Std dev	8.14	11.28
Std err	3.64	1.99
Thickness		
Max	70	79
Median	65	49.5
Min	51	37
Mean	62.40	51.00
Std dev	7.54	10.39
Std err	3.37	1.84
Circumference	(N=4)	(N=22)
Max	332	335
Median	306.5	274
Min	283	98
Mean	307.00	267.32
Std dev	21.46	50.65
Std err	10.73	10.80
Weight		(N=26)
Max	900	978
Median	818	504
Min	570	185
Mean	754.4	526.35
Std dev	149.03	211.45
Std err	66.65	41.47

* including 2 *percuteurs de concassage* from Unconformity B

damage, which render them unsuitable for this purpose. Nor is nut cracking a plausible function in light of the heavy damage marks on the artifacts, indicating the application of massive force and repeated blows. Such percussion is clearly unsuitable for the cracking of nuts, which requires a carefully measured application of force and dexterity. Woodworking has also been suggested as a possible function for *percuteurs de concassage* (Mora and de la Torre 2005). The GBY excavations did indeed uncover a preserved wood assemblage, including a worked (polished) wooden artifact (Belitzky et al. 1991), suggesting that woodworking was carried out at the site. However, the use of *percuteurs de concassage* for woodworking is not feasible in the context of GBY, as the wood assemblage consists mainly of soft types of wood and no signs of pounding damage have been identified on the logs and wood fragments from the site.

One possible function of the *percuteurs de concassage* could be the fragmentation of large bones. De Beaune (2000) noted that while percussors are often identified as flint knapping percussors, they were actually used for bone fragmentation. The rounded-flat morphology of the *percuteurs de concassage* and the dullness of their edges suggest they could have been used for recurrent battering during prolonged sessions of bone fragmentation (Célérier and Kervazo 1988 in de Beaune 2000:61). At GBY there is great variability in the frequencies of faunal remains in the different archaeological horizons, two of which are extremely rich and display the highest quantities, density, and taxonomic diversity at the site (Rabinovitch et al. 2008; Rabinovich et al. 2012). An additional faunal occurrence is an elephant skull that exhibits extensive pounding damage marks around its nasal

8.3 Typology and Technology of the Limestone Cores and Core Tools

Split percussors

This category (N=13) consists of split cobbles or large, thick and dorsally rounded flakes. Interestingly, they occur only in the different levels of Layer II-6, although limestone percussors occur throughout the cultural sequence. The splitting of the cobbles and the detachment of these flakes are interpreted as resulting from the use of percussors in extensive percussive activities on durable material involving the application of massive force. The ventral surface of these artifacts is generally rough and uneven and they usually lack the regular characteristics of conchoidal fracture such as identifiable impact points and clear bulbs of percussion. The dorsal surfaces regularly exhibit indicative battering signs, affirming the item's original use as a percussor. The artifacts in this category may result also from the splitting of knapping percussors.

All the artifacts in this category are complete (Table 8.10), their cortical coverage distribution is very similar to that of the percussors and their mean scar count is less than 3, making them closer to the group of knapping percussors. Their lengths and widths are similar to the combined group of percussors, while their thickness is significantly smaller, with a mean value of less than 4 cm (Table 8.11). It is interesting to note, though, that in circumference they are completely identical to percussors, thus strengthening their association with this group of artifacts.

Fig. 8.9 Limestone *percuteur de concassage* (#9545 Layer II-6 Level 7)

Fig. 8.10 Limestone *percuteur de concassage* (#7330 Layer II-6 Level 1)

foramen. The skull bears signs of pounding and cracking for marrow extraction (e.g. Goren-Inbar et al. 1994: pl. II, 102) and is associated with numerous small skull fragments (Goren-Inbar et al. 1994). Thus, the presence of fragmented bones of large mammals (e.g. elephants, rhinoceros, bovids, hippos) at GBY suggests that *percuteurs de concassage* were probably used in the cracking of large-mammal bones. Another possible function of the *percuteurs de concassage* could be related to the processing of vegetal foods. The presence of edible underground storage organs in various archaeological horizons at the site (Melamed et al. 2016), and the identification of basalt thin anvils that also likely served for this type of percussive activity (Section 7.4.3), may indicate that the *percuteurs de concassage* also played a role in their processing.

8.3.2 Modified Artifacts

Modified artifacts are cobbles with a maximum of three flake scars and no working edges and/or platforms (Fig. 8.11). This is the largest category of CCT in the limestone assemblage, alongside percussors (N=35; Table 8.8). Their vast majority

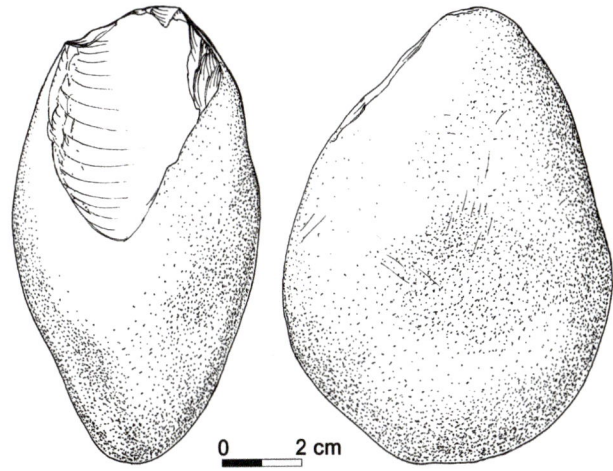

Fig. 8.11 Limestone modified artifact with scar resulting from spontaneous shattering (#7354 Trench II)

(80%) originated in the levels of Layer II-6. As discussed above, modified artifacts also exhibit pits or percussion damage (N=14) on their edges and/or surfaces, suggesting their use in percussive activities. In light of this high frequency, it is likely that the remaining modified artifacts had also been used as percussors and that spontaneous shattering of flakes occurred during their use, erasing the percussion damage signature.

The modified artifacts are mostly complete, with only a few artifacts showing lateral breakages (Table 8.10). Although most of them have a cortical coverage of 75–100%, they show a more variable pattern than the percussors, with some artifacts being completely devoid of cortex. They have low scar counts with a mean value of 2 scars per artifact, which is natural given their definition. In terms of size they are smaller and lighter than the percussors, having mean length, width, and thickness values of 7.1, 5.3, and 3.6 cm respectively (Table 8.11).

8.3.3 Chopping Tools

Chopping tools comprise some 9.8% of the limestone CCT assemblage (N=11; Table 8.8). Only one unifacial chopping tool (i.e. chopper) was recorded, while the remaining chopping tools exhibit a classic working edge with two platforms (Fig. 8.12). The majority (82%) originate in the levels of Layer II-6. Five of the 12 chopping tools have a second typological assignment associated with either percussors or pitted stones.

All of the artifacts in this category are complete, and all but two have a cortical coverage of more than 50% (Table 8.10). Their mean scar count is close to 7, substantially higher than that of modified artifacts and most of the percussors (excluding the *percuteurs de concassage*). In terms of size they are somewhat smaller than the percussors in most dimensions but larger than the modified artifacts (Table 8.11).

8.3.4 Pitted Stones

Pitted stones (N=19), which are defined on the basis of indicative pitting damage (Section 7.4.2), occur only as a secondary typology (Table 8.9). They are identified on both FFT (N=1) and CCT (N=18). Although they are present throughout the sequence, their highest occurrence is in Layer II-6 (84%). Pitted stones of the CCT category occur mostly on percussors (N=14), though they were also found on a modified artifact, a chopping tool, an amorphous core, and a heavy-duty scraper (Fig. 8.13). They are mostly cortical, with a single artifact bearing no cortex and the majority (N=14) exhibiting cortical coverage on 76–100% of their surface. Their size and shape fall within the relatively large category of percussors (mean dimensions in mm: length 89.95, width 76.25, thickness 53.00, circumference 271.11). Given that pitting damage is associated mostly with the cracking of nuts, its appearance in the limestone assemblage further highlights the key role of limestone in various percussive activities.

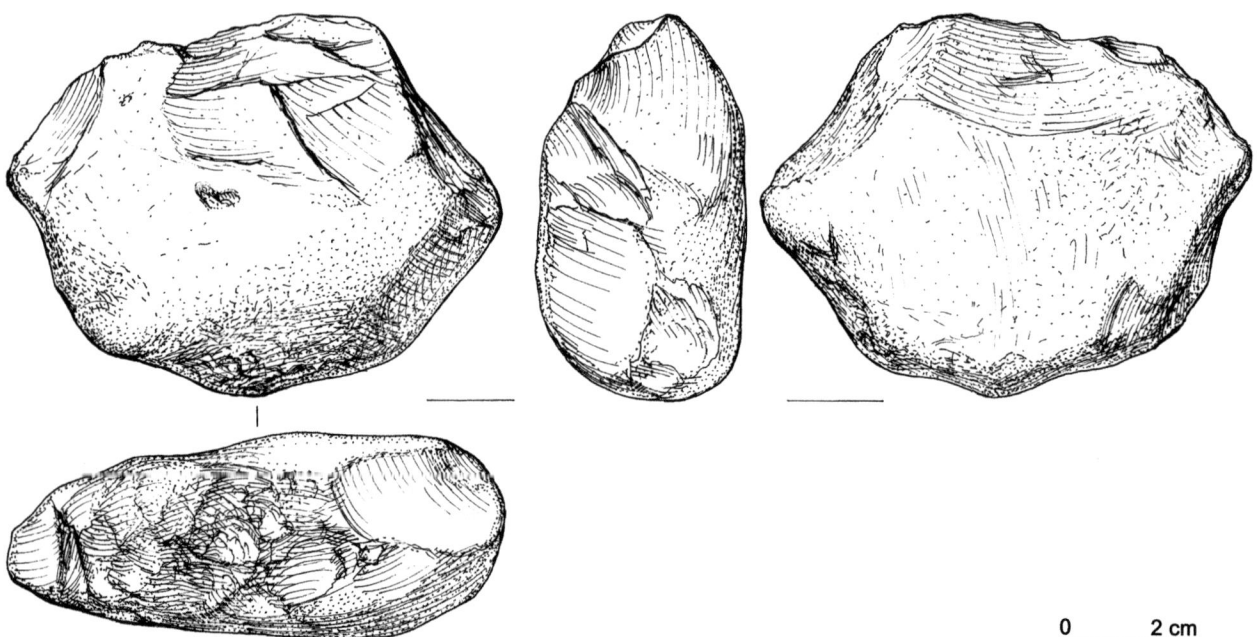

Fig. 8.12 Limestone chopping tool; note pecking/battering signs on the proximal end (#6038 Layer II-6 Level 4b)

8.3 Typology and Technology of the Limestone Cores and Core Tools

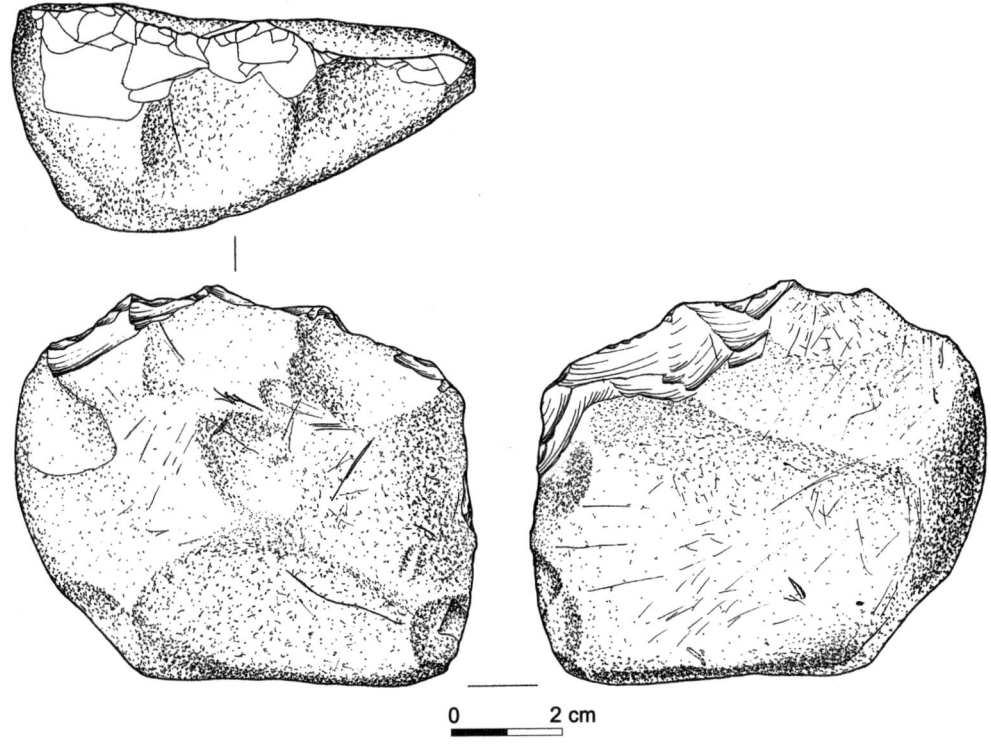

Fig. 8.13 Limestone pitted stone on massive scraper (#663 Layer II-6 Level 1)

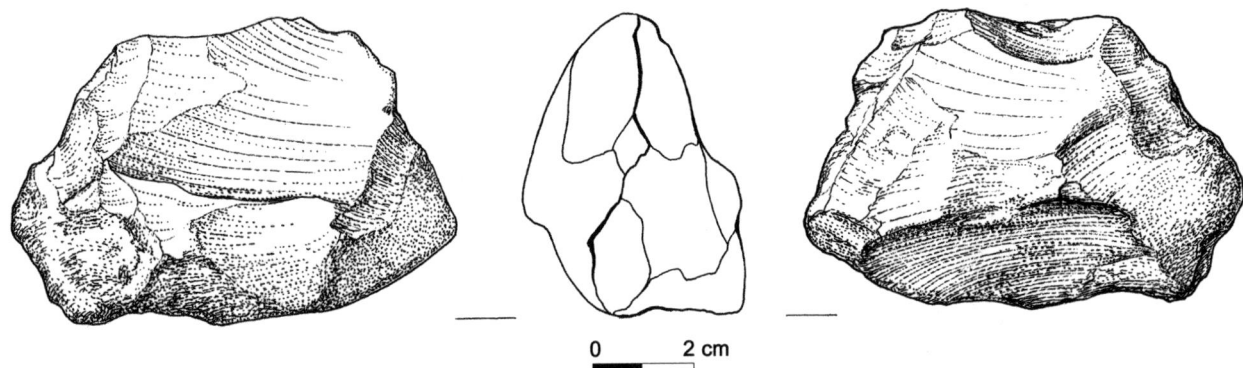

Fig. 8.14 Limestone core varia (#891 Layer II-6 Level 4)

8.3.5 Cores and Other Core Tools

The category of cores consists of six artifacts classified as varia cores (Fig. 8.14) and a single amorphous core. These are cores that have an identifiable working edge and organization of scars, but do not conform typologically to any of the formal core classes (Section 6.4). The single amorphous core also displays pitting damage, indicating its use in percussive activities. All the limestone cores are complete (Table 8.10) and present the lowest values of cortical coverage amongst the CCT, with more than 50% of the artifacts having coverage of less than 25%. Their scar count is relatively high, with more than 6 scars per artifact on average, but still somewhat lower than that observed on the chopping tools. However, there is a substantial variability in this attribute, as can be seen from the standard deviation. In terms of size this is the smallest category in the values of length, width, and circumference, while in thickness the mean value is similar to those of the split percussors and modified artifacts (Table 8.11). The scarcity of limestone cores, both in absolute numbers and in comparison to other limestone CCT categories, indicates that the reduction of cores for the production of limestone flakes was not a regular and systematic strategy of the GBY hominins. This conclusion is strengthened by the scarcity of core waste products, which consist of six angular fragments and a single core waste item (Table 8.8).

Alongside the cores, some isolated core tools were identified within the limestone component. These include two end-notched pieces and a single massive scraper. It should be noted that, while flaked limestone core tools were clearly not a systematic component at GBY, the presence of one biface preform, *éclats de taille de biface*, and ten bifaces (nine handaxes and a single cleaver; Section 7.5.2.1), as well as the abundance of microartifacts in some levels, indicate that the production of limestone bifacial tools did take place at the site, albeit very infrequently.

The limestone bifaces are presented and discussed together with the flint and basalt ones. The decision to present the bifacial component as an inclusive section derives from the relatively small number of the limestone bifaces and in order to avoid extensive repetition of identical typological classifications, description of measurements, and other attributes.

8.4 Discussion and Summery of the Limestone Component

In comparison to the other raw materials used by the GBY hominins, the limestone assemblage exhibits a different approach and mode of utilization. While basalt and flint consistently show higher frequencies of FFT than CCT, as well as varying ratios of these two categories throughout the archaeological sequence, the utilization of limestone seems to be fairly uniform throughout the sequence, with generally low frequencies of limestone artifacts and a CCT/FFT ratio closer to 2. Even before systematic examination of the limestone component, these patterns clearly indicate a particular exploitation mode in which FFT were not the main target products.

Within the examined sample, the overall number of limestone FFT (N=229) is very similar to the number of observed scars on the CCT (N=261). The pattern is variable among the archaeological horizons: in some assemblages the number of observed scars on CCT is smaller than the number of FFT, and in others it is larger. However, the overall similarity indicates that there is no shortfall of flakes that could have originated in the knapping of CCT. Furthermore, along with the presence of limestone microartifacts throughout the sequence, it suggests that the limestone FFT were the product of *in situ* knapping and utilization. If the limestone reduction sequence were oriented toward the production of flakes and flake tools, one would expect their number to exceed the observed number of scars on the CCT. However, considering the great similarity of the two values, the characteristics of the FFT (i.e. cortical, non-retouched, small-sized), and the dominance of percussors in these assemblages, it is most probable that the limestone flakes originate primarily in battering and accidental shattering as a result of the use of limestone percussors. Percussors and modified artifacts form the majority of limestone core tools, and these categories display similar patterns of breakage and cortical coverage, with a majority of complete artifacts and a particularly high cortical coverage. In terms of metric dimensions, however, there is a general similarity in size between percussors and chopping tools. This similarity may suggest that limestone nodules were selected for their particular size (fist size) and form and were transported to the site, as they form a foreign component of the sedimentological record of the lake margin (Section 3.5.2). The slight size gradient in several metric characteristics of these two categories enables a general reconstruction of the life-history of the limestone nodule – first as a percussor (larger, more cortical, fewer scars) and then, possibly after spontaneous breakage of the utilized percussor, modification into a chopping tool (slightly smaller, less cortical, more scars). Finally, the concluding stage in this limestone reduction sequence could be the transformation of a chopping tool into a core (smallest and least cortical). Fig. 8.15 illustrates the size distribution of all limestone components, including the small and unmodified limestone nodules that are a natural component of the sedimentological record. It demonstrates that 1) the natural component is much smaller than the percussors and their derived core tools; 2) the FFT, derived accidentally from the usage of the percussors, are also generally larger than the natural component; and 3) even though split percussors are classified as flakes, they are more similar in size to the percussors than to the FFT component.

A classic reduction sequence is determined by a succession of procedures, which generally begins with procurement of the raw material and terminates with the production of the desired end product, its use, and finally its discard. The GBY limestone component, however, presents a dynamic reduction sequence in which the desired end product (a percussor) is actually the starting point of the sequence (Fig. 8.16). Such a flexible and dynamic operational sequence is an indication of long-term planning and hence advanced cognitive abilities. These are expressed in an intentional selection of the limestone raw material involving decisions about the future function, which dictated the size and morphology of the selected nodule. From the very first stages of the limestone reduction sequence, the mobility of artifacts from their sedimentary source (fluvial terraces) necessitated means of transporting the limestone nodules – means that were different from those used for the basalt slabs (Goren-Inbar 2011b; Chapter 9).

The hominins of GBY apparently had a deep understanding of the association between the characteristics of raw materials and their particular mode of exploitation in the different reduction sequences. Since basalt percussors occur in GBY at even higher frequencies than those of limestone, one may ask what were the specific qualities of limestone that made its use desirable. Clearly, the basalt percussors differ in qualities from their limestone

8.4 Discussion: The Limestone Reduction Sequence

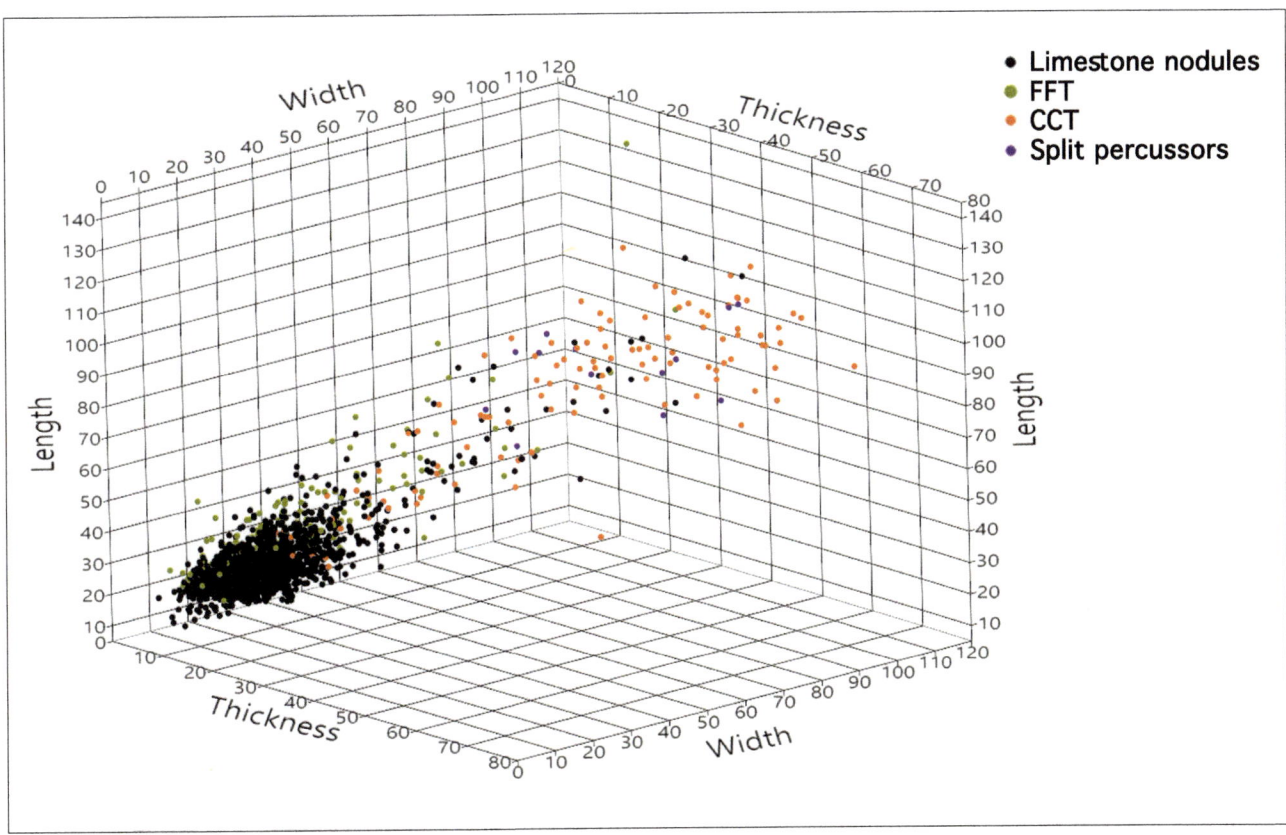

Fig. 8.15 Length, width, and thickness of limestone artifacts (nodules, CCT, split hammers, FFT)

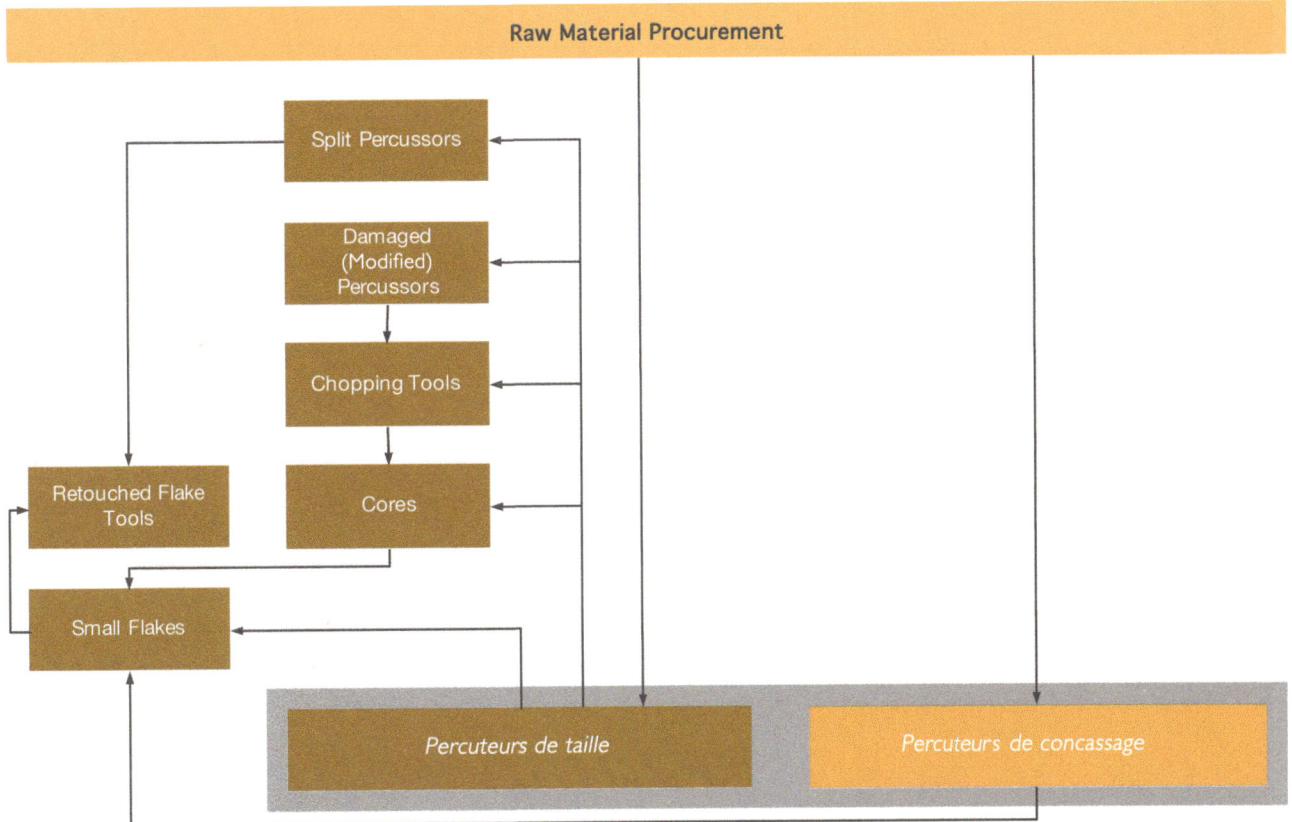

Fig. 8.16 Proposed scheme for the limestone reduction sequence

counterparts, the latter being a softer stone. Studies concerned with the qualities of limestone percussors illustrate the fact that their use resembles soft hammer techniques (Pelegrin 2000), techniques that are indeed observed at GBY, as demonstrated by the presence of their traits on basalt and even more clearly on flint flakes (Sharon and Goren-Inbar 1999; Goren-Inbar and Sharon 2006; Section 6.5.1.1). An antler fragment with evident marks of its use as a percussor was discovered at GBY (Goren-Inbar 2011a; Chapter 7). Both limestone percussors and that of antler were most probably instrumental in the final stages of biface modification. Furthermore, the identification of *percuteurs de concassage* and the occurrence of pitting damage on limestone artifacts indicates that they played a role in percussive activities other than mere knapping.

Chapter 9
Discussion and Conclusions

Abstract This chapter summarizes the results of the analyses of the lithic assemblages of Gesher Benot Ya'aqov and their behavioral, social, and cognitive implications. We discuss the place of these lithic assemblages within the Acheulian record and present a summary of the reduction sequences of each of the raw materials. We provide a comparative view of the reduction sequences, emphasizing typological and technological characteristics as well as issues of variability and conservatism along the cultural sequence. Finally, we present our conclusions on the behavioral patterns and cognitive abilities of the hominins that occupied the site, including issues of mobility, division of labor, group size, preferred locale, home base, and modern cognitive traits.

Keywords Cognition • conservatism and variability • division of labor • Gesher Benot Ya'aqov • group size • home base • mobility • reduction sequences

This chapter integrates the variety of information that was presented in previous parts of this volume, all concerned with the lithic assemblages of the Acheulian site of Gesher Benot Ya'aqov (GBY), dated to MIS 18–20. In Chapter 1 we provided comprehensive information on the geography, climate, vegetation, and fauna of the Upper Jordan Valley. A review of the history of research (Chapter 2) was followed by a detailed account of the geography, geology, stratigraphy, and available rock types at the site (Chapter 3). Chapter 3 also discusses the chronology of the site, whose stratigraphic archive overlies thick basalt flows dated to 1.2/1.3 Ma. Although the top of the sediment package was eroded, north of the site a basalt intrusion into the Benot Ya'akov Formation provided a minimal age estimate of 0.658 Ma. This volume also discusses the variety of excavation procedures and the laboratory methodologies used for the analyses of the lithic assemblages (Chapter 4) and provides a detailed account of each of the archaeological horizons and their inventories (Chapter 5). The results of the detailed analyses of the lithic assemblages, carried out individually for each of the three exploited raw materials, are presented in Chapters 6–8. The present chapter provides an integration between the three and furnishes discussions and conclusions touching on typological and technological as well as cultural and behavioral observations.

Our discussion begins with an attempt to place the GBY cultural record within the global Acheulian manifestations. Nowhere else in the Levant do we find an archaeological site that resembles GBY in the scope of its finds and its particular Acheulian tradition. Lithic assemblages that do resemble those of GBY were found thousands of kilometers away, making any detailed comparisons arbitrary. Thus, we focus here on discussing and summarizing different comparable aspects that emerge from the lithic analyses of the GBY assemblages. These aspects are all intertwined, making a straightforward topic-by-topic description an impossible task in view of the complexity expressed by the Acheulian knapping products. In view of the chronological timeframe (Chapter 3) and the techno-typological nature of the industry of GBY, we refrain from considering its assemblages "Early Acheulian" in the sense of Beyene et al. (2013), but see it as a somewhat younger entity within the Levantine Acheulian record.

The lithic assemblages of GBY are a classical example of the Large Flake Acheulian (LFA) (Sharon 2007). This tradition, a component of the Acheulian Technocomplex, is known for its hallmark handaxes and cleavers, made primarily on flakes. At GBY, this cultural tradition is documented along a continuous sedimentary lakeshore record, which is estimated to have lasted more than 100 ka at the same locality. The cleavers astonished Gilead (1970a) in their similarity to the corresponding African tools and indeed, after seventy years of research, the GBY entity still displays the greatest similarity to an array of African Acheulian sites and their assemblages.

Within the Levantine Corridor there is no site resembling GBY. Nowhere among the ca. 360 Acheulian find spots known in Israel (Alperson-Afil and Goren-Inbar n.d.) is there an entity that resembles GBY. Although isolated items (e.g. at Ma'ayan Barukh: Stekelis and Gilead 1967) resemble the GBY cleavers, these are usually made on flint. At most sites, identification of biface blanks is very difficult when they are extensively covered by flake scars (Chapter 7), but even when flake blanks

© Springer International Publishing AG 2018
N. Goren-Inbar et al., *The Acheulian Site of Gesher Benot Ya'aqov Volume IV*, Vertebrate Paleobiology and Paleoanthropology, https://doi.org/10.1007/978-3-319-74051-5_9

are identifiable they differ in technological and stylistic features from those of GBY. No other Acheulian site in the Levant shows the use of giant cores for the production of large flake blanks for the modification of bifaces. Therefore, GBY bifaces differ in the morphology of their blank, their size, and their particular type of striking platform and its systematic thinning (Chapter 7). Where cleaver-like small handaxes occur in younger Acheulian assemblages, such as that of Tabun (Acheulo-Yabrudian), they are entirely different from those of GBY (Matskevich 2006). The same holds true for some sites in Jordan, particularly 'Ain Soda, where thousands of flint "cleavers" were reported (Rollefson et al. 2006 and references therein). However, these are all oval flint handaxes that do not resemble the GBY cleavers in their typology, technology, or raw material (Sharon 2007). Further away in the Levant, none of the Syrian sites exhibit similarities, and nor do the Nile Valley sites. Perhaps the most similar tools are some bifaces reported over the years from Saudi Arabia (Whalen et al. 1983; Whalen et al. 1984; Whalen et al. 1986; Petraglia 2003, 2005; Petraglia et al. 2009; Shipton et al. 2014; Jennings et al. 2015). Slimak (Slimak et al. 2008) reported the presence of a large flake industry at Kaletepe Deresi 3 in Turkey that includes bifaces made on volcanic rocks. More recently, Galanidou et al. (2013) discussed Acheulian bifaces (including cleavers) from Greece, but these finds are technologically very different from the GBY assemblages.

Apart from isolated Acheulian occurrences, it is evident that GBY is the only site of its type in the Levantine Corridor and the eastern Mediterranean. However, it is beyond doubt that many other similar assemblages (and different types of sites) once existed, and that it is only due to various postdepositional processes, primarily tectonic activity (erosion caused by uplifting and accretion by subsidence) and other natural agents, that they have either remained hidden or been destroyed. The GBY Acheulian site remains a unique phenomenon in western Asia until future discoveries are made.

Although there are many Acheulian sites in Mediterranean Europe, such as Notarchirico (Piperno 1999), they do not resemble GBY. The Iberian Acheulian has a significant large flake component (Santonja and Villa 2006; Santonja and Pérez-González 2010; Santonja et al. 2016; Sharon and Barsky 2016). The chronology of these sites is debatable, but the pronounced presence of cleavers and the use of coarse-grained raw materials clearly place them as LFA. Such tool assemblages also occur in the Garonne and Tarn Valleys in France and at Rosaneto in Italy (Santonja and Villa 2006), but almost none have ever been reported from Europe beyond the Pyrenees (see Sharon and Barsky 2016 for discussion and references). The similarities to the African Acheulian in general and to the LFA in particular have been explained as resulting from cultural connections or from the diffusion of ideas, and perhaps also of hominins, from North Africa to Europe (e.g. Alimen 1975). In northern latitudes of Europe, where the LFA tradition is absent, some cleavers on flakes do occur, but these are usually negligible and are considered unintentional forms (e.g. White 2006).

Asia provides interesting and intriguing information. Some sites in Georgia may be considered to pertain to the LFA (e.g. Lioubin and Beliaeva 2004). But nothing matches the wealth of information originating in India, which is of prime importance in the presence of particular traditions and in regard to the large number of sites and the variability (morpho-techno-typological) of their assemblages (Corvinus 1981, 1983; Petraglia et al. 1999; Sharon 2007; Gaillard et al. 2010; Pappu et al. 2011).

In our attempt to collect the most recent data on the LFA, we found descriptions of several new and extremely promising areas rich in Acheulian sites, such as Buia (Martini et al. 2004) and Aalat (Ghinassi et al. 2015) in Eritrea, but were unable to find any recently published data pertaining to in-depth techno-typological research (Goren-Inbar and Sharon 2006; Sharon 2006; but see for example Gallotti et al. 2010; Diez-Martín et al. 2014). Clearly, this lack of data does not reflect the geographical distribution of this Acheulian tradition but rather demonstrates the absence of new and modern lithic studies.

Numerous genetic, sociological, and demographic modeling studies have shown that the survival of communities depends on a minimum group size and social relationships with other groups (see Section 9.2.2.2). Thus, the extensive intercontinental distribution of LFA assemblages testifies, in our view, to successful and resilient cultural survival. Considering the fact that lifestyles could only have been analogous to those of hunter-gatherers, the dearth of archaeological evidence should not be viewed as a realistic picture of past distribution. The paucity of data has frequently led to attempts to compare distantly located Acheulian entities. Clark (1966), however, warned against drawing conclusions that are based on such comparisons of cultural entities that are distantly located and are in different geographical zones. An example is the debate on the site of Bose in China (Hou et al. 2000), which, despite its repertoire of typical Acheulian artifacts (tools that are clearly handaxes, picks, and unifacials), is not accepted by many scholars as Acheulian (e.g. Corvinus 2004). In sum, the GBY assemblages are LFA, providing evidence for their close cultural affinities with the African Acheulian record. The GBY assemblages are thus a milestone of a distinct wave of hominin diffusion out of Africa that can be traced to India and possibly to China to the east and as far as Spain to the west.

9.1 The GBY Lithic Assemblages

This volume has provided a detailed description of the lithic assemblages produced from three different raw materials. In the

9.1 The GBY Lithic Assemblages

following sections we first summarize the reduction sequences of each of the raw materials (Section 9.1.1). Following this summary, we present a holistic synthesis in which we compare the different reduction sequences of the individual raw materials and describe typological and technological traits that are characteristic of the GBY lithic assemblages, regardless of rock type (Section 9.1.2). Although the lithic analyses touch upon a variety of behavioral, social, and cognitive issues, these are discussed separately in a later section of this chapter (Section 9.2)

In general, the GBY lithic assemblages include classic Acheulian tool types known from other LFA sites. The assemblages include unflaked/unmodified artifacts (e.g. percussors and anvils), cores and core tools (CCT), flakes and flake tools (FFT) and, the hallmark of the Acheulian Technocomplex, bifaces (handaxes and cleavers). These different components occur in varying frequencies in all the archaeological horizons, and the issues of homogeneity and variability are discussed in detail below (Section 9.1.3).

Although the GBY assemblages are typically Acheulian, certain components are missing. Spheroids and subspheroids were not recorded within the excavated sequence of GBY, although basalt and limestone spheroids were found north of the excavations (Sharon et al. 2010).

9.1.1 The Different Raw Materials and Their Reduction Sequences

The three types of raw material that were exploited at GBY appear in all archaeological horizons where the sample size is larger than 50 artifacts. The lithic analyses (Chapters 6–8) show that each of the raw materials was exploited in a similar manner throughout the stratigraphic sequence and that each of the morphotypes (the predetermined tools/artifacts) was destined to be produced on a particular raw material. The selection of the raw material encompasses more than a simple identification of the rock type; it includes recognition of the characteristics of the fracture mechanics and traits of the rock and its dimensions and planned (predetermined) mode of use. These repetitive patterns display a remarkable consistency that was maintained along the temporal trajectory of the GBY stratigraphic sequence. There is an impressive adherence to the sets of reduction sequences identified in each of the raw materials in each and every archaeological horizon.

We consider the patterns to result from a continuous tradition rooted in earlier Acheulian manifestations, such as 'Ubeidiya. Such selection of different types of raw materials seems to be an archaic behavioral trait that is almost completely lacking in assemblages younger than the GBY Acheulian and is entirely absent from the Levantine cave sequences of the Middle Paleolithic (Goren-Inbar 2011; Alperson-Afil 2015; but see, for a minimal appearance, Rosenberg et al. 2015).

In the Levant, the selection of multiple rock types (flint, basalt, and limestone) is documented at 'Ubeidiya (Bar-Yosef and Goren-Inbar 1993) and Latamne (Clark 1968). Thus, 'Ubeidiya, Latamne, and GBY display a somewhat similar mode of raw material exploitation. However, despite the similarities in the raw materials used, the mode of manufacture, technology, abundance, and typology of each of these sites' assemblages are entirely different, and only GBY, the youngest of the three, is assigned to the LFA.

The exploitation of multiple types of raw material is a well-known phenomenon and is reported from many Acheulian sites. Of particular interest are the African sites and their assemblages, as they show many similarities to GBY. All the notable African Acheulian sequences, such as Olduvai Gorge, Kalambo Falls, Olorgesailie, and Isimila, to mention but a few, are characterized by the exploitation of multiple types of raw materials. Furthermore, as at GBY, in many of the assemblages of these sites distinct reduction processes are associated with particular raw materials (e.g. Melka Kunture: Gallotti 2013; Olduvai Gorge: Santonja et al. 2016). The differences of the geological formations exposed in Africa and in the Levant do not permit us to establish direct comparisons. Nevertheless, a preference for coarse-grained raw materials for the production of bifacial tools is recognized as an LFA feature worldwide (e.g. the Arabian Peninsula: Petraglia 2003; for other sites see Sharon 2007, 2008).

A summary of the individual reduction sequences of GBY is provided below, highlighting the principal target products for each raw material and the associated reduction strategies.

9.1.1.1 Flint

Flint is the predominant raw material used at GBY in terms of frequencies rather than mass. The common denominator of the flint assemblages is the small size of the artifacts, an outcome of the original small size of the flint nodules. Among the available nodules, a few larger ones were utilized for the production of bifaces. This reduction sequence was limited and resulted in small numbers of flint handaxes. Nonetheless, products of this sequence (i.e. *éclats de taille de biface*) occur even when the bifaces themselves are missing, a phenomenon related to mobility (discussed in Section 9.2.1).

The main reduction sequence of flint (Fig. 9.1a) consists of the exploitation of small nodules for the production of FFT. The GBY knappers were capable of working with these small artifacts and reducing the volume of very small cores, including the transformation of flakes into cores (cores on flakes). This ability to handle small artifacts is also manifested in the fact that small retouched flakes were the desired target products of the

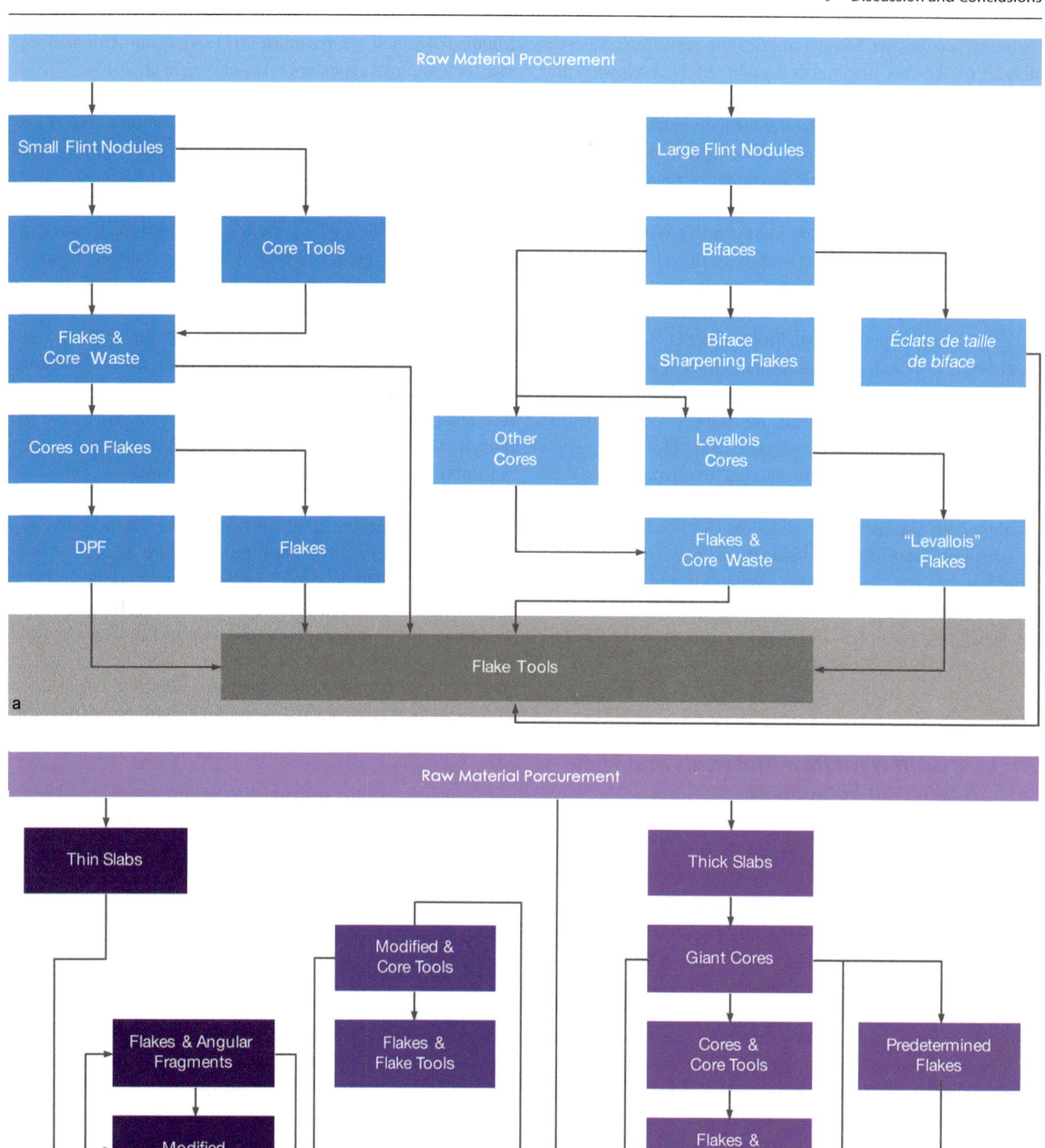

Fig. 9.1 Schematic illustration of the reduction sequences of a) flint, b) basalt, and c) limestone

9.1 The GBY Lithic Assemblages

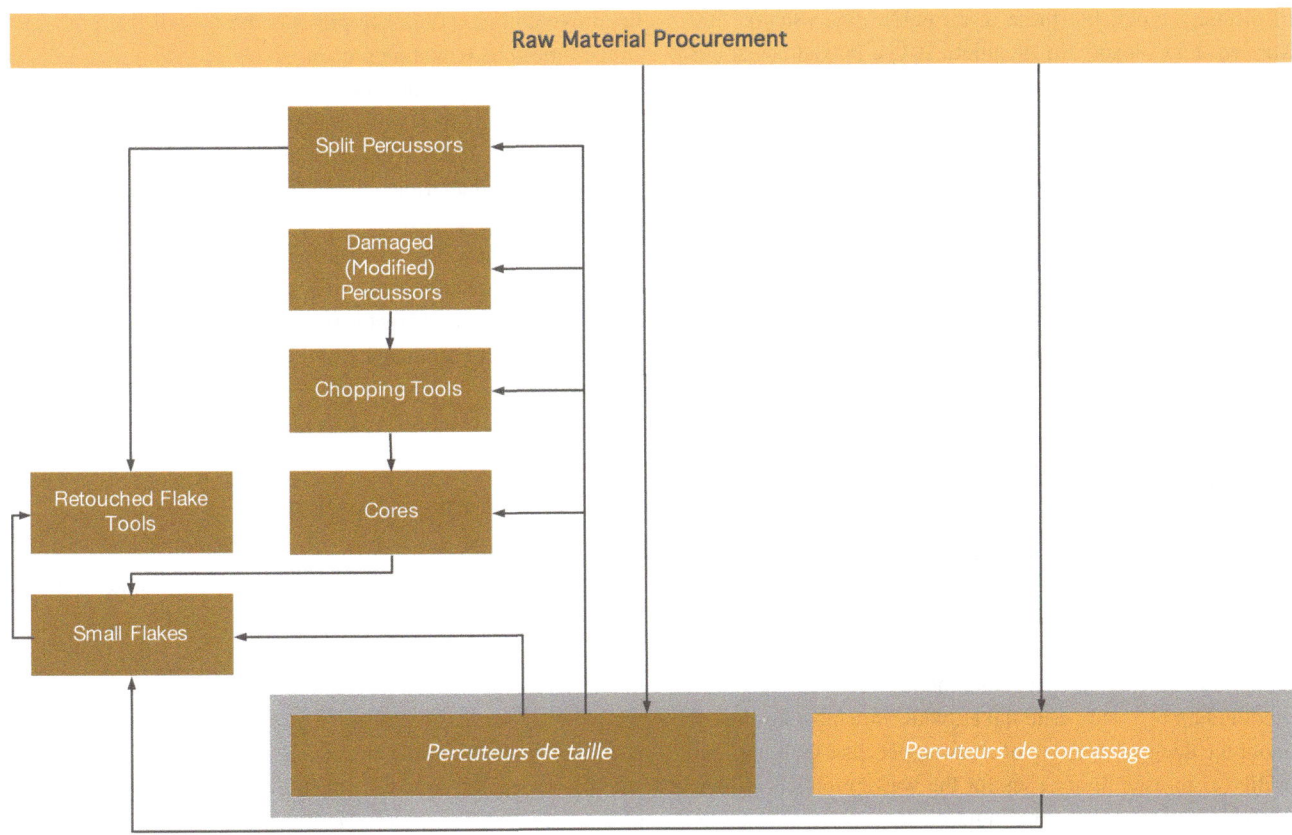

c

flint reduction sequence. Flint flakes are extensively retouched and occur in particularly high frequencies (37.9% tools out of the FFT assemblage of Layers V-5, V-6, II-6), a phenomenon that is also expressed by the relatively high frequencies of multiple tools (19.89% out of all tools).

One of our most remarkable observations is the fact that the striking platform of many of the retouched flakes was purposely removed. The difficult task of precisely removing the bulb of percussion and the striking platform was aimed at acquiring a flat, thin surface. The fact that these manipulations, requiring precision and dexterity, resulted in proximal ends of diverse forms implies that their purpose was to produce morphologies suitable for hafting of these flakes. Although hafting is clearly not an established characteristic of the Lower Paleolithic, the results of our analyses strongly support this interpretation (Chapter 6; Alperson-Afil and Goren-Inbar 2016 and references therein).

9.1.1.2 Basalt

Basalt is the predominant raw material at GBY in terms of mass and was used for knapping and percussive activities. These two functions necessitated acquisition of basalt items of distinctly different size and morphology (slabs and nodules; see below). The different reduction sequences associated with the diverse morphologies are independent from one another (Fig. 9.1b).

Percussive activities involved the use of unmodified basalt items of two different morphologies – thin slabs (anvils) and rounded cobbles (percussors). The assignment of percussive functions to these items necessitated great expertise and knowledge of their morphology, compactness, and weight properties on the part of the hominins. The group of percussive tools includes the markedly homogeneous thin anvils and diverse percussors that vary in size to suit different knapping requirements. The use of percussive tools may have resulted in the formation of other artifact categories, as well as in accidentally broken pieces that provided blanks for additional reduction sequences. Other percussive tools are pitted stones, which may occur on basalt CCT and FFT.

The knapping of basalt artifacts revolves primarily around the reduction sequence of the bifaces, the longest reduction process encountered in the lithic assemblages of GBY. It involves the identification of thick slabs, their extraction, their fragmentation, and their modification into giant cores.

Giant cores in Acheulian contexts are usually found in quarries and workshops, (e.g. Isampur: Petraglia et al. 1999; Vaal River: Kuman 2001) and are extremely rare in other types of occupations. They are likely more abundant in sites of different types, but these await publication (Sharon 2007; Mourre and Cologne 2011; Gallotti 2013). In contrast, at GBY giant cores are always found in association with bifaces, but they are not present in every Acheulian archaeological horizon that includes these tools.

At GBY the reduction process of giant cores involved several distinct methods, all aiming at systematic production of large flakes suitable for the modification of bifaces. The use of large flakes for the production of bifacial tools is one of the most characteristic technological features of the basalt tools of GBY. Large flakes underwent a typical modification during their transformation into handaxes and cleavers, often expressed by the removal of the bulb of percussion and the minimal modification of the edges. The basalt inventory of GBY includes artifacts typical of various stages of a long and complex reduction sequence (e.g. fragmented slabs, giant cores, giant core waste, large flakes of biface morphology, biface thinning flakes, *éclats de taille de biface*, etc.). While the aim of the long basalt reduction sequence was to produce bifaces, several other tools are by-products of the main reduction sequence. Of particular interest are massive scrapers modified on rejected waste flakes, which are most informative about the decision-making of the knappers and their particular ability to foresee future uses for the waste products (Chapter 7). Although there are small and medium-sized basalt flakes at GBY, it is clear from our analyses that these are not the products of an additional reduction sequence but rather the by-products of the biface reduction sequence. The paucity of other basalt tool types, and the minimal presence of basalt cores, could be the result of the dominant role of basalt in the production of bifaces.

9.1.1.3 Limestone

The limestone component is the smallest in each of the GBY lithic assemblages. Chapter 8 describes in detail the typology, technology, and frequencies of the limestone assemblages. These detailed analyses revealed the systematic collection of suitable limestone nodules for a single overall purpose – the use of these unmodified items as percussors. We have further identified two types of percussors, differing in their characteristics, which clearly derive from a particular use as knapping percussors and as *percuteurs de concassage*. While the first is only one type within a wider group of percussors made of flint, basalt, and antler, the second is known only from the limestone inventory. Based on the attribute analysis of the limestone artifacts, we conclude that the entire inventory (both tools and waste products) results from the unintentional fracturing of the knapping percussors. The ability to manipulate unpredictable situations caused by the breakage of the percussors provides additional insight into the hominins' mental and manual abilities and hence reflects their cognitive abilities. Fig. 9.1c presents the overall interpretation of the limestone reduction sequence.

The selection of limestone percussors for different functions may derive from their wide spectrum of traits, ranging from softer (Pelegrin 2000) to harder stone, making them suitable for knapping or for battering.

While we are now able to describe the reduction sequences of each of the raw materials and the particulars of their morpho-typo-technological characteristics, we are still far from a comprehensive understanding of the particular reasons for these selections.

9.1.1.4 A Comparative View

The results emerging from the lithic analyses show that the reduction sequences of the three raw materials differ drastically from one another. Two major factors contribute to the differences observed in the reduction sequences. The first is associated with the different properties (e.g. size and morphology) of each raw material, while the second involves the target tools and predetermined artifacts, which dictated and patterned each of the reduction sequences. The limestone reduction sequence began with the selection of target tools (percussors) and continued in a short, initially unplanned reduction, improvising in order to take advantage of the broken by-products of the limestone percussors. This stands in contrast with the reduction sequences of basalt and flint, which began with the selection of raw material and ended with the modification of the target tools. The reduction sequence of basalt is the longest (multiple stages) and most complex one and, though bifaces were the main target tools, hominins also used the various waste products (e.g. for massive scrapers).

The reduction sequences differed not only in their length, or the number of their stages, but also in their particular structures. In the basalt reduction sequence there was a general linear process from one stage to the next, starting with slabs of homogeneous morphology and continuing to the systematic production of flakes. For flint there was a different scenario, in which small nodules of diverse morphologies were transformed in a short process into cores, providing relatively few small flakes per core (Table 9.1).

Table 9.1 emphasizes the differences between the exploitation modes of the three raw materials. Cores are significantly more frequent in flint than in any of the other raw materials, although they seem to contribute similar frequencies of products (debitage, flake tools, and others). Despite this, the focus of the flint reduction sequence on the production of retouched flakes resulted in very high frequencies of flake tools and relatively low frequencies of unretouched flakes in comparison to basalt and limestone (Fig. 9.2).

9.1 The GBY Lithic Assemblages

Table 9.1 Counts, frequencies (%)[#], and and ratios of cores to different artifact categories[*] in Layer II-6 by raw material

	Basalt			Flint			Limestone		
	N	%	Core ratio	N	%	Core ratio	N	%	Core ratio
Cores	55	1.00	-	448	10.23	-	4	1.43	-
Debitage	4,274	77.94	1:77.71	1,853	42.33	1:4.14	200	71.43	1:50
Flake tools	303	5.53	1:5.51	1,282	29.28	1:2.68	20	7.14	1:5
Handaxes	371	6.77	1:6.75	21	0.48	1:0.05	8	2.86	1:2
Cleavers	148	2.70	1:2.69	-	-	-	1	0.36	1:0.25
Core tools	11	0.20	1:0.2	101	2.31	1:0.23	12	4.29	1:3
Others	322	5.87	1:5.85	673	15.37	1:1.50	35	12.50	1:8.75
Subtotal	5,484	100	-	4,378	100	-	280	100	-
Microartifacts	22,475	-	1:408.31	420,884	-	1:939.47	14,031	-	1:3,507.75

[#] calculated from the subtotal

[*] cores: Levallois, discoidal, globular, prismatic, pyramidal, varia, formless, on flake, on handaxe, giant; debitage: flake, blade, *édtdb*, biface sharpening flake; flake tools: retouched tool, massive scraper; core tools: chopping tool, disc, subspheroid, retouched CCT excluding massive scrapers; others: angular fragment, core waste, modified

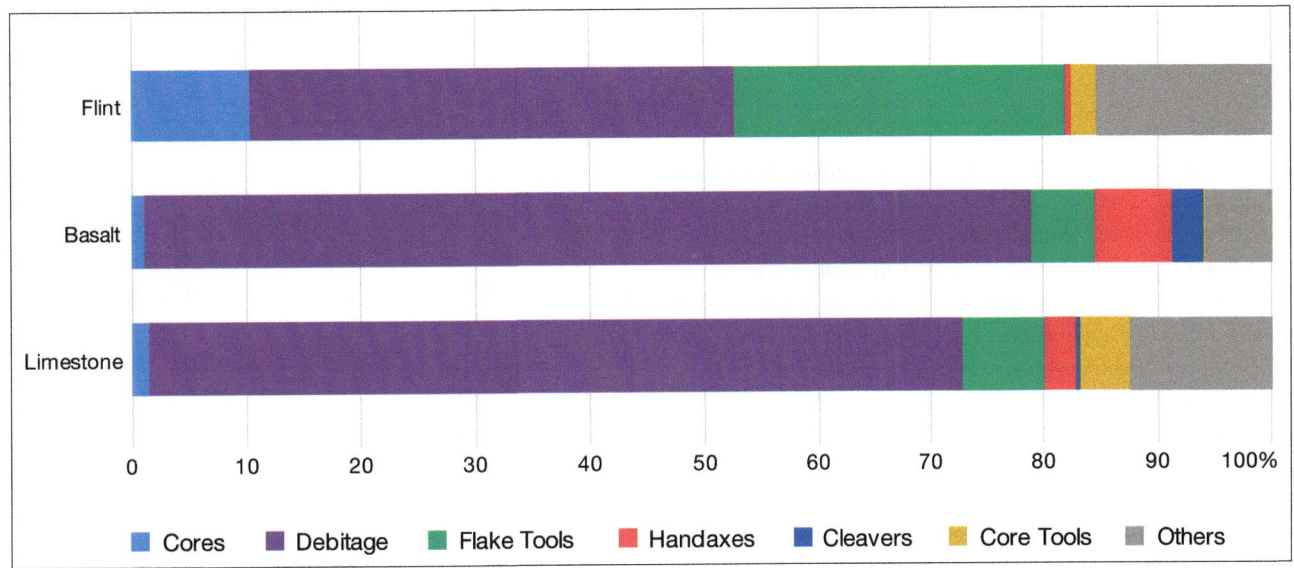

Fig. 9.2 Frequencies of different artifact categories in Layer II-6 by raw material

Bifaces are modified on basalt significantly more frequently than on limestone. Nonetheless, the number of limestone microartifacts suggests that knapping of limestone tools, including bifaces, took place at the site, and so their low occurrence may be related to artifact mobility (Section 9.2.1). A similar explanation may apply to the frequencies of flint bifaces.

The presentation of data in Table 9.1, though it does not include all lithic classes (e.g. natural slabs and percussors), provides an insight into the major characteristics of the particular reduction sequence of each raw material. The individual traits are so clear that even without the thorough analyses of Chapters 6–8 it is clear that basalt was exploited for biface manufacture, flint for flake tools, and limestone for core tools.

9.1.2 Typological and Technological Characteristics of the GBY Lithic Assemblages

9.1.2.1 Typological Characteristics

Handaxes and cleavers, particularly those modified on basalt, are the most prominent tool types of GBY, assigning the assemblages to the LFA. However, the most frequent typological component in all assemblages is small flint flake tools. In addition, many of the flint flake tools are multiple ones and a fraction of the flint cores was also modified into tools, providing yet another measure of the intensive effort that was invested in these small artifacts (Chapter 6). Such high frequencies of small flint retouched tools are definitely not a typical trait of the LFA. Small retouched tools were described for Oldowan assemblages (e.g. Gallotti and Mussi 2015) and for the Acheulian of 'Ubeidiya (Bar-Yosef and Goren-Inbar 1993), but nowhere do the frequencies resemble those observed at GBY.

Among the small flint flake tools of GBY, the dominant types are (in descending order) those with notched and denticulated edges, retouched flakes, borers, and side-scrapers. These morphologies are also the predominant ones subjcted to the removal of the striking platform (Section 9.1.2.4). While basalt flake tools are scarce, they comprise similar morphotypes and the dominant types are (in descending order) those with notched and denticulated edges, massive scrapers and side-scrapers, and retouched flakes. The few limestone flake tools of GBY comprise similar morphotypes, though in significantly lower frequencies.

Tools at GBY were modified on a variety of blanks, including cores, core trimming elements, and *éclats de taille de biface*.

An interesting typological phenomenon is the high frequency of multiple tools within the flint flake assemblage (Fig. 9.3). While double tools may occur on basalt and limestone flakes, they are much more common within the flint component. Triple tools occur (in meager frequencies) only in the flint component.

The bifaces of GBY lack discernible morphotypes that would allow the application of conventional typologies, and hence we differentiate only between handaxes and cleavers. In this we follow Isaac (1977), who suggested that LFA assemblages cannot be subdivided into typological categories beyond this dichotomy.

The typological distribution of the GBY handaxes reflects a high level of homogeneity, which conforms to the technological, morphological, and stylistic conservatism seen in handaxe production. The most common typologies in the thick form are core-like and amygdaloid, while in the thin form cordiform, ovate, and discoid compose the majority of artifacts. This reflects the relatively rounded planform of the GBY handaxes.

The bifaces of GBY, handaxes and cleavers alike, were modified on predetermined large flakes originating from giant cores. The use of large flakes allowed the knappers in most cases to avoid the initial "roughing-out" stage, as the flake already had the desired morphology, and thus "thinning and shaping" and "finishing" were minimally applied (Newcomer 1971).

Bifacial retouch and thinning are the most common modification modes applied on both faces of basalt handaxes, while thinning is more common on the ventral face. This pattern is consistent throughout the cultural sequence. For cleavers the dominant mode is bifacial retouch on the dorsal face and thinning on the ventral face. The dorsal faces of both handaxes and cleavers are more extensively covered with scars, while the ventral faces show consistently less modification. This pattern is generally more pronounced in cleavers than in handaxes.

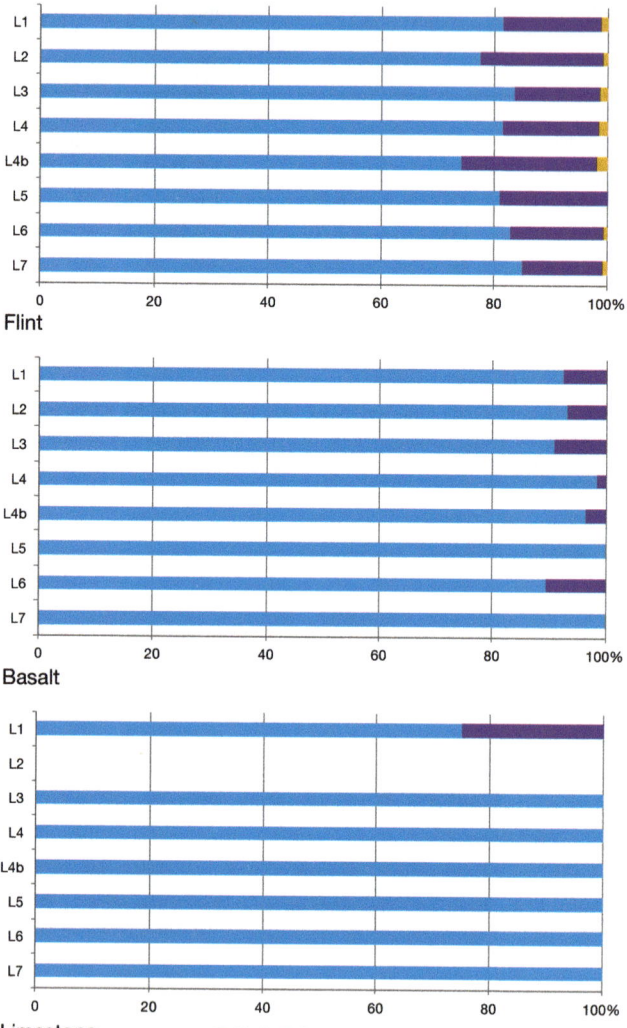

Fig. 9.3 Frequencies of single, double, and triple tools on flint, basalt, and limestone

Regardless of the characteristics of the selected flake, the basalt handaxes of GBY exhibit homogeneous morphologies. Cleavers, in contrast, are more morphologically diverse. Generally, the basalt bifaces of GBY follow an ideal morphological "mental template", although this template included a range of acceptable morphological variations (Herzlinger 2014). The common morphology of basalt handaxes displays several characteristics: the maximal width is close to the proximal end, the lateral edges range from straight to rounded, the distal end is pointed, and the cross-section is flat to biconvex. This morphology shows relative homogeneity among the different archaeological levels. However, alongside this inter-assemblage homogeneity, there is intra-assemblage variability in certain morphological attributes. In addition, flint handaxes generally conform to the morphology of those made on basalt, though with a significant difference in their thicker cross-sections.

Cleavers, like handaxes, show inter-assemblage homogeneity

with intra-assemblage variability. The morphology of cleavers consists of a straight, convex, or pointed proximal end, straight to convex lateral edges, and straight, convex, concave, or diagonal working edge. In cross-section their shapes are usually lenticular, trapezoid, or parallelogram.

Handaxes outnumber cleavers in most assemblages and, unlike cleavers which are made solely on basalt (except a single tool made on limestone), they are modified on three different raw materials. Flint and limestone handaxes occur in significantly smaller frequencies than those of basalt and they are smaller in size.

Percussive tools form a substantial typological component of the GBY lithic assemblages. They are unmodified blanks of distinct morphologies (knapping percussors, *percuteurs de concassage*, and thin anvils), except for the category of pitted stones. While thin anvils are made on basalt slabs and *percuteurs de concassage* on limestone nodules, the rest of the percussive tools occur on the three different types of raw materials. Clearly, percussive tools were involved in a variety of tasks, making them an important element of the Acheulian tool kit.

Interestingly, some tool types that are typical of Acheulian assemblages are altogether missing from GBY. These include polyhedrons, spheroids and discoids. Where chopping tools were identified, they are clearly not the typical tools that characterize Acheulian sites. Spheroids and polyhedrons, defined by Kleindienst (1962) and M.D. Leakey (1971) (see also Willoughby 1985 for an overview), are found at pre-Acheulian Developed Oldowan sites in Africa (Sahnouni et al. 1997; Sahnouni 2006) and the Levant (Le Tensorer et al. 2011). They are clearly part of assemblages assigned to the earlier phases of the Acheulian at sites like Olduvai Gorge, Sterkfontein, and 'Ubeidiya (Willoughby 1985; Bar-Yosef and Goren-Inbar 1993; Jones 1994; Leakey and Roe 1994; Mora and de la Torre 2005). Spheroids were also reported from the final part of the Lower Paleolithic at the Acheulo-Yabrudian site of Qesem Cave (Barkai and Gopher 2015) and hence appear throughout the entire Acheulian Technocomplex.

LFA assemblages are no exception, and spheroids appear in many of its sites and across its geographical range. Spheroids are reported from sub-Saharan African sites (e.g. Olduvai Gorge: Leakey and Roe 1994; Isimila: Howell et al. 1962: table 2; Olergesailie: Isaac 1977) and from North Africa (STIC, Casablanca, and Morocco: Biberson 1961). In addition, spheroids were reported from India (Hunsgi V: Paddayya 1982: table 3, Chirkei: Corvinus 1983).

Spheroids and polyhedrons are practically absent from the rich excavated lithic assemblages of GBY; only a single subspheroid was observed. It should be noted that no spheroids have ever been found among the hundreds of stone artifacts collected from the Jordan River banks in the vicinity of the GBY excavation (Goren-Inbar et al. 1992). However, nine spheroids and three subspheroids were collected from spoils dug by heavy machinery in proximity to the Acheulian locality north of the Benot Ya'aqov Bridge (the NBA site: Sharon et al. 2002, 2010: figs. 14–15; Sharon 2011). Six of these spheroids are modified on limestone and the rest on basalt. The presence of spheroids at NBA may conform with observations that they are often found concentrated in specific localities at Acheulian sites (e.g. Biberson 1961; Barkai and Gopher 2015). Their presence in the proximity of the NBA locality suggests, however, that spheroids were part of the lithic repertoire of the LFA Acheulian industry at GBY.

9.1.2.2 Percussive Techniques

Both ethnographic observations (e.g. Jones and White 1988) and experimental studies indicate that knappers invest great thought and effort in the selection of percussors. Prehistoric and recent knappers import their percussors into the site, sometimes from remote locations, and carry them with them out of the site when leaving (Jones and White 1988; Pétrequin and Pétrequin 1993; Jones 1994; Whittaker 1994). Knappers select their preferred percussors from a large variety of available shapes and sizes (Jones and White 1988; Pétrequin and Pétrequin 1993) and use them differentially in their knapping tasks. Such studies also demonstrate that various percussors are used alternately during different stages of the reduction sequence of a single tool; thus particular stages of the reduction sequence necessitate the use of percussors differing in size, morphology, and hardness (Jones 1994; Sharon and Goren-Inbar 1999; Madsen and Goren-Inbar 2004; to mention but a few).

The GBY knappers were familiar with the properties of the available materials suitable for use as knapping percussors. At GBY we have recovered a large variety of percussors made on basalt (Chapter 7), flint (Chapter 6), and limestone (Chapter 8). In addition, a cervid (*Dama* sp.) antler found at the site shows damage marks indicative of its use as a knapping percussor (Fig. 6.78).

Although we have been able to reconstruct the different reduction sequences of each of the raw materials used at GBY, we cannot determine which types of percussors were used. Nevertheless, it is evident that the biface reduction sequence involved several distinctly different percussors. Basalt percussors of very large dimensions were used for the "opening" of large basalt slabs, their fragmentation, and the handling of giant cores. It is likely that the modification of large flakes was carried out with smaller percussors, as well as soft (antler) percussors. The flint reduction sequence involved the use of small percussors, dictated by the small size of the knapped nodules. This difference between the very large percussors of the basalt reduction sequence and the very small percussors of the flint reduction sequence demonstrates the knappers' ability to suit the percussive technique to the task at hand, illustrating the cognitive flexibility and the dexterity of the GBY knappers.

Our knowledge of the percussive techniques used at GBY is not only based on the presence of percussors at the site, but is also supported by the knapping products and their characteristics. Soft percussors may have been used in several stages of the reduction sequences of basalt, flint, and limestone (Sharon and Goren-Inbar 1999). However, we can demonstrate only that soft percussors were applied in the shaping and finishing stages of bifaces, based on the technological characteristics of FFT (Chapters 6–7).

9.1.2.3 Core Reduction Methods

The results of the analyses of the GBY lithic assemblages demonstrated that different core reduction methods were in use (Chapters 6–8). Among the variety of core reduction methods, some were restricted to a particular raw material, while others were applied to more than one raw material. Core reduction methods at GBY include well-known technological strategies such as Levallois and Kombewa, while others are more obscure or undefined (e.g. those cores classified as varia, formless, globular, and amorphous). Interestingly, there are clear differences between the frequencies of defined methods identified on flint and basalt cores (undefined methods form 71.38% of the 559 flint cores and only 35% of the 60 basalt cores; Chapters 6–7). This difference reflects the single objective of the basalt reduction sequence, which was aimed at the reduction of giant cores in order to produce large flakes of particular morphologies that required minimal further modification for their transformation into bifaces. The flint cores, however, were reduced in order to produce small flakes of various morphologies that were intensively retouched and transformed into a variety of flake tools and multiple tools, many with additional proximal modifications in preparation for their hafting.

Another interesting observation is the integration of several core reduction methods in the handling of giant cores. Even though the procured basalt slabs were morphologically similar, different methods, including Levallois, Kombewa, slab slicing, and bifacial knapping, coexisted in their reduction. Another example of a similar complexity is the reduction method of cores on flakes. We assume that particular flakes were selected on the basis of their specific properties to be modified into cores, and thus we consider cores on flakes to be a distinct method. The reduction of cores on flakes sometimes also integrated the Levallois method. These cases illustrate greater complexity, since they involve a combination of core reduction methods.

Clearly, the core reduction methods identified at GBY do not represent the full range of methods used by the knappers, and some remain invisible. An example of such invisible methods can be seen in particular giant core waste products identified at the site, which are not associated with any of the defined methods (Chapters 7).

The Levallois Method

The GBY lithic analyses resulted in the identification of Levallois flint cores and a single Levallois giant basalt core. Cores assigned to the Levallois method are characterized by two hierarchical surfaces, one more intensively flaked than the other. These cores are characterized by a continuous acute angle of the working edge. In Chapter 6 (Section 6.5.1.3) we detailed the particular traits of these cores and concluded that they bear characteristics of the Levallois method, although they differ from the classical Middle Paleolithic products in lacking predetermined removals. While the scar patterns on the flint cores show similarities to the classical Levallois pattern, the artifacts differ in the extent to which individual flake removals penetrated the debitage surfaces. Despite these differences, we consider the cores as having been produced by a Levallois method, although they are clearly technologically inferior to Levallois cores from Middle Paleolithic contexts.

Within the FFT component, we were unable to identify artifacts that resemble classical Levallois products, although there are a variety of scar patterns that could be considered Levallois. The difficulty stems from the fact that products of the shaping and finishing of bifaces (e.g. *éclats de taille de biface*), which are abundant at GBY, are very similar to products of the Levallois method in their convexity, thinness, and dorsal scar patterns (Copeland 1995). Despite the paucity of true Levallois FFT, we consider those items that we classified as Levallois cores to be markers of this method, although they are not reduced by the conventional Levallois method, and its classical waste products (e.g. *débordant*, *outrepassé*) rarely occur.

The assemblage of cores on flakes, a well-defined core reduction method at GBY, includes artifacts that we classified as Levallois cores. Their presence provides further support for the intentionality of Levallois production. We conclude that the cognitive ability to conceptualize the production and exploitation of items that feature convexity is a component of the emerging Levallois technology. Moreover, we associate this cognitive ability with those that enabled the production of bifaces. The two are clearly associated, although handaxes that were transformed into cores that occur at many later Acheulian sites are minimally represented at GBY (e.g. de Bono and Goren-Inbar 2001; Marder et al. 2006) (Chapter 7). Nevertheless, the systematic production of handaxes, exploiting the feature of convexity, is indicative of an entrenched knowledge of this aspect. In later Acheulian assemblages, the transformation of handaxes into hierarchical cores attained high frequencies, for example, at the Acheulian sites of Tabun (Shimelmitz et al. 2016) and Revadim (Marder et al. 2006).

The Levallois component of the GBY assemblages provides evidence that the emergence of the Levallois method was not a sudden innovation that spread immediately and swiftly over a large area as a complete package. Rather, it was a slow, progressive process that initiated with the emergence of the lenticular morphology of bifaces and with the introduction of cores on flakes. Both reflect the ability to exploit biconvex surfaces by using the sharp angle of a continuous edge between two convex surfaces. The GBY assemblages present several components supportive of this concept – bifacial tools, cores on flakes, and Levallois cores – which are not identical with the Middle Paleolithic Levallois cores but exhibit the ability to exploit cores by the Levallois method (for the presence of these cores in the Acheulo-Yabrudian see Zaidner and Weinstein 2016; and in East Africa e.g. Tryon et al. 2005).

Our results contradict the views of Adler et al. (2014) on Armenia and White and Ashton (2003) on northwestern Europe, both of whom see the co-appearance of handaxes and Levallois technology as a local phenomenon. We have demonstrated that cores that can be considered Levallois and are similar to those described by Adler et al. (2014) appear as early as MIS 18 at GBY, much earlier than the Armenian examples (MIS 9). Our perspective envisages a slow process of adaptation of inventions and their geographical spread from centers of invention by diffusion and dispersal. Furthermore, highly sophisticated core methods that resemble the Levallois method and even show more advanced technological abilities are well documented in the Acheulian of South Africa (Sharon & Beaumont 2006) and India (Sharon 2007).

The Kombewa Method

The use of the Kombewa method for the production of flakes is minimally represented at GBY and is recorded only within the basalt component. It is likely that the Kombewa method was in fact more commonly used, but the combination of our rigorous classification and in-situ weathering has resulted in the underrepresentation of recorded Kombewa products. In addition, small flakes bearing two ventral faces were identified as dorsally plain flakes (DPF). These occur on both flint and basalt and are likely the products of core on flake knapping and the modification of bifaces on large flakes.

Among the cores, only one giant basalt core displays the use of this method. It is a giant flake exhibiting a flake removal on its ventral face (Fig. 7.132). Apart from this, the use of the Kombewa method is recorded in the presence of basalt Kombewa flakes in the FFT component, as well as in the assemblages of handaxes and cleavers.

Among the basalt FFT we have identified over 150 Kombewa flakes, 86% of which are unretouched. Kombewa flakes provide the ideal blank for biface modification, presenting the desired lenticular morphology. Nevertheless, although we would expect Kombewa flakes to co-occur with bifaces, no association was found between the frequencies of Kombewa flakes and those of bifaces. Kombewa blanks were identified in the assemblage of basalt handaxes (N=11; 2.82%) and cleavers (N=31; 19.74%). This difference stems either from the intensive scar cover of handaxes, which masks the method of flake production, or from the particular preference of Kombewa flakes for cleaver production.

9.1.2.4 Removal of Striking Platform

Analyses of the flint component have demonstrated that the removal of the striking platform is a frequent typo-technological feature of flint flake tools. A comparison between FFT with a striking platform and those in which the striking platform was removed demonstrates that the latter are significantly more often double patinated, are significantly more frequently retouched, exhibit significantly more radial scar pattern, bear significantly higher frequencies of removals on the ventral face, and are significantly more distally broken (Section 6.5.1.4). These observations led us to the conclusion that removal of the striking platform is in fact a typological feature, a product of special attention within the flint reduction process. It is our conclusion that these products, the target products of the flint reduction sequence, were proximally modified in order to facilitate their hafting (Alperson-Afil and Goren-Inbar 2016). The removal of the striking platform resulted in reduction of the thickness, and possibly the length, of these flakes. Another means of reducing the thickness of flakes is through thinning of the bulb of percussion. Although at GBY this feature is recorded on only three flint flakes, the thinning of the bulb is clearly much more frequent when we consider the numerous cases in which the striking platform was entirely removed. In the basalt component, on the other hand, the thinning of the bulb is a dominant shaping procedure (Goren-Inbar and Saragusti 1996), as at many other LFA sites (e.g. Isaac 1977; Sharon 2007; Gallotti 2013), and is one of the most characteristic technological features defining these assemblages. In fact, the removal of the striking platform of flint flakes and the thinning of the bulb area of basalt flakes share a common concept, as the aim of both was to thin the natural high relief formed by the hertzian cone. This concept, both typological and to some extent (see below) technological, is applied differently to flint and basalt. Within the basalt assemblage the thinning of the bulb occurs more frequently on large flakes and bifaces. This clearly contrasts with the very small flint FFT.

The production of basalt bifaces consistently uses a system of removal of the large bulb of percussion from the ventral face (and sometimes also the dorsal face, as in the Kombewa examples). These removals are very indicative and are observed on many of the bifaces, both handaxes and cleavers. These areas of removals, or thinning of the striking platform, have been noted in many of the LFA assemblages, whether in Africa or in Asia (e.g. Isaac 1977; Sharon 2007; Gallotti 2013). It seems obvious that such a well-executed process with the aim of achieving a well-proportioned and symmetrical cross-section for bifaces would also be applied to large unretouched flakes, including some artifacts that have the shape of a biface (see above). Thus, the knowledge and practice were applied to both raw materials, despite the difference in size between flint and basalt bifaces. Furthermore, for the first time, as far as we know, we can point to a common denominator that links the very different reduction sequences of flint and basalt – a shared expertise used for different purposes on different materials. We suggest that the purpose of removing and thinning the proximal area of flint flakes was to enable their hafting. As for the thinning of basalt flakes and bifaces, we envisage several options relating to concepts of symmetry, ease of handling, and possibly hafting. The ability of the GBY hominins to apply similar procedures to artifacts of very different sizes, types, and raw materials demonstrates advanced cognitive abilities (Section 9.2.4).

9.1.3 Cultural Variability and Conservatism

The cultural sequence of GBY is composed of a series of superimposed archaeological horizons. Each horizon is a well-defined stratigraphic unit that represents a discrete hominin occupation. The entire cultural sequence, from the lowermost to the uppermost occupation, is assigned to the Acheulian culture and is estimated to represent ca. 100 ka. However, the major part of this cultural sequence (Layer II-6 to Layer V-5), which is the focus of the detailed lithic analyses presented in this volume, is estimated at ca. 50 ka. It is impossible to determine the exact duration of occupation of each of the archaeological horizons but, as established by the paleobotanical identifications (Melamed et al. 2016), for most of them the minimal exposure time was four seasons (year round). The maximal duration could not be determined, but the preservation of faunal and botanical remains suggests that the exposure of each occupation was short.

The chronologically defined long cultural sequence of GBY provides a unique opportunity to explore the mechanisms of continuity and change over time. The lithic analyses demonstrated that certain typological and technological characteristics persist throughout the occupational sequence in conjunction with variations.

A recent attempt to discuss the issue of variability and conservatism at GBY (Sharon et al. 2011) focused on the basalt reduction sequence. This first attempt illustrated the co-existence of variability and consistency among typological and technological characteristics. In the following section these characteristics will be instrumental in an attempt to use the extensive data, results, and discussions provided earlier in this volume to enlarge the discussion of aspects of variability and conservatism at GBY.

9.1.3.1 Variability

One of the most striking differences between the various archaeological layers is seen in the frequencies of the raw materials exploited. Fig. 9.4 shows that there are intra-layer raw material preferences for the production of different lithic categories. In addition, when specific lithic categories are examined inter-layer variations are clearly evident, exhibiting a comb-like pattern of random fluctuations. An exception is the biface category, in which inter-layer variability is minimal and influenced primarily by the rare occurrence of bifaces made on flint and limestone (Chapter 7). Other lithic categories display much higher variability in raw material exploitation. In some assemblages (e.g. Layers V-5 and V-6) the exploitation of flint is more pronounced than in others, but even there the picture is more complex, as the frequencies of flint CCT drop meaningfully lower than those of flint FFT. Another example is Layer II-6 Level 4; here one would expect to find many cores on basalt due to the abundance of basalt flakes, yet the number of flint cores is higher than expected (Fig. 9.4). As for the exploitation of limestone in the different layers, it is consistently low and at times absent, and yet appears to be variable in each of the lithic categories.

Other patterns of variability can be recognized when the entire lithic assemblages are organized by lithic category and raw material. Fig. 9.5 presents the frequencies of lithic categories by raw material along the stratigraphic sequence. It provides insight into the observed variability in the exploitation of raw material, which is principally governed by the individual characteristics and target products of the three distinct reduction sequences of basalt, flint, and limestone.

Cultural variability is expressed in the fact that, with the exception of bifaces (particularly cleavers), the archaeological horizons differ strikingly from one another in the frequencies of each of the lithic categories. Particularly interesting are the patterns emerging from the comparison between the categories of debitage and flake tools. We would expect to find a similarity in the frequencies of both, so that when basalt debitage is dominant the basalt retouched tools are too. However, for flint the pattern is reversed. This pattern, however, is not homogeneous, as some assemblages (Layers V-5 and V-6) exhibit a dominance of flint in both debitage and flake tools.

In addition to the technological aspects of the observed variability, typological aspects are also encountered. Although the overall typological composition of the tool kit is similar throughout the cultural sequence, clear differences are evident in the frequencies of flint tool types (Fig. 9.6). While the category of retouched flakes (included within "other flake tools") is the most common in some layers/levels, in others it is the "notches & denticulates" category. The latter is outstandingly dominant in the assemblage of Layer V-5, while in Layer V-6 it is the category of side-scrapers that is predominant (Fig. 9.6). Again, no temporal pattern characterizes the cultural sequence, and

9.1 The GBY Lithic Assemblages

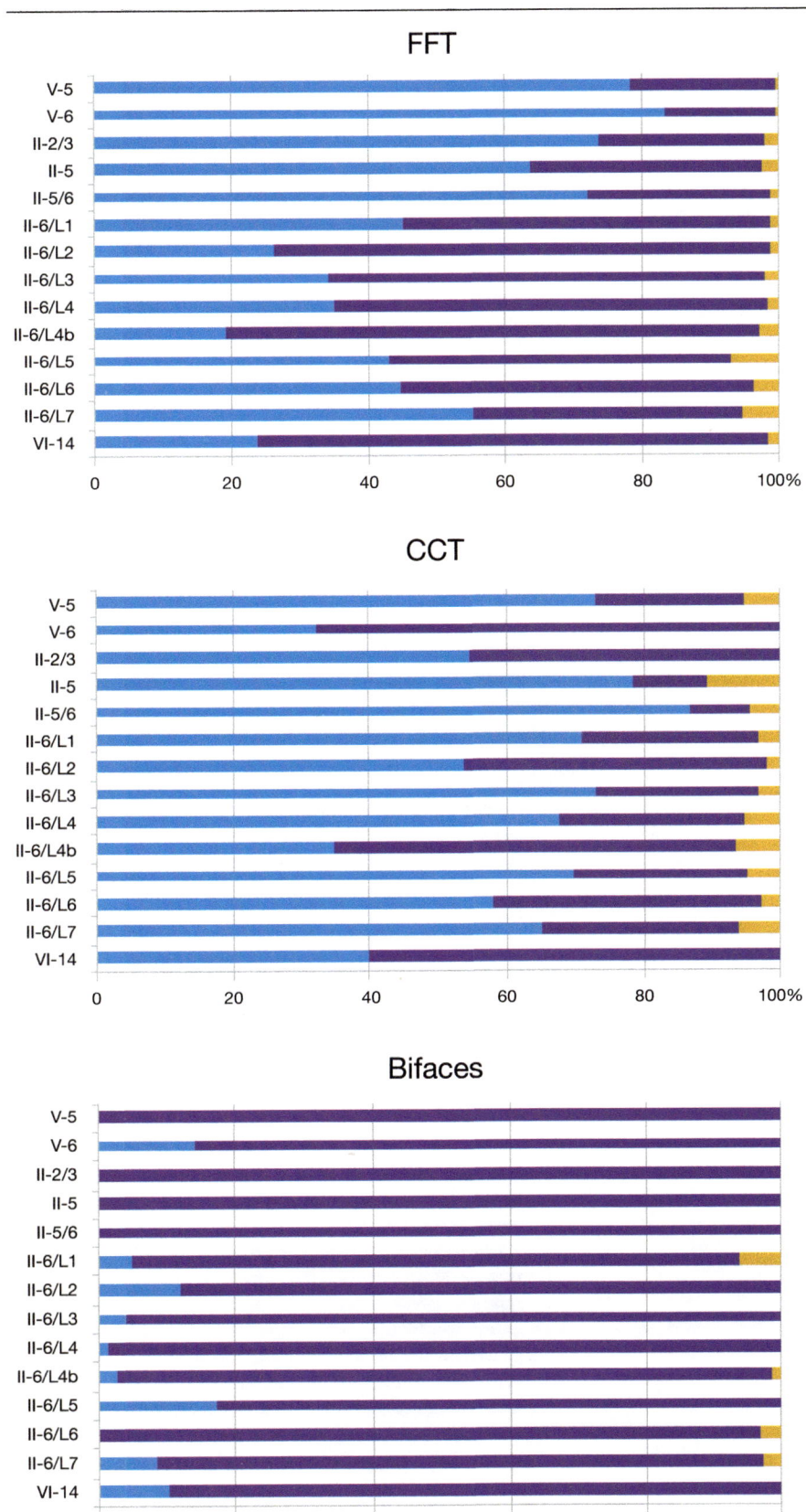

Fig. 9.4 Frequencies of FFT, CCT, and bifaces throughout the cultural sequence by raw material

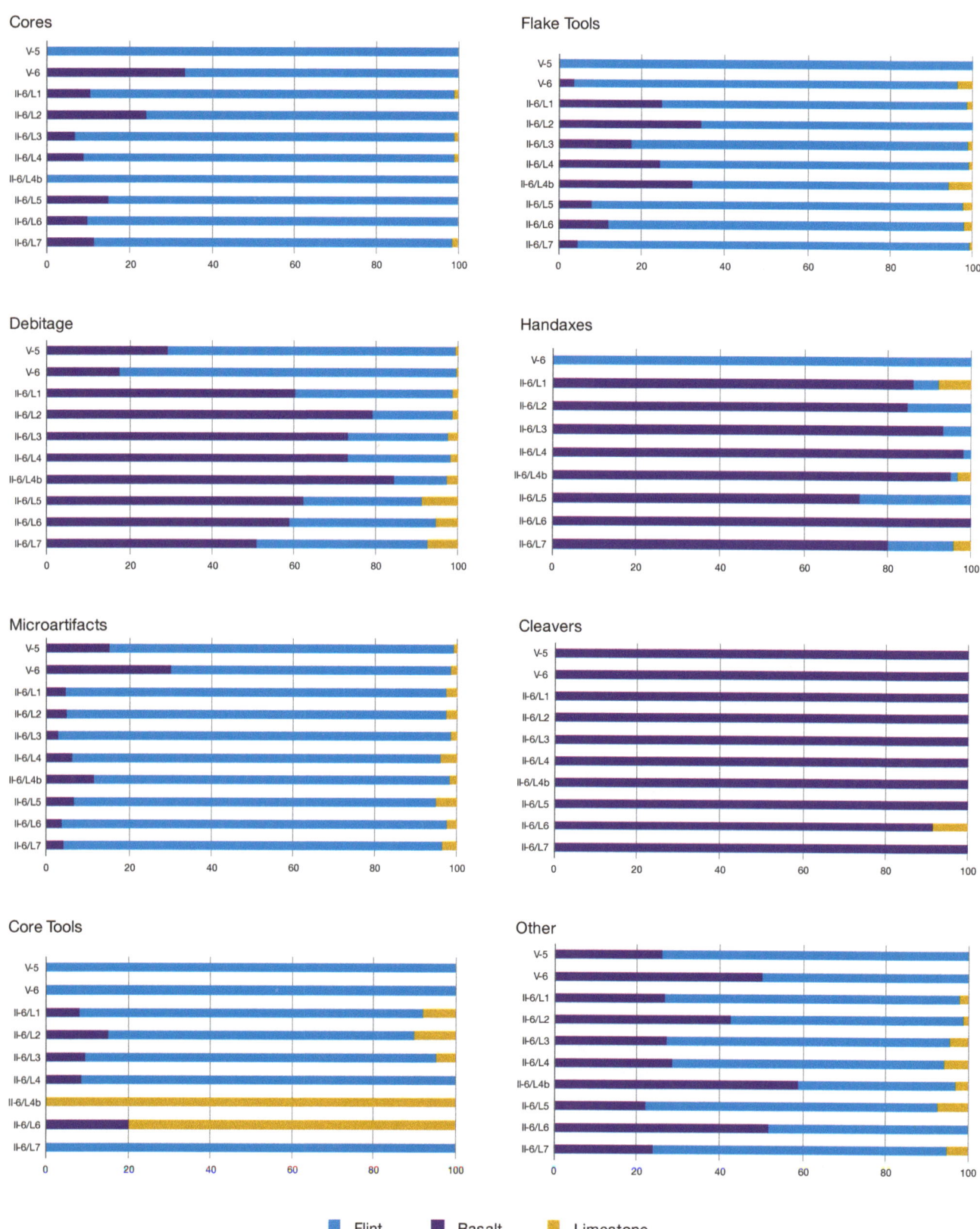

Fig. 9.5 Frequencies of different artifact categories throughout the cultural sequence by raw material

9.1 The GBY Lithic Assemblages

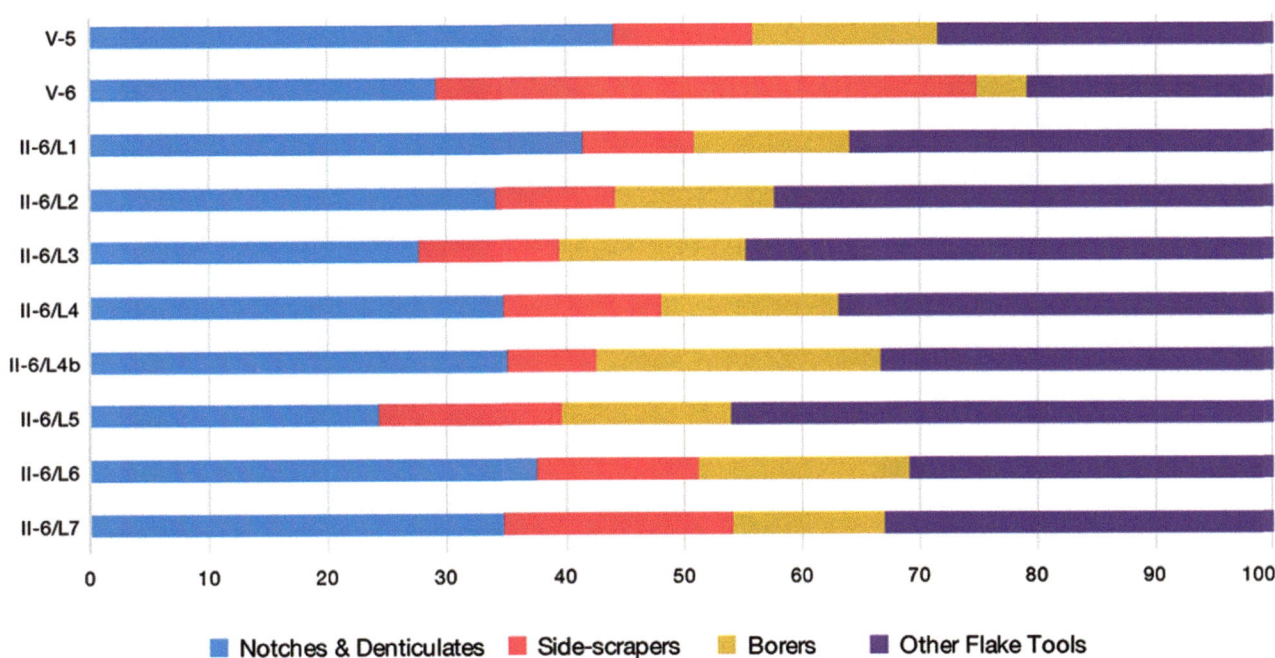

Fig. 9.6 Frequencies of particular types of flint flake tools throughout the cultural sequence

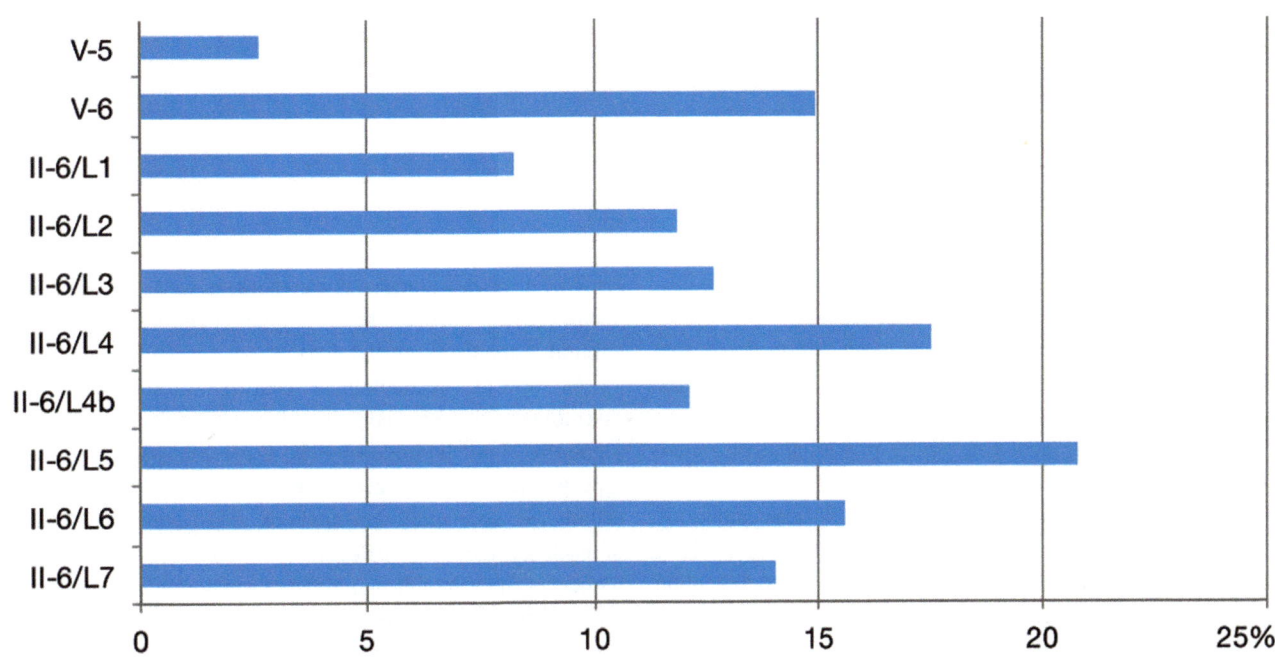

Fig. 9.7 Frequencies of flint FFT with removal of striking platform throughout the cultural sequence

each layer/level displays a different distribution of flint flake tools.

Additional typological variability is observed when the removal of the striking platform is examined (Fig. 9.7). Although this typological feature occurs along the entire stratigraphic sequence, there are pronounced variations in its frequency in each of the lithic assemblages. While the removal of the striking platform of flint flakes reaches only 2.6% in Layer V-5, in Layer II-6 Level 5 it characterizes 20% of the FFT (Fig. 9.7).

These patterns of variability suggest that at GBY, despite the relatively long duration of the cultural sequence, there is no trend in the pattern of variability along the time trajectory. Possible explanations for the observed variability are discussed in detail below and include mobility (Section 9.2.1.1) and site function (Section 9.2.3).

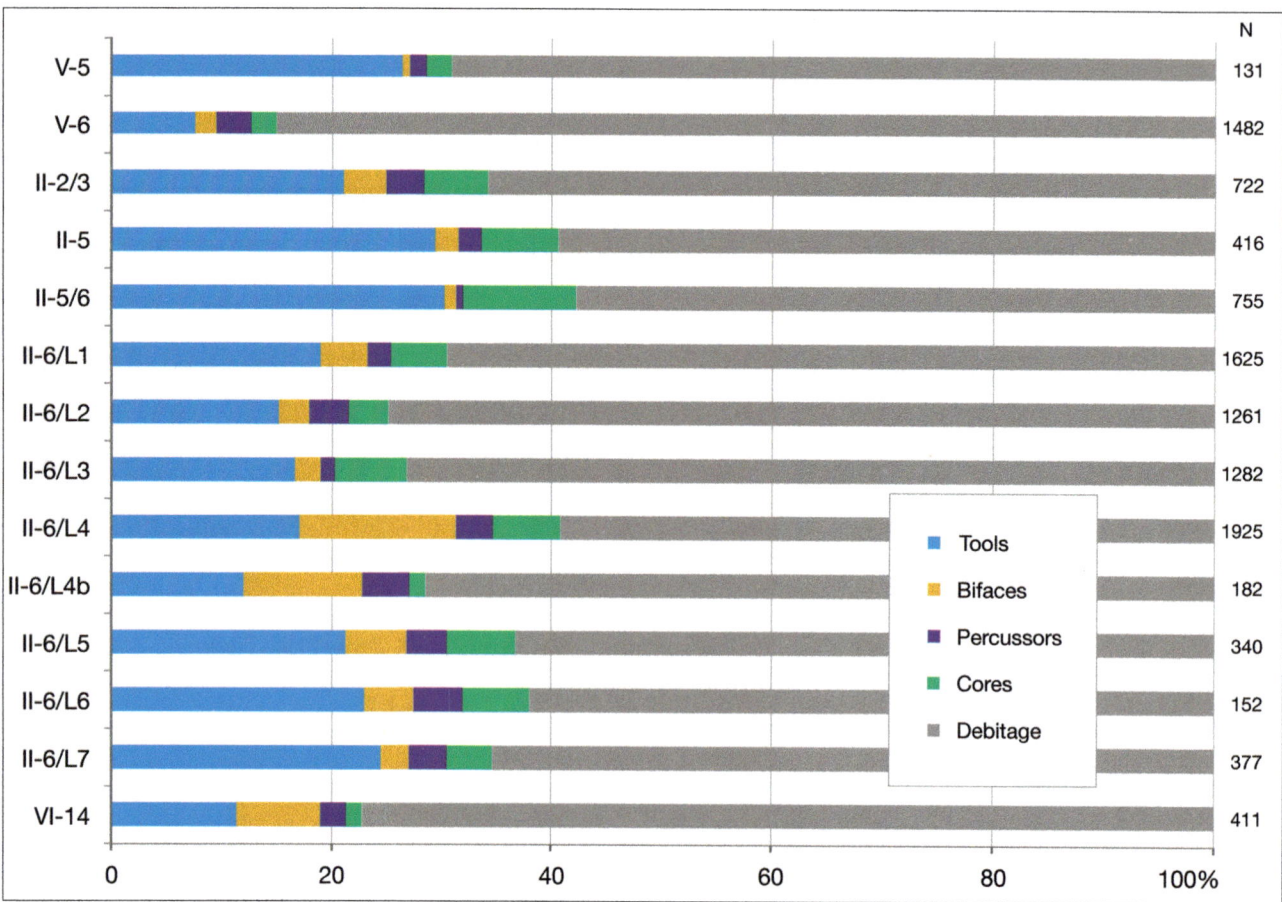

Fig. 9.8 Frequencies of main lithic categories of all raw materials throughout the cultural sequence

9.1.3.2 Conservatism

Analyses of the lithic assemblages have demonstrated that the typological and technological components of each of the assemblages are similar along the entire sequence, with debitage forming the most frequent category (Fig. 9.8). It is also clear that a certain degree of variation is evident and that components such as cores, bifaces, percussors, and tools vary randomly along the stratigraphic sequence (Section 9.1.3.1). The long duration of occupation at GBY exhibits a typical Acheulian typo-techno-morphological knapping tradition. As indicated above, any observed variation is related to the frequencies rather than presence/absence of artifacts.

Conservatism is expressed even in the initial stage of production, in which the acquisition of raw material involved particular sizes and morphologies that characterize the entire cultural sequence. This similarity in the traits of large basalt slabs, thin basalt slabs, and flint and limestone nodules implies a repetitive pattern of exploitation of the same sources.

We have further demonstrated that where raw materials are concerned, all three rock types appear in each of the archaeological horizons and their associated target artifacts are always similar. Furthermore, for each of the target artifact categories the raw material preferences, as well as their relative occurrence on each of the rock types, persist throughout the cultural sequence (Figs. 9.4–9.5).

As for the technological characteristics of the lithic assemblages, these are consistent throughout the entire cultural sequence. This homogeneity is expressed in a variety of aspects, including the sizes and types of percussors, percussive techniques, core reduction methods, and relative frequencies of each of the lithic categories. In addition, the reduction sequences of each of the exploited raw materials involve similar procedures and conserve the same *chaînes opératoires* along the entire cultural sequence. The predetermined products, particularly basalt bifaces and flint retouched tools, show a consistent typological composition, as well as relatively similar frequencies of particular tool categories, throughout the sequence (Fig. 9.6).

To sum up, the reduction sequences of basalt, flint, and limestone, as well as the characteristic tool kits, point to a distinct, well-established, and continuous tradition. This continuity is an indication of a homogeneous behavioral pattern and is the essence of the GBY cultural sequence. The patterns of variability and conservatism of the lithic assemblages described above were the outcome of diverse hominin abilities, a fact

9.2 Hominin Behavior at GBY

9.2.1 Mobility

Artifact mobility is one of the earliest hominin behavioral traits that have been recognized. Evidence for transportation of various rock types over the landscape is known as early as the Early Pleistocene (e.g. Féblot-Augustins 1990; Harmand 2009), and particular patterns of transportation have emerged from the analysis of Acheulian sites in Africa and beyond (Goren-Inbar 2011). A common strategy at 'Ubeidiya, for example, was initial biface production at the outcrops of raw material, with bifaces being transported to the site for final modification and use (Bar-Yosef and Goren-Inbar 1993; see also Shimelmitz et al. 2016 for the final Acheulian). The same pattern is observed at most of the African Acheulian sites and beyond, for example in the Arabian Peninsula (Petraglia 2005) and India (Sharon 2007; Akhilesh and Pappu 2015). Many studies of lithic Acheulian assemblages have demonstrated that the expected large numbers of primary flaking products of biface production are absent from the sites, and hence that the bifaces were imported to the site as finished tools (Bar-Yosef and Goren-Inbar 1993; Akhilesh and Pappu 2015). Other studies describe quarry or workshop sites where the nature of the assemblage is completely different, comprising numerous cores and waste products and smaller numbers of bifacial tools (Petraglia et al. 1999; Paddayya 2006; Barkai et al. 2009). In England, two different studies have been dedicated to tool transportation and discard patterns that resulted in differential tool and debitage frequencies at different Acheulian localities (Pope 2004; Hallos 2005). Both studies are based on refitting of lithic assemblages, to which the sites of Beeches Pits and Boxgrove have made a major contribution. In both cases the conclusion of the studies was that the configuration revealed by the excavation is a result of hominin behavior rather than postdepositional processes. Pope (2004) viewed the differential pattern of biface discard as being related to the exploitation of various environmental resources such as raw material, freshwater bodies, or game. Hallos (2005) stressed that the fragmented nature of the reduction sequence observed at many Acheulian sites is also a result of tool transportation and discard patterns, and that this is underlain by advanced cognitive capacities in the field of planning. It should be noted, however, that detailed examination of the actual biface quantities in relation to the excavated areas of the European sites reveals that their densities are far lower than those reported from the African sites or GBY.

At GBY, the issue of mobility is multifacetted and involves a variety of materials, including lithics, fauna, and flora. Hominins invested time and energy in transporting these materials into and out of the site, and the evidence for this is discussed below.

9.2.1.1 Lithics

The issue of artifact mobility at GBY is based primarily on the sedimentological nature of the lake margin at the site, which is typical of lakeshore depositional environments. This type of landscape is devoid of medium to large clasts, and hence lacks raw materials suitable for knapping. Examination of the lithic components of the site from the perspective of mobility is based on two classes of finds: those that were introduced to the archaeological horizons and those that were taken out. Based on our analyses (Chapter 6–8), we conclude that all the lithics exposed during excavations were introduced into the archaeological occupations; they are simply too large in size to be an integral part of the depositional environment of the lake margin.

We have demonstrated that the three raw materials were procured away from the site and that flint, limestone, and basalt were introduced into the archaeological horizons, each according to particular preferences in accordance with their particular reduction sequences (Chapters 6–8). Raw materials were selected, quarried, collected, and transported to be used at the site. In addition to the differences between the different raw materials, we have encountered mobility patterns within the same raw material. The most striking pattern emerges from the examination of the products of the basalt reduction sequence (Chapter 7). The basalt lithics transported to the site were intended to be used mainly as percussors and as blanks, the latter transformed into cores intended to be knapped at the site. The sedimentological record of the site furnishes indisputable evidence that giant artifacts, of a clast size foreign to the lake margin sediments, were transported to the site by hominins from basalt outcrops in the vicinity of the lake. Very large basalt slabs, fragmented slabs, and a variety of boulders are also present in the GBY archaeological horizons, but these occur only in a single layer. Clearly weight was not an obstacle for the hominins, who were certainly the only agent that could transport such materials into the site. Fig. 9.9 presents the weight of only three types of basalt artifacts, excluding FFT, percussors, and other categories, clearly illustrating the great mass that was carried into the site by the hominins.

Although the hominins of GBY transported natural raw material to be used or knapped at the site, we have evidence that other lithic elements were also brought into the site. On-site knapping could not be responsible for the entire lithic inventory, for several reasons. First, the basalt biface reduction sequence is

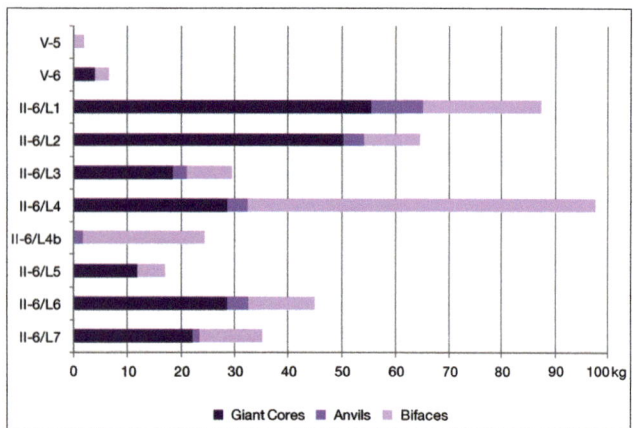

Fig. 9.9 Weights (kg) of different categories of basalt artifacts throughout the cultural sequence

based on the knapping of giant cores, documented at the site by their presence and their characteristic waste products. However, examination of all components of this reduction sequence indicates that the number of giant cores is low in comparison with the wealth of bifaces, suggesting that the majority of the products were introduced to the site as flakes or bifaces. Based on the data from GBY, only a few of the biface blanks were produced at the site, and many were brought in either as preforms or as complete tools. Although we do not know the location of the quarries or the knapping stations/workshops away from the site, we can assume that there were additional cores at these locations. These assumptions are supported by experimental work focused on knapping of giant cores, which concluded that in Layer II-6 Level 1 "…only 8–10% of the expected by-products from 68 basalt bifaces are actually present. This percentage indicates that many of the bifaces were not produced in the excavated area…" (Madsen and Goren-Inbar 2004:40). The basalt by-products in Layer II-6 Level 1 are extremely meager in comparison with those of the experimental work (812 vs. a minimum of 8,000 expected flakes). Hence, while over 60 bifaces were found in this level, "The on-site production is estimated to be in the range of 10–15 bifaces, depending on how much the imports were re-sharpened" (Madsen and Goren-Inbar 2004:40).

Table 9.2 presents the counts of basalt bifaces and FFT and their number of scars. The total number of scars on the bifaces and FFT provides an estimate of the number of flakes expected to be found. Subtracting the actual number of FFT from the expected number (i.e. the total number of scars) enables us to evaluate the number of missing flakes. Assuming that each scar should have a corresponding flake, the deficit of flakes ranges between 141 (Layer V-5) and more than 5,000 (Layer II-6 Level 4). However, some issues should be taken into consideration. First, FFT are by definition ≥ 2 cm in size, and scars measuring ≥ 2 mm were registered. It is possible that some of the products represented by the scars on FFT are actually present among the microartifacts. However, for the GBY biface assemblage (unlike later Acheulian bifaces) the vast majority of scars correspond to macroartifacts, as fine retouch on the edges is rare. It should also be remembered that the number of scars is an underestimate, as in many cases there are missing observations due to taphonomic issues. If we use an index of the average number of scars per face to assess the number of missing scars (i.e. flakes), the deficit of flakes is even higher. To conclude, from Table 9.2 we see that in all likelihood much of the biface shaping was carried out off-site. The table further shows the individual nature of each of the archaeological horizons and the fact that the mobility signature of each is very different.

Another striking example of the discrepancy between the number of basalt bifaces and the products of their manufacture is the "biface pavement" in Layer II-6 Level 4, from which some 225 basalt bifaces were retrieved (Chapter 5), while only one giant core and 788 basalt FFT were excavated from this level. The detailed analysis of the basalt component of Layer II-6 Level 4 (Chapter 7) clearly demonstrated that the large number

Table 9.2 Counts and scar counts of basalt bifaces and FFT and estimated counts of missing flakes throughout the cultural sequence

Layer	N of bifaces	N of scars on bifaces#*	N of FFT	N of scars on FFT#	Total scars^	Missing flakes^^
V-5	2	89 (N=4)	84	136 (N=63)	225	141
V-6	6	156 (N=12)	57	87 (N=41)	243	186
II-6 Level 1	71	1,572 (N=124)	912	1,699 (N=648)	3,271	2,359
II-6 Level 2	29	791 (N=58)	838	1,620 (N=581)	2,411	1,573
II-6 Level 3	24	406 (N=38)	715	1,270 (N=500)	1,676	961
II-6 Level 4	225	3,993 (N=375)	788	1,814 (N=583)	5,807	5,019
II-6 Level 4b	76	1,349 (N=120)	496	911 (N=314)	2,260	1,764
II-6 Level 5	19	349 (N=30)	179	236 (N=84)	585	406
II-6 Level 6	30	555 (N=45)	317	424 (N=144)	979	662
II-6 Level 7	32	555 (N=52)	527	661 (N=238)	1216	689

* the total number of scars is the sum of scars on the two faces of all bifaces; note that in some cases the scars of only a single face were registered
\# discrepancies in counts (N) result from cases where the preservation state did not allow registration of number of scars and from the cases of fully cortical flakes; scars ≥ 2 mm are registered
^ the total number of scars on bifaces and on FFT
^^ calculated by subtracting the number of FFT from the total number of scars

of bifaces excavated there could not all have been manufactured *in situ*, due to the deficit of giant cores that could have produced enough blanks for the bifaces. This is also expressed by the debitage: a simple calculation shows that many more debitage pieces would have been found if the entire reduction process, from giant core to biface, had taken place *in situ*. Experimental knapping suggested that the *façonnage* of a biface produces 110 flakes on average, and consequently the knapping of the 225 bifaces of Layer II-6 Level 4 should have resulted in 24,750 flakes. Our conclusion that numerous flakes originating from the shaping of bifaces are missing from the archaeological horizon is thus supported by two distinctly different avenues – archaeological and experimental. Both strongly point to the fact that off-site site knapping took place and that hominins transported preforms and finished bifaces into the site.

Another observation supportive of off-site knapping is the relatively small number of basalt microartifacts in all levels of Layer II-6. Fig. 9.10 shows the relationship between the amount of microartifacts and artifacts in each of the raw materials. High ratios represent numerous microartifacts and very few knapped artifacts, suggestive of on-site knapping and transportation of artifacts away from the site (particularly flint and limestone). Low ratios are suggestive of off-site knapping and transportation of artifacts into the site. The latter case is characteristic of the basalt component in all levels of Layer II-6, resulting from the introduction of complete or partial basalt bifaces into the site.

In addition to the transportation of lithics into the site, the mobility of artifacts at GBY is also expressed in the export of artifacts from the site. Indications of this are meager and more difficult to identify, and are evident mostly within the flint assemblage. A representative case is that of the paucity of flint bifaces in Layers V-5 and V-6 and the abundance of products associated with their manufacture (Sharon and Goren-Inbar 1999; Goren-Inbar and Sharon 2006) (Chapter 6). While the shaping and thinning of many flint bifaces took place in these layers, as evidenced by the very large amount of *éclat de taille de biface*, there is only one flint handaxe. The paucity of flint bifaces is interpreted as resulting from their removal and subsequent use and discard away from the site. Thus, these particular layers serve to illustrate the complexity of lithic artifact mobility at the site.

Another feature suggesting mobility of artifacts out of the site is the frequencies of microartifacts. Fig. 9.10 illustrates the high frequencies of limestone microartifacts in relation to artifacts. In the levels of Layer II-6 (excluding Levels 4b and 6) it is evident that such an abundance of microartifacts could not result from the production and modification of the few limestone artifacts. A similar pattern is observed in the basalt component of Layers V-5 and V-6, where microartifacts are abundant and artifacts are scarce. These two cases are both the result of on-site knapping of artifacts that were later taken away from the site. Another possibility is that these artifacts were not taken away but are still located beyond the excavated surface. However, as these patterns are repetitive and characterize a variety of levels, regardless of their individual nature, we consider mobility to be the agent responsible.

The patterns of mobility discussed above are extremely complex, and a simple explanation is inadequate for deciphering the products of the reduction sequences of the different raw materials. In addition, although each archaeological horizon at GBY is different, they all portray an intrinsically complex picture of mobility, illustrating the varied behavioral patterns of their makers. Our classification of mobility into and out of the site is clearly an oversimplification of the dynamic systems of the past, frozen by their discard and our limited powers of observation.

9.2.1.2 Plant and Animal Resources

The preservation of rich faunal and floral assemblages adds additional parameters to the issue of mobility. The hominins of GBY exploited a variety of resources at the site, such as the variety of biological components of the lake and its margins (e.g. water nuts: Goren-Inbar et al. 2014; fish: Zohar and Biton 2011; Zohar et al. 2014; driftwood: Goren-Inbar et al. 2002b). In addition, a wealth of resources was transported to the site from farther away. Carcasses of small and medium-sized animals were introduced to the site and were processed and consumed there. As for the largest of the mammals, the elephants, these are viewed as having been driven to the lake edge (Goren-Inbar et al. 1994), as their huge weight and size probably did not permit them to be carried. This may also be the case for hippos, large bovids, and rhinos, although their meager skeletal presence rules out credible suggestions. However, fallow deer, whose skeletal remains are the most abundant at the site, were clearly transported to and processed at the site (Rabinovitch et al. 2012).

The skeletal representation of fallow deer suggests that complete carcasses were introduced to the site, including the

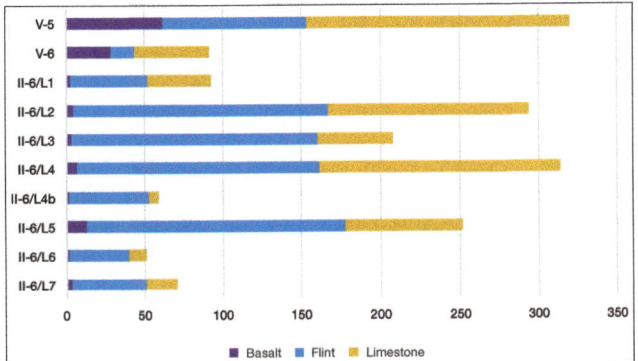

Fig. 9.10 Ratios of microartifacts to artifacts throughout the cultural sequence by raw material

skull and antlers. The latter were exploited for use as soft percussors.

Analyses of the plant material suggest that some plants grew in the immediate vicinity of the site, while others were brought from greater distances. These include acorns, olives, pear, almonds, etc. (Melamed et al. 2016).

The hominins of GBY exploited natural resources not only for nutritional purposes but also for other subsistence requirements. These include the collecting of wood for making fire (Alperson-Afil and Goren-Inbar 2010), which was available at times on the lake margin in the form of driftwood (Goren-Inbar et al. 2002b).

The lithic, faunal, and botanical remains of GBY provide insight into patterns of human mobility at the site. The behavior involving mobility was not limited to nutritional or subsistence requirements but likely involved other aspects. Unfortunately, this is documented at the site only by two bead-like crinoid columnal segments. These Jurassic fossils, probably from Mt. Hermon (25–30 km from the site), were not naturally transported (Goren-Inbar et al. 1991b) but were intentionally brought to the site by the hominins.

In light of the above discussion, which illustrates the great extent of mobility to and from the site, one should consider how and by what means hominins transported lithic, faunal, and botanical items. It is evident that the means of mobility differed in accordance with the objects under consideration. Clearly large slabs, fragmented slabs, some giant cores, and extremely large percussors were items that were handled individually due to their high weight (Fig. 9.9). We cannot determine the particular means of transportation, but carrying or rolling are two possible options for the heaviest basalt components, such as fragmented slabs and giant cores. Drawing from the ethnographic work of Pétrequin and Pétrequin (1993), it is possible that smaller artifacts, such as preforms, bifaces, and anvils, were tied together with thin vegetal (e.g. roots) or faunal (e.g. tendons) material.

The issue of mobility and the transportation of resources raises the possibility that more sophisticated means were used in the form of baskets, pouches, or a variety of containers. Such means are particularly efficient for the transportation of small materials, such as flint nodules and percussors as well as water nuts, acorns, olives, etc.

9.2.2 Social Aspects

The results of the lithic analyses add up to a wealth of information on the variety of available resources and their modes of exploitation. The accumulation of data on the hominins of GBY allows us to explore issues that may shed light on the social structure of the hominin groups that occupied the lake margin during Acheulian times. In the following discussion we focus on two main aspects – division of labor and group size.

Once again, our views on this aspect are derived from ethnographic studies of present-day hunter-gatherers (e.g. Kelly 1995).

9.2.2.1 Division of Labor

Our assumption, which clearly emerges from ethnographic studies, and can be applied to the Acheulian context, is that gender differences impose division of labor. Females are limited by their young offspring while males are at liberty to carry out longer and more arduous tasks. Several results of the GBY multidisciplinary analyses and their integration provide indications of the social structure, particularly division of labor. These indications derive from the lithic, faunal, and floral assemblages. Within the realm of lithics, we cannot rule out the possibility that knapping was carried out by both genders, though in present-day hunter-gatherer societies it is commonly a male activity. At GBY, the biface reduction sequence provides indications of a possible division of labor. The initial stages of the biface reduction process required great force, both in the extraction of the large basalt slabs and in their fragmentation, followed by their transformation into giant cores. These preliminary stages were carried out away from the site and necessitated great force and the use of very heavy percussors. In particular, one should consider the energy that was invested in the removal of the shoulder flakes, which required an immense input of energy in comparison to all other stages of giant core reduction (Chapter 7). The absence of human remains makes it difficult to hypothesize the identity of the GBY hominins, but the extreme energy needed for such blows and for the transportation of the giant slabs/cores suggests that the hominins were very muscular, similar to the descriptions of the fossil remains of *Homo ergaster*, *H. erectus* (Bodo type), and *H. heidelbergensis*, which have previously been described as possessing extreme muscle attachments (Anton 2003).

The best ethnographic illustration of the biface *chaîne opératoire* is drawn from the seminal study of Pétrequin and Pétrequin in Irian Jaya (1993). There, quarrying expeditions for obtaining volcanic rocks are organized and carried out entirely by men. The transportation of basalt slabs, their fragmentation, the preparation of roughouts, and all other stages of modification are carried out by men. Women are banned from participating in the quarrying expeditions and the subsequent activities, and their role is limited to providing food supplies. The extremely difficult extraction of the heavy basalt slabs at GBY, their fragmentation, and the expertise needed in order to produce high-quality basalt bifaces seems to be analogous to the reduction sequence of the Iriyan Jaya knappers. Indeed, the difference is in what the archaeological data do not provide:

the size of the expeditions for retrieval of the raw material, the differences between the work of experts who specialize in particular types of raw materials, and the continuous exchange of views during the knapping sessions (Pétrequin and Pétrequin 1993). Interestingly, not only are men involved in the production of bifaces, but they are the bearers and users of these tools. Australian aborigines similarly restrict the extraction of raw material to men, as this task is taboo for women (Rhys and White 1988).

Another activity that is limited to men in modern hunter-gatherer societies is the killing of very large mammals, particularly elephants, bovids, rhinos, and hippos. The most common faunal remains in all layers at GBY are elephants, hippos, and fallow deer (Rabinovich and Biton 2011; Rabinovich et al. 2012). Carcass processing of elephants is ethnographically known to be carried out by many individuals (see the discussion of group size below) (e.g. Crader 1983; Marks 2005). However, only one man, or a few men, do the hunting, while women are banned from participation. Such cases are reported from many groups, primarily in East and Central Africa (e.g. Janmart 1952). The GBY faunal assemblages also include small animals (fish, crabs, and birds) and an array of plants, with a significant frequency of edible species. We have described in detail the role of nuts in the Acheulian diet at GBY, including nuts of terrestrial (Goren-Inbar et al. 2002a) and aquatic (Goren-Inbar et al. 2002a; Melamed et al. 2011; Goren-Inbar et al. 2014) plants. The processing of these has been spatially associated at GBY with the presence of hearths, pitting stones, and basalt thin anvils, among other finds (Alperson-Afil et al. 2009; Goren-Inbar et al. 2015). Of the array of over 50 edible plant species and their abundant remains that were found at GBY, it is the nuts that are emerging as an important nutritional resource, and they appear in relatively large quantities (Melamed et al. 2016). Ethnographically, and in many societies in different geographical zones, women, children, and the aged collect and process small animals and edible plant materials. Hence we suggest that the collecting of crabs, fish, and other small animals could have been done by these members of the group.

Of particular interest is the presence of several species of edible plants producing underground storage organs (USO) in the archaeological horizons (as above). In present-day societies this component is commonly handled by women, as it requires collecting/gathering activities and also necessitates proximity to fire and water, both needed for the processing of these plants. The duration of processing of these resources is somewhat longer, due to the need for detoxification. Despite the view of Kuhn and Stiner (2006), the eastern Mediterranean, as demonstrated at GBY (Melamed et al. 2016), does have an important component of these plants, though it is restricted to particular ecological niches. The presence of these species on the lake margin, and the processes that they require, further support our view of the role of gender.

Based on the data from GBY and the supporting ethnographic

Table 9.3 Suggested generalized division of labor at GBY

Task	Female[#]	Male
Basalt biface reduction sequence		+
Transportation of large basalt slabs		+
Hunting large mammals		+
Carcass processing	+	+
Collecting aquatic resources	+	+
Gathering plants	+	
Collecting firewood	+	
Nut cracking	+	
Cooking/roasting	+	+

[#] ethnographically, female activities are accompanied by children and the aged

analogies, we conclude that during the Acheulian there was a distinct division of labor (Table 9.3). Our conclusion contradicts that of other scholars who reject the notion of division of labor even in later periods (e.g. Kuhn and Stiner 2006 for the Middle Paleolithic). Considering the variety of data from GBY, we maintain our view that the antiquity of the division of labor is much older than the current view, hence supporting the view of Isaac (1978), who associated behavior of this kind with much earlier hominins than the Acheulians of GBY.

9.2.2.2 Group Size

Estimates of prehistoric population size, particularly group size, have long been a topic of interest, as they throw light on the social life of the human genus. Clearly, hominin evolutionary success depended on intra-group social interactions and group cohesion. A common assumption in archaeology, derived from population genetics and ethnography, is that the minimum size of modern hunter-gatherer groups is 30–50 individuals. We present here preliminary data that have enabled new insight into the issue of group size during the Lower Paleolithic, suggesting that non-modern human group size may have been different. In general, judging from multidisciplinary studies of the Lower Paleolithic (archaeology, paleontology, paleobotany, and geology), the exceptional richness of some archaeological sites seems to indicate the presence of groups comprising more members than those suggested above. Such groups may at times have aggregated, similarly to ethnographic open-ended systems.

Surveys and excavations carried out in different parts of the African continent have yielded large collections of material culture and faunal remains of the Acheulian Technocomplex. Within the material culture records, bifaces display different technologies and styles and at times were found in association with skeletal remains of the local *H. erectus* (*H. ergaster*). One of the most striking traits of the African Acheulian is the immense quantities of bifaces, which occur in large numbers both at excavated sites and in surface scatters (e.g. Gadeb: Clark and Kurashina 1976; Isimila: Cole and Kleindienst 1974; Olorgesailie: Isaac 1977; Olduvai Gorge: Leakey and Roe 1994; Isenia: Roche and Texier 1991; Kilombe: Gowlett 1991; Kalambo Falls: Clark 2001; and Melka Kunture: Chavaillon and Piperno 2004, to mention but a few). In some parts of East Africa the surface density is so impressive that the paleontologist Jon Kalb referred to it as a reg when describing a survey in Ethiopia, where the team drove over 20 km and saw tens of thousands of bifaces (Kalb 2001). In many of the sites there are lithics of other classes, also in immense concentrations, as recently illustrated at the site of Kathu in South Africa (Walker et al. 2014). However, in-depth discussion of early hominin group size was not included in the discourse, as the available data contributed little information to this issue. Moreover, the model most frequently used for estimation of the population of prehistoric groups was that of extant hunter-gatherers. These communities are characteristically considered to consist of five or six nuclear families that share the same camp and move together from place to place (see references above).

Concentrations of bifaces have frequently been believed to result from the involvement of high-velocity water flow, powerful enough to transport bifaces to a new location and to redeposit them where they were found, or alternatively less powerful but strong enough to transport away all the smaller artifacts, producing a winnowing effect (Isaac 1971; Schick 1986). This view, which emerged from studies of site formation processes, provided further insight into issues of taphonomy and concluded that the abundance and organization of the bifaces are a result of taphonomic processes. However, these studies sometimes failed to use modern excavation methodologies and particularly sieving, hence missing a substantial component of the life-history of the sites. Moreover, experimental archaeology (e.g. Jones 1994) suggests that the production of bifaces may be an easy and fast procedure once the knapper is an expert, so that a vast number of bifaces does not necessarily represent many knappers. Thus, many scholars consider the observed vast quantities of bifaces to represent time-averaging processes (Isaac 1977; Schick 1986, 2001) rather than group size. This view has been so influential that it has overshadowed other potential explanations for the huge concentrations of bifaces as well as other "anomalous" concentrations of finds.

The massive presence of a single tool type is an Acheulian trait that was encountered in the lithic assemblages of 'Ubeidiya. At this site, where a very extensive sequence of occupations was revealed, such occurrences were reported for bifaces and spheroids. Each of these types was made on a different raw material (basalt and limestone respectively) and could occur in large numbers (Bar-Yosef and Goren-Inbar 1993). Thus, this particular behavioral trait is a component of the Early as well as more recent Acheulian. The Acheulian layers (sites) of 'Ubeidiya were located in diverse habitats and ecological niches that are documented by their sedimentary and paleontological remains. The possibility that the abundance of a particular tool type may be associated with functions related to particular paleoenvironmental conditions has been rebutted (Bar-Yosef and Goren-Inbar 1993). Thus, the abundance of a particular tool type in a particular location (and its minimal representation in others) remains to be elucidated, but clearly belongs to the behavioral sphere. The discrepancy between the high number of bifaces in Africa and the Jordan Valley and those revealed in Europe may suggest that the European Acheulian groups were smaller and hence produced smaller quantities of bifacial tools.

At GBY there are abundant indications of various kinds that the archaeological horizons underwent minimal postdepositional disturbances and were rapidly sealed. Thus, a different explanation for the abundance of a single tool type is required. The best example of striking abundance of a single tool type (basalt bifaces) is encountered in Layer II-6 Levels 4 and 4b (Chapter 5, Figs. 5.46–5.47). The densities there are indeed exceptional; for example, in Layer II-6 Level 4 there are 13.88 bifaces per 1 m^2 and in Level 4b there are 5.99 bifaces per 1 m^2. As the data are of high resolution, no taphonomic interpretations can be considered. First, it is possible to examine in depth the frequencies and size composition of both macroartifacts and microartifacts. Our conclusion is that in all archaeological horizons, apart from the gravely uppermost and lowermost layers, the observed spatial arrangement has undergone minimal change due to taphonomic processes (Alperson-Afil et al. 2009; Alperson-Afil and Goren-Inbar 2010; Rabinovich et al. 2012: Chapter 7; Feibel personal observations and additional studies under way). Second, the entire range of tool type composition occurs repetitively in *similar* lake margin environments and most probably at similar distances from the sources of the raw materials. While all the sites of GBY are located on the lake margin, relatively close to the raw material sources (Chapters 6–8), they display a substantial amount of variability, a phenomenon that is sometimes explained by varying distances from the source of the raw material. Clearly, the proximity of raw material cannot explain the variability encountered at GBY. Thus, the GBY data rule out the European interpretation of the presence of biface-rich assemblages (e.g. Pope 2004). An abundance of bifaces, therefore, cannot be interpreted as resulting from diverse environments or varying distances from raw material resources. Rather, it falls within the general framework of behavior and hominin thought processes, and perhaps the particular social structure of the group. Thus, we should seek alternative explanations for the phenomenon of tool concentrations.

Our modest suggestion is to view the concentrations of a single tool type as the outcome of the activities of a large group. Several characteristics of the biface-rich horizons may provide insight into this phenomenon. The absence of large wood pieces from these layers rules out woodworking, one of the most common functional explanations for the presence of bifaces; in fact, only one sign of such activity was identified in the GBY wood assemblage (Belitzky et al. 1991; Goren-Inbar et al. 2002b). Another possible explanations is a suggestion made by Potts (1984) in an attempt to explain abundance of stone tools in Early Stone Age East African sites. Potts viewed lithic accumulations as caches, focal points in the environment where hominins accumulated artifacts for future use. Food collected by hominins was transported to these focal points for processing away from the threat of predators. Later, Potts (1994) suggested that no single model can explain hominin behavior and that a variety of mechanisms could be responsible for the diversified nature of the sites.

The concept of caching has direct implications for human cognition, suggesting that hominins pre-planned their activities and prepared stone tools for use in the future (a trait also observed for apes: Bräuer and Call 2015). Although we support this concept, we prefer to relate the abundance of bifaces to the activities of the members of a large group who stockpiled tools for future hunts and carcass processing. At GBY, in contrast to interpretations of early East African sites, predators were not a threat due to the hominins' control of fire, and all the sites are embedded within a similar lake margin environment. Thus, we interpret biface-rich layers as representing a component of a long-term plan to exploit very large mammals (elephants, hippos, rhinos) that were driven to the lake margins, killed, and processed there (Section 5.4.9). Equipping is an essential stage in preparation for the hunting of large mammals. The opportunity to acquire an enormous amount of fat, marrow, and meat requires preparation and complex planning. It necessitates the acquisition of raw material, its transportation, and its modification into suitable tools, required not only for the hunt itself but for the processing of the large carcass. It is impossible to carry out the entire chain of production of the tools simultaneously with the hunt; this has to take place earlier and the tools need to be placed in proximity to the intended location of the kill, as the large carcass would be difficult to move. We suggest that the number of tools reflects the number of people involved in such a procedure. Ethnographic studies associate an individual with one or two bifacial tools for the duration of several years (Pétrequin and Pétrequin 1993). We find it difficult to accept that hundreds of Acheulian bifacial tools represent a group size similar to those studied ethnographically (i.e. 5–6 nuclear families), and suggest that the group of hominins was much larger.

Unfortunately, the size of the social group at GBY is a topic for which direct evidence is lacking. Nevertheless, we can gain some insight into the issue through examination of data acquired from various analyses of the lithic and faunal assemblages.

Evidence of large animal kills is documented in many archaeological horizons at the site (Rabinovich and Biton 2011: table 1). The repeated presence of the remains of elephants and hippos, among other large mammals, indicates that the community successfully hunted and processed these animals. Furthermore, elephants and hippos are the most common species of the largest fossil faunal assemblage at the site. While a single individual can carry out the hunt, processing, as documented in ethnographic studies, is handled by a large group and usually with the participation of the entire community, for which there are multiple reasons (the need to avoid decomposition, division among members of the community, etc.). Thus, the remains of large animals in each of the archaeological horizons testify to a large number of individuals in the group.

Additional support is found in the wealth of skeletal remains of fallow deer (*Dama*) and other medium-sized animals (Rabinovich et al. 2012). The detailed taphonomic study of the faunal remains indicates that systematic butchery of *Dama* sp. took place (Rabinovich et al. 2008). Furthermore, the diversity of species is not matched at any other Levantine Lower Paleolithic site. This unprecedented wealth, coupled with the original spatial distribution of "phantom hearths" (Alperson-Afil and Goren-Inbar 2010) and the rich lithic assemblage, testifies to a very short exposure before sealing (as for the other organic materials in all GBY layers) and implies the involvement of a large hominin group. It is unlikely that such extensive hunting and processing of fish, crabs, birds, and small to large mammals was carried out by a very small social group. The great numbers of carcasses clearly necessitated many hands at work. We have previously showed in a number of publications that the paleoclimate at GBY during the time of occupation was very similar to the present one (based on various proxies of flora, fauna, and isotopes). In such conditions the carcasses had to be consumed instantaneously or over a very few hours after the kill, since otherwise they would become a health hazard.

Another example supportive of a larger group size is what seems to be the involvement of many individuals in acquiring numerous very large fish over 1.5 m in length (more than 30,000 fish remains at the site), particularly in the complex of Layer II-6. Although only Level 2 has so far been described in detail (Alperson-Afil et al. 2009), Layer II-6 in general provides indications of extensive utilization of these very large fish, and the remains testify to their processing with fire (Zohar et al. 2014; Zohar et al. 2016). Clearly, such a large amount of fish, present in each of the levels, indicates consumption by many individuals.

To sum up, the multidisciplinary evidence from GBY of concentrations of numerous bifaces, large animal kills, the wealth of edible plants, and processing of diverse animal species, including great numbers of very large fish, suggests that the activities at GBY were carried out by a large number of individuals. Living in a large group implies cognitive abilities, since "Managing an increased number of social relationships would require several cognitive abilities, among them increased facility in facial recognition (for allies and enemies), memory for reciprocity in things like food sharing, and an expanded ability to manage simultaneous relationships" (Coolidge et al. 2015:183).

9.2.3 Home Base – The Preferred Locale

The co-occurrence of animal bones of various species and sizes, together with dense scatters of stone tools, in the Plio-Pleistocene deposits of the FLK Zinj floor at Olduvai Gorge yielded a new interpretation of living floors or home bases (Leakey 1971). The term "home base" was developed soon afterwards by Glynn Isaac into the "food-sharing hypothesis" or "central-place foraging hypothesis" (e.g. Isaac 1978, 1983) and revolutionized our understanding of the behavior of ancient hominins. As the "home base hypothesis" attributed various adaptive behaviors to ancient hominins, such as delayed consumption of food, gender division of labor, and food sharing, Isaac's hypotheses initiated a debate that is still ongoing (e.g. Binford 1984; Potts 1984, 1994; Bunn and Kroll 1986; Sept 1992; Rose and Marshall 1996; Rolland 2004).

However, Isaac used the "home base hypothesis" or "central place hypothesis" to interpret sites of very early age (Plio-Pleistocene), and the hypothesis was not thoroughly examined against the wealth of East African Early–Middle Pleistocene Acheulian sites. Reasons for this, in retrospect, were perhaps the lack of dates for these sites and the fact that many of these occurrences were taphonomically disturbed. The latter refers to assemblages that have been redeposited, such as the Late Acheulian of Olduvai Gorge (Leakey 1975; Kleindienst 1973) and Olorgesailie (Potts 1989).

We believe that the wealth of data at GBY, expressed in the wide spectrum of exploited resources and their modes of exploitation, enables us to adopt and adjust Isaac's hypotheses, suggesting that the archaeological horizons of GBY were indeed, as in Isaac's initial suggestion, "home".

9.2.3.1 GBY as a Preferred Locale

Regarding the archaeological horizons (sites) of GBY as "home bases" has certain implications. A home base is by definition a place to which a variety of resources are transported, suggesting that it was not an ephemeral occupation. Clearly, such a locale provided suitable conditions and resources in its immediate vicinity.

The southern shores of paleo-Lake Hula provided the GBY hominins with optimal conditions. This extremely small geographical zone (Chapter 3), bounded on the east by a desert and on the west by the Galilee Mountains, provided plentiful resources, essential for the existence and survival of the Acheulian communities. It furnished a year-round availability of fresh water (lake, rivers, and artesian springs) and a variety of habitats (the lake, wetlands, rivers, dry land, and diverse ecotones) that provided rich floral and faunal dietary resources. In addition, this locality offered easy access to the three lithic raw materials used along the entire cultural sequence. We suggest that the presence of laterally fissured mid-section basalt flows in particular served as *a focal point of attraction*, as basalt was preferred over other raw materials. This we relate to the powerful tradition that tied the production of bifaces with volcanic rocks (Alperson-Afil 2015). This phenomenon is well recorded in Africa, but additional information is currently being gained elsewhere (e.g. the Arabian Peninsula: Jennings et al. 2015).

It is evident that hominin communities frequented the margin of paleo-Lake Hula for generations (ca. 100 ka). The selection of this particular point in the landscape and the establishment of the lake margin as a focal point for diverse activities contradicts some views that consider the proximity of water and the dense vegetation to be disadvantageous (e.g. muddy surfaces, insects). However, what emerges from the multidisciplinary analyses of GBY is a scenario of a series of occupations that were located at the immediate water edge. This habitat of diverse ecological niches was extremely rich in biomass and hence food resources, both animals and plants, and together with the availability of raw materials formed an exceptional focal point in the Jordan Valley.

9.2.3.2 The Diversified Nature of Home

The attempts to interpret and classify Plio-Pleistocene sites provided various categorizations of sites according to the varying frequencies of the faunal versus lithic components. Isaac differentiated six such categories, including factory (Type A), butchering (Type B), and camp (Type C) sites. Such classifications suggest that different types of sites are devoted to different activities/tasks (Isaac and Crader 1981; Isaac 1983). The only type of site where a variety of activities is carried out is the camp site or home base (as above).

Our study demonstrates that it is unsatisfactory to lump sites with high frequencies of lithics into a "type" of site. The analyses of the GBY lithics demonstrate that their concentrations could be a result of different activities (e.g. extensive knapping, selected mobility, stocking). The same holds true for the bones, as their occurrences differ widely in nature (e.g. focusing on a single species or a wide spectrum of species). Any attempt to classify the GBY sites into strict categories will mask the observed diversity of lithics, bones, and botanical remains. This is encountered in each of the archaeological horizons, suggesting that hunting, gathering, food processing, and fire making (Alperson-Afil 2007; Alperson-Afil et al. 2009; Alperson-Afil and Goren-Inbar 2010) were all integral components of them.

The hunting and processing of small, medium-sized, and large mammals, birds, and fish and the consumption of their carcasses are documented by the abundant fragmentary bone remains and by a variety of bone damage types. While there is no direct evidence for hunting technology, the accumulation of different sizes and species, the intentional fragmentation of the bones for the extraction of marrow, and treatment with fire of the fish all testify to the ability of the GBY hominins to exploit the faunal niche extensively (Goren-Inbar et al. 1994; Rabinovich et al. 2008; Rabinovich and Biton 2011; Zohar and Biton 2011; Rabinovich et al. 2012). Gathering is also very well documented by the presence of over 55 edible plant species, with an emphasis on terrestrial and water nuts and many other foodstuff, including plants producing USO (Melamed 1997, 2003; Goren-Inbar et al. 2000; Goren-Inbar et al. 2002a; Goren-Inbar et al. 2002b; Goren-Inbar et al. 2004; Melamed et al. 2016). There are also some indications that crabs and turtles were collected and consumed (Alperson-Afil et al. 2009).

The archaeological horizons described in this volume show that hunting, fishing, meat and bone processing, nut and fruit gathering, and fire making were present throughout the cultural sequence of GBY. The co-occurrence of all of the above aspects with a conservative tool kit is found in each of the archaeological horizons of GBY in exactly the same locale – the Benot Ya'aqov Embayment. Despite the homogeneity of the components of the archaeological horizons, remarkable differences are evident when their frequencies are compared. The distribution maps of the different layers (Chapter 5) illustrate the extent of the cultural diversity, expressed in the changing frequencies and scatter patterning of the lithic, faunal, and floral components. We interpret the observed cultural diversity as reflecting different activities that were carried out in each of the occupations. If we had focused on the most conspicuous find, then Layer II-6 Level 1 would be categorized as an elephant kill site, Layer II-6 Levels 4–4b as biface stocking sites, and Layers V-5–6 as carcass processing sites. Consideration of all components (lithic, paleobotanical, and paleontological) of the archaeological horizons, including the presence of hearths in all horizons, points to repeated use of the same geographical location, a focal point, that provided advantageous conditions for the performance of different activities.

The scope of the activities carried out at GBY in the preferred locales was far more diverse than what we can suggest. Many functions remain unknown. What were the uses of some of the flint tools? If some of these were indeed hafted, as suggested here, how were they used and for what purpose? How were the basalt thin anvils used? And why do they exhibit diverse intensities and multitasking? Such unknown functions are present in each of the archaeological horizons and testify to the variety of activities carried out, in addition to the evident ones. Thus, despite the diversity, the archaeological horizons share common components illustrating that the wealth of resources available in this particular locale was always exploited, but in differing amplitudes.

The diverse activities that took place on each of the archaeological horizons contributed rock and bone mass to the extent that it actually changed the landscape on the lake margin. Although these horizons were covered by rapid processes of accumulation, it is still likely that such dense and extensive depositions influenced the topography (e.g. the presence of giant cores in several archaeological horizons) and the sedimentology, and probably marked the previous presence of hominins. The possibility that hominins observed these accumulations may have been instrumental in their decision to reoccupy that particular location. Each of these repeated visits resulted in the formation of an archaeological horizon, providing a cultural archive of occupation at the preferred locale for at least 100 ka.

This scenario provides an elaboration of Isaac's home base model. It is similarly concerned with a preferred locale to which various resources are transported and exploited, but it emphasizes the attractiveness of the locale, the stability of its conditions, and the fact that it witnessed continual repeated occupations. The botanical results indicate a mixture of all seasons in each of the rich archaeological horizons, suggesting year-round occupations (Melamed et al. 2016). We suggest that we should return to Isaac's initial proposal and acknowledge that such accumulations of activities, coupled with the data on group size and social structure, actually imply that the preferred locale served as "home" for the Acheulian hominins.

9.2.4 Cognition

In the previous sections, we have presented data supporting the notion that the archaeological remains at GBY represent at times the activities of large groups with a distinct division of labor that repeatedly occupied the preferred locale or "home" of the paleo-Lake Hula (Sections 9.2.2–9.2.3). These social implications, a direct outcome of the lithic analyses, are clearly a reflection of the cognitive abilities of the hominins of GBY. Correlations between indices of social complexity and indices of brain volume in primates have been recognized by several scholars (e.g. Sawaguchi and Kudo 1990; Dunbar 1992, 1998; Barton and Dunbar 1997; Lewis 2001) and are involved in cognitive mechanisms known as the "social brain" or "social cognition". Social cognition is concerned with the ability to acknowledge that other individuals have a mind that controls their behavior (synonymous with the theory of mind or ToM: Tomasello and Call 1998). The social brain hypothesis follows this notion and emphasizes the uniqueness of anthropoid primates in the complexity of their social systems and in the associated cognitive demands of these systems. Thus, to put it simply, a cognitive requirement for a large group is a larger brain, as suggested by the quantitative relationship between social group size and relative brain size (Dunbar and Shultz 2007; Shultz and Dunbar 2007, 2010; Dunbar et al. 2010). Following this line of reasoning, the assumption that the hominins of GBY operated at times in large groups likely involved other cognitively related aspects that may be revealed through the remains of their material culture. Our understanding of the reduction sequences of different lithic artifacts throughout the long occupational sequence at GBY enables us to address the use of social transmission of information, an issue that involves social learning as well as other required cognitive traits.

In the following section we will explore possible indications of the cognitive characteristics of the GBY hominins. Wynn and Coolidge (2011) suggest that a valid cognitive archaeological argument must have three components: the investigated cognitive ability must be well defined by cognitive science, the activities that require that cognitive ability must be identified, and finally attributes (preserved in the archaeological record) that can reliably stand for the activity must be determined. Furthermore, it is essential to "...use a variety of different activities if we wish to have reasonable confidence in our assessment" (Wynn and Coolidge 2011:4).

Following these suggestions, we correlate between particular archaeological data, our interpretation of the activities associated with them, and the cognitive implications deriving from them. To ensure that the cognitive abilities are well defined, the terminology used in our attempt to evaluate the cognitive abilities of the GBY hominins is based principally on the research of Coolidge and Wynn (Coolidge and Wynn 2001, 2007, 2009; Coolidge et al. 2015; Wynn 2002). As we endeavor to use a large range of activities, and in order to facilitate the discussion and minimize possible overlapping of different cognitive traits, we have subdivided the GBY cultural evidence into the smallest possible categories. Table 9.4 follows this outline by pointing to specific cognitive traits and their associated activity. The data on which the reconstructed activities are based were thoroughly discussed in previous chapters of this volume.

Table 9.4 clearly illustrates the very complex modes of survival and adaptation and the great technological sophistication expressed in the activities of the hominins of GBY. These archaeologically derived data contribute directly to our understanding of the cognitive abilities of these hominins. Furthermore, as the activities are recorded throughout the long occupational sequence of GBY the cognitive abilities required in order to carry out this vast range of activities are repetitive, indicating an established, developed cognition.

The data in Table 9.4 demonstrate the complex reduction sequences recorded at GBY. They provide direct evidence for the ability of the hominins to carry out tasks that necessitated a chain of steps, performed at different times in different places, requiring them to arrange their memories in sequential time (i.e. sequential memory). Furthermore, it is evident that the GBY hominins employed diverse core methods, often using alternating reduction strategies. This strongly suggests that they possessed a long-term working memory (Wynn and Coolidge 2004), which enabled them to achieve a remarkably high expertise in stone knapping. This ability is expressed in the technological flexibility recorded in all stages of the different reduction sequences aimed at the production and modification of lithic artifacts. Flexibility is also evident in the selection of different raw materials, as well as in the application of percussors of different sizes during a single reduction sequence and, similarly, shifting from hard to soft percussors. At GBY, in contrast to other Acheulian sites, flexibility is also evident in the different methods of giant cores employed for the production of large flakes. Another aspect of flexibility emerges from the study of the orientation (direction of blow) of basalt flakes, whether modified into bifaces or left unmodified. The diversity of directions indicates that the final goal could be achieved regardless of the direction and the obstacles that had to be overcome (the removal of thick striking platforms located at different places on the flakes). Another aspect of technological flexibility is seen within the biface reduction sequence and is expressed in the designation of particular products (apart from the large flakes) to be modified into a different morphotype – massive scrapers, a type that is not a component of the biface *chaîne opératoire*.

Technological flexibility, in fact, is frequently indicative of expertise, as only experts are capable of adopting alternative ways of resolving problems during lithic reduction and modification. Flexibility in the reduction sequence is a component of a decision-making process based on a complex system of repeated evaluation-action modes. The ability to shift strategies to overcome unforeseen problems during knapping is clearly an embedded cognitive trait recorded in the lithic assemblages of GBY, indicative of the long-term working memory of the hominins.

In addition, the hominins transported a variety of resources (lithic, faunal, and floral) into their preferred locale/home, so that immediate gratification was inevitably delayed (i.e. inhibition), and they were able to acknowledge these resources and their products even when not physically present (i.e. displaced reference). The acquired knowledge of the spatial location, advantages and disadvantages, quality, and availability of these diverse resources was essential for the adaptation and survival of these hominins. Such knowledge supported subsistence based on a wide breadth of diet (gathering of floral and aquatic foods, hunting large mammals) enhanced by roasting/cooking, enabled the knapping and hafting of tools, and was facilitated by enhanced executive functions. It required the simultaneous integration of multiple sources of information and necessitated the construction of different models of the environment that could be alternated to anticipate and plan future behaviors. Moreover, such accumulated knowledge was acquired through the ability of these hominins to "move" subjectively and fluidly between past and future events; for example, the memory of past seasonal availability of foodstuffs enables the construction of future gathering plans. Such "sequential processing" (i.e. autonoesis) is of major importance as, together with response inhibition and planning over distance and time, it is considered a reliable indicator of a modern executive reasoning ability (Coolidge et al. 2015).

The cognitive modernity of the GBY hominins is further expressed in the transformative technologies (i.e. bringing together

9.2 Hominin Behavior at GBY

Table 9.4 Activities recognized at GBY and their suggested cognitive traits

GBY cultural evidence	Sequential memory	Displaced reference	Fluid intelligence	Episodic memory and autonoesis	Inhibition	Innovation and adaptation	Transformative technology	Social cognition
Lithics								
Raw material procurement								
Acquisition of basalt, flint, and limestone raw material	+	+		+	+	+		+
Acquisition of thick and thin basalt slabs	+	+		+	+	+		+
Acquisition of percursors of different materials and sizes	+	+		+	+	+		
Core reduction								
Integrating off-site and on-site knapping	+			+				
Diverse ways of opening giant basalt slabs	+		+	+		+		
Production of large basalt flakes by a variety of core reduction methods	+		+	+		+		
Changing percursors of different sizes, materials, and hardness within and between reduction sequences	+		+	+		+		
Manipulating an extremely wide range of sizes	+		+	+		+		
Applying the same volumetric concept (Levallois) to giant basalt cores and small flint cores	+		+	+		+		
Manipulating convex surfaces in bifaces and cores on flakes	+		+	+		+		
Starting from the use of the target artifact and shifting to core reduction (in limestone)	+		+	+				
Expertise	+							
Blank selection								
Flexibility in selecting blanks for biface modification	+	+	+	+		+		
Selection of basalt flakes for future modification of massive scrapers	+	+		+		+		
Tool modification								
Modification of handaxes is not constrained to basalt; handaxes of similar morpho-typology are produced on flint and limestone	+	+	+	+		+		
Manipulating large flakes to adjust them to a required shape and volume (bifaces) despite their different morphologies (dictated by direction of blow)	+	+	+	+		+		
Removal of striking platform of basalt bifaces and of flint small flakes	+	+		+		+		
Modification of extremely small flint tools and removal of their striking platform (dexterity)	+	+		+		+		
Tool use								
Using the same artifact for different tasks (basalt thin anvils)	+	+	+	+		+		
Designating similar morphologies for different tasks (*concassage*)	+	+	+	+		+		
Hafting flint tools	+	+	+	+	+	+	+	
Fauna								
Exploiting and transporting a large range of faunal resources	+	+		+	+	+		+
Preference of particular animals within the large range				+				
Thorough anatomical knowledge (fallow deer)	+	+		+				
Exploitation of aquatic resources with preference of particular fish	+	+		+		+		
Marrow extraction from diverse species	+			+				
Flora								
Gathering and transporting a large range of edible plants	+			+	+			+
Collecting and processing water nuts	+	+		+	+	+		+
Fire								
Making fire	+	+	+	+	+	+	+	+
Processing nuts and USO	+	+	+	+	+	+		+
Processing fish	+	+						
Spatial patterning of activities around the fire	+							+

different materials and changing them to create something new; Coolidge and Wynn 2001, 2007; Wadley 2013) of tool hafting and the making, controlling, and using of fire, which appear to be Acheulian cultural innovations. Most studies describe inventions and innovations in relation to anatomically modern humans and Neanderthals in the Middle and Upper Paleolithic (e.g. Belfer-Cohen and Hovers 2010; McBrearty and Brooks 2000), and the common view is that inventions become innovations only when adopted by a large number of individuals (e.g. Hovers and Belfer-Cohen 2006 and references therein; for a different view, see Collard er al. 2016). A component of the cognitive abilities of the GBY hominins is concerned with inventions, innovations, and adoptions. As we are unable to identify *inventions* at GBY, we refer below only to *innovations* (following the terminology cited in Hovers and Belfer-Cohen 2006) that were used by the GBY hominins. For example, the controlled use of fire is undoubtedly an innovation, and one that has direct cognitive implications with regard to the maintenance and provisioning of fires (Twomey 2013). Evidence of fire is present in all the archaeological horizons at GBY. Yet hominins most probably mastered fire in earlier times, as is emerging from the African data (Gowlett and Wrangham 2013). If indeed that was the sequence of events, when did this innovation become a cultural component of the Acheulian? The same holds true for the innovation of the use of soft percussors in the production and modification of bifaces. Again, it is clearly present at GBY, but perhaps this innovation was deeply rooted within the set of "traditional" cultural knowledge of the Acheulian and reached the Levantine Corridor as a component of the Acheulian culture. Indeed, we lack evidence for the timing and dispersal of innovations that were brought into the Mediterranean landscape by diffusion from Africa. In addition to fire and the use of soft percussors, we now add the issue of hafting as proposed for the lithic assemblages of GBY. Again, our interpretation of particular lithic features views hafting as part of the cultural package, but the timing and diffusion mechanism remain unknown.

The data from GBY can thus be correlated with diverse activities that in turn are associated with a variety of cognitive traits. Even if we question each of the individual traits, we cannot ignore their accumulation (Table 9.4). Similarly, our reconstruction of the cognitive abilities of the GBY hominins does not derive from an individual task or the particular cognitive trait associated with it, but rather is based on our comprehensive understanding of the array of activities carried out by these Acheulian hominins. These activities enabled them to adapt, survive, and succeed throughout a long duration on the shores of paleo-Lake Hula.

9.3 Final Words

It is nearly thirty years since we started the excavation project of the Acheulian site of GBY. Despite the small scale of the field project and the minimal excavated volume and surface, the excavation yielded an incredible wealth of information on diverse domains of Pleistocene research that has broadened our knowledge of the Acheulian culture. Yet, even though the study continued without intermission through all these years, it seems that we have only scraped the surface of the knowledge buried in this site. Many questions and issues regarding the culture and behavior of the GBY hominins remain unsolved. What was the function of the tools? Why, despite the availability of driftwood, was only a single wooden tool found? Where are the contemporary sites in the Levantine Corridor? And finally, what were the mechanisms that led to the disappearance of the rich, complex, and continuous Acheulian culture?

These are only a few selected questions that remain unanswered out of an array of possible research avenues. Some research topics that were not addressed in this volume will soon be studied. Among these is the spatial distribution of the finds along the GBY cultural archive (Alperson-Afil, additional studies under way). This, like other studies, will be feasible due to the diverse methodologies that were applied during excavation and analysis. Moreover, some of the topics discussed in this volume could be studied further. An example is a renewed attempt to investigate the source of the basalt, a topic that is even more relevant after we have studied in depth and interpreted the assemblages of basalt artifacts. Continuing the multidisciplinary research approach could enhance our interpretations of the Acheulian communities that occupied the margins of paleo-Lake Hula and further reveal their behavioral patterns.

We initiated this study with a minimal understanding of the Acheulian of GBY, but we now feel that we have gained substantial knowledge of the hominins and their culture, behavior, and cognition. We have been constantly astounded by the package of diverse skills and abilities possessed by these hominins, which included memory, planning, knowledge, and social and cognitive traits. Our results suggest that the continuous culture documented at GBY not only required a fully developed language, but was entrenched in a distinctive, consistent, continuous, and homogeneous tradition, transferred through many generations and expressing an extremely diverse and impressive culture. We have demonstrated that these cultural expressions were not, as is frequently claimed, features of a typically stagnant Acheulian culture, but on the contrary a highly variable culture with multiple material expressions, *some* of which can be interpreted and understood as parallels to "modern behavior". We hope that many of the unexplained patterns described in this volume will be deciphered in future studies of the Acheulian culture.

In one of his many seminal papers, Isaac (1976) presented his two-part model of human evolution. The earlier developmental phase (Functional Complex I) included bipedal locomotion, carrying objects, meat eating, food gathering and sharing, home base, division of labor, tools, and equipment, and perhaps pair bonding. His model for the later phase (Functional Complex II,

9.3 Final Words

starting around 1 Ma and ending around 0.1 Ma) included the elements of Functional Complex I with the addition of speech and language cognition, social coordination, cultural rule systems, technical ingenuity, facility with symbols, insight, and cunning. With the possible exception of "facility with symbols", we believe that we have provided both data and insight on the presence of the components of both parts of Isaac's model, enabling us to turn Isaac's original model into an archaeological reality.

Finally, a few more words on the issue of artistic behavior ("facility with symbols"). Years ago, two bead-like fossils, each having two flat surfaces and a hole in the middle, were found during our excavations at GBY. These were identified as silicified crinoid columnal segments and were considered natural components of the sediments in which they were bedded (Goren-Inbar et al. 1991b). Currently, some 25 years later, our views have changed and our better understanding of the sediments and their accumulative history has proved us wrong. We now regard these two objects as ornamental beads that can be related to the artistic world of the GBY hominins, like similar objects reported from much younger European sites. The unfortunate meagerness of such art objects in the early to late Middle Pleistocene is well known (e.g. Goren-Inbar 1986; Goren-Inbar 1990), adding yet another enigma to the history of hominin abilities and culture.

This volume, finally ending, has dealt with a single aspect, the lithic assemblages, though providing insight into many others. It is due to the unique geological structure of GBY that we have been able to carve out a small "window" into the Pleistocene and gain a few glimpses into many spheres of the mostly unknown cultural world of ancient hominins.

References

Adler, D.S., Wilkinson, K.N., Blockley, S., Mark, D.F., Pinhasi, R., Schmidt-Magee, B.A., et al. (2014). Early Levallois technology and the Lower to Middle Paleolithic transition in the Southern Caucasus. *Science, 345* (6204), doi: 10.1126/science.1256484.

Alimen, M.H. (1975). Les 'isthmes' hispano-marocain et siculo-tunisien aux temps acheuléens. *L'Anthropologie, 79*, 399–436.

Alperson-Afil, N. (2007). Ancient flames: Controlled use of fire at the Acheulian site of Gesher Benot Ya'aqov, Israel. Ph.D. Dissertation, Hebrew University of Jerusalem.

Alperson-Afil, N. (2008). Continual fire-making by hominins at Gesher Benot Ya'aqov, Israel. *Quaternary Science Reviews, 27*, 1733–1739, doi: 10.1016/j.quascirev.2008.06.009.

Alperson-Afil, N. (2015). Forgotten knowledge: Basalt tool-making in the Lower Palaeolithic Levant. In G. Galizia & D. Shulman (Eds.), *Forgetting: An interdisciplinary conversation* (pp. 116–133). Jerusalem: Magnes Press.

Alperson-Afil, N., & Goren-Inbar, N. (2010). *The Acheulian site of Gesher Benot Ya'aqov, volume II: Ancient flames and controlled use of fire.* (Vertebrate Paleobiology & Paleoanthropology Series). Dordrecht: Springer.

Alperson-Afil, N., & Goren-Inbar, N. (2015). Scarce but significant: The Limestone component of the Acheulian lithic assemblages of Gesher Benot Ya'aqov, Israel. In M. N. Haidle, N. Conard, & M. Bolus (Eds.), *The Nature of Culture* (pp. 41-56, Vertebrate Paleobiology and Paleoanthropology Series. Dordrecht: Springer.). New York: Springer.

Alperson-Afil, N., & Goren-Inbar, N. (2016). Acheulian hafting: Proximal modification of small flint flakes at Gesher Benot Ya'aqov, Israel. *Quaternary International*, doi: 10.1016/j.quaint.2015.12.068.

Alperson-Afil, N., & Goren-Inbar, N. (n.d.). Database of the Acheulian spatial distribution in Israel. Unpublished raw data.

Alperson, N., Sharon, G., Zohar, I., Biton, R., Melamed, Y., Kislev, M.E., et al. (2009). Spatial organization of hominin activities at Gesher Benot Ya'aqov, Israel. *Science, 326*, 1677–1680.

Andrefsky, W. (2005). *Lithics: Macroscopic Approaches to Analysis.* Cambridge: Cambridge University Press.

Anton, S.C. (2003). Natural history of *Homo erectus*. *Yearbook of Physical Anthropology, 46*, 126–170, doi: 10.1002/ajpa.10399.

Ashkenazi, S. (2004). Human paleoecology in the Levantine Corridor. In N. Goren-Inbar & J.D. Speth (Eds.), *Human paleoecology in the Levantine Corridor* (pp. 167–190). Oxford: Oxbow Books.

Ashkenazi, S., & Mienis, H. (2005). The taxonomy of the mollusc assemblage of the Pleistocene site of Gesher Benot Ya'aqov (GBY) (pp. 12). Department of Evolution, Systematics and Ecology, Hebrew University of Jerusalem.

Ashkenazi, S., Motro, U., Goren-Inbar, N., Bitton, R., & Rabinovich, R. (2005). New morphometric parameters for assessment of body size and population structure in the freshwater fossil crab assemblage from the Pleistocene site of Gesher Benot Ya'aqov (GBY), Israel. *Journal of Archaeological Science, 32*, 675–689, doi: 10.1016/j.jas.2004.12.003.

Ashkenazi, S., Klass, K., Mienis, H.K., Spiro, B., & Abel, R. (2010). Fossil embryos and adult Viviparidae from the Early–Middle Pleistocene of Gesher Benot Ya'aqov, Israel: Ecology, longevity and fecundity. *Lathaia, 43*, 116–127, doi: 10.1111/j.1502-3931.2009.00178.x.

Ashton, N., & White, M. (2003). Bifaces and raw materials: Flexible flaking in the British Early Paleolithic. In M. Soressi & H. Dibble (Eds.), *Multiple approaches to the study of bifacial technology* (University Museum Monograph 115, pp. 109–124). Philadelphia: University of Pennsylvania Museum of Archaeology and Anthropology.

Bar-Yosef, O., & Goren-Inbar, N. (1993). *The lithic assemblages of 'Ubeidiya* (Vol. 34, Qedem). Jerusalem: Institute of Archaeology, Hebrew University.

Barkai, R., & Gopher, A. (2015). On anachronism: The curious presence of spheroids and polyhedrons at Acheulo–Yabrudian Qesem Cave, Israel. *Quaternary International*, doi: 10.1016/j.quaint.2015.08.005.

Barkai, R., Lemorini, C., Shimelmitz, R., Lev, Z., Stiner, M.C., & Gopher, A. (2009). A blade for all seasons? Making and using Amudian blades at Qesem Cave, Israel. *Human Evolution, 24*, 57–75, doi: o.1007/s10437-010-9083z.

Barkai, R., Lemorini, C., & Gopher, A. (2010). Palaeolithic cutlery 400 000–200 000 years ago: Tiny meat-cutting tools from Qesem Cave, Israel. *Antiquity, 84*.

Barnea, I. (2009). Hula monitoring report 2008. *The Hula Project.* Kiryat Shmona: KKL and Water Authority (in Hebrew).

Barton, R.A., & Dunbar, R.I.M. (1997). Evolution of the social brain. In A. Whiten & R.W. Byrne (Eds.), *Machiavellian intelligence II* (pp. 240–263). Cambridge, UK: Cambridge University Press.

Beaune, S.A. de (2000). *Pour une archeologie du geste.* Paris: Éditions CNRS.

Belfer-Cohen, A., & Hovers, E. (2010). Modernity, enhanced working memory, and the Middle to Upper Paleolithic record in the Levant. *Current Anthropology, 51*(Supplement 1), S167–S175, doi: 10.1086/649835.

Belitzky, S. (1987). Tectonics of the Korazim Saddle. M.Sc. Thesis, Hebrew University of Jerusalem.

Belitzky, S. (2002). The structure and morphotectonics of the Gesher Benot Ya'aqov area, Northern Dead Sea Rift, Israel. *Quaternary Research, 58*, 372–380, doi: 10.1006/qres.2002.2347.

Belitzky, S., Goren-Inbar, N., & Werker, E. (1991). A Middle Pleistocene wooden plank with man-made polish. *Journal of Human Evolution, 20*, 349–353, doi: 10.1016/0047-2484(91)90015-N.

Beyene, Y., Katoh, S., WoldeGabriel, G., Hart, W.K., Uto, K., Sudo, M., et al. (2013). The characteristics and chronology of the earliest Acheulean at Konso, Ethiopia. *Proceedings of the National Academy of Sciences, 110*, 1584–1591, doi: 10.1073/pnas.1221285110.

Biberson, P. (1961). *La cadre paléogéographique de la Préhistoire du Maroc atlantic* (Vol. 16). Rabat: Publications du Service des Antiquités du Maroc.

Binford, L.R. (1979). Organization and formation processes: Looking at curated technologies. *Journal of Anthropological Research, 35*, 255–273.

Binford, L.R. (1984). Butchering, sharing, and the archaeological record. *Journal of Anthropological Archaeology, 3,* 235–257, doi: 10.1016/0278-4165(84)90003-5.

Biton, R., Geffen, E., Vences, M., Cohen, O., Bailon, S., Rabinovich, R., et al. (2013). The rediscovered Hula painted frog is a living fossil. *Nature Communications,* Article 1959, doi: 10.1038/ncomms2959.

Blumenschine, R.J., Masao, F.T., Tactikos, J.C., & Ebert, J.I. (2007). Effects of distance from stone source on landscape-scale variation in Oldowan artifact assemblages in the Paleo-Olduvai Basin, Tanzania. *Journal of Archaeological Science.*

Boëda, E. (1995). Levallois: A volumetric construction, methods, a technique. In H.L. Dibble & O. Bar-Yosef (Eds.), *The definition and interpretation of Levallois technology* (Vol. 23). Madison Wisconsin: Prehistory Press, Monographs in World Archaeology Series.

Bordes, F. (1961). *Typologie du Paléolithique ancien et moyen.* Bordeaux: Imprimeries Delmas.

Brande, S., & Saragusti, I. (1996). A morphometric model and Landmark Analysis of Acheulian handaxes from northern Israel. In L.F. Marcus, M. Corti, A. Loy, G.J.P. Naylor, & D.E. Slice (Eds.), *Advances in morphometrics* (pp. 423–435). New York: Plenum Press.

Braüer, J., & Call, J. (2015). Apes produce tools for future use. *American Journal of Primatology,* 77, 254-263, doi:10.1002/ajp.22341.

Bunn, H.T., & Kroll, E.M. (1986). Systematic butchery by Pliocene/Pleistocene hominids at Olduvai Gorge, Tanzania. *Current Anthropology, 7,* 431–452.

Champault, B. (1966). L'Acheuléen évolué au Sahara occidental: Notes sur l'homme au Paléolithique ancien. Ph.D. Dissertation, Université de Paris.

Chavaillon, J. (1979). Essai pour une typologie du matériel de percussion. *Bulletin de la Societé Préhistorique Française, 76,* 230–233.

Chavaillon, J., & Piperno, M. (2004). *Studies on the Early Paleolithic site of Melka Kunture, Ethiopia.* Florence: Istituto Italiano di Preistoria e Protostoria.

Clark, J.D. (1966). Acheulian occupation sites in the Middle East and Africa: A study in cultural variability. *American Anthropologist, 68,* 202–229, doi: 10.1525/aa.1966.68.2.02a001010.

Clark, J.D. (1968). Further excavation (1965) at the Middle Acheulian occupation site at Latamne, northern Syria. *Quaternaria, 10,* 1–76.

Clark, J.D. (2001). *Kalambo Falls* (Vol. III). Cambridge: Cambridge University Press.

Clark, J.D., & Kurashina, H. (1976). New Plio-Pleistocene archaeological occurrences from the Plain of Gadeb, Upper Webi Shebele Basin, Ethiopia, and a statistical comparison of the Gadeb sites with other Early Stone Age assemblages. In J. Clark & G. Isaac (Eds.), *Les plus anciennes industries en Afrique* (pp. 158–216). Nice: UISPP IX Congrès.

Cole, G.H., & Kleindienst, M.R. (1974). Further reflections on the Isimila Acheulian. *Quaternary Research, 4,* 346–355, doi: 10.1016/0033-5894(74)90021-0.

Collard, M., Vaesen, K., Cosgrove, R., & Roebroeks, W. (2016). The empirical case against the 'demographic turn' in Palaeolithic archaeology. *Philosophical Transactions of the Royal Society B, 371*(1698).

Coolidge, F.L., & Wynn, T. (2001). Executive functions of the frontal lobes and the evolutionary ascendancy of *Homo sapiens. Cambridge Archaeological Journal, 11,* 255–260, doi: 10.1017/S0959774301000142.

Coolidge, F.L., & Wynn, T. (2007). The working memory account of Neandertal cognition: How phonological storage capacity may be related to recursion and the pragmatics of modern speech. *Journal of Human Evolution, 52,* 707–710, doi: 10.1016/j.jhevol.2007.01.003.

Coolidge, F.L., & Wynn, T. (2009). *The rise of Homo sapiens: The evolution of modern thinking.* Chichester, UK: Wiley-Blackwell.

Coolidge, F.L., Wynn, T., Overmann, K.A., & Hicks, J.M. (2015). Cognitive archaeology and the cognitive sciences. In E. Bruner (Ed.), *Human paleoneurology* (pp. 177–208). Cham, Switzerland: Springer.

Copeland, L. (1995). Are Levallois flakes in the Acheulean the results of biface preparation? In H.L. Dibble & O. Bar-Yosef (Eds.), *The definition and interpretation of Levallois technology* (pp. 171–183). Madison, Wisconsin: Prehistory Press.

Corvinus, G. (1981). *A survey of the Pravara River system in Western Maharashtra, India: The stratigraphy and geomorphology of the Pravara River system.* Tübingen: Verlag Archaeologia Venatoria.

Corvinus, G. (1983). *A survey of the Pravara River system in Western Maharashtra, India: The excavations of the Acheulian site of Chirki-on-Pravara, India* (Vol. 2). Tübingen: Verlag Archaeologia Venatoria.

Corvinus, G. (2004). *Homo erectus* in East and Southeast Asia, and the question of the age of the species and its association with stone artifacts, with special attention to handax-like tools. *Quaternary International, 117,* 141–151, doi: 10.1016/S1040-6182(03)00124-1.

Crader, D.C. (1983). Recent single-carcass bone scatter and the problem of 'butchery' sites in the archaeological record. In J. Clutton-Brock & C. Grigson (Eds.), *Hunters and their prey* (Vol. 163, pp. 107–141). Oxford: BAR International Series.

Dag, D., & Goren-Inbar, N. (2001). An actualistic study of dorsally plain flakes: a technological note. *Lithic Technology,* 26, 105-117.

deBono, H., & Goren-Inbar, N. (2001). Note on a link between Acheulian and Levallois technologies. *Journal of the Israel Prehistoric Society, 31,* 9–23.

Delage, C. (2007). Chert procurement in the Acheulean of Gesher Benot Ya'aqov (Israel): Preliminary assessments. In J. Hedges & E. Hedges (Eds.), *Chert availability and prehistoric exploitation in the Near East* (Vol. 1615, pp. 240–257). Oxford: BAR International Series.

Diez-Martín, F., Yustos, P.S., Rúa, D.G. de la, González, J.Á.G., Luque, L. de, & Barba, R. (2014). Early Acheulean technology at Es2-Lepolosi (ancient MHS-Bayasi) in Peninj (Lake Natron, Tanzania). *Quaternary International, 322–323,* 209–236, doi: 10.1016/j.quaint.2013.08.053.

Dimentman, C., Bromley, H.J., & Por, F.D. (1992). *Lake Hula: Reconstruction of the fauna and hydrobiology of a lost lake.* Jerusalem: Israel Academy of Sciences and Humanities.

Dunbar, R.I.M. (1992). Neocortex size and group size in primates: A test of the hypothesis. *Journal of Human Evolution, 28,* 287–296, doi: 10.1006/jhev.1995.1021.

Dunbar, R.I.M. (1998). The social brain hypothesis. *Evolutionary Anthropology, 6,* 178–190, doi: 10.1002/(SICI)1520-6505(1998)6:5<178::AID-EVAN5>3.0.CO;2-8.

Dunbar, R.I.M., & Shultz, S. (2007). The evolution of the social brain: Anthropoid primates contrast with other vertebrates. *Proceedings of the Royal Society, 274B,* 2429–2436, doi: 10.1098/rspb.2007.0693.

Dunbar, R.I.M., Gamble, C., & Gowlett, J.A.J. (Eds.) (2010). *Social brain, distributed mind* (Vol. 158, Proceedings of the British Academy). London: British Academy.

Ellenblum, R., Marco, S., Agnon, A., Rockwell, T., & Boas, A. (1998). Crusader castle torn apart by earthquake at dawn, 20 May 1202. *Geology, 26*(4), 303–306, doi: 10.1130/0091-7613(1998 026<0303:CCTABE>2.3.CO;2.

Eshel, G. (2008). Hula monitoring report 2007. *The Hula Project.* Kiryat Shmona: KKL and Water Authority (in Hebrew).

Feblot-Augustins, J. (1990). Exploitation des matières premières dans l'Acheuléen d'Afrique: Perspectives comportementales. *PALEO, 2,* 27–42.

Feibel, C.S. (2001). Archaeological sediments in lake margin environments. In J.K. Stein & W.R. Farrand (Eds.), *Sediments in archaeological contexts* (pp. 127–148). Salt Lake City: University of Utah Press.

Feibel, C.S. (2004). Quaternary lake margins of the Levant Rift Valley. In N. Goren-Inbar & J.D. Speth (Eds.), *Human paleoecology in the Levantine Corridor* (pp. 21–36). Oxford: Oxbow Books.

Friedman, E., Goren-Inbar, N., Rosenfeld, A., Marder, O., & Burian, F. (1995). Hafting during Mousterian times: Further indication. *Mitekufat Haeven, 26,* 8–31.

Friedman, H. (2010). The parting of the sea: How volcanoes, earthquakes

References

and plagues shaped the story of Exodus. *Journal of Archaeological Science, 37*, 224, doi: http://dx.doi.org/10.1016/j.jas.2009.09.032.

Gaillard, C., Singh, M., Rishi, K.K., & Bhardwaj, V. (2010). Atbarapur (Hoshiarpur District, Punjab), the Acheulian of the Siwalik Range within the South Asian context. *Comptes Rendus Palevol, 9*, 237–243, doi: 10.1016/j.crpv.2010.06.001.

Galanidou, N., Cole, J., Iliopoulos, G., & McNabb, J. (2013). East meets West: The Middle Pleistocene site of Rodafnidia on Lesvos, Greece. *Antiquity, 087*(336).

Gallotti, R. (2013). An older origin for the Acheulean at Melka Kunture (Upper Awash, Ethiopia): Techno-economic behaviours at Garba IVD. *Journal of Human Evolution, 65*, 594–620.

Gallotti, R., & Mussi, M. (2015). The unknown Oldowan: ~1.7-million-year-old standardized obsidian small tools from Garba IV, Melka Kunture, Ethiopia. *PloS ONE, 10*, 1–23, doi: 10.1371/journal.pone.0145101.

Gallotti, R., Collina, C., Raynal, J.-P., Kieffer, G., Geraads, D., & Piperno, M. (2010). The Early Middle Pleistocene site of Gombore II (Melka Kunture, Upper Awash, Ethiopia) and the issue of Acheulean bifacial shaping strategies. *African Archaeological Review, 27*, 291–322, doi: 10.1007/s10437-010-9083-z.

Garfunkel, Z. (1981). Internal structure of the Dead Sea leaky transform (rift) in relation to plate kinematics. *Tectonophysics, 80*, 80–108.

Gat, Z., & Paster, Z. (1974). *Agroclimate of the Golan Heights: The rain*. Bet Dagan: Israel Meteorological Service (in Hebrew).

Geraads, D., & Tchernov, E. (1983). Fémurs humains du Pleistocene moyen de Gesher Ya'aqov (Israël). *L'Anthropologie, 87*, 138–141.

Ghinassi, M., Oms, O., Papini, M., Scarciglia, F., Carnevale, G., Sani, F., et al. (2015). An integrated study of the *Homo*-bearing Aalat stratigraphic section (Eritrea): An expanded continental record at the early–middle Pleistocene transition. *Journal of African Earth Sciences, 112*, 163–185, doi: 10.1016/j.jafrearsci.2015.09.012.

Gilead, D. (1968). Gesher Benot Ya'aqov. *Hadashot Arkheologiyot, 27*, 34–35.

Gilead, D. (1970). Handaxe industries in Israel and the Near East. *World Archaeology, 2*, 1–11.

Gilead, D. (1970a). Early Paleolithic cultures in Israel and the Near East. Ph.D. Dissertation, Hebrew University of Jerusalem.

Gilead, D. (1973). Cleavers in the early Palaeolithic industries in Israel. *Paléorient, 1*, 73–86.

Goldstein, M. (2011). Taphonomic processes in lake margin environments: Microvertebrate remains at Gesher Benot Ya'aqov (Area A), Israel: A case study. M.A. Thesis, Hebrew University of Jerusalem.

Gorbushina, A., Krumbein, W.E., Rosenfeld, A., & Goren-Inbar, N. (1996) On the microbiology of flint tools and silica skins. In *International Union of Microbiological Societies, 8th International Congress of Bacteriology, and Applied Microbiology Division, Jerusalem, August 18–23 1996* (p. 103).

Goren, N. (1981). The Lower Palaeolithic in Israel and adjacent countries. In P. Sanlaville & J. Cauvin (Eds.), *Préhistoire du Levant* (pp. 193–205). Lyon: C.N.R.S., Colloque 598.

Goren-Inbar, N. (1985). The lithic assemblages of the Berekhat Ram Acheulian site, Golan Heights. *Paléorient, 11*, 7–28.

Goren-Inbar, N. (1986). A figurine from the Acheulian site of Berekhat Ram. *Mitekufat Haeven, 19*, 7–12.

Goren-Inbar, N. (1988). Too small to be true? A reevaluation of cores on flakes in Levantine Mousterian assemblages. *Lithic Technology, 17*, 37–44.

Goren-Inbar, N. (1990). *Quneitra: A Mousterian site on the Golan Heights* (Vol. 31, Qedem). Jerusalem: Institute of Archaeology.

Goren-Inbar, N. (2011b). Culture and cognition in the Acheulian industry: A case study from Gesher Benot Ya'aqov. *Philosophical Transactions of the Royal Society B, 366*, 1038–1049, doi: 10.1098/rstb.2010.0365.

Goren-Inbar, N. (2011a). Behavioral and cultural origins of Neanderthals: A Levantine perspective. In S. Condemi & G.-C. Weniger (Eds.), *150 years of Neanderthal discoveries: Continuity and discontinuity* (pp. 89–100, Vertebrate Paleobiology and Paleoanthropology Book Series). Dordrecht: Springer.

Goren-Inbar, N., & Belitzky, S. (1989). Structural position of the Pleistocene Gesher Benot Ya'aqov site in the Dead Sea Rift zone. *Quaternary Research, 31*, 371–376, doi: 10.1016/0033-5894(89)90043-4.

Goren-Inbar, N., & Saragusti, I. (1996). An Acheulian biface assemblage from the site of Gesher Benot Ya'aqov, Israel: Indications of African affinities. *Journal of Field Archaeology, 23*, 15–30, doi: 10.1179/009346996791974007.

Goren-Inbar, N., & Sharon, G. (2006). Invisible handaxes and visible Acheulian biface technology at Gesher Benot Ya'aqov, Israel. In N. Goren-Inbar & G. Sharon (Eds.), *Axe Age: Acheulian tool-making from quarry to discard* (pp. 111–135). London: Equinox.

Goren-Inbar, N., & Speth, J.D. (Eds.). (2004). *Human paleoecology in the Levantine Corridor*. Oxford: Oxbow Books.

Goren-Inbar, N., Lewy, Z., & Kislev, M.E. (1991b). The taphonomy of a Jurassic bead-like fossil from an Acheulian occupation at Gesher Benot Ya'aqov. *Rock Art Research, 8*, 133–136.

Goren-Inbar, N., Zohar, I., & Ben-Ami, D. (1991). A new look at old cleavers: Gesher Benot Ya'aqov. *Mitekufat Haeven, 24*, 7–33.

Goren-Inbar, N., Belitzky, S., Goren, Y., Rabinovitch, R., & Saragusti, I. (1992a). Gesher Benot Ya'aqov – the "bar": An Acheulian assemblage. *Geoarchaeology, 7*, 27–40, doi: 10.1002/gea.3340070103.

Goren-Inbar, N., Belitzky, S., Verosub, K., Werker, E., Kislev, M., Heimann, A., et al. (1992b). New discoveries at the Middle Pleistocene Gesher Benot Ya'aqov Acheulian site. *Quaternary Research, 38*, 117–128, doi: 10.1016/0033-5894(92)90034-G.

Goren-Inbar, N., Lister, A., Werker, E., & Chech, M. (1994). A butchered elephant skull and associated artifacts from the Acheulian site of Gesher Benot Ya'aqov, Israel. *Paléorient, 20*, 99–112.

Goren-Inbar, N., Feibel, C.S., Verosub, K.L., Melamed, Y., Kislev, M.E., Tchernov, E., et al. (2000). Pleistocene milestones on the Out-of-Africa Corridor at Gesher Benot Ya'aqov, Israel. *Science, 289*, 944–974.

Goren-Inbar, N., Sharon, G., Melamed, Y., & Kislev, M. (2002a). Nuts, nut cracking, and pitted stones at Gesher Benot Ya'aqov, Israel. *Proceedings of the National Academy of Sciences, 99*, 2455–2460, doi: 10.1073 pnas.032570499.

Goren-Inbar, N., Werker, E., & Feibel, C.S. (2002b). *The Acheulian site of Gesher Benot Ya'aqov, volume I: The wood assemblage*. Oxford: Oxbow Books.

Goren-Inbar, N., Alperson, N., Kislev, M.E., Simchoni, O., Melamed, Y., Ben-Nun, A., et al. (2004). Evidence of hominin control of fire at Gesher Benot Ya'aqov, Israel. *Science, 304*, 725–727.

Goren-Inbar, N., Sharon, G., Alperson-Afil, N., & Laschiver, I. (2008). The Acheulian massive scrapers of Gesher Benot Ya'aqov: An associated product of the large cutting tools *chaîne opératoire*. *Journal of Human Evolution, 55*, 702–712, doi: 10.1016/j.jhevol.2008.07.005.

Goren-Inbar, N., Kislev, M., Melamed, Y., Rabinovich, R., Zohar, I., Biton, R., et al. (2009). The effect of climate change on the environment and hominins of the Upper Jordan Valley between ca. 800Ka and 700Ka ago as a basis for prediction of future scenarios. Hebrew University of Jerusalem (Israel Science Foundation report).

Goren-Inbar, N., Grosman, L., & Sharon, G. (2011). The technology and significance of the Acheulian giant cores of Gesher Benot Ya'aqov, Israel. *Journal of Archaeological Science, 38*, 1901–1917, doi: 10.1016/j.jas.2011.03.037.

Goren-Inbar, N., Kislev, M., Melamed, Y., Rabinovich, R., Zohar, I., Biton, R., et al. (2012). The effect of climate change on the environment and hominins of the Upper Jordan Valley between ca. 800Ka and 700Ka ago as a basis for prediction of future scenarios. Jerusalem: Hebrew University (Israel Science Foundation grant).

Goren-Inbar, N., Melamed, Y., Zohar, I., Akhilesh, K., & Pappu, S. (2014). Beneath still waters: Multistage aquatic exploitation of Euryale ferox (Salisb.) during the Acheulian. *Internet Archaeology, 37*, doi: 10.11141/ia.37.1.

Goren-Inbar, N., Sharon, G., Alperson-Afil, N., & Herzlinger, G. (2015). A new type of anvil in the Acheulian of Gesher Benot Ya'aqov, Israel.

Philosophical Transactions of the Royal Society B, doi: 10.1098/rstb.2014.0353.

Gowlett, J.A.J. (1991). Kilombe: Review of an Acheulian site complex. In J.D. Clark (Ed.), *Cultural beginnings* (pp. 129–136). Bonn: Dr. Rudolf Habelt Gmbh.

Gowlett, J.A.J., & Wrangham, R.W. (2013). Earliest fire in Africa: Towards the convergence of archaeological evidence and the cooking hypothesis. *Azania: Archaeological Research in Africa, 48*, 5–30, doi: 10.1080/0067270X.2012.756754.

Green, J., & Short, N.M. (Eds.). (1971). *Volcanic landforms and surface features: A photographic atlas and glossary*. Berlin: Springer-Verlag.

Grosman, L., Smikt, O., & Smilansky, U. (2008). On the application of 3-D scanning technology for the documentation and typology of lithic artifacts. *Journal of Archaeological Science, 35*, 31013110, doi: 10.1016/j.jas.2008.06.011.

Grosman, L., Sharon, G., Goldman-Neuman, T., Smikt, O., & Smilansky, U. (2011). Studying post depositional damage on Acheulian bifaces using 3-D scanning. *Journal of Human Evolution*, 60, 398-406.

Grosman, L., Karasik, A., Harush, O., & Smilansky, U. (2014). Archaeology in three dimensions: Computer-based methods in archaeological research. *Journal of Eastern Mediterranean Archaeology and Heritage Studies, 2*, 48–64.

Hallos, J. (2005). ''15 minutes of fame'': Exploring the temporal dimension of Middle Pleistocene lithic technology. *Journal of Human Evolution, 49*, 155–179, doi: 10.1016/j.jhevol.2005.03.002.

Harash, A., & Bar, Y. (1988). Faults, landslides and seismic hazards along the Jordan River gorge, northern Israel. *Engineering Geology, 25*, 1–15, doi: 10.1016/0013-7952(88)90015-4.

Harmand, S. (2009). Raw materials and techno-economic behaviors at Oldowan and Acheulean sites in the West Turkana region, Kenya. In B. Adams & B.S. Blades (Eds.), *Lithic materials and Paleolithic societies* (pp. 3–14). Oxford: Wiley-Blackwell.

Hartman, G. (2004). Long-term continuity of a freshwater turtle (*Mauremys caspica rivulata*) population in the Northern Jordan Valley and its paleoenvironmental implications. In N. Goren-Inbar & J.D. Speth (Eds.), *Human paleoecology in the Levantine Corridor* (pp. 61–74). Oxford: Oxbow Books.

Heimann, A. (1990). *The development of the Dead Sea Rift and its margins in northern Israel during the Pliocene and the Pleistocene* (Report GSI/28/90). Jerusalem: Geological Survey of Israel (in Hebrew).

Heimann, A., & Ron, H. (1993). Geometric changes of plate boundaries along part of the northern Dead Sea Transform: Geochronologic and paleomagnetic evidence. *Tectonics, 12*, 477–491, doi: 10.1029/92TC01789.

Heimann, A., & Steinitz, G. (1988). K-Ar ages of basalts from the western slopes of the Golan Heights. In *Current Research* (Vol. 6, pp. 29–32). Jerusalem: Geological Survey of Israel.

Herzlinger, G. (2014). A method of geometric morphometric shape analysis and its application for the identification of individual knappers. M.A. Thesis, Hebrew University of Jerusalem.

Herzlinger, G., Pinsky, S., & Goren-Inbar, N. (2015). A note on handaxe knapping products and their breakage taphonomy: An experimental view. *Journal of Lithic Studies, 2*, 65–82, doi: 10.2218/jls.v2i1.1295.

Hooijer, D.-A. (1959). Fossil mammals from Jisr Banat Yaqub, south of Lake Hule, Israel. *Bulletin of the Research Council of Israel, G8*, 177–179.

Hooijer, D.A. (1960). A stegodon from Israel. *Bulletin of the Research Council of Israel, G8*, 104–107.

Horowitz, A. (1973). Development of the Hula Basin, Israel. *Israel Journal of Earth Sciences, 22*, 107–139.

Horowitz, A. (1979). *The Quaternary of Israel*. New York: Academic Press.

Horowitz, A. (1989). Palynological evidence for the Quaternary rates of accumulation along the Dead Sea Rift, and structural implications. *Tectonophysics, 164*, 63–71.

Horowitz, A. (2001). *The Jordan Rift Valley*. Lisse: A.A. Balkema Publishers.

Hou, Y.M., Potts, R., Yuan, B.Y., Guo, Z.T., Deino, A., Wang, W., et al. (2000). Mid-Pleistocene Acheulean-like stone technology of the Bose basin, South China. *Science, 287*, 1545–1700.

Hovers, E. (2007). The many faces of cores-on-flakes: A perspective from the Levantine Mousterian. In S.P. McPherron (Ed.), *Tools versus cores: Alternative approaches to stone tool analysis* (pp. 42–74). Newcastle, UK: Cambridge Scholars Publishing.

Hovers, E. (2009). *The lithic assemblages of Qafzeh Cave*. Oxford: Oxford University Press.

Hovers, E., & Belfer-Cohen, A. (2006). "Now you see it now you don't": Modern human behavior in the Middle Paleolithic. In E. Hovers & S. Kuhn (Eds.), *Transitions before the transition: Evolution and stability in the Middle Paleolithic and Middle Stone Age* (pp. 295–304). New York: Springer.

Howell, F.C., & Clark, D.J. (1961). Isimila: A Paleolithic site in Africa. *Scientific American, 205*, 118–129, doi: 10.1038/scientificamerican1061-118.

Howell, F.C., Cole, G.H., Kleindienst, M.R., & Haldemann, E.G. (1962). Isimila: An Acheulian occupation site in the Iringa highlands, Southern Highlands Province, Tanganyika. In G. Mortelmans & J. Nenquin (Ed.), *Actes du IVe Congrès Panafricain de Préhistoire et de l'Etude du Quaternaire* (Vol. 40, pp. 43–80). Tervuren, Belgique: Musée Royal de l'Afrique Centrale.

Inbar, M. (2002). A geomorphic and environmental evaluation of the Hula Drainage Project, Israel. *Australian Geographical Studies, 40*, 155–166, doi: 10.1111/1467-8470.00171.

Inizan, M.-L., Redouron-Ballinger, M., Roche, G., & Tixier, J. (1999). *Technology and terminology of knapped stone*. Nanterre: CREP.

Isaac, G.L. (1968). The Acheulian site complex at Olorgesailie, Kenya: A contribution to the interpretation of Middle Pleistocene culture in East Africa. Ph.D. Dissertation, University of Cambridge.

Isaac, G.L. (1969). Studies of early culture in East Africa. *World Archaeology, 1*, 1–28, doi: 10.1080/00438243.1969.9979423.

Isaac, G.L. (1976). Early hominins in action: A commentary on the contribution of archaeology to understanding the fossil record in East Africa. *Yearbook of Physical Anthropology*, 19–35.

Isaac, G.L. (1977). *Olorgesailie, archaeological studies of a Middle Pleistocene lake basin, Kenya*. Chicago: University of Chicago Press.

Isaac, G.L. (1978). The food sharing behavior of proto-human hominids. In B. Isaac (Ed.), *The archaeology of human origins: Papers by Glynn Isaac* (pp. 289–311). Cambridge: Cambridge University Press.

Isaac, G.L. (1981). Emergence of human behaviour patterns: Archaeological test of alternative models of early hominid behaviour: Excavation and experiments *Philosophical Transactions of the Royal Society B, 292*, 177–188.

Isaac, G.L. (1983). Bones in contention: Competing explanations for the juxtaposition of Early Pleistocene artefacts and faunal remains. In B. Isaac (Ed.), *The archaeology of human origins: Papers by Glynn Isaac* (pp. 37–103). Cambridge: Cambridge University Press.

Isaac, G.L., (1971). The diet of early man: aspects of archaeological evidence from Lower and middle Pleistocene sites in Africa. *World Archaeology, 2*, 278–298.

Isaac, G.L., & Crader, D.C. (1981). To what extent were early hominids carnivorous? An archaeological perspective. In R.S.O. Harding & G. Teleki (Eds.), *Omnivorous primates: Gathering and hunting in human evolution* (pp. 37–103). New York: Columbia University Press.

Janmart, J. (1952). Elephant hunting as practised by the Congo pygmies. *American Anthropologist, 54*, 146–147, doi: 10.1525/aa.1952.54.1.02a00440.

Jennings, R.P., Shipton, C., Breeze, P., Cuthbertson, P., Bernal, M.A., Wedage, W.M.C.O., et al. (2015). Multi-scale Acheulean landscape survey in the Arabian Desert. *Quaternary International 382*, 58–81, doi: 10.1016/j.quaint.2015.01.028.

Jones, P.R. (1994). Results of experimental work in relation to the stone industries of Olduvai Gorge. In M.D. Leakey & D.A. Roe (Eds.), *Olduvai Gorge excavations in Beds III, IV, and the Masek*

Beds 1968–1971 (Vol. 5, pp. 254–298). Cambridge: Cambridge University Press.

Kalb, J. (2001). *Adventures in the bone trade: The race to discover human ancestors in Ethiopia's Afar Depression*. New York: Copernicus Books.

Karcz, I. (1995). *Development of a geodetic system for monitoring of recent crustal movements along the Dead Sea rift*. Jerusalem: Earth Sciences Administration, Ministry of Energy and Infrastructure, Israel, (report #TR-GSI/7/95).

Karmon, Y. (1953). The settlement of the Northern Huleh Valley since 1838. *Israel Exploration Journal, 3*, 4–25.

Karmon, Y. (1960). The drainage of the Huleh swamps. *Geographical Review, 50*, 169–193.

Karmon, Y. (1973). *The Land of Israel: The geography of the land and regions*. Tel Aviv: Yavneh (in Hebrew).

Kelly, R. (1995). *The foraging spectrum: Diversity in hunter-gatherer lifeways*. Washington, D.C.: Smithsonian Institution Press.

Kleindienst, M.R. (1961). Variability within the Late Acheulian assemblage in East Africa. *South African Archaeological Bulletin, 16*, 35–52, doi: 10.2307/3886868.

Kleindienst, M.R. (1962). Components of the East African Acheulian assemblage: An analytic approach. In C. Mortelmans & J. Nenquin (Eds.), *Actes du IVe Congrès Panafricain de Préhistoire et de l'Etude du Quaternaire* (pp. 81–105). Tervuren, Belgique: Musée Royal de l'Afrique Centrale.

Kleindienst, M.R. (1973). Excavation at site JK2, Olduvai Gorge, Tanzania, 1961–1962: The geological setting. *Quaternaria, XVII*, 145–208.

Krumbein, W.C. (1941). Measurement and geological significance of shape and roundness of sedimentary particles. *Journal of Sedimentary Petrology, 11*, 64–72.

Kuhn, S.L., & Stiner, M.C. (2006). What's a mother to do? The division of labor among Neandertals and Modern Humans in Eurasia. *Current Anthropology, 47*, 953–980, doi: 10.1086/507197.

Kuman, K. (2001). An Acheulean factory site with prepared core technology near Taung, South Africa. *South African Archaeological Bulletin, 56*, 8–22.

Leakey, M.D. (1971). *Olduvai Gorge, excavations in Beds I and II, 1960–1963* (Vol. 3). Cambridge: Cambridge University Press.

Leakey, M.D. (1975). Cultural patterns in the Olduvai sequence. In K.W. Butzer & G.L. Isaac (Eds.), *After the Australopithecines: Stratigraphy, ecology and culture change in the Middle Pleistocene* (pp. 477–493). The Hague: Mouton.

Leakey, M.D., & Roe, D. (Eds.). (1994). *Olduvai Gorge excavations in Beds III, IV and the Masek Beds 1968–1971* (Vol. 5). Cambridge: Cambridge University Press.

Lepre, C.J., Roche, H., Kent, D.V., Harmand, S., Quinn, R.L., Brugal, J.-P., et al. (2011). An earlier origin for the Acheulian. *Nature, 477*, 82–85.

Lewis, K.P. (2001). A comparative study of primate play behaviour: Implications for the study of cognition. *Folia Primatologica, 71*, 417–421, doi: 10.1159/000052740.

Light, J.P. (2001). Non-destructive XRF sourcing of basaltic artifacts from Gesher Benot Ya'aqov, Israel: Implications for hominid behavior. M.Sc. Thesis, University of California, Davis.

Light, J.P., Verosub, K.L., & Goren-Inbar, N. (1999). Sourcing of basaltic artifacts from the Gesher Benot Ya'aqov archaeological site, Israel: Raw material selection and hominid behavior. Paper presented at the INQUA XV, Durban, South Africa,

Lioubin, V.P., & Beliaeva, E.V. (2004). Acheulian industry of the Kudaro I cave: Raw material diversity and characteristics of the assemblage. In M. Toussaint, C. Draily, & J.-M. Cordy (Eds.), *Human origins and the Lower Palaeolithic* (Vol. 1272, pp. 21–27). Oxford: BAR International Series.

Machin, A., Hosfield, R., & Mithen, S. (2007). Why are some handaxes symmetrical? Testing the influence of handaxe morphology on butchery effectiveness. *Journal of Archaeological Science, 34*, 883–393, doi: 10.1016/j.jas.2006.09.008.

Madsen, B., & Goren-Inbar, N. (2004). Acheulian giant core technology and beyond: An archaeological and experimental case study. *Eurasian Prehistory, 2*, 3–52.

Malinsky-Buller, A., Grosman, L., & Marder, O. (2011). A case of techno-typological lithic variability & continuity in the late Lower Palaeolithic. *Before Farming, 1*(article 3), 1–32, doi: 10.3828/bfarm.2011.1.3.

Marco, S., Agnon, A., Ellenblum, R., Eidelman, A., Basson, U., & Boas, A. (1997). 817-year-old walls offset sinistrally 2.1m by the Dead Sea transform, Israel. *Journal of Geodynamics, 24*, 11–20, doi: 10.1016/S0264-3707(96)00041-5.

Marder, O., Milevski, I., & Matskevich, Z. (2006). The handaxes of Revadim Quarry: Typo-technological considerations and aspects of intra-site variability. In N. Goren-Inbar & G. Sharon (Eds.), *Axe Age: Acheulian tool-making from quarry to discard* (pp. 223–242). London: Equinox.

Marks, S.A. (2005). *Large mammals and a brave people: Subsistence hunters in Zambia*. New Brunswick, NJ: Transactions Publishers.

Martínez-Navarro, B., & Rabinovich, R. (2011). The fossil Bovidae (Artiodactyla, Mammalia) from Gesher Benot Ya'aqov, Israel: Out of Africa during the Early-Middle Pleistocene transition. *Journal of Human Evolution, 60*, 375–386, doi: 10.1016/j.jhevol.2010.03.012.

Martínez-Navarro, B., Belmaker, M., & Bar-Yosef, O. (2012). The bovid assemblage (Bovidae, Mammalia) from the Early Pleistocene site of 'Ubeidiya, Israel: Biochronological and paleoecological implications for the fossil and lithic bearing strata. *Quaternary International, 267*, 78–97, doi: 10.1016/j.quaint.2012.02.041.

Martini, F., Libsekal, Y., Filippi, O., Ghbre/Her, A., Kashay, H., Kiros, A., et al. (2004). Characterization of lithic complexes from Buia (Dandiero Basin, Danakil Depression, Eritrea). *Rivista Italiana di Paleontologia e Stratigraphia, 110*(Supplement), 99–132.

Matskevich, Z. (2006). Cleavers in the Levantine Late Acheulian: the case of Tabun Cave. In N. Goren-Inbar & G. Sharon (Eds.), *Axe Age: Acheulian tool-making from quarry to discard*. London: Equinox.

McBrearty, S., & Brooks, A.S. (2000). The revolution that wasn't: A new interpretation of the origin of modern human behavior. *Journal of Human Evolution, 39*, 453–563, doi: 10.1006/jhev.2000.0435.

McNabb, J., & Cole, J. (2015). The mirror cracked: Symmetry and refinement in the Acheulean handaxe. *Journal of Archaeological Science: Reports, 3*, 100–111, doi: http://dx.doi.org/10.1016/j.jasrep.2015.06.004.

Melamed, Y. (1997). Reconstruction of the landscape and the vegetarian diet at Gesher Benot Ya'aqov archaeological site in the Lower Paleolithic period. M.Sc. Thesis, Bar-Ilan University (in Hebrew).

Melamed, Y. (2003). Reconstruction of the Hula Valley vegetation and the hominid vegetarian diet by the Lower Palaeolithic botanical remains from Gesher Benot Ya'aqov. Ph.D. Dissertation, Bar-Ilan University. (in Hebrew).

Melamed, Y., Kislev, M.E., Weiss, U., & Simchoni, O. (2011). Extinction of water plants in the Hula Valley: Evidence for climate change. *Journal of Human Evolution, 60*, 320–327, doi: 10.1016/j.jhevol.2010.07.025.

Melamed, Y., Kislev, M., Geffen, E., Lev-Yadun, S., & Goren-Inbar, N. (2016). Hominin plant foodstuff and diet at the Acheulian site of Gesher Benot Ya'aqov, Israel *PNAS USA, 113*, 14674–14679.

Mercier, N., Valladas, H., Valladas, H.G., Reyss, J.-L., Jelinek, A.J., Meignen, L., et al. (1995). TL dates of burnt flints from Jelinek's excavations at Tabun and their implications. *Journal of Archaeological Science, 22*, 495–509.

Mienis, H., & Ashkenazi, S. (2006). Remains of egg capsules from *Theodoxus* on 780,000- and 14,000-year old shells from Gesher Benot Ya'aqov. *Haasiana, 3*, 70–71.

Mienis, H.K., & Ashkenazi, S. (2011). Lentic Basommatophora molluscs and hygrophilous land snails as indicators of habitat and climate in the Early–Middle Pleistocene (0.78 Ma) site of Gesher Benot Ya'aqov (GBY). *Journal of Human Evolution, 60*, 328–340, doi: 10.1016/j.jhevol.2010.03.009.

Mora, R., & Torre, I. de la (2005). Percussion tools in Olduvai Beds I and II (Tanzania): Implications for early human activities. *Journal* of Anthropological Archaeology, 24, 179–192, doi: 10.1016/j.jaa.2004.12.001.

Moshkovitz, S., & Magaritz, M. (1987). Stratigraphy and isotope records of Middle and Late Pleistocene mollusks from a corehole in the Hula Basin, northern Jordan Valley, Israel. *Quaternary Research, 28*, 226–237, doi: 10.1016/0033-5894(87)90061-5.

Mourre, V. (2003). *Implications Culturelles de la Technologie des Hachereaux*. PH.D., Université de Paris X, Nanerre.

Mourre, V., & Colonge, D. (2011). La question du débitage de grands éclats à l'acheuléen. *PALEO, numéro spécial, 35*–48.

Newcomer, M.H. (1971). Some quantitative experiments in handaxe manufacture. *World Archaeology, 3*, 85–94, doi: 10.1080/00438243.1971.9979493.

Oron, T. *The Israel Nature and Parks Authority annual report 2012*. Jerusalem: Israel Nature and Parks Authority (in Hebrew).

Owen, W.E. (1938). The Kombewa culture, Kenya Colony. *Man, December*, 203–205, doi: 10.2307/2791552.

Paddayya, K. (1982). *The Acheulian culture of the Hunsgi Valley (Peninsular India): A settlement system perspective*. Pune: Deccan College.

Paddayya, K. (2006). Evolution within the Acheulian in India: A case study from the Hunsgi and Baichbal Valleys, Karnataka. *Bulletin of the Deccan College Research Institute, 66/67 (2006–2007)*, 95–111.

Pappu, S., Gunnell, Y., Akhilesh, K., Braucher, R., Taieb, M., Demory, F., et al. (2011). Early Pleistocene presence of Acheulian hominins in South India. *Science, 331*, 1596–1599.

Pelegrin, J. (2000). Les techniques de débitage laminaire au Tardiglaciaire: Critères de diagnose et quelques réflexions. In B. Valentin, P. Bodu, & M. Christensen (Eds.), *L'Europe centrale et septentrionale au tardiglaciaire. Confrontation des modèles régionaux de peuplement* (Vol. 7, pp. 73–86). Nemours: Memoires du Musée de Préhistoire d'Ile-de-France. A.P.R.A.I.F.

Petraglia, M.D. (2003). The Lower Paleolithic of the Arabian Peninsula: Occupations, adaptations, and dispersals. *Journal of World Prehistory, 17*, 141–179.

Petraglia, M.D. (2005). Hominin responses to Pleistocene environmental change in Arabia and South Asia. In M.G. Head & P.L. Gibbard (Eds.), *Early–Middle Pleistocene transitions: The land–ocean evidence* (Vol. Geological Society London, Special Publications 247, pp. 305–319). London: Geological Society.

Petraglia, M.D., Porta, P. L., & Paddayya, K. (1999). The first Acheulian Quarry in India: stone tool manufacture, bifaces morphology, and behaviors. *Journal of Anthropological Research, 55*, 39-70.

Petraglia, M.D., Drake, N., & Alsharekh, A. (2009). Acheulian landscapes and large cutting tool assemblages. In M.D. Petraglia & J.I. Rose (Eds.), *The evolution of human population in Arabia: Paleoenvironments, prehistory and genetics* (pp. 103–116, Vertebrate Paleobiology & Paleoanthropology Series). Dordrecht: Springer.

Pétrequin, P., & Pétrequin, A.-M. (1993). *Écologie d'un outil: La hache de pierre en Irian Jaya (Indonésie)*. Paris: Éditions CNRS.

Picard, L. (1952). The Pleistocene peat of Lake Hula. *Bulletin of the Research Council of Israel, G 2*, 147–156.

Picard, L. (1963). The Quaternary in the northern Jordan Valley. *Proceedings, Israel Academy of Sciences and Humanities, 1*(4), 1–34.

Picard, L. (1965). The geological evolution of the Quaternary in the central-northern Jordan Graben. *Bulletin of the Research Council of Israel, 84*, 337–366.

Piperno, M. (Ed.). (1999). *Notarchirico: Un sito del Pleistocene medio iniziale nel bacino di Venosa*. Venosa: Edizioni Osanna.

Piperno, M.D. (2001). *Melka Kunture*. Napoli: Dipartimento Discipline Storiche "Ettore Lepore", Università di Napoli "Federico II".

Pope, M.I. (2004). Behavioural implications of biface discard: Assemblage variability and land-use at the Middle Pleistocene site of Boxgrove. In E.A. Walker, F.W. Smith, & F. Healy (Eds.), *Lithics in action*. Oxford: Oxbow Books.

Potts, R. (1984). Home bases and early hominids. *American Scientist, 72*, 338–347.

Potts, R. (1989). Olorgesailie: New excavations and findings in Early and Middle Pleistocene contexts, southern Kenya Rift Valley. *Journal of Human Evolution, 18*, 477–484.

Potts, R. (1994). Variables versus models of early Pleistocene hominid land use. *Journal of Human Evolution, 27*, 7–24, doi: 10.1006/jhev.1994.1033.

Prodehl, C., Fuchs, K., & Mechie, J. (1997). Seismic refraction studies of the Afro-Arabian rift system: A brief review. *Tectonophysics, 278*, 1–13, doi: 10.1016/S0040-1951(97)00091-7.

Rabinovich, R., & Biton, R. (2011). The Early–Middle Pleistocene faunal assemblages of Gesher Benot Ya'aqov: Taphonomy and paleoenvironment. *Journal of Human Evolution, 60*, 357–374, doi: 10.1016/j.jhevol.2010.12.002.

Rabinovich, R., Gaudzinski-Windheuser, S., & Goren-Inbar, N. (2006). Site formation processes: The role of hominin and natural agents in the formation of striations and cut marks on bones at the Acheulian site of Gesher Benot Ya'aqov, Israel. Jerusalem: German-Israeli Foundation for Scientific Research and Development report.

Rabinovich, R., Gaudzinski, S., & Goren-Inbar, N. (2008). Systematic butchering of Fallow deer (*Dama*) at the early Middle Pleistocene Acheulian site of Gesher Benot Ya'aqov, (Israel). *Journal of Human Evolution, 54*, 134–149, doi: 10.1016/j.jhevol.2007.07.007.

Rabinovich, R., Gaudzinski-Windheuser, S., Kindler, L., & Goren-Inbar, N. (2012). *The Acheulian site of Gesher Benot Ya'aqov, volume III: Mammal taphonomy – The assemblages of Layers V-5 and V-6*. (Vertebrate Paleobiology & Paleoanthropology Series). Dordrecht: Springer.

Rhys, J., & White, N. (1988). *Point blank: Stone tool manufacture at the Ngilipitji Quarry, Arnhem Land, 1981*. Canberra: Dept. of Prehistory, Research School of Pacific Studies, Australian National University.

Rink, W.J., & Schwarcz, H.P. (2005). Short contribution: ESR and uranium series dating of teeth from the Lower Paleolithic site of Gesher Benot Ya'aqov, Israel: Confirmation of paleomagnetic age indications. *Geoarchaeology, 20*, 57–66.

Roche, H., & Texier, P.-J. (1991). La notion de complexité dans un ensemble lithique: Application aux séries acheuléen d'Isenya (Kenya). In *25 ans d'études technologiques en préhistoire*. Juan-les-Pins: APDCA.

Roe, D.A. (1964). The British Lower and Middle Palaeolithic: Some problems, method of study, and preliminary results. *Proceedings of the Prehistoric Society 30*, 245–267.

Roe, D.A. (1968). British Lower and Middle Palaeolithic handaxe groups. *Proceedings of the Prehistoric Society, 34*, 1–82.

Roe, D.A. (1994). A metrical analysis of selected sets of handaxes and cleavers from Olduvai Gorge. In M.D. Leakey & D.A. Roe (Eds.), *Olduvai Gorge excavations in Beds III, IV, and the Masek Beds 1968–1971* (Vol. 5, pp. 146–234). Cambridge: Cambridge University Press.

Rolland, N. (2004). Was the emergence of home bases and domestic fire a punctuated event? A review of the Middle Pleistocene record in Eurasia. *Asian Perspectives, 43*, 248–280.

Rollefson, G.O. (1978). Quantitative and qualitative typological analysis of bifaces from the Tabun excavations, 1967–1972. Ph.D. Dissertation, University of Arizona, Tucson.

Rollefson, G.O., Quintero, L.E., & Wilke, P.J. (2005). The Acheulian industry in the al-Jafr Basin of southern Jordan. *Journal of the Israel Prehistoric Society, 35*, 53–68.

Rollefson, G.O., Quintero, L.A., & Wilke, P.J. (2006). Late Acheulian variability in the southern Levant: A contrast of the western and eastern margins of the Levantine Corridor. *Near Eastern Archaeology, 69*, 61–72.

Rose, L., & Marshall, F. (1996). Meat eating, hominid sociality, and home bases revisited. *Current Anthropology, 37*, 307–338.

Rosenberg, D., Shimelmitz, R., Gluhak, T.M., & Assaf, A. (2015). The geochemistry of basalt handaxes from the Lower Palaeolithic site of Ma'ayan Baruch, Israel: A perspective on raw material selection. *Archaeometry, 57*(Supplement S1), 1–19, doi: 10.1111/arcm.12096.

Rufo, M.A., Minelli, A., & Peretto, C. (2009). The limestone industry of the Paleolithic site of Isernia La Pineta: An interpretative model of the behavioural strategies. *L'Anthropologie, 113*, 78–95, doi: 10.1016/j.anthro.2009.01.001.

Sahnouni, M., Schick, K., & Toth, N. (1997). An experimental investigation into the nature of faceted limestone "spheroids" in the Early Palaeolithic. *Journal of Archaeological Science*, 24, 701–713., doi:

Sahnouni, M. (2006). The North African Early Stone Age and the sites at Ain Hanech, Algeria. In N. Toth & K. Schick (Eds.), *The Oldowan: Case studies into the earliest Stone Age* (Vol. 1, pp. 77–11). Gosport, IN: Stone Age Institute Press.

Santonja, M., & Villa, P. (1990). The Lower Paleolithic of Spain and Portugal. *Journal of World Prehistory, 4*(1), 45–94.

Santonja, M., & Villa, P. (2006). The Acheulian of western Europe. In N. Goren-Inbar & G. Sharon (Eds.), *Axe Age: Acheulian tool-making from quarry to discard* (pp. 429–478). London: Equinox.

Santonja, M., & Pérez-González, A. (2010). Mid-Pleistocene Acheulean industrial complex in the Iberian Peninsula. *Quaternary International, 223–224*, 154–161, doi: 10.1016/j.quaint.2010.02.010.

Santonja, M., Pérez-González, A., Panera, J., Rubio-Jara, S., & Méndez-Quintas, E. (2016). The coexistence of Acheulean and Ancient Middle Palaeolithic techno-complexes in the Middle Pleistocene of the Iberian Peninsula. *Quaternary International*, 411, Part B, 367-377, doi:http://doi.org/10.1016/j.quaint.2015.04.056.

Saragusti, I. (2002). Changes in the morphology of handaxes from Lower Paleolithic assemblages in Israel. Ph.D. Dissertation, Hebrew University of Jerusalem

Saragusti, I., & Goren-Inbar, N. (2001). The biface assemblage from Gesher Benot Ya'aqov, Israel: Illuminating patterns in "Out of Africa" dispersal. *Quaternary International 75*, 85–89.

Saragusti, I., Sharon, I., Katzenelson, O., & Avnir, D. (1998). Quantitative analysis of the symmetry of artefacts: Lower Paleolithic handaxes. *Journal of Archaeological Science, 25*, 817–825, doi: 10.1006/jasc.1997.0265.

Saragusti, I., Karasik, A., Sharon, I., & Smilansky, U. (2005). Quantitative analysis of shape attributes based on contours and section profiles in artifact analysis. *Journal of Archaeological Science, 32*, 841–853, doi: 10.1016/j.jas.2005.01.002.

Sawaguchi, T., & Kudo, H. (1990). Neocortical development and social structure in primates. *Primates, 31*, 283–289, doi: 10.1007/BF02380949.

Schattner, U., & Weinberger, R. (2008). A mid-Pleistocene deformation transition in the Hula basin, northern Israel: Implications for the tectonic evolution of the Dead Sea Fault. *Geochemistry Geophysics Geosystems, 9*, doi: 10.1029/2007GC001937.

Schick, K.D. (1986). *Stone Age sites in the making: Experiments in the formation and transformation of archaeological occurrences* (Vol. S319). Oxford: BAR International Series.

Schick, K.D. (2001). An examination of Kalambo Falls Acheulian site B5 from a geoarchaeological perspective. In J.D. Clark (Ed.), *Kalambo Falls prehistoric site: The earlier cultures: Middle and Earlier Stone Age* (pp. 463–480). Cambridge: Cambridge University Press.

Schur, N. (2002). *The history of the Golan*. Tel Aviv, Israel: Effi Meltzer (in Hebrew).

Sept, J., King, B., McGrew, W., Moore, J., Paterson, J., Strier, K., et al. (1992). Was there no place like home? A new perspective on early hominid archaeological sites from the mapping of chimpanzee nests. *Current Anthropology, 33*, 187–207.

Sharon, G. (2006). Acheulian large flake industries: Technology, chronology, distribution and significance. Ph.D. Dissertation, Hebrew University of Jerusalem.

Sharon, G. (2007). *Acheulian large flake industries: Technology, chronology, and significance* (Vol. 1701). Oxford: BAR International Series.

Sharon, G. (2008). The impact of raw material on Acheulian large flake production. *Journal of Archaeological Science, 35*, 1329–1344, doi: 10.1016/j.jas.2007.09.004.

Sharon, G. (2009). Acheulian giant cores technology: A worldwide perspective. *Current Anthropology, 50*, 335–367, doi: 10.1086/598849.

Sharon, G. (2011). New Acheulian locality north of Gesher Benot Ya'aqov: Contribution for the study of the Levantine Acheulian. In J.-M.L. Tensorer, R. Jagher, & M. Otte (Eds.), *The Lower and Middle Palaeolithic in the Middle East and neighboring regions* (pp. 25–33). Basel: ERAUL.

Sharon, G., & Barsky, D. (2016). The emergence of the Acheulian in Europe – A look from the east. *Quaternary International*, 411, Part B, 25-33, doi:http://doi.org/10.1016/j.quaint.2015.11.108.

Sharon, G., & Beaumont, P. (2006). Victoria West: a highly standardized prepared core technology. In N. Goren-Inbar, & G. Sharon (Eds.), *Axe Age: Acheulian tool-making from quarry to discard* (pp. 181-199). London: Equinox.

Sharon, G., & Goren-Inbar, N. (1999). Soft percussor use at the Gesher Benot Ya'aqov Acheulian site? *Mitekufat Haeven, 28*, 55–79.

Sharon, G., & Goring-Morris, N. (2004). Knives, bifaces, and hammers: A study in technology from the southern Levant. *Eurasian Prehistory, 2*, 53–76.

Sharon, G., Feibel, C., Belitzky, S., Marder, O., Khalaily, H.M., & Rabinovich, R. (2002). 1999 Jordan River drainage project damages Gesher Benot Ya'aqov: A preliminary study of the archaeological and geological implications. In Z. Gal (Ed.), *Eretz Zafon: Studies in Galilean archaeology* (pp. 1*–19*). Jerusalem: Israel Antiquities Authority.

Sharon, G., Feibel, C., Alperson-Afil, N., Harlavan, Y., Feraud, G., Ashkenazi, S., et al. (2010). New evidence for the northern Dead Sea Rift Acheulian. *PaleoAnthropology*, 79–99, doi: 10.4207/PA.2010.ART35.

Sharon, G., Alperson-Afil, N., & Goren-Inbar, N. (2011). Cultural conservatism against variability in the Acheulian sequence of Gesher Benot Ya'aqov, Israel. *Journal of Human Evolution, 60*, 387–397, doi: 10.1016/j.jhevol.2009.11.012.

Shimelmitz, R., Bisson, M., Weinstein-Evron, M., & Kuhn, S.L. (2016). Handaxe manufacture and re-sharpening throughout the Lower Paleolithic sequence of Tabun Cave. *Quaternary International*, doi: 10.1016/j.quaint.2015.12.076.

Shipton, C., Parton, A., Breeze, P., Jennings, R., Groucutt, H.S., White, T.S., et al. (2014). Large flake Acheulean in the Nefud Desert of northern Arabia. *PaleoAnthropology*, 446–462, doi: 10.4207/PA.2014.ART85.

Shoshani, J., Goren-Inbar, N., & Rabinovich, R. (2001). A stylohyoideum of *Palaeoloxodon antiquus* from Gesher Benot Ya'aqov, Israel: Morphology and functional inferences. In *The World of Elephants – Proceedings of the 1st International Congress*, Rome, 2001 (pp. 665–667).

Shultz, S., & Dunbar, R.I.M. (2010). Species differences in executive function correlate with hippocampus volume and neocortex ratio across non-human primates. *Journal of Comparative Psychology, 124*, 252–260, doi: 10.1037/a0018894010.

Simmons, T. (2004). "A feather for each wind that blows": Utilizing avifauna in assessing changing patterns in paleoecology and subsistence at Jordan Valley archaeological sites. In N. Goren-Inbar & J.D. Speth (Eds.), *Human paleoecology in the Levantine Corridor* (pp. 191–205). Oxford: Oxbow Books.

Siret, T. (1933). Le coup de burin Moustérien. *Bulletin de la Société Préhistorique Français*, 120–127.

Slimak, L., Kuhn, S.L., Roche, H., Mouralis, D., Buitenhuis, H., Balkan-Atlı, N., et al. (2008). Kaletepe Deresi 3 (Turkey): Archaeological evidence for early human settlement in Central Anatolia. *Journal of Human Evolution, 54*, 99–111, doi: 10.1016/j.jhevol.2007.07.004.

Solodenko, N. (2010). On tools and elephants: An analysis of a lithic assemblage from Area B of the Late Acheulian Site Revadim Quarry. M.A. Thesis, Tel Aviv University (in Hebrew).

Stekelis, M. (1960). The Paleolithic deposits of Jisr Banat Yaqub. *Bulletin of the Research Council of Israel, G9*, 61–87.

Stekelis, M. (1966). *Archaeological excavations at Ubeidiya, 1960–1963*. Jerusalem: Israel Academy of Sciences and Humanities.

Stekelis, M., & Gilead, D. (1967). Ma'ayan Barukh: A Lower Palaeolithic site in Upper Galilee. *Mitekufat Haeven, 8*, 1–22.

Stekelis, M., & Picard, L. (1936). Jisr Banat Yaqub. *Quarterly of the Department of Antiquities in Palestine, 6*, 214–215.

Stekelis, M., Picard, L., & Bate, D.M.A. (1937). Jisr Banat Ya'qub. *Quarterly of the Department of Antiquities in Palestine, 6*, 214–215.

Stout, D., Apel, J., Commander, J., & Roberts, M. (2014). Late Acheulean technology and cognition at Boxgrove, UK. *Journal of Archaeological Science, 41*, 576–590, doi: 10.1016/j.jas.2013.10.001.

Tchernov, E. (1973). On the Pleistocene molluscs of the Jordan Valley. *Proceedings, Israel Academy of Sciences and Humanities, 11*, 1–46.

Tchernov, E. (1986). *Les mammifères du pleistocène inférieur de la Vallée du Jourdain à Oubeidiyeh* (Mémoires et Travaux du Centre de Recherche Français de Jérusalem). Paris: Association Paléorient.

Tchernov, E. (1996). Rodent faunas, chronostratigraphy and palaeobiogeography of the southern Levant during the Quaternary. *Acta Zoologica Cracoviensia, 39*, 513–530.

Tchernov, E., & Yom-Tov, Y. (1988). Zoogeography of Israel. In Y. Yom-Tov & E. Tchernov (Eds.), *The zoogeography of Israel: The distribution and abundance at a zoogeographical crossroad* (pp. 1–6). Dordrecht: Dr. W. Junk Publishers.

Tensorer, J.-M.L., Falkenstein, V. von, Tensorer, H.L., Schmid, P., & Muhesen, S. (2011). Etude préliminaire des industries archaïques de faciès oldowayen du site de Hummal (El Kowm, Syrie centrale). *Anthropologie, 115*, 247–266, doi: 10.1016/j.anthro.2011.02.006.

Tepper, Y., & Tepper, Y. (2003). The "Horses Barid" dated to the era of Sulten Mamluke Baybars. *Jerusalem and Eretz-Israel, 1*, 123–152 (in Hebrew).

Tixier, J. (1957). Le hachereau dans l'Acheuléen nord-africain: Notes typologiques. In *Congrès préhistorique de France: Compte rendu de la XVe session, Poitiers-Angoulême, 1956* (pp. 914–923). Paris: Société Préhistorique Française.

Tomasello, M., & Call, J. (1998). *Primate cognition*. New York: Academic Press.

Torre, I. de la (2011). The Early Stone Age lithic assemblages of Gadeb (Ethiopia) and the Developed Oldowan/Early Acheulean in East Africa. *Journal of Human Evolution, 60*, 768–812, doi: 10.1016/j.jhevol.2011.01.009.

Tryon, C.A., McBrearty, S., & Texier, P.J. (2005). Levallois lithic technology from the Kapthurin formation, Kenya: Acheulian origin and Middle Stone Age diversity. *African Archaeological Review, 22*, 199–229.

Twomey, T. (2013). The cognitive implications of controlled use of fire. *Cambridge Archaeological Journal, 23*, 113–128, doi: 10.1017/S0959774313000085.

Wadley, L. (2013). Recognizing complex cognition through innovative technology in stone age and palaeolithic sites. *Cambridge Archaeological Journal*, 23, 163-183, doi:10.1017/S0959774313000309.

Walker, S.J.H., Lukich, V., & Chazan, M. (2014). Kathu Townlands: A high density Earlier Stone Age locality in the interior of South Africa. *PlosONE, 9*, e103436, doi: 10.1371/journal.pone.0103436.

Weinstein, Y., Navon, O., Altherr, R., & Stein, M. (2006). The Role of Lithospheric Mantle Heterogeneity in the Generation of Plio-Pleistocene Alkali Basaltic Suites from NW Harrat Ash Shaam (Israel). *Journal of Petrology, 47*, 1017-1050.

Whalen, N., Sindi, H., Wahida, G., & Siraj-Ali, J.S. (1983). 1 – Excavation of Acheulean sites near Saffaqaj in ad-Dawadami 1402-1982. *Atlal: The Journal of Saudi Arabian Archaeology, 7*, 9–16.

Whalen, N., Siraj-Ali, J.S., & Davis, W. (1984). 1 – Excavation of Acheulean sites near Saffaqaj, Saudi Arabia, 1403 A H 1983. *Atlal: The Journal of Saudi Arabian Archaeology, 8*, 9–17.

Whalen, N., Ali, J.S., Sindi, H.O., & Pease, D.W. (1986). A Lower Pleistocene site near Shuwayhitiyah in northern Saudi Arabia. *Atlal: The Journal of Saudi Arabian Archaeology, 10*, 94–101.

White, M. (2006). Axeing cleavers: Reflections on broad-tipped large cutting tools in the British earlier Paleolithic. In N. Goren-Inbar & G. Sharon (Eds.), *Axe Age: Acheulian tool-making from quarry to discard* (pp. 365–386). London: Equinox.

White, M.J., & Ashton, N.M. (2003). Lower Palaeolithic core technology and the origins of the Levallois method in north-western Europe. *Current Anthropology, 44*, 598–609, doi: 10.1086/377653.

Whittaker, J. (1994). *Flintknapping: Making and understanding stone tools*. Austin: University of Texas Press.

Willoughby, P. (1985). Spheroids and battered stones in the African Early Stone Age. *World Archaeology, 17*, 44–60, doi: 10.1080/00438243.1985.9979949.

Wynn, T. (2002). Archaeology and cognitive evolution. *Behavioral and Brain Sciences, 25*, 389–402, doi: 10.1017/S0140525X02000079

Wynn, T., & Coolidge, F. (2004). The expert Neandertal mind. *Journal of Human Evolution, 46*, 467–487.

Wynn, T., & Coolidge, F.L. (2011). The implications of the Working Memory Model for the evolution of modern cognition *International Journal of Evolutionary Biology, 2011*, 1–12, doi: 10.4061/2011/741357.

Yom-Tov, Y., & Mendelssohn, H. (1988). Changes in the distribution and abundance of vertebrates in Israel during the 20th century. In Y. Yom-Tov & E. Tchernov (Eds.), *The zoogeography of Israel: The distribution and abundance at a zoogeographical crossroad* (pp. 515–547). Dordrecht: Dr. W. Junk Publishers.

Zaidner, Y. (2010). The lithic assemblage of Bitzat Ruhama: Lower Palaeolithic site in southern coastal plain, Israel. Ph.D. Dissertation, University of Haifa.

Zaidner, Y., & Weinstein-Evron, M. (2016). The end of the Lower Paleolithic in the Levant: The Acheulo-Yabrudian lithic technology at Misliya Cave, Israel. *Quaternary International*.

Zaidner, Y., Yeshurun, R., & Mallol, C. (2010). Early Pleistocene hominins outside of Africa: Recent excavations at Bizat Ruhama, Israel. PaleoAnthropology, 162–195, doi: doi:10.4207/PA.2010.ART38.

Zohar, I. (1993). A new approach to the analysis of cleavers: A topographical perspective. *Mitekufat Haeven, 25*, 103–119.

Zohar, I., & Biton, R. (2011). Land, lake, and fish: Investigation of fish remains from Gesher Benot Ya'aqov (paleo-Lake Hula). *Journal of Human Evolution, 60*, 343–356, doi: 10.1016/j.jhevol.2010.10.007.

Zohar, I., Goren, M., & Goren-Inbar, N. (2014). Fish and ancient lakes in the Dead Sea Rift: The use of fish remains to reconstruct the ichthyofauna of paleo-Lake Hula. *Palaeogeography, Palaeoclimatology, Palaeoecology, 405*, 28–41, doi: 10.1016/j.palaeo.2014.04.006.

Zohar, I., Ovadia, A., & Goren-Inbar, N. (2016). The cooked and the raw: A taphonomic study of cooked and burned fish. *Journal of Archaeological Science: Reports, 8*, 164–172, doi.org/10.1016/j.jasrep.2016.06.005.

Zohary, M. (1959). *Geobotany*. Merhavia: Sifriat Hapoalim (in Hebrew).

Zohary, M. (1960). The maquis of *Quercus calliprinos* in Israel and Jordan. *Bulletin of the Research Council of Israel, 9 D*, 51–72.

Zohary, M. (1962). *Plant life of Palestine: Israel and Jordan*. New York: Roland Press Company.

Zohary, M. (1966). *Flora Palaestina* (Vol. 1). Jerusalem: Israel Academy of Sciences and Humanities.

Zohary, M., & Orshansky, G. (1947). The vegetation of the Huleh Plain. *Palestine Journal of Botany (Jerusalem Series), 4*, 90–104.

Appendix 1

The Hula Reclamation Project and Its Implications for the Gesher Benot Ya'aqov Site

Tamar Nahmias-Lotan

This appendix includes three parts, which are concerned with 1) the history of research of the Hula Valley from the 19th to early 20th centuries, 2) the main drainage plans and works in the Hula Valley between 1837 and 2000 and 3) the relevant references. The data presented below are relevant to the history of the site, as all the developments that have taken place in this region have affected and damaged the Acheulian site of Gesher Benot Ya'aqov.

1. History of Research on the Hula Valley from the 19th to Early 20th Centuries

Most of the information documented up to the beginning of the 20th century derives from expedition and travel reports of researchers and surveyors (travelers and pilgrims to the Holy Land). Their impressions and scientific studies include data on the geography of the landscape and its inhabitants and a few reports on fauna and flora.

Starting at the end of the 19th century and continuing into the 1940s, numerous reports and studies were published, including taxonomic identifications of fauna and flora as well as abundant geophysical and some geological descriptions. During this period many observations of the lake were carried out; they were mainly biological in nature, but a variety of measurements, including a few chemical and physical ones, were also taken (Goren 1999; Gophen 2000).

Northern Galilee was infrequently traveled by Westerners at the beginning of this period. Few traveled north from Lake Kinneret toward Damascus and crossed the bridge of Benot Ya'aqov, which at the time was an impressive stone structure and an important crossing point. This is one of the bridges on the Jordan River between the Galilee and the Golan Heights, situated about 15 km north of Lake Kinneret and east of Rosh Pinna at an elevation of 70 m above modern sea level.

The bridge is situated along the Via Maris, the ancient road connecting Egypt and Syria (Karmon 1973:179). During the 12th century CE, the nearby convent of St. James, located in Safed, gained its main income from fees paid by those crossing the bridge. Hence the bridge's current name: the nuns (daughters = *benot* in Hebrew) of the convent of St. James (Jacob/Ya'aqov in Hebrew). It is otherwise known as *Filiarum Pontus Iacobis* (Latin), Jisr Banat Yaqub (Arabic), Pont de Jacob (French), Bridge of Jacob's Daughters (English) and Brucke am Jordan (German) (Karmon 1973:179).

In 1806 Seetzen reported that crossing the bridge was impossible from the eastern side, and walked along the western bank of the Jordan River. He reported that the basalt stone bridge was in good condition and that the river was 35 steps wide. He describes the vegetation as wild and the area as melancholic, a similar description to that given by Buchhardt in 1812 (Goren 2001:60). Johannes Rudolph Roth surveyed the area several times during his expedition with Von Schubert in 1836–1837, taking measurements of elevations along the Jordan Valley. This expedition was the first to note that the level of the Dead Sea was lower than that of the Mediterranean (Goren 1994:192, 1999:107). In 1843 Russegger surveyed the area, taking measurements using a mercury barometer. The elevation of Gesher Benot Ya'aqov was given as 123 m and that of Lake Kinneret as 203 m below sea level (Goren 1999:219). The first map of the area, published in 1846, shows the lake's shape along with the layout of the Jordan River's sources. Drawn by the English officer Frederick Holt Robb, the map does not include elevations; however, the topographic features are shaded, giving a clear impression of the area. In 1846 Louis von Wildenbruch published elevation measurements between Gesher Benot Ya'aqov and Tel Dan (Goren 1999:157, 2001:61).

From the second half of the 19th century, increasing numbers of Europeans traveled in the area. In 1846 the Rev. W.M. Thomson described the Hula Valley as:

"a plain broken by wild downs, covered with countless herds of cattle – chiefly bulls and buffaloes. The uppermost lake of the Jordan – about seven miles long and in its greatest width six miles broad, the mountains slightly compressing it at either extremity, surrounded by an impenetrable jungle of weeds abounding in wild fowl. This lake, now called Huleh, in old times bore the name of Merom, and afterwards Samachon, both probably from its upland situation… I asked an Arab if I could not reach the lake through the swamp. He regarded me with surprise for some time, as if to ascertain whether I was in earnest, and then, lifting his hand, swore by the Almighty, the Great, that not even a wild boar could get through" (Stanley 1877:390–391).

Francis Lindt's expedition in 1848–1852 produced a relatively accurate account of Lake Kinneret, the Hula Valley, and the sources of the Jordan River. Additional scholars who studied the area are F. Linch (visiting in 1849), H.B. Tristam (visiting in 1856), and V. Guérin (visiting in 1875) (Gophen 2000:12). The first to note that the area was hazardous to the health was the French researcher De Saulcy (visiting in 1860). His report also includes a detailed description of the fauna and flora. In 1853, Arthur Stanley surveyed the Jordan River sources (Goren 2001:62).

Of special interest is the report of John MacGregor (1869), who carried out the first modern mapping and was the first to suggest the idea of draining Lake Hula (Dimentman et al. 1992:9, Figs. 1, 2; Gophen 2000:12). In his own words, "I think that by cutting 400 yards long, and twenty feet deep, at the end of the Hooleh Lake, the whole of the marsh and lake would be made dry in a year, and an enormous tract of land would become productive and salubrious" (MacGregor 1869:305). He also describes in detail the Benot Ya'aqov Bridge: "bridge of black basalt… from the end of the lake this bridge was distant 650 paddle strokes, that is 3523 yards, or three yards over two miles" (MacGregor 1869:306).

> "The bridge is about 60 feet long, has three arches and no parapet. At the west end is an ugly tower, and a khan is over the river. The current is very trifling until quite close to the bridge… The bridge itself has been most likely built since the Crusades but the spot selected suggest that a ford was here, for it is just where the deep water ends, and before the high banks of the torrent begin… The channel soon becomes a mere torrent bed, wherein a white-foamed bursting of water hurries between rock thick set with oleanders, which often meet across the stream not a dozen feet in width. Before the river settles down into a thorough-going mill-race speed, it takes a sweep or two to right and left and its stream is divided by islands and large rocks. About a mile below the bridge are several imposing ruins… The distance between the Hooleh and the Gennesareth lake is not more than ten miles in a straight line, and the river has only a few long bends between them, which probably add not more than three miles to its course… yet the descent is very rapid here for it falls in these ten miles 700 feet" (MacGregor 1869:309).

In 1880 a research project carried out by the British Corps of Engineers was published. This survey was carried out for the Palestine Exploration Fund in 1871–1887 and includes detailed maps, sites, hydrology, fauna, flora, and more (Goren 2001:66).

In 1906, Treidel conducted research on the Hula Valley and its surroundings. This is a thorough report, covering the hydrology, topography, climate, population, petrography, and irrigation issues of the area (Treidel 1910).

2. The Drainage of the Hula Valley: Main Events and Impact

The Jordan River exits Lake Hula in the south, bordered by basalt rocks on the east and limestone on the west. The river's slope is gentle and the flow is slow due to the clogging caused by the basalt bedrock. South of the bridge, the slope and flow increase significantly. Consequently, any solution for the inundation of Lake Hula and the formation of swamps had to consider deepening the course of the river south of the lake as the first step (Karmon 1956:29).

Several drainage projects have been carried out in the Hula Valley, and additional plans were proposed but were not carried out for various reasons. All solutions were based on the assumption that the erosion around the Benot Ya'aqov Bridge had a "dam effect" that clogged and slowed the water flow to such an extent that the valley to the north was inundated (Zitrin 1987:32). The general solution would have to include deepening and widening the southern Jordan River along a 4.5 km stretch up to the Benot Ya'aqov Bridge (Avurmed 1983:22).

The drainage proposals can be divided into three major planning strategies:

A. *The southern solution*

Topographical and hydrological measurements (all cited in Ben Porat and Mintzker 1996) were carried out by engineers from Britain (Conder and Kitchener 1877), France (Le Clerce 1908), and Germany (Zoller and Weindner 1916) were the basis for the various drainage plans. All these had a common denominator – the deepening of the southern Jordan River from the lake's exit to the Crusader fortress of *Vadum Iacob* (French *Le Chastellet*, Hebrew Meitzad 'Ateret) – while ignoring the problem of the larger northern swamp (Ben Porat 2001:1–2; Ben Porat and Minzker 1996:44).

Between 1837 and 1887, under Ottoman rule, the southern Jordan River was deepened four times, the results of the work lasting for several years each time until erosion filled and blocked the river bed once again. The river flow capacity was 31 m^3/sec and the assumed level of the rock base at the Benot Ya'aqov Bridge was 69.5 m.

1837: Drainage of the southern Jordan River, exposing acres of swamp land in the southern basin.

1857: An American missionary reports "effendis" blowing up rocks and boulders in the southern Jordan River.

1861–1863: Drainage works in the southern Jordan River. Cotton was planted on the newly exposed lands, adjacent to the swamp.

1887: An Ottoman attempt at draining and reclamation of the Hula Basin. Turkish engineers concluded that the accumulated sediments in the vicinity of the Benot Ya'aqov Bridge were preventing the water from flowing freely, thus flooding the valley. A fourth arch was added to the Benot Ya'aqov Bridge, and the deepening of the river improved its flow capacity to 39 m^3/sec. Sediments were removed, lowering the water level of the lake by 1 m and exposing thousands of fertile acres around its southern end. Ancient irrigation channels testifying to agricultural activities in the valley were uncovered. Eventually, this initiative failed, as malaria outbreaks recurred due to the remaining areas of swamps (Karmon 1956:70; Zitrin 1987; Ben Porat and Minzker 1996:44).

1900: A French company, Société Agricole de Houlea, finances the deepening of the Jordan River, increasing the river's capacity from 26 to 38 m^3/sec (Ben Porat 2001).

1901–1903: Sultan Abdul Hamid II decided to turn the area between the southern end of Lake Hula and Lake Kinneret into the "Hula Concession". The intention was to transfer the area to an economic entity for development on the basis of a financial agreement. The concession was first granted to the Jewish settlers of Yesud Hama'alah, who decided to drain 3000 dunams of swamp adjacent to their settlement and solve the problem of malaria. Their intention was to deepen the Jordan River and drain the swamp by means of broad channels. With the aid of the JCA (Jewish Colonization Association), the channels were dug and fertile lands were exposed. This project came to a halt, and eventually failed, due to interference by Arab neighbors (Avurmed 1983:8).

1904–1905: Another Ottoman project was an attempt to lower the level of the water table by deepening the course of the Jordan River north of the swamps. The basic problem was not solved and the concession was given to a private company.

1906: Treidel was the first to determine the basic problem of the drainage of Lake Hula. In his report he asserted that the problem lay in the difference in the flow capacities of the water sources north and south of the lake. According to his measurements, the lake received water from its various origins at a rate of 52 m^3/sec, whilst the maximum flow capacity at the southern exit was 38 m^3/sec. The difference of 14 m^3/sec was the reason for the lake's existence. Treidel also noted that the Jordan River at the exit from the lake was 40 m wide and 1.7 m deep, with a flow velocity of 0.54 m^3/sec. South of the Benot Ya'aqov Bridge, the flow velocity reached 2.3 m^3/sec. He concluded that there was an urgent need to replace the bridge in order to carry out large-scale changes in the valley. This could be done only by the Hula concessionaries (Treidel 1910).

1908: The concession was given to a French company for five years, for the exploitation of papyrus reeds. A French engineer (Le Clerce) planned the deepening of the southern Jordan River and the diversion of the Dishon and Hazor streams south of the lake in order to improve the southern drainage of the Hula (Ben Porat and Minzker 1996:44; Ben Porat 2001:2).

1910: The concession was granted to two Syrian families with the obligation to drain the lake and swamps (Avurmed 1983).

1914: The concession was granted to the Syro-Ottoman Agricultural Society. Minor actions for the deepening of the river were carried out.

1916: Zoller and Weindner, German engineers, took measurements of the lake and swamps. Their report is missing, but their plans for the southern Jordan River and the replacement of the Benot Ya'aqov Bridge have survived. They recommended replacing the stone bridge with a suspension bridge (Ben Porat and Minzker 1996:51) and diverting the Dishon, Hazor, and Gilboa streams (the water sources of the southern Lake Hula) to the deepened southern Jordan River (Ben Porat 2001).

1918: British soldiers arrived at the bridge during their conquest of the region in World War I. The retreating Turkish soldiers blew up the bridge, which was subsequently rebuilt.

1919: The Cantor Report, compiled by a French engineer, proposed the deepening of the southern Jordan River along 3.4 km. This would achieve a gradient of 0.05% and a profile that would be able to transport 12 and 60 m^3/sec (in the lower level and the flooded level, respectively) (Ben Porat and Minzker 1996:44)

B. *The bypass solution*

1920: A bypass solution was suggested by Rothenberg and later by Werber (1934, cited in Ben Porat and Minzker 1996:44-45).

1921: Rothenberg was granted the Electricity Concession and the Jordan River Concession. According to the Rothenberg and Ettinger Plan (submitted in 1920 to the World Zionist Organization and the British Government), an Eastern Hula Bypass was proposed for draining the swamp and the future building of a hydroelectric station in the lake near *Vadum Iacob* ('Ateret Fortress) for the combined purpose of irrigation, electrical supply and drainage. Rothenberg planned a diversion channel dug east of the Hula Valley in the slopes of the Golan Heights, in which the water from the Banias and Dan Rivers would flow. This plan is the basis for the international border between Israel and Syria (Ben Porat 2001:4).

1929: The Syro-Ottoman company was granted the Hula Concession for eight years (until 1937) with the condition that drainage works be carried out immediately (Avurmed 1983:11; Ben Porat and Minzker 1996:45).

1930: The Benot Ya'aqov Bridge was replaced by a suspension bridge and the Jordan River was deepened in this area, causing a substantial fall of 1.5 m in the level of Lake Hula (from 70 m above msl to 68.5 m above msl). The river's flow capacity reached 46 m³/sec. During these works, prehistoric remains were revealed (Ben Porat 2001:5).

Several attempts were made by the Jewish settlers to gain the Hula Concession, finally succeeding in 1934, when it was handed over to the Israel Land Development Company (Hachsharat HaYeshuv).

1935: A detailed plan was prepared, based on a hydro-climatic survey of the basin. Amongst the conclusions was the need to redefine the water rights of the Electric Company and the Hula Concession. In order to avoid a legal confrontation with the Electric Company, a compromise (partial drainage) was suggested. This would involve the deepening of the southern Jordan River by an additional 2 m (from 68.5 m above msl to 66.5 m above msl), which would cause the lake's contraction to an area of 5000–7000 dunams at 67 m above msl. This suggestion was rejected, as partial drainage would not solve the problem of malaria in the area.

1936: The R.P.T. (Rendel, Palmer and Triton) Engineering Company proposed a detailed plan for the drainage and irrigation of the Hula Basin, based on the deepening of the Hazbani and Jordan Rivers and the straightening of the river's bends. The width of the suggested riverbed was 40 m, in order to allow a record flow capacity of 320 m³/sec. The lake's highest water level was 70.73 m above msl in 1935, two meters lower than in 1929, due to the replacement of the Benot Ya'aqov Bridge with a Bailey bridge (Karmon 1956:28). Two main drainage channels were suggested, in order to allow winter flows channeling directly to the Jordan south of the peat area. This plan predicted the subsidence of the surface: the "transit peat" would contract by 0.5 m and the deep peat by 1 m. The ground water of the swamp was planned to keep the deep peat permanently moist (Ben Porat and Minzker 1996:46).

The R.P.T. survey and its conclusions revealed that the Hula Valley Drainage project was not economically feasible. This was due to the Concession's condition that exposed lands suitable for agriculture were to be handed over to local farmers.

C. A combined solution: deepening in the south and regulating the Jordan's delta to the north

1937: Werber proposed alternatives: total drainage of the basin and swamps and the exploitation of the reeds (Ben Porat and Minzker 1996:47).

1942: Kovlanov proposed an overall plan for irrigation and drainage of the northern Hula Valley. This plan included a combined solution for the entire valley and total drainage of the lake (Ben Porat and Minzker 1996:47). The plan was to turn the valley into a major site for settlements. The need to absorb the Jewish immigration in the 1930s and the lack of land reserves was the motive for the acquisition of the Hula Concession. Between 1935–1947 a comprehensive hydro-climatic study of the valley was carried out. The term "drainage" referred mainly to the swamp area, as the Palestine Electricity Company was opposed to any action in the lake itself. This research pointed to the limited agricultural potential of the soils – both peat and lacustrine (limnic) sediments – and the marginal value of the peat.

1945: Werber concluded that, as a result of the basin's drainage, Lake Kinneret's water flow capacity would increase by 100 million m³. Therefore, there was no basis for the claim of the Palestine Electric Company that the total drainage of Lake Hula would damage the water capacity of Lake Kinneret. Werber's calculation of the water evaporation from Lake Hula over 20 years (100 million m³ per year) was one of the main arguments in favor of total drainage. In 1965, the engineers of TAHAL (Water Planning for Israel) estimated the water vaporization at 50 million m³ per year – 50% of the original estimate (Ben Porat 2001:6–7).

Prior to the establishment of the State of Israel, the Benot Ya'aqov Bridge was demolished on June 17th 1946 during the military operation of the Hagana known as *Leil HaGsharim* ("The Night of the Bridges"), which involved the demolition of several major bridges. It was attacked again on January 11th 1948, resulting in severe damage.

Budgetary problems and two world wars postponed the drainage of the valley, which was eventually carried out after the State of Israel was established. The final cost was 20 times higher than the original budget, with delays in execution throughout.

1948: The "First Hula Committee" commissioned plans for drainage of the Hula from the engineers Kovlanov and Werber. The Committee recommended a detailed plan for the deepening of the southern Jordan River, taking into consideration the emptying of Lake Hula.

1949: Due to lack of basic information regarding the exploitation of the peat, the Committee recommends preserving the peat as a land reserve of 8500 dunams ("the reserved peat"), which would remain waterlogged for future use.

1949: Picard planned 50 m deep drillings in the peat in order to check its potential economical feasibility (Ben Porat 2001:9).

1949: Kovlanov prepared plans for the east and west channels (Ben Porat 2001:10). The Committee accepted Kovlanov's plans, which represented an update of those of 1942. The alterations to the original plan were carried out with total disregard of the recommendations of researchers and experts, as well as the geopolitical changes (Ben Porat and Minzker

Appendix 1: The Hula Reclamation Project and Its Implications for the Gesher Benot Ya'aqov Site

1996:42). During the mid 1950s the Hula settlement plan was for 17 new settlements on land that was already subsiding as a result of the drainage.

The Kovlanov plan was based on deepening the depression at the valley's southern end, thus creating a route for carrying and draining water. The southern tip of the lake was used for temporary accumulation of flood waters while the deep peat area ("the reserved peat") was waterlogged.

1950: The Hula Concession was passed to HaKeren HaKayemet LeIsrael (KKL). Eventually, the swamp and lake were totally drained by KKL between 1951 and 1958.

The first stage of works began with the outlining of the channel north of the basin and the construction of a regulating dam in the southern Jordan which would enable the deepening of the original river bed. The first step planned was the widening and deepening of the southern Jordan River, thus creating a river channel whose bed would be 10 m wide, along 4.5 km. This channel would enable a flow of 80 m³/sec. This flow velocity was appropriate for absorbing flood water. The current channel enabled a flow of 20 m³/sec (CZA/KKL/1951).

"The work that Keren Kayemet is planning to begin in the near future is only part of a bigger plan. The goal is to deepen the Jordan River from the southern tip of Lake Hula up to a point situated about 1 kilometer south of the Benot Ya'aqov Bridge, during which an estimated half a million cubic meters of soil and stones will be excavated. This part alone consists of 15–20% of the drainage works" (CZA/Brachyahu/1951a).

According to the contractors, rocks alone comprised a volume of 100,000 m³. The duration of the excavations was expected to be at least nine months (CZA/Diskin/1951).

1951: Work started at the Benot Ya'aqov Bridge. Bedrock was reached. Excavation outlines were marked on both sides of the river. The transfer of heavy equipment to the eastern bed was denied by the Syrians, and excavations began adjacent to the bridge, working northward (Ben Porat 2001:10).

The Israel Department of Antiquities addressed KKL with regard to the archaeological remains around the Benot Ya'aqov Bridge that might be damaged during excavation work in the river, and an acceptable solution for the conservation of these finds was reached (CZA/Aharoni/1951). It was agreed that KKL would cover the expenses involved in "scientific examination" of the site, and that the area around the ancient Benot Ya'aqov Bridge would remain untouched (CZA/Yeivin/1951).

Stekelis, who had directed archaeological excavations near the bridge, was granted permission to enter no-man's-land for this purpose (CZA/Brachyahu/1951b).

The work schedule was reassessed, the objective being to achieve a free-flowing current along 4.1 km from Lake Hula's southern exit on schedule, before the approaching rainy season. The goal was to deepen the river just enough to lower the water level of the lake, though not as much as originally planned (CZA/Brachyahu/1951).

1951–1952: The first stage consisted mainly of the removal of basalt rocks blocking the river flow around the Benot Ya'aqov Bridge (CZA/KKL/1953). This work was carried out in difficult conditions along 4.5 km of the Jordan River, which was straightened (Dimentman et al. 1992:116), ending 1.48 km south of Gesher Benot Ya'aqov. Since this entire area was no-man's-land, the excavations were restricted to the right bank. The first stage of the project ended in 1953, but due to political, security and technical problems was only partially concluded. However, the deepening and widening of the river enabled the partial drainage of Lake Hula. The river's capacity reached 50–60 m³/sec. Overall, 300,000 m³ were excavated (CZA/KKL/1962:2).

The lake had receded to an elevation of 66.5 m above msl and to half its original area. Kovlanov planned the second stage (drying the swamps), first defining the flooded areas and the "reserved peat" area of 8500 dunams, surrounded by channels and levees. The second stage was divided into two stages, land and lacustrine excavation, and took longer than anticipated. With the deepening of the Jordan River's course, the river's water level was 6 m lower than that of the lake, a level that would enable the total drainage of the lake.

Excavation in the lake and swamps was carried out for two years with the assistance of two floating dredgers. The main work was on the east channel. In order to regulate the water level in the lake and prevent sweeping, two steel blockages were built, the southern one known as the Pkak Bridge (*Gesher HaPkak*). The lake was retained for another two and a half years.

1955: Against the advice of experts, who warned against drying the peat lands and the consequent subsidence of the surface soil level, a decision was made to dry out the deep peat lands for agricultural use. This decision was taken by KKL alone.

1957: Despite the official completion of the Hula Reclamation Project, the lake refused to dry up, remaining in a limited area of 8 km² as a consequence of sediment accumulation in the east channel. The channel was planned to reach a depth of 3–4 m and a width of 7–10 m at the bottom and 30–50 m at the top. Mud accumulated on the embankment situated a mere 65 m from the channel and its erosion filled the excavations. Eventually, in the course of two years, the channels were clogged.

1958: Accumulated sediments were cleared from all channels (40 km), and as a result the lake was entirely drained. Kovlanov's plan had been partially carried out. The drainage problems of the center of the Hula Basin, however, remained unsolved.

1959: Renewed excavations were carried out in the Jordan River, south of Lake Hula. The river was to be deepened, according to the plan, in order to enable a future water flow velocity of 265–300 m³/sec.

"The work intended is to deepen the river at its narrowest point – kilometer 2,080, 300 m north of the Benot Ya'aqov Bridge. If the work succeeds, the current velocity will reach 100–120 cubic meters per second. In order to allow a stronger current without further deepening the river, it is suggested to build higher embankments along the river" (CZA/Hammer/1959).

Steep slopes were arranged in order to prevent erosion and to ease future maintenance work. These works were carried out, as in the past, only on the west bank, while the east bank continued to collapse and clog the river.

The height of the west bank was 27 m. The excavation included basalt and other rocks that had to be transferred hundreds of meters away from the area. The estimate for the quarried material was over one million cubic meters.

An additional stage, which began in 1952 and ended in 1961, was the excavation of channels along 80 km of the eastern border in no-man's-land. The purpose was to transport all flows for irrigation (CZA/KKL/1962:3; Dimentman et al. 1992:118).

1960: Two committees, one from KKL and the other from the Water Authority (the "Doron Committee"), examined the drainage status of the Hula Basin. Prior to the drainage of the lake, it was assumed that the calcareous soil was inferior in quality and it was therefore designated for the collection of floodwater, thus avoiding the difficult excavation of the bedrock of the southern river. Since the soil later proved to be fertile, it was agreed that the main reservoir of floodwater would be the Jordan River itself, between the Benot Ya'aqov Bridge and the Pkak Bridge. Measurements of the bed of the drained Lake Hula showed subsidence of 50 cm in two years (Ben Porat and Minzker 1996:42).

1965: Western embankments (3 km long) were built along the Jordan River from the Benot Ya'aqov Bridge to the Pkak Bridge.

1967: The I.D.F. Engineering Corps erected a new Bailey bridge south of the British one. This new bridge was once again replaced in 2008.

The National Water Planning Committee reported that the regulation of the Jordan River south of the Pkak Bridge, which had begun with the Hula Reclamation Project, had not been completed due to the security situation in the area up to the Six Days War. Alternative solutions for the basin's drainage and its regulation were proposed by the committee, who divided the area from the Pkak Bridge to south of 'Ateret Fortress into three sections of approximately with a total length of 4 km.

In sum – to the south of the Benot Ya'aqov Bridge the Jordan River bed was too high to absorb a stronger water flow without flooding, and the southern river therefore had to be regulated in order to prevent continuous accumulations around the Benot Ya'aqov Bridge. The solution proposed was to install a profile 23 m wide, with a gradient of 1:3. As the right bank was much higher than the left one, the new profile should be on the left bank. The planned longitudinal slope was 0.2%. In order to save costs, it was advised to dispose of the excavated soil no farther away than 100 m; the excavated debris would form an elevated strip along the left bank, 1–2 m high and up to 120 m wide (CZA/Shoham/1967).

1969: A record-breaking flow at the Pkak Bridge (114 m^3/sec) caused an alluvial fan of the Jordan River in Lake Kinneret. Most probably, the eroded soil originated in the embankment spoils built in 1965 west of the southern Jordan River in order to prevent flooding (Ben Porat and Minzker 1996:45).

1971: The Hula Reclamation Project was completed according to the plan of TAHAL (Israel Water Planning); at the Pkak Bridge the river bed level was 59.2 m above msl and the flow capacity of the southern Jordan River was 126 m^3/sec. Two million cubic meters of soil and rocks had been removed in the excavations in the southern Jordan River in 1951–1971, as opposed to the 0.7 million originally planned by Kovlanov. The reasons for this huge difference were:

- The need to increase the southern Jordan River flow capacity from 80 to 126 m^3/sec by widening the river.
- The disposal of huge amounts of sediment load since the drainage works had begun.
- The fact that the project had been carried out in various phases.

For the second time, there was a failure to accomplish the goals: the rocks in the river's bed were not excavated to the planned depth. This was revealed by a measurement taken in 1994 (Ben Porat and Minzker 1996:45). The river was clogged by stones and cement along 200 m north of the Benot Ya'aqov Bridge accumulating 2.5 m higher than the bottom level measured in 1967. This current river bed level was a result of work carried out to support the left column of the bridge. New excavations in the Jordan River were essential, according to TAHAL (CZA/Natif/1972).

1973–1992: Several peat specialists concluded that the plantation of fodder crops would reduce the subsidence of the ground level.

1973: The river channel was regulated according to plans of the Israel Water Planning Committee. The original plan was to deepen and regulate the river at a longitudinal gradient of 0.06%. Topographical measurements taken in 1994 show that this plan was never completed (CZA/Minzker/2000:9). After the Yom Kippur War a second Bailey bridge was built in 1973. These two bridges stood parallel to one another, one for east-west traffic and the other for west-east traffic.

1983: The east and west banks were raised in order to prevent winter flooding (Ben Porat and Minzker 1996:45).

1985: A thorough soil survey of the Hula showed that the ground level had sunk by 3.5 m. The water table had maintained its level (Ben Porat and Minzker 1996:46).

1988: Accumulated measurements of the subsiding surface level showed that by 2005 an area of 5,000 dunams would be permanently inundated, as its surface would be lower than the surrounding area (62 m above msl in the center). This area would be 3 m above the summer water table, ensuring its continuing subsidence.

1993–1996: Consequential to the Hula Reclamation Project, an ecological restoration project – the Peat/Lignite Restoration Project – was suggested.

Advantages: Substantial improvement of agricultural drainage; diminishing frequency of flooding events; prevention of water loss from swamps and wadis.

Disadvantages: Peat drained of water will cause subsidence; dried peat is combustible; wind erosion effects on nearby agricultural plantations.

In order to prevent erosion of the peat, the surface water level would be maintained with the aid of vegetation. An alternative agro-tourist solution was chosen – a solution combining touristic and ecological objectives with agricultural development of the area (Ben Porat and Minzker 1996:46).

"After the destruction of a wealth of environments and of life forms, after causing unforeseen impact on downstream Lake Kinneret and River Jordan, after the lessons learned from the best possible use of the drained swamplands and after some preservation experience, the beds of the Hula Lake will be flooded again. The knowledge acquired of the past environments and the reasonably full picture of the original biodiversity of Lake Hula and of its swamp can supply the yardsticks for restoration work" (Dimentman et al. 1992:1).

1995: Proposal by the Kinneret Drainage Authority for the regulation of the Jordan River between the Pkak Bridge and 'Ateret Fortress in order to prevent flooding of the Hula Valley. The river's course between the Pkak Bridge and Lake Kinneret was divided into three parts, differing in topography and soil.

1999: According to the above restoration plans, approved by the Israel Water Authority in 1995, and the deepening of the river section as suggested, the gradient to be reached was 0.02555% (25.5 cm per 1 km), along a total length of 3.8 km. This regulation was deemed marginal in view of the flooding problems. The planners favored restoration works (deepening the river and widening the slopes) along 4.5 km as far as 'Ateret Fortress, but the authorities limited the extent of the work in consideration of the archaeological sites in the vicinity.

In total disregard of these agreed limitations, heavy excavation works were carried out by the Israel Water Authority in December 1999, beginning at 'Ateret Fortress and ending 1 km north of the Benot Ya'aqov Bridge. The works caused catastrophic damage to the Benot Ya'akov Formation and the excavation site. Hundreds of stone tools and bone remains were found in the debris mounds. The immense damage caused to the archaeological and geological strata is inconceivable and irreversible. Eventually, the work carried out was only minimally effective. Due to the restrictions, the existing width of the river bed (10–15 m) remained unchanged.

This regulation was considered the only possible solution in resolving the problem created by the Hula Reclamation Project (CZA/Minzker/2000:7–9).

2007: The two parallel Bailey bridges were replaced by a concrete bridge located where the southernmost bridge used to be, while one of the Bailey bridges was retained for emergency crossing.

Hydrological Summary and Consequences of the Hula Reclamation Project

The history of the Jordan River in the past 160 years can be summarized in terms of the flow capacity at the "bottleneck" and the river level at the Benot Ya'aqov Bridge.

During the 15,000 years of Lake Hula's existence, its level dropped gradually from 70 m to 69 m above msl, i.e., the river cut the basalt rock at an average pace of 0.07 mm per year (Ben Porat and Minzker 1996:48).

Over the past 70 years the river's bed has been excavated to an accumulated depth of 7.3 m. The "bottleneck" at the bridge has been deepened eight times in the past 160 years, accumulating debris and erosion between each excavation (Ben Porat and Minzker 1996:48). The deepening of the river bed (which was the basis for the drainage of Lake Hula) from 69.5 m to 58.3 m above msl in a span of 160 years proved to be insufficient, as the peat/lignite is still subsiding.

"The simple thought that the drainage problems of the Hula Valley were a result of neglect contributed to the general concept of arriving at a unique, one-time solution: the removal of the 'bottleneck'. A delicate balance exists between two fundamental factors in the Hula Valley: flooding and river erosion. This situation necessitates constant maintenance and deepening of the water channels. This reality, though common knowledge in deltas around the world, is yet to be recognized and accepted in Israel" (Ben Porat and Minzker 1996:49).

"No time was given to the scientists to tightly evaluate the impact of the project and work out proposals for correct management" (Dimentman et al.1992:1).

References

The following references are from the Central Zionist Archives (CZA) and include author (when available)/institution of the document/catalogue entry/date.

Aharoni, Y./KKL5/19014/18.1.1951 (in Hebrew).
Brachyahu, A./KKL5/18198/5.2.1951a (in Hebrew).
Brachyahu, A./KKL5/18198/24.8.1951b (in Hebrew).
Brachyahu, A./KKL5/18198/1.8.1959 (in Hebrew).
Diskin, S./KKL5/18198/14.3.1951 (in Hebrew).
Hammer, M./KKL5/25.12.1959 (in Hebrew).
Treidel, Y./Z2/642/1010 (translation) (in Hebrew).
Yeivin, S./KKL5/18197/1.8.1951 (in Hebrew).
KKL/KKL5/17395/2.5.1951 (in Hebrew).
KKL/KKL5/20723/1953 (in Hebrew).

KKL/A246/438/1962 (in Hebrew).

The following references are from the Kinneret Drainage Authority Archive and include author (when available)/institution of the document/catalogue entry/date.

Minzker, N./Kinneret Drainage Authority/1995 (in Hebrew).
Minzker, N./Kinneret Drainage Authority/2000 (in Hebrew).
Natif, A./Kinneret Drainage Authority/4-39-001/318/1971 (in Hebrew).
Shoham, D./Kinneret Drainage Authority/1967 (in Hebrew).

Avurmed, S. (1983). The Hula reclamation project. Seminar paper, Haifa University (in Hebrew).
Ben Porat, A. (2001). Main events in the Hula Valley 1837–2000. (pp. 1–13). Kibbutz Amir (in Hebrew).
Ben Porat, A., & Minzker, N. (1996). The Hula reclamation project: A history of the past 160 years. *Water and Irrigation (Mayim VeHashkaya), 361*, 42–51 (in Hebrew).
Dimentman, C., Bromley, H.J., & Por, F.D. (1992). *Lake Hula: Reconstruction of the fauna and hydrobiology of a lost lake*. Jerusalem: Israel Academy of Sciences and Humanities.
Gophen, M. (2000). History of research of the Kinneret and the drainage basin. *Agamit, 10*, 10–13 (in Hebrew).
Goren, H. (1994). J.R. Roth: A researcher of the Land of Israel. *Ariel, 102–103*, 190–201 (in Hebrew).
Goren, H. (1999). *Go view the land: German study of Palestine in the nineteenth century*. Jerusalem: Yad Ben Zvi (in Hebrew).
Goren, H. (2001). *Travelers descriptions of the Eastern Galilee and the Hula Valley from the 18th and 19th Centuries*. Tel Hai: Tel Hai Academic College (in Hebrew).
Karmon, Y. (1956). *The northern Hula Valley: Its natural and cultural landscape*. Jerusalem: Magness Press (in Hebrew).
Karmon, Y. (1960). The drainage of the Huleh swamps. *Geographical Review, 50*, 169–193.
Karmon, Y. (1973). *The Land of Israel: The geography of the land and regions*. Tel Aviv: Yavneh (in Hebrew).
MacGregor, J. (1869). *The Rob Roy on the Jordan, Nile, Red Sea & Gennesareth*. London: C.J. Murray.
Stanley, A.P. (1877). *Sinai and Palestine in connection with their history*. London: J. Murray.
Zitrin, Y. (1987). History of the Hula concessions. Ph.D. Dissertation, Bar-Ilan University (in Hebrew).

Appendix 2
Typological List

No.#	Type (modified after Bordes 1961*)	Typological group (this volume)	Class (this volume)
9	Simple straight side-scraper	Side-scrapers	Flake tools
10	Simple convex side-scraper	Side-scrapers	Flake tools
11	Simple concave side-scraper	Side-scrapers	Flake tools
12	Double straight side-scraper	Side-scrapers	Flake tools
13	Double straight-convex side-scraper	Side-scrapers	Flake tools
14	Double straight-concave side-scraper	Side-scrapers	Flake tools
15	Double convex-convex side-scraper	Side-scrapers	Flake tools
16	Double concave-concave side-scraper	Side-scrapers	Flake tools
17	Double convex-concave side-scraper	Side-scrapers	Flake tools
18	Convergent straight side-scraper	Side-scrapers	Flake tools
19	Convergent convex side-scraper	Side-scrapers	Flake tools
20	Convergent concave side-scraper	Side-scrapers	Flake tools
21	Offset side-scraper	Side-scrapers	Flake tools
22	Transversal straight side-scraper	Side-scrapers	Flake tools
23	Transversal convex side-scraper	Side-scrapers	Flake tools
24	Transversal concave side-scraper	Side-scrapers	Flake tools
25	Side-scraper on ventral face	Side-scrapers	Flake tools
26	Abruptly retouched side-scraper	Side-scrapers	Flake tools
27	Side-scraper with thinned back	Side-scrapers	Flake tools
28	Bifacially retouched side-scraper	Side-scrapers	Flake tools
29	Alternately retouched side-scraper	Side-scrapers	Flake tools
30	Typical end-scraper	End-scrapers	Flake tools
31	Atypical end-scraper	End-scrapers	Flake tools
32	Typical burin	Burins	Flake tools
33	Atypical burin	Burins	Flake tools
34	Typical borer	Borers	Flake tools
35	Atypical borer	Borers	Flake tools
36	Typical backed knife	Backed knives	Flake tools
37	Atypical backed knife	Backed knives	Flake tools
38	Naturally backed knife	Backed knives	Flake tools
42	Notch	Notches	Flake tools
54	End-notch	Notches	Flake tools
43	Denticulate	Denticulates	Flake tools
51	Denticulated point	Denticulates	Flake tools
40	Truncated flake	Truncations	Flake tools
106	Retouched flake	Retouched flakes	Flake tools
45	Retouch on ventral face	Retouched flakes	Flake tools
39	Raclette	Retouched flakes	Flake tools
114	Massive scraper	Massive scrapers	Flake tools
8	Limace	Varia	Flake tools
44	Alternately retouched bec	Varia	Flake tools

No.#	Type (modified after Bordes 1961*)	Typological group (this volume)	Class (this volume)
46–47	Abrupt and alternate retouch (thick)	Varia	Flake tools
48–49	Abrupt and alternate retouch (thin)	Varia	Flake tools
53	Pseudo-microburin	Varia	Flake tools
57	Tanged point	Varia	Flake tools
58	Tanged flake	Varia	Flake tools
62	Varia	Varia	Flake tools
87	Handaxe	Handaxes	Core tools
90	Handaxe varia	Handaxes	Core tools
91	Partial handaxe	Handaxes	Core tools
120	Handaxe with a "cleaver-like" edge	Handaxes	Core tools
123	Biface preform	Biface preforms	Core tools
128	Amorphous biface	Amorphous bifaces	Core tools
84	Cleaver	Cleavers	Core tools
85	Cleaver on flake	Cleavers	Core tools
59	Chopper	Chopping tools	Core tools
60	Chopper inverse	Chopping tools	Core tools
61	Chopping tool	Chopping tools	Core tools
116	Subspheroid	Subspheroids	Core tools
115	Discoid	Discoids	Core tools
95	Disc	Discs	Core tools
96	Percussor	Percussors	Core tools
127	Pitted stone	Pitted stones	Core tools
112	Anvil	Anvils	Core tools
100	Flake	Flakes	Flakes
108	Burin spall	Flakes	Flakes
101	Blade	Blades	Flakes
105	Dorsally plain flake	Dorsally plain flakes	Flakes
104	*Éclat de taille de biface*	*Éclat de taille de biface*	Flakes
124	Biface sharpening flake	Biface sharpening flakes	Flakes
99	Core waste	Core waste	Flakes
97	Angular fragment	Angular fragments	Non-flakes
122	Indeterminate waste	Indeterminate waste	Non-flakes
129	Too broken/rolled to classify	Indeterminate waste	Non-flakes
64	Levallois core for flakes	Levallois cores	Cores
65	Levallois core for points	Levallois cores	Cores
66	Levallois core for blades	Levallois cores	Cores
67	Discoidal cores	Discoidal cores	Cores
68	Globular core	Globular cores	Cores
69	Prismatic core	Prismatic cores	Cores
70	Pyramidal core	Pyramidal cores	Cores
107	Core on flake	Cores on flake	Cores
121	Core on handaxe	Cores on handaxes	Cores
71	Core varia	Core varia	Cores
72	Amorphous core	Amorphous cores	Cores
1000	Giant core	Giant cores	Cores
113	Modified	Modified	Modified
102	Microartifact	Microartifacts	Microartifact

\# After Bordes' typological list (1961) and with the authors' modifications (Chapter 4)

Appendix 3
List of Attributes for Flakes and Flake Tools

RM – Raw material
   1. flint
   2. limestone
   3. basalt
PS – Preservation
   1. fresh
   2. slightly abraded
   3. abraded
   4. heavily abraded
   5. exfoliated
PT – Patination
   1. no patina
   2. patinated
   3. double patinated
CB – Breakage
   1. complete
   2. distal
   3. lateral
   4. proximal
   5. lateral & distal
   6. proximal & distal
   7. fragment
   8. proximal & lateral
   9. indeterminate
   10. split (percussor)
LEV – Levallois
   1. Levallois
   2. not Levallois
   3. possibly Levallois
LEN – Flaking length
WID – Flaking width
THI – Thickness
MLEN – Maximum length
BR – Burned
   1. burned
   2. possibly burned
   3. not burned
CO – Cortical cover
   1. no cortex
   2. 0–25 %
   3. 26–50 %
   4. 51–75 %
   5. 76–100 %
   6. indeterminate
ASP – Angle of striking platform (interior)
D – Direction of blow
   1. indeterminate
   2. end-struck
   3. side-struck
   4. special side-struck
DOR – Scar pattern
   1. indeterminate
   2. cortical
   3. plain (Kombewa)
   4. simple (along axis)
   5. parallel
   6. convergent
   7. opposed
   8. radial
   9. ridged
   10. side
   11. along axis & side
   12. along axis & opposed
   13. opposed & side
   14. along axis & radial
NS – Number of dorsal scars
SP – Type of striking platform
   1. indeterminate
   2. cortical
   3. punctiform
   4. plain
   5. dihedral
   6. facetted
   7. removed
   8. missing
   9. crushed
LRA – Location of retouch A; similar for **LRB** (additional retouch)

1. distal
2. proximal
3. truncation
4. left edge
5. right edge
6. both edges
7. convergent
8. distal & both edges
9. circumference
10. distal & right
11. distal & left
12. proximal & both edges
13. indeterminate
14. proximal & left
15. proximal & right

TRA – Type of retouch A; similar for **TRB** (additional retouch)
1. signs of use
2. regular
3. end-scraper
4. side-scraper
5. notch/denticulate
6. indeterminate
7. irregular (not continuous)
8. semi-Quina (half Quina)
9. Quina
10. raclette (delicate)
11. invasive
12. bifacial
13. abrupt
14. Nahr Ibrahim
15. thinning
16. semi-abrupt (half abrupt)
17. bipolar
18. mixed (on the same edge)
19. burin blow

WFA – Form of retouched edge A; similar for **WFB** (additional retouch)
1. straight
2. convex
3. concave
4. convergent
5. wavy
6. denticulate
7. one tooth (awl-like)
8. indeterminate
9. oblique

ARA – Angle of retouch A; similar for **ARB** (additional retouch)
LR2 – Face of retouch A; similar for **LR2B** (additional retouch)
1. dorsal
2. ventral
3. dorsal & ventral
4. side

OU – Technological attributes 1
1. *outrepassé*
2. hinge
3. *débordant*
4. Kombewa
5. possibly Kombewa
6. ventrally curved
7. steps
8. Siret (*brut*)
9. hinge + *débordant*
10. Siret + *débordant*
11. possibly Kombewa + *débordant*
12. steps + *débordant*
13. hinge + Kombewa
14. Kombewa + *débordant*
15. Kombewa + Siret
16. Siret + steps

VARIA – Technological attributes 2
1. ventral removals
2. cone
3. large cone
4. thinning of bulb of percussion
5. bulbar scar
6. cones + ventral removals
7. cones + large cone

TYP – Typology following Bordes (1961) with additional types to suit the assemblage (Appendix 2)

LIP – Lipping
1. lipped
2. not lipped
3. possibly lipped

FLAKE TYPE (only for basalt flakes)
1. handaxe-shaped flake
2. biface thinning flake
3. roughing-out flake
4. wedge flake
5. cleaver-shaped flake
6. shoulder flake
7. working accident
8. waste of giant core

Appendix 4
List of Attributes for Cores and Core Tools

RM – Raw material
    1. flint
    2. limestone
    3. basalt
PS – Preservation
    1. fresh
    2. slightly abraded
    3. abraded
    4. heavily abraded
    5. exfoliated
PT – Patination
    1. no patina
    2. patinated
    3. double patinated
CB – Breakage
    1. complete
    2. distal
    3. lateral
    4. proximal
    5. lateral & distal
    6. proximal & distal
    7. fragment
    8. proximal & lateral
    9. indeterminate
    10. split percussor
LEV – Levallois
    1. Levallois
    2. not Levallois
    3. possibly Levallois
CO – Cortical cover
    1. no cortex
    2. 0–25 %
    3. 26–50 %
    4. 51–75 %
    5. 76–100 %
    6. indeterminate
TR – Type of retouch A
    1. signs of use
    2. regular
    3. end-scraper
    4. side-scraper
    5. notch/denticulate
    6. indeterminate
    7. irregular (not continuous)
    8. semi- Quina (half Quina)
    9. Quina
    10. raclette (delicate)
    11. invasive
    12. bifacial
    13. abrupt
    14. Nahr Ibrahim
    15. thinning
    16. semi-abrupt (half abrupt)
    17. bipolar
DOR – Scar pattern
    1. simple dominant
    2. horseshoe
    3. simple (along axis)
    4. parallel
    5. convergent
    6. opposed
    7. radial
    8. along axis & opposed
    9. along axis & side
    10. bipolar
    11. indeterminate
BR – Burned
    1. burned
    2. possibly burned
    3. not burned
SC – Cross-section
    1. triangular
    2. flat
    3. indeterminate
NS – Number of scars
ANG – Angle between platforms
WF – Form of working edge
    1. straight

2. convex
　　3. concave
　　4. convergent
　　5. wavy
　　6. denticulate
　　7. one tooth (awl-like)
　　8. indeterminate
　　9. circumferential
NW – Number of working edges
NP – Number of platforms
NS – Number of scars
MLEN – Maximum length
LEN – Length
WID – Maximum width
THI – Maximum thickness
CIR – Circumference
SLEN – Scar length; for cores; of dominant last scar
SWID – Scar width; for cores, of dominant last scar
WLEN – Length of working edge
TYP – Typology following Bordes (1961) with additional types to suit the assemblage (Appendix 2)
ANALYSES OF PITTED STONES
TYPE OF PIT (if more than one pit, measure the dominant one)
　　1. Flat
　　2. Easily defined and deep
　　3. Battered surface
　　4. Large and deep
　　5. With a step
　　6. Cluster of pits
LOCATION OF PIT ON ARTIFACT (see illustration)
　　1. Middle
　　2. Margins
　　3. Side
　　4. More than one pit
　　5. Indeterminate
SHAPE OF PIT
　　1. Round
　　2. Elliptical
　　3. Indeterminate
　　4. Indeterminate
　　5. Cluster of pits
INSIDE OF PIT
　　1. Smooth
　　2. Rough
　　3. Battered
PIT MAXIMUM LENGTH
NUMBER OF PITS

Appendix 5
List of Attributes for Handaxes

RM – Raw material
 1. flint
 2. limestone
 3. basalt

PS – Preservation
 1. fresh
 2. slightly abraded
 3. abraded
 4. heavily abraded
 5. exfoliated

PT – Patination
 1. no patina
 2. patinated
 3. double patinated

CB – Breakage
 1. complete
 2. distal
 3. lateral
 4. proximal
 5. lateral & distal
 6. proximal & distal
 7. fragment
 8. proximal & lateral
 9. indeterminate
 10. exfoliated
 11. slightly broken tip

CO – Cortical cover
 1. no cortex
 2. 0–25 %
 3. 26–50 %
 4. 51–75 %
 5. 76–100 %
 6. indeterminate

FC – Blank type
 1. flake
 2. chunk
 3. indeterminate
 4. Kombewa flake
 5. possibly Kombewa flake
 6. transversal flake
 7. possibly flake

D1 – Direction of blow (face 1); similar for **D2** (face 2)
 1 – 8

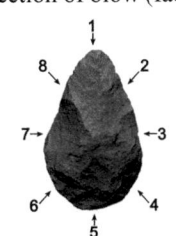

 9. indeterminate

SC – Cross-section
 1. lenticular
 2. plano-convex
 3. thick
 4. angular
 5. flat lenticular
 6. backed
 7. indeterminate
 8. triangular
 9. flat triangular

SP – Type of striking platform
 1. indeterminate
 2. cortical
 3. punctiform
 4. plain
 5. dihedral
 6. facetted
 7. removed
 8. missing

L1 – Location of retouch (face 1); similar for **L2** (face 2)
 1. distal
 2. proximal
 3. left edge
 4. right edge
 5. both edges
 6. convergent
 7. distal & both edges
 8. distal & left

9. distal & right
10. circumference
11. proximal & both edges
12. proximal & left
13. proximal & right
14. entirely covered
15. indeterminate
16. distal & proximal & left
17. distal & proximal & right
18. convergent & proximal

Q1 – Extent of retouch (face 1); similar for **Q2** (face 2)
1. 0–25 %
2. 26–50 %
3. 51–75 %
4. 76–100 %

TR1 – Type of retouch (face 1); similar for **TR2** (face 2)
1. flat and limited
2. scraper-like
3. thinning

6. indeterminate

12. bifacial

15. thinning & bifacial

NS1 – Number of scars (face 1)
NS2 – Number of scars (face 2)

ANG – Angle between face 1 and face 2

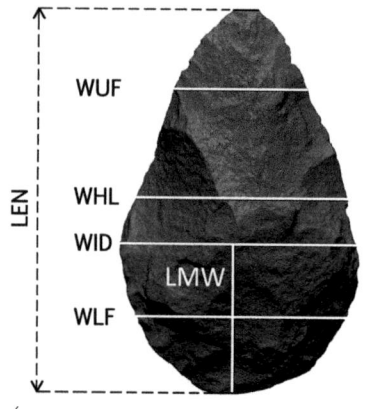

LEN – Length
WID – Width
THI – Thickness
WHL – Width at half length
LMW – Location of maximum width
LMT – Location of maximum thickness
WUF – Width at upper fifth
WLF – Width at lower fifth
THL – Thickness at half length
TUF – Thickness at upper fifth
TYP – Typology after Bordes (1961) with additional types to suit the assemblage (Appendix 2)

Appendix 6
List of Attributes for Cleavers

RM – Raw material
 1. flint
 2. limestone
 3. basalt

PS – Preservation
 1. fresh
 2. slightly abraded
 3. abraded
 4. heavily abraded
 5. exfoliated

PT – Patination
 1. no patina
 2. patinated
 3. double patinated

CB – Breakage
 1. complete
 2. distal
 3. lateral
 4. proximal
 5. lateral & distal
 6. proximal & distal
 7. fragment
 8. proximal & lateral
 9. indeterminate
 10. exfoliated
 11. slightly broken tip

CO – Cortical cover
 1. no cortex
 2. 0–25 %
 3. 26–50 %
 4. 51–75 %
 5. 76–100 %
 6. indeterminate

FC – Blank type
 1. flake
 2. chunk
 3. indeterminate
 4. Kombewa flake
 5. possibly Kombewa flake
 6. transversal flake

D1 – Direction of blow (face 1); similar for **D2** (face 2)
 1–8
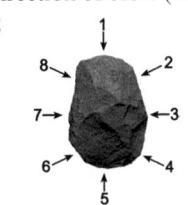
 9. indeterminate

SC – Cross-section
 1. rectangle
 2. trapeze
 3. parallelogram
 4. lenticular
 5. convex
 6. triangular
 7. rhombus
 8. indeterminate
 9. flat triangular

SP1 – Type of striking platform; similar for **SP2** (in case of two striking platforms)
 1. indeterminate
 2. cortical
 3. punctiform
 4. plain
 5. dihedral
 6. facetted
 7. removed
 8. missing

SCR – Scar pattern (face 1); similar for **SCR2** (face 2)
 1. indeterminate
 2. cortical
 3. plain (Kombewa)
 4. simple (along axis)
 5. parallel
 6. convergent
 7. mopposed
 8. radial
 9. ridged

10. side
11. along axis & side
12. along axis & opposed
13. opposed & side

L1 – Location of retouch (face 1); similar for **L2** (face 2)
1. distal
2. proximal
3. left edge
4. right edge
5. both edges
6. convergent
7. distal & both edges
8. distal & left
9. distal & right
10. circumference
11. proximal & both edges
12. proximal & left
13. proximal & right
14. entirely covered
15. indeterminate
16. distal & proximal & left
17. distal & proximal & right

TR1–Type of retouch (face 1); similar for **TR2** (face 2)
1. flat and limited
3. thinning
6. indeterminate
12. bifacial
15. thinning & bifacial
16. bifacial & scraper-like

WS – Shape of working edge
1. straight
2. convex
3. concave
4. pointed
5. indeterminate
6. oblique

WE – Design of working edge
1. before flake removal
2. after flake removal
3. indeterminate

SU – Type of retouch on working edge
1. signs of use
2. notch
3. wavy
4. slightly damaged

NS1 – Number of scars (face 1)
NS2 – Number of scars (face 2)
LEN – Length
MLEN – Maximum length
WID – Width
THI – Thickness

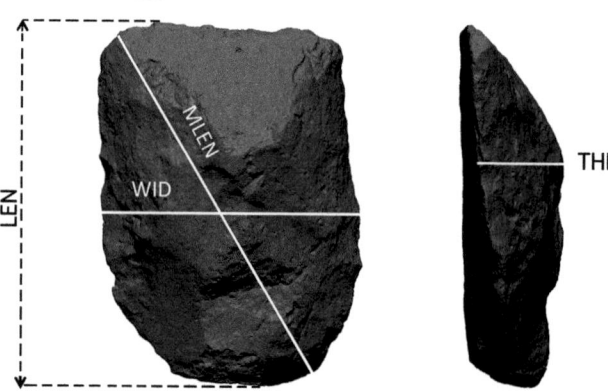

AWL – Angle between working edge and left edge
AWR – Angle between working edge and right edge
LWE – Length of working edge
LSC – Length of dominant scar
WSC – Width of dominant scar
AS1 – Angle of striking platform
AS2 – Angle of additional striking platform
TYP – Typology after Bordes (1961) with additional types to suit the assemblage (Appendix 2)

Appendix 7
A Comparison of Length of Flint Flake Tools

The following tables present the detailed results of the comparisons of the mean length of pairs of flint flake tool types. The comparison was performed for each pair using Student's t-test. The first table presents the frequenfy and mean length (in mm) of the tool groups along with the standard error and upper and lower 95% of observations. The second table presents the results of the statistical test for each pair of tool types, including the difference between means, the t-ratio, and the p-value of a two-tailed test. Significant differences are those in which the p-values equal 0.05 or lower (and are highlighted in bold). Significant differences indicate that Category I is significantly longer than Category II; degrees of freedom equal 2545.

Category	N	Mean length	Std err	Lower 95%	Upper 95%
Notch	237	27.90	0.59	26.74	29.06
Denticulate	120	28.46	0.83	26.83	30.09
Side-scraper	149	30.51	0.75	29.05	31.97
Borer	179	25.36	0.68	24.03	26.70
End-scraper	77	26.99	1.04	24.95	29.02
Knife	15	47.60	2.35	42.99	52.21
Truncation	42	29.26	1.40	26.51	32.02
Burin	8	39.38	3.22	33.07	45.68
Retouched flake	231	26.17	0.60	25.00	27.34
Unretouched flake	1497	25.73	0.24	25.27	26.19

Category I	Category II	Difference	t Ratio	p-Value
Knife	**Borer**	22.24	8.03	**> 0.0001**
Knife	**Unretouched flake**	21.87	9.26	**> 0.0001**
Knife	**Retouched flake**	21.43	8.84	**> 0.0001**
Knife	**End-scraper**	20.61	8.03	**> 0.0001**
Knife	**Notch**	19.70	8.13	**> 0.0001**
Knife	**Denticulate**	19.14	7.68	**> 0.0001**
Knife	**Truncation**	18.34	6.70	**> 0.0001**
Knife	**Side-scraper**	17.09	6.93	**> 0.0001**
Burin	**Borer**	14.01	4.26	**> 0.0001**
Burin	**Unretouched flake**	13.65	4.23	**> 0.0001**
Burin	**Retouched flake**	13.21	4.04	**0.0001**
Burin	**End-scraper**	12.39	3.66	**0.0003**
Burin	**Notch**	11.47	3.51	**0.0005**
Burin	**Denticulate**	10.92	3.29	**0.0010**
Burin	**Truncation**	10.11	2.88	**0.0040**
Burin	**Side-scraper**	8.86	2.68	**0.0073**
Knife	**Burin**	8.23	2.06	**0.0391**
Side-scraper	**Borer**	5.15	5.10	**> 0.0001**
Side-scraper	**Unretouched flake**	4.78	6.12	**> 0.0001**
Side-scraper	**Retouched flake**	4.34	4.54	**> 0.0001**
Truncation	**Borer**	3.90	2.50	**0.0125**

Category I	Category II	Difference	t Ratio	p-Value
Truncation	**Unretouched flake**	**3.54**	**2.48**	**0.0131**
Side-scraper	**End-scraper**	**3.52**	**2.76**	**0.0059**
Denticulate	**Borer**	**3.10**	**2.88**	**0.0040**
Truncation	**Retouched flake**	**3.09**	**2.03**	**0.0428**
Denticulate	**Unretouched flake**	**2.73**	**3.16**	**0.0016**
Side-scraper	**Notch**	**2.61**	**2.74**	**0.0062**
Notch	**Borer**	**2.54**	**2.82**	**0.0049**
Denticulate	**Retouched flake**	**2.29**	**2.24**	**0.0255**
Truncation	End-scraper	2.27	1.30	0.1926
Notch	**Unretouched flake**	**2.18**	**3.42**	**0.0006**
Side-scraper	Denticulate	2.05	1.84	0.0662
Notch	**Retouched flake**	**1.73**	**2.06**	**0.0394**
End-scraper	Borer	1.62	1.31	0.1905
Denticulate	End-scraper	1.47	1.11	0.2683
Truncation	Notch	1.36	0.89	0.3725
End-scraper	Unretouched flake	1.26	1.19	0.2359
Side-scraper	Truncation	1.25	0.79	0.4325
Notch	End-scraper	0.92	0.77	0.4430
End-scraper	Retouched flake	0.82	0.68	0.4945
Retouched flake	Borer	0.81	0.89	0.3740
Truncation	Denticulate	0.80	0.49	0.6224
Denticulate	Notch	0.56	0.54	0.5860
Retouched Flake	Unretouched flake	0.44	0.69	0.4914
Unretouched flake	Borer	0.36	0.50	0.6141

Index

A

Aalat, 398
Aborigines, 417
Abrupt and alternate retouch, 154
Acheulian, 1–2, 11, 14, 18, 19, 26, 36, 54, 89, 98, 116, 117, 134, 181, 214, 232, 234, 237, 246, 251, 281, 299, 321, 328, 331–332, 367, 368, 371, 374, 378, 397, 399, 401, 402, 404–408, 412–414, 416–422, 424
    Early Acheulian, 397, 399
    Large Flake Acheulian (LFA), 2, 19, 287, 371, 397–399, 403–405, 407
    Middle Acheulian, 14, 331
Acheulo-Yabrudian, 398, 405, 407
Adaptation, 1, 2, 19, 422, 423
    of inventions, 407
'Ain Soda, 398
Africa (*see also* Out of Africa), 1, 2, 3, 5, 14, 19, 24, 36, 54, 98, 246, 251, 331, 397–399, 405, 407, 413, 417–420, 424
Alternately retouched bec (beak), 154, 166–167, 248
Anatomically modern humans, 424
Angular fragments, 393
    basalt, 251, 321–323, 372
    flint, 182, 201, 219, 221, 229
Anvils, 58, 73, 77, 78, 91, 96, 98, 102, 103, 106, 108, 111, 114, 237, 298, 303, 310, 311, 372, 373, 374, 399, 401, 416
    pitted anvils, 69, 77, 78, 82, 84, 96, 102, 111, 115, 159, 306, 310, 387
    thin anvils, 251, 298, 299, 310, 313, 315–319, 372, 373, 374, 378, 391, 401, 405, 417, 421, 423
Apes, 419
    Chimpanzees, 307
Armenia, 407
Artistic behavior, 425
Attribute analysis, 43, 53–57, 332, 402
    attribute lists, 54–55
Asia, 1, 2, 398, 407
Autonoesis, 422, 423
Ayyelet Hashahar, 24

B

Backed knives, 152, 163, 185, 247
Basalt
    boulders and cobbles, 12, 13, 16, 29, 30, 44, 81, 115, 413
    flows, 27, 29, 34, 36, 38, 303, 324, 331, 372, 374, 397, 420
    nodules, 299, 306, 333, 372, 401
    slabs, 38, 57, 81, 83, 85, 274, 276, 277, 284, 303, 306, 311, 313, 315, 316, 318, 324, 325, 326, 331, 332, 372, 373, 374, 375, 394, 401, 402, 405, 406, 412, 413, 416
        slicing, 328–329
    reduction sequence, 237, 251, 259, 274, 287, 297, 298, 323, 344, 371–378
    taphonomy (*see* Taphonomy)
    technology
        bifaces, 337–346, 374–378
        cores, 323–331
        flakes, 250–298
    typology
        bifaces, 346–360, 362–369, 370–371
        core tools, 298–323
        flake tools, 241–250
Beads, 416, 425
Beeches Pits, 413
Ben-Ami, D., 15, 16
Benot Ya'aqov Formation (BYF), 6, 7, 16, 18, 19, 21–23, 26–30, 32, 34–37, 39, 41, 43, 44, 62, 63, 70, 71, 79, 81, 116, 332
Benot Ya'aqov Bridges, 6, 7–15, 17–19, 21–23, 27, 29
Benot Ya'aqov Embayment, 26–29, 37, 421
Biface (*see also* Cleaver, Handaxe), 2, 10, 13, 14, 58, 59, 117, 331–378, 398–399, 402, 413, 414, 417
    analyses, 55–58, 332, 337
    basalt, 172, 190, 225, 237, 251, 269, 272, 331–371, 401–402, 404, 406, 408, 416, 422
    blank, 251, 263, 269, 274, 287, 289, 293, 324, 326, 332, 338–339, 341–344, 376–377, 397, 398, 404, 407, 418
    caching/stocking, 419, 420, 421
    chaîne opératoire, 416, 422
    design/modification, 120, 187, 195, 201, 214, 226, 232, 264, 287, 298, 323, 346–348, 377–378, 414, 424
    direction of blow, 341–342
    façonnage, 269, 415
    finishing, 195, 287, 404, 406
    flint, 119, 181, 187, 195, 197, 223, 225, 226, 237, 272, 332–333, 399, 402, 415
    frequencies, 264, 272, 331, 418, 419
    limestone, 237, 272, 332, 337, 339, 381, 394, 396, 402
    morphology, 264, 287, 289, 293, 297, 348, 353–365, 377–378, 407
    pavement, 98, 414, 418
    raw material, 332, 398, 403, 420
    reduction sequence, 228, 237, 251, 269, 297, 332, 371, 372, 373, 374–378, 405, 416, 422
    retouch, 348–353
    sharpening flakes, 58, 130, 141, 147, 201, 214, 228, 230, 269
    striking platform, 342–346
    taphonomy, 239, 333–337, 339
    technology, 263, 332, 337–346
    thinning, 274, 301, 402
    typology, 58, 332, 358, 365–370, 404
    variability, 408, 412
Blades, 162, 185, 190, 192, 232

Bones, 2, 9, 10, 11, 12, 13, 15, 16, 18, 23, 29, 30, 34, 41, 45, 47, 48, 49, 50, 51, 62, 64, 65, 69, 73, 74, 75, 81, 83, 91, 94, 100, 102, 108, 111, 113, 120, 300, 303, 373, 420
   Conjoinable, 89
   cut marks, 19, 71
   fragmentation of, 19, 374, 390, 391
   marrow exploitation, 19, 71, 374, 391, 419, 421
Bordes, F., 54, 55, 56, 122, 136, 154, 332, 354, 355, 359, 365, 367, 368, 369, 377
Borers, 54, 117, 122, 132, 133–139, 140, 141, 162, 164, 167, 172, 176, 177, 182, 233, 247, 404
Bose, 398
Boxgrove, 413
Brain
   volume, 421
   social brain, 421
Buia, 398
Bulb of percussion, 55, 260, 304, 343, 388, 391
   diffused, 187, 226
   double cone, 54, 187
   pronounced, 187
   scar on bulb, 187, 193
   thinning/removing of, 56, 168, 172, 370, 377, 378, 401, 402, 407
Burins, 141, 146, 153–154, 182, 247, 248, 320
By-products, 195, 223, 228, 374, 384, 402, 414

C

Caching (*see also* Stocking), 419
Calcite, 85
Casablanca, 405
Central-place foraging hypothesis (*see* Home base)
Chaîne opératoire (*see* also Reduction sequence), 223, 379, 412, 416, 422
Charcoal, 36
China, 398
Chirkei, 405
Chopping tool, 54, 181, 182, 183, 202, 320, 379, 385, 392, 393, 394, 405
Cleavers, 1, 2, 15, 36, 43, 50, 51, 54, 59, 116, 172, 237, 332, 397, 398, 399, 402, 403, 405, 407, 408
   analysis, 55–57, 332, 362
   basalt, 13, 29, 79, 89, 94, 302, 326, 328, 331, 335, 336, 337, 342, 343, 346, 351, 352, 353, 373
   basalt flakes of cleaver morphology, 289, 293, 296–298, 324, 376
   blanks, 338–339, 376, 378
   distal configuration of, 358, 362, 364, 371
   flint, 15, 29, 332, 398
   limestone, 332, 394
   morphology, 354–355, 362–365, 377, 378
   predetermination, 370–371
   typology, 370, 404
Cognition, 1, 419, 421–424, 425
   cognitive abilities, 19, 378, 394, 397, 402, 406, 408, 413, 419, 422, 424
   cognitive flexibility, 373, 378, 405
   cognitive modernity, 422
   cognitive traits, 397, 422, 424
   social cognition, 421
Concretions, 53, 85, 95
Conservatism, 338, 364, 368, 373, 377, 378, 397, 404, 408, 412
Coolidge, F.L., 422
Core management pieces, 185
   basalt, 274, 323
   flint, 201, 225, 228
Core trimming elements, 201, 404
Core waste
   basalt, 274, 373, 393, 402, 406
   flint, 201, 219, 221, 223
Cores, 38, 53, 54, 55, 119, 165, 182, 183, 201, 203–218, 221, 223, 225, 227–230, 258, 269, 274, 298, 323–331, 344, 370, 371, 372, 379, 385, 392, 393, 394, 399, 402, 403, 404, 408, 412, 413
   discoidal, 212, 214, 231
   exhausted, 325, 326
   formless, 214, 216–218, 228
   giant cores, 39, 57, 59, 89, 91, 94, 232, 237, 251, 298, 301, 303, 311, 323–331, 332, 398, 401, 402, 405, 406, 414, 415, 416, 421
      reduction methods/strategies, 326–331, 374–376
      waste products, 276–287, 373
   globular, 212, 323
   Levallois, 91, 119, 205–212, 214, 230–232, 323, 329, 406, 407
   on a biface, 214, 231, 232
   on biface sharpening flakes, 230, 231
   on flakes, 195, 200, 212, 214, 223, 228, 263, 323, 406, 407
   prismatic, 214
   pyramidal, 214
   reduction methods, 406–407, 412, 422
   varia, 214, 216–218, 223, 228, 393, 406
Crinoid, 416, 425
Crusader fortress of Ateret (Vadum Iacob), 15, 18, 23, 26, 27, 29, 332

D

Dead Sea Rift, 2, 3, 5, 19, 24, 63
Decision-making, 370, 372, 376, 402, 422
Débordant flakes, 193, 201, 228, 232, 255, 406
Denticulates, 408
   basalt, 241, 244–245, 246, 298, 320, 373
   denticulated points, 132–133
   flint, 122, 126, 128–133, 152, 162, 176, 182, 223
   limestone, 382
Desired products (*see* Target artifacts)
Diet, 2, 19, 417, 422
   edible plant species, 6, 19, 91, 113, 391, 417, 419, 421
Dip, 43, 44, 46, 47, 48, 49, 51, 64, 72, 75, 79, 82, 83, 86, 89, 94, 113
Discoids, 231, 405
Displaced reference, 422
Division of labor, 397, 416–417, 420, 421, 424
Dorsally plain flakes (DPF), 407
   basalt, 263–264
   flint, 195–200, 228
Drilling (drilled cores), 18, 19, 29, 37, 41
   GBY #2, 27, 34, 36
   Eshel Ya'aqov (EY), 27, 34, 36
   of the Hula Basin, 38, 39

E

Éclat de taille de biface (*édtdb*), 58, 404
   basalt, 250, 269–272, 402
   flint, 53, 117, 119, 162, 185, 187, 190–195, 201, 226, 232, 399, 406, 415
   limestone, 381, 394
End-scrapers
   basalt, 248
   flint, 122, 143–146, 162, 164, 176, 182
   limestone, 382
England, 413
Epi-Paleolithic, 18
Ethiopia, 418
Ethnography, 307, 405, 416, 417, 419
Europe, 1, 2, 3, 331, 367, 368, 377, 398, 407, 413, 418, 425
Executive functions, 422
Eynan, 6

F

France, 10, 398
Fauna, 1, 3, 5, 6, 11, 13, 19, 26, 29, 32, 34, 53, 61, 62, 65, 67, 69, 72, 84,

Index

85, 87, 95, 100, 101, 106, 298, 390, 397, 408, 413, 415, 416, 417, 419, 420, 421, 422
antler, 57, 75, 227, 299, 396, 402, 405, 416
birds, 2, 3, 5, 6, 53, 75, 89, 417, 419, 420
    swan skull, 69
crabs, 53, 65, 69, 73, 75, 89, 113, 374, 417, 419, 421
fish (see Fish)
herpetofauna, 85, 87, 89
    amphibians, 5, 6, 65, 73, 75
    reptiles, 2, 5, 6, 65, 73, 75
laboratory analyses, 53
mammals, 2, 5, 6, 19, 53, 65, 72, 75, 80, 82, 89, 95, 96, 100, 101, 102, 301, 391, 415, 417, 419, 422
    Bos sp., 75, 100
    Bovidae gen. et sp. indet., 84, 100, 103, 106
    bovids, 303, 391, 415, 417
    Bovini indet., 100, 106
    Bovini sp., 103
    Caprini, 106
    carnivores, 5, 75, 87, 89, 95, 100, 101, 103, 106
    Cervidae sp., 100, 101, 103, 106
    cervids, 11, 75, 405
    Cervus cf. *elaphus*, 85, 89, 100, 101, 106
    Dama sp., 75, 82, 84, 87, 91, 95, 96, 100, 101, 103, 106, 405, 419
    Dicerorhinus hemitoechos, 103, 106
    Elephant, 9, 10, 11, 73, 89, 300, 303, 391, 415, 417, 419
        kill site, 421
        Palaeoloxodon antiquus, 75, 82, 84, 85, 87, 91, 95, 96, 100, 101, 103, 106
        skull, 89, 91, 94, 303, 329, 374, 390
        tusk, 9, 10, 11, 15
        vertebra, 73
    Equus cf. *africanus*, 106
    Equus sp., 101, 103, 106
    fallow deer, 19, 415, 417, 419
    Gazella cf. *G. gazella*, 82, 91, 101, 103, 106
    Hippopotamus amphibious, 75, 82, 84, 87, 89, 96, 100, 101, 103, 106
    hippos, 113, 303, 391, 415, 417, 419
    Megalocerus sp., 103, 106
    rhinos, 391, 415, 417, 419
    Sus scrofa, 5, 75, 95, 100, 101, 103
    Vulpes sp., 5, 89, 100, 103
micromammals, 34, 36, 53, 65, 73, 83, 84, 87, 89, 91, 95, 96, 100, 101
mollusks (see Mollusks)
turtles, 6, 421
Fire, 2, 19, 89, 374, 416, 417, 419, 420, 421, 424
Fish, 2, 5, 6, 19, 53, 75, 80, 85, 87, 89, 415, 417, 419, 420
    treatment with fire, 421
    fishing, 421
Flexibility, 19, 353, 373, 376, 377, 378, 379, 405, 422
Flint
    nodules, 29, 39, 57, 119, 183, 219, 221, 223, 225, 227, 228, 232, 338, 399, 405, 416
    reduction sequence, 165, 172, 221, 223–236, 399–401, 405
    taphonomy (see Taphonomy)
    technology
        cores, 201–218
        flakes, 185–201
    typology
        bifaces, 360–361, 369–370
        core tools, 181–185
        flake tools, 121–181, 232–236
Flora, 1, 3, 5, 6, 19, 26, 29, 61, 87, 113, 413, 415, 416, 419, 420, 421, 422
    acorns, 416
    almonds, 416

bark, 16, 49, 50, 52, 53, 71, 75, 113
Euryale ferox, 19
fruits, 2, 6, 19, 53, 65, 75, 89, 102, 374, 421
laboratory analyses, 53
nuts, 2, 6, 19, 36, 65, 82, 91, 374, 389, 390, 392, 415, 416, 417, 421
olives, 416
pear, 416
pollen, 34, 75
Quercus calliprinos, 5, 91
Quercus ithaburensis, 5, 113
Sallix, 5, 79
seeds, 2, 53, 65, 75, 89, 102
Trapa natans, 19
underground storage organs (USO), 2, 6, 19, 374, 391, 417, 421
wood, 6, 16, 19, 47, 48, 49, 50, 51, 52, 53, 65, 71, 73, 75, 79, 80, 81, 82, 83, 85, 87, 89, 91, 94, 100, 102, 103, 106, 113, 416, 418, 419
    driftwood, 415, 416, 424
    polished plank, 79, 390
    wooden log, 91, 94, 374
    wood working, 390, 418
Food-sharing hypothesis (see Home base)
FTIR, 38, 85

G

Gadeb, 2, 418
Gadot Formation, 16, 26, 38
Gardner, E. W., 9, 10
Garonne Valley, 398
Garrod, D. A. E., 7, 9, 10, 23
Gathering, 417, 420, 421, 422, 424
Gender, 416, 417, 420
Georgia, 398
Gesher Benot Ya'aqov site
    age, 34–37
    as a preferred locale, 420
    archaeological layers, 61–117
    comparisons, 397–398, 399
    drill cores (see Drilling)
    excavation areas, 18, 32, 43, 45, 47
        Area A, 74–78
        Area B, 78–118
        Area C, 62–74
        The Bar, 16–18, 29–32, 62
        Jordan Bank (JB), 32, 45, 69–73
        The Enclosure, 45, 73–74
    geography, 21–24
    geology, 24–28
        geological survey, 15–16
    history of archaeological research, 8–19
    hominin bones, 15
    methodology, 43–59
        conservation, 52–53
        cross-sections, 51
        drawing/drafting, 49–51
        elevations, 46–47
        grid, 43–44
        sieving, 51–52
    sediments
        black mud, 41, 71, 74, 75, 83, 84, 86, 87, 103, 111
        clay, 10, 11, 12, 13, 29, 30, 32, 51, 63, 64, 65, 71, 74, 75, 77, 79, 81, 82, 84, 89, 94, 101, 103,
        coarse-grained, 32, 84
        coquina, 26, 29, 62, 63, 64, 65, 71, 74, 75, 76, 79, 82, 83, 84, 85, 87
        fine-grained, 84
        gypsum, 85

    gravel, 12, 13, 15, 22, 26, 37, 38, 62, 87, 101, 102, 103, 115, 418
    sand, 13, 22, 26, 29, 62, 64, 65, 75, 87, 101, 102
        limonite, 85
  sediment sorting, 53–54
  sedimentology, 79, 421
  pollen, 34, 75
  stratigraphy, 28–34
  trenches, 44–45
  Unconformity, 36, 45, 48, 50, 51, 63, 64, 65, 74, 78, 81–82, 387
Gilead, D., 7, 13, 14, 36, 331, 397
Golan Heights, 5, 8, 21, 24, 26, 29, 37, 45, 53
Greece, 398
Group size, 397, 398, 416, 417–419, 421

H

Hafting, 119, 234, 236, 401, 406, 407, 408, 422, 424
Handaxes (*see also* Bifaces), 1, 9, 11, 12, 13, 15, 29, 36, 50, 51, 53, 54, 57, 59, 79, 86, 89, 94, 115, 116, 172, 185, 197, 223, 225, 228, 230, 231, 232, 301, 324, 326, 328, 377, 378, 394, 397, 398, 399, 402, 403, 404, 405, 406, 407, 415
  analysis, 55–57, 332
  amygdaloid, 367, 404
  basalt flakes of handaxe morphology, 284, 289, 293–296, 297, 298, 375
  blanks, 329, 376–377
  cleaver-like, 358, 398
  cordiform, 368, 404
  core-like, 367, 369, 404
  discoid, 368, 404
  elongation index, 354, 358
  finishing, 120, 187, 190, 195, 287, 404, 406
  flatness ratio, 354, 357, 358, 365
  ovate, 332, 368, 404
  pointedness ratio, 354, 355, 358, 368
  roughing-out, 404
  thinning and shaping, 120, 187, 269, 287, 378, 404, 406, 414, 415
  typology, 366–370, 404–405
Hearth (*see* Fire)
Hinge, 193, 255
Home base (*see also* Preferred locale), 397, 420–421, 424
Hominin
  adaptation, 1, 2
  behavior, 354, 413–424
  bones, 15
  cognitive abilities (*see* Cognition)
  diet (*see* Diet)
  group size (*see* Group size)
  inhibition, 422
  subsistence, 19, 420–421
Homo erectus, 1, 416
Homo ergaster, 416
Homo heidelbergensis, 416
Homogeneity, 72, 136, 140, 141, 172, 185, 223, 286, 318, 325, 337, 338, 357, 358, 360, 369, 371, 375, 377, 378, 383, 399, 404, 412, 421
Horowitz, A., 5, 6, 34
Hula
  Drainage, 3, 5, 7, 9, 12
  Lake, 3, 6, 7, 8, 12
  Nature Reserve
  paleo-Lake, 2, 19, 27, 29, 37, 420, 421, 424
  Valley, 3–6
Hunsgi V, 281, 328, 405
Hunter-gatherers (*see also* Hunting, Gathering), 19, 398, 416, 417, 418
Hunting, 94, 417, 419, 420, 421, 422

I

Iberian Peninsula, 1, 398
India, 281, 398, 405, 407, 413
Inhibition, 422
Innovations, 406, 424
Invention, 407, 424
Isaac, G. L., 332, 376, 404, 417, 420, 421, 424, 425
Isenia, 2, 418
Isimila, 399, 405, 418
Italy, 398

J

Jordan, 332, 398
Jordan River, 3, 6, 7, 8, 9, 10, 11, 13, 16, 18, 21, 22, 23, 26, 27, 29, 32, 36, 37, 38, 44, 45, 49, 53, 62, 63, 64, 69, 71, 73, 75, 78, 81, 82, 115, 405
  bottleneck, 29
  drainage, 7, 9, 11, 13, 15, 18, 29, 36, 63
  floodplain, 16, 22, 26, 32, 36, 44, 48, 75, 81, 82
Jordan Valley, 1, 3, 10, 39, 52, 374, 397, 418, 420

K

Kalambo Falls, 2, 181, 399, 418
Kalb, J., 418
Kaletepe Deresi 3, 398
Kathu, 418
Kilombe, 2, 287, 418
Kleindienst, M. R., 251, 324, 332, 405
Knapping, 120, 165, 168, 187, 223, 232, 237, 240, 255, 258, 260, 300, 301, 302, 303, 304, 319, 321, 323, 324, 325, 331, 342, 344, 373, 374, 375, 385, 389, 390, 391, 394, 396, 397, 401, 402, 403, 405, 406, 407, 412, 413, 414, 416, 417, 420
  dexterity, 234, 302, 376, 401, 405
  experimental, 54, 57, 227, 269, 299, 302, 332, 415
  expertise, 408, 416, 422
  off-site, 415
  on-site, 185, 225, 413, 415
Knowhow, 333, 372'
Kokiselei, 2,
Kombewa, 56, 342
  basalt, 263–269, 329, 338, 339, 370
  flint, 195
  method, 329, 375, 406, 407
Konso, 2
Korazim Saddle, 3, 21, 24, 26, 27, 36, 37

L

Language, 424, 425
Latamne, 399
Leakey, M.D., 54, 310, 319, 332, 405
Levallois, 10, 12
  basalt, 91, 323, 329
  flint, 119, 205, 206, 207, 212, 214
  method, 201, 230–232, 329, 375, 406–407
  emergence of, 406
  Middle Paleolithic, 232, 406, 407
Limace, 154
Limestone, 11, 12, 13, 37, 38–39, 41, 50, 53, 79, 100, 181, 203, 222, 227, 237, 272, 298, 299, 300, 302, 307, 321, 325, 332, 335, 337, 338, 339, 351, 371, 374
  nodules, 379, 394, 402, 405, 412
  reduction sequence, 57, 382, 383, 384, 394–396
  spheroids (*see* Spheroids)
  taphonomy (*see* Taphonomy)
  technology
    cores, 393
    flakes, 381–384

typology
    core tools, 385–392, 394
    flake tools, 381–384
Lip, 54, 55, 187, 195, 226, 255, 259
Lithics
    3-D scanning, 3, 59
    basalt (*see* Basalt)
    flint (*see* Flint)
    illustration, 59
    laboratory analyses, 53–57
    limestone (*see* Limestone)
    photography, 59

M

Ma'ayan Baruch, 15, 214
Massive scrapers, 241, 246–247, 261, 272, 320, 326, 372, 373, 378, 385, 394, 402, 404, 422
Melka Kunture, 2, 399, 418
Mental template, 377, 404
Microartifacts, 19, 53, 69, 89, 100, 119, 190, 195, 197, 225, 272, 379, 394, 403, 414, 415, 418
Middle Paleolithic, 232, 399, 406, 407, 417
Mishmar HaYarden Formation, 26, 38
Mobility, 57
    of lithics, 394, 397, 399, 403, 411, 413–415, 420
    of plant and animal resources, 415–416
Modern
    behavior, 424
    cognitive modernity (*see* Cognition)
    executive reasoning ability, 422
Modified artifacts
    basalt, 81, 319, 320, 372, 378
    flint, 182, 219–221
    limestone, 385, 391–392, 393, 394
Mollusks
    Bellamya, 62
    Melanopsis, 11, 13, 29, 76
    Theodoxus, 29, 76
    Unio, 29, 67
    Viviparid, 67, 71
        fossil embryos
        Vivipara sp., 11, 13
        Viviparus apameae, 6, 15, 26, 29, 76
        Viviparus apameae galileae, 34, 62
Morocco, 405
Mousterian, 1, 14, 18, 54, 181
Multiple tools, 54, 117, 136, 141, 146, 149, 168, 181, 250, 378, 401, 403, 404, 406
    triple tools, 146, 181, 385, 404

N

Nahal Mahanaim Outlet (NMO), 6
Neanderthals, 424
Nile Valley, 398
North of Bridges Acheulian (NBA), 18, 36, 116, 405
Notarchirico, 398
Notch
    basalt, 241, 242–244, 245, 272, 298, 373
    Clactonian notch, 122, 128
    end-notches, 122, 125, 126, 394
    flint, 122–128, 132, 133, 136, 140, 141, 144, 146, 152, 162, 168, 172, 176, 181, 182, 183, 223, 233, 404, 408
    limestone, 382, 385
Nuts (*see* Flora)
Nutting stones (*see also* Pitted stones), 307

O

Oldowan, 404
    Developed Oldowan, 405
Olduvai Gorge, 2, 54, 399, 405, 418, 420
    FLK Zinj floor, 420
Olorgesailie, 2, 181, 287, 399, 418, 420
Out of Africa, 1, 2, 14, 398
Outrepasseé flake, 201, 232, 255, 406

P

Paleoclimate, 419
Paleoclimatic cycle, 62
    Milankovitch cycles, 32
    MIS, 2, 32, 62, 116, 397, 407
    OIS, 79
Paleomagnetism, 16, 36
    Matuyama/Bruhnes chron boundary (M/B Chron boundary), 27, 36, 117
Patina (Patination), 39, 41, 54, 117, 119, 120, 134, 140, 155, 162, 212, 216, 237, 239, 240, 333, 335, 380
    double patina, 117, 162, 206, 212, 219, 221, 236
Percussive activity, 237, 298, 306, 373, 385, 389, 391, 392, 393, 396, 401
    battering, 182, 221, 299, 300, 304, 307, 311, 318, 385, 387, 388, 390, 391, 394, 402
    pounding, 221, 307, 388, 390, 391
Percussive technique (*see* Percussors)
Percussive tools, 237, 298–319, 371, 373, 374, 385, 401, 405
    anvils (*see* Anvils)
    pitted stones (*see* Pitted stones)
    percussors (*see* Percussors)
    reduction sequence of, 373–374
Percussors, 38, 54, 57, 81, 94, 237, 399, 401, 405–406, 412, 416, 422
    basalt, 298–306, 310, 311, 315, 316, 318, 319, 320, 372, 373, 374, 375, 388, 401, 413
    flint, 182, 219, 221, 299
    hard, 187, 193, 225, 232
    limestone, 79, 227, 299, 379, 385–391, 392, 394, 396, 402
    percussive technique, 225–227
    Percuteurs de taille, 385
    Percuteurs de concassage, 385, 387–391, 392, 396, 402, 405
    potential percussors, 304–306
    soft, 187, 193, 195, 225, 232, 406, 416, 424
        made of antler, 57, 227, 405
    split, 54, 251, 272, 303–304, 391, 385, 393
Picard, L., 6, 10, 13, 26, 34, 37
Picks, 332, 398
Pitted anvils (*see* Anvils, Pitted stones)
Pitted stones, 54, 237, 405
    basalt, 250, 251, 286, 298, 306–313, 315, 372, 373, 374, 378, 401
    limestone, 385, 292
Pleistocene, 1, 5, 6, 7, 16, 34, 37, 38, 39, 44, 45, 46, 48, 74, 81, 420, 424, 425
    Early Pleistocene, 2, 6, 37, 116, 117, 413, 420
    Early–Middle Pleistocene, 18, 19, 22, 32, 36, 61, 78
    Middle Pleistocene, 15, 16, 18, 19, 29, 34, 36, 37, 62, 65
    Late Pleistocene, 26
Polyhedrons, 405
Potts, R., 419
Preferred locale, 420–421, 422
Preservation, 2, 29, 38, 39–41, 52, 54, 67, 117, 332, 333, 335, 408, 415
    of basalt, 48, 80, 237–240, 248, 260, 299
    of flint, 119–120, 123, 128, 134, 140, 144, 162
    of limestone, 380, 383
Primates, 421

Q

Qesem Cave, 405

Quadrihedrals, 332
Quarrying, 324, 374, 413, 416

R

Radiometric dates, 15, 16, 19, 34, 36–37
    radiocarbon dating, 82
Raw material, 2, 3, 14, 15, 37–41, 50, 53, 54, 57, 62, 72, 89, 115, 119, 203, 214, 221, 223, 225, 237, 306, 307, 311, 320, 331, 332, 333, 335, 336, 337, 338, 339, 354, 361, 371, 372, 373, 374, 376, 378, 379, 380, 383, 394, 397, 398, 399, 401, 402, 403, 405, 406, 408, 412, 413, 415, 417, 418, 419, 420, 422
    basalt (*see* Basalt)
    chert, 12, 13, 53
    conservatism (*see* Conservatism)
    exploitation mode, 402
    flint (*see* Flint)
    limestone (*see* Limestone)
    manuports, 49, 71, 374
    natural nodules, 54, 57–58
    preservation (*see* Preservation)
    quartz, 53
    selection of (*see* Selection)
    variability (*see* Variability)
Reduction sequence, 57, 399, 405, 406, 407, 408, 412, 413, 414, 415, 416, 422
    comparative view, 399–403
    of basalt, 237, 259, 269, 274, 287, 297, 298, 323, 332, 344, 371, 372, 373–374, 378, 401–402, 408
    of bifaces (*see* bifaces)
    of flint, 165, 172, 221, 223–236, 399–401
    of limestone, 382, 383, 384, 394, 402
    reduction strategies, 374, 376, 399, 422
Retouched flakes
    basalt, 241, 246, 248–250, 310, 373
    flint, 117, 122, 125, 126, 132, 136, 140, 141, 146, 152, 154, 155–162, 172, 176, 233, 399, 401, 402, 404, 408
    limestone, 382
Revadim, 134, 214, 320, 406
Roe, D., 54, 55, 332, 354, 358, 359, 377
Rosaneto, 398
Rosh Pinnah River, 27, 29, 37, 38
Ruman basalt, 26

S

Saudi Arabia, 398
Scar pattern, 55, 58, 165, 185, 190, 195, 207, 212, 217, 218, 221, 232, 233, 234, 259, 262, 269, 272, 274, 278, 286, 287, 289, 296, 297, 321, 370, 383, 384, 406, 407
Selection, 38, 130, 140, 141, 162, 164, 176, 183, 223, 225, 233, 242, 264, 310, 325, 339, 372, 373, 375, 376, 377, 383, 384, 394, 399, 402, 405, 422
Sequential memory, 422
Sequential processing (*see* Autonoesis)
Side-scrapers, 54, 117, 404, 408
    basalt, 245–246, 247, 248, 298
    flint, 122, 136–143, 144, 152, 165, 168, 175, 176, 182, 233
    limestone, 382
    Mousterian, 176
Signs of use
    on CCT, 223, 316
    on FFT, 122, 152, 154, 159, 182
Siret flakes, 54, 260–262, 264
Site formation processes, 95, 418
    postdepositional processes, 39, 41, 57, 62, 67, 69, 80, 89, 94, 238, 240, 258, 299, 301, 335, 373, 380, 398, 413, 418
Site types, 420
    quarry site (*see* Quarry)
Social
    brain (*see* Brain)
    cognition (*see* Cognition)
    complexity, 421
    learning, 422
    structure, 416, 418, 421
Spain, 398
Spatial distribution, 19, 57, 69, 72, 94, 95, 417, 418, 419, 424
Spheroids, 399, 405, 418
    Subspheroids, 399
Stekelis, M., 2, 6, 10, 11, 12, 13, 14, 23, 36, 331
Steps, 193, 195, 226, 255
Sterkfontein, 405
STIC, 405
Stocking (*see also* Caching), 419, 420, 421
Strike, 18, 32, 43, 44, 46, 47, 48, 49, 50, 63, 64, 72, 75, 78, 79, 81, 82, 83, 86, 87, 89, 94, 95, 101, 113, 115
Striking platform, 55, 57, 58, 59, 185, 187, 192, 201, 205, 219, 232, 251, 255, 257, 259, 260, 262, 264, 269, 274, 277, 284, 286, 318, 319, 321, 329, 373, 375, 377, 383, 385
    abraded, 187, 226
    crushed, 187, 192, 195, 226
    facetted, 165, 187, 192, 195, 226, 232, 233, 286, 326
    lipped (*see* Lip)
    of bifaces, 337, 341, 342–346, 348, 370, 377, 378, 398, 407–408
    plain, 165, 187, 192, 264, 269, 274, 278, 281, 286, 287, 296, 297, 385
    punctiform, 192, 195
    removal of, 125, 134, 165–178, 233–236, 281, 293, 297, 384, 401, 404, 407–408, 411, 422

T

Tabun Cave, 214, 331, 398, 406
Taphonomy (*see also* Preservation), 45, 57, 62, 79, 89, 94, 95, 102, 113, 116, 117, 332, 414, 418, 420
    bone taphonomy, 19, 41, 53, 111, 419
    of basalt, 237–240, 241, 242, 248, 251, 257, 258, 262, 264, 272, 277, 286, 319, 371
    of flint, 119–120, 126, 162, 182
    of limestone, 41, 379, 380, 382, 387
Target artifacts, 57, 176, 200, 232, 236, 298, 326, 331, 372, 378, 382, 384, 394, 399, 402, 407, 408, 412
Tarn Valley, 398
Tchernov, E., 5, 15, 34, 53
Technology, 57, 119, 134, 223, 237, 398, 399, 406, 407
    basalt (*see* Basalt), 250–298, 323–331, 337–346, 371, 374–378
    biface (*see* Biface), 263, 332, 337–346, 377, 378
    flint (*see* Flint), 185–218
    limestone (*see* Limestone), 381–384, 393, 402
Tectonic activity, 24, 26, 27, 28, 34, 37, 43, 47, 62, 79, 398
Theory of mind (ToM), 421
Tilted beds, 13, 15, 18, 21, 22, 23, 26, 27, 29, 32, 34, 43, 44, 46, 47, 48, 49, 50, 51, 62, 64, 65, 71, 74, 78, 79, 81, 82, 94
Tradition, 1, 2, 287, 320, 374, 378, 397, 398, 399, 412, 420, 424
Transformative technology, 422
Transportation of lithics (*see* Mobility)
Trihedrals, 332
Truncations, 122, 146, 149, 153, 154, 168, 172, 175, 176
Turkey, 3, 398

U

'Ubeidiya, 2, 14, 43, 47, 134, 153, 237, 320, 332, 371, 399, 404, 405, 413, 418
Unretouched flakes, 402
    basalt, 248, 250, 251–298, 310, 373, 408
    flint, 123, 134, 140, 141, 162, 164, 165, 168, 172, 175, 176, 185–201, 212, 219, 225, 233

Index

limestone, 381–385
Upper Paleolithic, 424

V

Variability, 1, 38, 79, 80, 103, 251, 390, 398, 408–411, 418, 412
  technological, 274, 272, 264, 258, 243, 221, 212, 207, 202, 195, 187, 393, 383, 376, 358, 338, 323, 319, 304, 296, 286, 284, 278, 277
  typological, 122, 141, 143, 165, 168, 176, 182, 183, 241, 242, 244, 247, 302, 310, 318, 337, 364, 368, 373, 377, 404
Ventral removals
  basalt, 272–273
  flint, 154, 200–201
Volcanic activity, 7, 9, 19, 24, 36, 37, 38

W

Wynn, T., 422

Y

Yarda basalt, 6, 26, 34
Yehudiya Block, 24

PGMO 04/13/2018